Handbook of Applied Mycology

Volume 1: Soil and Plants

Handbook of Applied Mycology

Series Editor
Dilip K. Arora

Banaras Hindu University
Varanasi, India

Volume 1 Soil and Plants, edited by *Dilip K. Arora, Bharat Rai, K. G. Mukerji*, and *Guy R. Knudsen*

Volume 2 Humans, Animals, and Insects, edited by *Dilip K. Arora, Libero Ajello*, and *K. G. Mukerji*

Volume 3 Foods and Feeds, edited by *Dilip K. Arora, K. G. Mukerji*, and *Elmer H. Marth*

Volume 4 Fungal Biotechnology, edited by *Dilip K. Arora, Richard P. Elander, and K. G. Mukerji*

Volume 5 Mycotoxins in Ecological Systems, edited by *Deepak Bhatnagar, Eivind B. Lillehoj, and Dilip K. Arora*

Handbook of Applied Mycology

Volume 1: Soil and Plants

edited by

Dilip K. Arora

Banaras Hindu University
Varanasi, India

Bharat Rai

Centre of Advanced Study in Botany
Banaras Hindu University
Varanasi, India

K. G. Mukerji

University of Delhi
Delhi, India

Guy R. Knudsen

University of Idaho
Moscow, Idaho

MARCEL DEKKER, INC. New York • Basel • Hong Kong

Library of Congress Cataloging-in-Publication Data

Soil and Plants / edited by Dilip K. Arora ... [et al.].
p. cm. -- (Handbook of applied mycology; v. 1)
Includes bibliographical references and index.
ISBN 0-8247-8380-8
1. Mycorrihizae. 2. Phytopathogenic fungi--Biological control. 3. Wood-decaying fungi. 4. Fungal diseases of plants. I. Arora, Dilip K. II. Series.
QK918.S64 1991
589.2--dc20 90-25706
CIP

This book is printed on acid-free paper.

MARCEL DEKKER, INC.
270 Madison Avenue, New York, New York 10016

Current printing (last digit):
10 9 8 7 6 5 4 3 2 1

PRINTED IN THE UNITED STATES OF AMERICA

HANDBOOK INTRODUCTION

Applied mycology is one of the most stimulating and rapidly evolving areas of the biological sciences. It encompasses many facets of agricultural, industrial, pharmacological, medical, and food sciences, all of which include organisms or processes that are subject to biotechnological manipulation. However, published information on the practical significance of fungi is fragmented in many specialized publications. There is an obvious need for a comprehensive treatment of the basic principles, methods, and applications of mycology as an applied, integrated, and multidisciplinary subject. The potential usefulness of such a handbook is suggested by the practical significance of mycological research, by the uniqueness and increasing coherence of the field's conceptual directions, and by the international scope of the subject. These five volumes of the *Handbook of Applied Mycology* attempt to fulfill this need. These volumes present and collate the major aspects of applied mycology and are designed as a modern standard reference for the work in this field.

In developing these volumes we decided that the treatment of the subject should be comprehensive but concise enough to enable completion within a set of five volumes published closely together rather than a continuous encyclopedic series taking several years to complete. However, this does not preclude the possibility of publishing supplementary volumes, as appropriate, to communicate new information. Further, we wanted to develop a treatise that would offer a detailed overview of applied aspects of mycology in terms of its theoretical, methodological, and empirical contributions. By emphasizing the interdisciplinary aspects of the subject, we hope that the handbook will highlight recent prospective linkages among diverse research paradigms.

Since the information presented in these volumes is expected to serve students as well as veterans in the field, both basic and advanced specialized materials are included. In addition to addressing researchers in mycology, the handbook will prove useful to a broad spectrum of professional groups, including biotechnologists, medical scientists, molecular biologists, food scientists, pharmacologists, soil microbiologists, microbial ecologists, and plant pathologists. Each chapter is intended to provide a balanced view of the current state of knowledge within an expanding field of interest. The selection of the chapters and overall organization of the book were guided by several insightful colleagues from around the world. These volumes were authored by a large team of international experts. Every chapter within the five volumes has been edited by a distinguished group of scientists. As a result, nine editors and over 200 authors representing 20 countries have contributed to this project.

Because of the multidisciplinary nature of the subject, judgment had to be made about the relative importance and significance of the hundreds of potential topics related to applied mycology. Also, a balance had to be struck between comprehensive coverage and intellectual appeal of the field. Readers should find this work useful for in-depth information and as a route to additional information contained in its critical discussions and wide range of references provided. Each of the five volumes is intended to be self-contained. Therefore, some degree of duplication of materials, especially basic principles, is inevitable.

No work of this magnitude can be accomplished without the support and contributions of many individuals. The planning and eventual production of this handbook were a collaborative effort. I am deeply indebted to the group of coeditors and many colleagues who have assisted me throughout all stages of the project. I also appreciate the dedication and hard work of the chapter authors, who cooperated persistently in the production of these volumes. I acknowledge with gratitude the support of M. V. Wiese, N. W. Schaad, S. Stotzsky, J. S. Singh, Bharat Rai, R. S. Upadhyay, V. S. Jaiswal, B. Thompson, Jon Olson, A. K. Pandey, Rejeev Gaur, and Shusma Gupta, whose interest and cooperation have been invaluable to me. I am grateful to the University of Idaho for providing assistance in the form of staff support and facilities, and the Banaras Hindu University, India, for granting leave during the period I edited these volumes. I appreciate the encouraging support provided by Ms. Sandra Beberman of Marcel Dekker, Inc., at all stages of production of this book. My gratitude is expressed to my teacher, J. L. Lockwood, and to my parents, whose example has always been my guiding principle. Finally, to Meenakshi, Sidhartha, and Gautam for their patience, encouragement, and willingness to forgo other interests so that the handbook might be completed, I am grateful.

Dilip K. Arora

PREFACE

Mycology, the study of fungi, has in some respects only recently emerged from the shadows of natural history. More and more, mycologists are embracing studies in such diverse areas as physiology, genetics, and evolution, and finding new applications for this knowledge. The development of a conceptual basis for understanding the ecology and physiology of fungi in nature requires transcendence of narrow scientific disciplines. Our intent for this, the first volume in the *Handbook of Applied Mycology*, was to bring together scientists who are using diverse approaches to study, understand, and manipulate fungi. We focus on fungi associated with soil and plants; of necessity, we have had to narrow our focus still further, and concentrate on four general areas in which mycological research is rapidly generating practical applications.

The first three parts of this book cover a wide range of topics within three categories: mycorrhizal symbionts, fungi as biological control agents against plant pathogens and weeds, and fungi as degradative agents in nature. The fourth and final part concentrates on applications of mathematical modeling in mycology. Since modeling focuses on understanding systems and making predictions about them, it serves as a link between basic and applied mycology. Although the scope of this volume makes omission of a few topics inevitable, our aim was to represent the current progress in these areas as comprehensively as possible.

We wish to thank the scientists who have contributed their time and energy to this endeavor. Numerous colleagues and members of the Marcel Dekker, Inc., editorial staff provided comments and suggestions for improvement of the chapters, or reviewed and commented on manuscripts. We are much obliged to them all for their contributions, criticisms, and suggestions.

Dilip K. Arora
Bharat Rai
K. G. Mukerji
Guy R. Knudsen

CONTENTS

CONTRIBUTORS

Dilip K. Arora Department of Botany, Banaras Hindu University, Varanasi, India

Charles W. Bacon Toxicology and Mycotoxin Research Unit, Russell Research Center, Agricultural Research Station, United States Department of Agriculture, Athens, Georgia

D. Joseph Bagyaraj Department of Agricultural Microbiology, University of Agricultural Sciences, Bangalore, India

Ralph Baker Department of Plant Pathology and Weed Science, Colorado State University, Fort Collins, Colorado

Victoria Barrett Department of Botany and Microbiology, Auburn University, Auburn, Alabama

Michael J. Bazin Department of Microbiology, King's College, University of London, London, England

Lynne Boddy School of Pure and Applied Biology, College of Cardiff, University of Wales, Cardiff, Wales

John A. Buswell* Department of Biology, Paisley College of Technology, Paisley, Renfrewshire, Scotland

Jose De Battista Department of Agronomy, University of Georgia, Athens, Georgia

John C. Dodd Biological Laboratory, The University of Kent, Canterbury, Kent, England

---

**Present affiliation*: Department of Biology, The Chinese University of Hong Kong, Shatin, Hong Kong.

Vern J. Elliott Department of Plant Pathology, Agricultural Research Station, United States Department of Agriculture, and North Carolina State University, Oxford, North Carolina

Donna M. Gerten Agricultural Research Station, National Clonal Germplasm Repository, United States Department of Agriculture, Corvallis, Oregon

R. K. Gupta Department of Environmental Sciences, College of Basic Sciences and Humanities, G. B. Pant University of Agriculture and Technology, Pantnagar, India

Suresh K. Hasija Department of Biological Sciences, Rani Durgavati University, Jabalpur, India

Bushan L. Jalali Department of Plant Pathology, Haryana Agricultural University, Haryana, India

Indu Jalali Department of Plant Pathology, Haryana Agricultural University, Haryana, India

Wojciech Janisiewicz Appalachian Fruit Research Service, Agricultural Research Service, United States Department of Agriculture, Kearneysville, West Virginia

Peter Jeffries Biological Laboratory, The University of Kent, Canterbury, Kent, England

Guy R. Knudsen Department of Plant, Soil and Entomological Sciences, University of Idaho, Moscow, Idaho

Robert G. Linderman Horticultural Crops Research Laboratory, Agricultural Research Service, United States Department of Agriculture, Corvallis, Oregon

Mohan S. Manocha Department of Biological Sciences, Brock University, St. Catharines, Ontario, Canada

Paul Markham Department of Microbiology, King's College, University of London, London, England

K. G. Mukerji Department of Botany, University of Delhi, Delhi, India

Eric B. Nelson Department of Plant Pathology, Cornell University, Ithaca, New York

David M. O'Keefe* Soil Science Department, University of Florida, Gainesville, Florida

---

**Present affiliation*: Department of Agricultural Engineering.

Timothy C. Paulitz* Horticultural Crops Research Laboratory, Agricultural Research Service, United States Department of Agriculture, Corvallis, Oregon

James I. Prosser Department of Molecular and Cell Biology, Marischal College, University of Aberdeen, Aberdeen, Scotland

K. Raghu Plant Pathology and Pesticide Residues Section, Nuclear Agriculture Division, Bhabha Atomic Research Centre, Bombay, India

Douglas I. Rouse Department of Plant Pathology, University of Wisconsin–Madison, Madison, Wisconsin

N. Sethunathan Laboratory of Soil Microbiology, Central Rice Research Institute, Cuttack, India

U. D. Singh Division of Plant Pathology, Central Rice Research Institute, Cuttack, India

Pradeep K. Singhal Department of Biological Sciences, Rani Durgavati University, Jabalpur, India

James P. Stack† Department of Plant Pathology and Microbiology, Texas A&M University, College Station, Texas

Suvercha Department of Botany, University of Delhi, Delhi, India

David M. Sylvia Soil Science Department, University of Florida, Gainesville, Florida

C. Gerald Van Dyke Department of Botany, North Carolina State University, Raleigh, North Carolina

M. V. Wiese Department of Plant, Soil and Entomological Sciences, University of Idaho, Moscow, Idaho

---

Present affiliations:
*Department of Plant Sciences, Macdonald College of McGill University, Ste. Anne de Bellevue, Quebec, Canada.

†EcoScience Laboratories, Inc., Amherst, Massachusetts.

Part I

MYCORRHIZAE AND ENDOPHYTIC FUNGI

1

ECOLOGY OF VESICULAR-ARBUSCULAR MYCORRHIZAE

D. JOSEPH BAGYARAJ *University of Agricultural Sciences, Bangalore, India*

I. INTRODUCTION

Plant roots provide an ecological niche for many of the microorganisms that abound in soil. German Botanist Albert Bernard Frank in 1885 [1] introduced the greek word *mycorrhiza*, which literally means "fungus root," to scientific terminology. In natural ecosystems much of the root system can be colonized by mycorrhizal fungi. Colonization is restricted to the root cortex and does not enter the vascular cylinder. The symbiosis is so well balanced that, although many of the host cells are invaded by the fungal endophyte, there is no visible tissue damage and under certain conditions it enhances the growth and vigor of the host plant. Because most economically important plants form mycorrhizae, the subject is currently attracting much attention in agricultural, horticultural, and forestry research. Interest in this symbiosis has escalated dramatically in recent years, partly because of what we have learned about the benefits of mycorrhizae and partly because of economic and geopolitical events [2].

There are different types of mycorrhizae. The main types of mycorrhizae and the host plants in which they are commonly associated are given in Table 1.

II. VESICULAR-ARBUSCULAR MYCORRHIZAE

Of the five types of mycorrhizae listed in Table 1, vesicular-arbuscular (VA) mycorrhizae are geographically ubiquitous and occur over a broad ecological range. They are commonly found in association with agricultural crops. They are formed by zygomycetous fungi. The presence of vesicles and arbuscules is the diagnostic criterion for identifying VA mycorrhizal fungi in a root (Fig. 1). In this review, VA mycorrhizae alone is considered.

III. THE HOSTS

VA mycorrhizae have the widest host range and distribution of all the mycorrhizal associations. It is estimated that about 90% of vascular plants

TABLE 1. Different Types of Mycorrhizae and Their Hosts

Mycorrhiza type	Host plant/families
Ecto	Pinaceae, Fagaceae, Betulaceae, Leguminaceae, mostly caesalpinoid legumes, Salicaceae, Tiliaceae, Rosaceae, Juglandaceae, etc.
Vesicular-arbuscular (VA)	Majority of the plants including those important in agriculture, horiticulture, pastures, and tropical forests
Ericoid	Ericaceae and Epacridaceae
Orchidaceous	Orchidaceae
Arbutoid	*Arbutus*, *Monotropha*

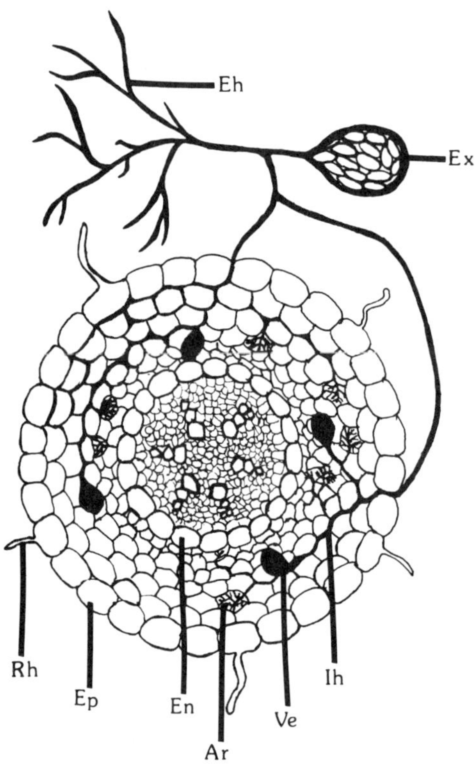

FIGURE 1. Diagrammatic representation of the transverse section of a root colonized by VA mycorrhizal fungus. Ar = arbuscle; Eh = external hyphae; En = endodermis; Ep = epidermis; Ex = extramatrical chlamydospore; Ih = internal hyphae; Rh = root hair; Ve = vesicle.

normally establish mutualistic relationship with VAM fungi. VA mycorrhizae have been observed in 1000 genera of plants representing some 200 families. There are at least 300,000 receptive hosts in the world flora [2], and there are about 120 species of VA mycorrhizal fungi [3]. If the hosts are divided up evenly among the fungi with no overlap in host range, each fungus would have more than 2500 potential partners. We know that host ranges overlap extensively, suggesting that some individual VAM fungi may well have access to thousands of hosts [2].

According to Gerdemann [4], it is easier to list most plant families that do not form VA mycorrhizae than to list those that do. Families not forming VA mycorrhizae include Pinaceae, Betulaceae, Orchidaceae, Fumariaceae, Commelinaceae, Urticaceae, and Ericaceae. Families that rarely form VA mycorrhizae include the Brassicaceae, Chenopodiaceae, Polygonaceae, and Cyperaceae. Families that form both ectomycorrhizae and VA mycorrhizae include Juglandiaceae, Tiliaceae, Myrtaceae, Salicaceae, Fagaceae, and Caesalpiniaceae [4]. Occasionally, immune species may occur within a normally susceptible family, e.g., *Lupinus consentinii* [5], which has a thick fringe of long root hairs resembling the proteoid roots of members of the Proteaceae that are also immune. Important crops with VA mycorrhiza include wheat, maize, all millets, potatoes, beans, soybeans, tomatoes, apples, oranges, grapes, banana, castor, tobacco, tea, coffee, cocoa, sugar cane, mango, asparagus, rubber, cardamom, pepper, etc. Most of the tropical rain forest trees are VA mycorrhizal [6]. Harley [7] has listed the gymnosperms in which VA mycorrhizae have been observed. They are found in Pteridophytes [8] and Bryophytes [9]. Recently, VA mycorrhizal colonization has been reported in floating [10] and submerged aquatic plants [11]. Usually VA mycorrhizae are confined to the roots; they have been reported in diverse structures such as the modified leaves of water fern *Salvinia cucullata* [10], fruiting peg of peanut [12], and modified scalelike leaves and rhizomes of ginger and canna [13]. They are generally not found in proteoid roots and root nodules [14].

It is not understood why VAM fungi have such broad host ranges. The ability of a single VAM isolate to infect roots of unrelated plants such as strawberry, ash, various grasses and legumes, and even hazel (which is also ectomycorrhizal) suggests that its nutritional requirements are not very specific. Nevertheless, an isolate with such a broad host range will not grow in pure culture on any nutrients tested so far. It needs a susceptible root. In addition, there is recent evidence that plant species within a single family can vary in their susceptibility as much as plant species in very distinct families. Even varieties or cultivars of the same species can be colonized to different extents. As yet, there is no well-substantiated hypothesis to explain these phenomena.

IV. THE FUNGI

VA mycorrhizal fungi, being obligate biotrophs, do not grow on synthetic media and hence are classified according to morphological characteristics of the spores formed in the soil. One hundred and twenty species have been described by 1987 [3], and they are all grouped in a single family, the

Endogonaceae. The family is considered to be in an order of its own, the Endogonales tentatively placed in the class Zygomycetes in the subdivision Zygomycotina [15]. The seven genera, *Endogone, Glomus, Sclerocystis, Entrophospora, Acaulospora, Scutellospora* and *Gigaspora*, presently constitute the Endogonaceae. Of these, only *Endogone* forms zygospores characteristic of Zygomycetes, while others probably lack sexual reproduction. Species of *Endogone* do not form VA mycorrhizal association. The other six genera form VA mycorrhizal association.

The main features of VA mycorrhizal fungi are summarized by Trappe and Schenck [16] and Hall [17]. Hall and Abbott [18] compiled photographic transparencies illustrating diagnostic features of the Endogonaceae. Walker [19] has suggested a standardized terminology for use in description of endogonaceous spore walls involving a stylized graphic representation of the wall structure ("murograph"). This concept is recommended for future descriptions of new species. Berch [20] reviewed the taxonomy, specificity, fossil record, and phylogeny of Endogonaceous fungi. Recently, Schenck and Perez [3] published a manual for the identification of VA mycorrhizal fungi.

Glomus, the most common genus of VAM fungi, has over 50 species which form globose, ellipsoid or rather irregularly shaped spores that range from 20–400 μm. These spores are thick-walled (up to 30 μm), and hyaline, yellow, red–brown, brown or black. Spores are attached to a single subtending hypha (or to two, three, or more in exceptional species) and produced in the soil near plant roots, at the soil surface, or occasionally in roots, and they may form singly, in groups of a few to many, or in large aggregates called sporocarps. Sporocarps are most commonly found in undisturbed forest communities with perennial plants and accumulated organic material. Sporocarpic species, when grown on plants in pot culture, usually produce spores singly or in small aggregates. Some *Glomus* species (e.g., *Glomus lacteum*) are not known to form spores in sporocarps, while others form spores outside the plant root only infrequently (e.g., *Glomus intraradices*). The spores of most species germinate by emergence of a germ tube directly through the subtending hypha.

Sclerocystis species produce chlamydospores that are very similar to those of *Glomus*, except that they tend to be clavate rather than globose. All *Sclerocystis* species form spores only in sporocarps up to 700 μm in diameter. The spores, which can be up to 125 μm long, develop in a single layer, radially arranged around a central plexus of sterile and sporogenous hyphae. The sporocarps may group together in masses up to several cm in diameter on the soil surface, on leaves, twigs, and mosses. There is no published information on the germination of *Sclerocystis* spores.

Spores of the genus *Acaulospora* appear to form laterally on the neck of a small thin-walled saccule, called variously a mother spore, a vesicle, a hyphal terminus, a sporiferous saccule, or a sporogenous saccule [21]. When mature, the spore does not display a subtending hypha because the vesicular structure that gives rise to it loses its cytoplasm and collapses during development of the spore. The spores are globose or ellipsoid, from <100 to >400 μm in diameter, and hyaline, yellow, or reddish-brown. The surface of the sporewall may be ornamented with pits, projections of various shapes, folds, spines, or reticulations, and the wall is up to 12 μm thick. One species of *Acaulospora* is now known to produce sporocarps, but in other species, spores are normally found in the soil of the rhizosphere. *Entrophospora* is more or less similar to *Acaulospora* except that the spore is formed inside the parent hypha just below the vesicle.

Scutellospora spores can be greater than 600 μm in diameter, and the smallest ones are just under 200 μm in diameter. Spores develop terminally on a bulbous hyphal suspensor, which remains attached at maturity and may bear short lateral projections. Spores are usually globose but can be ovoid, obovoid, piriform, or irregular. Spores are hyaline, white, yellow, pinkish, grey-green, or brown. Spore wall structure is composed of at least two wall groups, with one or more flexible membranous or coriaceous walls in the inner group or groups. The genus is characterized in particular by its mode of spore germination. Germination is by means of one or more germ tubes produced near the spore base from a germination shield formed within or upon a flexible inner wall. Thin walled, knobby, or broadly papillate auxiliary cells (vesicles) are borne in soil on straight or coiled hyphae, formed singly or in clusters. *Gigaspora* produces spores which are superficially similar to those of *Scutellospora* spores. The striking difference is the germination of spores, which is by direct growth of one or more germ tubes through the spore wall. It is possible that this genus evolved from *Scutellospora* [21].

Another fungus, or group of fungi, that forms endomycorrhizae has been referred to in the literature as "the fine endophyte," since its hyphae are <4 μm wide, compared to 10-20 μm for typical VAM fungi. This anomalous endophyte has been described as *Glomus tenue*, though it does not resemble the other species of *Glomus*. Its spores are 10-15 μm in diameter, and when young are hyaline and indistinguishable from the vesicles that form in the roots; when mature, however, they are pigmented. Because of their small size as compared to those of other VAM fungi, the spores of *Glomus tenue* are not often recorded in nature, though mycorrhizae of this type are common, particularly in dry regions.

V. DISTRIBUTION

VA mycorrhizae, in addition to their widespread distribution throughout the plant kingdom, are also geographically ubiquitous and occur in plants growing in arctic, temperate, and tropical regions [22]. As an explanation for their remarkably widespread distribution, Trappe [23] proposed that VAM fungi were disseminated intercontinentally prior to continental drift. The supercontinent Gondwanaland is thought to have begun to break apart and drift north about 125 million years ago. Fossil records of plants containing VAM-like structures have occurred as Trappe suggested. In general, VAM population is more in cultivated soil compared to virgin soil [24]. They are mostly seen in the top 15-30 cm of soil, and their numbers decrease markedly below the top 15 cm [25]. They are normally not found in depths beyond the normal root range of plants [22]. Although VA mycorrhizal fungi are ubiquitous in soils, the patterns of distribution of individual species have not been fully understood. Studies on the distribution of species have either covered large geographical areas [24,26] or have been limited to smaller regions [27]. The distribution of species of VA mycorrhizal fungi varies with climatic and edaphic environment as well as with land use. For example, *Acaulospora laevis* is common in western Australia [27] and New Zealand [24] but occurs less frequently in soils of eastern Australia [24]. *Glomus* spp.

appear to have widest distribution. *Gigaspora* and *Sclerocystis* spp. are more common in tropical soils. *Acaulospora* seems to be better adapted to soils with pH <5.0. In fact, certain VAM fungi have been linked to particular kinds of soil: *Glomus mosseae* with fine textured, fertile, high pH soils; *Acaulospora laevis* with coarse-textured, acid soils; and *Gigaspora* species with sand dune soils [2].

A fundamental problem in studies of the distribution of VA mycorrhizal fungi lies in identifying the fungi. It relies almost entirely on spore morphology [26], which may change with the spore age [26,28]. Description of many of taxa have been based on single collections of uncertain age. This, coupled with the possibility of finding undescribed taxa, can make the identification of spores collected from field samples very difficult. Another difficulty in determining the distribution of VA mycorrhizal fungi is that spores of all species are not equally easy to recover from soil. Soil often contain spores of more than one species of VA mycorrhizal fungus [27]. Moreover, the same root may become colonized by more than one species of VAM fungus [8,29]. Some VAM fungi develop characteristic morphological features within plant roots [8,28]. This has allowed quantitative estimates to be made of the total amounts of coarse or fine VA mycorrhizal hyphae within plant roots [30,31]. It is also possible to assess the extent of infection formed by different species of VAM fungi with coarse hyphae occurring in the same soil [32].

Management practices can influence the types of VA mycorrhizal fungi found in agricultural soils. In Australia, *G. monosporum* was present in some cultivated soils but absent in virgin soils [27]. The population of VA mycorrhizal fungi has been changed by long-term fertilizer applications in some instances [24]. Crush [33] suggested that pasture development will change the VA mycorrhizal fungal population. It is clear that there are many species of VA mycorrhizal fungi, and that the factors that control their distribution are poorly understood except in a few cases.

VI. DISPERSAL

Dispersal of VAM fungi is usually by spread from one living root to another, spread of spores and other propagules in water and wind-blown particles, and spread by animal vectors. These can be divided into two groups: (a) active dissemination (growth of mycelium through soil), and (b) passive dissemination, where VAM fungi are moved by wind, water, or animal vectors.

A. Active Dissemination

Active dispersal occurs as mycelia grow through the soil, although it is effective over a limited range. Powell [34] determined that an efficient mycorrhizal fungus would advance 65 meters in 150 years, or 0.43 meters per year under greenhouse conditions. This slow extension limits the potential importance of this mode of dispersal. Powell further demonstrated that the introduced fungus spread more readily through soils which already contain low populations of VAM fungi, particularly if the population has been depressed by cultivation to nonhost crops. More recently, it has been shown that plant species and root density may significantly influence the rate of VAM fungus spread [35]. In clover, the greatest rate of spread of *Glomus fasciculatum* was 1 cm/week, whereas in fescue, *G. fasciculatum* spread at a rate of only 0.7 cm/week. These experiments showed that root density is most critical when plants are young and root density is low. In fact, supraoptimal root density was achieved in fescue (a grass that develops an extensive root system), and rate of fungal spread was reduced as plant size increased. A similar supraoptimal root density was not achieved in clover because of the less extensive root system.

The previous experiments were conducted in the greenhouse in fumigated soil, and it is difficult to project the rate of mycelial growth and VAM fungus spread through field soils. Colonization by morphologically similar indigenous VAM fungi in nonsterile field soils make this type of study difficult. However, using sporulation by nonindigenous VAM fungi as proof of spread, Mosse et al. [36] demonstrated that *Glomus caledonicum* was able to move 7 – 13 cm from an inoculation point after 13 weeks. No correlation was observed between rate of spread and plant size, but spread rate was greater in nonsterilized plots than in those receiving formalin treatments. In these experiments, host species also significantly affected spread rate.

Whether VAM fungi grow in a directed way, i.e., towards a root stimulus or randomly in soil, has been debated. Directed growth would most likely make optimum use of energy supplies in the spore and would increase the number of infective hyphal strands which reach a host [37]. Powell [38] using the buried slide technique in partially sterilized soil, demonstrated little or no attraction of VAM hyphae to root until random contact occurred, except with hyphae from honey-colored spores (*Acaulospora laevis*), which frequently grow toward the roots. Koske [39] has demonstrated chemotactic attraction of hyphae of *Gigaspora margarita* to host roots suspended above germinating spores; the attractant is probably a volatile substance. Whether such chemotactic substances are produced under field conditions and can direct mycelial growth in the field has not been studied.

B. Passive Dissemination

Passive dissemination can occur through biotic agents like rodents, worms, insects, and birds or through abiotic agents like wind and water.

Dissemination Through Biotic Agents

VA mycorrhizal fungi sporulate singly in soil and also in hypogeous or epigeous sporocarps. A wide variety of animals are known to have VA mycorrhizal spores in their gut tracts or feces. These include a few arthropod and nonarthropod invertebrates, although no systematic search of these animals has been conducted. As early as 1922, endogonaceous spores were observed in the digestive tracts of millipedes [40], and more recently they have been found in grasshoppers and crickets [41] and in earthworm and ant casts [42]. Endogonaceous spores remain viable following earthworm ingestion since earthworm casts give rise to typical VAM colonization [42]. Recently, we found that VAM propagules can survive in earthworm cast stored for a period of 11 months (K. M. Harinikumar and D. J. Bagyaraj, unpublished). In this group, millipedes, earthworms, and sphecid wasps are potential vectors because they directly use soil in one way or another. Grasshoppers and crickets, however, are less likely candidates since they get spores from plant surfaces. The major contribution of some of these vectors may be that soil containing spores or spores themselves are brought to the soil surface, thus favoring further dispersal by wind if that occurs. Soils containing spores can also be brought to soil surface by the activity of mud dauber wasps, robins, or sparrows in whose nests spores have been found. Spores from swallow nests were also able to initiate typical VAM colonizations. In Kuwait, birds have been observed to feed on sporocarps of certain desert truffles and may be important spore vectors.

Spores of VA mycorrhizal fungi have been observed in the gut tracts or feces of 42 different species of mammals [43]. Fogel and Trappe [44] observed a relationship between mammal size and the size of sporocarps ingested. Shrews, jumping mice, mice, rats, lemmings, voles, and pikas are the usual vectors. Despite the large spore size ($\leqslant$400 μm), these spores pass through the rodent digestive system and remain germinable. Furthermore, the defecated spores are capable of initiating typical VAM fungal infections. Mycophagy remains the only known dispersal method for spores of Endogonaceae formed in hypogenous sporocarps.

The importance of digging and burrowing animals that do not themselves ingest fungi but bring soil to the surface has recently been illustrated by MacMahon [45]. In strip-mined areas, rodent dissemination of VAM fungal propagules has been shown to increase the rate of revegetation. At Mount St. Helens, burrowing and digging animals that bring inoculum to the surface were equally vital to revegetation [46]. When animals consume sporocarps, they must be "dusted" with spores, which adhere to their bodies. This seems to be reported only for the meadow jumping mouse, *Zapus hudsonicus* [47]. This method of dispersal, while rarely observed, may ultimately be shown to have significance. There is no requirement in this case for digestion-resistant spores. Additionally, the spores may be deposited over a broad area rather than in a discrete fecal package.

Dissemination Through Abiotic Agents

VA mycorrhizal propagules brought to the surface by a variety of vectors can be disseminated by abiotic agents like wind, water, etc. Ponder [48] reported VAM fungal spores in grasshopper and rabbit droppings. While occasional bits of soil might be ingested by these animals, they are primarily leaf feeders. The presence of VAM fungal spores in their digestive tracts or feces, therefore, implies that spores were present on leaves prior to feeding, probably as a result of wind dispersal. The presence of viable VAM fungal spores in rabbit droppings certainly implies that the spores can survive wind dissemination [48]. Taber [49] has observed VAM fungal spores in *Portulca* seed capsules, which are oriented on the plant toward the wind. VAM fungal spores appear trapped on the mucilaginous surface of seeds with the capsule and were probably wind disseminated. MacMahon and Warner [43] collected airborne VA mycorrhizal spores from modified sticky traps, thus giving direct evidence of wind dispersal of VAM spores. Recently, wind dispersal of spores up to 2 kilometers was demonstrated [50].

Surface flow of water that would cause mass flow of soil might also move spores. Rain flows as surficial sheet erosion—an ideal scenario for movement of spores at or near the soil surface. Suggestive of this is

Powell's [51] study dealing with the mycorrhizal infectivity of eroded soils in which spore density decreased with increasing soil erosion. Especially in open environments like deserts and tundras with sparse vegetation, wind and water may be more important dispersal agents than are animals.

VII. ASSOCIATION

Successful mycorrhizal formation depends on the presence of appropriate host, fungus, and environment. The stimulation of spore germination and the initial attraction of mycorrhizal fungi to plant roots appears to be nonspecific, as this can occur in the rhizosphere of nonhosts. Spores in the soil germinate even in the absence of a suitable partner, and often in the absence of any apparent stimulus, but usually in conditions that are also appropriate for plant seed germination and root growth. Powell [38] demonstrated that VAM fungal spores germinated similarly whether or not onion roots were present. That spores will germinate well in nonsterile soil in the absence of host roots is well documented [52]. Daniels and Trappe [52] observed that germination occurred equally well in the presence of nonhost or ectomycorrhizal plant roots, which could lead to reduced populations of VAM fungi in soils. If a receptive root is not encountered, the cytoplasm may be retracted from the germ tube and the spore may retain the capacity to germinate on several subsequent occasions. This multiple germination may reduce the vulnerability of the fungus to grazing by mycophagous soil microfauna [53], as well as increase the likelihood that the fungus will successfully colonize a plant.

The interaction of fungus and plant may begin well before the hypha and the root make physical contact. Germ tubes of *Gigaspora gigantea* were attracted through the air toward roots of bean or corn, presumably by volatile substances that were active over at least 10 mm [54]. Response of the VAM fungus to volatile attractants emitted by roots would increase the probability of eventual mycorrhiza formation. In addition, germinating spores of *Glomus mosseae* liberated substances having the same kind of biological activity as gibberellins and cytokinins [55]. The role of these substances in recognition and colonization under natural conditions is not known and deserves further investigation.

It is now considered that increase in root exudation associated with phosphorus deficiency are not related to the infection process [56]. It is possible, but unproven, that pectic enzymes are produced to assist penetration of the root, but we do not know if VAM fungi have such constitutive enzymes or if they are induced by root exudates or by cell wall contact. There appear to be no enzyme studies so far on external hyphae to act as controls for settling this question.

It is usually considered that if the spore germinates but does not encounter a root, it retracts cytoplasm from the hyphae and resumes the quiescent state [57]. However, hyphae from germinating spores of *Glomus caledonicum* and *Acaulospora laevis* can retain their infectivity for up to 4 months [58], but decreasing infectivity of hyphae after 4–6 weeks correlated with loss of ability to form prepenetration and penetration structures on the root. This could indicate some transient factor associated with hyphae which enhances infection.

The lack (or sparsity) of infection of Chenopodiaceae, Brassicaceae, Caryophyllaceae, and a few other families could be due to numerous reasons, ranging from a physical barrier of the cell wall, to an absence of essential nutrients, to production of toxins by the plant. An absence of essential nutrients is not likely to be the major reason, for VA endophytes can grow in the rhizosphere of nonhost plants [52,59–61]. However, in studies by Tommerup [60], spore germination and hyphal growth from spores of *G. caledonicum* placed with *Brassica napus* were much slower than when placed with *Trifolium subterraneum*. This may be due to a partial lack of nutrients from *B. napus* or production of mild toxins. Inhibition of mycorrhiza formation on susceptible plants by plantings of Brassicaceae has been recorded [61–63], lending further weight to a toxin theory. The occurrence of a wide range of S compounds in Brassicaceae and of betalins (related to phenol and anthocyanins) in Chenopodiaceae, both of which have fungistatic activity, might be important [64]. Glenn [65] suggested that these families lack functional mycorrhizae because of the presence of glucosinolates and their hydrolysis products, isothiocyanates, in and around their roots.

If the germ tube or any network of hyphae ramifying through the soil encounters a receptive root or root hairs, the hypha swells to form an appresorium [26]. The infecting hyphal branch originating from the appresorium penetrated into or between epidermal cells [66], in the zone of differentiation and elongation of the fine feeder roots. The root tips are not normally infected [67].

Within the epidermal and outermost cortical cells, the VAM fungus may form loops of hyphae, or it may traverse these cells to penetrate the next deeper layer of cells. Subsequently, the hyphae grow between or within the cortical cells, although the meristematic cells and the endodermal cells are never entered. Specialized branches enter individual cortical cells and form finely arborescent structures called arbuscules, which are apparently the sites of exchange between the fungus and the plant. The arbuscules gradually break down and their fine branches disappear. It was formerly assumed that the plant was digesting the fungal structure, and that phosphorus passed to the plant as a result of this digestion. However, Cox and Tinker [68] demonstrated that digestion of the arbuscule could not provide the plant with more than 0.065% of the phosphorus that enters the mycorrhiza [69]. This seems to indicate that there must be an active mechanism transferring phosphorus to the plant throughout the duration of the mycorrhizal relationship. Polyphosphate granules involved in P transport within the fungal cytoplasm have been seen in vacuoles in the inter- and intracellular hyphae [70] but are no longer observed in the finest branches of the arbuscules [71]. Those branches have been found to contain acid and alkaline phosphatases [72].

Individual arbuscules seem to function for a period of 4–15 days [73], and the cytoplasm of the colonized host cell remains alive, and seems in fact to be greatly stimulated, throughout this period. The nucleus, in particular, of the arbuscule-containing cells is much larger than that of noncolonized cells, and the volume of the cytoplasm is also greatly increased. Ultimately, retraction septa form in the arbuscule as the cytoplasm is drawn back, beginning with the fine branches and progressing toward the main trunk [74]. The fine branches then collapse and the cytoplasm of the root cell eventually returns to normal, though traces of the fungal wall may persist long after the active fungus has decamped.

Many, though not all, endomycorrhizal fungi that are characterized by the formation of arbuscules also form terminal and/or intercalary vesicles within the mycorrhiza. These are thin-walled, expanded structures which are not delimited by a septum and often contain large quantities of lipids. Up to 500 vesicles per cm of root in leek, resulting in the root cortex taking on the appearance of an almost solid mass of vesicles, has been observed [64]. This occurs when the root zone concerned is no longer as actively involved in absorption as in the earlier stages of colonization, and it seems probable that the cortex is no longer functional as a pathway for nutrient uptake by the plant. Since the stele is not colonized, it can, of course, still translocate substances to and from the active feeder roots. Vesicles, which may be involved in temporary storage of modified photosynthates received from the plant, are not formed by the species of *Gigaspora*.

The progress of VAM colonization in a single root system commonly follows a sigmoid curve. These sigmoid progressions have been observed in controlled environments as well as in the field [75] and show them to be an intrinsic property of root systems and not solely the result of seasonal changes. The lag phase presumably reflects the slow initial establishment of colonization, but the sharp rise in colonization is not understood. It could be due to the attainment of a critical root density allowing inter-root infection [76]. The dynamics of spread of infection within a root system have been approached in two ways [64]:

1. Mathematical modeling based on progressive measurement of infection and infection points on the growing root and certain assumptions [77]. The modeling has highlighted the need for more detailed experimental study of the biology of individual infections.
2. Detailed experimental study of particular questions, e.g., the rate of spread of individual infections, the mode of spread, the susceptibility of different parts of the root, and ages of root to infection, etc. [59].

The former approach has tended to dominate till now.

In conjunction with the development of the fungus inside the root, which is generally described as the intra-matrical phase, the fungus proliferates in the soil outside the root, this being described as the extra-matrical phase. A network of extramatrical hyphae ramifies through the soil, extending as far as 8 cm from the root [78]. This means that a mycorrhizal plant can exploit several times the volume of soil available to a nonmycorrhizal plant, since active translocation of minerals along the extra-matrical hyphae to the host plant has been convincingly demonstrated [78]. It has now been established that these fungi differ in the amount of external hyphae they produce [79]. Effectiveness of a fungus can be correlated with its ability to produce more external hyphae [80]. Two basic factors probably affect the production of external hyphae: (a) the availability of the substrate in the root [56,79] and (b) the various soil factors [81]. More studies to determine the effect of soil factors on extraradical growth are needed.

VIII. FACTORS AFFECTING THE DEVELOPMENT OF MYCORRHIZAE

A. Physical Factors

Temperature

Temperature has been shown to have a significant influence on colonization and sporulation by VAM fungi under greenhouse conditions. Higher temperatures generally result in greater root colonization and increased sporulation [82]. Studying the effects of temperature on VAM establishment, Schenck and Schroder [83] observed that maximum arbuscule development occurred near 30°C but that mycelial colonization of the root surface was greatest between 28 and 34°C. Daniels and Trappe [52] observed that optimum temperature for the germination of *Glomus* and *Acaulospora* spores is around 20–25°C, whereas *Gigaspora* had a much higher optima. These studies suggest that increased soil temperature hastens the development of VA mycorrhizae. This may explain the slow development of infection in agricultural crops in temperate soils [76] where soil temperatures are low compared to tropical soils. Since most species of VAM fungi exist worldwide, it is possible that strains and species may be temperature adapted. This is suggested by the work of Schenck et al. [84], who found that two isolates of *Glomus* from Florida germinated best at 34°C, whereas one from Washington had an optimum of 20°C.

Light

VA mycorrhizal fungi obtain their carbon source from the host plant and thus rely on both the photosynthetic ability of the plant and the translocation of photosynthates to the root. Hence, light can strongly affect the development of mycorrhizae. The stimulatory effect of light on the development of VA mycorrhizae has been shown by Furlan and Fortin [85]. Shading not only reduces root colonization and spore production but also the plant response to VA mycorrhiza [86]. Redhead [87] postulated that day length may plan an important role in VA mycorrhiza development, and this was confirmed by Daft and El-Giahmi [88]. However, the effect of light on VA mycorrhizae seems to depend on the photosensitivity of the species of host plant [87]. In fact, a photoperiod of 12 hr or more may be more important than light intensity in providing high levels of root colonization [83], but if suitable day length is provided, increased light intensity may still increase colonization [88]. It could be particularly interesting to examine this aspect in plantation crops like coffee, cardamom, etc., which are cultivated under shade trees in the tropics.

Water

A continual decline of germination of *Gigaspora gigantea* in polyethylene glycol from 80% at –0 MPa to 15% at –10 MPa was reported by Wilson [89]. *Glomus caledonicum*, *Gigaspora calospora*, and *Acaulospora laevis* spores showed high germination from –0 to 1.4 MPa, but delayed germination at –2.2 MPa [90]. These data suggest that more spore germination can probably occur at soil moisture levels at which roots do not normally grow [91].

VAM occur over a wide range of soil water contents. Colonization has been found in arid regions in xerophytes [92], very wet soils of marshes [93], and also in free-floating [10] and submerged aquatic plants [11]. It was believed earlier that under saturated conditions, O_2 concentration can inhibit VAM spore germination and root colonization [94]. Le Tacon et al. [95] observed for spores germinated in air that their subsequent growth was little affected by lack of O_2 until the oxygen tension was below 3%. The recent observations on mycorrhizae in aquatic plants suggest that perhaps in water there is adequate dissolved O_2, and concentrations of toxic substances such as Mn, H_2S, organic acids, etc., developing under anaerobic conditions are probably absent. Soil water may select for certain species of VAM fungi, which adapt to that environment [96].

B. Chemical factors

pH

Studies conducted in the laboratory on agar media suggest that good germination of VAM spore occurs between pH 6 and 7, although there are cases of fungi germinating at pH 5 and below as well as pH 8 and above [97]. Optima for germination of *Gigaspora corraloidea*, *Gigaspora heterogama*, and *Glomus mosseae* on the agar have been recorded at pH 5, 6, and 7, respectively [98].

In soil, effects of pH are difficult to evaluate since many chemical properties of soil vary with changes in pH. In soil, more than 40% germination of *Glomus epigeum* spores was found over the pH range 4.8–8.0, the optimum being 7.0 [52]. Sparling and Tinker [99] found no obvious effect of pH on infection in three grassland sites at pH 4.9, 5.9, and 6.2; however, such infection may have been due to different endophyte species. There is some evidence for differences in adaptation of strains and species of VAM fungi to pH. Lambert and Cole [100] reported that six isolates of *Glomus tenue* differed in their ability to form mycorrhizae at low pH. Daft et al. [101] reported considerable VAM colonization in plants growing in a bituminous mine spoil of pH 2.7. Natural soils of the world cover the pH range 2.8 to >10.0 [102]. Soil pH may affect the distribution of mycorrhizae in a subtle way.

Phosphorus

The addition of phosphorus affects VAM colonization of roots [103,104]. Conclusive recommendations for specific soil P levels for mycorrhizal production cannot be made. There are several reasons for this. First, it is not soil P per se that regulates mycorrhizal colonization, but rather the amount of P absorbed by the host plant [104]. Second, methods for evaluating available soil P often differ greatly, and plant tissue analysis is a far more reliable method for determining available soil P than most methods that analyze soil. Finally, since host plants vary in their ability to absorb P and mycorrhizal fungi vary in their response to P, each plant–soil–VAM symbiont system must be evaluated separately [103,104]. For example, in one sandy soil, maximum spore production by *Glomus fasciculatum* on sour orange occurred at 50 ppm added P, but this amount of P applied to a different citrus variety (*Troyer citrange*) did not encourage sporulation [104].

Tissue P is also not always a good estimate for mycorrhizal colonization because the mycorrhizae themselves influence that factor. It is thought that P influences VAM colonization by affecting concentrations of root carbohydrates [103] or the amount of root exudates [105] and for that reason the effects of P concentration may be partially overcome by other factors such as light intensity [106]. The best indicator for identifying a soil that will provide good VAM colonization appears to be the percentage of P in plants at the time of VAM colonization [103]. Recently, Sreenivasa and Bagyaraj [107] observed that rock phosphate applied at 100 ppm P level resulted in more of infective propagules of *Glomus fasciculatum* compared to bone meal and superphosphate fertilizers [108]. Long-term application of superphosphate fertilizer (15 years of 150 kg P/ha/yr) was found to result in a population of VA mycorrhizal endophytes that was little affected by subsequent additions of P [109]. It appeared that P-tolerant strains have developed during the 15-year period of superphosphate application. More work in isolating P-tolerant strains is needed, as they can be used for inoculating high-yielding varieties of field crops and horticultural crops demanding application of heavy doses of fertilizers.

Nitrogen

Hayman [110] showed that N fertilizers (188 kg N/ha as Nitro-chalk) had a large negative effect on the mycorrhizal population. This was confirmed later by Jenson and Jakobsen [111]. Alexander and Fairley [112] reported reduction in mycorrhizal colonization following application of 300 kg N/ha as $(NH_4)_2SO_4$. Menge [113] noted that daily fertilization of citrus with more than 100 ppm N as a mixture of NO_3^- and NH_4^+ retarded mycorrhizal development. Davis and Young [114] reported NO_3^- salts to be more inhibitory to VAM development than NH_4^+ salts. Comparing calcium nitrate, urea, and calcium ammonium nitrate at different levels, we found that calcium ammonium nitrate applied at 80 ppm N levels produced a maximum number of infective propagules of *Glomus fasciculatum* in association with Rhodes grass (M. N. Sreenivasa and D. J. Bagyaraj, unpublished). Levels greater than 80 ppm decreased the number of infective propagules. These results indicate that nitrogen contents of soils could greatly influence the distribution and abundance of VAM.

Micronutrients

Micronutrients such as manganese and zinc inhibit spore germination of mycorrhizal fungi [115]. Zinc and copper was shown to inhibit mycorrhizal colonization in clover, onion, maize, soybean, and pinto bean [22,116]. Zinc, copper, and manganese applied at the rate of 12, 2.5, and 40 ppm, respectively, along with Ruakura solution improved mycorrhizal root colonization, sporulation, and the most probable number of infective propagules of *Glomus fasciculatum* [117].

Acid soils contain much soluble Al, Mn, and Fe, and in neutral soils Mn and Fe can be released in large quantities when reducing conditions prevail after waterlogging [118]. Many soils are grossly polluted with heavy metals as a result of mining and smelting. Gildon [119] investigated the reclaimed site of an old lead mine at Shipham, England, and found considerable colonization in white clover growing on soil heavily contaminated with Zn and Cd. This suggests that VA endophytes can develop strains adapted to particular soil conditions. Hence, one of the prime strategies for reclamation of mine tailings should be the selection, introduction, and maintenance of suitable mycorrhizal fungi.

Pesticides

Trappe et al. [120] have reviewed the effect of pesticides on mycorrhizal fungi and mycorrhizae. Most research with pesticides has been undertaken with soil fumigants, fungicides, and nematicides which exhibit a range of activity toward VAM fungi. Fumigation of soil with biocides such as methyl bromide, chloropicrin, formaldehyde, Mylone, Vapam, and

Vorlex effectively kill endophytes in the treatment zone. Fortunately, these fungi reinvade most fumigated soils within several years. Unlike the general biocides, most nematicides such as 1,3-dichloropene (1,3-D), 1,2-dibromoethane (EDB), 1,2-dibromo-3-chloropropane (DBCP), and some of the organophosphates (Bhenomiphos) and organocarbamates (aldicarb) at recommended rates exhibit no or a slight inhibitory effect. Interestingly, some nematicides, such as DBCP and 1,3-D, can stimulate root infection in host plants. This type of response is believed to be due to control of competitive and plant-pathogenic microflora, to possible stimulation of exudates from host plant roots, or to other factors.

The systemic fungicides like thiobendazole, benomyl, and triadimefon are more toxic to these fungi. Pentachloronitrobenzene, which is not systemic, is also highly toxic. Many fungicides are probably only fungistatic to these fungi [121]. Limited research has been done on the effect of herbicides and insecticides on VAM. Paraquat and simazine are toxic to these fungi, whereas trifluralin, bromacil, and diuron are nontoxic. The insecticides Metasystox and aldrin differ in their activity on VAM fungi, the former being less toxic than the latter.

Many pesticides contain heavy metals, and the presence of such metals in soils may be responsible for poor germination of these fungi [122]. In highly acid soils, especially those of the tropics, manganese and aluminum may be more soluble and reach concentrations that limit growth of these fungi. Selecting pesticides with no adverse effect on VAM but reducing contaminants in pot cultures will be useful in the maintenance and mass production of these fungi [113; M. N. Sreenivasa and D. J. Bagyaraj, unpublished].

Salinity

Sodium and chloride ions inhibit spore germination of VAM fungi [116]. Bowen [123] found typical VAM colonization and spores in a soil with more than 5000 ppm chloride. There are several reports of VAM in maritime salt marshes [124]. It is not known whether the endophytes involved show a special adaptation to the saline conditions. Bowen [123] hypothesized that VAM fungi could counteract some forms of soil toxicity by absorbing elements harmful to plants or assisting plant tolerance of high alkalinity or high salinity in tropical soils.

Organic Matter

Organic matter influences soil structure, pH, nutrient profile, and water-holding capacity, all of which may directly and/or indirectly influence VAM development and efficiency. According to Sheikh et al. [125], endogonaceous spore population seems to be closely correlated with the level of organic matter content in Pakistani soils. Maximum spore numbers were recovered from soils containing 1–2% organic matter, and spores were sparse in soils with below 0.5% organic matter. No such correlation has been observed in temperate soils with higher (2–13%) organic matter contents, although organic manures often enhance mycorrhizal development in tropical soils [126].

An aspect of the study of organic matter that deserves attention is the impact of mycorrhizal root residues themselves on the ecology of VAM fungi in soil. Numerous mycorrhizal plants are annuals, and mycorrhizal root systems are thus continuously being incorporated into soils and degraded by soil microorganisms. Almost nothing is known regarding the fate of the endophyte mycelia outside and inside the root tissues.

Redhead [25] suggested that seasonal dieback of Sudan and Sahel savanna grasses could stimulate endogonaceous spore production, thus increasing spore populations, as observed when arable crops such as maize, barley, and wheat are harvested. Mycorrhizal root debris in soil can also be an important reservoir of inoculum. Daft et al. [127] claimed that most infections of *Endymion non-scriptus* in nature arose directly from the decaying old root systems through which the new roots grew, and Warner and Mosse [128] indicated some saprophytic ability of VAM fungi in soil that would enable them to establish a base (possibly in particles of organic material) from which they could infect a host plant. Many authors have emphasized that spores are not important for maintaining infection when colonized roots are present especially in the case of natural plant communities when VAM fungi may be nonsporing or poorly sporing types [99]. Rives et al. [129] also suggested that in areas with low annual rainfall, contact between colonized root debris and roots of uninfected plants may constitute the most efficient mode of mycorrhizal spread.

C. Biological Factors

Host Plant

The presence or absence of a host plant obviously plays an important role in whether or not colonization and subsequent sporulation will occur. Nonhost plants such as Chenopodiaceae and Brassicaceae species can become minimally colonized by VAM fungi [130], particularly when grown in the presence of host plants [61]. The influence of nonhost plants on the colonization of host plants has been studied with contradictory results. For example, the presence of nonmycorrhizal plants has resulted in reduced colonization of mycorrhizal host plants [131] possibly because of toxic nonhost exudates [65,131]. In contrast, Ocampo et al. [61] detected no reduced colonization of mycorrhizal plants cropped together with nonmycorrhizal hosts. In fact, onions became more colonized when grown with nonhost swedes than when grown alone, and similar results were observed in barley cropped together with rape. These experiments were conducted in the greenhouse. If planting host and nonhost species together results in more rapid utilization of soil nutrients because there is more root per pot, an increase in root colonization might not be unexpected. The presence of a so-called nonhost seems to be better for a mycorrhizal fungus than no plant at all [63], as VAM hyphae have been observed colonizing the root surfaces of nonhosts [59,61], thereby procuring an ecological niche in which to survive in the absence of a host root. Not only does this indicate that these nonhost roots are not releasing toxic factors into the surrounding soil, it also raises the possibility of some saprophytic growth in rhizosphere soil by VAM fungi.

That the crop species itself can exert a selective effect on which VAM species in a mixed indigenous population become predominent is well illustrated in a study by Schenck and Kinloch [132]. They found marked differences in populations of VAM fungi between different crops grown for 7 years in monoculture. Three species of *Gigaspora* were most prevalent in association with soybean roots, in contrast to cotton and peanut, where two species of *Glomus* were most numerous. The largest number of VAM species was found with sorghum.

Although VAM fungi have extremely wide host ranges [133], the existence of host preference has been suggested by many researchers [134, 135]. The preferential association between certain plant and fungal species can be evaluated with respect to combinations that provide the greatest plant growth stimulation. At present, we do not have a good explanation for the variation in mycorrhizal dependency of different host plants. One possibility is that plants with coarse and relatively fewer hairs are more dependent on mycorrhiza compared to those plants with fine roots and long root hairs [136]. Recent investigations suggest that mycorrhizal colonization could be treated as a genetic trait [137]. Like resistance or susceptibility of crop plants to infection by fungal pathogens, Krishna et al. [137] have shown that the extent of root colonization is a plant-heritable trait. There are no reports at present of exclusive colonizations between a VAM fungus and a host plant. The factors that determine host–symbiont affinities have not been studied and no doubt are of considerable importance. Reduction in plant size by pruning and defoliation of plants [138] can decrease mycorrhizal root colonization and sporulation.

VAM Fungal Efficiency

VAM fungi are not always equally infective to any one plant species, and they certainly vary in their physiological interactions with different plants and hence in their effects on plant growth. Species and strains of VAM fungi have been shown to differ in the extent to which they increase nutrient uptake and plant growth [139]. These observations have led to introduction of the term "efficient" or "effective" strains [140]. Generally those fungi that infect and colonize the root system more rapidly are considered to be "efficient" strains [141]. The ability to form extensive external hyphae and the ability to absorb P from soil solution and improve plant growth are also important in determining the effectiveness of VAM endophytes [134,142; J. Reena and D. J. Bagyaraj, unpublished]. Comprehensive studies are needed with as many species and strains as possible to see whether "better" ones exist in nature or can perhaps be created by techniques like genetic manipulation.

VAM Fungal Dormancy

Spore dormancy in VAM fungi was first reported by Tommerup [90]. *Acaulospora laevis* spores are dormant for about 6 months, *Gigaspora calospora* for 6–12 weeks, and *Gigaspora decipiens, Glomus caledonicum*, and *G. monosporum* for a few days. Dormancy periods were much shorter in dry soils than in moist soils [90]. Dormancy can be broken by desiccation or cold treatment [143]. Tommerup [90] suggests that short dormancy periods could prevent spores from germinating immediately after formation around the root; such germination would thwart their roles as infection agents for later crops. Long dormancy may protect spores against early false breaks in a season; selection pressure may have operated to select such a population of VAM fungi. More studies on dormancy of VAM fungi and its ecological implications are needed.

Other Soil Microorganisms

Of the various microorganisms colonizing the rhizosphere, VAM fungi occupy a unique ecological position, as they are partly inside and partly outside the host. The part of the fungus within the root does not encounter competition with other soil microorganisms. This advantage enables them to achieve a more functional biomass in intimate contact with the root and thus increases their chances of exerting a greater effect on plants. The different aspects of biological interactions with VAM fungi have been reviewed (see Chapter 4) [144,145]. Since aspects of this subject are covered in detail in other chapters of this book by different authors, the topic is not reviewed here except for bringing out certain general conclusions.

VAM markedly improve nodulation and nitrogen fixation by legume bacteria, mainly by providing the high phosphorus requirement for fixation process. Mycorrhizal colonization also allows introduced populations of beneficial soil organisms like *Azotobacter* and phosphate-solubilizing bacteria to maintain higher numbers than around nonmycorrhizal plants and to exert synergistic effects on plant growth. In general, VAM decrease the severity of root diseases. The various mechanisms of suppression of pathogens by VAM has also been reviewed [144]. Secilia and Bagyaraj [146] recently suggested one more possible mechanism of suppression of root pathogens by VAM fungi, i.e., stimulation of antagonistic actinomycetes in the rhizosphere. Research on parasites and predators of VAM fungi should be intensified, as they will probably play an important role in the success or failure of VAM inoculum trials in the field, as should research on survival and persistence of indigenous and introduced endophytes. Most of the biological interaction studies have been conducted in pots, and hence there is a need to exploit this potential under field conditions. To begin with, experiments could be done with plants important in agriculture, horticulture, and forestry, which are usually raised in nursery and then planted out in the field. As only small quantities of the mycorrhizal inoculum would be required, the method could be applied immediately to practical farming without much difficulty.

IX. SURVIVAL OF VAM FUNGI AND COMPETITION BETWEEN SPECIES

A. Survival

The sheer abundance of VAM fungi, in spite of their dependence on higher plants, is indicative of an impressive capacity for survival. VA mycorrhizal spores are the largest fungal spores known; they are full of nutrient reserves and their individual survival potential is high, which perhaps compensates for the relatively small numbers produced compared to many other fungi. Both spores of VAM fungi and colonized root pieces can be effective propagules initiating typical VAM colonization in host plants [35,38,147]. The greater ability of spore inoculum to withstand environmental conditions has been suggested by Hall [148]. The large thick-walled resting spores probably survive in soil for long periods even in extreme environments. The absence of a large decrease in spore numbers in soil cropped with the nonhost swede [110] suggests that they can survive at least 1 year in the field, although populations fell during the second year with a nonhost or during a fallow [76]. In semi-arid tropical soil we found significant reduction in the number of infective propagules after leaving the land fallow for one season or cultivating a nonmycorrhizal host [149]. Tommerup [150] has indicated that spores of VAM fungi can probably survive for at least 20 years in dry and 2 years in moist soil and can return to dormancy after germination if the germ tube does not encounter a suitable host [22]. They may germinate again up to 10 or more times [53] if the original germ tube is removed. In an experiment, which is still in progress, we found that VAM inoculum stored at 45°C for 2 years is still viable.

Colonized root pieces remained viable for 6 months when stored in dry conditions [147] but rapidly lost viability in moist soils [148]. Viability of mycorrhizal root fragments up to 3 years in a surface mined top soil was reported by Rives et al. [129]. Survival of VAM propagules in an Australian soil reaching 70°C during summer was recently observed by McGee et al. [151]. Survival of propagules is probably directly influenced by hyperparasites, unsuccessful attempts of colonization, proximity of host plants, etc. [46]. It is interesting to note that VAM fungi frequently sporulate in protected areas in the soil. For example, *Glomus fasciculatum* spores form in old seed coats or even within nematode cysts. *Glomus deserticola* has also been observed to sporulate flatly around the inner cap surface of small plastic ultracentrifuge tubes [46]. Perhaps these sites provide protection from parasitic microorganisms and predatory larvae.

B. Competition Between Fungal Species

Many attempts have been made over the years to determine the influence of mixed VAM fungal species on root colonization and sporulation. Whether VAM fungal species have synergistic or competitive effects on each other has been difficult to determine because of the morphological similarity of mycorrhizal structures formed by the various VAM endophytes. Only with *Glomus tenue* and a few other such species, which produce morphologically distinct structures, have such studies been possible.

Because of the experimental difficulties, the effect of multiple VAM fungal species colonizations has only been assessed in terms of the resulting plant growth response. There are numerous reports of increased plant growth from inoculation with nonindigenous VAM species even into non-sterile soils containing indigenous species. The competitive ability of *Glomus tenue* was demonstrated by Powell and Daniel [152], who observed increased plant growth when *Glomus tenue* was added to host pot cultures already colonized by other VAM fungi. Further evidence for competition between VAM fungal species came from Ross and Ruttencutter [153] who found that less colonization occurred in *Glomus macrocarpum* inoculated with peanuts and soybeans than when these hosts were inoculated with *Gigaspora gigantea* and *Glomus macrocarpum* together. The percentage of *G. macrocarpum* colonization, as measured by vesicle formation, was reduced when combined with *Gigaspora gigantea*, suggesting the better competitive ability of *G. gigantea*.

The ability of a particular fungus to successfully compete with another fungus for an infection site depends partly on the aggressiveness of the fungus. An aggressive isolate or fungus is one that is capable of maintaining infectivity in a mixed inoculum relative to its infectivity from a single inoculum, better than the fungus with which it is competing [89]. The intensity of infection depends on inoculum density, along with the strategies used for spread of infection, and is sensitive to all of the host, soil, and climatic factors affecting fungal interactions. Prior occupation of a root by one fungus also can reduce infection by a second colonizer, but the ability to exclude a competitor is fungus dependent. The host also can influence the success of a fungus by inhibiting an increase in secondary entry points if they would be in close proximity to a primary colonized site; thus, the host can prevent potentially viable propagules from forming infection points. Use of techniques like gel electrophoresis [154], enzyme-linked immunosorbent assay (ELISA) [155], and fluorescent antibody technique [156] will help to understand the competition between different VAM fungi in colonizing a root and to evaluate mixed fungal inoculations versus inoculation with a single fungus.

X. NATURAL ECOSYSTEMS AND VAM

The great majority of research on mycorrhizal ecology has been autecological. Research on the synecological roles of mycorrhizae has been reported only in the last few years [157]. Mycorrhizae or lack of mycorrhizal fungi or hosts can affect the composition of plant communities. The ability of plant species to grow at a particular site is governed by climatic and edaphic factors. Among the latter, nutrient availability is one of the most important, and the ability to form mycorrhiza improves the growth potential in nutritionally marginal sites. The ability of a species with a relatively restricted root system, e.g., many trees and legumes [158], to establish in such environments in competition with less demanding species or ones with more extensive root systems, e.g., many grasses, can be contingent on mycorrhizal infection. Bowen [123] discusses ways in which competition for scarce nutrients among species in a mixed community can be reduced. Thus, in a mixed community, the more sparsely rooted plants may derive more benefit from the mycorrhizal infection and thus improve their competitive ability [159]. Antagonistic and synergistic relationships can occur between different plant species and their mycorrhizal fungi in mixed communities [22]. Although mycelial connections between disparate plant species presumably occur, since it is possible to infect one from the other, it is exceedingly difficult to demonstrate this in practice, and the frequency of such connections remains to be established [160].

The influence of mycorrhizae on plant succession may depend on the response of individual species. The dependence of plants on VA mycorrhizal infection forms a continuum with species that are obligately mycotrophic, facultatively mycotrophic, and nonmycotrophic [161]. Some tropical forest species die without their fungal symbionts and thus are obligately mycotrophic [162], whereas temperate grassland species are mainly facultatively mycotrophic or nonmycotrophic [161]. Individual facultative mycotrophs show different degrees of response to infection [163]. Examining the response to infection by species from different stages of succession may further implicate mycorrhizae as a mechanism of succession. Mycorrhizae is seen to play an important role in plant establishment succession on disturbed sites [164]. For more information on VAM and natural ecosystems, readers may consult Janos [6,161], Allen [164], and Miller [165].

XI. INFLUENCE OF MAN ON VAM

Most of man's activities on the land, ranging from organized agriculture and forestry to such disruptive processes as mining and road building, will have effects on the extent of mycorrhizal infection and on the population of mycorrhizal fungi. It is likely that development of agriculture has greatly modified the distribution of mycorrhizal fungi. Relatively little is known about the changes in VA endophyte populations when virgin soil is brought into cultivation. Janos [162] found that a tropical rain forest was established with difficulty in swampy areas that had been colonized mainly by *Carex* (a nonmycorrhizal genus). He attributed the difficulty in establishment to lack of VA endophytes in the soil. Crush [33] studied the effectiveness of endomycorrhizal populations during pasture development, and he concluded that there is a rapid increase in the effectiveness of soil endophyte populations as pasture development proceeds.

Prolonged monoculture of agricultural crops can lead to a build up of VAM or to a decline according to host species and fertilizer treatment, etc. [130]. Continuous wheat farming in western Nebraska shifted species dominance from *G. fasiculatum* in native grass land to *G. mosseae* [166]. Fallow tillage, intercropping, crop rotations, soil fumigation, irrigation, use of pesticides, breeding for disease resistance, etc., can alter mycorrhizal populations. Man himself, or using of agricultural implements, domesticated animals, etc., can aid dispersal of VAM fungi. Major disturbances of the surface soil involving removal and storage, as in open cast mining, gravel extraction, and road building, will reduce mycorrhizal populations, at least temporarily [167].

Man also deliberately introduces mycorrhizal fungi through inoculation to improve plant growth. These introductions will alter the native VA mycorrhizal composition. There is very little information available on the competitive ability and persistence of these introduced fungi to agricultural and forest ecosystem. It has been suggested that a mixed inoculum containing fungi with differing ecological strategies might give more consistent and permanent results in promoting plant growth [168]. Though commercial production of VAM fungi is not that easy, methods are available to mass-produce VAM fungi [113,169; D. J. Bagyaraj, unpublished]. In view of the widespread interest in VA mycorrhiza that has arisen in recent years, it is likely that methodology, understanding, and biotechnological application of this mutalistic symbiont will continue to improve in the coming years. Man is trying to use these beneficial fungi to increase the productivity of food, fuel, and fiber. Without a clear understanding of the ecology of VA mycorrhizal fungi, man's ability to manipulate the mycorrhizal symbiosis for maximizing productivity will be severely limited.

XII. CONCLUSIONS

It is now established that VA mycorrhizae improve plant growth, mainly through phosphorus nutrition; other beneficial effects are in the biological control of root pathogens, biological nitrogen fixation, hormone production, and greater ability to withstand water stress. Because of their unique ability to increase the uptake of phosphorus by plants, mycorrhizal fungi can be utilized as practical substitutes for phosphatic fertilizers. Several workers have stressed the potential for commercial utilization of VAM fungi in crop production. One of the impediments in conducting field experiments is the inability to grow VAM fungi on laboratory media. Another impediment to field experimentation is the fragmentary information available on the ecology of these fungi. The factors that influence their survival, germination, root colonization, sporulation, spread, and survival are only partially understood. The information gathered so far is mostly from laboratory and pot experiments. It is dangerous to draw ecological inferences from glass house trials. Hence, there is a greater need for field experiments to understand their distribution and function.

Principally there could be two strategies for managing VAM fungi. The first strategy is to develop inoculation techniques with efficient VAM fungi, adapted to the crop plant and the environment. Considerable work is being undertaken at present on this aspect. The second strategy is to manage the indigenous VAM fungi by agricultural practices in such a way that efficient native fungi are enhanced and inefficient fungi are depressed. A good deal of research is necessary before we formulate some general recommendations for this management method for each soil and crop. Therefore, manipulation of this symbiotic association to attain its full ecological and economic potential should be the goal for future studies.

ACKNOWLEDGMENTS

I thank K. M. Harinikumar, M. S. Byra Reddy, B. C. Mallesh, M. Vasanthakrishna, J. Secilia, J. Reena, and B. R. Mohan for their help in the preparation of this manuscript.

REFERENCES

1. Frank, A. B. Über die auf Werzelsymbiose beruhende Ernährung gewisser Bäume durch unterirdische Pilze. *Ber. Dtsch. Bot. Ges.*, *3*: 128–145 (1885).
2. Kendrick, B. and Berch, S. Mycorrhizae: Applications in agriculture and forestry. In *Comprehensive Biotechnology*, Vol. 4 (C. W. Robinson, ed.), Pergamon Press, Oxford, 1985, pp. 109–150.
3. Schenck, N. C. and Perez, Y. *Manual of the Identification of VA Mycorrhizal fungi*, University of Florida, Gainesville, FL, 1987, pp. 1–245.
4. Gerdemann, J. W. Vesicular-arbuscular mycorrhizae. In *The Development and Function of Roots* (J. G. Torrey and D. T. Clarkson, eds.), Academic Press, London, 1975, pp. 576–591.
5. Trinick, M. J. Vesicular arbuscular infection and soil phosphorus utilization in *Lupinus* spp., *New Phytol.*, *78*:297–304, 1977.
6. Janos, D. P. Tropical mycorrhizas. In *Tropical Rain Forest: Ecology and Management* (S. L. Sutton, T. C. Whitmore, and A. C. Chadwick, eds.), Blackwell Scientific Publications, Oxford, 1983, pp. 327–345.
7. Harley, J. L. *The Biology of Mycorrhiza*, 2nd ed., Leonard Hill, London, 1969.
8. Cooper, K. M. A field survey of mycorrhizas in New Zealand ferns, *New Zealand J. Bot.*, *14*:169–181, 1976.
9. Parke, J. L. and Linderman, R. G. Association of vesicular-arbuscular mycorrhizal fungi with the moss *Funaria hygrometrica*, *Can. J. Bot.*, *58*:1898–1904, 1980.
10. Bagyaraj, D. J., Manjunath, A., and Patil, R. B. Occurrence of vesicular arbuscular mycorrhizas in some tropical aquatic plants, *Trans. Br. Mycol. Soc.*, *72*:164–167, 1979.

11. Clayton, J. S. and Bagyaraj, D. J. Vesicular arbuscular mycorrhizas in submerged aquatic plants of New Zealand, *Aquatic Botany, 19*: 251–262, 1984.
12. Graw, D. and Rehm, S. Vesikular-arbuskular Mykorrhiza in den Fruchttragern von *Arachis hypogea, L. Z. Acker-Und Pflanzenbau., 144*:75–78, 1977.
13. Selvaraj, T., Kannan, K., and Lakshminarasimhan, C. Vesicular-arbuscular mycorrhizal fungi in roots and scale-like leaves of *Canna indica* L. (Cannaceae). *Curr. Sci., 55*:728–730, 1986.
14. Lamont, B. The morphology and anatomy of proteoid roots in the genus *Hakea*. *Aust. J. Bot., 20*:155–174, 1972.
15. Benjamin, R. K. Zygomycetes and their spores. In *The Whole Fungus* (B. Kendricks, ed.), National Museums of Canada, Ottawa, 1979, pp. 573–616.
16. Trappe, J. M. and Schenck, N. C. Taxonomy of the fungi forming endomycorrhizae. A. Vesicular-arbuscular mycorrhizal fungi (Endogonales). In *Methods and Principles of Mycorrhizal Research* (N. C. Schenck, ed.), American Phytopathological Society Publication, St. Paul, MN, 1982, pp. 1–10.
17. Hall, I. R. Taxonomy of VA mycorrhizal fungi. In *VA Mycorrhiza* (C. L. Powell and D. J. Bagyaraj, eds.), CRC Press, Boca Raton, FL, 1984, pp. 57–94.
18. Hall, I. R. and Abbott, L. K. Photographic slide collection illustrating features of Endogonaceae. *Bull. Br. Mycol. Soc., 17*:54–55, 1983.
19. Walker, C. Taxonomic concepts in the Endogonaceae: Spore wall characteristics in species description. *Mycotaxon., 18*:443–455, 1983.
20. Berch, S. M. Endogonaceae: Taxonomy, specificity, fossil record, phylogeny. In *Frontiers in Applied Microbiology*, Vol. 2 (K. G. Mukerji, N. C. Pathak, and V. P. Singh, eds.), India Print House, Lucknow, 1986, pp. 161–186.
21. Walker, C. Current concepts in the taxonomy of the Endogonaceae. In *Mycorrhizae in the Next Decade: Practical Applications and Research Priorities* (D. M. Sylvia, L. L. Hung, and J. H. Graham, eds.), University of Florida Publication, Gainesville, FL, 1987, pp. 300–302.
22. Mosse, B., Stribley, D. P., and Le Tacon, F. Ecology of mycorrhizae and mycorrhizal fungi. In *Advances in Microbial Ecology*, Vol. 5 (M. Alexander, ed.), Plenum Press, New York, 1981, pp. 137–209.
23. Trappe, J. M. Biogeography of hypogeous fungi: Trees, mammals, and continental drift, Abstr. 2nd Int. Mycol. Congr. Inc. (H. E. Bigelow and E. G. Simmons, eds), University of South Florida, Tampa, 1977, p. 675.
24. Mosse, B. and Bowen, G. D. The distribution of *Endogone* spores in some Australian and New Zealand soils and in the experimental field soil at Rothamsted. *Trans. Br. Mycol. Soc., 51*:485–492, 1968.
25. Redhead, J. F. Endotrophic mycorrhizas in Nigeria: Species of the Endogonaceae and their distribution. *Trans. Br. Mycol. Soc., 69*:275–280, 1977.

26. Hall, I. R. Species and mycorrhizal infections of New Zealand Endogonaceae. *Trans. Br. Mycol. Soc.*, *68*:341–356, 1977.
27. Abbott, L. K. and Robson, A. D. The distribution and abundance of vesicular arbuscular endophytes in some western Australian soils. *Aust. J. Bot.*, *25*:515–522, 1977.
28. Abbott, L. K. and Robson, A. D. A quantitative study of the spores and anatomy of mycorrhizas formed of *Glomus*, with reference to its taxonomy. *Aust. J. Bot.*, *27*:363–375, 1979.
29. Mosse, B. Plant growth responses to vesicular-arbuscular mycorrhiza. X. Responses of *Stylosanthus* and maize to inoculation in unsterile soils. *New Phytol.*, *78*:277–288, 1977.
30. Sparling, G. P. and Tinker, P. B. Mycorrhizal infection in Penning grassland. II. Effects of mycorrhizal infection on growth of some upland grasses on irradiated soil. *J. Appl. Ecol.*, *15*:951–958, 1978.
31. Powell, C. L. Inoculation of white clover and ryegrass seed with mycorrhizal fungi. *New Phytol.*, *83*:81–85, 1979.
32. Abbott, L. K. and Robson, A. D. Infectivity and effectiveness of five endomycorrhizal fungi: Competition with indigenous fungi in field soils. *Aust. J. Agric. Res.*, *32*:621–630, 1981.
33. Crush, J. R. Changes in effectiveness of soil endomycorrhizal fungal populations during pasture development. *New Zealand J. Agric. Res.*, *21*:683–685, 1978.
34. Powell, C. L. Spread of mycorrhizal fungi through soil. *New Zealand J. Agric. Res.*, *22*:335–339, 1979.
35. Warner, A. and Mosse, B. Factors affecting the spread of vesicular-arbuscular mycorrhizal fungi in soil. I. Root density. *New Phytol.*, *90*:529–536, 1982.
36. Mosse, B., Warner, A., and Clarke, C. A. Plant growth response to vesicular-arbuscular mycorrhiza. XIII. Spread of an introduced VA endophyte in the field and residual growth effects of inoculation in the second year. *New Phytol.*, *90*:521–528, 1982.
37. Wallace, H. R. Dispersal in time and space: Soil pathogens. In *Plant Disease: An Advanced Treatise*, Vol. 2 (J. G. Horsefall and E. B. Cowling, eds.), Academic Press, New York, 1978, pp. 181–202.
38. Powell, C. Development of mycorrhizal infections from *Endogone* spores infected root segments. *Trans. Br. Mycol. Soc.*, *66*:439–445, 1976.
39. Koske, R. E. Attraction of germtubes of *Gigaspora gigantea* to plant roots by a volatile mechanism. *5th North Am. Conf. Mycorrhizae*, 18, 1981.
40. Thaxter, R. A revision of the Endogonaceae. *Proc. Am. Acad. Sci.*, *57*:291–351, 1922.
41. Hansen, R. M. and Uckert, P. N. Dietary similarity of some primary consumers. *Ecology*, *51*:641–680, 1970.
42. McIlveen, W. P. and Cole, H., Jr. Spore dispersal of Endogonaceae by worms, ants, wasps and birds. *Can. J. Bot.*, *54*:1486–1489, 1976.
43. MacMahon, J. A. and Warner, A. Dispersal of mycorrhizal fungi: Processes and Agents. In *VA Mycorrhizae and Reclamation of Arid and Semi Arid Lands* (S. E. Williams and M. F. Allen, eds.), University of Wyoming Publication, Laramie, 1984, pp. 24–41.

44. Fogel, R. and Trappe, J. M. Fungus consumption (Mycophagy) by small animals. *Northwest Sci.*, *52*:1–31, 1978.
45. MacMahon, J. A. Mount St. Helens revisited. *Nat. Hist.*, *91*:14, 1982.
46. Hetrick, B. D. Ecology of VA mycorrhizal fungi. In *VA Mycorrhiza* (C. L. Powell and D. J. Bagyaraj, eds.), CRC Press, Boca Raton, FL, 1984, pp. 35–55.
47. Whitaker, J. O., Jr. *Endogone, Hymenogaster* and *Melanogaster* as small mammal foods. *American Midland Naturalist*, *67*:152–156, 1962.
48. Ponder, F., Jr. Rabbits and grasshoppers: Vectors of endomycorrhizal fungi on new coal mine spoil. *Forest Service Res. Note.* Washington, D.C. Spl. Issue No. 1, 1980.
49. Taber, R. A. Occurrence of *Glomus* spores in weed seeds in soil. *Mycologia*, *74*:515–520, 1982.
50. Warner, N. J., Allen, M. F., and MacMahon, J. A. Dispersal agents of vesicular-arbuscular mycorrhizal fungi in a disturbed arid ecosystem. *Mycologia*, *19*:721–729, 1987.
51. Powell, C. L. Mycorrhizal infectivity of eroded soils. *Soil Biol. Biochem.*, *12*:247–250, 1980.
52. Daniels, B. A. and Trappe, J. M. Factors affecting spore germination of the vesicular-arbuscular mycorrhizal fungus, *Glomus epigaeus*. *Mycologia*, *72*:457–471, 1980.
53. Koske, R. E. Multiple germination by spores of *Gigaspora gigantea*. *Trans. Br. Mycol. Soc.*, *76*:328–330, 1981.
54. Koske, R. E. Evidence for a volatile attractant from plant roots affecting germ tubes of a VA mycorrhizal fungus. *Trans. Br. Mycol. Soc.*, *79*:305–310, 1982.
55. Barea, J. M. and Azcon-Aguilar, C. Production of plant growth regulating substances by the vesicular-arbuscular mycorrhizal fungus *Glomus mosseae*. *Appl. Environ. Microbiol.*, *43*:810–813, 1982.
56. Schwab, S. M., Menge, J. A., and Leonard, R. T. Comparison of stages of vesicular-arbuscular mycorrhiza formation in sudangrass grown at two levels of phosphorus nutrition. *Am. J. Bot.*, *70*: 1225–1232, 1983.
57. Mosse, B. Vesicular-arbuscular mycorrhiza research for tropical agriculture. *Res. Bull. Hawaii Inst. Trop. Agric. Hum. Resources*, 194, 1981.
58. Rovira, A. D. and Bowen, G. D. The effects of microorganisms upon plant growth. II. Detoxication of heat sterilized soils by fungi and bacteria. *Pl. Soil.*, *25*:129–142, 1966.
59. Bevege, D. I. and Bowen, G. D. *Endogone* strain and host plant differences in development of vesicular-arbuscular mycorrhizas. In *Endomycorrhizas* (F. E. Sanders, B. Mosse, and P. B. Tinker, eds.), Academic Press, London, 1975, pp. 77–86.
60. Tommerup, I. C. Development of infection by a vesicular-arbuscular mycorrhizal fungus in *Brassica napus* L. and *Trifolium subterraneus* L. *New Phytol.*, *98*:487–495, 1984.
61. Ocampo, J. A., Martin, J., and Hayman, D. S. Influence of plant interactions on vesicular-arbuscular mycorrhizal infections. I. Host and nonhost plants grown together. *New Phytol.*, *84*:27–35, 1980.
62. Powell, C. L. Effect of kale and mustard crops on response of white clover to VA mycorrhizal inoculation in pot trial. *New Zealand J. Agric. Res.*, *25*:461–464, 1982.

63. Harinikumar, K. M. and Bagyaraj, D. J. Effect of crop rotation on native vesicular-arbuscular mycorrhizal propagules in soil. *Pl. Soil.*, *110*:77–80, 1988.
64. Bowen, G. D. The biology and physiology of infection and its development. In *Ecophysiology of VA Mycorrhizal Plants* (G. R. Safir, ed.), CRC Press, Inc., Boca Raton, FL, 1987, pp. 27–58.
65. Glenn, M. G. VA mycorrhizal infection in *Brassica*, a nonhost. *Phytopathology*, *72*:951, 1982.
66. Gianinazzi-Pearson, V., Morandi, D., Dexheimer, J., and Gianinazzi, S. Ultrastructural and ultracytochemical features of a *Glomus tenuis* mycorrhiza. *New Phytol.*, *88*:633–640, 1981.
67. Hayman, D. S. Endomycorrhizae. In *Interactions Between Non-Pathogenic Soil Microorganisms and Plants* (Y. R. Dommergues and S. W. Krupa, eds.), Elsevier, Amsterdam, 1978, pp. 401–442.
68. Cox, G. and Tinker, P. B. Translocation and transfer of nutrients in vesicular-arbuscular mycorrhizas. I. The arbuscule and phosphorus transfer: A quantitative ultrastructural study. *New Phytol.*, *77*:371–378, 1976.
69. Sanders, F. E. and Tinker, P. B. Phosphate flow into mycorrhizal roots. *Pestic. Sci.*, *4*:383–395, 1973.
70. Cox, G., Sanders, F. E., Tinker, P. B., and Wild, J. A. Ultrastructural evidence relating to host-endophyte transfer in a vesicular arbuscular mycorrhiza. In *Endomycorrhizas* (F. E. Sanders, B. Mosse, and P. B. Tinker, eds.), Academic Press, London, 1975, pp. 297–312.
71. Callow, J. A., Capaccio, L. C. M., Parish, G., and Tinker, P. B. Detection and estimation of polyphosphate in vesicular-arbuscular mycorrhizas. *New Phytol.*, *80*:125–134, 1978.
72. Gianinazzi, S. V., Gianinazzi-Pearson, V., and Dexheimer, J. Enzymatic studies on the metabolism of vesicular-arbuscular mycorrhiza. III. Ultrastructural localization of acid and alkaline phosphatase in onion roots infected by *Glomus mosseae* (Nicol. and Gerd.). *New Phytol.*, *82*:127–132, 1979.
73. Carling, D. E. and Brown, M. F. Anatomy and physiology of vesicular-arbuscular and non-mycorrhizal roots. *Phytopathology*, *72*:1108–1114, 1982.
74. Kaspari, H. Fine structure of the host-parasite interface in endotrophic mycorrhiza of tobacco. In *Endomycorrhizas* (F. E. Sanders, B. Mosse, and P. B. Tinker, eds.), Academic Press, London, 1975, pp. 325–334.
75. Saif, S. R. The influence of host development on vesicular-arbuscular mycorrhizae and endogonaceous spore population in field-grown vegetable crops. I. Summer grown crops. *New Phytol.*, *79*:341–348, 1977.
76. Black, R. and Tinker, P. B. The development of endomycorrhizal root systems. II. Effect of agronomic factors and soil conditions on the development of vesicular-arbuscular mycorrhizal infection in barley and on the endophyte spore density. *New Phytol.*, *83*:401–413, 1979.
77. Buwalda, J. G., Stribley, D. P., and Tinker, P. B. The development of endomycorrhizal root systems. V. The detailed pattern of

development of infection and the control of infection level by host in young leek plants. *New Phytol.*, *96*:411–427, 1984.

78. Rhodes, L. H. and Gerdemann, J. W. Phosphate uptake zones of mycorrhizal and non-mycorrhizal onions. *New Phytol.*, *75*:555–561, 1975.
79. Abbott, L. K. and Robson, A. D. Formation of external hyphae in soil by four species of vesicular-arbuscular mycorrhizal fungi. *New Phytol.*, *99*:245, 1985.
80. Scheltema, M. A., Abbott, L. K., Robson, A. D., and De'Ath, G. The spread of *Glomus fasciculatum* through roots of *Tribolium subterraneum* and *Lolium rigidum*. *New Phytol.*, *100*:105–114, 1985.
81. Rovira, A. D., Bowen, G. D., and Foster, R. C. The significance of rhizosphere microflora and mycorrhizas in plant nutrition. In *Inorganic Plant Nutrition. Encyclopedia of Plant Physiology: New Series*, Vol. 15A (A. Lauchli and R. L. Bieleski, eds.), Springer-Verlag, Heidelberg, 1983, pp. 61–78.
82. Furlan, V. and Fortin, J. A. Formation of endomycorrhizae by *Endogone calospora* on *Allium cepa* under three different temperature regimes. *Nat. Can.* (Quebec), *100*:467–477, 1973.
83. Schenck, N. C. and Schroder, V. N. Temperature response of *Endogone mycorrhiza* on soybean roots. *Mycologia*, *66*:600–605, 1974.
84. Schenck, N. C., Graham, S. O., and Green, N. E. Temperature and light effects on contamination and spore germination of vesicular-arbuscular mycorrhizal fungi. *Mycologia*, *67*:1189–1192, 1975.
85. Furlan, V. and Fortin, J. A. Effect of light intensity on the formation of vesicular-arbuscular endomycorrhizas on *Allium cepa* by *Gigaspora calospora*. *New Phytol.*, *79*:335–340, 1977.
86. Gerdemann, J. W. Vesicular-arbuscular mycorrhiza and plant growth. *Annu. Rev. Phytopathol.*, *6*:397–418, 1968.
87. Redhead, J. F. Endotrophic mycorrhizas in Nigeria: Some aspects of the ecology of the endotrophic mycorrhizal association of *Khaya grandifolia* C.D.D. In *Endomycorrhizas* (F. E. Sanders, B. Mosse, and P. B. Tinker, eds.), Academic Press, London, 1975, pp. 447–459.
88. Daft, M. J. and El-Giahmi, A. A. Effect of arbuscular mycorrhiza on plant growth. VIII. Effect of defoliation and light on selected hosts. *New Phytol.*, *80*:365–372, 1978.
89. Wilson, J. M. Comparative development of infection by three vesicular-arbuscular mycorrhizal fungi. *New Phytol.*, *97*:413–426, 1984.
90. Tommerup, I. C. Spore dormancy in vesicular-arbuscular mycorrhizal fungi. *Trans. Br. Mycol. Soc.*, *81*:37–45, 1983.
91. Bowen, G. D. Tree roots and the use of soil nutrients. In *Nutrition of Plantation Forests* (G. D. Bowen and E. K. S. Nambiar, eds.), Academic Press, London, 1984, pp. 147–167.
92. Khan, A. H. The occurrence of mycorrhizas in halophytes and xerophytes and of *Endogone* spores in adjacent soils. *J. Gen. Microbiol.*, *81*:7–14, 1974.
93. Dowding, E. S. Ecology of *Endogone*. *Trans. Br. Mycol. Soc.*, *42*: 449–457, 1959.

94. Saif, S. R. The influence of soil aeration on the efficiency of vesicular-arbuscular mycorrhizae. I. Effect of soil oxygen on the growth and mineral uptake of *Eupatorium odoratum* L. inoculated with *Glomus macrocarpus*. *New Phytol.*, *88*:649–659, 1981.
95. Le Tacon, F., Skinner, F. A., and Mosse, B. Spore germination and hyphal growth of a vesicular-arbuscular mycorrhizal fungus, *Glomus mosseae* (Gerdemann and Trappe), under decreased oxygen and increased carbon dioxide concentration. *Can. J. Microbiol.*, *29*: 1280–1285, 1983.
96. Manjunath, A., Mohan, R., Raj, J., and Bagyaraj, D. J. Vesicular-arbuscular mycorrhizas in cultivars of rice. *J. Soil. Biol. Ecol.*, *1*: 1–4, 1981.
97. Siqueira, J. O., Hubbell, D. H., and Schenck, N. C. Spore germination and germ tube growth of a vesicular-arbuscular mycorrhizal fungus *in vitro*. *Mycologia*, *74*:952–959, 1982.
98. Green, N. E., Graham, S. O., and Schenck, N. C. The influence of pH on the germination of vesicular-arbuscular mycorrhizal spores. *Mycologia*, *68*:929–934, 1976.
99. Sparling, G. P. and Tinker, P. B. Mycorrhizal infection in Pennine grassland. I. Levels of infection in the field. *J. Appl. Ecol.*, *15*: 943–950, 1978.
100. Lambert, D. H. and Cole, H., Jr. Effects of mycorrhizae on establishment and performance of forage species in mine spoil. *Agron. J.*, *72*:257–260, 1980.
101. Daft, M. J., Hacskaylo, E., and Nicolson, T. H. Arbuscular mycorrhizas in plants colonizing coal spoils in Scotland and Pennsylvania. In *Endomycorrhizas* (F. E. Sanders, B. Mosse, and P. B. Tinker, eds.), Academic Press, London, 1975, pp. 561–580.
102. Baas Becking, L. G. M., Kaplan, I. R., and Moore, D. Limits of the natural environment in terms of pH and oxidation reduction potentials. *J. Geol.*, *68*:243–284, 1960.
103. Jasper, P. A., Robson, A. D., and Abbott, L. K. Phosphorus and the formation of vesicular-arbuscular mycorrhizas. *Soil Biol. Biochem.*, *11*:501–505, 1979.
104. Menge, J. A., Steirle, D., Bagyaraj, D. J., Johnson, E. L. V., and Leonard, R. T. Phosphorus concentrations in plants responsible for inhibition of mycorrhizal infection. *New Phytol.*, *80*:575–578, 1978.
105. Graham, J. H., Leonard, R. T., and Menge, J. A. Membrane-mediated decrease in root exudation responsible for phosphorus inhibition of vesicular arbuscular mycorrhiza formation. *Plant Physiol.*, *68*:548–552, 1981.
106. Graham, J. H., Leonard, R. T., and Menge, J. A. Interaction of light intensity and soil temperature with phosphorus inhibition of vesicular arbuscular mycorrhiza formation. *New Phytol.*, *91*: 683–690, 1982.
107. Sreenivasa, M. N. and Bagyaraj, D. J. Suitable form and level of phosphorus for mass production of the VA mycorrhizal fungus, *Glomus fasciculatum*. *Zbl. Mikrobiol.*, *144*:34–36, 1989.
108. Clarke, C. and Mosse, B. Plant growth response to vesicular-arbuscular mycorrhiza. XII. Field inoculation responses of barley at two soil levels. *New Phytol.*, *87*:695–703, 1981.
109. Porter, W. M., Abbott, L. K., and Robson, A. D. Effect of rate of application of superphosphate on populations of vesicular-

arbuscular endophytes. *Aust. J. Exp. Agric. Anim. Husb., 18*: 573–578, 1978.

110. Hayman, D. S. The occurrence of mycorrhiza in crops as affected by soil fertility. In *Endomycorrhizas* (F. E. Sanders, B. Mosse, and P. B. Tinker, eds.), Academic Press, London, 1975, pp. 495–509.
111. Jensen, A. and Jakobsen, I. The occurrence of vesicular-arbuscular mycorrhiza in barley and wheat grown in some Danish soils with different fertilizer treatments. *Pl. Soil., 55*:403–414, 1980.
112. Alexander, I. J. and Fairley, R. I. Effects of nitrogen fertilization on populations of fine roots and mycorrhizas in spruce humus. *Pl. Soil., 71*:49–53, 1983.
113. Menge, J. A. Inoculum production. In *VA Mycorrhiza* (C. L. Powell and D. J. Bagyaraj, eds.), CRC Press, Inc., Boca Raton, FL, 1984, pp. 188–199.
114. Davis, E. A. and Young, J. L. Endomycorrhizal colonization of glasshouse grown wheat as influenced by fertilizer salts when banded or soil mixed. *Can. J. Bot., 63*:1196–1203, 1985.
115. Hepper, C. M. Germination and growth of *Glomus caledonius* spores: The effects of inhibitors and nutrients. *Soil. Biol. Biochem., 11*: 269–277, 1979.
116. Gildon, A. and Tinker, P. B. Interaction of vesicular-arbuscular mycorrhiza infection and heavy metals in plants. I. The effect of heavy metals on the development of vesicular arbuscular mycorrhizas. *New Phytol., 95*:247–261, 1983.
117. Sreenivasa, M. N. and Bagyaraj, D. J. Suitable level of zinc, copper and manganese for mass production of the vesicular-arbuscular mycorrhizal fungus, *Glomus fasciculatum*. *Proc. Indian Acad. Sci., 98*:135–138, 1988.
118. Ponnamperuma, F. N. The chemistry of submerged soils. *Adv. Agron., 24*:29–96, 1972.
119. Gildon, A. *The influence of vesicular-arbuscular fungus on the host plant uptake of heavy metals*. Ph.D. thesis, University of London, 1981.
120. Trappe, J. M., Molina, R., and Castellano, M. Reactions of mycorrhizal fungi and mycorrhiza formation to pesticides. *Ann. Rev. Phytopathol., 22*:331–359, 1984.
121. Nemec, S. VA mycorrhizae in horticultural systems. In *Ecophysiology of VA Mycorrhizal Plants* (G. R. Safir, ed.), CRC Press, Inc., Boca Raton, FL, 1987, pp. 193–211.
122. Hepper, C. M. and Smith, G. A. Observations on the growth of *Endogone spores*. *Trans. Br. Mycol. Soc., 66*:189–194, 1976.
123. Bowen, G. D. Mycorrhizal roles in tropical plants and ecosystems. In *Tropical Mycorrhiza Research* (P. Mikola, ed.), Oxford University Press, Oxford, 1980, pp. 165–190.
124. Mason, E. Note on the presence of mycorrhiza in the roots of salt marsh plants. *New Phytol., 27*:193–195, 1928.
125. Sheikh, N. A., Saif, S. R., and Khan, H. G. Ecology of *Endogone*. II. Relationship of *Endogone* spore population with chemical soil factors. *Islamabad J. Sci., 2*:6–9, 1975.
126. Johnson, C. R. and Michelini, S. Effect of mycorrhizae on container grown *Acacia*. *Proc. of the Florida State Hort. Soc., 87*:520–522, 1974.

127. Daft, M. J., Chilvers, M. T., and Nicolson, T. H. Mycorrhizas of the Liliflorae. I. Morphogenesis of *Endymion nonscriptus* (L.) Garcke and its mycorrhizas in nature. *New Phytol.*, *85*:181–189, 1980.
128. Warner, A. and Mosse, B. Independent spread of vesicular-arbuscular mycorrhizal fungi in soil. *Trans. Br. Mycol. Soc.*, *74*: 407–410, 1980.
129. Rives, C. S., Bajwa, M. I., and Liberta, A. E. Effects of topsoil storage during surface mining on the viability of VA mycorrhiza. *Soil Sci.*, *129*:253–257, 1980.
130. Kruckelmann, H. W. Effects of fertilizers, soils, soil tillage and plant species on the frequency of *Endogone* chlamydospores and mycorrhizal infection in arable soils. In *Endomycorrhizas* (F. E. Sanders, B. Mosse, and P. B. Tinker, eds.), Academic Press, London, 1975, pp. 511–525.
131. Iqbal, S. H. and Qureshi, K. S. The effect of vesicular-arbuscular mycorrhizal associations on growth of sunflower (*Helianthus annus* L.) under field conditions. *Biologia*, *23*, 189–196, 1976.
132. Schenck, N. C. and Kinloch, R. A. Incidence of mycorrhizal fungi on six field crops in monoculture on a newly cleared woodland site. *Mycologia*, *72*:445–455, 1980.
133. Mosse, B. The role of mycorrhiza in phosphorus solubilization. In *4th Int. Conf. Global Impacts Appl. Microbiol.* (J. S. Furtado, ed.), Sau Paulo, Brazil, 1973, pp. 543–561.
134. Bagyaraj, D. J., Reddy, M. S. B., and Nalini, P. A. Selection of an efficient inoculant VA mycorrhizal fungus for leucaena. *Forest Ecol. and Management*, *27*:791–801, 1988.
135. Mosse, B. Specificity in VA mycorrhizas. In *Endomycorrhizas* (F. E. Sanders, B. Mosse, and P. B. Tinker, eds.), Academic Press, London, 1975, pp. 469–484.
136. St. John, T. V. Root sizc, root hairs and mycorrhizal infection. A re-examination of Baylis's hypothesis with tropical trees. *New Phytol.*, *84*:483–487, 1980.
137. Krishna, K. R., Shetty, K. G., Dart, P. J., and Andrews, D. J. Genotype dependent variation in mycorrhizal colonization and response to inoculation pf pearl millet. *Pl. Soil.*, *86*:113–135, 1985.
138. Sreenivasa, M. N. and Bagyaraj, D. J. Effect of culture age and pruning on mass production of the VA mycorrhizal fungus, *Glomus fasciculatum*. *Pertanika*, *11*:143–145, 1988.
139. Powell, C. L., Metcalfe, D. M., Buwalda, J. G., and Waller, J. E. Phosphate response curves of mycorrhizal and nonmycorrhizal plants. II. Responses to rock phosphate. *New Zealand J. Agric. Res.*, *23*:477–482, 1980.
140. Abbott, L. K. and Robson, A. D. Infectivity and effectiveness of vesicular-arbuscular mycorrhizal fungi: Effect of inoculum type. *Aust. J. Agric. Res.*, *32*:631–639, 1981.
141. Munns, D. N. and Mosse, B. Mineral nutrition of legume crops. In *Advances in Legume Science* (R. J. Summerfield and A. H. Bunting, eds.), University of Reading Press, Reading, 1980, pp. 115–125.
142. Abbott, L. K. and Robson, A. D. The role of vesicular-arbuscular mycorrhizal fungi in agriculture and selection of fungi for inoculation. *Aust. J. Agric. Res.*, *33*:389–408, 1982.

143. Gemma, J. G. Physiology and ecology of vesicular arbuscular mycorrhizal fungi in a sand dune system. Ph.D. thesis, University of Rhode Island, Kingston, 1987.
144. Bagyaraj, D. J. Biological interactions with VA mycorrhizal fungi. In *VA Mycorrhiza* (C. L. Powell and D. J. Bagyaraj, eds.), CRC Press, Inc., Boca Raton, FL, 1984, pp. 131–154.
145. Linderman, R. G. Mycorrhizal interactions with the rhizosphere microflora: The mycorrhizosphere effect. *Phytopathology, 78*: 366–371, 1988.
146. Secilia, J. and Bagyaraj, D. J. Bacteria and Actinomycetes associated with pot cultures of VA mycorrhizas. *Can. J. Microbiol., 33*: 1069–1073, 1987.
147. Tommerup, I. C. and Abbott, L. K. Prolonged survival and viability of VA mycorrhizal hyphae after root death. *Soil. Biol. Biochem., 13*:431–433, 1981.
148. Hall, I. R. Soil pellets to introduce vesicular-arbuscular mycorrhizal fungi into soil. *Soil. Biol. Biochem., 11*:85–86, 1979.
149. Harinikumar, K. M. and Bagyaraj, D. J. Effect of different seasons of the year on vesicular arbuscular mycorrhizal population in soil. *Arid Soil Res. and Rehabil., 2*:139–143, 1988.
150. Tommerup, I. C. Physiology and ecology of VAM spore germination and dormancy in soil. In *Mycorrhizae in the Next Decade: Practical Applications and Research Priorities* (D. M. Sylvia, L. L. Hung, and J. H. Graham, eds.), University of Florida Publication, Gainesville, 1987, pp. 175–177.
151. McGee, P., Smith, S. E., and Smith, F. A. The survival of VA mycorrhizal propagules in arid environments. In *Mycorrhyzae in the Next Decade: Practical Applications and Research Priorities* (D. M. Sylvia, L. L. Hung, and J. H. Graham, eds.), University of Florida Publication, Gainesville, 1987, p. 159.
152. Powell, C. L. and Daniel, J. Mycorrhizal fungi stimulate uptake of soluble and insoluble phosphate fertilizer from a phosphate-deficient soil. *New Phytol., 80*:351–358, 1978.
153. Ross, J. P. and Ruttencutter, R. Population dynamics of two vesicular-arbuscular mycorrhizal fungi and the role of hyperparasitic fungi. *Phytopathology, 67*:490–496, 1977.
154. Hepper, C. M. Gel electrophoresis for identification of VAM fungi. In *Mycorrhizae in the Next Decade: Practical Applications and Research Priorities* (D. M. Sylvia, L. L. Hung, and J. H. Graham, eds.), University of Florida Publication, Gainesville, 1987, pp. 308–310.
155. Aldwell, F. E. B. and Hall, I. R. Monitoring spread of *Glomus mosseae* through soil infested with *Acaulospora laevis* using serological and morphological techniques. *Trans. Br. Mycol. Soc., 87*: 131–134, 1986.
156. Kough, J., Malajczuk, N., and Linderman, R. G. Use of indirect immunofluorescent techniques to study the vesicular-arbuscular fungus *Glomus epigeum* and other *Glomus* species. *New Phytol., 94*: 57–62, 1983.
157. Trappe, J. M. Mycorrhizae and productivity of arid and semi arid rangelands. In *Advances in Food Producing Systems for Arid and Semi Arid Lands* (J. T. Manassah and E. J. Briskey, eds.), Academic Press, New York, 1981, pp. 581–599.

158. Barley, K. P. The configuration of the root system in relation to nutrient uptake. *Adv. Agron.*, *22*:159–201, 1970.

159. Christie, P., Newman, E. I., and Campbell, R. The influence of neighbouring grassland plants on each other's endomycorrhizas. *Soil. Biol. Biochem.*, *10*:521–527, 1978.

160. Newman, E. I. Mycorrhizal links between plants: Their functioning and ecological significance. *Adv. Ecol. Res.*, *18*:243–270, 1988.

161. Janos, D. P. Mycorrhizae influence tropical succession. *Biotropica*, *12*:56–64, 1980.

162. Janos, D. P. Vesicular-arbuscular mycorrhizae affect lowland tropical rainforest plant growth. *Ecology*, *61*:151–162, 1980.

163. Graw, D., Moawad, M., and Rehm, S. Untersuchungen zur Wirts- und Wirkungsspezifitat der VA-Mykorrhiza. *Z. Acker. Pflanzenbau.*, *148*:85–98, 1979.

164. Allen, M. F. Ecology of vesicular-arbuscular mycorrhizae in an arid ecosystem: Use of natural process promoting dispersal and establishment. In *Mycorrhizae in the Next Decade: Practical Applications and Research Priorities* (D. M. Sylvia, L. L. Hung, and J. H. Graham, eds.), University of Florida Publication, Gainesville, 1987, pp. 133–135.

165. Miller, R. M. The management of VA mycorrhizae in semi-arid environments. In *Mycorrhizae in the Next Decade: Practical Applications and Research Priorities* (D. M. Sylvia, L. L. Hung, and J. H. Graham, eds.), University of Florida Publication, Gainesville, 1987, pp. 139–141.

166. Allen, M. F. and Boosalis, M. G. Effects of two species of vesicular-arbuscular mycorrhizal fungi on drought tolerances of winter wheat. *New Phytol.*, *93*:67–76, 1983.

167. Loree, M. A. J. and Williams, S. E. Vesicular-arbuscular mycorrhizae and severe land disturbance. In *VA Mycorrhizae and Reclamation of Arid and Semi Arid Lands* (S. E. Williams and M. F. Allen, eds.), University of Wyoming Publication. Laramie, 1984, pp. 1–14.

168. Hayman, D. S. The physiology of vesicular-arbuscular endomycorrhizal symbiosis. *Can. J. Bot.*, *61*:944–963, 1983.

169. Wood, T. Commercial production of VA mycorrhiza inoculum: Axenic versus non-axenic techniques. In *Mycorrhizae in the Next Decade: Practical Applications and Research Priorities* (D. M. Sylvia, L. L. Hung, and J. H. Graham, eds.), University of Florida Publication, Gainesville, 1987, p. 274.

2

MECHANISMS OF THE VESICULAR-ARBUSCULAR MYCORRHIZAL PLANT-GROWTH RESPONSE

DAVID M. O'KEEFE and DAVID M. SYLVIA *University of Florida, Gainesville, Florida*

I. INTRODUCTION

Plant responses to vesicular-arbuscular mycorrhizal fungus (VAMF) infection have received several explanations; nonetheless, improved phosphorus (P) uptake has consistently proved to be the primary factor providing a benefit. The mechanism of P uptake is, therefore, the focus of this review. The principles examined may also be applied to other poorly mobile elements in the soil, such as copper and zinc. The relationships among VAMF infection, micronutrients, and P uptake are discussed below. Nonnutritional plant responses to VAMF infection, such as disease resistance, improved water relations, salinity tolerance, and hormone production, have been proposed; however, these will not be discussed in this review, except in the sense that most of these responses are secondary effects of improved plant nutrition [1].

In 1969, Harley [2] listed the following hypotheses to account for plant growth responses to VAMF infection: (a) assimilation of gaseous nitrogen; (b) conduction of material from the soil to the host; (c) secretion of substances into the soil, which affect the availability of nutrients; (d) changes in the nature or area of the absorbing surface; and (e) production of vitamins or growth stimulators by the fungus within plant tissues.

By 1973 the search for the primary mechanism of the mycorrhizal growth response had been narrowed to improved phosphorus nutrition. Sanders and Tinker [3] listed the following hypotheses to account for the underlying mechanism of plant growth promotion: (a) morphological changes in the plant root system; (b) the ability of infected roots to absorb phosphate more rapidly than noninfected roots; (c) the ability of infected roots to absorb phosphate from sources not available to noninfected roots; and (d) uptake of phosphate by the external mycelium from labile or other sources and transport of phosphate to the root by the fungus.

Sanders and Tinker's [3,4] work supported primarily the fourth hypothesis. In addition, these authors noted the importance of the distribution of external hyphae beyond the P-depletion zones of roots and root hairs for enhancement of P uptake by VAMF-infected roots. They went on to explain that acceptance of the fourth hypothesis was contingent on the demonstration of uptake of P and its translocation to the host by VAMF hyphae. The capacity of VAMF hyphae to transport P to the plant was demonstrated by Hattingh, Gray, and Gerdemann [5] and by Rhodes and Gerdemann [6]. Sanders and Tinker [3] also stated that VAMF hyphae must have P fluxes high enough to account for any P increase in the mycorrhizal versus nonmycorrhizal plant. Based on calculated P-inflow rates, the number of hyphal entry points to the root, and the average diameters of these entry points, Sanders and Tinker [3] calculated P flux in the hyphae at the entry points to be 3.8×10^{-8} mol cm^{-2} s^{-1}. In a subsequent experiment, Pearson and Tinker [7] measured maximum P fluxes of 10^{-9} mol cm^{-1} s^{-1} in external hyphae. Tinker [8] concluded that the difference between these values was reasonable, based on the difference in experimental conditions and distance from the root where the flux was estimated.

The above-mentioned work supports the fundamental theory for the mycorrhizal plant growth response, as paraphrased from Sanders and Tinker [4] and Mosse [9], that external hyphae form a better-distributed surface for abosorbing P from the soil solution than do roots alone. The ability of these fungi to increase the solubility of P in the rhizosphere, as suggested by Harley's third hypothesis, is still plausible. Sanders and Tinker's second hypothesis, that the uptake kinetics of the mycorrhizal root system are different from those of the root alone, also deserves further consideration.

Silberbush and Barber's [10] work, using a mathematical model to investigate P-uptake processes, demonstrates the importance of distribution of the absorbing surface for P-uptake. However, additional parameters were identified in their model, such as the growth (extension) rate and radius of the abosorbing organ. The role of radii was implicit in Sanders and Tinker's [3,4] and Tinker's [8,11] papers. Silberbush and Barber [10], however, more fully detailed this relationship.

While the fundamental hypothesis that the VAMF-infected root system is a "better distributed surface for absorbing P from the soil solution" is undoubtedly correct, Harley's fourth hypothesis [2], stating that "a change in the nature or area of the absorbing surface occurs," is more flexible. The latter hypothesis does not exclude the possibilities of (a) rhizosphere/hyphasphere alterations, (b) differences in extension rates between roots and hyphae, and (c) extended longevity of the P-absorbing capacity of mycorrhizal roots relative to nonmycorrhizal roots. Also, it would be beneficial to expand on the concept of a better-distributed hyphal network since little is known about the distribution of external hyphae.

Mechanisms that have been contemplated for improved P uptake by VAMF-infected roots since Tinker's [8,11] reviews include: (a) better distribution of the abosorbing network; (b) more favorable geometry (smaller radii) of hyphae relative to roots; (c) greater surface area; (d) different uptake kinetics; (e) faster extension rate; (f) chemical alteration of rhizosphere/hyphasphere; (g) increased functional longevity; (h) mineralization of organic phosphorus; (i) exploration of smaller pore spaces; (j) greater carbon-use efficiency of VAMF hyphae vs. root hairs; and (k) differences in rhizosphere populations. This review will summarize current knowledge relevant to these hypotheses and will also attempt to identify pertinent areas for future research.

II. TEMPORAL CONSIDERATIONS

In order to understand the mechanism of the VAMF plant-growth response, the timing of the response must be determined [12]. Growth increase in young plants may persist throughout their lives, therefore, multiple harvests [12,13] or repeated, nondestructive measurements are necessary to identify the onset of the VAMF plant-growth response.

To emphasize the importance of timing, the VAMF plant-growth response can be considered in two time frames. The simplest case would be a short-term, glasshouse or growth-chamber study. The other extreme would be the long-term response in natural environments. The underlying mechanism may or may not be the same for long- and short-term responses. In the short term, the VAMF response may be the result of either a transitory event or a continuous phenomenon. In the case of natural environments, there is evidence that the VAMF effect is transitory and varies with the degree of drought stress [14] or with the season [15].

A chronology of events leading to the VAMF plant-growth response, adapted from Gueye et al. [16] and Snellgrove et al. [17], might be as follows: (a) infection and proliferation of external hyphae; (b) elevation of tissue P levels; and (c) growth increase. While this sequence of events is simplistic and perhaps obvious, it points to the need to take measurements prior to the time the growth response becomes apparent in the shoot.

The response of the cowpea to VAMF infection appears to follow this proposed chronology (Fig. 1). The proliferation of external hyphae at 20–25 days after planting is in agreement with observations by Owusu-Bennoah and Wild [18], who observed that changes in P-depletion zones between mycorrhizal and nonmycorrhizal onion roots occurred within 21 days. Sanders et al. [13] reported that P inflow was higher in some VAMF-inoculated treatments than for noninoculated controls after 17 days, followed by shoot dry weight differences between 30 and 40 days. Interestingly, the authors noted that increased P inflow occurred while colonization

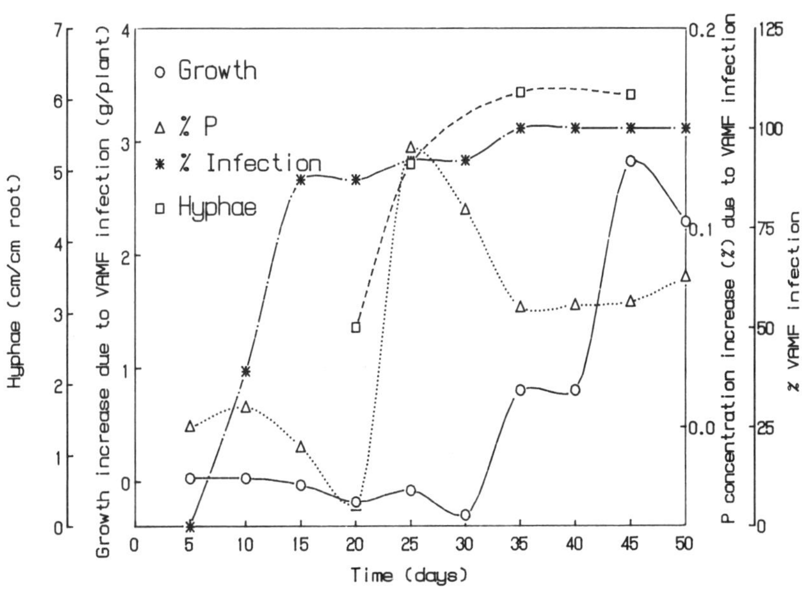

FIGURE 1. Chronology of the mycorrhizal response in cowpea. (From Ref. 16.)

intensity and external hyphae proliferation were relatively low. This suggests that even small increases in external hyphae can increase P inflow. In their experiments, as well as in the experiments of Sanders and Tinker [3], P inflow into mycorrhizal roots exceeded that into nonmycorrhizal roots throughout the experiment, suggesting that, at least in this case, the mycorrhizal response may be a continuous phenomenon. The mycorrhizal response occurs when P inflow is greater for the mycorrhizal plant than for the nonmycorrhizal plant, and also when P inflow exceeds the diffusion limits to the root alone [3].

III. MECHANISTIC HYPOTHESES FOR IMPROVED P UPTAKE

A. Comparative Geometries of Roots and Hyphae

Nye [19] theorized that, when a nutrient is deficient in the soil solution, the critical root parameter for its uptake is surface area and not volume. Factors that increase the surface area of the root, such as root hairs and mycorrhizal hyphae, will be advantageous to the absorption of nutrients present in low concentration in the soil solution. This relationship is demonstrated in the model of Silberbush and Barber [10], which predicts nutrient uptake based on nutrient supply characteristics of soil and several root parameters (Table 1). A sensitivity analysis of their model demonstrated that root growth rate (K) and root radius (r_0) are the two most important root parameters for P uptake (Fig. 2). Phosphorus uptake increases with r_0 because surface area increases with r_0. However, increasing surface area by increasing root radii is not an efficient strategy for uptake of nutrients having low diffusion coefficients because the steepness of the diffusion gradient is inversely related to the radius of the absorbing unit, and consequently the soil solution should be less depleted at the surface of a VAMF hyphae compared to the surface of a root. These authors calculated the P gradients around mycorrhizal hyphae, root hairs, grass, and corn roots (Fig. 3), concluding that the relatively high concentration of P at the hypha surface was due to the effect of the radial geometry on nutrient flux to the root. As Barber noted [20], an absorbing structure of larger radius will have greater uptake per unit length than an absorbing structure of smaller radius, but the narrower absorbing unit will be more efficient on a surface area basis. Mycorrhizal hyphae should, therefore, be less P-stressed than roots under identical soil conditions. For this reason the relative uptake kinetics of the absorbing surfaces may be of only secondary importance to nutrient uptake. The reader should refer also to our subsequent section on uptake kinetics.

The smaller diameter of hyphae allows them to explore smaller pore spaces than roots and root hairs. This affords the mycorrhizal root system two advantages over the nonmycorrhizal root system. First, the mycorrhizal root system can explore a greater volume of soil. Second, the physical environment of the smaller pore spaces may be different from the larger pore spaces accessible to roots and root hairs. For instance, it is likely that volumetric water content will be higher in small pores than in large pores. Higher volumetric water contents will lead to increased P-diffusion coefficients.

The advantage of a long, thin absorptive system is diminished considerably if soil volume is held constant, because depletion zones then will

TABLE 1. Plant and Soil Parameters for Simulation of P Uptake – Definition and Initial Values for Williams Soybeans Growing on Raub Silt Loam

Symbol	Parameter	Initial value
D_e	Effective diffusion coefficient	2.3×10^{-9} cm^2 s^{-1}
b	Buffer power	163
C_{li}	Initial conc. in soil soln.	0.0136 mol cm^{-3}
v_0	Water influx to root	5.0×10^{-7} cm^{-3} cm^{-2} s^{-1}
r_0	Root radius	0.015 cm
r_1	Half-distance between roots	0.2 cm
I_{max}	Maximal influx rate	6.43×10^{-7} mol cm^{-2} s^{-1}
C_{min}	Minimal conc. where $I_n = 0$	0.2×10^{-4} mol cm^{-3}
K_m	Conc.-C_{min} when $I_n = 1/2\ I_{max}$	5.45×10^{-3} mol cm^{-3}
L_0	Initial root length	250 cm
k	Root growth rate	0.03 cm s^{-1}

I_n = net influx at r_0.
Source: Ref. 10.

begin to overlap [10,21]. Therefore, the mycorrhizal effect should decrease with pot size. For example, in a glasshouse experiment we found that mycorrhizal dependency (dry weight mycorrhizal plant/dry weight nonmycorrhizal plant) was reduced from 8.13 to 1.80 and from 5.67 to 1.72 using *Glomus etunicatum* and *Aculospora rugosa*, respectively, when pot size was reduced from 11,350 cm^3 to 574 cm^3 (D. M. O'Keefe and D. M. Sylvia, unpublished results). Similarly, in an experiment using field beans, Kucey and Janzen [22] concluded that P concentration and dry matter production due to VAMF infection decreased with pot size. That restricted root growth will diminish the mycorrhizal growth response should be kept in mind when interpreting data from pot studies.

Due to the advantages in P uptake inherent in narrower absorbing structures, the possibility that VAMF strains that have narrow hyphae may be more effective in promoting P uptake than strains with wide hyphae is intriguing, but has yet to be demonstrated. In fact, Lopez-Aguillon and Mosse [23] reported that *Glomus tenue*, with a hyphal diameter of <2 μm, was less effective in promoting growth of clover and sorghum than *G. fasciculatum* and *Gigaspora margarita*, which had hyphal diameters of 2–5 and 5–10 μm, respectively.

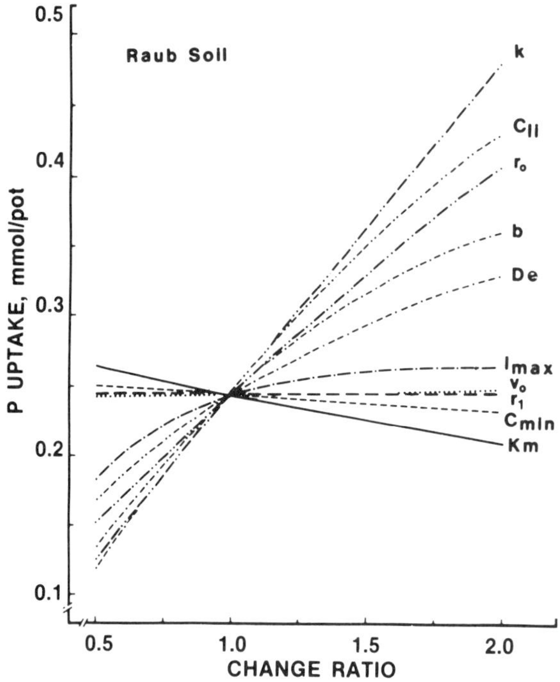

FIGURE 2. Results of a sensitivity analysis for P uptake showing the relative effect of changing each model parameter separately, from 0.5 to 2.0 times the initial value, while leaving all other parameters at the initial levels shown in Table 1. (Used with permission from Ref. 10.)

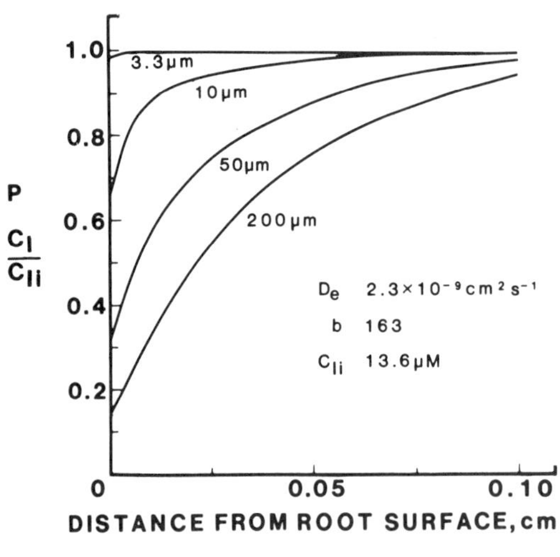

FIGURE 3. The effect of root radius on depletion of the relative P concentration in the soil next to the root surface after 10 days of uptake. Soil and plant parameters, other than root radius, were held constant. See Table 1 for definitions. (Used with permission from Ref. 10.)

B. Surface Area

Hyphae clearly increase the surface area of a VAMF-infected root system. The reported increase in surface area due to VAMF hyphae, assuming average diameters of 8 μm and 250 μm for hyphae and roots, respectively, range from >1800% to <1% (Table 2). The proportion of hyphae that were metabolically active in these studies range from 2 to 96%. Metabolically active hyphae may be a more relevant measure than total hyphae for understanding P uptake, since P uptake by fungi is believed to be an active process [24]. The wide range in the proportion of active external hyphae suggests that the proportion of VAMF external hyphae involved in P uptake can vary considerably. This should be considered when using values for total hyphal length in nutrient-uptake models. The P-inflow values reported by Sanders and Tinker [3] allow comparison of the relative increase in surface area due to VAMF hyphae to the relative increase in P inflow; a mean increase of 477% in P inflow occurred with a mean increase in surface area of only 3%. The large increase in P inflow due to a small increase in surface area indicates that there is more to the mycorrhizal growth response than a simple increase in absorptive surface area.

The high surface-to-volume ratio of VAMF hyphae compared to roots should make mycorrhizal uptake a more carbon-efficient mechanism for P uptake. Assuming the same radii for hyphae and root used above (8.0 μm and 250.0 μm for the hyphae and root, respectively), the surface to volume ratios are 25:100 for hypha and 8:1000 for roots. While the carbon cost of symbiosis may be high [25], allocation of carbon on a surface-area basis is certainly lower for VAMF hyphae than for roots.

The portion of the metabolically active hyphae that absorb P is not known. It is possible that only the growing hyphal tips actively take up P. If this is the case, an area for future research would be the role of hyphal branching in determining the effectivity of VAMF isolates.

C. Distribution of VAMF Hyphae

If mycorrhizae are to be most effective in increasing P uptake, their external hyphae should be distributed beyond the P-depletion zone around roots and root hairs. Interestingly, a model for nutrient uptake by root hairs developed by Itoh and Barber [26] demonstrates that root hair length has a large effect on P uptake. If VAMF hyphae can be considered as extended root hairs, it follows from their model that VAMF hyphae will increase P uptake considerably. Within the root-hair cylinder, i.e., the zone around the root occupied by root hairs, P is depleted uniformly and root hairs compete for P [27]. Within the root-hair cylinder it is not clear whether VAMF hyphae would add to P uptake (see Section IV. D). Using the quantitative autoradiographic technique of Bhat and Nye [27], Owusu-Bennoah and Wild [18] found that VAMF hyphae increased the measurable P-depletion zone around the roots of onion from 0.1 to 0.2 cm. The authors point out that the method used could not detect differences in P concentration of less than 10–20%, and consequently the P-depletion zone may have extended further than 0.2 cm from the root. If a hyphal system is very diffuse, it is probable that depletion gradients around roots due to hyphae will not be detected with current methods. The authors also commented on the possible role of soil physical properties in affecting the extent of hyphal spread from the root. For example, increased bulk density could restrict the spread of external hyphae.

TABLE 2. Increase in Nutrient-Absorbing Surface Area Due to VAMF Hyphae, Assuming Radii of Hyphae and Roots to be 8 μm and 250 μm, Respectively

			Surface area increase (%)		
Rhizoplane hyphae (cm/cm IRL[a])	Soil hyphae (cm/cm RL[b])	Active hyphae (%)	96% Hyphae active	2% Hyphae active	Ref.
	592.0		1,818.6	37.9	33
32.0		96-2	98.3	2.1	71
6.0			18.4	0.4	72
	30.0		92.2	1.9	73
	9.0		27.7	0.6	74
	1.3		4.0	0.1	75
	0.8		2.5	0.1	3
		96-15			76
	0.5		0.15	0.003	16

[a]Infected root length.

[b]Root length.

TABLE 3. P-Uptake Kinetics Data for Mycorrhizal and Nonmycorrhizal Soybean and Tomato Plants

	Soybean[a]		Tomato[b]	
	K_m (μM)	V_{max} (nmol sec^{-1} m^{-2})	K_m μM	V_{max} (μM P g^{-1} fresh weight × h)
Mycorrhizal	20.0	58.0	1.6	0.10
Nonmycorrhizal	3.5	19.0	3.9	0.10
Difference (%)	−83.0	−67.0	+144.0	0.00

Solution P ranges for soybean and tomato were 79−30 μM KH_2PO_4 and 79−20 μM KH_2PO_4, respectively.

[a]*Source*: Ref. 32.

[b]*Source*: Ref. 31.

In artificial systems, P uptake by VAMF hyphae at distances much greater than 0.2 cm have been demonstrated [5,6,28]. Furthermore, Warner and Mosse [29] found that VAMF hyphae could extend up to 2.0 cm from an infected root system and thereby spread to another host plant.

The above discussion assumes that the soil is homogeneous and that hyphae will grow out from the root in a uniform cylinder. However, St. John and coworkers [30] found that VAMF hyphae proliferated at nutrient-rich sites and, in effect, provided the root system with an efficient strategy for exploiting soil-heterogeneity. Their experiment examined heterogeneity of organic matter distribution; however, P distribution in soils is also heterogeneous and hyphae may proliferate at sites of higher P concentration. In a discussion of mycorrhizal response to rock phosphate application, Tinker [11] proposed that VAMF hyphae proliferated at sites of low-solubility sources of P such as rock phosphate grains.

D. Uptake Kinetics

The central question concerning P-uptake kinetics of mycorrhizal roots compared to nonmycorrhizal roots is, do mycorrhizal roots have more absorbing surface area, higher P uptake per unit surface area, or a combination of the two? If V_{max} (the maximum uptake rate) is larger for mycorrhizal roots than for nonmycorrhizal roots, then one may conclude that the mycorrhizal roots have more absorbing sites. If, on the other hand, K_m (the phosphate ion concentration at which the uptake rate is half the maximum rate) for mycorrhizal roots is lower than for nonmycorrhizal roots, then one may conclude that mycorrhizal roots have greater uptake per unit surface area. Kinetics data from two research reports are presented in Table 3. Cress et al. [31] reported that mycorrhizal roots of tomato had similar V_{max} values, but lower K_m values than nonmycorrhizal roots. They concluded that the mycorrhizal effect was due, at least in part, to a greater affinity of sites for P uptake in mycorrhizal roots, and that the increased P-absorbing surface due to external hyphae was not a major contributing factor to P uptake. Working with soybeans, Karunaratne et al. [32] came to the opposite conclusion. These authors suggested that the discrepancy between results was due to the fact that Cress et al. [31] used plants which were physiologically impaired due to age (124 days) and to severe P starvation. Karunaratne et al. [32] observed that P efflux at low P solution concentrations was lower for mycorrhizal than for nonmycorrhizal roots. This suggests that mycorrhyzal roots can remove P from soil solutions having lower P concentrations than can nonmycorrhizal roots. In effect, a mycorrhizal root would have a lower C_{min} value (the solution concentration of an ion at which efflux equals influx) in Silberbush and Barber's [10] model. The C_{min} may be important for understanding uptake kinetics and mycorrhizal response. A lower C_{min} value for mycorrhizal roots, as compared to nonmycorrhizal roots, may explain the finding of Owusu-Bennoah and Wild [18] that P concentrations in depletion zones close to mycorrhizal roots were lower than for nonmycorrhizal roots. Mycorrhizal roots also had wider depletion zones than did nonmycorrhizal roots. If C_{min} values are identical for mycorrhizal and nonmycorrhizal roots, then the zones of depletion in Owusu-Bennoah and Wild's [18] study which were attributed to uptake by the root alone (i.e., a 0.1 cm cylinder surrounding the root) should have had equal concentrations of P in the mycorrhizal and nonmycorrhizal treatments, but they did not.

E. Hyphal Growth Rates

As mentioned earlier, nutrient uptake will increase with an increase in the extension rate of the abosorbing unit [10]. With rapid growth through the soil, there will be less time available for formation of depletion zones around the root apex. The faster the absorbing unit elongates, the longer a portion of the absorbing surface will be in undepleted or less severely depleted soil. While there are no reported measurements of actual extension rates for VAMF hyphae in soil, we suggest from the data presented in Table 2 that the rate of hyphal extension ranges from 59,000 to 5 times the rate of root growth. Excluding the exceptionally high value reported by Sylvia [33], the average hyphal extension rate would be 823 times the extension rate of roots. This substantial increase in the extension rate of the absorbing unit should have a significant effect on P uptake. The importance of VAMF hyphae extension rates needs further investigation.

F. Chemical Alteration of the Rhizosphere/Hyphasphere

It is generally accepted that plants can increase P uptake by chemically modifying the rhizosphere [34–36]. Basic mechanisms by which this occurs include: (a) anion exchange of organic acids for adsorbed phosphate anions; (b) formation of Fe– and Al–organic acid complexes (chelation); and (c) pH alteration through excretion of H^+ and HCO_3^-. The extent to which mycorrhizal and nonmycorrhizal roots differ with regard to these mechanisms is not known. However, as a general group, fungi possess the capacity to alter P availability in the soil by the same mechanisms outlined above for higher plants [24].

There is extensive evidence that mycorrhizal and nonmycorrhizal plants draw from the same pools of soil P, such as organic vs inorganic P [3,9, 37–39]. This suggests that mycorrhizal and nonmycorrhizal root systems do not differ with regard to the mechanisms of solubilization discussed above. However, these scientists all compared L values between inoculated and noninoculated plants to arrive at their conclusions. An L value is a measurement of isotopic dilution or specific activity of ^{32}P in a plant grown on ^{32}P-amended soil [40]. If L values are similar between mycorrhizal and nonmycorrhizal plants, then the plants are thought to be drawing from the same pool of soil P. Bolan et al. [41] argued that forms of P that may differ in terms of availability to mycorrhizal and nonmycorrhizal plants may still be equally labeled by applied ^{32}P. The authors demonstrated that, of five mechanisms for the increased uptake of P by mycorrhizal plants, only solubilization of organic P would result in different specific activities between mycorrhizal and nonmycorrhizal plants (Fig. 4). Therefore, the conclusion that mycorrhizal and nonmycorrhizal root systems are drawing from the same pool of P may not be valid [42].

Tinker [11] provided a thorough discussion of the various hypotheses used to explain the differences between mycorrhizal and nonmycorrhizal plants in terms of uptake from insoluble P sources. One hypothesis was the chemical alteration of the rhizosphere by mycorrhizae; however, this ability has yet to be demonstrated. Exudation of organic acids, presence of specific chelators, and pH changes attributed to VAMF infections each need to be demonstrated in order to change current thinking on the role of mycorrhizae in enhanced P uptake through chemical alteration of the rhizosphere.

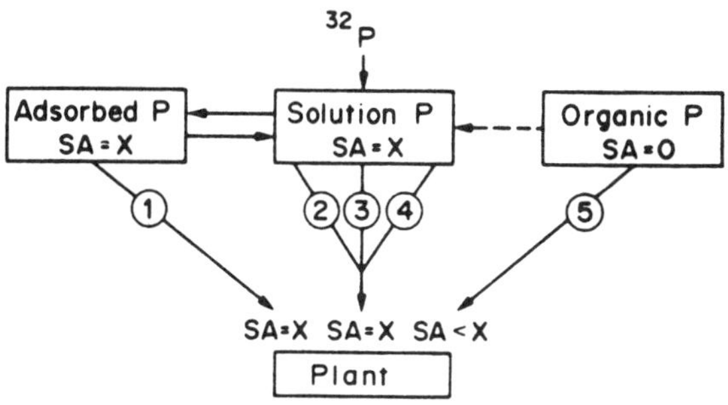

FIGURE 4. Mechanisms for the increased uptake of P by mycorrhizal plants and their effects on specific activity (SA) of P in the plants. (Used with permission from Ref. 41.)

The role of mineralization of organic phosphorus by the phosphatases of mycorrhizae is uncertain. Mycorrhizal fungi do produce phosphatases, and there have been repeated, measurable differences between mycorrhizal and nonmycorrhizal root systems of the same plant species. MacDonald and Lewis [43], using a cytochemical method, demonstrated that VAMF have acid phosphatases both inside and outside the roots. Azcon et al. [44] found that VAMF infection reduced root-surface phosphatase activity of lavender compared to nonmycorrhizal plants, but that VAMF infection had no effect on wheat phosphatase activity. Gianinazzi-Pearson and Gianinazzi [45] found that VAMF had no effect on root-surface acid-phosphatases of clover. Dodd et al. [46] reported that acid-phosphatase activity of wheat and onion varied with VAMF species, whereas phosphatase activity increased over the control for two fungi (*Glomus geosporum* and *G. mosseae*) and was not different from the control for a third fungus (*G. monosporum*). The authors concluded that increased phosphatase activity in the rhizosphere of colonized plants was not due to excretion by the fungal hyphae, but rather to increased root-surface phosphatase activity and subsequent enhanced diffusion away from the root. The degree to which these phosphatases may be important in increasing P uptake has not been assessed quantitatively; nonetheless, it appears that VAMF infection has little effect on the uptake of organic P.

G. Increased Longevity of the Absorbing Unit

Bowen [47,48] discussed the possibility that ectomycorrhizae increased the time during which a section of root functions in ion uptake. He explained that, as the radius affected by P depletion increases with time, the volume of soil supplying phosphorus would increase, allowing for continued P uptake. The root must retain its capacity to take up nutrients over an extended period in order to take advantage of this expanding zone of P supply. Presumably, the presence of hyphae promotes extended uptake capacity. It is reasonable to assume that VAMF also increase the longevity of root-uptake capacity, though there is as yet no direct experimental evidence for this. Extended uptake capacity would become important especially as plants reach maturity and as root extension slows.

H. Role of VAMF-Altered Rhizosphere Populations

Mycorrhizal infection has been shown to affect rhisozphere populations [49, 50]. Furthermore, rhizosphere organisms have been shown to affect plant growth through hormone production and changes in nutrient availability [51–53]. It is uncertain, though, if VAMF-mediated changes in rhizosphere populations have a significant effect on P uptake by plants.

There are two basic mechanisms by which VAMF infection may affect nutrient uptake through effects on rhizosphere populations. Firstly, hormone production by rhizosphere bacteria can alter root morphology [54]. However, there is no evidence to suggest that VAMF increase the total number or proportion of hormone-producing bacteria in the rhizosphere. Second, it is well established that rhizosphere bacteria can solubilize soil phosphorus. A possible synergistic effect on P uptake among mycorrhizal fungi and P-solubilizing bacteria has been demonstrated [55,56], but in another study synergism was not reported [57]. However, research to date has not demonstrated that VAMF alter populations or activities of P-solubilizing bacteria. Meyer and Linderman [58] found that a plant-growth–promoting bacterium (*Pseudomonas putida*) proliferated equally in rhizoplanes of both mycorrhizal and nonmycorrhizal clover.

The increase in nutrient-absorbing surface area (Table 2) due to VAMF hyphae should increase potential surfaces for bacterial colonization, thus supporting a larger rhizosphere population on a per-plant basis. It is possible that a larger rhizosphere population would increase the total amount of organic acids exuded from mycorrhizae and associated microorganisms into the soil. The degree to which this influences the VAMF plant response would be difficult to quantify. On the other hand, Rozycki and Strelczyk [59], working with ectomycorrhizae, found that *Streptomyces* spp. extracted from the mycorrhizosphere of pine actually produced less organic acid than did populations isolated from the rhizosphere or bulk soil.

IV. OTHER POORLY MOBILE ELEMENTS

Several micronutrients have been implicated in the VAMF plant response. Like P, these micronutrients have low diffusion coeffecients and low soil-solution levels (Table 4), so uptake of these ions could be diffusion-limited, and VAMF infection should aid in their uptake. Tinker and Gildon [60] reviewed the literature on this subject and concluded that there was clear evidence that VAMF take up Zn [61] and Cu [62]. Munjunath and Habte [63] demonstrated the role of VAMF in increasing the concentration of Cu and Zn in *Leucaena leucocephala*. Pacovsky [64] found that VAMF increased Cu and Zn concentrations in soybeans, but actually decreased Mn and Fe concentrations relative to P-fertilized, nonmycorrhizal plants. Kucey and Janzen [22] measured P, Zn, Cu, Fe, and Mn uptake by wheat

TABLE 4. Diffusion Coefficients for Phosphorus and Selected Micronutrients

Ion	Medium	Diffusion coefficient ($cm^2 s^{-1}$)	Ref.
$H_2PO_4^-$	Asv	$10^{-8}-10^{-11}$	20
Z^+	Asv	$10^{-7}-10^{-11}$	
Mn^{2+}	Asv	$5.0 \times 10^{-7}-1 \times 10^{-9}$	
Fe	Asv	3.0×10^{-10}	
Fe^{2+}, Cu^{2+}			77
Zn^{2+}, Mn^{2+}	M	1.0×10^{-6}	
Fe^{3+}	M	2.5×10^{-7}	
Zn^{2+}, Cu^{2+}	W	7.2×10^{-6}	78
H_2PO_4	W	8.9×10^{-6}	79

Asv = average soil value; M = montmorillonite; W = water.

and field beans. The VAMF infection increased uptake of P, Zn, Cu, and Fe by field beans and of P and Zn by wheat. Raju et al. [65] reported increased total uptake of P, Mn, Fe, Cu, and Zn by sorghum; however, they did not report tissue concentrations of the micronutrients. Therefore, the increase in these nutrients may have been a secondary effect of increased P uptake. A recalculation of Raju's data [65], on a concentration basis, indicates that Cu and Zn concentrations in shoots were increased by VAMF infection, while Mn and Fe concentrations were decreased.

It is likely that the mechanisms responsible for increased P uptake also apply to uptake of other diffusion-limited micronutrients. Using ^{65}Zn, Swaminathan and Verma [66] found that mycorrhizal and nonmycorrhizal roots utilized the same pool of soil Zn (labile, soil-solution Zn), as is the case with P. In addition, Lambert et al. [67] demonstrated that, in cases where high soil P decreased mycorrhizal development, Cu and Zn concentrations in corn and soybean were reduced correspondingly. Singh et al. [68] demonstrated a similar effect with Zn deficiency for wheat grown in the field. In their study, plants from high P plots contained high concentrations of P but a significant decrease in Zn concentration compared to non-P-fertilized plants. Roots from the high P plots had reduced VAMF colonization compared to the low P plots. In addition, Zn concentration in the shoot was correlated positively ($r = 0.84-0.85$) with VAMF colonization.

Increased uptake of Fe due to VAMF, on the other hand, may be due in part to the production of specific Fe chelators (siderophores). Cress and coworkers [69] found siderophore activity associated with VAMF infections of four species of fungi but no activity in nonmycorrhizal roots of the host plant, *Hilaria jamesii.*

V. CONCLUSIONS

Since Tinker's reviews [8,11], a better understanding of the effects of VAMF infection on the uptake kinetics of plant nutrients has been achieved. However, additional study, especially on relative C_{min} values between VAMF-infected and noninfected roots is required. The question of how VAMF hyphae are distributed around roots is still unclear, but new methods for studying external hyphae should improve our understanding (see Ref. 70). Data are needed on external hyphae: their distribution, lengths, growth rates, viability, radii, and the physiology of P uptake. Mycorrhizal research will be more meaningful if considered in light of the P chemistry of soil. To this end, diffusion coefficients and soil-solution levels of relevant nutrients should be reported, along with references to the methods employed. In addition to putting VAMF research data in a proper physical and chemical perspective, it is also necessary to understand the temporal framework of the VAMF effect [12].

At present, work concerning VAMF-mediated alteration of rhizosphere chemistry and microbial populations is inconclusive. This research should benefit considerably from more quantitative considerations of hyphal length, absorbing surface area, and chemical changes. Attempts should be made to incorporate these values into current nutrient-uptake models. Evaluation of rhizosphere changes with respect to nutrient stresses is required to evaluate the relevance of data concerning organic acid production, phophatase activity, and qualitative and quantitative shifts in microbial populations caused by VAMF infections.

A relatively unexplored area is the mechanism of VAMF function in field situations where soil properties are heterogeneous. While this will undoubtedly prove to be difficult, it may suggest new mechanistic hypotheses for the VAMF growth response, such as decreased P availability due to water stress, salinity, or aluminum toxicity. Sufficient groundwork has been laid in artificial systems to begin applying mechanistic concepts to the understanding of VAMF effects in the field.

ACKNOWLEDGMENTS

This work was supported partially by the Gas Research Institute as part of a cooperative program of research with the Institute of Food and Agriculture Sciences, University of Florida, Gainesville. This review is published as Florida Agricultural Experiment Station journal series no. 9372.

REFERENCES

1. Graham, J. H. Non-nutritional benefits of VAM fungi – do they exist? In *Mycorrhizae in the Next Decade: Practial Applications and Research Priorities. Proceedings of the 7th North American Conference on Mycorrhizae* (D. M. Sylvia, L. L. Hung, and J. H. Graham, eds.), University of Florida Press, Gainesville, 1987, pp. 237–239.

2. Harley, J. L. *The Biology of Mycorrhizae*. Leonard Hill, London, 1969.
3. Sanders, F. E. and Tinker, P. B. Phosphate flow into mycorrhizal roots. *Pestic. Sci.*, *4*:385–395, 1973.
4. Sanders, F. E. and Tinker, P. B. Mechanism of absorption of phosphate from soil by endogone mycorrhizas. *Nature*, *233*:278–279, 1971.
5. Hattingh, M. J., Gray, L. E., and Gerdemann, J. W. Uptake and translocation of ^{32}P-labeled phosphate to onion roots by endomycorrhizal fungi. *Soil Sci.*, *16*:383–387, 1973.
6. Rhodes, L. H. and Gerdemann, J. W. Phosphate uptake zones of mycorrhizal and non-mycorrhizal onions. *New Phytol.*, *75*:555–561, 1975.
7. Pearson, V. and Tinker P. B. Measurement of phosphorus fluxes on the external hyphae of endomycorrhizas. In *Endomycorrhizas* (F. E. Sanders, B. Mosse, and P. B. Tinker, eds.), London, Academic Press, 1975, pp. 277–287.
8. Tinker, P. B. Effects of vesicular-arbuscular mycorrhizas on higher plants. *Symp. Soc. Exp. Biol. No.*, *29*:325–350, 1975.
9. Mosse, B. Advances in the study of vesicular-arbuscular mycorrhizae. *Annu. Rev. Phytopathol.*, *11*:171–196, 1973.
10. Silberbush, M. and Barber, S. A. Sensitivity of simulated phosphorus uptake to parameters used by a mechanistic-mathematical model. *Pl. Soil*, *74*:93–100, 1983.
11. Tinker, P. B. Soil chemistry of phosphorus and mycorrhizal effects on plant growth. In *Endomycorrhizas* (F. E. Sanders, B. Mosse, and P. B. Tinkers, eds.), Academic Press, London, 1975, pp. 353–371.
12. Abbott, L. K. and Robson, A. D. The effect of mycorrhizae on plant growth. In *VA Mycorrhiza* (C. L. Powell and D. J. Bagyaraj, eds.), CRC Press, Boca Raton, FL, 1984, pp. 113–130.
13. Sanders, F. E., Tinker, P. B., Black, R. L. B., and Palmerley, S. M. The development of endomycorrhizal root systems: I. Spread of infection and growth-promoting effects with four species of vesicular-arbuscular endophyte. *New Phytol.*, *78*:257–268, 1977.
14. Fitter, A. H. Effect of benomyl on leaf phosphorus concentration in alpine grasslands: A teat of mycorrhizal benefit. *New Phytol.*, *103*:767–776, 1986.
15. Douds, D. D. and Chaney, W. R. The effect of high nutrient addition upon seasonal patterns of mycorrhizal development, host growth, and root phosphorus and carbohydrate content in *Fraximus pennsylvanica* Marsh. *New Phytol.*, *103*:91–106, 1986.
16. Gueye, M., Diem, G. H., and Dommergues, Y. R. Variation in N_2 fixation, N and P content of mycorrhizal *Vigna unguiculata* in relation to the progressive development of extraradical hyphae of *Glomus mosseae*. *MIRCEN J.*, *3*:75–86, 1987.
17. Snellgrove, R. C., Splittstoesser, W. E., Stribley, D. P., and Tinker, P. B. The distribution of carbon and the demand of the fungal symboint in leek plants with vesicular-arbuscular mycorrhizas. *New Phytol.*, *92*:75–87, 1982.
18. Owusu-Bennoah, E. and Wild, A. Autoradiography of the depletion zone of phosphate around onion roots in the presence of vesicular-arbuscular mycorrhiza. *New Phytol.* *82*:133–140, 1979.
19. Nye, P. H. The relation between the radius of a root and its nutrient-absorbing power (α). *J. Exp. Bot.*, *24*:783–786, 1973.

20. Barber, S. A. *Soil Nutrient Bioavailability.* John Wiley and Sons, Inc., New York, 1984.
21. Baldwin, J. P., Tinker, P. B. H., and Nye, P. H. Uptake of solutes by multiple root systems from soil. II. The theoretical effects of rooting density and pattern of uptake of nutrients from soil. *Pl. Soil, 36*:693–708, 1972.
22. Kucey, R. M. N. and Janzen, H. H. Effects of VAM and reduced nutrient availability on growth and phophorus and micronutrient uptake of wheat and field beans under greenhouse conditions. *Pl. Soil, 104*:71–78, 1987.
23. Lopez-Aguillon, R. and Mosse, B. Experiments on competitiveness of three endomycorrhizal fungi. *Pl. Soil, 97*:155–170, 1987.
24. Beever, R. E. and Burns, D. J. W. Phosphorus uptake, storage and utilization by fungi. *Adv. Bot. Res., 8*:127–219, 1980.
25. Pang, P. C. and Paul, E. A. Effects of vesicular-arbuscular mycorrhiza on ^{14}C and ^{15}N distribution in nodulated Faba beans. *Can. J. Soil Sci., 60*:241–250, 1982.
26. Itoh, S. and Barber, S. A. A numerical solution of whole plant nutrient uptake for soil-root systems with root hairs. *Pl. Soil, 70*: 403–413, 1983.
27. Bhat, K. K. S. and Nye, P. H. Diffusion of phosphate to plant roots in soil. I. Quantitative autoradiography of the depletion zone. *Pl. Soil, 38*:161–175, 1973.
28. Alexander, C., Alexander, I. J., and Hadley, G. Phosphate uptake by *Goodyera repens* in relation to mycorrhizal infection. *New Phytol., 97*:401–411, 1984.
29. Warner, A. and Mosse, B. Spread of vesicular-arbuscular mycorrhizal fungi between separate root systems. *Trans. Br. Mycol. Soc., 80*: 353–354, 1983.
30. St. John, T. V., Coleman, D. C. and Reid, C. P. P. Growth and spatial distribution of nutrient absorbing organs: Selective exploitation of soil heterogeneity. *Pl. Soil, 71*:487–493, 1983.
31. Cress, W. A., Throneberry, G. O., and Lindsey, D. L. Kinetics of phosphorus adsorption by mycorrhizal and nonmycorrhizal tomato roots. *Plant Physiol., 64*:484–487, 1979.
32. Karunaratne, R. S., Baker, J. H., and Barker, A. V. Phosphorus uptake by mycorrhizal and nonmycorrhizal roots of soybean. *J. Plant Nutr., 9*:1303–1313, 1986.
33. Sylvia, D. M. Spatial and temporal distribution of vesicular-arbuscular mycorrhizal fungi associated with *Uniola paniculata* in Florida foredunes. *Mycologia, 78*:728–734, 1987.
34. Jungk, A. Soil-root interaction in the rhizosphere affecting plant available phosphorus. *J. Plant Nutr., 10*:1197–1204, 1987.
35. Marschner, H., Romheld, V., and Cukmak, I. Root-induced changes of nutrient availability in the rhizosphere. *J. Plant Nutr., 10*: 1175–1184, 1987.
36. Nye, P. H. and Kirk, G. J. D. The mechanism of rock phosphate solubilization in the rhizosphere. *Pl. Soil, 100*:127–134, 1987.
37. Hayman, D. D. and Mosse, B. Plant growth responses to vesicular-arbuscular mycorrhiza.III. Increased uptake of labile P from soil. *New Phytol., 71*:41–47, 1972.
38. Owus-Bennoah, E. and Wild, A. Effects of vesicular-arbuscular mycorrhiza on the size of the labile pool of soil phosphate. *Pl. Soil, 54*:233–242, 1980.

931441

39. Powell, C. L. I. Plant growth responses to vesicular-arbuscular mycorrhiza. VII. Uptake of P by onion and clover infected with different *Endogone* spore types in ^{32}P labelled soils. *New Phytol.*, *75*: 563–566, 1975.
40. Larsen, S. Soil phosphorus. *Adv. Agron.*, *19*:151–210, 1967.
41. Bolan, N. S., Robson, A. D., Barrow, N. J. and Aylmore, L. A. G. Specific activity of P in mycorrhizal and nonmycorrhizal plants in relation to the availability of P to plants. *Soil Biol. Biochem.*, *16*: 299–304, 1984.
42. Bolan, N. S., Robson, A. D., and Barrow, N. J. Effects of vesicular-arbuscular mycorrhiza on the availability of iron phosphates to plants. *Plant and Soil*, *99*:401–410, 1987.
43. MacDonald, R. M. and Lewis, M. The occurrence of some acid phosphatases and dehydrogenases in vesicular-arbuscular fungus *Glomus mosseae*. *New Phytol.*, *80*:135–142, 1978.
44. Azcon, R., Borie, F., and Barea, J. M. Extracellular acid phosphatase activity of lavender and wheat roots as effected by phytase and mycorrhizal inoculation. In *Les Mycorhizes: Biologie et Utilization*. Les collogues de I'INRA, Vol. 13, edited by Gianinazzi, S., V. Gianinazzi-Pearson, A. Trouvelot. Paris, INRA, 1982, pp. 83–85.
45. Gianinazzi-Pearson, V. and Gianinazzi, S. The physiology of improved phosphorus nutrition in mycorrhizal plants. In *Physiological and Genetical Aspects of Mycorrhizae. First European Symposium on Mycorrhizae* (V. Gianinazzi-Pearson and S. Gianinazzi, eds.), 1986, pp. 101–109.
46. Dodd, J. C., Burton, C. C., Burns, R. C., and Jeffries, P. Phosphatase activity associated with roots and rhizosphere of plants infected with vesicular-arbuscular mycorrhizal fungi. *New Phytol.*, *107*: 163–172, 1987.
47. Bowen, G. D. Mineral nutrition of ectomycorrhizae. In *Ectomycorrhiza: Their Ecology and Physiology* (G. C. Marks and T. T. Kozlowski, eds.), Academic Press, New York, 1973, pp. 151–197.
48. Bowen, G. D. and Theodoru, C. Studies on phosphate uptake by mycorrhizas. *Proc. Int. Union Forest Res. Organ. 14th Munich*, *5*: 116–138, 1967.
49. Ames, R. N., Reid, C. P. P., and Ingham, E. R. Rhizosphere bacterial population responses to root colonization by vesicular-arbuscular mycorrhizal fungi. *New Phytol.*, *96*:555–564, 1984.
50. Meyer, J. R. and Linderman, R. G. Response of subterranean clover to dual inoculation with vesicular-arbuscular mycorrhizal fungi and a plant growth-promoting bacterium, *Pseudomonus putida*. *Soil Biol., Biochem.*, *18*:185–190, 1986.
51. Bowen, G. D. and Rovira, A. D. Microbial colonization of roots. *Annu. Rev. Phytophathol.*, *14*:121–141, 1976.
52. Rovira, A. D. Interactions between plant roots and soil microorganisms. *Annu. Rev. Microbiol.*, *19*:241–266, 1965.
53. Tinker, P. B. The role of microorganisms in mediating the uptake of plant nutrients from soil. *Pl. Soil*, *76*:77–92, 1984.
54. Tien, T. M., Gaskins, M. H., and Hubbell, D. H. Plant growth substances produced by *Azospirillum brasilense* and their effect on the growth of pearl millet (*Pennisetum americanum L.*). *Appl. Environ. Microbiol.*, *37*:1016–1024, 1979.

55. Azcon, R., Barea, J. M., and Hayman, D. S. Utilization of rock phosphate in alkaline soils by plants inoculated with mycorrhizal fungi and phosphate solubilizing bacteria. *Soil Biol. Biochem.*, *8*:135–138, 1976.
56. Raj, J., Bagyaraj, D. J., and Manjunath, A. Influence of soil inoculation with vesicular-arbuscular mycorrhizae (*Glomus fasciculatum*) and a phosphate-dissolving bacterium (*Bacillus circulans*) on plant growth and ^{32}P-uptake. *Soil Biol. Biochem.*, *13*:105–108, 1981.
57. Azcon-Aguilar, C., Gianinazzi-Pearson, V., Fardeau, J. C., and Gianinazzi, S. Effect of vesicular-arbuscular mycorrhizal fungi and phosphate-solubilizing bacteria on growth and nutrition of soybean in a neutral-calcareous soil amended with ^{32}P-^{45}Ca-tricalcium phosphate. *Pl. Soil*, *96*:3–15, 1986.
58. Meyer, J. R. and Linderman, R. J. Selective influence on populations of rhizosphere or rhizoplane bacteria and actinomycetes by mycorrhizas formed by *Glomus fasciculatum*. *Soil Biol. Biochem.*, *18*:191–196, 1986.
59. Rozycki, H. and Strzelczyk, E. Organic acids production by *Streptomyces* spp. isolated from soil, rhizosphere and mycorrhizosphere of pine (*Pinus sylvestris L.*). *Pl. Soil*, *96*:334–345, 1986.
60. Tinker, P. B. and Gildon, A. Mycorrhizal fungi and ion uptake. In *Metals and Micronutrients, Uptake and Utilization by Plants* (D. A. Robb and W. S. Pierpoint, eds.), Academic Press, New York, 1983, pp. 21–32.
61. Cooper, K. M. and Tinker P. B. Translocation and transfer of nutrients in vesicular-arbuscular mycorrhizae. II. Uptake and translocation of phosphorus, zinc. and sulfur. *New Phytol.*, *81*:43–52, 1978.
62. Timmer, L. W. and Leyden, R. F. The relationship of mycorrhizal infection to phosphorus-induced copper deficiency in sour-orange seedlings. *New Phytol.*, *85*:15–24, 1980.
63. Manjunath, A. and Habte, M. Development of vesicular-arbuscular mycorrhizal infection and the uptake of immobile nutrients in *Leucaena leucocephala*. *Pl. Soil*, *106*:97–103, 1988.
64. Pacovsky, R. S. Micronutrient uptake and distribution in mycorrhizae of phosphorus-fertilized soybeans. *Pl. Soil*, *95*:379–388, 1986.
65. Raju, P. S., Clarck, R. B., Ellis, J. R., and Maranville, J. W. Vesicular-arbuscular mycorrhizal infection effects on sorghum growth, phosphorus efficiency, and mineral element uptake. *J. Plant Nutr.*, *10*:1331–1339, 1987.
66. Swaminathan, K. and Verna, B. C. Responses of three crop species to vesicular-arbuscular mycorrhizal infection on zinc-deficient Indian soils. *New Phytol.*, *82*:481–488, 1979.
67. Lambert, D. H., Baker, D. E., and Cole, H. The role of mycorrhizae in the interactions of phosphorus with zinc, copper and other elements. *Soil Sci. Soc. Am. J.*, *43*:976–980, 1979.
68. Singh, J. P., Karamanos, R. E., and Stewart, J. W. B. Phosphorus-induced zinc deficiency in wheat on residual phosphorus plots. *Agron. J.*, *78*:668–675, 1986.
69. Cress, W. A., Johnson, G. V., and Barton, L. L. The role of endomycorrhizal fungi in iron uptake by *Hilaria jamesii*. *J. Plant Nutr.*, *9*:547–556, 1986.
70. Sylvia, D. M. Distribution, structure, and function of external hyphae of vesicular-arbuscular mycorrhizal fungi. In *Rhizosphere Dynamics*

(J. E. Box and L. C. Hammond, eds.), Westview Press, Boulder, CO, in Press.
71. Sylvia, D. M. Activity of external hyphae of vesicular-arbuscular mycorrhizal fungi. *Soil Biol. Biochem.*, *20*:39–43, 1988.
72. Diem, H. G., Gueye, M., and Dommergues, Y. R. The development of extraradical hyphae in relation to the response of cowpea to VA mycorrhizal infection. In *Physiological and Genetical Aspects of Mycorrhizae. Proceedings of the 1st European Symposium on Mycorrhizae* (V. Gianinazzi-Pearson and S. Gianinazzi, eds.), INRA, Paris, 1986, pp. 227–232.
73. Abbott, L. K. and Robson, A. D. Formation of external hyphae in soil by four species of vesicular-arbuscular mycorrhizal fungi. *New Phytol.*, *99*:245–255, 1985.
74. Abbott, L. K., Robson, A. D., and De Boer, G. The effect of phosphorus on the formation of hyphae in soil by the vesicular-arbuscular mycorrhizal fungus, *Glomus fasciculatum. New Phytol.*, *97*:437–446, 1984.
75. Tisdall, J. M. and Oades, J. M. Stabilization of soil aggregates by the root segments of ryegrass. *Aust. J. Soil Res.*, *17*:429–441, 1979.
76. Schubert, A., Marzachi, C., Mazzitelli, M., Cravero, M. C., and Bonfante-Fasolo, P. Development of total and viable extraradical mycelium in the vesicular-arbuscular mycorrhizal fungus *Glomus clarum* Nicol. & Schenck. *New Phytol.*, *107*:183–190, 1987.
77. Wilkinson, H. F. Movement of micronutrients to plant roots. In *Micronutrients in Agriculture* (R. C. Dinauer, ed.), Soil Science Society of America, Inc., Madison, WI, 1972, pp. 139–169.
78. Parsons, R. *Handbook of Electrochemical Constants.* Academic Press, New York, 1959.
79. Drew, M. C. and Nye, P. G. The supply of nutrient ions by diffusion to plant roots in soil III. Uptake of phosphate by roots of onion, leek, and rye-grass. *Pl. Soil*, *33*:545–563, 1970.

3

DROUGHT RESPONSE IN FUNGI AND MYCORRHIZAL PLANTS

R. K. GUPTA *College of Basic Sciences and Humanities, G. P. Pant University of Agriculture and Technology, Pantnagar, India*

I. INTRODUCTION

Water deficit is one of the most important environmental factors responsible for productivity and geographical distribution of plant species. About one third of the world's potentially arable land suffers from an inadequate supply of water, and on most of the remainder crop yields are periodically reduced by drought [1]. Almost all plant physiological and biochemical parameters are affected by drought. Some physiological parameters such as cell growth, cell wall synthesis, and nitrate reductase are very sensitive to water deficit stress, being reduced by a fall in water potential of only a few bars, whereas stomatal opening, photosynthesis, and respiration become affected when water potential falls around −1.2 MPa [2].

There is a fundamental difference between water economy of lower and higher plants. The terms "homoiohydrous" and "poikilohydrous" were used by Walter [3] to describe the water economics of higher and lower plants as well as fungi and bacteria. The homoiohydrous plants have adaptations which restrict water loss or maintain a good water supply from a deep root system; their water content varies very little as the atmospheric vapor pressure decreases. The poikilohydrous plants readily lose water to the atmosphere along a diffusion gradient of decreasing vapor pressure. The physiological and ecological behavior of higher plants are, to some extent, independent of water supply in the atmosphere. In lower plants, on the other hand, poikilohydry means that when atmospheric vapor pressure deficit occurs, physiological activities are curtailed. In these plants, which Buch [4] has called "pallacauophytes," water is lost rapidly to dry air, and cytoplasm itself shows a high degree of tolerance.

According to Levitt [5], plant responses to environmental stresses may be considered in terms of "avoidance" or "tolerance." When considering the response of lower plants in these terms, the poikilohydrous water economy will restrict the type of drought avoidance that can occur. Since the tissues are unable to maintain a constant water status in the presence of a vapor pressure deficit, avoidance can only be shown by organisms which grow in aquatic or constantly damp environments. While many poikilohydrous plants live in such circumstances, there are others that freely colonize habitats in which vapor pressure deficits of various levels normally occur. For these species to survive, their tissues must be able to tolerate stress. Among the vascular plants, more than 100 species of poikilohydric resurrection plants from arid regions of South Africa and Australia are known [6,7]. Most of these plants dehydrate to air dry condition. In most of these plants the tissue water potentials range between −100 and −200 MPa when collected from the field. They are able to withstand exposure to a very dry atmosphere for long periods.

A number of review papers have been published regarding various physiological aspects in relation to water deficit stress in homoiohydrous plants [2,8] and more recently in poikilohydrous plants [9,10,43,45]. Nevertheless, drought responses in free-living fungi and fungi-infected plants (mycorrhizal plants) seem not to have been reviewed most extensively. In the present article such studies are, therefore, reviewed extensively. In addition, drought responses in lichenized fungi have also been included.

II. FREE-LIVING FUNGI

Responses of water deficit stress in various poikilohydrous and homoiohydrous plants have been studied in detail [2,9−12], however, little is known about the biochemical changes in fungi. Mahajan et al. [13] found that desiccation stress in a xorophilic thermophile *Humicola* species caused marked increase in glucose content. Such increase of glucose due to desiccation is probably associated with the hydrolysis of glycogen to glucose. Phospholipids also decreased considerably with the increasing desiccation stress. Thus, it is quite likely that desiccation stress may have affected membrane integrity in *Humicola* species. Like higher plants, proline content in *Humicola* species increased considerably with the increase of desiccation stress. The amount of sterols in *Humicola* species increased markedly with the increasing desiccation stress. It is known that with increased cholesterol content, the permeability of artificial membrane to water is decreased [14]. Thus, a change in sterol content due to drought stress will probably reflect alterations in membrane permeability. Further studies are needed to investigate response of drought stress on drought-sensitive fungal species to understand mechanisms of drought tolerance in thermophilic fungi.

The ability of some microorganisms to grow under conditions of reduced water availability has been investigated by many workers [13−15]. The xerophilic yeast *Saccharomyces rouxii* accumulates glycerol in response to low external water potential [15]. Ability to accumulate polyols in response to decrease in water potential of external mycelium is a key feature in the osmoregulation of sugar-tolerant yeast [16,17]. Little information is available about the osmoregularity mechanism in xerophilic filamentous fungi. It has been assumed [15] that microorganisms use polyols, mainly glycerol, as osmoregulator in the same way that some yeasts and algae do. Using natural abundance ^{13}C nuclear magnetic resonance spectroscopy, Hocking and Norton [18] showed that in four xerophilic and one nonxerophilic fungi glycerol, erythritol, arabitol, and mannitol acted as osmoregulatory solutes. Despite considerable research on thermotolerance in fungi [9−12], there exists an untraversed scope of research regarding the mechanism of drought tolerance. Lipid content of spores and mycelia of mesophilic mucorales was lower than that of closely related thermotolerant and thermophilic species, fatty acids of thermophilic fungi were more saturated than those of corresponding mesophilic species [22]. Since fungi are poikilohydrous organisms, drought tolerance must be due to protoplasmic characteristics. Further work is required to study various protplasmic characteristics under drought stress in tolerant and sensitive fungi. Such work has been reported in lower plants [9–12].

Some work is available on desiccated and rehydrated nongerminated spores. Florance et al. [23] made electron microscopic studies on *Aspergillus nidulans* after hydration of dry conidia. The cells of dry, dormant conidia are three layered, and the germtube walls are continuous with the innermost layer. Mitochondria and endoplasmic reticulum increase in number and amount, respectively, during hydration and germination, and vacuoles increase in size. The apex of the young germtube is multivesicular. Further investigations are needed to correlate these electron microscopic studies with biochemical ones. Dry spores of *Neurospora crassa* and *Rhizopus stolonifera* showed single ribosome and almost no polysomes [24, 25]. Ribosomes of dry spores of *Aspergillus oryzae* seem to be identical to those of hydrated spores [26]. Polysomes revealed an increase in number when dry spores of *Neurospora crassa* were hydrated. This clearly suggests that polysome integrity might be lost during spore drying and that they can reform rapidly on rehydration. Dry urediospores of *Puccinia graminis* resume respiration within 10 min of rehydration [27].

III. LICHENIZED FUNGI

A. Physiological Changes

The drought responses of the whole lichen thallus are, of course, a combination of algal (phycobiont) and fungal (mycobiont) components. Water deficit stress induces changes in photosynthesis, and these reflect on the metabolic state of the algal component which, although comprising 3–10% of the thallus by weight, is the fixer of carbon, transported subsequently as carbohydrates to a more massive fungal component [10]. The fungal component is the major site of respiration. A number of review articles have been written regarding response of drought on carbon assimilation in lichens [24,47,56,64,92]; readers should consult these for further detail.

A number of studies have been conducted on respiration in relation to drought stress in lichens [9–11,33]. As lichens show symbiosis between algae and fungi, the study of their respiration may elucidate something about symbiotic association. In each symbiont, respiration from glycolysis, the pentose phosphate pathway, and a variety of other catabolic processes are presumably included [28]. When water was added to a dry lichen, the rate of respiration rapidly rose for a time to a value higher than the controls (saturated plant), while the rate of CO_2 assimilation began at a low level, rising slowly to that of controls. The magnitude of this rewetting effect was much less in lichens from dry habitats than in those from moist ones [33,34]. Two distinct phenomena occur when water is added to dry lichen: (a) an immediate nonmetabolic release of a substantial volume of

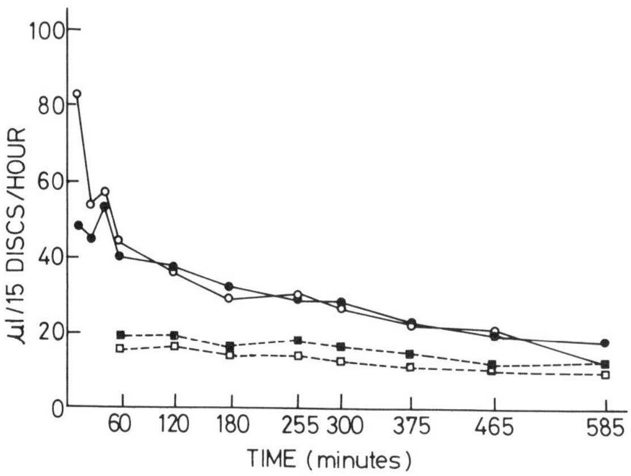

FIGURE 1. Progress of respiration of *Peltigera polydactyla* discs after addition of water to dry discs: (●), O_2 uptake; (○) CO_2 output. Control discs on water: (■) O_2 uptake; (□) CO_2 output. (From Ref. 37.)

gas containing CO_2 called as "wetting burst," and (b) a rapid rise of respiration to a rate well above control value known as "resaturation respiration" [35–37]. Wetting burst is purely a physical process, also exhibited by nonliving tissues and inert substances, and is of no significance to desiccation tolerance [35]. Resaturation respiration could be a symptom of a temporary loss of control over metabolic rates during recovery from drying, or it could supply energy vital to the rapid reestablishment of cell function after wetting [37]. The progress of resaturation respiration in *Peltigera polydactyla* is shown in Figure 1. After 15 min, CO_2 output was 5 times the rate in control discs, and oxygen uptake $2\frac{1}{2}$ times. Thereafter, the rate slowly declined but did not finally return to control levels until 9–10 hr after the start of experiment. The resaturation respiration persisted for about 9 hr in *Peltigera polydactyla* and 2 hr in *Xanthoria aureola* [37]. In *P. polydactyla* resaturation respiration occurred if the water of discs had fallen below approximately 40% of the maximum. *Xanthoria aureola*, which grows in drier habitats than *P. polydactyla*, was much more tolerant of cycles of wetting and drying and showed less intense resaturation respiration [37]. The magnitude of resaturation respiration depends upon the conditions of drying and wetting. If lichens are resaturated by absorbing moisture from air before being saturated with water, then no resaturation respiration takes place [38,39]. In *Chondropsis semiveridis*, following a mild desiccation treatment, the resaturation respiration was very low; basal respiration was obtained within an hour [4]. Brown et al. [41] further confirmed the physical nature of wetting burst by drying lichens in CO_2-free air, after which the burst was eliminated. A prior fast-drying rate produces higher resaturation respiration rate than does slow drying. Resaturation respiration is also markedly affected by seasonal change; it was very high in plant material collected from under snow in December and February and during snowmelt in April [41]. However, in experimental material collected in May following a dry period, the peak of resaturation respiration was completely absent; material collected following a 2-week period of wet weather in June revealed a partial recovery of resaturation respiratory activity. The experimental combination of length of thallus saturation with drying rate was also responsible for the amplitude of resaturation respiration. Unlike basal respiration, resaturation respiration is aside and cyanide sensitive in *P. polydactyla*; it is also DNP-sensitive in *H. physodes* [35]. In this paper, the effect of one inhibitor concentration was studied, but these inhibitors have different effects upon metabolism at different concentrations. It is not known whether the resaturation respiration is associated with the mitochondria. It will be interesting to know the relationships among respiration, ATP production, and respiratory substrates. The main respiratory substrates in lichens are polyols (e.g., mannitol and arabitol) plus nonpolyol sucrose [35,37,42]. Net loss of polyols from the thallus of *H. physodes* on rehydration exceeds that which can be accounted for by resaturation respiration alone; the balance is lost due to extensive leakage of metabolites during the first few minutes of rewetting of dry lichen [35]. Farrar [43] has hypothesized that the establishment of a sizeable and readily accessible pool of respirable substrate is an important "physiological buffer." Thus, in times of metabolic stress imposed by alternate wetting and drying cycles, substrate may be drawn from the pool, and hence degradation of insoluble cell components (e.g., carbohydrates and protein polymers) does not occur. The pool is replenished by photosynthesis when favorable conditions prevail. If the polyol pool is depleted below a critical level, recovery from desiccation is scarcely possible.

Brown et al. [41] discussed biochemical basis of resaturation respiration. This may result from a stimulation of basal respiration reflecting an increased cellular demand for ATP as a direct consequence of loss of membrane integrity during dehydration. Below critical water level, lipid bilayer degenerates into a leaky hexagonal phase displacing associated proteins. However, Simon [44] suggested that restoration of membrane configuration is a physical process and is not dependent on the source of ATP. Dhindsa and Matowe [45] correlated resaturation respiration with decline in superoxide dismutase activity following rewetting. This results in an increased level of free O_2 radicals and stimulation of free radical lipid peroxidation inducing increased oxygen uptake with no change in CO_2 production. Although this pattern has been found in *Tortula ruralis* [46], it is not seen in *P. polydactyla* [41].

Recently Cowan et al. [47] studied the anhydrobiotic metabolism in lichen species, *Ramalina celastri* and *P. polydactyla*, exposed to known atmospheric humidities of tritium-labeled water vapors. Quantitative and qualitative estimations of ^3H labeling of water-soluble cell constituents (amino acids, organic acids, sugars, sugar phosphates, and sugar alcohols), neutral and polar lipids, water-soluble macromolecules and insoluble cell residues have shown that diverse metabolic systems operated at different degrees of desiccation. TCA cycle reactions and related amino acids intercoversions were shown to be relatively desiccation resistant, whereas Calvin cycle metabolism appeared to be highly sensitive. Quite conversely, synthesis of sugar alcohols was found to be active beyond the limit of photosynthetic metabolism. Neither lipid nor macromolecule synthesis was shown to occur in the desiccated state to a significant degree.

B. Mechanisms

Determination of viability of a lichen either in the field or under controlled conditions is perhaps the most difficult factor in studies of survival under extreme environments. Each partner (mycobiont and phycobiont) may have different tolerance to desiccation [48]; a single test for viability does not obviously give accurate results. Owing to these difficulties, different kinds of measurements and observations have been used to determine lichen responses to drought. Since Höfmeister [49], it has visually been observed in many studies that lichens can tolerate desiccation for several months [50–52] or years [53]. In recent years viability of lichens has been determined mainly using a photosynthetic technique [9,11,29–31], which indicates drought survival of phycobiont rather than of mycobiont. However, in some cases respiration has been used to measure drought survival; this reflects drought tolerance in mycobiont.

Drought-tolerant species recover their physiological processes within a very short time after rewetting of desiccated plant material, but drought-sensitive species do not recover at all [9–11]. It will be absorbing to note why some species can tolerate drought whereas others do not. It seems that considerable amount of work has been done regarding response of water stress on various physiological processes. However, little information is available to explain mechanisms of drought tolerance in lichens. The presence of polyols in cells might be important during desiccation to protect macromolecules as water of hydration is lost [9], but there is no direct evidence to prove this statement in lichens. On the other hand, protective effects of various sugars against drought injury on cellular membranes have been studied extensively [49]. More work is needed to ascertain if polyols can protect lichens against drought injury or not. Very little work has been done to draw a clear conclusion as to the mechanism of drought tolerance in lichens. However, because of lichen's poikilohydrous nature, drought tolerance in lichens must be protoplasmic in nature rather than in mechanical. Further work is needed to investigate various biochemical and ultrastructural changes during desiccation and rehydration in drought-resistant and -sensitive lichens. Also, changes in membrane permeability, in terms of measurement of membrane components and leakage of solutes, due to drought stress in drought-tolerant and -sensitive species would prove useful. Such study has been extensively carried out in bryophytes [11,12].

IV. RESPONSE OF VA MYCORRHIZAE TO WATER RELATIONS OF THE HOST PLANT

Vesicular-arbuscular (VA) mycorrhizae, symbiotic associations of a fungus and plant root, can increase the growth of the host plant. The enhanced growth of plants is primarily due to the increase of phosphorus uptake by mycorrhizae [55–57]. In P-deficient soils mycorrhizal plants typically grow better than nonmycorrhizal ones, but the reverse may be true in soils well supplied with phosphate [58]. VA mycorrhizae can improve growth and survival of plants in numerous habitats [59]. In mesic to wet habitats, increased phosphate uptake as regarded to be the major cause [60], but in the semi-arid regions of the North American Great Plains, water is generally the major factor limiting production [61]. If VA mycorrhizae are capable of substantially altering production and survival in these habitats, they must improve water uptake and resistance in host plant. The improved water relations of the host plant may be suggested by the observations of Mosse and Hayman [62] that leaf tips of uninfected onion seedlings withered after 2 weeks of transpiration whereas those of mycorrhizal seedlings did not.

A. Water Transport

Safir et al. [63,64] were the first to demonstrate that mycorrhizal soybean plants had lower root resistance to water transport than nonmycorrhizal plants. Safir et al. [64] found that although resistance of whole soybean plants differed by 40%, there was no difference in the resistance of stem plus leaves, indicating that the major effect of mycorrhizae was to reduce the resistance of the root. Since the fungitoxicant *p*-chloronitrobenzene had no effect on resistance to water transport, reduced resistances probably were not caused by direct modification of the transport pathway by fungus. Addition of nutrients essentially eliminated difference in the resistance to water transport between mycorrhizal and nonmycorrhizal plants. Thus, lowered resistance of mycorrhizal roots growing in soil with low levels of nutrients probably resulted from enhanced nutrient status of the plant brought about by the fungus. However, the exact way in which enhanced nutrient status reduces resistance was not known. Sieverding [65] found a lower water consumption per unit shoot dry weight in VA mycorrhizal as compared with nonmycorrhizal *Eupatorium odoratum, Guizotia abyssimica, Sorghum bicolor*, and *Trigonella foenumgraecum*. Smaller differences were observed between best and worst water use values over a range of soil water regimes in mycorrhizal plants. On the argument that the findings of Safir et al. [63,64] could be due to larger size of their mycorrhizal plants, Levy and Krikun [66] compared VA mycorrhizal with similar sized uninfected rough lemon seedlings. From changes in leaf xylem pressure potential and transpiration during and after recovery from water stress, they concluded that root resistances were similar in two sets of plants. However, since the leaf conductances were consistently higher in the mycorrhizal plants, they postulated a mycorrhizal effect on stomatal regulation via changes in the root/shoot hormonal balance. Allen and coworkers [67,68] reported that the VA mycorrhizae increased transpiration, and the whole-plant resistance to water transport was reduced considerably. However, the nonmycorrhizal plants in these experiments were generally smaller than the VA-mycorrhizal plants, and possibly were less vigorous due to phosphorus deficiency. A number of workers studied water transport in terms of hydraulic conductivity of root. Safir et al. [64] and Levy and Krikum [66] determined hydraulic conductivity indirectly by measuring transpiration rate and leaf water potential and assuming soil water potential was nearly zero under well-watered conditions [69]. Hardie and Leyton [70] calculated root hydraulic conductivities directly by measuring the rates of water forced through detopped root systems using a pressure chamber. The hydraulic conductivities of the root systems were very much higher in mycorrhizal than in nonmycorrhizal P-deficient red clover plants. This could probably be attributed to the greater lengths and diameters and hence to the total surface area of mycorrhizal roots. Per unit length of root, the conductivities of the mycorrhizal roots were still two to three times higher, suggesting that this was mainly due to hyphal growth in the soil. However, altered size of root systems alone can change root conduction in a nonlinear fashion [71]. Allen [72] compared water transport between nonmycorrhizal *Boutelous gracilis* and VA mycorrhizal *B. gracilis* infected with *Glomus fasciculatus*. Mycorrhizal infection increased water uptake rates even with similar sized root systems and the estimated differences were within the range of evapotranspiration in other fungal species. These results tend to support a predominantly direct flow mechanism [70] as the reduced resistances might be attributable to the greater flow rates associated with increased absorptive surface area [73]. Levy et al. [74] used the pressure chamber technique to compare rough lemon seedlings of equal size and P nutrition and found that VA-mycorrhiza had either no effect or slightly reduced root conductivity under well-watered conditions. The slight decrease in root conductivity by VA mycorrhizae was attributed to root-limiting conditions in containers in which mycorrhizal plants developed lower leaf water potentials, extracted more water from soil, and experienced greater water-deficit stress. Since water stress alone reduces root conductivity of rough lemon [74,75], greater water use by VA mycorrhizal plants may have been responsible for the decrease in root conductivity. More recently, Graham and Syvertsen [76] compared root hydraulic conductivity of mycorrhizal and P-fertilized nonmycorrhizal citrus seedlings of equal size and P nutrition in a nonlimiting soil volume under well-watered conditions, with a view to ascertaining if enhancement of water uptake by mycorrhizal roots was attributable to improve host P nutrition and/or to a direct effect of hyphae which might provide a lower resistance pathway from soil to the root. Mycorrhizal seedlings of citrus root stocks with adequate P nutrition had more than twice the root hydraulic conductivity of P-deficient nonmycorrhizal plants of equal size. Greater water uptake by mycorrhizal citrus is comparable to the increases found for mycorrhizal soybean [64], *Bouteloua gracilis* [72], and red clover [70]. They found that flow of water to roots via hyphae alone could not account for the greater water uptake by mycorrhizal roots. However, P nutrition was primarily responsible for the greater conductivity of roots since no differences were found between root hydraulic conductivity of mycorrhizal and nonmycorrhizal plants of equal P status under well-watered conditions. Thus, this study confirms the nutritional hypothesis set forth by Safir et al. [60,64] that enhanced P nutrition by mycorrhiza reduces resistance to water transport in roots. Graham and Syvertsen [76] pointed out that their finding did not, however, rule out the possibility that mycorrhizal hyphae absorbed and transported significant amounts of water to the plant when moisture depletion zones formed around roots at low soil water potential. The reason for decrease of water uptake by roots of P-deficient citrus is not known. It is likely that P effects on root conductivity are morphological and/or physiological. A significant difference in the physiology of roots at low P level is an increase in root permeability [77]. This could result in greater leakage of metabolites from root cells that are involved in water uptake and transport. Mycorrhizal plants with better P nutrition had greater vascular development, but this was in comparison to smaller nonmycorrhizal plants; the root anatomy of similar size mycorrhizal and nonmycorrhizal plants of same and different P status should be explained to distinguish the effects of P nutrition from those of plant size on vascular development [78].

Levy et al. [74] studied that role of VA mycorrhizae on the hydraulic conductivity of roots in plant responses during and after recovery from drought stress. The hydraulic root conductivity of nonbagged plants (plants enclosed in polythene bags) was significantly decreased by previous water stress treatments and further reduced by VA mycorrhizae. Since water stress alone can reduce hydraulic conductivity [75], water could have been the determining factor for decrease of hydraulic conductivity of the mycorrhizal plants. It is interesting to note that hydraulic root conductivity responded to mycorrhizae and water stress in ways parallel to shoot: root ratios. Plants with the highest conductivity also had the highest shoot:root ratio. The longer total root length of mycorrhizal plants could have compensated for decreases in root hydraulic conductivity leading to higher transpiration rates. However, enclosing the pots in polythane bags apparently negated effects of various drought cycles and mycorrhizae on hydraulic conductivity of root. The bags may have had an effect on the mycorrhizal plant interaction, as nonbagged plants had generally higher root hydraulic conductivity. Bagging may have reduced the O_2 concentration in the root zone and thereby reduced root and VA mycorrhizae vigor. This effect should have been more pronounced, however, in the mycorrhizal plants since they tend to have higher root densities.

B. Transpiration and Associated Parameters

A large number of reports are available regarding influence of VA-mycorrhiza on transpiration and its associated parameters in host plants. Levy and Krikum [66] found that VA mycorrhizal infection increased transpiration. Mycorrhizal infection increased transpiration rates by over 100% with 5 – 70% lower leaf resistances to water vapor diffusion [68]. Mycorrhizal plants had 50% lower leaf resistance with no change in leaf or root water potentials; thus transpiration was increased 100% with 50% reduction in whole plant, soil to root, and root to leaf resistance. Hardie and Leyton [70], while studying the influence of mycorrhizae on water relations of red clover plants, reported that under an adequate water regime, the mycorrhizal plants were losing water more rapidly than the nonmycorrhizal. This must be explained by their greater size and leaf area, and the same applies to the increased transpiration of the nonmycorrhizal plants given additional phosphate. However, on a unit area basis, the mycorrhizal plants still showed higher transpiration rates. Levy and Krikum [66] found higher transpiration fluxes and lower diffusion resistance in infected rough lemon seedlings before and after water stress, but the differences decreased rapidly with the increasing water stress. In case of red clover plant, when watering was discontinued the mycorrhizal plants wilted much more rapidly than the nonmycorrhizal (2 – 13 days compared with 13 – 30 days, respectively, according to P supply); transpiration of mycorrhizal plants, 24 hr before wilting, though higher in absolute terms, was significantly lower than for nonmycorrhizal plants, on unit area basis [70].

Whittingham [79] reported greater stomatal resistance in mycorrhizal than in nonmycorrhizal *Plantago lanceolata* when under water stress. It appears that mycorrhizal infection can indeed influence stomatal opening by inducing greater water stress and hence lower leaf water potential [70]. Levy et al. [74] found that mycorrhizal infection increased significantly the mean daily transpiration rate of well-watered as well as previously stressed plants over that of noninoculated high phosphorus control plants (Fig. 2). Hardie and Leyton [70] showed VA mycorrhizae enhanced transpiration rates. Differences in nutritional level or leaf area could not have been responsible for the observed trends in transpiration rates. Within the mycorrhizal plants, there were no significant differences between the transpiration rates of well-watered and previously drought-stressed seedlings. Stomatal conductance were higher in mycorrhizal than in nonmycorrhizal

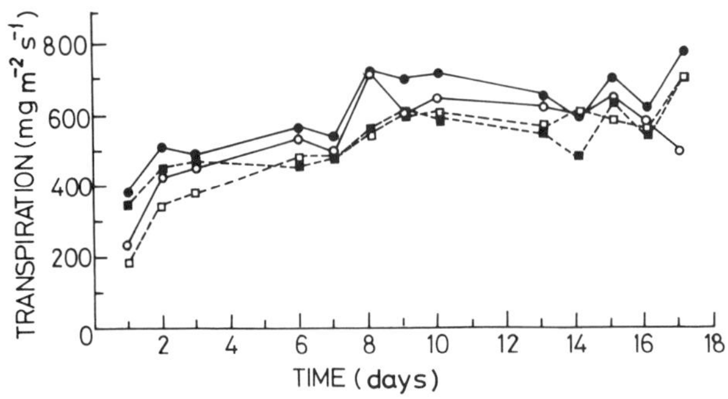

FIGURE 2. The effect of VAM and three prior drought cycles on whole plant transpiration rates of well-watered rough lemon seedlings. Each point is the mean of five plants calculated by averaging 24-hr weight loss data over the 12-hr photoperiod. (□) – water – VAM; (■) + water – VAM; (○) – water + VAM; (●) + water + VAM. (From Ref. 74.)

plants in both wet and dry treatments. Stomatal closure in mycorrhizal plants occurred at lower leaf water potentials and after greater desiccation than in nonmycorrhizal plants, but some leaves of *Glomus mosseae*-infected plants showed no stomatal response to drought and continued to transpire at water potential as low as −4.1 MPa [67]. Mycorrhiza significantly increased whole plant transpiration of *Carrizo citrange* (CC) but not of sour orange (SO). When transpiration was expressed on a unit root length basis to allow for the smaller size of mycorrhizal root systems, mycorrhizal SO as well as CC had greater transpiration rates than nonmycorrhizal plants. When both root stocks were considered together, transpiration rate, on a unit length basis, was found to be positively correlated with root hydraulic conductivity [76]. Bethlenfalvay et al. [80] studied the influence of VA mycorrhizae on soybean plant, both under water deficit stress and nonstress conditions. In stressed plants, transpiration and leaf conductance in mycorrhizal plants were significantly greater than in nonmycorrhizal P-fed plants during the first part of the stress cycle (Fig. 3). Transpiration rates and leaf conductance declined linearly with time for both mycorrhiza and P-fed treatments and converged towards the end of stress cycle. This decline occurred even during the early part of the drought cycle where moisture sensors were not sensitive enough to detect a change in soil water potentials (Fig. 4). The relationship between transpiration and leaf conductance was essentially linear towards the end of the stress cycle. However, during the first days of the cycle, while soil water potential was still high (Fig. 4), large changes in leaf conductance were accompanied by relatively small change in transpiration. Koide [81] also found that mycorrhizal condition enhanced transpiration and leaf conductance during an important portion of recurring drought cycles. Huang et al. [82] found that leaf conductance and transpiration rates per unit leaf area in mycorrhizal plants were often twice as great as in nonmycorrhizal plants. They also found that leaf conductance during the drier midday period appeared to be strongly coupled with the leaf of air vapor pressure deficit (LAVD) with mycorrhizal plants showing the more sensitive response. These results are not in complete consonance with the

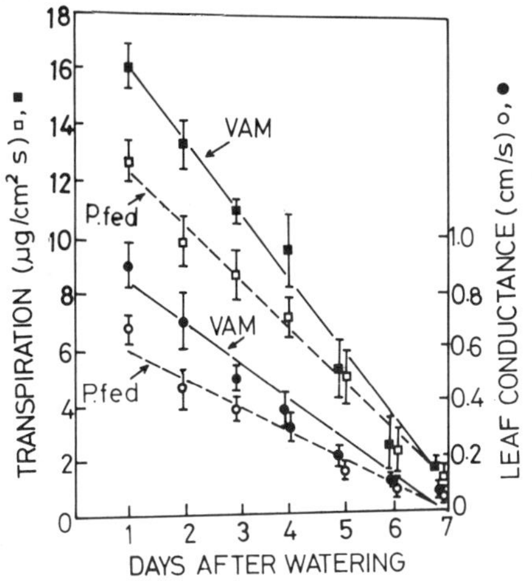

FIGURE 3. Changes in transpiration rates and leaf conductance in VAM and non-VAM, P-fed plants during the final stress cycle. Correlation coefficients were $r = -0.99$ (transpiration) and $r = -0.98$ (leaf conductance) for both VAM and non-VAM plants. (From Ref. 80.)

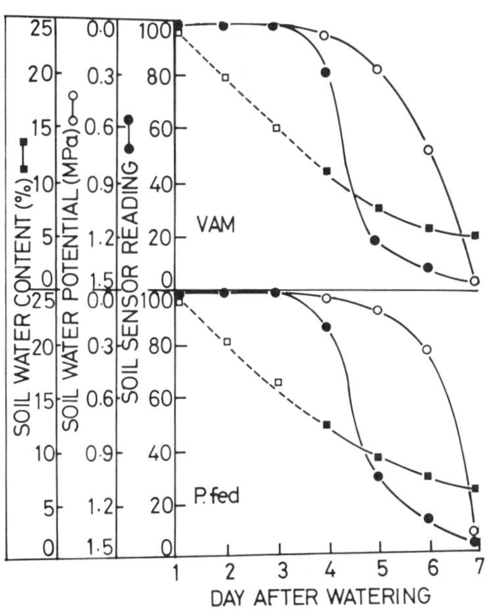

FIGURE 4. Soil water content, water potential, and moisture sensor relationships during the final stress cycle of VAM and non-VAM, P-fed soybean plants. Soil sensor data points reflect the average of six replications and are related to soil water data by calibration in a pressure plate apparatus. The dotted portion of the soil water content curve indicates an approximate position of the data points for days 2 and 3 of the cycle. Soil water content values were not measured, but calculated from a calibration curve based on soil sensor readings and the saturation value on day 1 (24%, w/w). Since soil sensors were not sensitive to change during the first 3 days, the curve was interpolated and the symbols placed at the appropriate position. (From Ref. 80.)

conclusions of Hardie and Leyton [70] and Whittingham [79]. In contrast to their findings for red clover, Huang et al. [82] found that mycorrhizal *Leucaena* plants tended to have greater xylem pressure potential than did nonmycorrhizal, especially during morning hours. Unlike other workers, Nelson and Safir [83] found that there were no significant differences in plant water relations (transpiration and water potentials) under either well-watered and drought-stressed conditions between mycorrhizal and nonmycorrhizal plants.

C. Water Potential

Some work has been done regarding the response of mycorrhizal infection on leaf/soil water potential of host plant. Allen [72] reported no change in leaf water potential due to mycorrhizal infection under well-watered conditions. No difference in leaf water potential was found between mycorrhizal and nonmycorrhizal plants under either water-deficit and well-watered conditions [68,84,85]. Levy et al. [74] reported that leaf water potential was nearly the same in mycorrhizal and nonmycorrhizal plants under well-watered conditions, but it was greater in mycorrhizal plants under water-deficit stress condition (Table 1). Hardie and Leyton [70] found that water potential of soil, and hence leaf, at which mycorrhizal plants wilted was significantly lower than in nonmycorrhizal plants. According to them, there are two, not mutually incompatible, explanations for this: (a) The soil water potentials quoted are bulk values, if, because of their smaller root systems, exploitation of the soil moisture of the nonmycorrhizal plants was not so effective; there may have been drier regions in the vicinity of the roots and wetter regions elsewhere. In other words, water potential of soil around the roots may be lower than the bulk value for the nonmycorrhizal plants but much nearer to it in the case of the mycorrhizal plants of their more extensive root system, including fungal hyphae. (b) In some way, mycorrhizal infection enables the leaf to develop a lower water potential, thus drying out the soil more efficiently. It is well established that when subjected to water stress, leaf turgor can be maintained by decreasing the osmotic potential of the vacuolar

TABLE 1 Mean Leaf Water Potential (MPa) Just Prior to the Light Period and 6 hr Later at the End of Third Drought Cycle

Time		Well-watered	Water-stressed
Prephotoperiod	−VAM	−0.22c	−0.76b
	+VAM	−0.21c	−1.72a
Mid-photoperiod	−VAM	−1.38b	−1.52b
	+VAM	−1.41b	−1.77a

Separation of four treatment means at each time period by Duncan's multiple range test ($p < 0.05$). a, b, c indicate that treatments having same superstate do not differ significantly.
Source: Ref. 74.

sap by an increase in inorganic/or organic ions, sugars, etc. In the case of inorganic ions, this is much more likely to be achieved by an increased uptake of, say, K rather than of P. There are indeed cases where endomycorrhizal infection has led to increased concentrations of K [86], but this response is not consistent [87]. Recently, Huang et al. [82] studied diurnal changes of xylem pressure potential and soil water potential for mycorrhizal and nonmycorrhizal *Leucaena leucocephala* (Lam.) de Wit. Soil water potential remained high (−0.3 MPa) in the nonmycorrhizal pots during the entire day compared to a comparatively large decrease in mycorrhizal pots (−0.3 to −2.2 MPa). Xylem pressure potential also decreased markedly for both mycorrhizal and nonmycorrhizal plants, with somewhat greater decrease for nonmycorrhizal plants during the morning. By late afternoon, a maximum xylem pressure potential of nearly −2.0 MPa occurred for both treatments, but the differences between the xylam pressure potential and the soil water potential was less for mycorrhizal plants (−0.1 vs. −1.6 MPa). Subsequent increase in xylem pressure potential and soil water potential of over −0.5 MPa occurred after about 1800 hr in pots with mycorrhizal plants.

V. PHOTOSYNTHESIS AND PHOTORESPIRATION

Very little work has been done regarding response of mycorrhizae to photosynthesis, photorespiration, and chlorophyll content under different drought conditions. Levy and Krikun [66] found that VA mycorrhizal infection increased $^{14}CO_2$ uptake in citrus upon recovery from drought. Recently Ramakrishnan et al. [85] studied the influence of introduced VAM fungus *Glomus caledonius* and that of indigenous population on photosynthesis, photorespiration, and chlorophyll content in osmotically stressed maize grown in sterile/unsterile soil, with or without P additions. There was no significant increase in photosynthesis under different treatments and soil conditions in maize plant at −0.3 MPa water potential. However, at severe stress (−1.2 MPa) there was a significant increase in photosynthetic rate in VAM (10.4%) and P-amended plants (23.9%) in sterile soil. In sterile soil some increase in chlorophyll −a and −b content was found in VAM, IVAM + GC, and P-amended plants over the non-VAM at zero osmotic potential. In nonstressed plants, evolution of $^{14}CO_2$ by decarboxylation of ^{14}C-glycine was considerably greater in VAM and IVAM + GC treatments than in non-VAM plants. This increase in photorespiratory $^{14}CO_2$ evolution may reflect a decrease in growth of VAM and IVAM + GC plants over non-VAM. At a mild water stress (−0.6 and −0.8 MPa) there was no significant change in photorespiration in sterile soil; at severe stress (−1.2 MPa), it was considerably more in VAM, IVAM + GC, and P-amended soil than in non-VAM plants.

VI. GROWTH PARAMETERS

Some work has been reported regarding the influence of mycorrhiza on growth of plants under different drought conditions. The water-stressed nonmycorrhizal onion plants, despite high levels of P fertilization, were only 23% as large as the stressed mycorrhizal onions grown at the low P level [75]. Under both nonstress and water-stress conditions, total root length and total leaf area were markedly greater in mycorrhizal than in

TABLE 2. Effect of VAM and Water Stress on Leaf Phosphorus, Total Leaf Area, Total Feeder Root Length, and Shoot:Root Ratio of 12-Month-Old Rough Lemon Seedlings

Treatment		Well-watered	Water-stressed
Leaf phosphorus (% dry wt)	−VAM (+P)	0.22ab	0.25a
	+VAM	0.19bc	0.17c
Total leaf area (cm^2)	−VAM	597	592 n.s.
	+VAM	677	632
Root length (m)	−VAM	31.3	35.0 n.s.
	+VAM	38.4	43.2
Shoot:root ratio (dry wt)	−VAM	2.31a	1.87b
	+VAM	2.09ab	1.83b

Each value is the mean of 8−10 leaves or plants from each of the four treatments. Separation of the four treatment means by Duncan's multiple range test ($p < 0.05$). a, b, c indicate that treatments having same superstate do not differ significantly. n.s., not significant.
Source: Ref. 74.

nonmycorrhizal plants (Table 2); however, mycorrhizal infection caused slight decrease in leaf phosphorus and shoot-root ratio [74]. Plants infected with *G. fasciculatus* had dry weight similar to nonmycorrhizal wheat plants; infection with *G. mosseae* significantly reduced growth of wheat under both wet and dry conditions. Dry weights of both nonmycorrhizal and *G. fasciculatus*-infected plants were significantly reduced (27 and 33%, respectively) by drought. Leaf dry weights were 34% lower in *G. mosseae*-infected plants with dry versus wet treatments, but there was no difference in root dry weights [67]. Nodule development and activity, leaf and root parameters, including fresh and dry weight, and N and P nutritions of VAM plants and of non-VAM, P-fed plants grown under the same control conditions were compared. All the parameters, except N content, were greater in VAM plants than the P-fed, non-VAM plants, when under water stress. The opposite was generally true in the unstressed comparisons [80].

VII. DROUGHT RESISTANCE

As mentioned earlier, mycorrhizal infection improves water relations and growth of plants, both under wet and dry conditions. However, there are few reports to ascertain the influence of mycorrhiza on drought resistance of host plant. VA mycorrhizae can improve survival of host plants in numerous habitats [59]. Nelsen and Safir [83] revealed that drought tolerance of mycorrhizal onion plants was associated with the improved P nutrition. Using a leaf osmotic potential as a criterion, infection by *G. fasculatus* appeared to increase wheat drought tolerance, whereas infection by

G. mosseae did not. *G. fasciculatus*-infected wheat plants had significantly lower leaf osmotic potential than both nonmycorrhizal and *G. mosseae*-infected plants.

It is known that a number of physio-biochemical factors are associated with drought tolerance in plants. Proline accumulation correlates with drought resistance in various plant species [42,89]. Recently Ramakrishnan et al. [90] found that mycorrhizal maize plants yielded considerably greater amounts of proline than the nonmycorrhizal plants at different water-deficit conditions. Thus, mycorrhizal infection can increase drought resistance in host plants. At the specific water stress treatment, there was more increase in membrane permeability in drought-sensitive plant tissues [11,12, 91–93]. Drought-resistant plant tissues revealed greater amounts of phospholipids [12,93,94]. Photosynthesis and respiration has also been correlated with drought tolerance in moss plants [91,95]. Chlorophyll stability index can be correlated with drought tolerance in lower [96] and higher plants [97]. Levitt [5] reported protein-SH content to be related to drought tolerance in plants. Further studies are required to investigate whether these physio-biochemical parameters can be correlated with the increase of drought tolerance due to mycorrhizal infection.

VIII. CONCLUSION

The various aspects of drought responses have been reviewed in free-living and lichenized fungi and mycorrhizal host plants. Unlike other poikilohydrous plants, very little work has been done in fungi regarding response of drought on the cellular activities. Various types of metabolites are affected due to drought in fungi. The exact mechanism of drought resistance in fungi is still not known. Ability to accumulate polyols (mainly glycerol) in response to decrease in leaf water potential may be the key feature in the osmoregulation of sugar-tolerant yeast. However, relatively little information is available about the osmoregularity mechanism in xerophyllic filamentous fungi. Free and lichenized fungi from drier habitats revealed greater tolerance to desiccation. By far the most studied metabolic process in lichenized fungi is respiration. Two distinct phenomena occur when water is added to an air-dry lichenized fungi: (a) an immediate nonmetabolic release of a substantial volume of gas containing CO_2, called "wetting burst," and (b) a rapid rise of respiration to a rate well above control value, known as "resaturation respiration." However, relatively very little work has been done at the cellular level of organization to determine the consequence of drought in lichenized fungi.

A number of reports are available regarding influence of VA mycorrhizae on water relations in plants. Generally, mycorrhizal infection increases water transport in host plants, which either may be related to enhanced nutrient status of mycorrhizal plants grown in low nutrient soils or may be due to larger size of mycorrhizal plants. However, the exact reason for greater water uptake in mycorrhizal plants is not known. Mycorrhizal infection also increases transpiration rates in host plants. This may be explained by the greater size and leaf area of mycorrhizal plants. Generally, no difference in leaf water potential was found between mycorrhizal and nonmycorrhizal plants, either under drought or well-watered conditions. Very little work has been done on the influence of mycorrhizae on photosynthesis, photorespiration, and photosynthetic pigments, either under well-watered and drought conditions. There is no significant increase in photosynthesis under different drought treatments and soil conditions in unstressed maize plant, but it increases significantly in VAM and P-amended stressed plants. Under nonstressed conditions photorespiration is considerably greater in VAM and IVAM + *Glomus caledonius* (GC)-treated plants over other treatments. Some increase of chlorophyll-a and -b has been found in VAM, IVAM + GC, and P-amended unstressed plants. Under nonstressed and water-stressed conditions, various growth parameters are increased markedly in mycorrhizal plants over nonmycorrhizal plants. Mycorrhizal infection may induce drought hardiness in host plants.

REFERENCES

1. Kramer, P. J. Drought stress, and the origin of adaptations. In *Adaptations of Plants to Water and High Temperature Stress* (N. C. Turner and P. J. Kramer, eds.), John Wiley and Sons, New York, 1980, pp. 7–20.
2. Hsiao, T. C. Plant responses to water stress. *Annu. Rev. Pl. Physiol.*, *24*:519–570, 1973.
3. Walter, H. Water economy and the hydrature of plants. *Annu. Rev. Pl. Physiol.*, *6*:239–252, 1955.
4. Buch, H. De l'adaptation des bryophytes, des algues et des lichens aux conditions de l'humidite et de la concentration de la solution nutritive dans les milieux naturets. *Rev. Bryol. Lichen*, *16*:54–73, 1947.
5. Levitt, J. Responses of plants to environmental stresses. Academic Press, New York, 1972.
6. Gaff, D. F. Desiccation tolerant vascular plants of Southern Africa. *Oecologia*, *31*:95–109, 1977.
7. Gaff, D. F. and Latz, P. The occurrence of resurrecting plants in the Australian flora. *Aust. J. Bot.*, *26*:485–492, 1978.
8. Iljin, W. S. Drought resistance in plants and physiological processes. *Annu. Rev. Plant Physiol.*, *8*:257–274, 1957.
9. Bewley, J. D. Physiological aspects of desiccation tolerance. *Annu. Rev. Plant Physiol.*, *30*:195–238, 1979.
10. Bewley, J. D. and Krochko, J. E. Desiccation—Tolerance. In *Physiological Plant Ecology 11, Encyclopedia of Plant Physiology*, Vol. 12 (O. L. Lang, P. S. Novel, C. B. Osmond, and H. Ziegler, eds.), Springer-Verlag, Heidelberg, 1982, pp. 325–370.
11. Gupta, R. K. Drought tolerance in poikilohydrous plants. *Indian Rev. Life Sci.*, *2*:189–212, 1982.
12. Gupta, R. K. Leakage phenomenon and plant cellular membranes under drought. *Indian Rev. Life Sci.*, *4*:69–86, 1984.
13. Mahajan, M. K., Johri, B. N., and Gupta, R. K. Influence of desiccation stress in a xerophilic thermophile *Himicola* sp. *Curri. Sci.*, 55:928–930, 1986.
14. Finkelstein, A. and Cass, A. Effect of cholesterol on the water permeability of thin lipid membranes. *Nature*, *18*:717–718, 1967.
15. Brown, A. D. Compatible solutes and extreme water stress in eukaryotic microorganisms. *Adv. Microbiol. Physiol.*, *17*:181–242, 1978.

16. Anand, J. G. and Brown, A. D. Growth rate patterns of so-called osmophilic and non-osmophilic yeasts in solutions of polyethylene glycol. *J. Gen. Microbiol.*, *52*:205–212, 1968.
17. Brown, A. D. and Simpson, J. R. Water relations of sugar-tolerant yeasts: The role of intracellular polyols. *J. Gen. Microbiol.*, *72*: 589–591, 1972.
18. Hocking, A. D. and Norton, R. S. Natural abundance ^{13}C under magnetic resonance studies on the internal solutes of xerophilic fungi. *J. Gen. Microbiology*, *129*:2915–2925, 1983.
19. Conney, D. G. and Emersion, R. *Thermophilic Fungi. An Account of Their Biology: Activities and Classification*, W. H. Freeman and Co., San Francisco, 1964, p. 188.
20. Alexandrov, V. Y., Lomagin, A. G., and Feldman, N. L. The responsive increase in thermostability of plant cells. *Protoplasma*, *69*: 417–458, 1970.
21. Sumner, J. L. and Morgan, K. D. The fatty acid composition of sporangispores and vegetative mycelium of temperature adapted fungi in the order mucorales. *J. Gen. Microbiol.*, *59*:215–221, 1969.
22. Bruszweski, T. E., Fergus, C. L., and Mumma, R. O. Thermophilic fungi. IV. The lipid composition of six species. *Lipids*, *7*: 695–698, 1972.
23. Florance, E. R., Denison, W. C., and Allen, T. C. Ultrastructure of dormant and germinating conidia of *Aspergillus nidulans*. *Mycologia*, *64*:115–123, 1972.
24. Mirkes, P. E. Polysomes, ribonucleic acid, and protein synthesis during germination of *Neurospora crassa* conidia. *J. Bacteriol.*, *117*: 196–202, 1974.
25. Van Etten, J. L., Dunkle, L. D., and Knight, R. H. Nucleic acids and fungal spore germination. In *The Fungal Spore—Form and Function* (D. J. Weber and W. M. Hess, eds.), Wiley, New York, 1976, pp. 243–300.
26. Kimura, K., Ono, T., and Yanagita, T. Isolation and characterization of ribosomes from dormant and germinating conidia of *Aspergillus oryzae*. *J. Biochem.*, *58*:569–576, 1965.
27. Maheshwari, R. and Sussman, A. S. Respiratory changes during germination of unrediospores of *Puccinia graminis* f. sp. tritici. *Phytopathology*, *60*:1357–1364, 1970.
28. Farrar, J. F. Lichen physiology: Progress and pitfalls. In *Air Pollution and Lichens* (D. W. Ferry, M. S. Baddeley, and D. L. Hawksworth, eds.), University of Toronto Press, Toronto, 1973, pp. 238–282.
29. Harris, G. B. Water content and productivity of lichens. In *Water and Plant Life, Problems and Modern Approaches*, Vol. 19 (O. L. Lang, L. Kappen, and E. D. Schulze, eds.), Springer-Verlag, Berlin, 1976, pp. 452–468.
30. Kappen, L. Response to extreme environments. In *The Lichens* (V. Ahmadijian and M. E. Hale, eds.), Academic Press, New York, 1973, pp. 311–380.
31. Lang, O. L., Schulze, E. D., Kappen, L., Buschbom, U., and Evenari, M. Adaptations of desert lichens to drought and extreme temperatures. In *Environmental Physiology of Desert Organisms*

(N. F. Hadley, ed.), Hutchinson and Ross, Stroudsburg, 1975, pp. 20–37.
32. Lang, O. L. and Kappen, L. Photosynthesis of lichen from Antarctica. In *Antarctic Research Series*, Vol. 20 (G. A. Llano, ed.), Methuen, London, 1972, pp. 83–95.
33. Ried, A. Nachwirkungen der Entquellung auf den Gaswechsel von Krustenflechten. *Biol. Zentralbl.*, *79*:657–678, 1960.
34. Ried, A. Stoffwechsel and Verbreitungsgrenzen von Flechten, Wasser und Assimilationshaushalt, Erustenflechten benachbarter Standorte. *Flora*, *149*:345–385, 1960.
35. Farrar, J. F. and Smith, D. C. Ecological physiology of the lichen *Hypogymnia physodes*. III. The importance of the rewetting phase. *New Physiol.*, *77*:115–125, 1976.
36. Smith, D. C. Symbiosis and the biology of lichenzed fungi. *Symbiosis Symp. Soc. Exp. Biol.*, *29*:373–407, 1975.
37. Smith, D. C. and Molesworth, S. Lichen physiology. XIII. Effect of rewetting dry lichens. *New Phytol.*, *72*:525–533, 1973.
38. Bertsch, A. Über den CO_2-Gaswechsel einiger Flechten nach Wasserdampfaufnahmne. *Planta*, *68*:157–166, 1966.
39. Lang, O. L. Ecophysiological investigation in lichens of the Nagev desert. CO_2 gas exchange of *Ramalina maciformis* (Del.) Bory under controlled conditions in the laboratory. *Flora*, *158*:324–359, 1969.
40. Ried, A. Physiologische Aspekte der Vertikalzonierung von Algen des marinen Litorals. *Ber. Deutt. Bot.*, *82*:127–140, 1969.
41. Brown, D., Macfarlane, J. D., and Kershaw, K. A. Physiological–environmental interactions in lichens. XVI. A re-examination of resaturation respiration phenomenon. *New Phytol.*, *93*:237–246, 1983.
42. Smith, D. C. The biology of lichen thalli. *Biol. Rev.*, *37*:537–570, 1962.
43. Farrar, J. F. The lichen as an ecosystem: Observation and experiment. In *Lichenology: Progress and Problems* (D. H. Brown, D. L. Hawksworth, and R. H. Bailey, eds.), Academic Press, New York, 1976, pp. 385–406.
44. Simon, K. W. Phospholipids and plant membrane permeability. *New Phytol.*, *73*:377–420, 1974.
45. Dhindsa, R. S. and Matowe, W. Drought tolerance in two mosses: Corelated with enzymatic defence against lipid peroxidation. *J. Expl. Bot.*, *32*:79–91, 1981.
46. Krochko, J. E., Winner, W. E., and Bewley, J. D. Respiration in relation to adenosine triphosphate content during desiccation and rehydration of a desiccation-tolerant and a desiccation-intolerant moss. *Plant Physiol.*, *64*:13–17, 1979.
47. Cowan, D. A., Green, T. G. A., and Wilson, A. T. Lichen metabolism. 1. The use of tritium labelled water in studies of anhydrobiotic metabolism in *Ramalina celastri* and *Peltrigera polydactyla*. *New Phytol.*, *82*:489–503, 1979.
48. Lang, O. L. Hitze- und Trochenresistenz der Flechten in Beziehung zu ihrer Verbreitung. *Flora*, *140*:39–97, 1953.
49. Höfmeister, W. *Allgemeine Morphologie der Gewaschse*. Engelmann, Leipzig, 1868.

50. Becquerel, P. Reviviscence du *Xanthoria parientina* desseche avec sa faune, a – 189°C. Ses consequences biologiques. *C. R. Acad. Sci.*, *226*:1413–1414, 1948.
51. Ewart, A. J. On the power of withstanding desiccation in plants. *Proc. Liverpool Biol. Soc.*, *11*:151–159, 1897.
52. Schroeder, G. Über die Austrocknungsfähigkeit der Pflanzen. *Unters. Bot. Inst. Tübingen*, *2*:1–52, 1986.
53. Thomas, H. H. Some observations on plants in the Lybian desert. *J. Ecol.*, *2*:75–88, 1921.
54. Heber, U. and Santarius, K. A. Water stress during freezing. In *Ecological Studies, Analysis and Synthesis* (O. L. Lang, L. Kappen, and K. D. Schulze, eds.), Springer-Verlag, Heidelberg, 1976, pp. 253–267.
55. Daft, M. J. and Nicholson, T. H. Effect of endogone mycorrhiza on plant growth. *New Phytol.*, *65*:343–350, 1966.
56. Gerdemann, J. W. The effect of mycorrhiza on the growth of maize. *Mycologia*, *56*:342–349, 1964.
57. Gray, L. E. and Gerdemann, J. W. Influence of vesicular-arbuscular mycorrhizas on the uptake of phosphorus-32 by *Liriodendron tukipifera* and *liquidambar styraciflua*. *Nature*, *213*:106–107, 1967.
58. Fitter, A. H. and Hay, R. K. M. *Environmental Physiology of Plants*, Academic Press, New York, 1981, pp. 110–114.
59. Heyman, D. S. Mycorrhiza and crop production. *Nature*, *287*: 487–488, 1980.
60. Sanders, F. E., Tinker, P. B., Black, R. L. B., and Palmerley, S. M. The development of endomycorrhizal root systems. I. Spread of infection and growth-promoting effects with four species of vesicular-arbuscular endophyte. *New Phytol.*, *78*:257–268, 1977.
61. Lauenroth, W. K., Dodd, J. L., and Sims, P. L. The effect of water and nitrogen induced stress on plant community structure in a semi-arid grassland. *Oecologia*, *36*:211–222.
62. Mosse, B. and Hayman, O. J. Plant growth responses to vesicular-arbuscular mycorrhiza. II. In unsterile field soil. *New Phytol.*, *70*: 29–34, 1971.
63. Safir, G. R., Boyer, J. S., and Gerdemann, J. W. Mycorrhizal enhancement of water transport in soybean. *Science*, *172*:581–583, 1971.
64. Safir, G. R., Boyer, J. S., and Gerdemann, J. W. Nutrient status and mycorrhizal enhancement of water transport in soybean. *Plant Physiol.*, *49*:700–703, 1972.
65. Sieverding, E. Einfluss der Bodenfeucht auf die Effektivitat der VA-mykorrhiza. *Angewandta Botanik*, *53*:91–98, 1979.
66. Levy, I. and Krikun, J. Effect of vesicular-arbuscular mycorrhiza on *Citrus jambhiri* water relations. *New Phytol.*, *85*:25–32, 1980.
67. Allen, M. F. and Boosalis, M. G. Effect of two species of VA-mycorrhizal fungi on drought tolerance of winter wheat. *New Phytol.*, *93*:67–76.
68. Allen, M. F., Smith, W. K., Moore, T. S., Jr., and Christensen, M. Comparative water relations and photosynthesis of mycorrhizal and non-mycorrhizal *Bouteloua gracilis* (H.B.K.) Lag Ex steud. *New Phytol.*, *88*:683–693, 1981.

69. Boyer, J. S. Water transport in plants: Mechanism of apparent changes in resistance during absorption. *Planta, 117*:187–201, 1974.
70. Hardie, K. and Leyton, L. The influence of vesicular-arbuscular mycorrhiza on growth and water relations of red clover. *New Phytol., 89*:599–608, 1981.
71. Fiscus, E. L. and Markhardt, A. H. Relationships between root system water transport properties and plant size in *Phaseolus*. *Plant Physiol., 64*:770–773, 1979.
72. Allen, M. F. Influence of vesicular-arbuscular mycorrhizae on water movement through *Bouteloua gracilis* (H.K.B.) Lag Ex Steud. *New Phytol., 91*:191–196, 1982.
73. Fiscus, E. L. Determination of hydraulic and osmotic properties of soybean root systems. *Plant Physiol., 59*:1013–1020, 1977.
74. Levy, Y., Syvertsen, J. P., and Memec, S. Effect of drought stress and vesicular-arbuscular mycorrhiza on citrus transpiration and hydraulic conductivity of roots. *New Phytol., 93*:61–66, 1983.
75. Ramos, C. and Kaufmann, M. R. Hydraulic resistance of rough lemon roots. *Physiol. Plantarum, 45*:311–314, 1979.
76. Graham, J. H. and Syvertsen, J. P. Influence of vesicular-arbuscular mycorrhiza on the hydraulic conductivity of roots of two citrus root stocks. *New Phytol., 97*:277–284, 1984.
77. Graham, J. H., Leonard, R. T., and Menge, J. A. Membrane-mediated decrease in root exudation responsible for phosphorus inhibition of vesicular-arbuscular mycorrhiza formation. *Plant Physiol., 68*:548–552.
78. Daft, M. J. and Okusanya, B. O. Effect of endogone mycorrhiza on plant growth. VI. Influence of infection on the anatomy and reproductive development in four hosts. *New Phytol., 72*:1333–1339, 1973.
79. Whittingham, J. The biology of mycorrhizae with special reference to the ecology of semi-natural limestone grasslands. Ph.D. thesis, University of Sheffield, U.K., 1980.
80. Bethlenfalvay, G. J., Brown, M. S., Mishra, K. L., and Stafford, E. Glycine Glomus-Rhizobium symbiosis. *Plant Physiol., 85*:115–119, 1987.
81. Koide, R. The effect of VA mycorrhizal infection and phosphorus status on sunflower hydraulic and stomatal properties. *J. Exp. Bot., 36*:1087–1098, 1985.
82. Huang, R.-S., Smith, W. K., and Yost, R. S. Influence of vesicular-arbuscular mycorrhiza on growth, water relations, and leaf orientation in *Leucaena leucocephala* (Lam.) de Wit. *New Phytol., 99*:229–243, 1985.
83. Nelsen, C. E. and Safir, G. R. Increased drought tolerance of mycorrhizal onion plants caused by improved phosphorus nutrition. *Planta, 154*:407–413, 1982.
84. Nelsen, C. E. and Safir, G. R. The water relations of well-watered mycorrhizal and non-mycorrhizal onion plants. *J. Amer. Soc. Hort. Sci., 107*:271–274, 1982.
85. Ramakrishnan, R., Johri, B. N., and Gupta, R. K. Effect of vesicular-arbuscular mycorrhizal fungus on photosynthesis and

photorespiration in water-stressed maize. *Photosynthetica, 22*: 443–447, 1988.
86. Mosse, B. Growth and chemical composition of mycorrhizal and non-mycorrhizal apples. *Nature, 179*:922–924, 1957.
87. Schultz, R. C., Kormanik, P. P., Bryan, W. C., and Brister, G. H. Vesicular-arbuscular mycorrhizae influence growth but not mineral concentrations in seedlings of eight sweetgum families. *Can. J. For. Res., 9*:218–223, 1979.
88. Singh, T. N., Aspinall, D., and Paleg, L. G. Proline accumulation and varietal adaptability to drought in barley: A potential metabolic measure of drought resistance. *Nature, 236*:188–190, 1972.
89. Aspinall, D. and Paleg, L. G. Proline accumulation: Physiological aspects. In *The Physiology and Biochemistry of Drought Resistance in Plants* (L. G. Paleg and D. Aspinall, eds.), Academic Press, 1981, Australia, 1981, pp. 206–240.
90. Ramakrishnan, R., Johri, B. N., and Gupta, R. K. Influence of VAM-fungus *Glomus calendonius* on free proline accumulation in water-stressed maize. *Curri. Sci., 57*:1082–1084, 1988.
91. Gupta, R. K. The study of photosynthesis and leakage of solutes in relation to the desiccation effects in bryophytes. *Can. J. Bot., 55*: 1186–1194, 1977.
92. Gupta, R. K. Effects of various rates of desiccation on photosynthesis and photosynthates leakage in young and aged tissues of the moss *Fissidens adianthoides* (Hedw.). *Photosynthetica, 15*:347–350, 1981.
93. Gupta, R. K. Drought injury and membrane integrity in young and aged tissues of a moss, *Neckera crispa* (Heds.). *Trop. Plant Sci. Res., 1*:259–261, 1983.
94. Gupta, R. K. Morphological and physiological studies in bryophytes: A comparative study of cell size, soluble sugars, proteins and phospholipids in resistant and non-resistant liverwort. *Indian J. Exp. Biol., 15*:695–697, 1977.
95. Gupta, R. K. An artefact in studies of the responses of respiration of bryophytes to desiccation. *Can. J. Bot., 55*:1195–1200, 1977.
96. Gupta, R. K. The physiology of the desiccation resistance in bryophytes: Effect of desiccation on water status and chlorophyll-a and b in bryophytes. *Indian J. Exp. Biol., 16*:354–356, 1978.
97. Kaloyereas, S. A. A new method of determining drought resistance. *Plant Physiol., 33*:232–234, 1958.

4

MYCORRHIZAL INTERACTIONS WITH SOIL ORGANISMS

TIMOTHY C. PAULITZ* and ROBERT G. LINDERMAN *Horticultural Crops Research Laboratory, Agricultural Research Service, United States Department of Agriculture, Corvallis, Oregon*

I. INTRODUCTION

Mycorrhizal fungi occur in nearly all soils on earth and form a symbiotic relationship with roots of most terrestrial plants. The mycorrhizal morphology is the basis for grouping them into two major groups: ectomycorrhizae and endomycorrhizae; the latter group is further divided into ericoid, orchid, and the ubiquitous and large group of vesicular-arbuscular (VA) mycorrhizae. A few plant families are facultative and can form either ecto- or endomycorrhizae, and some form no mycorrhizae. Endomycorrhizal fungi penetrate root cortical cells intracellularly and form special haustoria-like structures that interface with the host cytoplasm, separated only by the host plasma membrane and the fungal cell wall. Ectomycorrhizal fungi form an external fungal mantle and an intercellular hyphal network in the cortex called a Hartig net. These fungal structures in the cortex provide increased surface area for metabolic exchanges between the host and fungal partners.

Mycorrhizal fungi also interface directly with the surrounding soil at the mantle (ectomycorrhizae only) and by means of hyphae, hyphal strands, or rhizomorphs that extend into the soil and increase the nutrient and water absorption potential of the root system and contribute greatly to the improvement of soil texture for better aeration and water percolation.

Colonization of short roots by ectomycorrhizal fungi induces drastic changes in root branching due to production of phytohormones in the colonized tissues, resulting in greatly increased surface area for nutrient absorption. Formation of endomycorrhizae does not change root morphology other than by loss of root hairs. When plants become mycorrhizal, their physiology significantly changes. For example, there are changes in concentrations of growth-regulating compounds such as auxin, cytokinin, gibberellins, and ethylene. Photosynthetic rates increase, and there is a change in the partitioning of photosynthate to shoots and roots. The nutritional composition of the host tissues changes in response to altered uptake of minerals from the soil. Altered nutrient composition in turn can change structural and biochemical aspects of root cells that can alter

---

**Present affiliation*: Macdonald College of McGill University, Ste. Anne de Bellevue, Quebec, Canada.

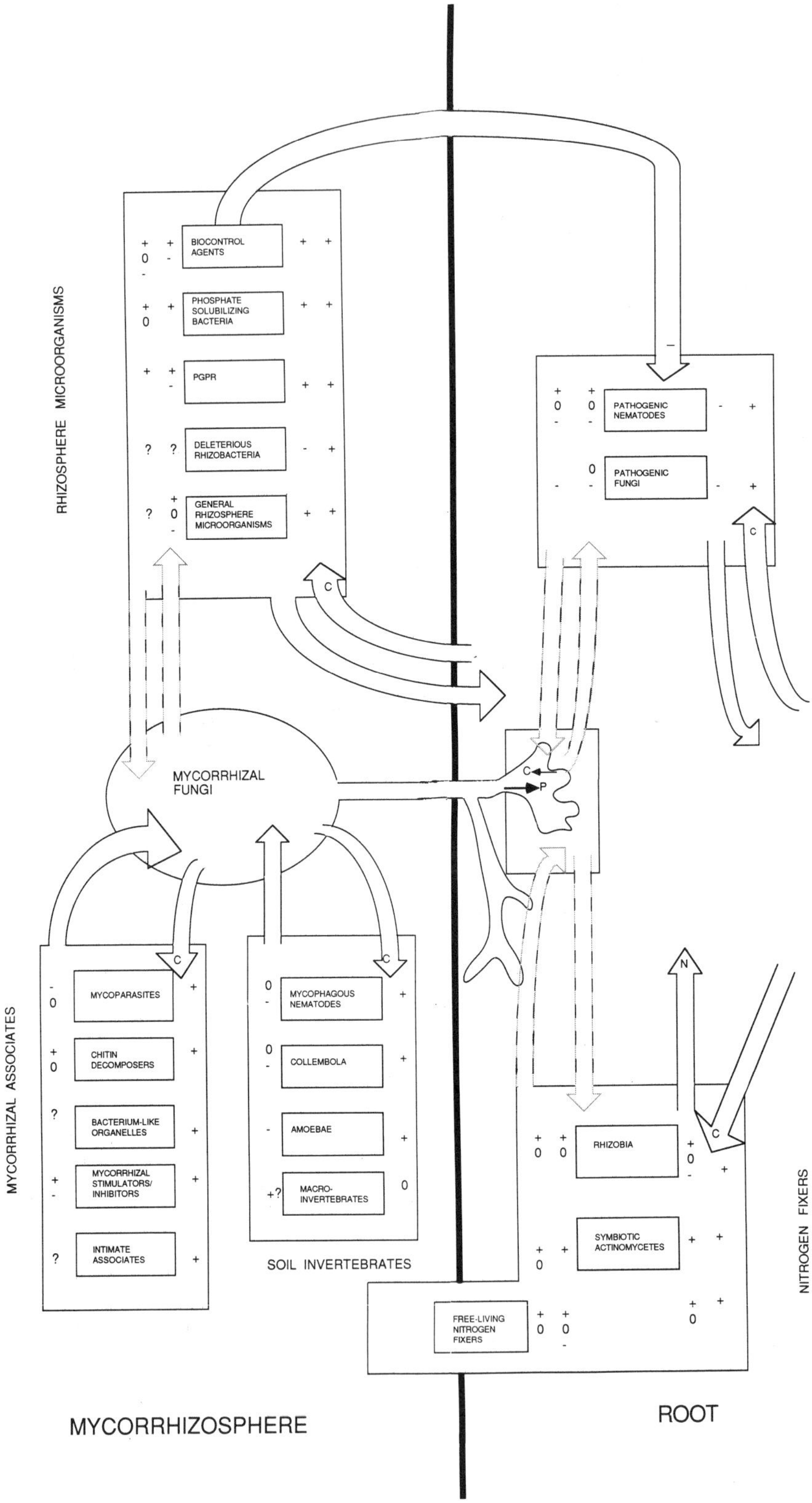
RHIZOSPHERE MICROORGANISMS
BIOCONTROL AGENTS
PHOSPHATE SOLUBILIZING BACTERIA
PGPR
DELETERIOUS RHIZOBACTERIA
GENERAL RHIZOSPHERE MICROORGANISMS
PLANT PATHOGENS
PATHOGENIC NEMATODES
PATHOGENIC FUNGI
MYCORRHIZAL FUNGI
MYCORRHIZAL ASSOCIATES
MYCOPARASITES
CHITIN DECOMPOSERS
BACTERIUM-LIKE ORGANELLES
MYCORRHIZAL STIMULATORS/ INHIBITORS
INTIMATE ASSOCIATES
MYCOPHAGOUS NEMATODES
COLLEMBOLA
AMOEBAE
MACRO-INVERTEBRATES
SOIL INVERTEBRATES
NITROGEN FIXERS
RHIZOBIA
SYMBIOTIC ACTINOMYCETES
FREE-LIVING NITROGEN FIXERS
MYCORRHIZOSPHERE
ROOT

membrane permeability and thus the quality and quantity of root exudation. These changes are likely to be more or less permanent, and thus may induce changes in the composition of microorganisms in the rhizosphere soil. It is appropriate, therefore, to refer to the rhizosphere around mycorrhizae as the "mycorrhizosphere" [1,2]. The purpose of our review is to discuss those changes and the ways that mycorrhizal fungi and mycorrhizae influence or are influenced by other occupants of the mycorrhizosphere.

When mycorrhizal fungi form a symbiotic association with plant roots, they may interact directly with other organisms in the soil, or they may influence those organisms indirectly by their effects on the host plant physiology that could change root morphology, physiology, and thus patterns of exudation into the mycorrhizosphere. Similarly, other root symbionts, such as nodule-forming bacteria, can indirectly influence the behavior of mycorrhizal fungi by changes in the host physiology. The kinds of direct and indirect interactions between mycorrhizal fungi and other soil organisms are shown in Figure 1. Mycorrhizosphere reactions are also influenced by edaphic and host factors. For example, soil pH, moisture, nutrient content, organic matter content, and physical properties can greatly influence the microbial interactions that occur, because each may influence the behavior of a specific organism in some way. Similarly, those edaphic factors, as well as environmental and climatic factors, can also influence the growth and physiology of the host plant. In addition, applications of various chemicals to the host plant can alter its physiology, which in turn can affect rhizosphere reactions. And, of course, changing the genotype of the host can change the nature of the mycorrhizal symbiosis too, as was recently demonstrated by Krishna et al. [3]. Thus, our discussion of the interactions that may occur between mycorrhizae and soil organisms must include consideration of the host genotype and physiology as well as influences of the soil and environmental factors on all mycorrhizosphere processes.

II. MYCORRHIZA–SOIL ORGANISM INTERACTIONS

A. Interactions with General Rhizosphere Microflora

Induced Rhizosphere Changes: Mycorrhizosphere Effect

Implicit in the concept of "mycorrhizosphere" is the hypothesis that mycorrhizae exert a strong influence on the microflora of the rhizosphere. Mycorrhizal fungi might alter the physiology and exudation of the root, resulting in a change in composition of microbial populations that depend on roots for a source of energy. External hyphae of mycorrhizal fungi might provide a physical or nutritional substrate for bacteria. Ectomycorrhizal fungi excrete growth-promoting substances and antimicrobial compounds that might alter the microbial makeup of the mantle. Many of the

---

FIGURE 1. Types of interactions that can occur between mycorrhizal fungi and other soil organisms. Interactions, depicted by direction of arrows, are either direct between any two types of organisms (solid arrows) or indirect and mediated by the host plant (broken arrows). The nature of the interaction is positive (+), negative (−), neutral (o), or unknown (?). Some arrows indicate the direction of internal flow of carbon (C) and nitrogen (N) within the host root.

papers on interactions with specific microbial groups, such as free-living nitrogen fixers, provide evidence for these hypotheses. The work cited in this section will deal with more general groups of microbes not covered in the latter categories.

In rhizosphere analysis of mycorrhizal and nonmycorrhizal tomato plants, Bagyaraj and Menge [4] showed that greater populations of bacteria and actinomycetes occurred in the mycorrhizosphere, compared to the noninoculated control. Newman et al. [5] used a multivariate statistical approach to study mycorrhizae and root-surface microorganisms on *Plantago*. They reported a weak correlation ($p < 0.02$) between mycorrhizal infection and bacterial cover. Another approach to documenting bacterial interactions in the rhizosphere was reported by Ames et al. [6]. Five bacterial species isolated from blue grama were added back to noninoculated and VAM-inoculated pot-grown plants. In the rhizosphere soil, populations of an actinomycete, a fluorescent pseudomonad, and a gram-positive nonmotile rod were lower in the treatments inoculated with *Glomus mosseae*; however, the only difference that was statistically significant was with the gram-positive bacterium, when expressed as cfu/g of whole root. On the other hand, a species of *Corynebacterium* and a gram-negative rod were more numerous in the mycorrhizal treatment. Based on the small number of bacterial strains studied, the authors suggested that the species composition of the rhizosphere microflora may be altered by the formation of mycorrhizae. This hypothesis was supported by Meyer and Linderman [7], who demonstrated differences in populations of taxonomic and functional groups of bacteria associated with VAM and non-VAM plants. Populations of total bacteria were higher in the rhizoplane of mycorrhizal sweet corn and clover, compared to nonmycorrhizal plants. Chitinase-producers, *Streptomyces* spp., and fluorescent pseudomonads were more numerous in the rhizosphere of nonmycorrhizal plants. The authors concluded that the microbial equilibrium was altered by the formation of VA mycorrhizae.

The realization that VAM fungi can alter the microbial makeup of the rhizosphere has led to increased scrutiny of control treatments in mycorrhizal experiments. Ames et al. [8] refuted the assumption that similar microbial groups become established in VAM fungal and control treatments. They demonstrated that microbial populations differ between VAM and non-VAM plants, and that soil from non-VAM pot cultures cannot be used as a valid control. Populations of bacteria and actinomycetes were quantitatively and qualitatively analyzed in pot cultures of different VA mycorrhizal fungi by Secilia and Bagyaraj [9]. The populations of total bacteria and N fixers in the rhizosphere of guinea grass were greater in plants colonized with *Glomus fasciculatum*, *Gigaspora margarita*, and *Sclerocystis dussi* than in the non-VAM plants. Actinomycetes were stimulated in roots that were colonized by *G. fasciculatum* and *Acaulospora laevis*. Morphological and physiological groups of bacteria were also examined, including gram-positive and gram-negative bacteria, spore formers, urea hydrolyzers, and starch hydrolyzers. The frequency of bacteria in each group varied among VAM species. Especially notable were the high numbers of starch hydrolyzers found on the roots colonized by *G. margarita* and *Sclerocystis dussi*. Urea hydrolyzers were present in all the VAM plants, but were absent in the nonmycorrhizal plants. This study confirmed that VAM fungi influence the populations of bacteria in greenhouse pot cultures; and that this effect varies with each VAM fungus.

Volatile substances produced in the mycorrhizosphere might also have an influence on general microbial populations. The in vivo effect of volatiles from ectomycorrhizae on soil microbes was examined by Schisler and Linderman [10], who used a device that purged volatiles from the soil around ectomycorrhizal Douglas fir seedlings (donor), and introduced the volatiles into a second soil (receiver). Volatiles from the ectomycorrhizae of *Hebeloma crustuliniforme* and *Laccaria laccata* stimulated the total bacterial populations over the nonmycorrhizal control.

Mycorrhizae exude organic materials that are substrates for other soil microbes that colonize the mycosphere. The hyphal associates frequently produce polysaccharides that cause soil particles to adhere, creating small aggregates. These aggregates are present in most undisturbed soils and impart structural properties that allow for aeration and water percolation, as well as stability. Sutton and Sheppard [11] showed that hyphae of VA mycorrhizal fungi play a key role in the formation of aggregates that improve the stability of sand dunes. Forster and Nicolson [12] found that these aggregates contained a range of microorganisms. These microbial associates of mycorrhizal fungi may have profound effects on the further development of hyphae in soil. The microbial metabolic products may readily be absorbed by the hyphae and translocated to the host. The specific functional composition of these aggregates and the metabolic products produced therein are virtually unknown and should be the subject of much needed future research.

It is difficult to draw any general conclusions from the above studies, other than the fact that microbial populations in the rhizosphere are influenced by mycorrhizal fungi. The effects vary with time, the species of mycorrhizal fungus, the host, and other environmental parameters. These studies also point out the difficulty of determining biologically significant differences in population densities of microbes. Many of these studies claimed statistical differences, but one could question whether the proper statistical tests were used. Microbial population densities in the rhizosphere are extremely variable and tend to follow a log-normal distribution [13]. Statistical tests that assume a normal distribution of populations might overestimate the true mean of the population. The interpretation of plate-count data is difficult, and other techniques should be developed that measure the specific activity of an organism, rather than colony-forming units. For example, the measurement of specific functional groups that break down or produce a certain substance might give a clearer picture, rather than just measuring total numbers of bacteria or actinomycetes.

VA Mycorrhiza Stimulation/Inhibition

In most studies that claim an interaction between bacteria and mycorrhizal fungi, root colonization was the only fungal parameter measured. This final "snapshot" represents the culmination of a series of developmental events. These events include spore germination, growth of mycelia through soil, stimulation and attachment of infective hyphae to the root, penetration of the root, establishment of biotrophy, and secondary spread outside and inside the root. Most attempts to narrow in on these events have focused on these first two steps – spore germination and germ tube growth – which can be easily examined in vitro. However, the latter steps are tied to a living plant, since VAM fungi are obligate biotrophs and cannot be grown independent of the plant. This makes it more difficult to study these subsequent events. Yet these final events occur within the rhizosphere, which contains active and dynamic populations of microorganisms that can affect the formation of the symbiosis, either by stimulating or inhibiting the formation of mycorrhizae.

Stimulators: Many researchers have observed that spores of VAM fungi germinated better on agar in the presence of contaminating bacteria. Further evidence of this phenomenon was provided by Daniels and Trappe [14] with spores of *Glomus epigaeus* (= *G. versiforme*) that did not germinate in autoclaved or gamma-irradiated soils, but germinated in nonsterile soils. The addition of autoclaved kaolin or activated charcoal to the autoclaved soil enhanced spore germination. The authors suggested that self-inhibitors were present in the spores of *G. epigaeus*, which were inactivated by soil microbes or absorbed and immobilized by substances with a high ion exchange capacity. Mayo et al. [15] showed that spores of *G. versiforme* that were completely surface disinfested and incubated on water agar germinated less than spores that were contaminated with bacteria. When antibiotics were added to the agar to inhibit bacterial growth, VAM spore germination was suppressed. Several of these bacteria, which belonged to genera including *Pseudomonas* and *Corynebacterium*, enhanced the germination of spores. VAM hyphae that developed from spores germinated in the presence of bacteria were longer and more extensively branched than those from spores without bacteria. Root colonization of alfalfa plants by *Glomus mosseae* in sterile soil was higher when nonsterilized spores were used as inoculum, as compared to surface-disinfested spores [16]. The addition of filtrates from natural soil that contained soil microorganisms increased VAM infection from surface-disinfested spores. The authors inferred that soil microorganisms influence the independent hyphal growth of preinfection stages of VAM formation. The presence of unidentified soil fungi on water agar hastened the germination of *G. mosseae* spores, when compared to the uninoculated control [17]. These fungi also enhanced VAM hyphal growth and the formation of vegetative spores. This stimulation was evident even though there was no contact between *G. mosseae* and the fungi, indicating that a volatile or diffusible substance might be affecting the VAM fungus. In a later study, cell-free supernatant from a culture of a stimulatory bacterium was applied to filter paper discs, which were placed on water agar containing spores of *G. mosseae* [18]. Both the cell-free extract and the complete bacterial culture enhanced hyphal growth and the formation of secondary vesicles. Washed bacterial cells gave no effect. This gives further evidence that some water-soluble, diffusible microbial metabolite is responsible for the effect.

Spores of the isolate of *G. mosseae* used by Mugnier and Mosse [19] did not germinate spontaneously on water agar. When the actinomycete *Streptomyces orientalis* was applied to the bottom of an agar overlay, 85–90% of the spores germinated. If the actinomycete was applied to one side of a divided petri dish, the same effect was seen, indicating the involvement of a volatile substance. The hypothesis that volatile substances produced by the plant or soil microbes might stimulate mycorrhizal infection was examined by St. John et al. [20], who used sterile clover seedlings inoculated with *Gigaspora margarita* and grown in closed containers. In containers with potassium permanganate, a substance that oxidizes a wide range of organic compounds, there was significantly less mycorrhizal infection. The authors speculated that volatiles produced by the plant or microbes stimulated VAM infection, and that these volatiles were removed by potassium permanganate. Koske [21] and Gemma and Koske [22] gave evidence that a volatile attractant from plant roots affected the germ tubes of *Gigaspora gigantea*. Germ tubes grew vertically through the air to contact bean or corn roots up to 11 mm away. This effect was seen in both sterile and nonsterile systems. Microbes might produce these same volatiles or influence their production and diffusion by the plant.

Soil microbial factors could also be involved in the stimulation or inhibition of the development of the extramatrical hyphae of VAM. Graham et al. [23] demonstrated that these hyphae do not necessarily develop in proportion to the extent of intraradical hyphal colonization. Some isolates or species of VA mycorrhizal fungi appear to be incompatible with certain soils and fail to develop extramatrical hyphae, even though spore germination and root colonization have occurred. One explanation might be the lack of specific microbes that stimulate hyphal development or suppress those that might inhibit it. Sutton and Sheppard [11] demonstrated that VA mycorrhiza formation on bean in sand dune soil was suppressed in sterilized soil but enhanced in nonsterile soil or sterile soil inoculated with nonsterile extracts, presumably containing a complex of microorganisms. Furthermore, sand aggregation, indicating development of extramatrical hyphae, was also enhanced by inoculation with *Glomus* plus soil extract. While the primary energy source for hyphal development is the host plant, the extent of the hyphal network appears to be strongly influenced by the soil microflora.

Inhibitors/Antagonists: Many VAM spores exist in a quiescent state and and will germinate in some soils but not in others [24]. If inhibitory soils are pasteurized, added spores can germinate. Therefore, soil microbes might also be involved in the inhibition of spore germination. Quiescent spores of some VAM species failed to germinate in soils under crops or pastures but did germinate in virgin forest soil [25]. Water-soluble, heat-labile compounds which prevented germination were isolated from inhibitory soils, but not from the forest soil. Self-inhibitors might also be present, since washing of the spores increased germination. The blockage of germination caused by sterile inhibitory soil extracts was relieved by adding VAM spore-contaminating bacteria. Tommerup [25] speculated that germination is regulated by a number of factors, including inhibitory soil compounds, self-inhibitors, microbial activity, and inducers within the spore. This inhibition of germination might have an ecological function, that is, to enhance survival during times when successful infection of plants is unlikely.

Other studies have indirectly suggested the involvement of microbes in mycorrhizal inhibition. The amendment of nonsterile soil to steamed soil reduced sporulation of *Glomus macrocarpus* [26]. Similarly, VAM sporulation was higher in fumigated than in nonfumigated microplots planted with soybean. Ross suggested that biotic factors in nontreated agricultural soils reduce VAM sporulation compared to fumigated soils, and this resulted in low inoculum density for the subsequent growing season.

Many authors have demonstrated that soil microorganisms in nonsterile soil may suppress spore germination and mycorrhizal growth response [27–30]. However, another explanation is that steam treatment of the soil removed minor pathogens, such as *Pythium* spp., or increased nutrient availability, which would mimic a mycorrhizal growth response. The lack of growth response in nonsterile soil could also be due to infection by indigenous VAM species that were not as effective as the introduced species.

Data from later research [31] puts the theory of microbial suppression of VAM into some doubt. Using the same VAM species and big bluestem, plants exposed to nonsterile soil were more extensively colonized and grew faster than plants in pasteurized soil. Only after 13−16 weeks did the colonization and dry weights of plants in the pasteurized treatment exceed those in the nonsterile. If reduced inoculum potential of the VAM fungus in nonsterile soil was responsible for growth suppression, then colonization and growth in the nonsterile soil should have lagged behind those in the pasteurized soil. The opposite result was found. Also, the suppression of growth and root colonization in nonsterile soil should have been overcome by increasing VAM inoculum dosages, but it was not. The authors then concluded that the loss of inoculum potential alone cannot fully explain the suppression of plant growth and root colonization in nonsterile soil.

Ectomycorrhiza Stimulation/Inhibition

Evidence for both microbial stimulation and inhibition of ectomycorrhizae has been reported. Interactions between bacteria and ectomycorrhizal fungi were tested on laboratory media and on the rhizoplane of *Pinus radiata* [32]. In vitro results were inconsistent, but in the rhizosphere the effects of the bacteria ranged from growth depression to stimulation. One fluorescent pseudomonad, strain FP4, reduced root colonization by all ectomycorrhizal fungi tested to less than 20% of the noninoculated control. This inhibition was not relieved by the addition of glucose, suggesting that antibiosis might be a possible mechanism. An isolate of *Bacillus* from the mycosphere of *Rhizopogon luteolus* stimulated colonization by up to 158% of the control. This work demonstrated that ectomycorrhizal fungi interact with soil microorganisms, and this might determine whether inoculated ectomycorrhizal fungi can be introduced and persist. However, the rhizoplane assay which Bowen and Theodorou [32] used was performed in sterile, autoclaved soil and probably would not reflect the natural situation, where individual bacteria have to compete with other microbes. In experiments performed under nonaxenic conditions, Garbaye and Bowen [33,34] avoided this artifact and isolated yeasts, actinomyetes, and bacteria from the mantle of *R. luteolus* on *Pinus radiata*. Growth of the ectomycorrhizal fungi in the soil and along the root were monitored in the presence and absence of these microbes. Out of the 25 significant interactions observed, 17 were positive. Seedlings of *P. radiata* were also grown in autoclaved soils that were inoculated with ectomycorrhizal fungi and amended with filtrates from four nonautoclaved soils. The microbial suspensions altered mycorrhizal infection in 14 out of 36 treatments, with stimulation occurring in eight cases and inhibition occurring in six cases. *Rhizopogon luteolus* formed mycorrhizae only when a filtrate from a nursery soil was added. The effect of the filtrates on ectomycorrhizae differed with each soil, suggesting that the soil type or plant-soil interaction played a role in the selection of a stimulatory microflora. In a similar experiment, McAfee and Fortin [35] grew jack pine and larch seedlings in autoclaved vermiculite−peat moss substrate inoculated with ectomycorrhizal fungi and amended with a soil suspension. The addition of the soil suspension significantly enhanced mycorrhiza formation by *Laccaria bicolor* and *Pisolithus tinctorius* on jack pine and increased the shoot length of the seedlings when compared to mycorrhizal seedlings without the suspension. On larch, the soil suspension had no influence on mycorrhizal formation, but did enhance shoot length of seedlings inoculated with *L. bicolor* and *P. tinctorius*. Strzelczyk and Kampert [36] showed that five bacterial isolates from the mycorrhizosphere of *Pinus sylvestris* had no effect or slightly stimulated the growth of four ectomycorrhizal fungi when grown together in liquid culture. In contrast, Summerbell [37] found that *Trichoderma viride* and *T. polysporum*, isolated from the mycorrhizosphere of black spruce, were strongly antagonistic to mycorrhizal colonization by *L. bicolor* when tested on axenically cultivated black spruce seedlings. *Tolypocladium inflatum* and *Trichosporon beigelii* also inhibited mycorrhizae formation, but not as much as the *Trichoderma* spp.

Despite these studies, there is still much to be learned about organisms that stimulate or inhibit mycorrhizal fungi. Most of the studies on spore germination and hyphal growth were done on water agar, a highly artificial system. These organisms need to be tested in competition with the background microflora to see if the observed effects still persist, or whether the observation is an artifact of the system. Axenic systems might still be useful for screening organisms, and the use of sterile sand or soil might overcome some of the nutritional artifacts of agar. Future research should look at effects on other stages besides spore germination and germ tube growth. There is very little literature on the involvement of microorganisms in the attachment and penetration of the root by the VAM fungus, in external secondary spread along the rhizoplane, and in sporulation. All of these events take place in the rhizoplane and rhizosphere, which contain a tremendous diversity and activity of microorganisms. Most of the work comparing sterile and nonsterile soil is difficult to interpret, since these treatments are so radical and change not only biological, but also physical, chemical, and nutritional properties.

B. Interactions with Rhizobacteria

Symbiotic Nitrogen Fixers

Rhizobium: Most leguminous plants are simultaneously symbiotic with nodule-forming *Rhizobium* and VA mycorrhizal fungi. *Rhizobium* fixes nitrogen, which benefits plant growth, and the mycorrhizal fungi benefit the host by increasing the efficiency of uptake of mineral elements and water from the soil as well as altering host metabolism and other physiological parameters. However, the tripartite relationship of host–*Rhizobium*–VAM represents a new dimension unlike either dipartite symbiosis. A wide range of leguminous host species including those used for forage and range (*Medicago*, subclover, sweetvetch, siratro, trifolium), as human food sources (soybean, faba bean, bean, cowpea, pigeon pea, chickpea), and some woody perennial tree species (*Leucaena*, *Hedysarum*) have been extensively studied to determine the nature of the relationship. These studies have been conducted under controlled conditions in containers using soilless media or field soil that had been sterilized or pasteurized or left natural, or were conducted under field conditions. Although experimental conditions had a major impact on the results and their interpretation, there were some rather consistent findings throughout all the studies. The subject has been reviewed by Tilak [38], Barea and Azcon-Aguilar [39], and Cluett and Boucher [40], so much of the detailed *Rhizobium*–mycorrhiza interaction reports will not be specifically mentioned.

When legumes are symbiotic with both *Rhizobium* and VA mycorrhizal fungi, plant growth is generally much greater than control plants symbiotic with either alone. Many reports [41–43] indicate that the VAM increased P uptake, which under P stress conditions would itself increase plant growth. Improved P nutrition in turn would favor the nitrogen fixation process by *Rhizobium*. The combination of effects resulted in further growth enhancement [44–49]. However, factors that influence each symbiont separately may also affect the tripartite relationship. All strains of *Rhizobium* or strains or species of VA mycorrhizal fungi do not affect their host plant in the same manner or to the same degree. Different edaphic factors influence the behavior and effectiveness of the symbionts, so would affect the combination as well [50,51]. Furthermore, it has been suggested in some studies [52] that interendophyte compatibility may also play an important role in the combined reaction on plant growth.

Mycorrhizal fungi and *Rhizobium* could directly interact in the rhizosphere of their host plant before either has formed a symbiotic relationship with the host. No experimental evidence exists to support that idea, however. Generally, VA mycorrhizal fungi are first to colonize the root, and their advance establishment seems to favor the process of nodulation by *Rhizobium* [53,54]. Increased numbers of nodules on mycorrhizal vs. nonmycorrhizal plants has been demonstrated in numerous studies. One study [55], however, demonstrated that advanced formation of VAM inhibited nodulation, but that may have been a case of interendophyte incompatibility or competition. Even when nodule number has not been significantly increased by VAM, the size and nitrogen-fixing activity of the nodules has been shown to be increased [56]. The explanation for this seems to be in the partitioning of photosynthate into the roots supporting both symbionts. The photosynthetic rate of mycorrhizal plants is greater than in nonmycorrhizal plants [57,58], and a greater percentage of the photosynthate goes to nodules of mycorrhizal than nonmycorrhizal plants [56, 58–61]. Furthermore, the P content of nodules of mycorrhizal plants is generally greater in VAM plants that non-VAM plants [48]. These factors could account for the increased size of the nodules as well as the increased energy for nitrogen fixation. In addition, VAM plants absorb more micronutrients from the soil which plays a rolc in the enhanced nitrogen fixation process [62,63].

Formation of VAM generally results in increased P levels in the plant tissues and thus changes in membrane permeability. The resultant changes in the quality and quantity of root exudates can induce changes in the rhizosphere populations, which could influence the competition between *Rhizobium* and other rhizobacteria. If bacteria selectively favored in the rhizoplane enhanced *Rhizobium* competitiveness, then nodulation would be favored. Grimes and Mount [64] showed, for example, that *Pseudomonas putida* markedly increased nodulation on bean, a report supported later by Meyer and Linderman [65] on subclover. Interestingly, the bacterium used in the bean study [64] also produced 2-ketogluconic acid, a P-solubilizing compound which might have increased the available P as nodulation occurred. Similarly, the bacterium used by Meyer and Linderman [65] was known to produce siderophores that could have limited the Fe available to microbes competing with the *Rhizobium*.

The fact that increasing available fertilizer P frequently has increased the rate of nitrogen fixation has led many to conclude that the role of VAM is to increase P nutrition of the plant. Recently, however, Ames and Bethlenfalvay [66] used a split-root system to demonstrate a localized, nonsystemic, non–P-mediated influence of VAM on cowpea root growth and nodule activity. The belief that the increased N fixation phenomenon with VAM was not P-mediated had previously been shown by Asimi et al. [67]

and others by fertilizing non-VAM plants with enough P to give equal P nutrition to VAM plants, thereby eliminating P growth effects without loss of enhanced N fixation.

Mycorrhizal fungi greatly assist plants in acquisition of mineral elements from soil, especially immobile elements such as P, Cu, Zn, and Mn. The extramatrical hyphae bridge the gap across the zone of nutrient depletion that forms near the root. Pacovsky [63] demonstrated that in soybean plants with and without VAM and fertilized to be of equal size, growth stage, and P status, that Cu and Zn concentrations, essential for nitrogen fixation, were always higher in VAM plants, while Fe and Mn were lower. Ames et al. [68] and later Barea et al. [69], using ^{15}N techniques, showed that both NH_4 and NO_3 sources of N can be readily absorbed by extramatrical hyphae of VAM and that growth enhancement of legumes by VAM can be attributed both to enhanced N fixation as well as improved N uptake from soil, especially NH_4 forms. Smith et al. [70] showed that VA mycorrhizal fungi are able to assimilate ammonium by means of their production of glutamine synthetase, an ability that would be important in uptake from soil as well as translocation within the plant. Ammonium released from nodules could then be absorbed by VAM hyphae, thereby improving the N economy of the plant. Similarly, adjacent nonleguminous plants that shared the same VAM symbiont and were connected by extraradical hyphae could benefit by transfer of nitrogen through the hyphal conduits [71,72]. VAM generally are not found in legume nodules, but such was recently reported for bean [73].

Nearly all leguminous plants with nodules formed by *Rhizobium* grow better if they are also colonized by VA mycorrhizal fungi. The tripartite association has been studied in containers with soilless media and in the field. Generally, the growth increase was explained by increased photosynthesis and increased carbon flow to the nodule and VAM sinks, resulting in more and larger nodules that fix more nitrogen for the plant. As with other VAM situations, however, the plant and the *Rhizobium* system benefit from the presence of VAM by the correction of that deficiency which is limiting the system. If P is limiting, then VAM enhancement of P uptake enhances the nitrogen-fixing process. If minor elements that are needed for maximum nitrogen fixation are limiting, then VAM can correct that deficiency. All the components of the tripartite system must function optimally, however, for maximum plant growth to occur. The *Rhizobium*–VAM partnership nearly always exists, but may not necessarily be optimized with the best combination of symbionts for the host species. The challenge to researchers is to characterize each tripartite relationship and optimize all components so that they function compatibly and in concert.

Symbiotic Actinomyetes (Frankia): Actinomycete endophytes in root nodules of certain shrubs and trees can fix nitrogen in situ. These actinomycetes, which are in the genus *Frankia*, form symbiotic associations with 15 genera of flowering plants, typically woody perennials growing in temperate areas, such as *Alnus*, *Cercocarpus*, *Purshia*, and *Ceanothus*. These host plants usually colonize nutrient-poor habitats such as forests, sand dunes, and arid soils. Relatively little is known about the ecology and survival of *Frankia* in soil, and even less is known about mycorrhizal–actinorhizal interactions.

In a survey of actinorhizal plants, both actinomycete nodules and VAM fungi were found on all 25 species examined [74]. These included species of *Alnus, Casuarina, Ceanothus, Hippophae, Cerocarpus,* and *Purshia. Glomus* spp. were the most commonly found VAM fungi and were associated with 16 of the 25 species. In another study, *Hippophae rhamnoides*, growing on sand dunes, was reported to be infected with *Glomus fasciculatum* [75]. *Purshia* and *Cowania*, rosaceous shrubs from arid regions of the western United States, were found to be both actinorhizal and mycorrhizal [76]. A triple symbiosis was observed on the tree *Casuarina equisetifolia*, on which an ectomycorrhizal fungus, a VAM fungus, and actinorhizal nodules were observed [77].

The tripartite association between actinomycetes and VAM fungi was examined in *Ceanothus* [78]. Plants with both actinomycete nodules and *Glomus gerdemannii* mycorrhizae had greater plant dry weights, number of nodules, nodule weight, and more nitrogenase activity than plants that were inoculated with the actinomycete only. Inoculation with the actinomycete also significantly enhanced the mycorrhizal infection, P content, and N content of plants. Gardner et al. [79] also reported enhanced plant dry weights, P content, and nitrogenase activity in *H. rhamnoides* inoculated with both actinomycete nodules and *G. fasciculatum*, compared to treatments with either alone. Because the plants were grown in sand with low levels of P and N, the authors attributed the increase in nitrogenase activity to better phosphorus nutrition. Diem and Gauthier [80] inoculated seedlings of *C. equisetifolia*, grown in sterile soil with low available P, with *G. mosseae, Frankia* nodules, or both. After 6 months, there were twice as many nodules on the VAM-infected actinorhizal plants, compared to the plants that were only actinorhizal. A similar trend was seen with nitrogen content in the shoots, but *Frankia* had no effect on VAM infection or P content.

Free-Living Nitrogen Fixers

Although symbiotic nitrogen-fixing bacteria account for a majority of the biologically fixed nitrogen, some free-living bacteria also can fix nitrogen. Free-living diazotrophic species are found in the genera *Azotobacter, Beijerinckia, Clostridium, Pseudomonas,* and *Azospirillum.* The ecology of nonsymbiotic dinitrogen fixers is reviewed by Subba Rao et al. [81].

Azospirillum spp. are ubiquitous soil organisms capable of colonizing the roots of tropical grasses and C_3 plants. Field trials have shown gains of 21 kg N/ha by inoculating grass seed with *Azospirillum* and increased yields in sorghum, maize, and barley [82]. Inoculation of barley and pearl millet plants with both *Azospirillum brasilense* and VAM fungi increased dry matter and grain yields, when compared to the VAM treatments without bacterial inoculation [83,84]. Inoculation with *A. brasilense* or VAM fungi alone did not significantly increase plant growth, and *A. brasilense* had no effect on VAM infection. These same effects were seen in a low-phosphorus soil fertilized with superphosphate or rock phosphate [85]. *Azospirillum brasilense* also stimulated VAM root infection in the noninoculated control (a nonsterile field soil) and in treatments inoculated with *G. margarita* or *G. fasciculatum*. *Azospirillum brasilense* stimulated the VAM infection of maize and ryegrass by *G. mosseae* [86]. Root and shoot dry weight of maize in the dually inoculated treatments were greater than in the treatment with *A. brasilense* + phosphorus and were similar to that of the noninoculated treatment fertilized with N and P. Sorghum plants, grown in a soil with low available N and P, were inoculated with *G. fasciculatum*, *A. brasilense*, or both [87]. The plants were fertilized with a nutrient solution without N or P. Both the VAM fungus and *A. brasilense* alone stimulated plant growth over noninoculated controls. The combination of the two symbionts enhanced plant growth over the single-symbiont treatments. The increases observed were additive, and no interaction was evident. The addition of *A. brasilense* to VAM treatments resulted in greater VAM colonization, biomass, and numbers of internal vesicles.

Later research in that laboratory examined carbon and nitrogen flow in the *Sorghum* – *Azospirillum* – *Glomus* association by use of $^{15}N_2$ and $^{14}CO_2$ labeling [88]. The presence of mycorrhizae resulted in a 60% decrease in N_2 fixation by *A. brasilense.* In a short-term assay, plants with *Azospirillum* alone fixed five times more $^{15}N_2$ than plants that were also mycorrhizal. The presence of *Azospirillum* and/or *G. fasciculatum* decreased net C fixation, and the presence of VAM increased the total allocation of C to the roots by 4–7%. Mycorrhizal plants were less colonized by *A. brasilense,* while *A. brasilense* increased VAM infection, suggesting that VAM fungi might compete with *Azospirillum* for carbon.

One difficulty in interpreting the results of many of these experiments is that of separating nutritional effects from other physiological effects. Pacovsky [89] circumvented this difficulty by fertilizing noninoculated sorghum plants with equivalent amounts of P and N to compensate for the input from *Azospirillum* and *Glomus.* The total dry weights of the fertilized, noninoculated and inoculated treatments were identical. However, the *Glomus*-inoculated plants contained less P, starch, sucrose, and more Cu and Zn than the P-fertilized plants. Plants inoculated with *Azospirillum* contained less N, glucose, glutamine, and more glutamate than N-amended plants. The mycorrhizal plants contained five unusual fatty acids not found in the noninfected roots. One of these fatty acids was correlated with internal vesicles [90]. The authors concluded that the response of sorghum to *Azospirillum* and *Glomus* cannot be accounted for solely as a N or P effect. These symbionts apparently alter the host physiology even under high-fertility regimes where nutrient input by the endophytes is negligible.

Interactions between VAM fungi and *Azotobacter*, another free-living diazotroph, were demonstrated by Bagyaraj and Menge [4]. The populations of *Azotobacter chroococcum* in the rhizosphere of tomato were enhanced by inoculation with *G. fasciculatum.* Inoculation with *Azotobacter* increased sporulation and root infection by *G. fasciculatum.* The combination of *Azotobacter* and *Glomus* together increased root dry weights over treatments with either used alone. In a recent report, Mohandas [91] found that inoculation of tomato plants in the field with *G. fasciculatum* or *Azotobacter vinelandii* alone or in combination increased plant growth, N content, P content, and yield over the uninoculated control.

Brown and Carr [92] reported that the simultaneous inoculation of roots of lettuce seedlings with *A. chroococcum* and VAM fungi produced larger plants than with either inoculum alone. The use of VAM fungi alone (a mixture of *G. mosseae* and *G. caledonius*) benefited plant growth in a partially sterilized, low-P soil but depressed growth in a soil with adequate P. *Azotobacter* did not influence the level of VAM infection, but the populations of *Azotobacter* in the rhizosphere were frequently lower in the presence of the VAM fungi.

Beijerinckia mobilis, another free-living diazotroph, stimulated spore production of *G. fasciculatum* but had no effect on root infection of onions [93]. The combination of *Glomus* and *Beijerinckia* increased root and shoot weight and nitrogen and phosphorus content of onions, when compared to plants inoculated with either organism alone.

Some free-living nitrogen fixers have been found closely associated with the thalli of mycorrhizal fungi. Tilak et al. [94] isolated *Azospirillum* sp. from surface-disinfested mycorrhizal roots of onion. *Azospirillum* spp. were also recovered from surface-disinfested spores of *G. fasciculatum*, *G. intraradices*, *G. mosseae*, and *Gigaspora gilmori* [95]. *Klebsiella pneumoniae* and two isolates of *Pseudomonas fluorescens*, all capable of fixing nitrogen, were isolated from the interior sporocarp tissue of the ectomycorrhizal fungi *Hymenogaster parksii*, *Suillus ponderosus*, and *Tuber melanosporum* [96]. Nitrogenase activity was detected within the sporocarps of *Laccaria laccata* and *Rhizopogon vinicolor*, which were incubated in nitrogen-free liquid medium under microaerophilic conditions [97,98]. *Clostridium* spp. and *Azospirillum* spp. were subsequently isolated from nitrogenase-active cultures. Li [99] also isolated diazotrophic bacteria, tentatively identified as *Azospirillum* spp., from the sporocarps of the ectomycorrhizal fungi *Barssia oregonensis*, *Hysterangium setchellii*, *Leccinum scabrum*, and *Rhizopogon parksii*.

This work suggests a close spatial association between mycorrhizal fungi and free-living diazotrophic bacteria. These bacteria might utilize P from mycorrhizal fungi, or mycorrhizal fungi might transport fixed nitrogen from the bacteria to the plant. Both possibilities should be investigated. *Azospirillum brasilense* has been isolated not only from the rhizosphere, but also from the root cortex (endorhizosphere) [100]. This opens up the possibility of a direct interaction between a VAM fungus and *Azospirillum* within the plant. Interactions might also occur via the production of plant hormones by free-living nitrogen fixers. Tien et al. [101] suggested that plant hormone production by *A. brasilense* might account for the increased growth of inoculated pearl millet. *Azotobacter vinelandii* and *A. beijerinckii* produced auxins, gibberellins, and cytokinins in culture [102]. Cell-free supernatants and whole cells of *A. vinelandii* and *A. beijerinckii* stimulated plant growth and VAM infection of lavender, tomato, and alfalfa inoculated with *G. mosseae* [103]. These same effects were mimicked by the addition of a combination of gibberellic acid, kinetin, and IAA. The involvement of plant hormones, especially cytokinins, in mycorrhizae has been suggested by many workers [104–106]. The production of cytokinins by both ectomycorrhizal and VA mycorrhizal fungi has been documented [107,108]. The stimulation of VAM infection and plant growth by some free-living nitrogen fixers could be due to a hormonal interaction, but more work is needed to definitively prove this hypothesis.

Plant Growth-Promoting Bacteria

Plant-growth promoting rhizobacteria (PGPR) are rhizosphere-inhabiting bacteria that have a beneficial effect on plant growth. This subject has been reviewed by Burr and Caesar [109]. The discovery that specific strains of the *Pseudomonas fluorescens-putida* group could rapidly colonize potato roots and produce significant yield increases [110] has stimulated research in this area. Exactly how beneficial rhizobacteria promote plant growth is not well understood, but they may produce antibiotics or siderophores that suppress minor plant pathogens or produce phytohormones or other growth-enhancing compounds that directly affect the plant. These rhizobacteria may also interact with mycorrhizal fungi to produce the growth-enhancing effects by as yet unknown mechanisms.

The involvement of phytohormones in both ecto and VA mycorrhizae has been suggested by many studies. These hormones may be produced by the mycorrhizal fungi themselves or by the plant in response to mycorrhizal colonization. Indole-3-acetic acid (IAA) production by ectomycorrhizal fungi was demonstrated conclusively by Ek et al. [111] using GC-MS analysis and by Madej and Haggblom [112] using an immunoassay technique. In most of these studies, the bacteria were grown in high concentrations of tryptophan, which might not be present in vivo. However, Mitchell et al. [113] found that the levels of IAA in short roots of *Pinus echinata* colonized with *Pisolithus tinctorius* were threefold higher than in nonmycorrhizal short roots. Many studies have suggested that dichotomous branching and other morphological changes that occur in mycorrhizal roots are induced by the production of auxin by the ectomycorrhizal fungus. However, recent work by Wilson and Field [114] and Rupp and Mudge [115] suggests that ethylene, possibly induced by endogenous auxin or auxin produced by the mycorrhizal fungus, was actually responsible for inducing dichotomous branching.

Ethylene production by ectomycorrhizal fungi was first demonstrated by Graham and Linderman [116]. Ethephon, an ethylene-releasing compound, stimulated lateral root formation of Douglas fir [117]. The authors proposed that ectomycorrhizal fungi may contribute to the production of low levels of ethylene in the rhizosphere, which might control mycorrhiza formation by triggering the development of short lateral roots. Wilson and Field [114] observed that ethylene stimulated dichotomous branching of short roots of *Pinus radiata*. Similarly, Rupp and Mudge [115] suggested that a close relationship exists between auxin and ethylene, and confirmed that ethylene could induce dichotomous branching. On the other hand, soil and foliar applications of ethephon (ethrel) depressed VAM formation in alfalfa and wheat [118].

Evidence for cytokinin production by ectomycorrhizal fungi was obtained by Ng et al. [107]. Strzelczyk et al. [119] confirmed that *Rhizopogon luteolus* and *Paxillus involutis* synthesized cytokinins. Barea and Azcon-Aguilar [108] extracted gibberellin and cytokinin-like compounds from water in which the spores of *G. mosseae* were germinated axenically. Axenically grown *Bouteloua gracilis* infected with *G. fasciculatum* had higher levels of cytokinins in the leaves and roots than in nonmycorrhizal plants [104]. The level of gibberellin-like substances was significantly increased in the leaves of infected plants but decreased in the roots [120]. All of this literature suggests that phytohormones and other growth-promoting substances might play some role in mycorrhizae formation and their effects on plant growth.

Many soil microbes have also been shown to produce phytohormones. Soil microorganisms commonly produce auxins in the presence of tryptophan. These bacteria can affect the morphology of roots in the same way as gibberellic acid and indoleacetic acid. Phytohormone production by bacteria can also have a deleterious effect on plant growth [121]. Plant-deleterious rhizobacteria also affect the plant through the production of toxins such as cyanide [122], which might have an effect on mycorrhizal formation.

Rhizobacteria alone can also enhance root exudation [123]. The presence of *Pseudomonas putida* in the rhizosphere of wheat caused a doubling of root exudation, when compared to axenic controls [124]. VAM and rhizobacteria might also interact via root exudation.

These morphological and physiological changes in the plant induced by bacteria-produced hormones might influence the formation of the mycorrhizal symbiosis. Growth-promoting substances might also act directly on the mycorrhizal fungi themselves. Is there any evidence for these interactions? Some work has implied that production of plant hormones by free-living nitrogen fixers and phosphate solubilizers is responsible for the observed growth increases. The interactions of mycorrhizal fungi with these organisms has been discussed in a previous section but may involve plant-growth–promoting substances. Cell-free supernatants from bacterial cultures of *Azotobacter* and a PSB (*Pseudomonas* sp.), known to contain growth substances, increased growth of plants colonized by *Glomus* spp. and enhanced VAM colonization [103]. The application of pure plant hormones mimicked these effects.

Microbes that produce plant-growth–promoting substances might have a close spatial association with the mycorrhizal fungi, residing on or around the fungal hyphae. Ectomycorrhizal fungi, bacteria, and actinomycetes were isolated from the mycorrhizosphere of *Pinus sylvestris* and were grown in tryptophan and tryptophan-free media [125]. Many of the bacteria and ectomycorrhizal fungi produced auxins and gibberellin-like substances. Some of these bacteria were capable of making cytokinin-like substances [126] and B vitamins. This work demonstrates the possibility that PGPR produce plant hormones and may be present in close association with the mycorrhizae. Soil microbes might also interfere with the production of hormones by the mycorrhizal fungi themselves. The synthesis of cytokinins by *Rhizopogon luteolus* in culture was reduced by the addition of pasteurized filtrates from a culture of *Streptomyces* [119].

More research is needed to determine the influence that PGPR and deleterious rhizobacteria have on mycorrhizae. In the future, the use of mutants deficient in the production of these growth substances can clarify the influence of these bacteria on mycorrhizae.

Phosphate-Solubilizing Bacteria

Many bacteria produce organic acids and solubilize inorganic and organic forms of phosphorus that are unavailable to plants. VAM fungi, on the other hand, do not take up insoluble P, but more efficiently utilize labile forms [127]. Some studies question this "central dogma" of VAM fungi and show that differences in the availability of soil P to mycorrhizal and nonmycorrhizal plants are not reflected by differences in ^{32}P labeling [128]. Nevertheless, VAM fungi do not seem to take up much P from insoluble sources.

Barea et al. [129] suggested the possibility of an interaction between mycorrhizal fungi and phosphate-solubilizing bacteria (PSB). They observed a synergistic increase in dry weight and P uptake in maize grown in a low-P soil and dually inoculated with *Endogone* (E3 type) and P-solubilizing *Pseudomonas* sp. and *Agrobacterium* sp. These results were duplicated in a later study, which also showed an enhancement of populations of PSB in the rhizosphere of VAM-infected plants, although the effect disappeared after 2 months [130]. Delorenzini et al. [131] found that the dry weight of clover dually inoculated with PSB and *G. mosseae* was significantly higher than with either treatment alone. Nitrogen and phosphorus levels were also enhanced by the combination of the two organisms. The PSB (a combination of *Pseudomonas* sp., *Agrobacterium* sp., and *Bacillus* sp.) increased the VAM infection of plants by indigenous VAM fungi. An interaction between PSB (*Bacillus megaterium* var. *phosphaticum* and *Pseudomonas fluorescens*) and *Glomus macrocarpum* on pepper was observed by Krone et al. [132]. The dry weight of plants in the VAM + PSB treatments was higher than in treatments with either alone. Further support for this hypothesis was given by Raj et al. [133], who examined the interactions between *Glomus fasciculatum* and *Bacillus circulans* on pearl millet. The study was conducted using ^{32}P-labeled superphosphate and insoluble tricalcium phosphate in a natural soil. Inoculation with PSB resulted in higher soil levels of available ^{32}P, suggesting that P was solubilized to an available form. The VAM treatments did not change the level of available P in the soil, although these treatments had higher dry weights and P concentrations. This supports the theory that VAM fungi do not increase P availability, but acquire more P by exploring greater volumes of soil. The PSB also increased ^{32}P activity in the plant and total P uptake. VAM enhanced the population densitites of PSB in the rhizosphere, but PSB had no effect on VAM spore production, the only VAM parameter measured.

In a similar study [134] in which soybeans were fertilized with ^{32}P-labeled tricalcium phosphate, the results were contrary to the above hypothesis. Inoculation with *G. mosseae* or *Glomus* sp. E3 improved growth and yield, but PSB (unidentified strains) did not. Mycorrhizal infection was significantly improved by PSB inoculation at all levels of added phosphate. Utilization of ^{32}P from the added insoluble phosphate was improved by the VAM fungi, but not by PSB. This work shows that PSB did not directly enhance plant P nutrition, although they did increase mycorrhizal infection. Another study by Lee and Bagyaraj [135] showed that PSB did not increase dry matter or P uptake in tomatoes when inoculated alone or in combination with *Gigaspora margarita*. The population of PSB declined until a nutrient solution was added to the soil, so the absence of growth increases might be due to inability of the added PSB to grow and multiply in the soil.

The interactions observed might be due to the production of plant hormones or vitamins by PSB, rather than to P solubilization. Barea et al. [136] demonstrated that PSB produced indoleacetic acid, gibberellins, and cytokinins. Cell-free supernatant from liquid cultures of a PSB (*Pseudomonas* sp.) enhanced the growth and VAM infection of tomato, alfalfa, and lavender, an effect mimicked by the addition of pure plant hormones [103]. In a similar experiment with the same PSB on alfalfa [137], culture filtrates enhanced VAM infection and dry weight, but washed cells did not. The authors hypothesized that plant hormones present in the supernatant enhanced VAM infection and plant growth.

PSB have also been shown to produce vitamins [138]. In general, PSB isolated from the root region produced greater amounts of vitamins than PSB from the nonrhizosphere soil and PSB produced greater amounts of vitamins than nonphosphate solubilizing bacteria. Although a correlation was found between vitamin production and phosphate solubilization, it is unknown whether vitamin production has any effect on plant growth or mycorrhizal fungi. Plants can synthesize their own vitamins and mycorrhizal fungi can probably satisfy their vitamin requirements from the plant. Nevertheless, competition for vitamins among rhizosphere microflora might occur, and this should be investigated.

The production of phosphatase enzymes by rhizosphere microbes might constitute another possible interaction. Acid phosphatase activity in the rhizoplane and rhizosphere of wheat and onion was increased by colonization with *G. mosseae* or *G. geosporum* [139]. Their results suggest that rhizosphere phosphatase activity is unrelated to the external hyphal network of VAM, since increased enzyme activity occurred when VAM infection was low (15%). Although the experiment was carried out in autoclaved growth medium irrigated with sterile nutrient solution, no attempt was made to keep the system sterile. Therefore, the possibility exists that a limited root microflora might have developed and contributed to phosphatase activity. Furthermore, the activity of these phosphatase-producing microbes might be enhanced by VAM infection, as suggested by Krishna and Bagyaraj [140].

Considerable controversy exists as to whether improved P nutrition by ectomycorrhizal fungi is due to changes in root morphology and exploration of relatively undepleted soil away from the root surface by hyphae and mycelial strands, or to solubilization of soil P not available to nonmycorrhizal plants. Recent experiments with labeled ^{32}P have supported the first hypothesis [141]. Nevertheless, ectomycorrhizal fungi in culture have been shown to solubilize large amounts of phosphorus from insoluble mineral phosphates such as calcium phosphate, iron phosphate, and phlogopite [142]. The interactions between ectomycorrhizal fungi and *Agrobacterium radiobacter*, a PSB, were examined in culture [143]. Both *A. radiobacter* and ectomycorrhizal fungi solubilized P from tricalcium phosphate. When added to 25-day-old cultures, *A. radiobacter* stimulated the growth of *Paxillus involutus* in the presence of iron phosphate. The total P solubilized in the presence of both *A. radiobacter* and *Laccaria laccata* or *Paxillus involutus* was two and five times greater, respectively, than if fungi were cultured alone. Leyval and Berthelin [144] investigated the interaction of ectomycorrhizal fungi and PSB in soil. Two PSB (*Agrobacterium* sp. and *A. radiobacter*) and/or *Laccaria laccata* were inoculated in the rhizosphere of beech and pine trees grown in greenhouse lysimeters in sand mixed with rock phosphate and phlogopite. After 2 years, both *A. radiobacter* and *L. laccata* increased beech growth and nutrient uptake. However, dual inoculation had no significant effect on plant growth and nutrition. Increased amounts of organic acids were detected in the effluents from plants inoculated with the bacterial inoculum, but no differences were seen between mycorrhizal and nonmycorrhizal plants.

Biological Control Agents

Biological control is the reduction of the amount of inoculum or disease-producing activity of a pathogen accomplished by or through one or more organisms other than man [145]. The use of microorganisms to control plant pathogenic fungi and bacteria is an active and promising area of research, and it is beyond the scope of this chapter to review the tremendous volume of literature. Instead, the reader is referred to some excellent books and reviews on the subject [145–148]. Despite all the research on biological control of plant pathogens, very few studies have examined the possible interactions with VAM and ectomycorrhizal fungi, which are ubiquitous components of most agricultural and forest ecosystems. If biocontrol agents are ever to become widely used in agriculture and horticulture, studies must be done on the interactions between these groups of beneficial organisms. Biocontrol agents might antagonize not only plant pathogens but also mycorrhizal fungi, which in terms of plant productivity would be self-defeating. Conversely, mycorrhizal fungi might enhance the activity of biocontrol agents, either directly or through an indirect effect on the microflora of the mycorrhizosphere. Biocontrol agents might also stimulate mycorrhizal spore germination and mycorrhiza formation through the production of growth-promoting substances or indirect effects on the host plant.

The first report of such an interaction on VAM was made by Krishna et al. [149], who observed that *Streptomyces cinnamomeous*, an actinomycete that inhibited the growth of plant pathogenic fungi, reduced the sporulation and infection of *Glomus fasciculatus* on finger millet. This reduction was most prominent when *S. cinnamomeous* was added 2 weeks prior to inoculation with *G. fasciculatum*. In a more recent study, dual inoculation of subterranean clover with *Pseudomonas putida* and an indigenous VAM species resulted in significantly greater shoot dry weights than inoculation with VAM alone, PGPR alone, or no inoculation [65]. Inoculation with *P. putida* increased VAM colonization of roots at 6 weeks, but no differences were seen at the 12-week harvest. At 6 weeks, the concentrations of Fe, Cu, Al and Zn were higher in plants inoculated with PGPR and VAM than in the treatment with either alone. However, this effect was not observed at the 12-week harvest. The combination of PGPR + VAM also enhanced nodulation by indigenous rhizobia in the nonsterile soil.

Mycorrhizae exert a selective pressure on populations of soil microorganisms. Those mycorrhizosphere microbes can have specific effects on root pathogens. For example, sporangium and zoospore production by the root pathogen *Phytophthora cinnamomi* was reduced in the presence of rhizosphere leachates from VAM-inoculated sweet corn and chrysanthemum, when compared to leachates from nonmycorrhizal plants [7]. Similarly, actinomycetes were isolated from VAM pot cultures on guinea grass and tested for their antagonistic activity against selected bacteria and fungi [9]. Pot cultures of *Glomus fasciculatum* harbored more actinomycetes antagonistic to *Fusarium solani* and *Pseudomonas solanacearum* than the non-VAM control or the other VAM species. Actinomycetes antagonistic to *Xanthomonas vignicola* were in higher numbers in pot cultures of *Gigaspora margarita*, while pot cultures of *Acaulospora laevis* were stimulatory to antagonists of *Fusarium solani*. This work suggests that VAM fungi exert an influence on rhizosphere organisms in pot cultures.

Inoculation of cassava with an isolate of *Pseudomonas putida*, which was antagonistic to *Diplodia* and *Pythium*, enhanced the sporulation of the VAM fungus *Acaulospora longula* [150]. The heights of plants colonized by *A. longula* were increased by inoculation with *P. putida*, although root colonization was unaffected.

The interactions between fluorescent pseudomonads and two species of VAM fungi in the rhizosphere of cucumber were examined by Paulitz and Linderman [151]. They treated seeds with rifampicin-resistant mutants of *Pseudomonas putida* and *P. fluorescens* and monitored the population density of the bacteria over time in the rhizosphere of mycorrhizal and nonmycorrhizal cucumber plants. The population density of most strains was significantly reduced on plants colonized by *Glomus intraradices*, compared to the nonmycorrhizal plants. This effect was not seen in plants colonized by *Glomus etunicatum*, a less aggressive colonizer. The authors proposed that changes in root exudation and competition for carbon induced by rapid infection with *Glomus intraradices* were responsible for the reduction of fluorescent pseudomonads. In a related series of greenhouse experiments, high populations of fluorescent pseudomonads added to the soil had no effect on the inoculum potential of either VAM species, as determined by most-probable-number and inoculum density–root colonization assays. However, two isolates of *P. fluorescens* temporarily delayed germination of *G. etunicatum* in raw soil.

Interactions between ectomycorrhizal and biological control organisms have been investigated in Australia. Actinomycetes were isolated from the rhizospheres of plants in western Australia and tested for their antifungal activity against *Phytophthora cinnamomi, P. cryptogea, P. nicotiana, Pythium proliferum*, and *Laccaria laccata* in vitro [152]. In general, actinomycetes isolated from root surfaces showed general activity against all the fungi tested. However, two groups of actinomycetes were isolated in higher numbers from the rhizosphere of ectomycorrhizal *Pinus radiata*, and these two groups showed a high degree of activity against the *Phytophthora* species tested. The microflora of mycorrhizal *Eucalyptus* roots was examined as part of a study to determine why some soils were suppressive to root rot caused by *Phytophthora cinnamomi* [153,154]. In the conductive soil, *E. calophylla*, which is more resistant to P. *cinnamomi*, was more heavily colonized by mycorrhizal fungi than the susceptible *E. marginata.* In the conducive soil, the total population densities of bacteria and actinomycetes in the mycorrhizosphere of the resistant species were higher than with the susceptible *Eucalyptus* sp., although no statistical analysis was performed. Qualitative differences in bacteria associated with different types of mycorrhizae were also seen in EM sections of *Eucalyptus* roots. The populations of bacteria and actinomycetes were four to five times higher in the mycorrhizosphere of *Eucalyptus* seedlings, compared to the rhizosphere of nonmycorrhizal roots. The authors concluded that different types of mycorrhizal roots and their associated bacterial microflora may contribute to differential susceptibility of the two species to infection by *P. cinnamomi* in conducive soils. In a later study using light and scanning electron microscopy [155], Malajczuk observed that zoospores of *P. cinnamomi* germinated and penetrated nonmycorrhizal jarrah roots grown in soil and roots colonized by B and H type ectomycorrhizal fungi. However, in the white ectomycorrhiza type, encysted zoospores were colonized by bacteria and actinomycetes, and lysis was evident. One type of fungus formed a constriction ring and "captured" the germ tube of the pathogen, reminiscent of nematode trapping fungi. The author suggests that mycorrhizosphere microbes were the primary agents that reduced the activity of *P. cinnamomi* zoospores and limited root infection; although there might also be a direct antagonism by the ectomycorrhizal fungus.

As biological control systems are developed for commercial applications, many questions about their interactions with mycorrhizae must be answered. At the same time, biocontrol agents should be developed that have neutral or positive effects on mycorrhizae, and vice-versa. By incorporating mycorrhizae into the development strategy of the biocontrol agent, either through selection and screening or genetic engineering, the maximum potential of the system to enhance plant growth can be realized.

Bacteria Associated with Mycorrhizal Fungi

Many of the bacteria described as intimately associated with mycorrhizal fungi either stimulate or inhibit some aspect of mycorrhizal formation and are discussed in a previous section. Some of the bacteria observed in close association with VAM fungi have a more neutralistic relationship. One example is the chitin-decomposing microorganisms, shown by Ames [156] to be associated with the spores of VAM fungi. Some of the actinomycetes significantly increased VAM root infection and the density of VAM hyphae in the rhizosphere [157].

Other organisms have been found inside the walls and membranes of VAM spores. Bacterium-like-organelles (BLOs) were detected in the VAM fungus *Glomus caledonius* by electron microscopy [158]. BLOs were observed in the cytoplasm of resting spores, intercellular hyphae, and arbuscule trunks. Bacteria were also associated with the eroded walls of the resting spores and were embedded within the wall layers. In addition, BLOs were observed in the spores of *Glomus mosseae*, *Gigaspora margarita*, and *Gigaspora heterogama* [159,160]. The authors proposed that BLOs could by prokaryotic cytobionts, which have established an obligate association within their fungal host or that they could be permanent, nonprokaryotic features of the cytoplasm or organelles of unknown function and origin. However, without the in vitro cultivation of BLOs, the first possibility could not be proven by the authors. Spiroplasma-like organisms were found in the cytoplasm of spores of a *Glomus* sp. that had been infected with a chytrid mycoparasite, *Phlyctochytrium kniepii* [161]. These spiroplasms were also associated with the thalli of the chytrid. Attempts to culture these helical organisms in cell-free media were not successful. By use of scanning electron microscopy, dividing bacterial cells were seen colonizing the inner surface of the spore walls of *Glomus macrocarpus* var. *macrocarpus* [162]. A number of bacteria were isolated from the crushed surface-disinfested spores, including *Pseudomonas* spp., *Streptococcus*, and an enteric type. Some doubt still exists as to whether the isolated bacteria were really the BLOs observed by previous workers, since the authors did not examine the cytoplasm with transmission electron microscopy. However, scanning electron microscopy has shown that the isolated bacteria are inhabitants of the lumen, inner walls, and membrane of VAM spores. The observation of cells undergoing binary fission also indicates that these bacteria are actively growing. These observations were confirmed in a later study, in which bacteria were seen on the inner spore membrane of *Glomus microcarpum* [163]. Anaerobic, fermentative, heterotrophic bacteria were isolated, primarily *Klebsiella* sp., *Clostridium* sp., and *Streptococcus lactis*.

An interesting intimate association between bacteria and orchid mycorrhizae was recently reported by Wilkinson et al. [164]. A variety of bacteria were isolated from mycorrhizal tissues of a variety of terrestrial orchids. Symbiotic germination of one orchid was enhanced by one of these bacteria. The bacteria were isolated directly from the mycorrhizal fungus hyphae, where they seemed to be very closely associated.

C. Interactions with Fungi

Saprophytes

Most studies that deal with mycorrhizal–fungal interactions focus on plant pathogens rather than saprophytes. Little is known about the effect of general rhizosphere fungi on mycorrhizal fungi. Bagyaraj and Menge [4], in their study on the interactions between a VAM fungus and *Azotobacter*, noted that the total fungal population in the rhizosphere of tomato was unaffected by inoculation with *Glomus fasciculatum*. Similar results were reported on barley plants inoculated with *Glomus macrocarpum* [8]. Secilia and Bagyaraj [165] examined the fungi associated with VAM pot cultures of *Glomus fasciculatum*, *Gigaspora margarita*, *Acaulospora laevis*, and *Sclerocystis dussi*. Total fungal populations in the root zone of guinea grass were not affected by inoculation with any VAM fungi except *S. dussi*, where populations were reduced compared to noninoculated plants. Qualitative changes in the fungal makeup among the mycorrhizospheres of different VAM species were also recorded.

The association between saprophytic fungi and ectomycorrhizae was reviewed by Davey [166]. More recently, Summerbell [167] enumerated populations of fungi from ectomycorrhizal roots of black spruce. The most predominant were sterile isolates, *Trichoderma* spp., *Penicillium* spp., *Fusarium* spp., and *Micromucor* spp. Most of these fungi had no effect on the formation of ectomycorrhizae by *Laccaria bicolor* on axenically cultivated black spruce seedlings. However, *Trichoderma viride* and *T. polysporum* were strongly antagonistic, while two other soil fungi, *Tolypocladium inflatum* and *Trichosporon beigelii*, had a small inhibitory effect [37]. The total fungal populations in the mycorrhizosphere of *Pinus kesiya* increased as the ectomycorrhiza developed [168]. Many of the fungal species were restricted to certain ectomycorrhizal development stages in the pine seedlings.

Plant Pathogens

A tremendous amount of research has been reported on the interactions between VAM fungi and soilborne fungal plant pathogens. (This topic is covered by another chapter in this book.) Several excellent reviews on the subject have been written, [169–173]. Rather than reviewing the recent literature in detail, we will synthesize the literature, discuss some general trends, and hypothesize on possible mechanisms of interaction.

Most of the literature suggests that VAM fungi reduce soilborne disease or the effects of disease caused by fungal pathogens. In a review by Dehne [172], disease damage was reduced in 17 out of the 32 reports cited. In 11 studies performed since 1982, 8 reported decreases in disease symptoms. But in many experiments, no effect of the VAM fungus on disease was observed [174–178] or the disease severity increased [179–181]. This makes it difficult to draw any general conclusions. Part of the reason for such varying results is that many different pathogens were used, all having different mechanisms of pathogenesis. Clearly, VAM fungi will not have the same effect on such diverse pathogens.

How do VAM fungi interact with fungal pathogens to reduce plant disease? One of the first questions that needs to be answered is whether disease symptoms are really reduced, or whether the results are confounded by a nutritional or phosphorus effect. In most experiments, no attempts were made to compensate for the P effect by supplying nonmycorrhizal plants with similar amounts of phosphorus. Such experiments would help clarify the role of P nutrition in suppression of disease. Studies using growth measurements should make comparisons relative to both mycorrhizal and nonmycorrhizal controls, rather than absolute comparisons. For example, Davis [178] showed that mycorrhizal citrus seedlings infected with *Thielaviopsis basicola* were larger than diseased nonmycorrhizal ones. But the relative reduction of growth of both mycorrhizal and nonmycorrhizal seedlings by *T. basicola* was similar.

In most experiments, plants were preinoculated with VAM fungi and challenged with the pathogen later. In experiments where the plants were simultaneously inoculated with both the pathogen and VAM fungus, no reduction in disease was seen [182,183]. This would limit the practical applications of VAM solely for disease control, since many root pathogens can germinate and grow faster than VAM fungi and attack the roots before they become mycorrhizal. Some pathogens, such as *Pythium* spp., selectively attack young root tips, tissue which is not normally mycorrhizal. However, VAM fungi could be ideally adapted to transplanted field and horticultural crops, where plants with preestablished mycorrhizae could be outplanted. This was demonstrated by Stewart and Pfleger [184] on poinsettia root rot (*Pythium* and *Rhizoctonia*) where preinoculation with VA mycorrhizal fungi gave disease control, but simultaneous inoculation with the pathogens did not. Preinoculation of rooted cuttings with mycorrhizal fungi is very feasible in the ornamental horticulture industry.

An effect observed in many studies is that root infection by the pathogen reduced mycorrhizal colonization of roots [174–176, 182]. This is not surprising, because VAM fungi are obligate symbionts, and any detrimental effects on roots would be reflected on the biotroph. On the other hand, with an obligate parasite, *Olpidium brassicae*, individual cells infected with either *O. brassicae* or *Glomus mosseae* existed side by side in the same root, but the two never occupied the same cell [173]. With obligate parasites, there might be a direct competition for susceptible cells.

Another observation, gained from split root experiments, is that resistance associated with VAM appears to be localized in some cases but systemic in others. Dry weights of nonmycorrhizal citrus roots inoculated with *P. parasitica* were not significantly different whether they were split from mycorrhizal or nonmycorrhizal roots, indicating a lack of systemic effects [180]. When both the pathogen and VAM fungus were inoculated onto the same root, mycorrhizal colonization was reduced. A similar lack of systemic effects was noted on pea roots infected with *Aphanomyces euteiches* and a VAM fungus [182]. On the other hand, evidence for a systemic effect was given in the same paper. When one side of the split root was inoculated with VAM fungus, oospore production by *A. euteiches* in the opposite non-VAM root was reduced, compared to plants with no VAM on either root [182]. In roots inoculated with *P. parasitica*, the percentage of the nonmycorrhizal roots that were healthy was significantly greater when these roots were split from mycorrhizal roots than when they were split from nonmycorrhizal roots [180]. A nutritional effect can also be systemic, i.e., the mycorrhizal half of the split root provided greater phosphorus nutrition to the nonmycorrhizal half. This possibility must be considered in split root experiments. Observation of a systemic effect is not concrete proof for induced resistance.

Most of the mechanisms of interaction that have been hypothesized assume that the pathogen and VAM fungus do not directly affect each other, but that there is an indirect effect caused by mycorrhizal-mediated changes in the host metabolism and physiology. This is unlike ectomycorrhizal fungi, which directly interact with the pathogen.

One hypothesis is that VAM fungi alter host physiology, thereby making the root more resistant to pathogens. VAM fungi influenced the phenol metabolism and lignification of tomato and cucumber root cells [185]. They speculated that deposition of lignin in the cell walls of the endodermis and stele increased resistance to Fusarium wilt. Similarly, Becker [186] found that localized sites on mycorrhizal onion roots appeared to be more resistant to pink root caused by *Pyrenochaeta terrestris*, apparently by preventing the internal development of the pathogen by cell wall thickenings (callose or lignitubers). Mycorrhizal plants showed increased chitinase activity, which might be effective against fungal pathogens [187]. Higher levels of arginine accumulated in mycorrhizal roots, which inhibited the sporulation of *Thielaviopsis basicola* [187]. Holly plants colonized by *Glomus mosseae* showed a 28% increase in induced wound-barrier formation in response to *Thielaviopsis basicola*, when compared to control treatments [188]. Mycorrhizal plants produced wound barriers faster than nonmycorrhizal plants; however, *T. basicola* was not excluded from mycorrhizal cells. They attributed the increase in wound-barrier formation to increased vigor of the mycorrhizal plants. Many other ultrastructural and metabolic changes have been documented in mycorrhizal infected roots, any one of which might influence pathogen infection [189–191].

Other studies suggest that reduction of disease in mycorrhizal plants is due to increased phosphorus nutrition provided by the VAM fungus, which might make the plant more resistant or able to compensate for the effects of the disease. Some studies indicate that the increased growth response of diseased plants colonized by VAM fungi is eliminated by adding phosphorus. P-deficient nonmycorrhizal wheat plants developed more take-all disease than mycorrhizal plants, but adding 50 μg P/g soil of superphosphate also reduced disease [192]. The authors speculated that increased P status of the plant led to a decrease in leakage of root exudates and reduced pathogen activity, although the latter hypothesis has never been confirmed. Roots of sweet orange grown in low phosphorus soil and infected with *Phytophthora parasitica* and *Glomus fasciculatum* were healthier than roots infected with *P. parasitica* alone [180]. However, the beneficial effect was eliminated by adding 56 μg P/g soil. The tolerance of mycorrhizal roots to *P. parasitica* was attributed to the ability of mycorrhizal roots to take up more phosphorus. In some cases, this increased level of phosphorus in mycorrhizal plants led to an increase in disease [179]. Graham and Egel [193] used mycorrhizal and nonmycorrhizal sweet orange seedlings of similar size of leaf P to examine the interactions between *Glomus intraradices* and *Phytophthora parasitica*. In that situation, the VAM fungus did not reduce root rot development. They concluded that the VAM fungus did not increase the resistance or tolerance of sweet orange to Phytophthora root rot, unless the VAM fungus gave the plant increased phosphorus nutrition over the nonmycorrhizal plant.

Some plant species are less responsive to additional phosphorus, so phosphorus may have little or no influence on disease. For example, an increase in phosphorus level in leaves and stems of tomato had no effect on the population of *Fusarium oxysporum* f. sp. *radicis-lycopersici* and root necrosis, but did reduce root weight at any of the P concentrations used, and the VAM fungus did not alter phosphorus levels in the leaves or stems. But in the *G. intraradices* treatment, root necrosis and population densities of the pathogen were reduced.

A third possible explanation for reduced disease in the presence of VAM is that mycorrhizae alter the makeup of microbial populations in the rhizosphere and enhance the populations of organisms antagonistic to the plant pathogen. Most soilborne pathogenic fungi must germinate and grow saprophytically in the rhizosphere and rhizoplane prior to establishing a parasitic relationship with the plant. Some bacteria and fungi in the rhizosphere are antagonistic to or can compete with fungal pathogens for carbon substrates. This premise is the basis for much of the research on biological control of plant pathogens. Meyer and Linderman [7] found higher populations of fluorescent pseudomonads in the rhizoplane of mycorrhizal subterranean clover when compared to nonmycorrhizal plants. They also demonstrated that fewer sporangia and zoospores of *Phytophthora cinnamomi* were formed in extracts from mycorrhizal roots of sweet corn when compared to extracts from nonmycorrhizal roots. Pot cultures of *Glomus fasciculatum* and *Acaulospora laevis* contained more actinomycetes that were antagonistic to *Fusarium solani* than non-VAM pot cultures [9]. Caron et al. [194–196] reported a reduction in the soil population density of *Fusarium oxysporum* f. sp. *radicis-lycopersici* on tomatoes colonized by *Glomus intraradices*. This indicates a mechanism operating external to the plant, since the populations were sampled from soil with the roots removed. This effect was observed even when mycorrhizal infection was low, indicating that it is probably not due to competition for root exudates. The authors suggest that the interaction might occur in the remote extramatrical phase of the VAM fungus. Therefore, mycorrhizosphere organisms might also be involved in this phenomenon. Despite these promising reports, to our knowledge, no other research has investigated this hypothesis.

The interaction of ectomycorrhizal fungi and fungal pathogens, resulting in the reduction of disease, has been recognized for over 40 years. Several reviews have been written on the subject [197,198]. Many of the mechanisms described for VAM fungi also apply to this group. Other mechanisms include the formation of a Hartig net and mantle, which provide a physical barrier to the pathogen [199], production of antibiotics [200,201], induction of phenolic compounds and other inhibitory substances in the mycorrhizal root [202], and a change in root exudates available to the pathogen. The establishment of a protective microbial rhizosphere population might be another mechanism. The selective influence of the mycorrhizosphere on microbial populations was recognized in many earlier works. More recent studies suggest that ectomycorrhizal fungi on jarrah enhanced microbes capable of reducing the activity of zoospores of *Phytophthora cinnamomi* [155,203]. A recent report by Zhao and Kuo [204] demonstrated that ectomycorrhizal fungi may also exhibit hyperparasitic capacity against root pathogens like *Rhizoctonia*.

Mycoparasites

Over the past 40 years, numerous observations of mycoparasitism of VAM fungi have been reported, but we will only review the literature of the last 10 years. To our knowledge, nothing has been published on mycoparasites of ectomycorrhizal fungi, except for mycoparasites of macroscopic fruiting bodies.

Tzean and Huang [205], in a report on the occurrence of VAM on citrus and maize in Taiwan, observed hyphae in the chlamydospores of *Glomus* sp. *Fusarium oxysporum* was isolated from the affected chlamydospores. Bhattacharjee et al. [206] noticed actinomycete-like hyphal growth on the inside surface of the azygospore of *Gigaspora*. Twenty percent of ruptured azygospore shells of *Gigaspora gigantea* from sand dune soil contained the spindle cells of *Labyrinthula* sp., an amoeboid organism classified in the division Myxomycota [207]. Azygospores of *Gigaspora* in weed seeds occassionally were found to be parasitized by unidentified filamentous organisms [208]. Tzean et al. [161] observed chlamydospores of *Glomus* sp. infected by the chytrid *Phlyctochytrium kniepii*. Chytridiaceous fungi were observed on the chlamydospores of *G. clarum* and *G. etunicatum* at wetter matric potentials [209]. *Fusarium* sp., *Penicillium* sp., and *Trichoderma* sp. contaminated the outside of VAM spores and survived surface disinfestation.

Some research has gone beyond a simple descriptive observation of mycoparasitism of VAM fungi. During a population dynamics study of two VAM fungi in field plots, Ross and Ruttencutter [210] found two mycoparasites that attacked *Glomus macrocarpum* var. *geosporum* at the end of the second season of the study. One was a species of *Phlyctochytrium*, and the other was a "Pythium-like" fungus. The presence of these mycoparasites was correlated with a decline in spore production by the VAM fungus on soybean. The authors postulated that mycoparasites might play a major role in limiting populations of VAM fungi. Daniels and Menge [211] isolated *Anguillospora pseudolongissima* and *Humicola fuscoatra* from parasitized spores of *Glomus fasciculatum* isolate 0-1 (= *G. deserticola*) and *G. epigaeum* (= *G. versiforme*) in greenhouse pot cultures. When the spores of *G. epigaeum* were added to water agar or autoclaved sand containing *A. pseudolongissima* or *H. fuscoatra*, a large percentage became parasitized. A third chytrid-like mycoparasite, identified as *Phlyctochytrium* sp., was isolated from spores of *G. fasciculatum*. This study also showed that spores of *G. epigaeum* were more susceptible to the mycoparasites when treated with KOH or oxidizing agents like hydrogen peroxide or sodium hypochlorite. The authors hypothesized that these substances decolor or deactivate melanin pigments in the spore wall, which might render the spore more susceptible to attack by the mycoparasites. Daniels [212] tested the effects of these mycoparasites added separately to citrus plants inoculated in the greenhouse with VAM fungi. After 7 months, none of the mycoparasites affected infection and growth response of *G. epigaeum* or *G. fasciculatum* isolate 92 but did inhibit *G. fasciculatum* 0-1. Sylvia and Schenck [213] showed that fenaminosulf, a fungicide active against oomycetes, reduced the percentage of *Gigaspora margarita* spores covered with sporangia of chytridiaceous fungi.

Interactions between the chytrid *Spizellomyces punctatum* (= *Phlyctochytrium punctatum*) and spores of *Gigaspora margarita* were extensively studied by Paulitz and Menge [214] (Fig. 2). They found that the chytrid mostly attacked nongerminated and probably nonviable VAM spores. The incidence of colonization of VAM spores by the chytrid was greater when the spores were first killed with heat or propylene oxide. Inoculation of the VAM spores with *S. punctatum* did not significantly reduce the incidence of VAM spore germination. In greenhouse trials, high populations of the chytrid did not reduce infection or growth response of cotton plants inoculated with *G. margarita*. The authors concluded that *S. punctatum* was primarily a saprophyte that attacked nonviable or stressed spores of *G. margarita*. Another mycoparasite, *Anguillospora pseudolongissima*, reduced root colonization by *Glomus deserticola* and growth response of onions grown in a low-phosphorus sand [215]. However, this effect was seen only in treatments where the VAM inoculum was diluted to 10^{-3} of the original inoculum. The authors postulated that under conditions of low VAM inoculum density and low phosphorus availability, a mycoparasite could indirectly reduce plant dry weight by adversely affecting the fungal mycorrhizal symbiont. However, this study was performed in autoclaved sand with high inoculum densities of the mycoparasite, so it is difficult to extrapolate these results to a natural soil.

Siqueira et al. [216] found *Stachybotrys chartarum* as a contaminant of spores of *Gigaspora margarita* isolated from pot culture. The spores of

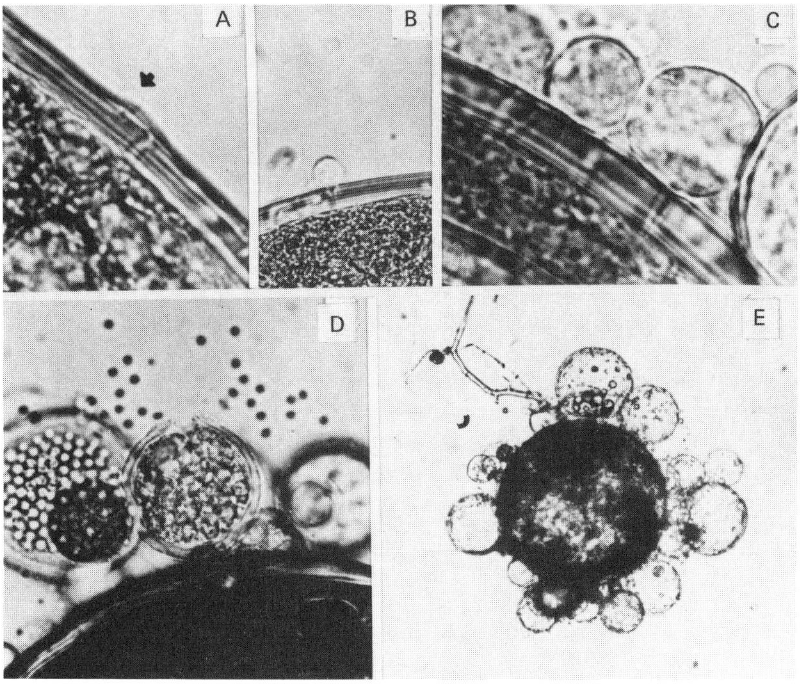

FIGURE 2. Sequence of events in the colonization of azygospores of *Gigaspora margarita* by the chytrid *Spizellomyces punctatum.* (A) 5 hr postinoculation. A zoospore of *S. punctatum* has encysted on the wall of an azygospore of *G. margarita* (arrow). The penetration tube runs laterally along a laminate layer before penetrating the wall. (B) 30 hr postinoculation. The thallus and penetration tube of *S. punctatum* is visible. (C) 46 hr postinoculation. The thalli of *S. punctatum* have developed into immature sporangia. (D) 54 hr postinoculation. Mature sporangia of *S. punctatum.* Zoospores are released through numerous exit pores in the walls of the sporangia. (E) One week postinoculation. Note the empty sporangia of *S. punctatum* surrounding the necrotic azygospore. (Parts C–E used with the permission of The New York Botanical Garden, Bronx, NY, copyright 1984.)

G. margarita were inoculated with spores of *S. chartarum* and transferred to membrane filters on water agar plates. After 10 days of incubation, there was a significant negative correlation between percent infection by *S. chartarum* and germination and germ tube growth of VAM spores. However, the experimental system was somewhat artificial, since the antagonist might have derived some nutrition from the water agar.

Despite this literature, there is still much to be learned about mycoparasites of mycorrhizal fungi. Most of the literature consists of observations of possible mycoparasitism. In very few cases has it been proven that these supposed "mycoparasites" are really mycoparasites, and not just saprophytes growing on necrotic or nonviable VAM spores. Most experiments were performed on water agar or on other sterile substrates. Very few greenhouse studies on the overall effect of the mycoparasites on the symbiosis have been performed and none of those were done with natural soils. One study, however, examined the distribution of mycoparasites of VAM fungi in natural soils [217]. Soils from citrus groves were baited with surface-disinfested spores of *Glomus deserticola.* Most of the fungi isolated from the spores were common soil saprophytes, such as *Fusarium* spp., *Humicola* sp., *Verticillium* sp., *Penicillium* sp., and *Pyrenochaeta* sp., which showed no mycoparasitic ability when further tested. Four fungi exhibited mycoparasitic ability: *Anguillospora pseudolongissima* and three unidentified nonsporulating isolates (Fig. 3). The mycoparasites occurred in all the soils tested, but the total populations were low. Despite this study, the impact of these mycoparasites on the ecology of VAM fungi in natural soils is unknown.

D. Interactions with Invertebrates

Plant Pathogenic and Mycophagous Nematodes

A large volume of information has been generated on the interactions between plant pathogenic nematodes and mycorrhizae, especially VAM. Both are present in the rhizosphere, and in the case of endoparasitic nematodes, both colonize the root cortex. Much of the interest in these interactions undoubtedly stemmed from the hope that mycorrhizae could reduce the deleterious effects of these nematodes on agricultural crops. The literature on this subject is reviewed, along with other plant pathogens, in the general references cited in Section II. C. In addition, Hussey and Roncadori [218] and Ingham [219] also assess this topic.

In a majority of the studies, nematode-susceptible plants colonized by VAM fungi were better able to tolerate plant pathogenic nematodes [220–222]. This avoidance of the plant growth reduction from nematodes is often accompanied by decreases in the population densities of the endoparasitic nematodes, expressed as number of galls, juveniles, or eggs per unit root length. Some examples are with *Meloidogyne incognita* [223,224], *M. hapla* [225], and *Rotylenchulus reniformis* [222]. However, in some cases, fungal colonization had no effect on nematode populations, as shown with studies on *Radopholus similis* [226], *M. incognita* [227], or *M. hapla* [221]. In some instances, the populations of the nematode were consistently higher in mycorrhizal plants, as reported with *M. arenaria* on grapes [228]. Other measures of nematode infection have been assessed relative to mycorrhizal infection. Larval penetration [229], nematode development [230], nematode reproduction [231], and number of giant cells [232] were all reduced by mycorrhizal colonization.

This reduction of symptoms and nematode effects might be dependent on the level and timing of VAM infection. Larvae and adults of *M. hapla* were absent from lucerne roots where mycorrhizal infection was more than 10% [221]. Hussey and Roncadori [233] observed that most of the *P. brachyurus* were in cotton roots that were not mycorrhizal and that nematodes were rarely seen in roots that were mycorrhizal. When less than 38% of cotton roots were colonized by *Glomus fasciculatum*, no reduction in number of larvae or eggs of *M. incognita* was observed [224]. However, when VAM infection exceeded 85%, both nematode parameters were lower. Similarly, 20–30% VAM infection of cotton was insufficient to suppress *M. incognita*, but nematode reproduction was reduced 54% on roots that were more than 50% colonized by *G. intraradices* [231].

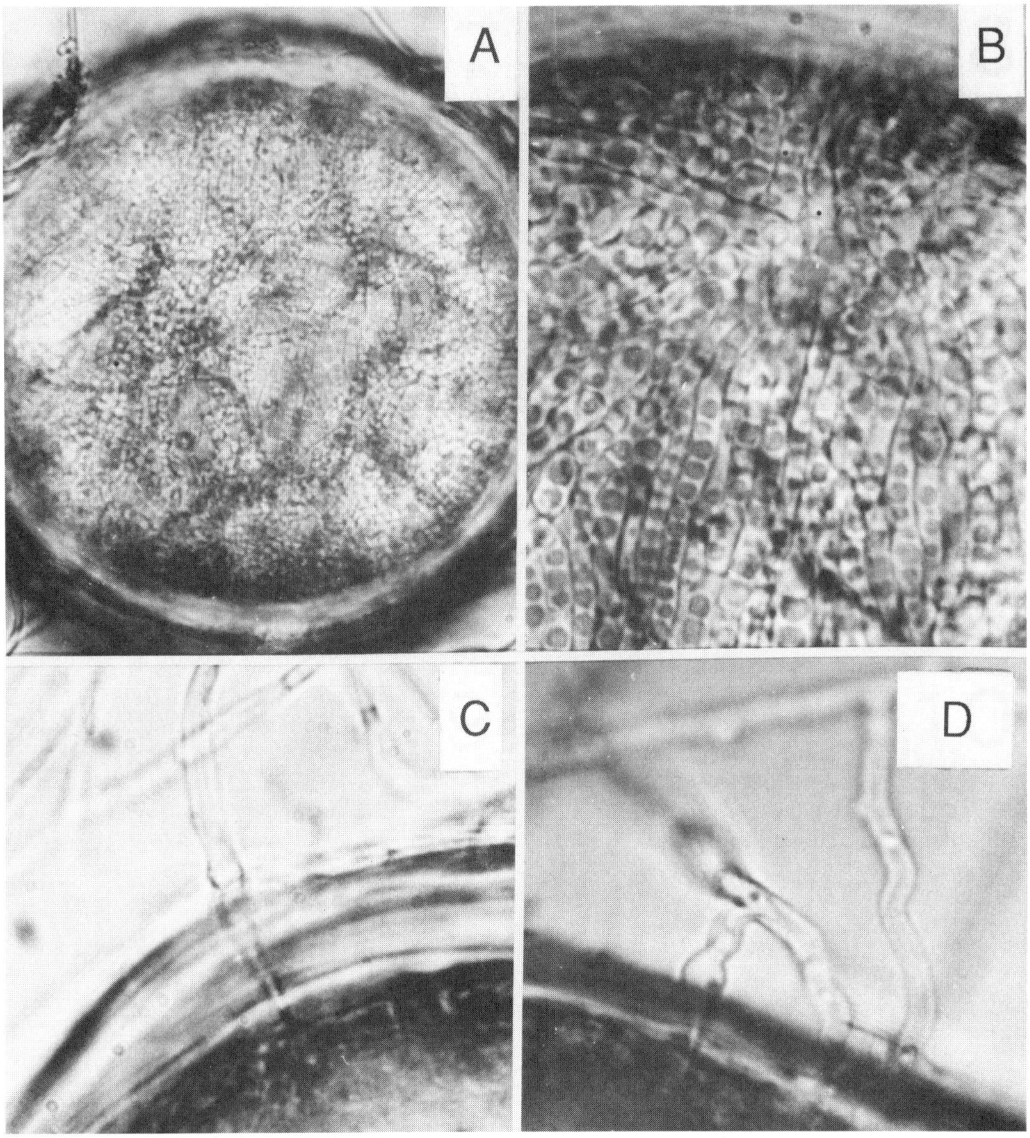

FIGURE 3. (A) Parasitism of chlamydospores of *Glomus clarum* by an unidentified mycoparasite (*G. c.* isolate). Note tightly packed hyphae of the mycoparasite filling the VAM spore. (B) Close-up of (A). Note large oil vacuoles in hyphae of mycoparasite. (C) and (D) Penetration and infection of chlamydospores of *Glomus clarum* by *G. c.* isolate. Hyphae of mycoparasite penetrating spore wall of *G. clarum*. In (C) note appressorium-like swellings of the hyphae proximal to wall penetration. The penetrating hyphae (1.5 μm diameter) are narrower than the vegetative hyphae (3 μm diameter).

The timing of VAM inoculation is also important. Plants preinoculated with VAM fungi 4 weeks before challenge with *M. incognita* and *M. hapla* had less nematode infection and development than plants inoculated simultaneously [225,234]. Mycorrhizal effects on *M. incognita* were more pronounced when cotton plants were inoculated with *G. intraradices* 28 days before the addition of the nematode [231].

Chlamydospores of VAM fungi have been detected in the cysts of the soybean cyst nematode *Heterodera glycines* [235]. In another report, chlamydospores occupied 1–24% of the cysts from field samples [236]. In some cases, eggs were infected with VAM fungi, but in other cases, VAM chlamydospores, healthy eggs, and juveniles existed in the same cyst. When soybean plants were preinoculated with an isolate of *Glomus fasciculatum* from cysts, the number of first-generation females was decreased 26%. However, there were more second-generation females in the mycorrhizal treatments, and the VAM fungus infected only 1% of the eggs. Direct infection of nematode eggs or cysts is unlikely to explain the reduction of nematodes by VAM fungi.

The response of the VAM fungus to plant pathogenic nematodes is variable. Spore production was decreased [228], not affected [237–240], or increased [220]. The response of mycorrhizal infection ranged from enhancement [223] to no effect [225,237] to inhibition [226].

Both VAM fungi and endoparasitic nematodes occupy the root cortex, and histological observations have been made to ascertain the spatial arrangement within that tissue. These studies have examined whether both organisms occupy the same cells and whether gall tissue or giant cells formed by the root-knot nematode can be colonized by VAM fungi. In general, VAM fungi are rarely inside the gall tissue but are adjacent to it [218]. VAM fungi have not been observed in the giant cells of *Meloidogyne*. However, Kellam and Schenck [237] found mycelia, arbuscules, and vesicles of *G. macrocarpus* in hypertrophied tissue within galls. But these mycorrhizal structures were atypical, characterized by lysing and disintegrating hyphae. They never observed appressorium formation and penetration of gall tissue by the VAM fungus. No VAM structures were observed in the gall tissue of lucerne infected with *M. hapla*, although hyphae frequently traversed the gall surface without penetration [221].

A few reports on interactions between plant pathogenic nematodes and ectomycorrhizae can be found in the literature. Both *Tylenchorhynchus claytoni* and *Helicotylenchus dihystera* were observed to penetrate and migrate through the fungal mantle formed by *Pisolithus tinctorius* on pine [241]. The structural integrity of the fungal mantle was disrupted by *H. dihystera*, increasing root rot caused by *Phytophthora cinnamomi*. They [241] speculated that the nematode caused breaks in the fungal mantle, providing entry points for the pathogen. In a study of *Pratylenchus penetrans* on seedlings of *Pinus radiata*, most of the nematodes were extracted from the dichotomously branched lateral roots [242]. Nematodes invaded mycorrhizal roots, moving freely within the cortex and completing their life cycle.

A number of mechanisms can be proposed to explain the interactions between nematodes and VA mycorrhizal fungi. First, VAM fungi can alter the physiology of the root, including the exudates responsible for chemotactic attraction of the nematode. Sikora [230] suggested that the attractiveness of the root system to *M. incognita* larvae was altered by the presence of *G. mosseae*. This and other studies [229,238] found decreased larval penetration and retarded nematode development in mycorrhizal roots.

Second, VAM fungi improve plant vigor, especially in low P soils, and this offsets yield loss normally caused by nematodes [218]. Several studies have attempted to look at the phosphorus component of this interaction. Typically, increased P results in greater root mass, and thus greater numbers of nematodes, when expressed on a per plant basis [225,227]. Additional P also stimulated nematode reproduction when nematodes were added at the time of planting, but not when the nematodes were added 28 days later [231]. This would suggest that the phosphorus effect was on the early stages of plant growth.

In some cases, phosphorus did not influence nematodes. Roncadori and Hussey [220] found no effect of high phosphorus fertilizer applications on eggs per plant or eggs per gram of root. The survival and reproduction of *M. hapla* were not affected by phosphorus applications [238].

However, in many of these studies, there were fewer nematodes per gram of root, and fewer eggs per egg mass in phosphorus-fertilized treatments, indicating some phosphorus effect. Hussey and Roncadori [233] suggested that this was due to a dilution effect from larger root systems induced by phosphorus fertilization. Cooper and Grandison [225] eliminated P as a variable by using VAM fungi that would infect tomato and white clover under high P conditions. Under high P fertilization, mycorrhizal infection reduced numbers of *M. hapla* in these plants, both when expressed on a per root and per root weight basis. Thus, even under high phosphorus conditions, VAM fungi reduced the level of nematodes. Similar results were obtained with *M. incognita* on tamarillo [234]. This led the authors to surmise that resistance was not due entirely to improved host nutrition, but that the physiology of the root system must have been altered. Strobel et al. [240] used a split root system with peach seedlings to evaluate this phenomenon. Phosphorus applied to a root opposite a nematode-infected root mimicked the effects of *Gigaspora margarita*, in that both treatments increased shoot and root dry weight. This suggested that increased vigor due to better P nutrition had overcome the effects of the nematode *M. incognita*. However, a mycorrhizal root had fewer eggs per gram than a comparable root treated with phosphorus fertilizer. If the phosphorus or *Gigaspora margarita* were applied to the root opposite the nematode-infected one, no effect was seen. This seems to be strong evidence that the nematode and VAM fungus have to be on the same root in order to get an antagonistic interaction, which cannot be duplicated by the addition of phosphorus. Similar results were obtained by Oliveira and Zambolim [243], who reported a reduction in the reproduction of *M. javanica* on roots co-inoculated with *Glomus etunicatum*.

Third, nematode development and reproduction are retarded in mycorrhizal plants due to decreases in available food. Indirect evidence for this is seen in studies that show reduced numbers of nematodes and eggs per root biomass [224]. More direct evidence is seen in the reduction of the size of the feeding nematodes. Suresh et al. [232] reported that nematodes in the mycorrhizal roots were smaller than in nonmycorrhizal roots, although the differences were not significant. The number of giant cells, on which developing larvae feed, were reduced in the mycorrhizal treatment. The body width of feeding females of *Rotylenchulus reniformis* in tomato was reduced by 9% in the mycorrhizal roots [222]. Nematode development and reproduction, which might be influenced by nematode nutrition, were delayed in mycorrhizal plants [238].

Fourth, there may be competition for space and feeding sites, as suggested by O'Bannon and Nemec [226]. Nematodes would infect roots well before a VAM fungus, but this mechanism might function in cases where the plant was preinoculated with the VAM fungus.

Fifth, physiological changes in VAM roots may result in resistance due to the production of antagonistic substances. Extracts from mycorrhizal tomato roots resulted in 44% mortality of *M. incognita* larvae after 3 days, compared to only 24% mortality in the extracts from nonmycorrhizal plants [232].

A sixth possible mechanism is that mycorrhizal fungi alter the microbial makeup of the rhizosphere, in turn affecting the survival of nematodes and the penetration of roots by nematodes. However, there is no direct evidence for this in the literature.

At present, the exact mechanism responsible for nematode reduction by VAM fungi is unclear. Part of the difficulty in interpreting the conflicting results of these studies is that the measured parameters, such as gall counts and numbers of nematodes in soil, do not accurately reflect nematode damage or actual numbers of nematodes in the root. The use of different host plant–nematode–VAM fungus combinations also makes it difficult to draw general conclusions. Still, the results to date are promising enough to warrant further research in this area. The option of chemical control of nematodes is becoming less viable because of cost and the threat of groundwater contamination. Mycorrhizae might confer additional benefits besides improvement of plant nutrition, and the study of nematode–VAM interactions is an important way to realize that goal.

Mycophagous, or fungus-feeding nematodes, are common inhabitants of the rhizosphere. Although these nematodes are thought to be important consumers of saprophytic fungi, they probably do not have any direct effects on plants. However, they might feed on the external hyphae of VA mycorrhizal fungi and the hyphal strands and mantle of ectomycorrhizae, leading some researchers to speculate that mycophagous nematodes might have adverse effects on these mycorrhizae. Shafer et al. [244] reported that the mycophagous nematode *Aphelenchoides bicaudatus* parasitized hyphae of an ericaceous endomycorrhizal fungus in vitro. The reduction of growth in culture, which was caused by these hyphal grazers, led the authors to speculate that these nematodes could indirectly reduce the nutrient supply to the plant by destroying external hyphae. The addition of the nematode *Aphelenchus avenae* to soybeans inoculated with *Glomus* sp. resulted in an antagonistic interaction, with a decrease in shoot and root weight in treatments with both VAM and the nematode [245]. Fewer nodules and *Glomus* spores were found in the combination treatment. The nematode had no effect on nonmycorrhizal plants. Although they did not quantify mycorrhizal infection, they speculated that the destruction of VAM hyphae by the nematodes reduced the P uptake of the plant, resulting in insufficient nodulation. The addition of *A. avenae* to cotton inhibited the growth enhancement by *Gigaspora margarita*, but only when very high inoculum densities of the nematode were added [246]. However, shoot growth enhancement by *Glomus etunicatum* was not inhibited at these same nematode levels. The nematodes had no effect on spore production or root colonization by either VAM fungus. Since such high levels of nematodes were needed to adversely affect the VAM fungus, the authors surmised that these fungus feeders would not make a major impact in a field situation. The application of the nematicide carbofuran increased native VAM infection of blue grama grass, from 0–2% in the control to 15% in the nematicide treatments [247]. The authors speculated that the reduction in fungus-feeding nematode populations lessened the predation pressure on the VAM fungi, allowing more colonization to occur.

However, the nematicide treatment also changed fungal and bacterial populations, which might have an indirect influence on VAM colonization.

The influence of mycophagous nematodes on ectomycorrhizae formation has been investigated, but to our knowledge there are no recent papers on the subject.

Soil Invertebrates, Microarthropods, and Protozoans

Soil fauna, particularly microarthropods, play an important role in decomposition and nutrient turnover in the soil. These detritivores can also feed on fungi, grazing on hyphae and spores. They have been implicated as regulators in soil nutrient cycling and below-ground food webs [248]. Mycophagous amoebae in the soil have been shown to reduce the inoculum potential of plant pathogenic fungi [249] and are frequent colonizers of the rhizosphere. Despite the importance of soil fauna and mycorrhizal fungi in nutrient cycling, very few studies have examined the interactions between these two ubiquitous groups of organisms.

In a factorial experiment on leeks with *Glomus fasciculatus* and the Collembola *Folsomia candida*, a strong interaction between the VAM fungus and the springtail was detected [250]. The Collembola reduced the mycorrhizal growth increase, and populations of Collembola were three times higher in the mycorrhizal treatments. The presence of springtails reduced the phosphorus content of mycorrhizal plants, although mycorrhizal infection was not affected. The authors concluded that the Collembola were feeding on the external hyphae of *G. fasciculatus*, thus reducing its effectiveness. Finlay [251] also demonstrated higher populations of another Collembola, *Onychiurus ambulans*, on mycorrhizal leek plants. Mycorrhizal stimulation was negated by the presence of Collembola, which had no effect on nonmycorrhizal plants. In field plots treated with the insecticide chlorfenvinphos, the populations of Collembola were significantly reduced, and the plants had higher shoot dry weights and P concentrations than plants in the noninsecticide plots. He suggested that mycorrhizal efficiency was higher when the Collembola densities were low. This was further demonstrated in a greenhouse experiment where *Folsomia candida* decreased VAM colonization of soybeans by 37% [252]. The addition of chlorfenvinphos to a field grassland plot reduced the density of Collembola and increased the shoot dry mass of *Holcus lanatus* [253]. During the final harvest interval, the inflow rate of P was three times higher in the insecticide treatment, which suggested that Collembola feeding on VAM external hyphae reduced P uptake. More direct evidence for Collembola feeding on VAM fungi was obtained by Moore et al. [254], who observed four species of Collembola feeding on the hyphae of axenically germinated VAM spores. The feeding preferences of Collembola on ectomycorrhizal fungi from pine plantations was studied by Shaw [255]. The Collembola *Onychiurus armatus* showed a pronounced preference for the mycorrhizal fungus *Lactarius rufus*. On the other hand, *Hebeloma crustuliniforme*, which contains muscarine and other toxins, was avoided.

Other larger arthropods have been shown to consume VAM spores. The digestive systems of sowbugs and crickets contained chlamydospores and azygospores of VAM fungi [256]. Larger soil arthropods might also function as vectors of VAM fungi. Grasshoppers or earthworms were added to fumigated surface mine soils, which were then planted with ryegrass, fescue, and clover [257]. After 8 weeks, 33% of the plants in the grasshopper treatment and 52% of the plants in the earthworm treatment were infected with endomycorrhizae, indicating that these invertebrates could move VAM fungi. The digestive systems of soil macro-invertebrates were examined for the ingestion of VAM fungal spores [258]. The guts of earthworms, sowbugs, ground beetles, scarab beetles, and millipedes contained VAM spores, including those of *Glomus constrictum*, *G. geosporum*, *G. etunicatum*, and *G. microaggregatum*. Eighty percent of the earthworms that were sampled contained VAM spores, and this group ingested a greater density and diversity of VAM spores than soil microarthropods. This work suggested that soil macro-invertebrates might play a role in the dispersal of VAM fungi. The spores of the VAM fungus *Acaulospora bireticulata* have also been found in the exoskeletons of oribatid mites [259]. The most likely explanation is that the fungus penetrated the decomposed exoskeleton and subsequently sporulated.

Mycophagous amoebae are also known to feed on mycorrhizal fungi. Old and Wong [260] showed that amoebae feed on VAM fungi, and two species of mycophagous amoebae, *Saccamoeba* sp. and *Gephyramoeba* sp., reduced the colonization of pine roots by *Rhizopogon luteolus* [261,262]. The authors concluded that the protozoans could influence the establishment of inoculated symbiotic fungi. Old and Patrick [263] also observed the giant Vamprellid amoeba, *Arachnula impatiens*, feeding on the basidiospores of the ectomycorrhizal fungi *Agaricus sylvaticus* and *Suillus granulatus*, leaving perforated holes in the basidiospores.

III. IMPLICATIONS AND PRACTICAL ASPECTS

Mycorrhizal fungi are known to benefit plant growth in a number of ways. The major way is to enhance nutrient uptake from soil by increasing the absorptive surface of the root system. The extramatrical hyphae of the mycorrhizal fungi can bridge the zone of nutrient depletion adjacent to the root, and thus make available immobile elements such as phosphorus, zinc, and copper by translocating them from more remote sites through hyphae to the host plant root. These hyphae are able to translocate mobile elements, such as nitrogen, as well. The result is that plants with mycorrhizae are better nourished than plants without, especially if the plants are growing in soil where nutrient deficiencies occur. Mycorrhizae are also known to reduce plant response to drought by increasing the water absorption capacity of the root system or by altering the host physiology in such a way as to increase drought tolerance. Mycorrhizae have been reported to increase plant tolerance to a number of stressors, including high salt levels, toxicities associated with mining operations, heavy metals, and mineral imbalances. They are known to reduce the incidence and/or severity of root diseases in some cases, in others not. Many of these responses are due to altered physiological processes and/or morphological changes induced by the mycorrhizal fungi. External to the root, mycorrhizal fungi can alter the chemical and physical properties of soil, largely due to the effects of the extramatrical hyphae or hyphal strands. These penetrate the soil and generate a mycosphere effect, which is a selective stimulation of certain other soil microbes supported by hyphal exudates. These microbes may have sticky surfaces which cause soil to aggregate and be bound together by the mycorrhizal fungal hyphae.

In light of the interactions between mycorrhizal fungi and other soil organisms discussed in this review, it seems appropriate to raise the question as to the likelihood that mycorrhizal fungi alone could account for all of the benefits to plants for which they are credited. More likely, these benefits should be credited to combinations of mycorrhizal fungi and other mycorrhizosphere microbial associates. And when benefits to plants are not apparent, perhaps the deleterious effects of some other microbes might be involved. In a similar fashion, we should reconsider the benefits attributed to other beneficial rhizosphere microbes in light of their probable interaction with mycorrhizal fungi. This is especially logical and likely since mycorrhizal fungi are nearly ubiquitous in all soils. Only exceptional families of plants and special circumstances or treatments of soils preclude mycorrhizal fungi from being a fundamental component of the rhizosphere and having a pronounced effect on the physiology and development of plants.

It seems apparent that many observed inconsistencies in field trials using beneficial soil microbes to stimulate plant growth, whether mycorrhizal fungi, free-living or symbiotic nodule-forming bacteria, biocontrol agents, or plant-promoting rhizobacteria, could be explained on the basis of the uncontrolled variable of other competing or associative microbes. Where benefits are observed, it could mean that a compatible additive or synergistic interaction between rhizosphere microbes has occurred. Where no effects or deleterious effects occur, it could mean that some incompatible reactions have occurred that have blocked some important step in one or more critical processes. If the edaphic conditions are such that they impose a stress on the plant, and the countering symbiont is not able to fully establish a relationship with its host, then the stressor will retard the growth and development of the plant. Since plants have undoubtedly co-evolved with mycorrhizal fungi, it is logical that other microbes could have evolved along with the mycorrhizal fungi.

Microbial associations with plant roots and with each other are normal in natural ecosystems, but in disturbed situations, such as the agroecosystem, sustained crop productivity is only possible with considerable inputs to compensate for the effects of the disturbance. Sustained plant growth in natural ecosystems occurs because of the balance that has developed over evolutionary time, a balance between the host plants, their microbial associates, and the macroenvironment. Disturbance to that balance decreases any chance for sustainability, but the balance could be restored by incorporation of the necessary microflora. We believe that reconstruction should begin with the establishment of appropriate populations of mycorrhizal fungi and as many of their beneficial associates as possible. This attempt to manage the rhizosphere microbiology could include inoculation with mycorrhizal fungi properly placed to cause early colonization, along with fungal and bacterial biocontrol agents with rhizosphere competence and compatibility with their mycorrhizal fungus associates. For legumes, appropriate strains of *Rhizobium* should be included as seed treatments. For nonlegumes, inoculation with free-living nitrogen fixing bacteria could be useful. Addition of growth-promoting rhizobacteria could give added benefit. While the utility of such combinations has not been demonstrated, the logic of their use as a crop management strategy seems apparent.

IV. CONCLUSIONS

With relatively few exceptions, plant roots form associations with mycorrhizal fungi. The degree of colonization of roots by mycorrhizal fungi and thus the effects of the symbiosis may vary, depending on total interaction between host, symbiont, and environment. In most cases, the evidence is strong that mycorrhizae may significantly change the host morphology and/or physiology. The net effect is that other microorganisms in the soil are affected by the change, and a new rhizosphere balance results. We are only beginning to understand just what changes do occur and what organisms are affected and how.

There can be no doubt that mycorrhizae affect other soil microbes, and other microbes affect mycorrhizae. These interactions can be either direct by metabolic exchange or indirect as mediated by changes in the host plant. Because these interactions are sure to occur, they must be studied in detail in order to increase the predictability of inoculations with mycorrhizal fungi or other beneficial microbes and thereby also increase their consistency of performance and benefit to plant growth and development.

Since many of the possible interactions between mycorrhizal fungi and other soil organisms are positive, it makes sense to think of managing the rhizosphere by applying combinations of organisms with multiple purposes. It would not only increase the cost effectiveness of inoculation, but would stabilize the rhizosphere with beneficial microbes that can at least dominate that zone for some critical period of time. We know very little about how long some of the organisms might survive, but the evidence suggests that mycorrhizal fungi are likely to survive indefinitely in the roots of perennial plants and throughout the season for annual plants. Most, if not all, mycorrhizal annuals form VA mycorrhizae, and the fungal symbionts are able to produce resting spores to survive in soil and thus carry over to the next crop. Their microbial associates may be more prone to seasonal and annual fluctuations, and it may be necessary to give "booster" applications to increase their populations. To maximize the combination effects, mycorrhizal fungi and synergistic associates should be in the rhizosphere to colonize roots early in the growing season and to the extent needed to influence plant growth and development.

The prospectus for the future is good for the application of mycorrhizal fungi and positive associative organisms for a number of reasons. We have the technology to produce inoculum of each of the desired microbial components, and the application of beneficial microorganisms is very timely in the face of lessened availability of chemicals to fertilize and protect crop plants. It is therefore incumbent on researchers to merge the technologies developed in the fields of mycorrhizal fungi, nitrogen fixation, biological control, and growth enhancement by growth-promoting and nutrient-cycling microbes. The time is right for the development of "inoculum packages" that are compatible with host and soil and that can benefit plants in ways not possible by chemical additives. In many parts of the world, the application of biologicals is the only way to grow crops successfully and regain or reach a level of sustained productivity that can meet the food and fiber needs of future generations.

REFERENCES

1. Rambelli, A. The rhizosphere of mycorrhizae. In *Ectomycorrhizae* (G. L. Marks and T. T. Kozlowski, eds.), Academic Press, New York, 1973, pp. 299–343.
2. Linderman, R. G. Mycorrhizal interactions with the rhizosphere microflora: The mycorrhizosphere effect. *Phytopathology*, *78*:366–371, 1988.

3. Krishna, K. R., Shetty, K. G., Dart, P. J., and Andrews, D. J. Genotype dependent variation in mycorrhizal colonization and response to inoculation of pearl millet. *Pl. Soil, 86*:113–125, 1985.
4. Bagyaraj, D. J. and Menge, J. A. Interactions between a VA mycorrhiza and *Azotobacter* and their effects on rhizosphere microflora and plant growth. *New Phytol., 80*:567–573, 1978.
5. Newman, E. I., Heap, A. J., and Lawley, R. A. Abundance of mycorrhizas and root-surface micro-organisms of *Plantago lanceolata* in relation to soil and vegetation: A multivariate approach. *New Phytol., 89*:95–108, 1981.
6. Ames, R. N., Reid, C. P. P., and Ingham, E. R. Rhizosphere bacterial populations responses to root colonization by a vesicular-arbuscular mycorrhizal fungus. *New Phytol., 96*:555–563, 1984.
7. Meyer, J. R. and Linderman, R. G. Selective influence on populations of rhizosphere or rhizoplane bacteria and actinomycetes by mycorrhizas formed by *Glomus fasciculatum*. *Soil Biol. Biochem., 18*:191–196, 1986.
8. Ames, R. N., Mihara, K. L., and Bethlenfalvay, G. J. The establishment of microorganisms in vesicular-arbuscular mycorrhizal and control treatments. *Biol. Fertil. Soils, 3*:217–223, 1987.
9. Secilia, J. and Bagyaraj, D. J. Bacteria and actinomycetes associated with pot cultures of vesicular-arbuscular mycorrhizas. *Can. J. Microbiol., 33*:1069–1073, 1987.
10. Schisler, D. A. and Linderman, R. G. The influence of volatiles purged from soil around Douglas fir ectomycorrhizae on soil microbial populations. *7th North American Conference on Mycorrhizae.* University of Florida, Gainesville, 1987, p. 218.
11. Sutton, J. C. and Sheppard, B. R. Aggregation of sand-dune soil by endomycorrhizal fungi. *Can. J. Bot., 54*:326–333, 1976.
12. Forster, S. M. and Nicolson, T. H. Aggregation of sand from a maritime embryo sand dune by microorganisms and higher plants. *Soil Biol. Biochem., 13*:199–203, 1981.
13. Loper, J. E., Suslow, T. V., and Schroth, M. N. Lognormal distribution of bacterial populations in the rhizosphere. *Phytopathology, 74*:1454–1460, 1984.
14. Daniels, B. A. and Trappe, J. M. Factors affecting spore germination of the vesicular-arbuscular mycorrhizal fungus, *Glomus epigaeus*. *Mycologia, 72*:457–471, 1980.
15. Mayo, K., Davis, R. E., and Motta, J. Stimulation of germination of spores of *Glomus versiforme* by spore-associated bacteria. *Mycologia, 78*:426–431, 1986.
16. Azcon-Aguilar, C. and Barea, J. M. Effect of soil micro-organisms on formation of vesicular-arbuscular mycorrhizas. *Trans. Br. Mycol. Soc., 84*:536–537, 1985.
17. Azcon-Aguilar, C., Diaz-Rodriguez, R. M., and Barea, J. M. Effect of soil micro-organisms on spore germination and growth of the vesicular-arbuscular mycorrhizal fungus *Glomus mosseae*. *Trans. Br. Mycol. Soc., 86*:337–340, 1986.
18. Azcon, R. Germination and hyphal growth of *Glomus mosseae* in vitro: Effects of rhizosphere bacteria and cell-free culture media. *Soil. Biol. Biochem., 19*:417–419, 1987.
19. Mugnier, J. and Mosse, B. Spore germination and viability of a vesicular-arbuscular mycorrhizal fungus, *Glomus mosseae*. *Trans. Br. Mycol. Soc., 88*:411–413, 1987.

20. St. John, T. V., Hays, R. I., and Reid, C. P. P. Influence of a volatile compound on formation of vesicular-arbuscular mycorrhizas. *Trans. Br. Mycol. Soc.*, *81*:153–154, 1983.
21. Koske, R. E. Evidence for a volatile attractant from plant roots affecting germ tubes of a VA mycorrhizal fungus. *Trans. Br. Mycol. Soc.*, *79*:305–310, 1982.
22. Gemma, J. N. and Koske, R. E. Pre-infection interactions between roots and the mycorrhizal fungus *Gigaspora gigantea*: Chemotropism of germ-tubes and root growth response. *Trans. Br. Mycol. Soc.*, *91*:123–132, 1988.
23. Graham, J. H., Linderman, R. G., and Menge, J. A. Development of external hyphae by different isolates of mycorrhizal *Glomus* spp. in relation to root colonization and growth of Troyer citrange. *New Phytol.*, *91*:183–189, 1982.
24. Tommerup, I. C. Spore dormancy in vesicular-arbuscular mycorrhizal fungi. *Trans. Br. Mycol. Soc.*, *81*:37–45, 1983.
25. Tommerup, I. C. Inhibition of spore germination of vesicular-arbuscular mycorrhizal fungi in soil. *Trans. Br. Mycol. Soc.*, *85*:267–278, 1985.
26. Ross, J. P. Effect of nontreated field soil on sporulation of vesicular-arbuscular mycorrhizal fungi associated with soybean. *Phytopathology*, *70*:1200–1205, 1980.
27. Wilson, G. T., Hetrick, B. A., and Kitt, D. G. Suppression of VA mycorrhizal fungus spore germination in nonsterile soil. In *7th North American Conference on Mycorrhizae*, University of Florida, Gainesville, 1987, p. 224.
28. Hetrick, B. A. D., Kitt, D. G., and Wilson, G. T. The influence of phosphorus fertilization, drought, fungal species, and nonsterile soil on mycorrhizal growth response in tall grass prairie plants. *Can. J. Bot.*, *64*:1199–1203, 1986.
29. Hetrick, B. A., Wilson, G. T., and Kitt, D. G. Suppression of mycorrhizae by soil microorganisms. In *7th North American Conference on Mycorrhizae*, University of Florida, Gainesville, 1987, p. 200.
30. Kitt, D. G., Hetrick, B. A. D., and Wilson, G. T. Sporulation of two vesicular-arbuscular mycorrhizal fungi in nonsterile soil. *Mycologia*, *79*:896–899, 1987.
31. Wilson, G. W. T., Hetrick, B. A. D., and Kitt, D. G. Suppression of mycorrhizal growth response of big bluestem by non-sterile soil. *Mycologia*, *80*:338–343, 1988.
32. Bowen, G. D. and Theodorou, C. Interaction between bacteria and ectomycorrhizal fungi. *Soil Biol. Biochem.*, *11*:119–126, 1979.
33. Garbaye, J. and Bowen G. D. Effect of mycorrhizospheric microorganisms on ectomycorrhizal infection of *Pinus radiata*. In *7th North American Conference on Mycorrhizae*, University of Florida, Gainesville, 1987, p. 196.
34. Garbaye, J. and Bowen, G. D. Effect of different microflora on the success of ectomycorrhizal inoculation of *Pinus radiata*. *Can. J. For. Res.*, *17*:941–943, 1987.
35. McAfee, B. J. and Fortin, J. A. Comparative effects of the soil microflora on ectomycorrhizal inoculation of conifer seedlings. *New Phytol.*, *108*:443–449, 1988.
36. Strzelczyk, E. and Kampert, M. Growth of mycorrhizal fungi cultured with bacteria. *Agnew. Botanik*, *61*:157–162, 1987.

37. Summerbell, R. C. The inhibitory effects of *Trichoderma* species and other soil microfungi on formation of mycorrhiza by *Laccaria bicolor in vitro*. *New Phytol.*, *105*:437–448, 1987.
38. Tilak, K. V. B. R. Interaction of vesicular-arbuscular mycorrhizae and nitrogen fixers. *Proc. Soil Biol. Symp.*, Hisar.: 219–226, 1985.
39. Barea, J. M. and Azcon-Aguilar, C. Mycorrhizas and their significance in nodulating nitrogen-fixing plants. *Adv. Agron.*, *36*:1–54, 1985.
40. Cluett, H. C. and Boucher, D. H. Indirect mutualism in the legume-Rhizobium-mycorrhizal fungus interaction. *Oecologia*, *59*:405–408, 1983.
41. Manjunath, A. and Bagyaraj, D. J. Response of pigeon pea and cowpea to phosphate and dual inoculation with vesicular arbuscular mycorrhiza and *Rhizobium*. *Trop. Agric.*, *61*:48–52, 1984.
42. Purcino, A. A. C., Lurlarp, C., and Lynd, J. Q. Mycorrhiza and soil fertility effects with growth, nodulation and nitrogen fixation of *Leucaena* grown on a typic eutrustox. *Commun. in Soil Sci. Plant Anal.*, *17*:473–489, 1986.
43. Bethlenfalvay, G. J. and Yoder, J. F. The *Glycine-Glomus-Rhizobium* symbiosis. I. Phosphorus effect on nitrogen fixation and mycorrhizal infection. *Physiol. Plant.*, *52*:141–145, 1981.
44. Manjunath, A., Bagyaraj, D. J., and Gopala Gowda, H. S. Dual inoculation with VA mycorrhiza and *Rhizobium* is beneficial to *Leucaena*. *Pl. Soil*, *78*:445–448, 1984.
45. Subba Rao, N. S., Tilak, K. V. B. R., and Singh, C. S. Dual inoculation with *Rhizobium* sp. and *Glomus fasciculatum* enhances nodulation, yield and nitrogen fixation in chickpea (*Cicer arietinum* Linn.) *Pl. Soil*, *95*:351–359, 1986.
46. Azcon-Aguilar, C. and Barea, J. M. Field inoculation of *Medicago* with V-A mycorrhiza and *Rhizobium* in phosphate-fixing agricultural soil. *Soil Biol. Biochem.*, *13*:19–22, 1981.
47. Redente, E. F. and Reeves, F. G. Interactions between vesicular-arbuscular mycorrhiza and *Rhizobium* and their effect on sweetvetch growth. *Soil Sci.*, *132*:410–415, 1981.
48. Kawai, Y. and Yamamoto, Y. Increase in the formation and nitrogen fixation of soybean nodules by vesicular-arbuscular mycorrhiza. *Plant Cell Physiol.*, *27*:399–405, 1986.
49. Ganry, F., Diem, H. G., and Dommergues, Y. R. Effect of inoculation with *Glomus mosseae* on nitrogen fixation by field grown soybeans. *Pl. Soil*, *68*:321–329, 1982.
50. Hicks, P. M. and Loynachan, T. E. Phosphorus fertilization reduces vesicular-arbuscular mycorrhizal infection and changes nodules occupancy of field-grown soybean. *Agron. J.*, *79*:841–844, 1987.
51. Bethlenfalvay, G. J., Brown, M. S., Mihara, K. L., and Stafford, A. E. *Glycine-Glomus-Rhizobium* symbiosis. V. Effects of mycorrhiza on nodule activity and transpiration in soybeans under drought stress. *Plant Physiol.*, *85*:115–119, 1987.
52. Bayne, H. G. and Bethlenfalvay, G. J. The *Glycine-Glomus-Rhizobium* symbiosis. IV. Interactions between the mycorrhizal and nitrogen-fixing endophytes. *Plant. Cell and Environment*, *10*:607–612, 1987.
53. Smith, S. E., Nicholas, D. J. D., and Smith, F. A. Effect of early mycorrhizal infection on nodulation and nitrogen fixation in *Trifolium subterraneum* L. *Aust. J. Plant Physiol.*, *6*:305–316, 1979.
54. Smith, S. E. and Daft, M. J. Interactions between growth, phosphate content and nitrogen fixation in mycorrhizal and non-mycorrhizal *Medicago sativa*. *Aust. J. Plant Physiol.*, *4*:403–413, 1977.

55. Bethlenfalvay, G. J., Brown, M. S., and Stafford, A. E. *Glycine-Glomus-Rhizobium* symbiosis. II. Antagonistic effects between mycorrhizal colonization and nodulation. *Plant Physiol.*, *79*:1054–1058, 1985.
56. Pacovsky, R. S., Fuller, G., Stafford, A. E., and Paul, E. A. Nutrient and growth interactions in soybeans colonized with *Glomus fasciculatum* and *Rhizobium japonicum*. *Pl. Soil*, *92*:37–45, 1986.
57. Allen, M. F., Smith, W. K., Moore, T. S., Jr., and Christensen, M. Comparative water relations and photosynthesis of mycorrhizal and nonmycorrhizal *Bouteloua gracilis*. *New Phytol.*, *88*:683–693, 1981.
58. Kucy, R. M. N. and Paul, E. A. Carbon flow, photosynthesis, and N_2 fixation in mycorrhizal and nodulated faba beans (*Vicia faba* L.). *Soil Biol. Biochem.*, *14*:407–412, 1982.
59. Pang, P. C. and Paul, E. A. Effects of vesicular-arbuscular mycorrhiza on ^{14}C and ^{15}N distribution in nodulated fababeans. *Can. J. Soi. Sci.*, *60*:241–250, 1980.
60. Harris, D., Pacovsky, R. S., and Paul, E. A. Carbon economy of soybean-*Rhizobium-Glomus* associations. *New Phytol.*, *101*:427–440, 1985.
61. Kosslak, R. M. and Bohlool, B. B. Suppression of nodule development of one side of a split-root system of soybeans caused by prior inoculation of the other side. *Plant Physiol.*, *75*:125–130, 1984.
62. O'Hara, G. W., Dilworth, N. J., Boonkerd, N., and Parkpian, P. Iron-deficiency specifically limits nodule development in peanut inoculated with *Bradyrhizobium* sp. *New Phytol.*, *108*:51–57, 1988.
63. Pacovsky, R. S. Micronutrient uptake and distribution in mycorrhizal or phosphorus-fertilized soybeans. *Pl. Soil*, *95*:379–388, 1986.
64. Grimes, H. D. and Mount, M. S. Influence of *Pseudomonas putida* on nodulation of *Phaseolus vulgaris*. *Soil Biol. Biochem.*, *16*:27–30, 1984.
65. Meyer, J. R. and Linderman, R. G. Response of subterranean clover to dual inoculation with vesicular-arbuscular mycorrhizal fungi and a plant growth-promoting bacterium, *Pseudomonas putida*. *Soil Biol. Biochem.*, *18*:185–190, 1986.
66. Ames, R. N. and Bethlenfalvay, G. J. Localized increase in nodule activity but no competitive interaction of cowpea rhizobia due to pre-establishment of vesicular-arbuscular mycorrhiza. *New Phytol.*, *106*:207–215, 1987.
67. Asimi, S., Gianinazzi-Pearson, V., and Gianinazzi, S. Influence of increasing soil phosphorus levels on interactions between vesicular-arbuscular mycorrhizae and *Rhizobium* in soybeans. *Can. J. Bot.*, *58*:2200–2205, 1980.
68. Ames, R. N., Reid, C. P. P., Porter, L. K., and Cambardella, C. Hyphal uptake and transport of nitrogen from two ^{15}N-labeled sources by *Glomus mosseae*, a vesicular-arbuscular mycorrhizal fungus. *New Phytol.*, *95*:381–396, 1983.
69. Barea, J. M., Azcon-Aguilar, C., and Azcon, R. Vesicular-arbuscular mycorrhiza improve both symbiotic N_2 fixation and N uptake from soil as assessed with a ^{15}N technique under field conditions. *New Phytol.*, *106*:717–725, 1987.
70. Smith, S. E., St. John, B. J., Smith, F. A., and Nicholas, D. J. D. Activity of glutamine synthetase and glutamate dehydrogenase in *Trifolium subterraneum* L. and *Allium cepal.*: effects of mycorrhizal infection and phosphate nutrition. *New Phytol.*, *99*:211–227, 1985.
71. Haystead, A., Malajczuk, N., and Grove, T. S. Underground transfer of nitrogen between pasture plants infected with vesicular-arbuscular mycorrhizal fungi. *New Phytol.*, *108*:417–423, 1987.

72. Van Kessel, C., Singleton, P. W., and Hoben, H. J. Enhanced N-transfer from a soybean to maize by vesicular arbuscular mycorrhizal (VAM) fungi. *Plant Physiol.*, *79*:562–563, 1985.
73. Kucy, R. M. N., Peron, S. C., Purtugal, E. P., and Saito, S. M. T. Occurrence of VAM within nodules of common bean. In *7th North American Conference on Mycorrhizae*, University of Florida, Gainesville, 1987, p. 371.
74. Rose, S. L. Mycorrhizal association of some actinomycete nodulated nitrogen-fixing plants. *Can. J. Bot.*, *58*:1449–1454, 1980.
75. Clelland, D. M., Gardner, I. C., and Scott, A. Occurrence of *Glomus fasciculatus*, a mycorrhizal endophyte, in the nitrogen-fixing non-legume *Hippophae rhamnoides*. *Microbios Lett.*, *24*:107–113, 1983.
76. Perry, C. B., Stutz, J. G., and Righetti, T. L. Endomycorrhizae in actinomycete nodulated nitrogen-fixing shrubs in the Rosaceae. In *6th North American Conference on Mycorrhizae*, Bend, OR., 1984, p. 312.
77. Ba, A. M., Sougoufara, B., and Thoen, D. The triple symbiosis of *Casuarina equisetifolia* in Senegal. In *7th North American Conference on Mycorrhizae*, University of Florida, Gainesville, 1987, p. 121.
78. Rose, S. L. and Youngberg, C. T. Tripartite associations in snowbrush (*Ceanothus velutinus*): Effect of vesicular-arbuscular mycorrhizae on growth, nodulation, and nitrogen fixation. *Can. J. Bot.*, *59*:34–39, 1981.
79. Gardner, I. C., Clelland, D. M., and Scott, A. Mycorrhizal improvement in non-leguminous nitrogen fixing associations with particular reference to *Hippophae rhamnoides*. *Pl. Soil*, *78*:189–199, 1984.
80. Diem, H. G. and Gauthier, D. Effet de l'infection endomycorhizienne (*Glomus mosseae*) sur la nodulation et la croissance de *Casuarina equisetifolia*. *C. R. Acad. Sc. Paris*, *294*:215–218, 1982.
81. Subba Rao, N. S., Tilak, K. V. B. R., and Singh, C. S. Ecology of nonsymbiotic dinitrogen fixers. In *Progress in Microbial Ecology* (K. G. Mukerji, V. P. Agnihotri and R. P. Singh, eds.), Lucknow, Print House (India), 1981.
82. Tilak, K. V. B. R. Recent developments in the field of *Azospirillum* and its interaction with vesicular-arbuscular mycorrhizal (VAM) fungi in increasing the crop yields. In *Crop Productivity* (H. L. Svivastava, S. Bhaskaman, and K. Menon, eds.), Oxford & IBH Book Publ. Co., New Delhi, 1985, pp. 287–305.
83. Subba Rao, N. S., Tilak, K. V. B. R., and Singh, C. S. Synergistic effect of vesicular-arbuscular mycorrhizas and *Azospirillum brasilense* on the growth of barley in pots. *Soil Biol. Biochem.*, *17*:119–121, 1985.
84. Subba Rao, N. S., Tilak, K. V. B. R., and Singh, C. S. Effect of combined inoculation of *Azospirillum brasilense* and vesicular-arbuscular mycorrhiza on pearl millet (*Pennisetum americanum*). *Pl. Soil*, *84*:283–286, 1985.
85. Tilak, K. V. B. R. and Singh, C. S. Response of pearl millet (*Pennisetum americanum*) to inoculation with vesicular-arbuscular mycorrhizae and *Azospirillum brasilense* with different sources of phosphorus. *Curr. Sci.*, *57*:43–44, 1988.
86. Barea, J. M., Bonis, A. F., and Olivares, J. Interactions between *Azospirillum* and VA mycorrhiza and their effects on growth and nutrition of maize and ryegrass. *Soil Biol. Biochem.*, *15*:705–709, 1983.

87. Pacovsky, R. S., Fuller, G., and Paul, E. A. Influence of soil on the interactions between endomycorrhizae and *Azospirillum* in sorghum. *Soil Biol. Biochem.*, *17*:525–531, 1985.
88. Harris, D., Pacovsky, R. S., and Paul, E. A. C-flow and N_2 fixation in Sorghum-*Azospirillum-Glomus* associations. In *7th North American Conference on Mycorrhizae*, University of Florida, Gainesville, 1987, p. 374.
89. Pacovsky, R. S. Influence of inoculation with *Azospirillum brasilense* and *Glomus fasciculatum* on sorghum nutrition. *Pl. Soil*, *110*:283–287, 1988.
90. Pacovsky, R. S. and Fuller, G. Mineral and lipid composition of *Glycine-Glomus-Bradyrhizobium* symbioses. *Physiol. Plant.*, *72*:733–746, 1988.
91. Mohandas, S. Field response to tomato (*Lycopersicon esculentum* Mill 'Pusa Ruby') to inoculation with a VA mycorrhizal fungus *Glomus fasciculatum* and with *Azotobacter vinelandii*. *Pl. Soil*, *98*:295–297, 1987.
92. Brown, M. E. and Carr, G. R. Interactions between *Azotobacter chroococcum* and vesicular-arbuscular mycorrhiza and their effects on plant growth. *J. Appl. Bacteriol.*, *56*:429–437, 1984.
93. Manjunath, A., Mohan, R., and Bagyaraj, D. J. Interaction between *Beijerinckia mobilis*, *Aspergillus niger* and *Glomus fasciculatus* and their effects on growth of onion. *New Phytol.*, *87*:723–727, 1981.
94. Tilak, K. V. B. R., Li, C. Y., and Ho, I. Occurrence of nitrogen-fixing *Azospirillum* on surface sterilized mycorrhizal roots of green onion (*Allium cepa*). In *7th North American Conference on Mycorrhizae*, University of Florida, Gainesville, 1987, p. 222.
95. Tilak, K. V. B. R., Li, C. Y., and Ho I. Recovery of nitrogen-fixing *Azospirillum* from spores of VA-mycorrhizal fungi. In *7th North American Conference on Mycorrhizae*, University of Florida, Gainesville, 1987, p. 223.
96. Li, C. Y. and Castellano, M. A. Nitrogen-fixing bacteria isolated from within sporocarps of three ectomycorrhizal fungi. In *6th North American Conference on Mycorrhizae*, Bend, OR, 1984, p. 264.
97. Li, C. Y. and Castellano, M. A. *Azospirillum* isolated from within sporocarps of the mycorrhizal fungi *Hebeloma crustuliniforme, Laccaria laccata* and *Rhizopogon vinicolor*. *Trans. Br. Mycol. Soc.*, *88*:563–565, 1987.
98. Li, C. Y. and Hung, L. L. Nitrogen-fixing (acetylene-reducing) bacteria associated with ectomycorrhizae of Douglas-fir. *Pl. Soil*, *98*:425–428, 1987.
99. Li, C. Y. Diazotrophic bacteria in sporocarps of ectomycorrhizal fungi *Barssia oregonensis, Hysterangium setchellii, Leccinum scarbum,* and *Rhizopogon parksii*. In *7th North American Conference on Mycorrhizae*. University of Florida, Gainesville,, 1987, p. 205.
100. Tilak, K. V. B. R. and Subba Rao, N. S. Association of *Azospirillum brasilense* with pearl millet (*Pennisetum americanum* (L.) Leeke). *Biol. Fertil. Soils*, *4*:97–102, 1987.
101. Tien, T. M., Gaskins, M. H., and Hubbell, D. H. Plant growth substances produced by *Azospirillum brasilense* and their effect on the growth of pearl millet (*Pennisetum americanum* L.). *Appl. Environ. Microbiol.*, *37*:1016–1024, 1979.
102. Azcon, R. and Barea, J. M. Synthesis of auxins, gibberellins and cytokinins by *Azotobacter vinelandii* and *Azotobacter beijerinckii*

related to effects produced on tomato plants. *Pl. Soil, 43*:609–619, 1975.

103. Azcon, R., Azcon-Aguilar, C., and Barea, J. M. Effects of plant hormones present in bacterial cultures on the formation and responses to VA endomycorrhiza. *New Phytol., 80*:359–364, 1978.
104. Allen, M. F., Moore, T. S., and Christensen, M. Phytohormone changes in *Bouteloua gracilis* infected by vesicular-arbuscular mycorrhizae: I. Cytokinin increases in the host plant. *Can. J. Bot., 58*:371–374, 1980.
105. Ho, I. Phytohormonal control of seedling growth by ectomycorrhizal fungi. In *7th North American Conference on Mycorrhizae*, University of Florida, Gainesville, 1987, p. 252.
106. Mudge, K. W. Hormonal involvement in ectomycorrhizal development. In *7th North American Conference on Mycorrhizae*, University of Florida, Gainesville, 1987, p. 228.
107. Ng, P. P., Cole, A. L. J., Jameson, P. E., and Mcwha, J. A. Cytokinin production by ectomycorrhizal fungi. *New Phytol., 91*:57–62, 1982.
108. Barea, J. M. and Azcon-Aguilar, C. Production of plant growth-regulating substances by the vesicular-arbuscular mycorrhizal fungus *Glomus mosseae*. *Appl. Environ. Microbiol., 43*:810–813, 1982.
109. Burr, T. J. and Caesar, A. Beneficial plant bacteria. *Critical Reviews in Plant Sciences, 2*:1–20, 1984.
110. Kloepper, J. W., Leong, J., Teintze, M., and Schroth, M. N. Enhanced plant growth by siderophores produced by plant growth-promoting rhizobacteria. *Nature* (London), *286*:885–886, 1980.
111. Ek, M., Ljungquist, P. O., and Stenstrom, E. Indole-3-acetic acid production by mycorrhizal fungi determined by gas chromatography-mass spectrometry. *New Phytol., 94*:401–407, 1983.
112. Madej, A. and Haggblom, P. Radioimmunoassay for determination of indole-3-acetic acid in fungi and plants. *Physiol. Plantarum, 64*:389–392, 1985.
113. Mitchell, R. J., Garrett, H. E., Cox, G. S., and Atalay, A. Boron and ectomycorrhizal influences on IAA levels and IAA oxidase and peroxidase activities of *Pinus echinata* roots. *Tree Physiol., 1*:1–8, 1986.
114. Wilson, E. R. L. and Field, R. J. Dichotomous branching in lateral roots of pine: The effect of 2-chloroethylphosphonic acid on seedlings of *Pinus radiata* D. Don. *New Phytol., 98*:465–473, 1984.
115. Rupp, L. A. and Mudge, K. W. Ethephon and auxin induce mycorrhiza-like changes in the morphology of root organ cultures of Mugo pine. *Physiol. Plant., 64*:316–322, 1985.
116. Graham, J. H. and Linderman, R. G. Ethylene production by ectomycorrhizal fungi, *Fusarium oxysporum* f. sp. *pini*, and by aseptically synthesized ectomycorrhizae and Fusarium-infected Douglas-fir roots. *Can. J. Microbiol., 26*:1340–1347, 1980.
117. Graham, J. H. and Linderman, R. G. Effect of ethylene on root growth, ectomycorrhizae formation, and Fusarium infection of Douglas-fir. *Can. J. Bot., 59*:149–155, 1981.
118. Azcon-Aguilar, C., Rodriguez-Navarro, D. N., and Barea, J. M. Effects of ethrel on the formation and responses to VA mycorrhiza in *Medicago* and *Triticum*. *Pl. Soil, 60*:461–468, 1981.

119. Strzelczyk, E., Kampert, M., and Michalski, L. Production of cytokinin-like substances by mycorrhizal fungi of pine (*Pinus sylvestris* L.) in culture with and without metabolites of actinomycetes. *Acta Microbiologica Polonica, 34*:177–186, 1985.
120. Allen, M. F., Moore, T. S., and Christensen, M. Phytohormone changes in *Bouteloua gracilis* infected by vesicular-arbuscular mycorrhizae. II. Altered levels of gibberellin-like substances and abscisic acid in the host plant. *Can. J. Bot., 60*:468–471, 1982.
121. Schroth, M. N., Loper, J. E., and Hildebrand, D. C. Bacteria as biocontrol agents of plant disease. In *Current Perspectives in Microbial Ecology* (M. J. Klug and C. A. Reddy, eds.), American Society for Microbiology, Washington, D.C., 1984.
122. Schippers, B., Bakker, A. W., Bakker, P. A. H. M., Weisbeek, P. J., and Lugtenburg, B. Plant growth-inhibiting and stimulating rhizosphere microorganisms. In *Microbial Communities in Soil* (V. Jensen, A. Kjoller, and L. H. Sorensen, eds.), Elsevier Applied Science Publishers, London, 1986.
123. Barber, D. A. and Martin, J. K. The release of organic substances by cereal roots into soil. *New Phytol., 75*:69–80, 1976.
124. Prikryl, Z. and Vancura, V. Root exudates of plants. IV. Wheat root exudation as dependent on growth, concentration gradient of exudates and the presence of bacteria. *Pl. Soil, 57*:69–83, 1980.
125. Strzelczyk, E. and Pokojska-Burdziej, A. Production of auxins and gibberellin-like substances by mycorrhizal fungi, bacteria and actinomycetes isolated from soil and the mycorrhizosphere of pine (*Pinus sylvestris* L.). *Pl. Soil, 81*:185–194, 1984.
126. Strzelczyk, E., Dahm, H., Kampert, M., Pokojska, A., and Rozycki, H. Activity of bacteria and actinomycetes associated with mycorrhiza of pine (*Pinus sylvestris* L.). *Agnew. Botanik, 61*:53–64, 1984.
127. Sanders, F. E. and Tinker, P. B. Mechanism of absorption of phosphate from soil by *Endogone* mycorrhizas. *Nature* (London), *223*:278–279, 1971.
128. Bolan, N. S., Robson, A. D., Barrow, N. J., and Aylmore, L. A. G. Specific activity of phosphorus in mycorrhizal and non-mycorrhizal plants in relation to the availability of phosphorus to plants. *Soil Biol. Biochem., 16*:299–304, 1984.
129. Barea, J. M., Azcon, R., and Hayman, D. Possible synergistic interactions between *Endogone* and phosphate-solubilizing bacteria in low-phosphate soil. In *Endomycorrhiza. Proc. Symp. Univ. Leads. July, 1974* (B. Mosse, F. E. Sanders, and P. B. Tinker, eds.), Academic Press, New York, 1975.
130. Azcon, R., Barea, J. M., and Hayman, D. S. Utilization of rock phosphate in alkaline soils by plants inoculated with mycorrhizal fungi and phosphate solubilizing bacteria. *Soil Biol. Biochem., 8*:135–138, 1976.
131. Delorenzini, G., Barea, J. M., and Olivares, J. Fertilizacion biologica (micorriza + *Rhizobium* + fosobacterias) de *Trifolium pratense* en diferentes condiciones de cultivo. *Rev. Lat-amer. Microbiol., 21*:129–134, 1979.

132. Krone, W., Bichler, B., Viebrock, E., and Moawad, A. M. Interactions between VA mycorrhiza and phosphate solubilizing bacteria. In *7th North American Conference on Mycorrhizae*, University of Florida, Gainesville, 1987, p. 328.
133. Raj. J., Bagyaraj, D. J., and Manjunath, A. Influence of soil inoculation with vesicular-arbuscular mycorrhiza and phosphate-dissolving bacterium on plant growth and ^{32}P-uptake. *Soil. Biol. Biochem.*, *13*:105–108, 1981.
134. Azcon-Aguilar, C., Gianinazzi-Pearson, V., Fardeau, J. C., and Gianinazzi, S. Effect of vesicular-arbuscular mycorrhizal fungi and phosphate-solubilizing bacteria on growth and nutrition of soybean in a neutral-calcareous soil amended with ^{32}P-^{45}Ca-tricalcium phosphate. *Pl. Soil*, *96*:3–15, 1986.
135. Lee, A. and Bagyaraj, D. J. Effect of soil inoculation with vesicular-arbuscular mycorrhizal fungi and either phosphate rock dissolving bacteria or thiobacilli on dry matter production and uptake of phosphorus by tomato plants. *New Zealand J. Agricult. Res.*, *29*:525–531, 1986.
136. Barea, J. M., Navarro, E., and Montoya, E. Production of plant growth regulators by rhizosphere phosphate-solubilizing bacteria. *J. Appl. Bact.*, *40*:129–134, 1976.
137. Azcon-Aguilar, C. and Barea, J. M. Effects of interactions between different culture fractions of 'phosphobacteria' and *Rhizobium* on mycorrhizal infection, growth, and nodulation of *Medicago sativa*. *Can. J. Microbiol.*, *24*:520–524, 1978.
138. Baya, A. M., Boethling, R. S., and Ramos-Cormenzana, A. Vitamin production in relation to phosphate solubilization by soil bacteria. *Soil Biol. Biochem.*, *13*:527–531, 1981.
139. Dodd, J. C., Burton, C. C., Burns, R. G., and Jeffries, P. Phosphatase activity associated with the roots and the rhizosphere of plants infected with vesicular-arbuscular mycorrhizal fungi. *New Phytol.*, *107*:163–172, 1987.
140. Krishna, K. R. and Bagyaraj, D. J. Phosphatases in the rhizosphere of mycorrhizal and non-mycorrhizal peanut. In *7th North American Conference on Mycorrhizae*, University of Florida, Gainesville, 1987, p. 372.
141. Thomas, G. W., Clarke, C. A., Mosse, B., and Jackson, R. M. Source of phosphate taken up from two soils by mycorrhizal (*Thelephora terrestris*) and nonmycorrhizal *Picea sitchensis* seedlings. *Soil Biol. Biochem.*, *14*:73–75, 1982.
142. Leyval, C. and Berthelin, J. Comparison between the utilization of phosphorus from insoluble mineral phosphates by ectomycorrhizal fungi and rhizobacteria. In *Physiological and Genetical Aspects of Mycorrhizae*. (V. Gianinazzi-Pearson and S. Gianinazzi, eds.), INRA, Paris, 1986.
143. Leyval, C. and Verthelin, J. Interactions between ectomycorrhizal fungi and phosphate-solubilizing bacteria: Phosphorus mobilization from different inorganic phosphates. In *Current Perspectives in Environmental Biogeochemistry* (G. Giovannozzi-Sermanni and P. Nannipieri, eds.), Rome, Italy, 1988.

144. Leyval, C. and Berthelin, J. Influence of acid-producing *Agrobacterium* and *Laccaria laccata* on pine and beech growth, nutrient uptake and exudation. In *2nd European Symposium on Mycorrhizae*, Prague, Czechoslovakia, 1988.
145. Cook, R. J. and Baker, K. F. *The Nature and Practice of Biological Control of Plant Pathogens.* The American Phytopathological Society, St. Paul, MN, 1983.
146. Baker, K. F. and Cook, R. J. *Biological Control of Plant Pathogens.* W. H. Freeman, San Francisco, 1974.
147. Deacon, J. W. *Microbial Control of Plant Pests and Diseases.* American Society for Microbiology, Washington, D. C., 1983.
148. Papavizas, G. C. *Biological Control in Crop Production.* Allanheld, Osmun Publishers, Granada, 1981.
149. Krishna, K. R., Balakrishna, A. N., and Bagyaraj, D. J. Interactions between a vesicular-arbuscular mycorrhizal fungus and *Streptomyces cinnamomeous* and their effects on finger millet. *New Phytol.*, *92*:401–405, 1982.
150. Oliveira, E., Sieverding, E., and Toro, S. Interaction between three species of VAM fungi and one isolate of *Pseudomonas putida* on cassava. In *7th North American Conference on Mycorrhizae*, University of Florida, Gainesville, 1987, p. 216.
151. Paulitz, T. C. and Linderman, R. G. Interactions between VA mycorrhizal fungi and fluorescent pseudomonads in the rhizosphere. *Phytopathology, 78*:1567, 1988.
152. Keast, D. and Tonkin, C. Antifungal activity of western Australian soil actinomycetes against *Phytophthora* and *Pythium* species and a mycorrhizal fungus, *Laccaria laccata. Aust. J. Biol. Sci.*, *36*:191–203, 1983.
153. Malajczuk, N. The microflora of unsuberized roots of *Eucalyptus calophylla* R. Br. and *Eucalyptus marginata* Donn ex Sm. seedlings grown in soils suppressive and conducive to *Phytophthora cinnamomi* Rands. II. Mycorrhizal roots and associated microflora. *Aust. J. Bot.*, *27*:255–272, 1979.
154. Malajczuk, N. and McComb, A. J. The microflora of unsuberized roots of *Eucalyptus calophylla* R. Br. and *Eucalyptus marginata* Donn ex Sm. seedlings grown in soils suppressive and conducive to *Phytophthora cinnamomi* Rands. I. Rhizosphere bacteria, actinomycetes, and fungi. *Aust. J. Bot.*, *27*:235–254, 1979.
155. Malajczuk, N. Interaction between *Phytophthora cinnamomi* zoospores and micro-organisms on non-mycorrhizal and ectomycorrhizal roots of *Eucalyptus marginata. Trans. Br. Mycol. Soc.*, *90*:375–382, 1988.
156. Ames, R. N. Chitin-decomposing microorganisms associated with spores of a VA mycorrhizal fungus. *Phytopathology, 76*:1110 (abstr.), 1986.
157. Ames, R. N. Mycorrhiza development in onion in response to inoculation with chitin-decomposing actinomycetes. *New Phytol.*, *112*:423–427, 1989.
158. MacDonald, R. M. and Chandler, M. R. Bacterium-like organelles in the vesicular-arbuscular mycorrhizal fungus *Glomus caledonius. New Phytol.*, *89*:241–246, 1981.

159. MacDonald, R. M., Chandler, M. R., and Mosse, B. The occurrence of bacterium-like organelles in vesicular-arbuscular mycorrhizal fungi. *New Phytol.*, *90*:659–663, 1982.
160. Sward, R. J. The structure of the spores of *Gigaspora margarita*. I. The dormant spore. *New Phytol.*, *87*:761–768, 1981.
161. Tzean, S. S., Chu, C. L. and Su, H. J. Spiroplasmalike organisms in a vesicular-arbuscular mycorrhizal fungus and its mycoparasite. *Phytopathology*, *73*:989–991, 1983.
162. Varma, A. K., Singh, K., and Lall, V. K. Lumen bacteria from endomycorrhizal spores. *Curr. Microbiol.*, *6*:207–211, 1981.
163. Singh, K. and Varma, A. K. Association of bacteria in the lumen of vesicular-arbuscular mycorrhizal spores extracted from the rhizosphere of xerophytes. *Trans. Mycol. Soc. Japan*, *26*:511–516, 1985.
164. Wilkinson, K., Dixon, K. W., and Sivasithamparam, K. Interaction of bacteria, fungi and seed in relation to germination of terrestrial orchids. *Proc. 5th International Congress of Plant Pathology*, Kyoto, Japan, 1988, p. 193.
165. Secilia, J. and Bagyaraj, D. J. Fungi associated with pot cultures of vesicular arbuscular mycorrhizas. *Trans. Br. Mycol. Soc.*, *90*:117–119, 1988.
166. Davey, C. B. Nonpathogenic organisms associated with mycorrhizae. In *Mycorrhizae, Proceedings of the First North American Conference on Mycorrhizae*, Misc. Publication 1189, U.S. Department of Agriculture, Forest Service, Washington, D.C., 1969.
167. Summerbell, R. C. Fungi in the mycorrhizosphere of black spruce in boreal Canada. In *6th North American Conference on Mycorrhizae*, Bend, OR, 1984, p. 267.
168. Sharma, G. D. Influence of mycorrhizal developmental stages and plant age on rhizosphere mycoflora of *Pinus kesiya* (Royale). *Acta Mycologia*, *17*:45–51, 1981.
169. Schenck, N. C. and Kellam, M. K. *The influence of vesicular arbuscular mycorrhizae on disease development*. Fla. Agric. Exp. Stn. Tech. Bull., 798, 1978.
170. Schenck, N. C. Can mycorrhizae control root disease? *Plant Dis.*, *65*:232–234, 1981.
171. Schenck, N. C. Vesicular-arbuscular mycorrhizal fungi and the control of fungal root diseases. In *Innovative Approaches to Plant Disease Control* (I. Chet, ed.), John Wiley and Sons, New York, 1987.
172. Dehne, H. W. Interactions between vesicular-arbuscular mycorrhizal fungi and plant pathogens. *Phytopathology*, *72*:1115–1119, 1982.
173. Schonbeck, F. Endomycorrhiza in relation to plant diseases. In *Soil-Borne Plant Pathogens* (B. Schippers and W. Gams, eds.), Academic Press, New York, 1979, pp. 271–280.
174. Zambolim, L. and Schenck, N. C. Reduction of the effects of pathogenic, root-infecting fungi on soybean by the mycorrhizal fungus, *Glomus mosseae*. *Phytopathology*, *73*:1402–1405, 1983.
175. Zambolim, L. and Schenck, N. C. Effect of *Macrophomina, Rhizoctonia, Fusarium* and the mycorrhizal fungus *Glomus mosseae* on nodulated and non-nodulated soybeans. *Fitopatologia Brasileira*, *9*:129–138, 1984.

176. Baath, E. and Hayman, D. S. Plant growth responses to vesicular-arbuscular mycorrhiza. XIV. Interactions with *Verticillium* wilt on tomato plants. *New Phytol.*, *95*:419–426, 1983.

177. Baath, E. and Hayman, D. S. No effect of VA mycorrhiza on red core disease of strawberry. *Trans. Br. Mycol. Soc.*, *82*:534–536, 1984.

178. Davis, R. M. Influence of *Glomus fasciculatus* on *Thielaviopsis basicola* root rot of citrus. *Plant Dis.*, *64*:839–840, 1980.

179. Davis, R. M., Menge, J. A., and Erwin, D. C. Influence of *Glomus fasciculatus* and soil phosphorus on Verticillium wilt of cotton. *Phytopathology*, *69*:453–456, 1979.

180. Davis, R. M. and Menge, J. A. Influence of *Glomus fasciculatus* and soil phosphorus on Phytophthora root rot of citrus. *Phytopathology*, *70*:447–452, 1980.

181. Davis, R. M. and Menge, J. A. *Phytophthora parasitica* inoculation and intensity of vesicular-arbuscular mycorrhizae in citrus. *New Phytol.*, *87*:705–715, 1981.

182. Rosendahl, S. Interactions between the vesicular-arbuscular mycorrhizal fungus *Glomus fasciculatum* and *Aphanomyces euteiches* root rot of peas. *Phytopath. Z.*, *114*:31–40, 1985.

183. Bartschi, H., Gianinazzi-Pearson, V., and Vegh, I. Vesicular-arbuscular mycorrhiza formation and root rot disease (*Phytophthora cinnamomi*) development in *Chamaecyparis lawsoniana*. *Phytopath. Z.*, *102*:213–218, 1981.

184. Stewart, E. L. and Pfleger, F. L. Development of poinsettia as influenced by endomycorrhizae, fertilizer and root rot pathogens *Pythium ultimum* and *Rhizoctonia solani*. *Florist's Rev.*, *159*:37,79–80, 1977.

185. Dehne, H. W. and Schonbeck, F. Untersuchungen zun Einfluss der Endotrophen Mycorrhiza auf Pflanzenkrankheiten. II. Phenolstoffwechsel und Lignifizierung. (The influence of endotrophic mycorrhiza on plant diseases. II. Phenolmetabolism and lignification.) *Phytopath. Z.*, *95*:210–216, 1979.

186. Becker, W. N. Quantification of onion vesicular-arbuscular mycorrhizae and their resistance to *Pyrenochaeta terrestris*. Ph.D. thesis, University of Illinois, Urbana, 1976.

187. Dehne, H. W., Schonbeck, F., and Baltruschat, H. Untersuchungen zum Einfluss der endotrophen Mycorrhiza auf Pflanzenkrankheiten. 3. Chitinase-Aktivitat und Ornithinzyklus. (The influence of endotrophic mycorrhiza on plant diseases. 3. Chitinase-activity and ornithine-cycle). *Z. Pflkrankh.*, *85*:666–678, 1978.

188. Wick, R. L. and Moore, L. D. Histology of mycorrhizal and nonmycorrhizal *Ilex crenata* 'Helleri' challenged by *Thielaviopsis basicola*. *Can. J. Plant Pathol.*, *6*:146–150, 1984.

189. Gianinazzi-Pearson, V., Morandi, D., Dexheimer J., and Gianinazzi, S. Ultrastructural and ultracytochemical features of a *Glomus tenuis* mycorrhiza. *New Phytol.*, *88*:633–639, 1981.

190. Gianinazzi, S., Gianinazzi-Pearson, V., and Dexheimer, J. Enzymatic studies on the metabolism of vesicular-arbuscular mycorrhiza. III. Ultrastructural localization of acid and alkaline phosphatase in onion roots infected by *Glomus mosseae* (Nicol. & Gerd.), *New Phytol.*, *82*:127–132, 1979.

191. Marx C., Dexheimer, J., Gianinazzi-Pearson, V., and Gianinazzi, S. Enzymatic studies on the metabolism of vesicular-arbuscular mycorrhizas. IV. Ultracytoenzymological evidence (ATPase) for active transfer processes in the host-arbuscule interface. *New Phytol.*, *90*:37–43, 1982.
192. Graham, J. H. and Menge, J. A. Influence of vesicular-arbuscular mycorrhizae and soil phosphorus on take-all disease of wheat. *Phytopathology*, *72*:95–98, 1982.
193. Graham, J. H. and Egel, D. S. Phytophthora root rot development on mycorrhizal and phosphorus-fertilized nonmycorrhizal sweet orange seedlings. *Plant Dis.*, *72*:611–614, 1988.
194. Caron, M., Fortin, J. A., and Richard, C. Effect of phosphorus concentration and *Glomus intraradices* on Fusarium crown and root rot of tomatoes. *Phytopathology*, *76*:942–946, 1986.
195. Caron, M., Fortin, J. A., and Richard, C. Influence of substrate on the interaction of *Glomus intraradices* and *Fusarium oxysporum* f. sp. *radicis-lycopersici* on tomatoes. *Pl. Soil*, *87*:233–239, 1985.
196. Caron, M., Fortin, J. A., and Richard, C. Effect of *Glomus intraradices* on infection by *Fusarium oxysporum* f. sp. *radicis-lycopersici* in tomatoes over a 12-week period. *Can. J. Bot.*, *64*:552–556, 1986.
197. Marx, D. H. Ectomycorrhizae as biological deterrents to pathogenic root infections. *Annu. Rev. Phytopathology*, *10*:429–454, 1972.
198. Marx, D. H. and Krupa, S. V. Mycorrhizae. A. Ectomycorrhizae. In *Interactions Between Non-Pathogenic Soil Microorganisms and Plants* (Y. R. Dommergues and S. V. Krupta, eds.), Elsevier Scientific Publishing Company, New York, 1978.
199. Perrin, R. and Garbaye, J. Influence of ectomycorrhizae on infectivity of Pythium-infested soils and substrates. *Pl. Soil*, *71*:345–351, 1983.
200. Sylvia, D. M. and Sinclair, W. A. Suppressive influence of *Laccaria laccata* on *Fusarium oxysporum* and on Douglas-fir seedlings. *Phytopathology*, *73*:384–389, 1983.
201. Duchesne, L. C., Peterson, R. L., and Ellis, B. E. Interaction between the ectomycorrhizal fungus *Paxillus involutus* and *Pinus resinosa* induces resistance to *Fusarium oxysporum*. *Can. J. Bot.*, *66*:558–562, 1988.
202. Sylvia, D. M. and Sinclair, W. A. Phenolic compounds and resistance in fungal pathogens induced in primary roots of Douglas-fir seedlings by the ectomycorrhizal fungus *Laccaria laccata*. *Phytopathology*, *73*:390–397, 1983.
203. Malajczuk, N. and Sanfelieu, C. L. Survival of *Phytophthora cinnamomi* zoospores on non-mycorrhizal and ectomycorrhizal roots of jarrah. In *6th North American Conference on Mycorrhizae*, Bend, OR, 1984. p. 336.
204. Zhao, Z. P. and Kuo, S. C. Study on hyphal hyperparasitic relationships between *Rhizoctonia solani* and ectomycorrhizal fungi. In *Proc. 5th International Congress of Plant Pathology*, Kyoto, Japan, 1988, p. 196.
205. Tzean, S. S. and Huang, Y. S. The occurrence and formation of vesicular-arbuscular mycorrhizae of citrus and maize. *Bot. Bull. Academia Sinica*, *21*:119–134, 1980.

206. Bhattacharjee, M., Mukerji, K. G., Tewari, J. P., and Skoropad, W. P. Structure and hyperparasitism of a new species of *Gigaspora*. *Trans. Br. Mycol. Soc.*, *78*:184–188, 1982.
207. Koske, R. E. *Labyrinthula* inside the spores of a vesicular-arbuscular mycorrhizal fungus. *Mycologia*, *73*:1175–1180, 1981.
208. Taber, R. A. *Gigaspora* spores and associated hyperparasites in weed seeds in soil. *Mycologia*, *74*:1026–1031, 1982.
209. Sylvia, D. M. and Schenck, N. C. Germination of chlamydospores of three *Glomus* species as affected by soil matric potential and fungal contamination. *Mycologia*, *75*:30–35, 1983.
210. Ross, J. P. and Ruttencutter, R. Population dynamics of two vesicular-arbuscular endomycorrhizal fungi and the role of hyperparasitic fungi. *Phytopathology*, *67*:490–496, 1977.
211. Daniels, B. A. and Menge, J. A. Hyperparasitism of vesicular-arbuscular mycorrhizal fungi. *Phytopathology*, *70*:584–588, 1980.
212. Daniels B. A. The influence of hyperparasites of VAM fungi on growth of citrus. *Phytopathology*, *71*:212–213, (Abstr.), 1981.
213. Sylvia, D. M. and Schenck, N. C. Soil fungicides for controlling chytridiaceous mycoparasites of *Gigaspora margarita* and *Glomus fasciculatum*. *Appl. and Environ. Microbiol.*, *45*:1306–1309, 1983.
214. Paulitz, T. C. and Menge, J. A. Is *Spizellomyces punctatum* a parasite or saprophyte of vesicular-arbuscular mycorrhizal fungi? *Mycologia*, *76*:99–107, 1984.
215. Paulitz, T. C. and Menge, J. A. The effects of a mycoparasite on the mycorrhizal fungus, *Glomus deserticola*. *Phytopathology*, *76*:351–354, 1986.
216. Siqueira, J. O., Hubbell, D. H., Kimbrough, J. W., and Schenck, N. C. *Stachybotrys chartarum* antagonistic to azygospores of *Gigaspora margarita in vitro*. *Soil Biol. Biochem.*, *16*:679–681, 1984.
217. Paulitz, T. C. Mycoparasitism of vesicular-arbuscular mycorrhizal fungi. Ph.D. dissertation, University of California, Riverside, 1984.
218. Hussey, R. S. and Roncadori, R. W. Vesicular-arbuscular mycorrhizae may limit nematode activity and improve plant growth. *Plant Dis.*, *66*:9–14, 1982.
219. Ingham, R. E. Interactions between nematodes and VA mycorrhizae. *Agriculture, Ecosystems, and Environment*, *24*:169–182, 1988.
220. Roncadori, R. W. and Hussey, R. S. Interaction of the endomycorrhizal fungus *Gigaspora margarita* and root-knot nematode on cotton. *Phytopathology*, *67*:1507–1511, 1977.
221. Grandison, G. S. and Cooper, K. M. Interaction of vesicular-arbuscular mycorrhizae and cultivars of alfalfa susceptible and resistant to *Meloidogyne hapla*. *J. Nematology*, *18*:141–154, 1986.
222. Sitaramaiah, K. and Sikora, R. A. Effect of the mycorrhizal fungus *Glomus fasciculatus* on the host-parasite relationship of *Rotylenchulus reniformis* in tomato. *Nematologica*, *28*:412–419, 1982.
223. Bagyaraj, D. J., Manjunath, A., and Reddy, D. D. R. Interaction of vesicular-arbuscular mycorrhiza with root knot nematodes in tomato. *Pl. Soil*, *51*:397–403, 1979.
224. Saleh, H. and Sikora, R. A. Relationship between *Glomus fasciculatum* root colonization of cotton and its effect on *Meloidogyne incognita*. *Nematologica*, *30*:230–237, 1984.

225. Cooper, K. M. and Grandison, G. S. Interaction of vesicular-arbuscular mycorrhizal fungi and root-knot nematode on cultivars of tomato and white clover susceptible to *Meloidogyne hapla*. *Ann. App. Biol.*, *108*:555–565, 1986.
226. O'Bannon, J. H. and Nemec, S. The response of *Citrus limon* seedlings to a symbiont, *Glomus etunicatus*, and a pathogen, *Radopholus similis*. *J. Nematology*, *11*:270–275, 1979.
227. Cason, K. M. T., Hussey, R. S., and Roncadori, R. W. Interaction of vesicular-arbuscular mycorrhizal fungi and phosphorus with *Meloidogyne incognita* on tomato. *J. Nematology*, *15*:410–417, 1983.
228. Atilano, R. A., Menge, J. A., and Van Gundy, S. D. Interaction between *Meloidogyne arenaria* and *Glomus fasciculatus* in grape. *J. Nematology*, *13*:52–57, 1981.
229. Sitaramaiah, K. and Sikora, R. A. Influence of the endomycorrhizal fungus *Glomus mosseae* on *Rotylenchulus reniformis*. Penetration and development on bush bean, cucumber and muskmelon. *Med. Fac. Landbuoww, Rijksuniv, Gent.*, *46*:695–702, 1981.
230. Sikora, R. A. Einfluss der Endotrophen Mykorrhiza (*Glomus mosseae*) auf das Wirt-Parasit-Verhältnis von *Meloidogyne incognita* in Tomaten. *Z. Pflkrankh.*, *83*:197–202, 1978.
231. Smith, G. S. Hussey, R. S., and Roncadori, R. W. Penetration and postinfection development of *Meloidogyne incognita* on cotton as affected by *Glomus intraradices* and phosphorus. *J. Nematology*, *18*:429–435, 1986.
232. Suresh, C. K., Bagyaraj, D. J., and Reddy, D. D. R. Effect of vesicular-arbuscular mycorrhiza on survival, penetration and development of root-knot nematode in tomato. *Pl. Soil*, *87*:305–308, 1985.
233. Hussey, R. S. and Roncadori, R. W. Interaction of *Pratylenchus brachyurus* and *Gigaspora margarita* on cotton. *J. Nematology*, *10*:16–20, 1978.
234. Cooper, K. M. and Grandison, G. S. Effects of vesicular-arbuscular mycorrhizal fungi on infection of tamarillo (*Cyphomandra betacea*) by *Meloidogyne incognita* in fumigated soil. *Plant Dis.*, *71*:1101–1106, 1987.
235. Willox, J. and Tribe, H. T. Fungal parasitism in cysts of *Heterodera*. I. Preliminary investigations. *Trans. Brit. Mycol. Soc.*, *62*:585–594, 1974.
236. Francl, L. J. and Dropkin, V. H. *Glomus fasciculatum*, a weak pathogen of *Heterodera glycines*. *J. Nematology*, *17*:470–475, 1985.
237. Kellam, M. K. and Schenck, N. C. Interactions between a vesicular-arbuscular mycorrhizal fungus and root-knot nematode on soybean. *Phytopathology*, *70*:293–296, 1980.
238. MacGuidwin, A. E., Bird, G. W., and Safir, G. R. Influence of *Glomus fasciculatum* on *Meloidogyne hapla* infecting *Allium cepa*. *J. Nematology*, *17*:389–395, 1985.
239. O'Bannon, J. H., Inserra, R. N., Nemec, S., and Vovlas, N. The influence of *Glomus mosseae* on *Tylenchulus semipenetrans*-infected and uninfected *Citrus limon* seedlings. *J. Nematology*, *11*:247–250, 1979.
240. Strobel, N. E., Hussey, R. S., and Roncadori, R. W. Interactions of vesicular-arbuscular mycorrhizal fungi, *Meloidogyne incognita*, and soil fertility on peach. *Phytopathology*, *72*:690–694, 1982.

241. Barham, R. O., Marx, D. H., and Ruehle, R. L. Infection of ectomycorrhizal and nonmycorrhizal roots of shortleaf pine by nematodes and *Phytophthora cinnamomi*. *Phytopathology*, *64*:1260–1264, 1974.
242. Marks, G. C., Winoto-Suatmadji, R., and Smith, I. W. Effects of nematode control on shoot, root and mycorrhizal development of *Pinus radiata* seedlings growing in a nursery soil infested with *Pratylenchus penetrans*. *Aust. For. Res.*, *17*:1–10, 1987.
243. Oliveira, A. A. R. and Zambolim, L. Interacao entro o fungo endomicorrizico *Glomus etunicatum* e o nematoide de galhas *Meloidogyne javanica* em feijoeiro com raiz partida. *Fitopatol. Bras.*, *12*:222–225, 1987.
244. Shafer, S. R., Rhodes, L. H., and Riedel, R. M. In-vitro parasitism of endomycorrhizal fungi of ericaceous plants by the mycophagous nematode *Aphelenchoides bicaudatus*. *Mycologia*, *73*:141–149, 1981.
245. Salawu, E. O. and Estey, R. H. Observations on the relationships between a vesicular-arbuscular fungus, a fungivorous nematode, and the growth of soybeans. *Phytoprotection*, *60*:99–102, 1979.
246. Hussey, R. S. and Roncadori, R. W. Influence of *Aphelenchus avenae* on vesicular-arbuscular endomycorrhizal growth response in cotton. *J. Nematology*, *13*:48–52, 1980.
247. Ingham, E. R., Trofymow, J. A., Ames, R. N., Hunt, H. W., Morley C. R., Moore, J. C., and Coleman, D. C. Trophic interactions and nitrogen cycling in a semi-arid grassland soil. II. System responses to removal of different groups of soil microbes or fauna. *J. Appl. Ecology*, *23*:615–630, 1986.
248. Moore, J. C., Walter D. E., and Hunt, H. W. Arthropod regulation of micro- and mesobiota in below-ground detrital food webs. *Annu. Rev. Entomol.*, *33*:419–439, 1988.
249. Chakraborty, S. and Warcup, J. H. Soil amoebae and saprophytic survival of *Gaeumannomyces graminis tritici* in a suppressive pasture soil. *Soil Biol. Biochem.*, *15*:181–185, 1983.
250. Warnock, A. J., Fitter, A. H., and Usher, M. B. The influence of a springtail *Folsomia candida* (Insecta,Collembola) on the mycorrhizal association of leek *Allium porrum* and the vesicular-arbuscular mycorrhizal endophyte *Glomus fasciculatus*. *New Phytol.*, *90*:285–292, 1982.
251. Finlay, R. D. Interactions between soil micro-arthropods and endomycorrhizal associations of higher plants. In *Ecological Interactions in Soil: Plants, Microbes, and Animals* (A. M. Fitter, D. Atkinson, D. J. Read, and M. B. Usher, eds.), Blackwell Scientific Publications, Oxford, 1985, pp. 319–331.
252. Kaiser, P. A. Collembolan effect on mycorrhizal colonization in soybeans. *Mycological Society of America Newsletter*, *39*:33 (Abstr.), 1988.
253. McGonigle, T. P. and Fitter A. H. Evidence that Collembola suppress plant benefit from vesicular-arbuscular mycorrhizas (VAM) in the field. In *7th North American Conference on Mycorrhizae*, University of Florida, Gainesville, 1987, p. 209.
254. Moore, J. C., St. John, T. V., and Coleman, D. C. Ingestion of vesicular-arbuscular mycorrhizal hyphae and spores by soil microarthropods. *Ecology*, *66*:1979–1981, 1985.
255. Shaw, P. J. A. Grazing preferences of *Onychiurus armatus* (Insecta: Collembola) for mycorrhizal and saprophytic fungi of pine plantations.

In *Ecological Interactions in Soil, Plants, Microbes and Animals*, (A. H. Fitter, D. Atkinson, D. J. Reed, and M. B. Usher, eds.), Blackwell Scientific Publications, Oxford, 1985, pp. 333–337.
256. Rabatin, S. C. and Stinner, B. R. Arthropods as consumers of vesicular-arbuscular mycorrhizal fungi. *Mycologia*, *77*:320–322, 1985.
257. Buckalew, D. W., Riley, R. K., Yoder, W. A., and Vail, W. J. Invertebrates as vectors of endomycorrhizal fungi and *Rhizobium* upon surface mine soils. *Proc. W. Va. Acad. Sci.*, *54*:1 (Abstr.), 1982.
258. Rabatin, S. C., Stinner, B. R., and Edwards, C. A. Ingestion and dispersal of vesicular-arbuscular mycorrhizal fungi by soil-inhabiting invertebrates. In *7th North American Conference on Mycorrhizae*, University of Florida, Gainesville, 1987, p. 163.
259. Rabatin, S. C. and Rhodes, L. H. *Acaulospora bireticulata* inside oribatid mites. *Mycologia*, *74*:859–861, 1982.
260. Old, K. M. and Wong J. N. F. Perforation and lysis of fungal spores in natural soils. *Soil Biol. Biochem.*, *8*:285–292, 1976.
261. Chakraborty, S., Theodorou, C., and Bowen, G. D. Mycophagous amoebae reduction of root colonization by *Rhizopogon*. In *6th North American Conference on Mycorrhizae*, Bend, OR, 1984, p. 275.
262. Chakraborty, S., Theodorou, C. and Bowen, G. D. The reduction of root colonization by mycorrhizal fungi by mycophagous amoebae. *Can. J. Microbiol.*, *31*:295–297, 1985.
263. Old, K. M. and Patrick Z. A. Giant soil amoebae: Potential biocontrol agents. In *Soil-Borne Plant Pathogens* (B. Schippers and W. Gams, eds.), Academic Press, London, 1979, pp. 617–628.

5

MYCORRHIZA IN PLANT DISEASE CONTROL

BUSHAN L. JALALI and INDU JALALI *Haryana Agricultural University, Haryana, India*

I. INTRODUCTION

Many ecosystems are characterized by a steady-state balance, an equilibrium which is maintained by homeostatic mechanisms [1]. The plant-microorganism system (including pathogenic and symbiotic microbial systems) behaves broadly in this way; nevertheless, microbial populations do not exhibit a perfectly stable equilibrium, but rather an ever-fluctuating one. The related microbial shifts are either short term, such as induced by root exudation rhythms, or long term and more progressive changes such as those linked to the plant growth cycle. The equilibrium of the plant-microorganism system can, however, be upset by biotic and abiotic stresses. The level and magnitude of stress required to upset the equilibrium varies according to the level of buffering of the plant-microorganism complex. Successful superimposition of selected mycorrhizal fungi in an existing microbial spectrum, including pathogens, is one such dimension to biological equilibrium in the plant-microorganism system.

Conceptually, root diseases and mycorrhizal associations have a distinct similarity in terms of parasitism. In both host-microbe systems, more or less, similar types of parasitism are involved. In the case of vesicular-arbuscular (VA) mycorrhizal fungi, the infection is triggered by the formation of an appressorium on the host root surface from which the epidermal cells are penetrated by hyphae, whereas with the ectomycorrhizal fungus, endophytes are stimulated by host roots, symbiotically infect and ultimately transform the feeder cortical cells enclosed in the Hartig net. These cells are isolated from direct contact with the soil by the fungal mantle. In a similar fashion, root pathogens like *Rhizoctonia, Fusarium, Verticillium, Sclerotinia, Pythium*, and *phytophthora* are stimulated by roots; they cause infection on the meristematic and primary cortex tissues, progressively invade and colonize surrounding tissues, and ultimately cause restricted or extensive necrosis. In advanced stages of root necrosis, the root pathogens spread into vascular tissues.

The mycorrhizal endophytes as well as root pathogens in the soil environment are part of a complex interacting system. The dynamics of such microbial systems in the rhizosphere exert measurable effects on plants by altering the morphology and physiology of the root system. Equally important is the significant change brought about in the chemical nature of soils through the influence exerted by root exudation. Such soil microbial systems also bring about discernable alterations in the permeability of root cells by damaging root tissues, by altering root metabolism, by selectively utilizing some root exudate components, or by excreting toxins. The level of such rhizosphere responses is influenced markedly by biological factors.

During the past decade, our knowledge of various aspects of mycorrhizae has expanded significantly. Valuable data on induced suppression of soilborne pathogens have opened up the prospects for their practical application. This subject has been reviewed by many "mycorrhizasts" [2–7], who have attempted to focus attention on the biocontrol potentialities of this microbial model system and thereby "raised their study to the level of a reputable pursuit" [8]. We intend to highlight the different directions that research on mycorrhizae have taken in recent years in terms of their impact on plant disease and the influences exerted by fungitoxicants on the mycorrhizal system. This may help us to build a comprehensive picture of various interacting biotic and abiotic parameters involved in these interactions and to assess their advantages, as well as limitations, in practical plant disease control strategy.

II. VA MYCORRHIZA AND PLANT DISEASE

A variety of pathogenic fungi, bacteria, viruses, and nematodes derive their energy from plant roots. Yet, in recent years several reports have indicated that a host plant previously inoculated with a VA mycorrhizal fungal symbiont exhibits increased resistance to several root diseases [2, 3,5]. However, in some cases VA mycorrhizal infection may promote the development of plant disease. Suppression or promotion of plant disease as affected by mycorrhizal development may ultimately be determined by several biotic and abiotic interacting forces involved in the pathogen–host–mycorrhiza system. Changes in the soil environment, particularly in terms of its nutrient and water status, significantly affect the germination, growth, development, and pathogenic bahavior of many soil microorganisms.

A. Fungal Pathogen–Mycorrhiza Interactions

Host plants previously inoculated with fungal symbionts have been shown to exhibit an increased resistance to fungal root rots and wilts [9–11]. Growth of mycorrhizal seedlings infected by *Phytophthora parasitica* was greater than that of infected nonmycorrhizal citrus seedlings [12], which is in contrast to earlier findings by the same authors where the beneficial response of citrus to mycorrhizal infection was significantly suppressed by *P. parasitica* at high inoculum density [13]. A lower initial inoculum density of the pathogen did not prevent mycorrhizal roots from absorbing more phosphorus and other nutrients than infected, nonmycorrhizal plants. With mycorrhiza still functioning in seedlings infected by *P. parasitica*, the growth rate of mycorrhizal seedlings was greater than that of nonmycorrhizal ones, which grew slowly because of their inability to absorb nutrients as effectively as mycorrhizal seedlings. It is possible that infection by VAM fungi elicits a resistance mechnism by the host which may inhibit the subsequent infection by fungal pathogens.

Root colonization by VA mycorrhizal fungi may decrease, increase, or exert no significant effect on the development of fungal root pathogens and consequently the severity of disease. The resultant impact may be due to the interaction between the fungal pathogens and the mycorrhizal endophyte. Zambolim and Schenck [14], while studying the interactions between *Glomus mosseae* and *Macrophomina phaseolina* in soybean, observed that the colonization of the host root by the mycorrhizal symbiont compensated for the effect of the pathogen without changing incidence of disease in the host. In a series of investigation by Caron et al. [15–17], root colonization by *Glomus* was not affected by the presence of pathogen *Fusarium oxysporum* f. sp. *radicis-lycopersici*. The number of propagules of this pathogen was consistently lower when the plants were inoculated with the mycorrhizal endophyte. The presence of the fungal symbiont *Glomus intraradicis* significantly inhibited root necrosis caused by the pathogen. There was a significant increase of root endomycorrhizal colonization (REC) index with time (Fig. 1). The relation between time and density of *Fusarium* propagules was quadratic; the shape of the curves (Fig. 2) was the same for both treatments (i.e., plants with pathogen in the presence or absence of fungal symbiont), showing a decrease in the number of pathogen propagules before 5-weeks followed by an increase. But the mean number of propagules was consistently lower when plants were also inoculated with *Glomus*. This leads to the conclusion that for better assessment of such interactions, quantification of responses between various components involved in the process must be made at different time intervals after inoculation with the pathogen;

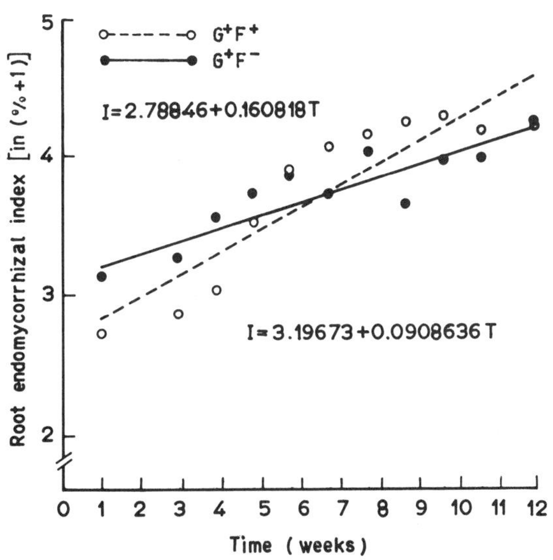

FIGURE 1. Observed and estimated effect of time (T) on the root endomycorrhizal colonization index (I) on tomato plants inoculated with *Glomus intraradices* in the presence (○ - - ○) or the absence (●——●) of *Fusarium oxysporum* f. sp. *radicis-lycopersici*. (From Ref. 16.)

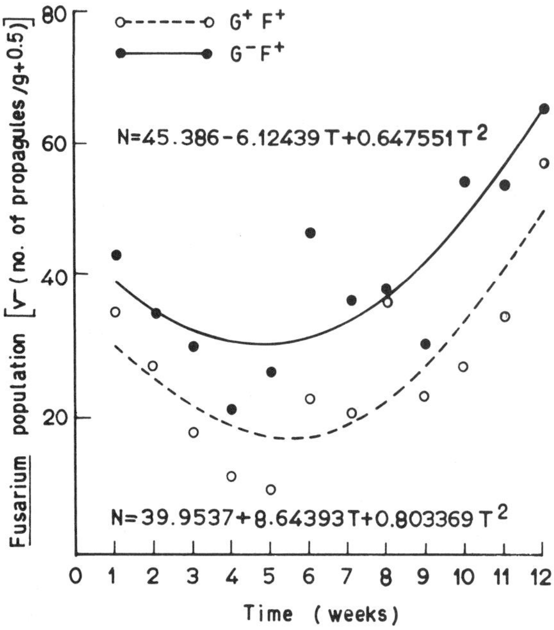

FIGURE 2. Observed and estimated effect of time (T) on the *Fusarium* population (N) in the substrate after inoculation of tomato plants with *Fusarium oxysporum* f. sp. *radicis-lycopersici* in the presence (○ - - ○) or the absence (●—●) of *Glomus intraradices*. (From Ref. 16.)

this will ensure that observations thus recorded are a representation of the interaction.

Since very few pathogen–mycorrhiza studies have investigated the role of phosphate nutrition, it may not be possible to conclude whether responses of mycorrhizal plants to pathogens would differ from nonmycorrhizal plants at a comparable nutritional status. Smith [18] reviewed this dimension and examined those studies that compared the influence of phosphate nutrition with VA mycorrhizal fungi on the host–pathogen relationship. Earlier Schönbeck [5] indicated that in the presence of VA mycorrhizae, fungal disease generally was inhibited in the root system, while the shoot system exhibited enhanced disease incidence. Gerdemann [19] raised the question whether altered disease resistance in mycorrhizal plants was due to altered plant nutrition or due to other, more direct, mechanisms. In their studies, Kaye et al. [20] reported that mycorrhizal root colonization was greater in plants grown in soils inoculated with *Pythium ultimum* at each density than in plants grown in soil without the pathogen inoculation. In soil inoculated with 39 colony forming units (CFU) per gram of *P. ultimum*, the relative reduction of shoot dry weight was 26 and 46%, respectively, for mycorrhizal and nonmycorrhizal plants. Similarly, height was reduced 18 and 28%, respectively, for mycorrhizal and nonmycorrhizal plants. In addition, enhanced phosphate uptake in mycorrhizal plants and reduction in the density of *P. ultimum* in the rhizosphere of mycorrhizal plants were recorded. Summarizing their observations thus obtained, it has been suggested that mycorrhizal plant's tolerance to *P. ultimum* could be attributed to factors other than improved plant nutrition alone. However, the possible role of pathogen–mycorrhizal interaction in nonrhizosphere soil raises several questions. For example, no significant reduction in the subsequent densities of *P. ultimum* added at 38 and 132 CFU/g in the presence of mycorrhizae occurred. In fact, in soil inoculated with 38 CFU/g, an increase in pathogen density was observed, whereas significant reduction was recorded in the presence of mycorrhizae in soil inoculated with 66 CFU/g of *P. ultimum*. Although explanations are difficult, this study suggests that *Glomus fasiculatus* acts as a biological deterrent of *P. ultimum* causing root rot in commercial poinsettia.

Jalali and Thareja [10], while studying the influence of a mycorrhizal fungus on infection of chickpea by the wilt pathogen *Fusarium oxysporum* f. sp. *ciceri*, indicated that inoculation with *Glomus* sp. resulted in extensive colonization of roots by mycorrhiza, and this was associated with significant reduction in the percentage of plants infected with *Fusarium* in those treatments which had been inoculated with pathogen (Fig. 3). Inoculation with *Glomus* also increased nodule number, total dry matter production, and phosphorus content in those plants not infected by the pathogen. This suggests that mycorrhizal colonization may provide effective protection against infection with *Fusarium* and perhaps other root pathogens.

Rosendahl [11] while observing that mycorrhizal infections were able to reduce root rot of peas caused by *Aphanomyces euteiches*, explained that this effect was dependent on the time lapse between the mycorrhizal inoculation and challenge inoculation with the pathogen. In order to avoid subjective quantification, percent infected root length was taken as an estimate of infection with the pathogen. A clear correlation was obtained between percent infected root length with the pathogen or VAM and the

FIGURE 3. Reduction of chickpea wilt (*Fusarium oxysporum* f. sp. *ciceri*) infection by addition of mycorrhizal endophyte. (From Ref. 10.)

disease index in several pilot tests. Early establishment of the mycorrhizal symbiont appeared to be vital if this microbial system was to be employed effectively to control root diseases. Baath and Hayman [21] carried out studies with mycorrhizae and *Verticillium* on tomato plants. Irrespective of the mycorrhizal status of the host plants, inoculation with *V. albo-atrum* decreased plant growth compared to noninoculated controls. Addition of extra phosphorus to the soil enhanced plant growth, but this enhancement was nullified by *Verticillium*. These studies indicate that mycorrhiza might not always be helping a plant to tolerate a root pathogen. In fact, they later found no interaction between *Phytophthora fragariae* and two VA mycorrhiza fungi in roots of strawberry [22]. This could be due to different modes of pathogen attack, since *P. fragariae* is a very aggressive pathogen and the test cultivar used was very susceptible to red-core disease. Prevalence and severity of sweet potato wilt at 32°C was highest when host plants were inoculated with *Fusarium oxysporum* f. sp. *batatas* alone, but at 21°C more wilt developed when plants were dually inoculated with the pathogen and the mycorrhizal fungi [23]. Wilt prevalence and severity increased with increasing phosphate levels.

Jalali [7] showed that mycorrhizal inoculation of mungbean resulted in a measurable reduction in infection by the root pathogen *Rhizoctonia solani*; this effect was more pronounced when seed pelleting with sporocarps of the endophyte was done (Fig. 4). This suppressive effect was directly correlated with the ability of the mycorrhizal symbiont to develop in the absence of available soil phosphate. These studies indicate that soil conditions favorable to mycorrhizal colonization were in many respects not conducive to the growth and establishment of the test pathogen. A wide range of studies, differing in methods and approach, have been carried out on various

FIGURE 4. Effect of mycorrhizal endophyte on root rot infection by *Rhizoctonia solani* in mungbean. (From Ref. 7.)

aspects of interactions between mycorrhizal endophyte and fungal pathogens [24–30].

B. Nematode–Mycorrhiza Interactions

Mycorrhizal fungi and plant parasitic nematodes are commonly found inhabiting the rhizosphere and colonizing roots of their host plants. These two groups of microorganisms exert a characteristic but opposite effect on plant health [31,32]. The potential role of mycorrhizal fungi as biocontrol agents for nematode infection has recently received considerable attention [33–37]. Generally, the severity of nematode diseases is reduced in mycorrhizal plants [33–35]. In several studies, mycorrhizal fungi have exhibited antagonistic influence on the population of plant parasitic nematodes [33,37, 39, 40]. Establishment of VA mycorrhiza in plants usually confers resistance to nematode parasitism [36,40] or adversely affects nematode reproduction [40]. However in some instances, root colonization by mycorrhizal fungi may have no effect or may even promote nematode numbers [41,42].

The expression of interaction response depends upon the species of mycorrhizal fungus, host plant, nematode, and nematode inoculum densities [43], the susceptibility of the host cultivar to nematode [41,44], and soil fertility [35,36,40,45]. Varietal differences in host plant efficiency and its sensitivity in response to nematode [46] and to mycorrhizal fungal infection [47] have been demonstrated. Parasitism of plant roots by VA mycorrhizal fungus may also change a plant's response to nematode penetration and development and hence influence the tolerance and resistance parameters of the host. Although the role of phosphorus in mycorrhiza–nematode interactions remains incompletely understood, phosphorus status of soil is known to influence plant growth and nematode reproduction favorably [41] and adversely [34] or may exert no effect [48]. Greenhouse testing to evaluate the influence of *Glomus mosseae* on *Rotylenchulus reniformis* penetration and development on bush bean, cucumber, and muskmelon revealed that reduction in the number of nematodes that penetrated mycorrhizal bush bean was 35 and 41% lower than the controls 4 and 8 days, respectively, after nematode inoculation [49]. A similar reduction in larval penetration was observed on mycorrhizal cucumber and muskmelon 8 days after nematode introduction. These studies have shown that the VA mycorrhizal fungus *Glomus mosseae* increases plant resistance to nematode infection. Preinoculation allowed the growing fungal symbiont to establish itself in the roots before exposure to *R. reniformis*. In further studies, inoculation of tomato transplants or seed beds with the fungal endophyte appreciably reduced juvenile penetration and development on mycorrhizal plants, compared with control plants [38]. The development of the gelatinous matrix was delayed and fewer eggs/egg sac were produced on plants with mycorrhiza. It has been demonstrated that several phases of the life of *R. reniformis* are affected adversely by *Glomus fasciculatus*.

Cooper and Grandison [35] showed that mycorrhizal plants exhibited more resistance to root knot nematode (*Meloidogyne hapla*) at all phosphate levels, and growth benefits were generally more significant in plants preinfected with mycorrhizal fungi. In mycorrhizal root systems, nematode numbers increased in the lower phosphate soils. However, at higher phosphate levels, nematode numbers were either unaffected or reduced in mycorrhizal treatments; mycorrhizal root length remained unaffected by nematodal inoculation. Similar responses were obtained in several other studies [34,37].

C. Virus-Mycorrhiza Interactions

Unlike the generalization that the incidence of diseases caused by soilborne fungal pathogens and nematodes is reduced in mycorrhizal plants, the diseases caused by viruses are more severe in the presence of VAM [5,50]. Whether viral infection was local [51] or systemic [52], mycorrhizal plants exhibited higher disease incidence than nonmycorrhizal plants. In mycorrhizal plants, the virus multiplied at a faster rate than in nonmycorrhizal plants [50]. Bondoux and Perrin [53] have reviewed effects of mycorrhiza on viruses, their mode of action, problems in using mycorrhizae against plant parasites,equilibrium of microorganisms in the host, and participation of mycorrhizae in the control of diseases of soil origin. Schönbeck and Spengler [54] demonstrated that the increased virus concentration in mycorrhizal plant roots is usually confined to the arbuscular stage of the endophyte. In electron microscopic studies, tobacco mosaic virus (TMV) was found to be restricted to plant cell cytoplasm [55,56]; however, it was more numerous in cells that contained arbuscules. Fungi have been demonstrated to vector plant viruses in a number of diseases [57]. The fungi acting as virus vectors usually depended upon the existence of an ectoplast-limited thallus that allows exchange of virus between vector and host [57]. In host plant-VA mycorrhizal associations, exchange of nutrients within the host occurs across the arbuscules, so these fungi are of particular interest in this regard. Such thalli, being membrane bound, also provide sites for virus adsorption and endocytotic uptake. However, it has also been demonstrated that there is no adsorption or aquisition of virus particles by arbuscular structures, and virus particles are not detected in hyphae of VAM fungi [56].

III. ECTOMYCORRHIZA AND PLANT DISEASE

The mycorrhizal root system is functional longer than the nonmycorrhizal root system. However, the success of mycorrhizal inoculation in inducing suppression of root diseases will largely depend upon the sequence of infection by root pathogen and mycorrhizal endophyte. If a pathogen infects a feeder root system prior to infection by a mycorrhizal symbiont, mycorrhizal development would likely be restricted. On the other hand, if the sequence of parasitic attack is reversed, i.e., the symbiont infects the feeder root system prior to infection by the pathogen, the damage caused by the pathogen is often drastically reduced. Following the sequence of events, a pathogen infecting a nonmycorrhizal root is initially confronted externally with succulent, thin-walled epidermal cells, and internally with cortex cells which have not undergone secondary cell wall thickening. But a pathogen attacking an ectomycorrhizal root system is initially confronted externally with the tightly interwoven network of hyphae that makes up the fungal mantle, and then internally with cortical cells whose cell walls are surrounded by Hartig net hyphae tissue [3]. It thus becomes apparent that a pathogen of a mycorrhizal root system must have the physical and chemical ability to penetrate these barriers.

Mycorrhizal tree seedlings have the ability to resist root infections by potential pathogens. While working with seedlings of loblolly pine under aseptic conditions, Wingfield [58] demonstrated that the mycorrhizal symbiont (*Pisolithus tinctorius*) increased survival of seedlings growing with the potential root pathogen *Rhizoctonia solani*. Nonmycorrhizal pine seedlings inoculated with the pathogen exhibited lower survival and vigor. Ross and Marx [59] showed that *Pinus clausa* seedlings were significantly protected against infection by *Phytophthora cinnamomi* by mycorrhizal colonization of *P. tinctorius*. Extensive root necrosis occurred in seedlings which were nonmycorrhizal and were infected by *P. cinnamomi*. After 2 months, only 40% of the nonmycorrhizal plants survived. Cortical tissues in the mycorrhizal roots were free from pathogen infection, thus confirming their role in resistance to attack by this pathogen. Marx [60] observed that the growth of shortleaf pine (*Pinus echinata*) seedlings with mycorrhizal infection were not reduced by *P. cinnamomi*. Nonmycorrhizal pine seedlings infected with the pathogen had significantly less root dry matter, as well as fewer new lateral roots than nonmycorrhizal seedlings without pathogen infection. Furthermore, the inoculum density of the pathogen in soil with the nonmycorrhizal seedlings at the start and at the end of the experiment did not change significantly.

Richard et al. [61] carried out elaborate studies on the interaction between the mycorrhizal endophyte *Suillus granulatus* and the root pathogen *Mycelium radicis atrovirens* on aseptic seedlings of *Picea mariana*. *Suillus granulatus*–inoculated seedlings grew better than nonmycorrhizal ones, while those inoculated with *M. radicis atrovirens* were chlorotic and extensively stunted. When dual inoculations with *S. granulatus* and the root pathogen were made, the disease expression was drastically inhibited.

Powell et al. [62] recorded a striking increase in mycorrhizal growth and development by *Scleroderma bovista*; this response was obtained after application of various biocides to the soil around pecan (*Carya illinoensis*) with symptoms of feeder root necrosis. Most of the toxicants did not affect the population densities of *Pythium* sp. or nematodes which were the casual agent of the root necrosis. They demonstrated convincingly that populations of competing soil microorganisms were suppressed by the toxicants which eventually caused stimulation in mycorrhizal development by *S. bovista*. In addition, increased mycorrhizal development enhanced nutrient transport and absorption by the host trees, and the mycorrhizal colonization functioned as an effective biocontrol tool. Earlier it was shown by Corte [63] that *Suillus granulatus*–inoculated seedlings appear to protect *Pinus excelsa* from *Rhizoctonia* root rot infections.

Sylvia [64] investigated the influence of *Laccaria laccata* on *Fusarium oxysporum* and its ability to induce resistance to the pathogen in the primary roots of Douglas fir seedlings. Root rot of Douglas fir by *F. oxysporum* can be lethal if the primary roots of very young seedlings become infected. The ectomycorrhizal fungus *L. laccata* protects young seedlings of Douglas fir by colonizing the primary roots shortly after germination [65]. Prior colonization of roots by *L. laccata* or exposure to its cell-free metabolites did not reduce either the rate at which the *F. oxysporum* approached the root or the extent of surface colonization. *Fusarium oxysporum* colonized 18% of the surface of the control roots compared to 43 and 25%, respectively, for roots exposed to *L. laccata* or its metabolites. The conclusion was that antibiosis was not important in root protection, although *L. laccata* had the capability of inhibiting *F. oxysporum* in vitro. However, cortical invasion of roots by the pathogen was reduced significantly in seedlings that had been exposed to *L. laccata* or its metabolites. *Fusarium oxysporum* occurred over 25% of the surface of control roots compared to 0 and

5%, respectively, for roots exposed to *L. laccata* or its metabolites. A highly significant inverse relationship was found between hyphal frequency and accumulation of osmiophilic material in cortical cells of control plants.

In dual cultures on media with pH less favorable for the growth of *F. oxysporum* than *L. laccata*, diffusable metabolites of a mycorrhizal strain of *L. laccata* inhibited growth and caused distortion of hyphae of the pathogen [66]. Cell-free fluid from cultures of *Laccaria* delayed germination of microconidia and chlamydospores of *F. oxysporum*. Only slight suppression occurred under conditions that favored rapid growth of the pathogen on agar media but culture filtrates did not affect spore germination. Antibiosis may be a partial basis of root protection by this mycorrhizal strain of *L. laccata*. On agar plates and in gnotobiotic system without soil, *L. laccata* or its cell-free metabolites suppressed root growth of Douglas fir. In sterile soil, however, no suppression occurred.

Recently, Perrin and Garbaye [67] carried out a bioassay to estimate soil infectivity. The results provide evidence that some ectomycorrhizal fungi like *Hebeloma crustuliniforme* decreased the infectivity of *Pythium*-infested soil to such a level that it protects beech seedlings from disease under field conditions. Striking differences were found between mycorrhizal fungi, and this may be of important ecological significance in influencing damping-off or root rot diseases. Their observations suggest that the decrease in soil infectivity is the result of a direct effect of *H. crustuliniforme* against the population of *Pythium* sp. The concept of soil infectivity and the methods used to measure it enables us to demonstrate microbial protection in the rhizosphere and to study the mechanism and its expression in nursery and forests.

When young pines growing in natural and autoclaved forest soils were inoculated with *Phytophthora cinnamomi*, the fungus had little effect on trees which had abundant mycorrhizal roots, whereas those growing in autoclaved forest soils were stunted and half were killed by the pathogen [68]. Inoculation of autoclaved forest soils with *Thelephora terrestris* improved tree growth in the absence as well as presence of *P. cinnamomi*. Condition for the formation of such ectomycorrhizae seem most favorable in rotten wood. Pot and field inoculation experiments showed that *Boletus* sp. and *Suillus grevillei* were highly effective in controlling pre- and post-emergence damping-off of pine (*Pinus tabulaeformis*) seedling caused mainly by *Rhizoctonia solani* and *Fusarium oxysporum* [69]. Seedling mortality in the plots was decreased by 21 – 35% in mycorrhizal seedlings as compared to nonmycorrhizal controls.

Chakravarty and Unestam [70] studied the differential influence of ectomycorrhiza on plant growth and disease resistance in *Pinus sylvestris* seedlings. Seedling growth and short-root formation were differentially stimulated, depending upon the mycobiont and the period of time. Disease severity and plant mortality were significantly reduced in the presence of mycorrhizal fungi, even before mycorrhizal formation. The protection continued throughout the year except with *H. crustiliniforme*, which protected the seedlings for only 3 months. Growth stimulation and mycorrhizal frequency were not always correlated with level of protection. However, the mycorrhizal fungi had little or no antagonistic effects on the pathogen in vitro. Mycorrhizal fungi could even prevent severe root damage in stressed pine seedlings [71]. The common rhizosphere fungus *Cylindrocarpon destructans* caused severe root damage, leading to mortality in light-starved, stunted *P. sylvestris* seedlings grown aseptically. This root damage was completely prevented by the presence of the mycorrhizal fungi *L. laccata* or *P. tinctorius* or by conifer soil containing several species of mycorrhizal fungi. Protection took place even before any visible mycorrhizae were found on the roots, thus suggesting that root protection by mycorrhizal fungi may be effective not only inside the root but also in the rhizoplane or rhizosphere.

Fungal isolates from the rhizosphere of black spruce (*Picea mariana*) were tested for their ability to inhibit mycorrhizal formation by *Laccaria bicolor* on axenically cultured black spruce seedlings [72]. *Trichoderma viride* and *T. polysporum* were strongly antagonistic. This antagonism was seen both in trials where *Trichoderma* and *L. bicolor* were inoculated simultaneously and in a trial designed to allow establishment of the mycorrhizal fungus in the rhizosphere prior to a challenge with *T. viride*. Despite its antagonism in flask trials, *T. viride* did not parasitize the hyphae of *L. bicolor* in agar cultures. *Tolypocladium inflatum* and *Trichosporus beigelii* also had small, but significant, inhibitory effects on mycorrhizal formation in flask trials. The other fungal isolates from the rhizosphere had negligible effect. It was concluded that *Trichoderma* spp. are likely to be antagonistic to mycorrhizal colonization in vitro, particularly in fumigated soils.

The effect of ectomycorrhizal fungi has been studied in case of fungal pathogens, but not much has been investigated with other root pathogens like bacteria, viruses, and nematodes. Barham et al. [73] inoculated intact ectomycorrhizae of short leaf pine with spiral or stunt nematodes (*Helicotylenchus dihystera* and *Tylenchorhynus claytoni*) and zoospores of *Phytophthora cinnamomi*. Both nematodes penetrated and migrated through the fungal mantle and Hartig net of ectomycorrhizae. The resultant disturbance in structural integrity of the fungus mantle by spiral nematodes created infection courts for *P. cinnamomi*, and intracellular hyphae of the pathogen were located in cortex cells surrounded by the Hartig net.

Malajczuk and McComb [74] reported intriguing results comparing rhizosphere populations around mycorrhizal and nonmycorrhizal eucalyptus seedlings grown in soils suppressive or conducive to root rot caused by *Phytophthora cinnamomi*. Microflora associated with mycorrhizal roots were distinctly different from that of nonmycorrhizal roots; high counts of bacteria were obtained throughout the fungal mantle, both within and between symbiont cells. Many bacteria were highly antagonistic to species of *Phytophthora* and *Pythium* causing root diseases.

IV. MECHANISMS OF OPERATION

Several hypotheses have been put forward to explain the mechanisms involved in protection of roots from pathogenic infection by mycorrhizal fungi. Subtle changes in the nutrient and water status of the soil can affect germination, growth, survival, and virulence of many soil microorganisms [75]. In pathogenic fungi, the quiescence of spores, lysis of mycelia, and formation of resistant structures can be induced by microbial competition for nutrients. The exogenous addition of nutrients containing energy sources can reverse the effect of competition. A mycorrhizosphere is nutritionally different than a nonmycorrhizosphere. Water status of the surrounding soil, the translocation characteristics of mycorrhizal hyphae as well as root exudation may enhance such differences. Propagules of plant pathogens are normally quiescent in the soil, away from the sources of energy-yielding nutrients. Root exudates may stimulate germination and provide the energy required for infection. Mycorrhizal fungi colonizing host roots would be expected to compete for such nutrients, reducing the quantity available to pathogens and restricting pathogenic infection.

The interactions between VA mycorrhizal endophytes and plant pathogens can be explained in two broad statements: (a) mycorrhizal fungi are able to retard pathogen development in the root system, and this influence is limited to the site of mycorrhizal establishment, and (b) mycorrhizae are able to increase disease incidence systemically, particularly in nonmycorrhizal plant parts. The systemic influence can be attributed to better nutrition, enhanced plant growth, and physiological stimulation in mycorrhizal plants. With increased concentration of assimilates, these plants can serve as better nutrient sources for plant parasitic organisms [76]. In contrast to this, there is a localized and specific impact of VA mycorrhizal symbionts on host physiology. It has been demonstrated that the mycorrhizal infection enhances lignification in the root tissues, particularly in stelar region [77]. This may be responsible for restricting the parasitic soilborne organisms that invade the host roots.

The symbiotic host–VAM fungus relationship is characterized by the formation of arbuscules which are successively degraded. This process may represent digestion of the fungus by the host [78]. Roots colonized by the mycorrhizal symbiont exhibit high chitinolytic activities. Similarly, the level of arginine in mycorrhizal roots was shown to be sixfold higher, due to an increased specific activity of enzymes in the ornithine cycle and blocking of the cycle by arginase enzyme system in such roots [79]. Mycorrhizal colonization induces strengthening of cell walls by enhanced production of polysaccharides and increased lignification. It has been postulated that the growth of root pathogens in such tissues is inhibited, as was shown with *Fusarium oxysporum* [79] and *Pyrenochaeta terrestris* [80]. A more extensive vascular system is likely to increase nutrient flow, impart greater mechanical strength and eventually reduce detrimental pathogenic effects. Similarly high chitinase activity in the mycorrhizal root tissues may also restrict the growth and the progression of root pathogens in host tissues [5].

In the ectomycorrhizae–pathogen system, several mechanisms have been postulated to explain disease protection of feeder roots on host trees [2,3, 81]. The mechanisms are as follows: (a) utilization of surplus carbohydrates in roots by mycorrhizal endophytes, eventually reducing the quantum of nutrients available to pathogen, (b) formation of mechanical barrier in the form of fungal mantle against ingress by the pathogen, (c) secretion of antibiotics inhibitory to the pathogens, (d) formation of a protective microbial rhizosphere population, and (e) production of metabolites released by mycorrhizal host cortical cells functioning as inhibitor to infection and spread of the pathogen.

Several investigators have demonstrated that the fungal mantle of ectomycorrhizal roots are formidable physical barriers that restrict penetration by several root pathogens, particularly *Phytophthora cinnamomi* [82,83]. This conclusion was drawn on the basis of histological studies of several ectomycorrhizal pine trees that had been inoculated with *P. cinnamomi*. The Hartig net surrounding the cortical cells may function as a supplementary physical barrier since progression of the pathogen was also blocked in this region. Krupa and Fries [84] suggested biochemical and cytological response of the cortical cells in ectomycorrhizal root system. They showed that the fungal endophyte *Suillus variegatus* liberated the fungistatic volatile compounds isobutanol, isobuteric acid, as well as other volatile compounds in pure culture. Volatile organic compounds were also extracted from intact seedling roots of *Pinus sylvestris* grown in pure culture with and without the fungal symbiont. Increased production and accumulation of volatile terpenes and sesquiterpenes was seen in mycorrhizal root systems compared with nonmycorrhizal roots. Several of these compounds are fungistatic, synthesized as a nonspecific response of the host cells to symbiont infection. In further studies, infection of feeder roots of *Pinus echinata* by the ectomycorrhizal fungus *Pisolithus tinctorius* resulted in a 40-fold increase in the level of 3-carene [85]. *Cenococcum graniforme* induced a 30-fold increase in β-phellandrene. Vapors of these volatile compounds affected the vegetative growth of different root pathogenic fungi and ectomycorrhizal endophytes.

Sylvia and Sinclair [81] showed that a mycorrhizal strain of *Laccaria laccata*, which can protect primary roots of Douglas fir seedlings from root rot infection by *Fusarium oxysporum*, induced an accumulation of osmiophilic material in cortical cells or primary roots. When host roots were challenged with fungal pathogens following incubation with *L. laccata*, invading hyphae were least abundant and often absent in tissue with the greatest accumulation of osmiophilic materials. Histochemistry of fresh sections reveal that the induced materials were primarily phenolic in nature. The authors presented direct evidence that the mycorrhizal strain of *L. laccata* can protect seedlings of Douglas fir from this root pathogen in the absence of other organisms. The potential role of other soil organisms in modifying root protection, however, appears large and deserves greater study. These authors stated that strains of mycorrhizal fungi which colonize the primary root rapidly, compete with other soil organisms aggressively, and induce a consistent deposition of phenolics in roots may provide biological control. Recently, the results obtained by Chakravarty and Unestam [71] suggest that root protection by mycorrhizal fungi may be effective not only inside the root but also in the rhizosphere and rhizoplane. It was shown further that disease severity and plant mortality were reduced in the presence of mycorrhizal fungi, even before the mycorrhizal formation took place.

V. IMPACT OF PESTICIDES ON MYCORRHIZAL DEVELOPMENT

The behavior of target and nontarget microorganisms at the soil–root interface is largely mediated by factors that affect host physiology and the quantitative/qualitative nature of root exudates [86]. Among the many factors, pesticidal applications are able to induce substantial changes in the zone of root influence. Not only are pathogens affected, but changes in the function of mycorrhizal fungi would exert measurable influences on the spectrum of root exudation [87], as well as the mineral nutrition of host plants with a resultant potential for predisposition to disease.

Pesticides are tailored to react with living cells. Although these toxicants are directed against target pathogenic organisms, a good number of these biocides have deleterious influences on nontarget organisms, including mycorrhizal fungi. Trappe et al. [88] and Menge [89], while reviewing the impact of pesticides on mycorrhizal fungi, indicated that a wide range of pesticides inhibit infection and resultant colonization of mycorrhizal fungi. In fumigated soils, citrus seedlings exhibited suppressed growth rates as well as reduced phosphate uptake [90]. In such soils, stunting and chlorosis were brought about by the killing of mycorrhizal fungi. Fumigation with methyl bromide or chloropicrin drastically suppressed mycorrhizal development [91]. As a result of this application, growth of citrus seedlings were restricted as compared to satisfactory growth of nonfumigated mycorrhiza-infected seedlings. Similarly, furrow application of pyroxychlor significantly reduced chlamydospore numbers of *Glomus fasciculatus* [92].

While quantifying the impact of soil fungitoxicants on the growth and development of VA mycorrhiza in wheat, Jalali [93] observed that soil application of four toxicants adversely affected the colonization of the mycorrhizal fungi in host roots. The effects were most pronounced in PCNB-treated soil, particularly when the test plants were grown beyond 60 days. The ability for increased phosphate uptake by mycorrhizal root was also lost when soils were treated with thiram or PCNB. Earlier, Jalali and Domsch [94] observed that seed, as well as foliar applications, with conventional and systemic fungitoxicants restricted the development of mycorrhizal endophytes on host roots. They postulated that since foliarly applied pesticides may not be translocated intact to the roots, the side effects on mycorrhizae may be brought about by changes in the spectrum of root exudates as a result of the stress exerted by the pesticides. In further tests the systemic fungicides triforine and tridemorph applied to the host foliage changed the pattern of amino acid exudation [87].

Formation of mycorrhiza in clover roots was prevented by soil drenches of benomyl and thiophenate methyl, and the spread of established infections was halted [95]. Immersion of fungal inoculum in suspension of the fungicides reduced infectivity. However, clover plants grown in benomyl-treated soil did not retain enough fungicide to affect the amount of infection after transplanting into benomyl-free soil. Similarly, while studying the effect of five fungicides on VA mycorrhizal symbiosis in onion grown in phosphate-deficient soils, Manjunath and Bagyaraj [96] showed that except for captan, all other test fungicides applied even at lower concentrations reduced plant growth and phosphate uptake.

The foliar application of the symplastic fungicide fosetyl-al at different concentrations to mycorrhizal leek (*Alium porrum*) plants significantly increased colonization by *Glomus intraradices*, the number of intramatrical vesicles and plant growth compared to inoculated but untreated plants [97]. These effects did not diminish with time. On the other hand, mycorrhizal infection and growth of inoculated leek seedlings were reduced by soil drenches of the systemic fungicide metalxyl. Inoculated but untreated plants had extensive mycorrhiza. The stimulation of *G. intraradices* by fosetyl-al could be direct effect of the toxicant on the plant or the mycorrhizal fungus, or both. Lack of significant difference among the slopes of the concentration response curves indicated that this toxicant has its greatest effect early in the colonization process. The mechanism by which fosetyl-al produces this effect seems to be related to its role in altering root exudation. This fungicide causes a significant increase in exudation of soluble sugar from mycorrhizal roots, especially during the first few days after treatment [98]. These studies provide support for the hypothesis that root exudates may be a predisposing factor for the colonization and the resultant spread of the mycorrhizal fungus within the host [99].

Population dynamics of mycorrhizal fungi in arable fields are inversely affected more by fungicides than by herbicides, nematicides, and insecticides. These chemicals usually decrease mycorrhizal infection and spore numbers, similar to the effects of benomyl applications. Benomyl, which decomposes in soil to yield carbendazim and butyl isocyanate, has been shown to reduce percentage mycorrhizal infection in roots samples [100]. Benomyl suspensions are toxic to several mycorrhizal endophytes on direct immersion and when mixed with irradiated or infested soil. This probably offers positive indications that soil drenched with benomyl is toxic to external mycelium from established infections so that new roots remain uninfected. Ectomycorrhizal development by artificially introduced *Pisolithus tinctorius* and naturally occurring fungi was significantly inhibited by three to four foliar applications with triadimefon applied with the initial objective of controlling fusiform rust on loblolly seedlings [101]. Basidiocarp production by ectomycorrhizal fungi in fungicide-treated plots was 3 to 10 times less than in the control.

Several investigators have demonstrated that several commonly used herbicides drastically affect the mycorrhizal fungi. However, Kelley and South [102] observed that with few exceptions, herbicide concentrations necessary to affect fungal growth were much higher than recommended doses. Furthermore, several herbicides are reported to stimulate growth of some ectomycorrhizal fungi in axenic cultures, usually at low concentrations [102,103]. However, this response does not correlate with any specific group of toxicants. Chlorotoluron, under certain conditions, has resulted in increased spore populations of mycorrhizal fungi in soil, but mycorrhizal formation in hosts remains either unaffected or suppressed [104–107]. Mycorrhizal growth and development were suppressed on citrus species [104], but mycorrhizal formation of *Chenopodium quinona* was enhanced [108]. Schwab et al. [108] postulated that mycorrhizal formation was promoted by a simazine-induced increase in the root exudation of sugars and amino acids. Selectivity against certain mycorrhizal fungi by herbicides has been demonstrated in soil [100,109,110]. In nurseries, herbicides that tend to promote mycorrhizal diversity might be preferred under certain conditions [109], but in others it might be possible to control weeds with a toxicant that at the same time is able to promote growth of specific, inoculated mycorrhizal fungi.

Much less is known about the impact of insecticides and nematicides on mycorrhizal infection and resultant colonization. However, there are indications that these compounds may influence the developmental processes of these fungi. The studies of Backman and Clark [111] offer indications that the nematicide carbofuron caused a significant decrease in mycorrhizal formation in peanut roots one month after planting. Mycorrhizal development returned to normal levels when field plots were reexamined 4 months after planting. However, aldicarb, dibromochloropropene, and dichloropropene appear to stimulate both spore production and mycorrhizal formation [105], which seems to be directly related to significant reduction in nematode populations [112].

By and large, the systemic pesticides as a group appear more damaging to mycorrhizal symbiosis than nonsystemic ones. Such toxicants affect spore germination and ultimately colonization of the mycorrhizal endophyte within the host root system. Since translocation is primarily upwards, systemics would be more damaging to mycorrhizal fungi when applied as soil drenches. These findings indicate that investigations on the impact of a pesticide on mycorrhizal colonization should consider whether infection level attained under pesticide use will benefit the host. In case of significant disturbance, the repeated use of soil pesticides in normal farming systems will, therefore, need careful consideration.

VI. CONCLUSIONS

In recent years significant advances have been made in understanding the influences exerted by mycorrhizal fungi in conferring resistance to plant disease in wide spectrum of host plants. Mycorrhizae have the potential to biologically suppress root pathogens and benefit the host plant in conditions that are deleterious to host root growth and development (e.g., compensation for disease and counteracting toxicities). Parameters that contribute to the efficiency of mycorrhizal fungi such as inoculum placement, inoculum density, spore germination, growth through the soil, competitive abilities, survival mechanisms, and effective infection capabilities are all part of the inoculum potential of a mycorrhizal isolate. Therefore, the selection of efficient strains of the endophyte is critical to the development of inoculant fungi for use in agricultural technology. Other factors including hyphal spread through the root alignments, efficiency of phosphate uptake, and resultant transfer may effect the intrinsic efficiency of the symbiont. Precise determinations of the factors governing mycorrhizal efficiency and resolving its relevance to suppression of soilborne pathogens need to be seriously attempted for selecting efficient isolates for meaningful use in farming systems.

Most studies on the role of mycorrhiza in biological control have been carried out under controlled glasshouse conditions, and the results thus obtained may not be demonstrable under natural conditions. We must ensure that potentially efficient strains of mycorrhizal symbiont establish and survive in the face of competition from less efficient and naturally occurring symbionts, as well as antagonism for resident root pathogens and other rhizosphere microorganisms, if we are to realize the potential of mycorrhiza as a meaningful biocontrol tool. We need to develop appropriate technologies to increase the suppressiveness of mycorrhizal endophyte to root diseases. Not much information has been generated about the epidemiological considerations of predicting the impact of known populations of crop diseases and in predicting situations in which measurable responses would be obtained by inoculation with mycorrhizal symbiont. This area of investigation needs attention.

Vesicular-arbuscular mycorrhizal fungi, being obligate symbionts, cannot be grown on synthetic culture media. However, in order to harness their potential in viable biocontrol, workable techniques for commercial production of VA mycorrhizal fungi need to be designed. For example, B'ecard and Fortin [113] developed an in vitro system using Ri T-DNA transformed roots and the VA mycorrhizal fungus *Gigaspora margarita* to probe into the initial events of mycorrhiza formation. Using a single spore as inoculum, they obtained colonization of a preselected point on the root and could time the colonization process. The system is appropriate for studying the triggering of the fungal biotrophy towards the root. The initial signal which switches the germinated spore from dependence on its reserves for growth and sporulation remains to be found. Fortunately, the situation is quite promising in ectomycorrhizal fungi. Commercial production of *Pisolithus tinctorius* has been successfully used for establishing mycorrhizae on pines.

The influences of mycorrhizal fungi on root pathogens have been postulated to have physical and physiological bases particularly in terms of dynamics of root growth and function, root exudation, nutritional effects, and competition for host photosynthates, space, or infection site. In most cases the primary influence of mycorrhizal symbiosis in host–pathogen relationships have been correlated with enhanced phosphate nutrition, since substitution of mycorrhizal endophytes with phosphate fertilizers has paralleled the mycorrhizal host reponse. Nevertheless, recent studies have offered convincing evidence that mycorrhizal fungi affect host–pathogen relationships differently than phosphate nutrition. In-depth studies on non-nutritional influences of fungal symbionts on plant pathogens may yield a more conclusive picture. Although early establishment of fungal symbionts is important if this microbial system is to be employed as a meaningful biocontrol tool, additional investigations are needed to confirm its validity and relevance under field conditions.

Mycorrhizal responses to different pesticidal applications have been variable. Recent literature has added to the list of contradictory results without significantly improving our understanding of why most pesticides affect mycorrhizal fungi in the way that they do. This regression might be expected to continue, because new pesticides are being formulated faster than mycorrhiza researchers can deal with them [88]. The application of systemic pesticides has been shown to change the host metabolism, which is likely to affect the survival of mycorrhizal fungi in soil. Under the impact of pesticides, sensitive symbiotic relationships between the host and the fungal symbiont tend to deviate from their normal pattern of development; this is of practical significance and should not be overlooked in our future strategy for meaningful plant protection systems.

Clearly, the interactions between mycorrhizal fungi, plant pathogens, and pesticidal applications are not all that simple. In any case, they constitute only a part of the complex pattern of relationships existing between various biotic and abiotic forces. It is time to use existing information on various dimensions of mycorrhizal associations and undertake appropriate research and development efforts to solve problems of practical importance. We must concentrate more mycorrhizal research on "Does it work in the field?" rather than on "How does it work in the laboratory?" [114].

REFERENCES

1. Odum, E. P. *Fundamentals of Ecology*. Saunders, Philadelphia, 1971.
2. Zak, B. Role of mycorrhizae in root disease. *Annu. Rev. Phytopath.*, *2*:377–392, 1964.
3. Marx, D. H. Mycorrhizae and feeder root diseases. In *Ectomycorrhizae* (G. C. Marks and T. T. Kozlowski, eds.), Academic Press, New York, 1973, pp. 351–382.
4. Marx, D. H. and Krupa, S. V. Ectomycorrhizae. In *Interactions Between Non-Pathogenic Soil Microorganisms and Plants* (Y. R. Dommergues and S. V. Krupa, eds.), Elsevier, Amsterdam, 1978, pp. 373–400.
5. Schönbeck, F. Endomycorrhizas in relation to plant diseases. In *Soil-borne Plant Pathogens* (B. Schippers and W. Gams, eds.), Academic Press, New York, 1979, pp. 271–280.

6. Dehne, H. W. Interaction between vesicular arbuscular mycorrhizal fungi and plant pathogens. *Phytopathology, 72*:1115–1119, 1982.
7. Jalali, B. L. Vesicular-arbuscular mycorrhiza: Current status. In *Vistas in Plant Pathology* (A. Varma and J. P. Varma, eds.), Malhotra Publ. House, New Delhi, 1986, pp. 437–450.
8. Mosse, B. Advances in the study of vesicular-arbuscular mycorrhizae. *Annu. Rev. Phytopath., 11*:171–196, 1973.
9. Schenck, N. C. and Kellam, M. K. The influence of vesicular-arbuscular mycorrhiza on disease development. *Florida Agric. Stn. Tech. Bull.*, 798, 1978.
10. Jalali, B. L. and Thareja, M. L. Suppression of *Fusarium* wilt of chickpea in vesicular-arbuscular mycorrhizal inoculated soils. *Int. Chickpea Newsletter, 4*:21–22, 1981.
11. Rosendahl, S. Interactions between the vesicular-arbuscular mycorrhizal fungus *Glomus fasciculatum* and *Aphanomyces euteiches* root rot of peas. *J. Phytopathol., 114*:31–40, 1985.
12. Davis, R. M. and Menge, J. A. *Phytophthora parasitica* inoculation and intensity of vesicular-arbuscular mycorrhizae in citrus. *New Phytol., 87*:705–715, 1981.
13. Davis, R. M., Menge, J. A., and Zentmyer, G. A. Influence of vesicular-arbuscular mycorrhizae on *Phytophthora* root rot of three crop plants. *Phytopathology, 68*:1614–1617, 1978.
14. Zambolim, L. and Schenck, N. C. Reduction of the effects of pathogenic, root-infecting fungi on soybean by the mycorrhizal fungus, *Glomus mosseae. Phytopathology, 73*:1402–1405, 1983.
15. Caron, N., Fortin, J. A., and Richard, C. Influence of substrate on the interaction of *Glomus intraradices* and *Fusarium oxysporum* f. sp. *radicis-lycopersici* on tomatoes. *Pl. Soil, 87*:233–239, 1985.
16. Caron, M., Fortin, J. A., and Richard, C. Effect of *Glomus intraradices* on infection by *Fusarium oxysporum* f. sp. *radicis-lycopersici* in tomatoes over a 12-week period. *Can. J. Bot., 64*:552–556, 1986.
17. Caron, M., Richard, C., and Fortin, J. A. Effect of preinfestation of the soil by a vesicular-arbuscular mycorrhizal fungus, *Glomus intraradices*, on *Fusarium* crown and root rot of tomatoes. *Phytoprotection, 67*:15–19, 1986.
18. Smith, G. S. The role of phosphorus nutrition in interactions of vesicular-arbuscular mycorrhizal fungi with soil borne nematodes and fungi. *Phytopathology, 78*:371–374, 1988.
19. Gerdemann, J. W. Vesicular-arbuscular mycorrhiza. In *The Development and Function of Roots* (J. G. Torry and D. T. Clarkson, eds.), Academic Press, New York, 1975, pp. 571–591.
20. Kaye, J. W., Pfleger, F. L., and Steward, E. L. Interaction of *Glomus fasciculatum* and *Pythium ultimum* on greenhouse grown poinsettia. *Can. J. Bot., 62*:1575–1579, 1984.
21. Baath, E. and Hayman D. S. Plant growth responses to vesicular-arbuscular mycorrhizas XIV. Interactions with *Verticillium* wilt on tomato plants. *New Phytol., 95*:419–426, 1983.
22. Baath, E. and Hayman, D. S. No effect of VA mycorrhiza on red core disease of strawberry. *Trans. Br. Mycol. Soc., 82*:534–536, 1984.
23. Ngeve, J. M. and Roncadori, R. W. Effect of vesicular-arbuscular mycorrhizae, temperature and phosphorus on *Fusarium* wilt of sweet

potato. In *Tropical Root Crops: Production and Uses in Africa.* International Development Research Centre, Ottawa, 1984, pp. 197–202.

24. Iqbal, S. H. and Mahmood, T. Occurrence of VA mycorrhizae and damping-off caused by *Rhizoctonia solani* in relation to the age of *Brassica napus* seedlings. *Bull. Mycol.*, *2*:41–58, 1982.
25. Krishna, K. R. and Bagyaraj D. J. Interaction between *Glomus fasciculatum* and *Sclerotium rolfsii* in peanut. *Can. J. Bot.*, *61*:2349–2351, 1983.
26. Wick, R. L. and Moore, L. D. Histology of mycorrhizal and nonmycorrhizal *Ilex crenata* Helleri challenged by *Thielaviopsis basicola. Can. J. Pl. Pathol.*, *6*:146–150, 1984.
27. Hegde, S. V. and Rai, P. V. Influence of *Glomus fasciculatum* on damping-off of tomato. *Curr. Sci.*, *53*:588–589, 1984.
28. Bisht, V. S., Krishna, K. R., and Nene, Y. L. Interaction between vesicular-arbuscular mycorrhiza and *Phytophthora drechsleri* f. sp. *cajani. Int. Pigeonpea Newsletter, 4*:63–64, 1985.
29. Melo, I. S., Costa, C. P. and Silveira, A. P. D. Effect of vesicular-arbuscular mycorrhizas on aubergine wilt caused by *Verticillium albo-atrum* Reinke & Berth. *Summa Phytopathologica, 11*:173–179, 1985.
30. Chakravarty, P. and Mishra, R. R. Influence of endotrophic mycorrhizae on the fusarial wilt of *Cassia tora* L. *J. Phytopath.*, *115*:130–133, 1986.
31. Hussey, R. S. and Roncadori, R. W. Vesicular-arbuscular mycorrhizae may limit nematode activity and improve plant growth. *Plant Dis.*, *66*:9–14, 1982.
32. Schenck, N. C. Can mycorrhizae control rot diseases? *Plant Dis.*, *65*:230–234, 1983.
33. Saleh, H. and Sikora, R. A. Relationship between *Glomus fasciculatum* root colonization of cotton and its effect on *Meloidogyne incognita. Nematologica, 32*:230–237, 1984.
34. Oliveira, A. A. R. and Zambolim, L. Interaction between the ectomycorrhizal fungus *Glomus etunicatum* and the gall nematode *Meloidogyne javanica* at different phosphorus levels on bean (*Phaseolus vulgaris*). *Fitopathologia Brasileira, 11*:217–216, 1986.
35. Cooper, K. M. and Grandison, G. S. Interaction of vesicular-arbuscular mycorrhizal fungi and root-knot nematode on cultivars of tomato and white clover susceptible to *Meloidogyne hapla. Ann. App. Biol.*, *108*:555–565, 1986.
36. Smith, G. S., Roncadori, R. W., and Hussey, R. S. Interaction of endomycorrhizal fungi, superphosphate, and *Meloidogyne incognita* on cotton in microplot and field studies. *J. Nematol.*, *18*:208–216, 1986.
37. Grandison, G. S. and Cooper, K. M. Interaction of vesicular-arbuscular mycorrhizae and cultivars of alfalfa susceptible and resistant to *Meloidogyne hapla. J. Nematol.*, *18*:141–149, 1986.
38. Sitaramaiah, K. and Sikora, R. A. Effect of the mycorrhizal fungus *Glomus fasciculatus* on the host-parasite relationship on *Rotylenchus reniformis* in tomato. *Nematologica, 28*:412–419, 1982.
39. Kellam, M. K. and Schenck, N. C. Interactions between a vesicular-arbuscular mycorrhizal fungus and root-knot nematode on soybean. *Phytopathology, 70*:293–296, 1980.

40. Hussey, R. S. and Roncadori, R. W. Interaction of *Pratylenchus brachyurus* and *Gigaspora margarita* on cotton. *J. Nematol.*, *10*:15–20, 1978.
41. Roncadori, R. W. and Hussey, R. S. Interaction of the endomycorrhizal fungus *Gigaspora margarita* and root-knot nematode on cotton. *Phytopathology*, *67*:1507–1511, 1977.
42. Sikora, R. A. Biological and chemical interactions in the rhizosphere. *Proceedings symposium*, Stockholm, 1981, pp. 115–136.
43. Atilano, R. A., Menge, J. A., and Gundy, S. D., van. Interaction between *Meloidogyne arenaria* and *Glomus fasciculatus* in grape. *J. Nematol.*, *13*:52–59, 1981.
44. Schenck, N. C., Kinloch, R. A., and Dickson, D. W. Interaction of endomycorrhizal fungi and root-knot nematode on soybean. In *Endomycorrhizas*, (F. E. Sanders, B. Mosse, and P. B. Tinker, eds.), Academic Press, London, 1975, pp. 607–617.
45. Strobel, N. E., Hussey, R. S., and Roncadori, R. W. Interaction of vesicular-arbuscular mycorrhizal fungi, *Meloidogyne incognita* and soil fertility on peach. *Phytopathology*, *72*:690–694, 1982.
46. Cook, R. Nature and inheritance of nematode resistance in cereals. *J. Nematol.*, *13*:52–57, 1974.
47. Harley, J. L. and Smith, S. E. *Mycorrhizal symbiosis*. Academic Press, London, 1983, p. 483.
48. Thomson, Casen, M. K., Hussey, R. S., and Roncadori, R. W. Interaction of vesicular-arbuscular mycorrhizal fungi and phosphorus with *Meloidogyne incognita* on tomato. *J. Nematol.*, *15*:410–417, 1983.
49. Sitaramaiah, K. and Sikora, R. A. Influence of the endomycorrhizal fungus *Glomus mosseae* on *Rotylenchulus reniformis* penetration and development on bush bean, cucumber and muskmelon. *Med. Fac. Landbouww, Rijksuniv. Gent.*, *46*:695–702, 1981.
50. Nemec, S. and Myhre, D. Virus-*Glomus etunicatum* interactions in citrus root stocks. *Plant Dis.*, *68*:311–314, 1984.
51. Schönbeck, F. and Schinzer, U. Investigations on the influence of endotrophic mycorrhiza on TMV lesion formation in *Nicotiana tabacum* L. Var. *Xanthine*. *Phytopath. Z.*, *73*:78–80, 1972.
52. Green, S. K. and Deng, T. C. Effect of ectomycorrhizal fungus *Glomus mosseae* on soybean mosaic virus in soybean. *Pl. Prot. Bull.*, Taiwan, *27*:353–358, 1985.
53. Bondoux, P. and Perrin, R. Mycorrhizas and plant protection. *Comptes Rendus des seances de I' Academie d' Agric. de France*, *68*:1162–1177, 1982.
54. Schönbeck, F. and Spengler, G. Detection of TMV in mycorrhizal cells of tomato by immunofluorescence. *Phytopath. Z.*, *94*:84–86, 1978.
55. Schönbeck, F. and Dehne, H. W. Untersuchungen zum Einfluss der endotrophen Mycorrhiza auf Pflanzenkrankheiten. 4. Pilzliche Sprossparasiten *Olpidium brassiae*, TMV. *Z. Pflanzen-krankh. Pflanzenschutz*, *86*:103–112, 1979.
56. Jabaji-Hare, S. H. and Stobbs, L. W. Electron microscopic examination of tomato roots coinfected with *Glomus* sp. and tobacco mosaic virus. *Phytopathology*, *74*:277–279, 1984.

57. Campbell, R. N. Fungal vectors of plant viruses. In *Fungal Viruses* (H. P. Molitoris, M. Hollings, and H. A. Wood, eds.), Springer-Verlag, New York, 1979, pp. 8–24.
58. Wingfield, E. B. Mycotrophy in loblolly pine. I. The role of *Pisolithus tinctorius* and *Rhizoctonia solani* in survival of seedlings. II. Mycorrhizal formation after fungicide treatment. *Ph.D. dissertation*, Virginia Polytechnic Institue, Blacksburg, 1968, p. 78.
59. Ross, E. W. and Marx, D. H. Susceptibility of sand pine to *Phytophthora cinnamomi*. *Phytopathology, 62*:1197–1200, 1972.
60. Marx, D. H. Growth of ectomycorrhizal and nonmycorrhizal shortleaf pine seedlings in soil with *Phytophthora cinnamomi*. *Phytopathology, 63*:18–23 1973.
61. Richard, C., Fortin, J. A., and Fortin, A. Protective effect of an ectomycorrhizal fungus against the root pathogen, *Mycelium radicis atrovirens*. *Can. J. For. Res., 1*:246–251, 1971.
62. Powell, W. M., Hendrix, F. F., and Marx, D. H. Chemical control of feeder root necrosis of pecans caused by *Pythium* species and nematodes. *Plant Dis. Rep., 52*:577–580, 1968.
63. Corte, A. Research on the influence of the mycorrhizal infection on the growth, vigor, and state of health of three *Pinus* species. *Arch. Bot. Biogeogr. Ital., 45*:1, 1969.
64. Sylvia, D. M. Role of *Laccaria laccata* in protecting primary roots of Douglas-fir from root rot. In *Tree Root Systems and Their Mycorrhizas* (D. Atkinson, K. K. S. Bhat, M. P. Coutts, P. A. Mason, and D. J. Read, eds.), Martinus Nijhoff, The Hague, 1983, pp. 299–302.
65. Sinclair, W. A., Sylvia, D. M., and Larsen A. O. Disease suppression and growth promotion in Douglas-fir seedlings by the ectomycorrhizal fungus *Laccaria laccata*. *For. Sci., 28*:191–201, 1982.
66. Sylvia, D. M. and Sinclair, W. A. Suppressive influence of *Laccaria laccata* on *Fusarium oxysporum* and on Douglas-fir seedlings. *Phytopathology, 73*:384–389, 1983.
67. Perrin, R. and Garbaye, J. Influence of ectomycorrhizae on infectivity of *Pythium*-infested soils and substrates. In *Tree Root Systems and Their Mycorrhizas* (D. Atkinson, K. K. S. Bhat, M. P. Coutts, P. A. Mason, and D. J. Read, eds.), Martinus Nijhoff, The Hague, 1983, pp. 345–351.
68. Bumbieris, M. Effect of mycorrhizas on susceptibility of *Pinus radiata* to *Phytophthora cinnamomi*. In *Biennial Rept. Waite Agric. Res. Inst.*, Adelaide. Univ., South Australia, 1982.
69. Kuo, S. C. and Tang, F. The protective effect of several mycorrhizal fungi, *Boletus* sp. and *Suillus grevillei* on the damping-off disease of seedlings of *Pinus tabulaeformis*. *Acta Bot. Yunnanica, 3*:359–366, 1981.
70. Chakravarty, P. and Unestam T. Differential influence of ectomycorrhizae on plant growth and disease resistance in *Pinus sylvestris* seedlings. *J. Phytopath., 120*:104–120, 1987.
71. Chakravarty, P. and Unestam, M. Mycorrhizal fungi prevent disease in stressed pine seedlings. *J. Phytopath., 118*:335–340, 1987.
72. Summerbell, R. C. Inhibitory effect of *Trichoderma* species and other soil microfungi on formation of mycorrhiza by *Laccaria bicolor* 'in vitro'. *New Phytol., 105*:437–448, 1987.

73. Barham, R. O., Marx, D. H., and Ruehle, J. L. Infection of ectomycorrhizal and nonmycorrhizal roots of shortleaf pine by nematodes and *Phytophthora cinnamomi*. *Phytopathology*, *64*:1260–1264, 1974.
74. Malajczuk, N. and McComb, A. J. The microflora of unsuberized roots of *Eucalyptus calophylla* R. Br. and *Eucalyptus margirata* Donn. ex. Sm. seedlings grown in soils suppressive and conducive to *Phytophthora cinnamomi* Rands.I. Rhizoplane bacteria, actinomycetes, and fungi. II. Mycorrhizal roots and associated microflora. *Aust. J. Bot.*, *27*:235–272, 1979.
75. Lockwood, J. L. Fungistasis in soils. *Biol. Rev. Camb. Phil. Soc.*, *52*:1–43, 1977.
76. Davis, R. M. Influence of *Glomus fasciculatus* on *Thielaviopsis basicola* root rot of citrus. *Plant Dis.*, *64*:839–840, 1980.
77. Dehne, H. W. Untersuchungen über den Einfluss der endotrophen Mycorrhiza auf die *Fusarium*-Wekle on Tomato and Gurke. Ph.D. dissertation, University of Bonn, 1977, p. 150.
78. Schönbeck, F. Endomycorrhiza: Ecology, function and phytopathological aspects. *Forum Microbial.*, *3*:90–96, 1980.
79. Dehne, H. W., Schönbeck, F., and Baltruschat, H. The influence of endotrophic mycorrhiza on plant diseases. 3. Chitanese activity and the ornithine-cycle. *Z. Pflanzenkrankh. Pflanzenschutz*, *85*:666–678, 1978.
80. Becker, W. N. Quantification of onion vesicular-arbuscular mycorrhizae and their resistance to *Pyrenochaeta terrestris*. Ph.D. dissertation, Duke University, Durham, NC, 1976, p. 147.
81. Sylvia, D. M. and Sinclair, W. A. Phenolic compounds and resistance to fungal pathogens induced in primary roots of Douglas-fir seedlings by the ectomycirrhizal fungus *Laccaria laccata*. *Phytopathology*, *73*:390–397, 1983.
82. Marx, D. H. and Davey, C. B. The influence of ectotrophic mycorrhizal fungi on the resistance of pine roots to pathogenic infections. IV. Resistance of naturally occurring mycorrhizae to infections by *Phytophthora cinnamomi*. *Phytopathology*, *59*:559, 1969.
83. Marx, D. H. Influence of ectotrophic mycorrhizal fungi on the resistance of pine roots to pathogenic infections. V. Resistance of mycorrhizae to infection by vegetative mycelium of *Phytophthora cinnamomi*. *Phytopathology*, *60*:1472, 1970.
84. Krupa, S. and Fries, N. Studies on ectomycorrhizae of pine. I. Production of volatile organic compounds. *Can. J. Bot.*, *49*:1425, 1971.
85. Krupa, S. Anderson, J., and Marx, D. H. Studies on ectomycorrhizae of pine. IV. Volatile organic compounds in mycorrhizal and nonmycorrhizal root systems of *Pinus echinata* Mill. *Eur. J. For. Pathol.*, *3*:194–200, 1973.
86. Hale, M. G. and Moore, L. D. Factors affecting root exudation II: 1970–78, *Adv. Agron.*, *31*:93–124, 1979.
87, Jalali, B. L. and Domsch, K. H. Effect of some fungitoxicants on the amino acid spectrum of root exudates in relation to endomycorrhizal development. *Phytopath. Z.*, *90*:22–27, 1977.
88. Trappe, J. M., Molina, R., and Castellano, M. Reaction of mycorrhiza formation to pesticides. *Annu. Rev. Phytopathol.*, *22*:331–359, 1984.

89. Menge, J. A. Effect of soil fumigants and fungicides on vesicular-arbuscular fungi. *Phytopathology, 72*:1125–1132, 1982.
90. Kleinschmidt, G. D. and Gerdemann, H. W. Stunting of citrus seedlings in fumigated nursery soils related to the absence of endomycorrhizae. *Phytopathology, 62*:1447–1453, 1972.
91. O'Bannon, J. H. and Nemec, S. Influence of soil pesticides on vesicular-arbuscular mycorrhizae in a citrus soil. *Nematropica, 8*:56–61, 1978.
92. Pfleger, F. L. and Stewart, E. L. The influence of fungicides on endemic populations of *Glomus* species in association with field grown peas. *Proc. Am. Phytopathol. Soc., 3*:274, 1976.
93. Jalali, B. L. Effects of soil fungitoxicants on the development of VA-mycorrhiza and phosphate uptake in wheat. In *Soil-Borne Plant Pathogens* (B. Schippers and W. Gams, eds.), Academic Press, London, 1979, pp. 525–530.
94. Jalali, B. L. and Domsch, K. H. Effect of systemic fungitoxicants on the development of endotrophic mycorrhiza. In *Endomycorrhizas* (F. E. Sanders, B. Mosse, and P. B. Tinker, eds.), Academic Press, London, 1975, pp. 619–626.
95. Boatman, N., Paget, D., Hayman, D. I., and Mosse, B. Effects of systemic fungicides on vesicular-arbuscular mycorrhizal infection and plant phosphate uptake. *Trans. Br. Mycol. Soc., 70*:443–450, 1978.
96. Manjunath, A. and Bagyaraj, D. J. Effect of fungicides on mycorrhizal colonization and growth of onion. *Pl. Soil, 80*:147–150, 1984.
97. Jabaji-Hare, S. H. and Kendrik, W. B. Response of an endomycorrhizal fungus in *Allium porrum* L. to different concentrations of systemic fungicides, metalaxyl (Ridomil) and fosetyl-Al (Aliette). *Soil Biol. Biochem., 19*:95–99, 1987.
98. Jabaji-Hare, S. H. and Kendrik, W. B. Effects of fosetyl-Al on root exudation and on composition of extracts of mycorrhizal and nonmycorrhizal leek roots. *Can. J. Pl. Pathol., 7*:118–126, 1985.
99. Graham, J. H., Leonard, R. T., and Menge, J. A. Membrane-mediated decrease in root exudation responsible for phosphorus inhibition of vesicular-arbuscular mycorrhiza formation. *Pl. Physiol., 68*:548–552, 1981.
100. Tommerup, I. C. and Briggs G. C. Influence of agricultural chemicals on germination of vesicular-arbuscular endophyte spores. *Trans. Br. Mycol. Soc., 76*:326–328, 1981.
101. Marx, D. H., Cordell, C. E., and France, R. C. Effects of triadimefon on growth and ectomycorrhizal development of loblolly and slash pines in nurseries. *Phytopathology, 76*:824–831, 1986.
102. Kelley, W. D. and South, D. B. Effects of herbicides on 'in vitro' growth of mycorrhizae of pine (*Pinus* spp.). *Weed Sci., 28*:599–602, 1980.
103. Lake, D. B., Ippoliti, D. J., Brandow, C. C., and Otrosina, W. J. Effect of herbicides on the growth of *Pisolithus tinctorius* and *Scleroderma aurantium* in pure culture. *5th N. Am. Conf. Mycorrhizae*, Quebec, Laval University, (Abstr.), 1981, p. 65.
104. Nemec, S. and Tucker, D. Effects of herbicides on endomycorrhizal fungi in Florida citrus (*Citrus* spp.) soils. *Weed Sci., 31*:427–431, 1983.

105. Ocampo, J. A. and Hayman, D. S. Effects of pesticides on mycorrhiza in field-grown barley, maize and potatoes. *Trans. Br. Mycol. Soc.*, *74*:413–416, 1980.
106. Smith, T. F., Noack, A. J., Cosh, S. M. The effect of some herbicides on vesicular-arbuscular endophyte abundance in the soil and on infection of host roots. *Pestic. Sci.*, *12*:91–97, 1981.
107. Pope, P. E. and Holt, H. A. Paraquat influences development and efficacy of the mycorrhizal fungus *Glomus fasciculatus*. *Can. J. Bot.*, *59*:518–521, 1980.
108. Schwab, S. M., Johnson, E. L. V., and Menge, J. A. Influence of simazine on formation of vesicular-arbuscular mycorrhizae in *Chenopodium guinona* Wilid. *Pl. Soil,* *69*:283–287, 1982.
109. Trappe, J. M. Effects of the herbicides bifenox, DCPA, and napropamide on mycorrhiza development of ponderosa pine and Douglas-fir seedlings in six western nurseries. *For. Sci.*, *29*:464–468, 1983.
110. Walker, C. and Mize, C. W. Populations of endogenaceous fungi at two locations in central Iowa. *Can. J. Bot.*, *60*:2518–2529, 1982.
111. Backman, P. A. and Clark, E. M. Effect of carbofuron and other pesticides on VA mycorrhizae in peanuts. *Nematropica,* 7:13–17, 1977.
112. Germani, G., Diem, H. G., and Dommergues, Y. R. Influence of 1,2 dibrome-3-chloropropane fumigation on nematode population, mycorrhizal infection, N_2 fixation and yield of field-grown groundnut. *Rev. Nematol.*, *3*:75–79, 1980.
113. B'ecard, G. and Fortin, J. A. Early events of vesicular-arbuscular mycorrhiza formation on Ri T-DNA transformed roots. *New Phytol.*, *108*:211–218, 1988.
114. Marx, D. H. and Schenck, N. C. Potential of mycorrhizal symbiosis in agricultural and forest productivity. In *Challenging Problems in Plant Health* (T. Kommedahal and P. H. Williams, eds.), American Phytopathological Society, Minnesota, 1983, pp. 334–347.

6

THE USE OF MYCORRHIZAL INOCULANTS IN FORESTRY AND AGRICULTURE

PETER JEFFRIES and JOHN C. DODD *The University of Kent, Canterbury, Kent, England*

I. INTRODUCTION

Mycorrhizae are mutualistic symbioses between certain groups of soil fungi and plant root systems. It is now accepted that in a natural ecosystem most plants will have a well-developed mycorrhizal association. Some critical nutrients that the plant requires will come via this network rather than directly into the roots. In most agricultural situations, the important role of mycorrhizae in promoting healthy plant growth has been overlooked. Suboptimal yields have resulted when plants have been grown in soils that have low nutrient availability if an effective mycorrhizal association has not been able to develop. In some cases this problem has been overcome by the routine use of large amounts of agricultural fertilizers or by the development of an efficient natural indigenous mycorrhizal population.

This capacity to overcome deficiencies in mycorrhizal infection has in many cases hindered or negated the need for artificial inoculation of soils with mycorrhizal fungi and has promulgated the idea that mycorrhizal studies are primarily of academic interest rather than of practical merit. The literature is full of exciting results where spectacular growth increases have followed mycorrhizal inoculation of important agricultural crops. On closer examination, however, it is evident that the noninoculated control plants have been grown in unnatural conditions perhaps lacking a natural phosphorus source or in a sterilized soil with no indigenous mycorrhizae. Alternatively, the amount of inoculum that has been added would be totally impractical in a large-scale field situation. In some cases it is apparent that inoculum would have to be freshly prepared in situ because of the difficulties experienced in producing inoculum that will retain viability during processes of preparation, storage, and distribution. It is hardly surprising, then, that many of these spectacular results have not been translated into real yield increases in true field situations.

Another factor influencing the lack of adoption of mycorrhizal inoculants in agriculture is the fact that much of the published mycorrhizal research has been carried out in "developed" countries which have high-input agricultural systems. This form of agro-industry will be the least suitable beneficiary of mycorrhizal inoculation in the short term, as fertilizer needs are provided in excess, crops are high yielding, and marginal soils are not thought worthy of cultivation. The impetus for development of mycorrhizal inoculants is thus lacking, especially when research budgets have become more focused towards cost-effectiveness on a parochial scale. In contrast, the level of mycorrhizal technical knowledge and exploitation is much lower in developing countries where lower input and subsistence level agriculture is prevalent. This type of agriculture is widespread; half the world's soils are acid, and crops grown in them suffer from phosphorus deficiency. Such situations are ideal for adapting mycorrhizal technology to local needs in order to improve farming efficiency. There is, however, a breakdown of communication between agriculturalists who have know-how but are in situations where they are unable to take advantage of it, and those that lack know-how and have the situations that most need it.

Technology transfer is possible, however, and there are several examples where successful field results have been obtained. In this review we will describe particular cases where beneficial growth responses have been achieved in poor agricultural soils. Our aim is to encourage a dialogue between scientists with a mycorrhizal background and those agriculturalists who face problems of soil infertility or limited fertilizer supplies. We wish to stimulate the development of strategies of mycorrhizal management for improvement and sustainability of yields. We consider that the plethora of data obtained in academic studies of mycorrhizae, coupled with the reluctance of investigators to enter into prolonged studies in field situations, has led to a negative attitude being taken towards mycorrhizal exploitation. This has tended to obscure the fact that there are many situations, albeit strictly defined, under which mycorrhizal manipulation can be of tremendous benefit in practical agriculture. For vesicular-arbuscular (VA) mycorrhizae we believe that most benefit will accrue in low-input agricultural systems on tropical acid-infertile soils. The aim will be to take advantage of the mycorrhizal symbiosis for the more efficient use of cheap and readily available fertilizers in order to achieve higher yields of food crops. Ectomycorrhizal inoculants, on the other hand, will benefit the forestry industry, not only on poor soils but also where modern approaches are being taken. In this chapter we wish to highlight these situations and describe some of the successful results that are appearing in the literature. We will go on to list some of the types and sources of inoculum currently on the market and then make some predictions about the most likely future benefits.

II. BENEFITS CONFERRED BY MYCORRHIZAE

A. Nutrient Uptake

There is no doubt that the growth of many plants can be substantially improved if they possess a well-developed mycorrhizal system. The relative benefit that a plant obtains may vary with different genera, and it is possible to rank species according to their relative mycorrhizal dependency (Table 1). Plenchette et al. [1] have devised a scheme for calculating a relative field mycorrhizal dependency index that can be useful in assessing the importance that mycorrhizae have to particular host plants. Dependence is apparently closely related to the extensiveness of the root system [2]. The fungus establishes an intimate relationship with the root system and then spreads out into the surrounding soil, thus increasing the volume accessible to the plant. Because of this, the principal benefit conferred by the mycorrhizal condition is the increased availability of plant nutrients of low diffusion coefficient.

TABLE 1. Mycorrhizal Dependency of Agricultural Crops in Low P Conditions

Strong	Medium	Weak
Cassava	Soya	Potato
Citrus	Wheat	Rice
Allium	Barley	Melon
Tropical forage	Cotton	Sunflower
legumes	Cowpea	Beans
Pinus (ectomycorrhizae)	Grain legumes	Maize
Peppers	Tomatoes	Sorghum
Many tropical fruits	Grapevine	
Brachiaria spp.		

Phosphorus is the most important of these nutrients, and much of the mycorrhizal response can be related to improved phosphorus uptake. This aspect has been extensively reviewed elsewhere [3] (see also Chapter 2). In some cases, however, mycorrhizae can act in an opposite way and help to protect the plant from the presence of toxic metal ions. VA mycorrhizae, for example, can ameliorate the effect of manganese toxicity toward soybean [4] and toward leek plants and mung bean, especially in sterilized soils where manganese levels tend to be elevated (D. P. Stribley, personal communication). Other ions such as magnesium and aluminum can also be accumulated by the fungi and excluded from the plant. VA mycorrhizal infection of grasses growing in zinc-polluted soils has been shown to protect the plants from the toxic effects of this metal [5]. Fruiting bodies of some ectomycorrhizal fungi are known to accumulate heavy metals from contaminated soils, and apparently ectomycorrhizal colonization can protect *Pinus* and *Picea* seedlings from heavy metal toxicity at low or intermediate soil concentrations of cadmium, copper, nickel, lead, and zinc [6]. Thus, the mycorrhizal fungus can have either a protective role or a scavenging role depending on the relative concentration of the particular ion in question. This property of mycorrhizal infection can become important when soil management strategies are being introduced that include the addition of sewage sludge [7] or similar additions to infertile soil. These amendments frequently contain high levels of certain metal ions, and when first added to unamended soils can prove toxic to the indigenous mycorrhizal population [8], resulting in poor growth responses when the site is revegetated. Mycorrhizal fungi with tolerance to high levels of toxic metals have already been reported from environments that have had prolonged exposure to the toxicant [9]. For example, strains of mycorrhizal fungi have been found to grow well at zinc concentrations of 1 mg/g [7]. It is thus important to consider whether or not mycorrhizal inoculation with tolerant strains ought to accompany this type of soil amendment.

Although most benefits of mycorrhizal infection have been traced to a phosphorus effect, other relatively immobile nutrients, such as zinc [10], have also been implicated. Improvement in Zn nutrition leading to growth responses in field crops following mycorrhizal infection have been reported [11]. Improved growth leads inevitably to a more rapid uptake of all nutrients, and it is important to establish which is the cause for improvement. This brings us back to the phosphorus effect and highlights the difficulties faced by physiologists investigating the mechanics of plant growth stimulation following mycorrhizal infection. In general, adding phosphorus alone must mirror the mycorrhizal response to show it to be the growth-limiting substrate. Definitive studies should compare the phosphorus response curves for mycorrhizal and nonmycorrhizal plants. Yost and Fox [12], for example, have done this in field plots on tropical oxisols, using methyl bromide fumigation of soil maintained at different levels of P. They found that the soil solution P level at which mycorrhizal infection ceased to have an effect varied between 0.1 and 1.6 mg P/L, depending upon species. In a more focused example, Menge et al. (13) have suggested that citrus trees, which have high mycorrhizal dependency, would benefit from the symbiosis in soils containing less than 34 mg/L available phosphorus, 12 mg/L available zinc, or 27 mg/L available manganese. These conditions apparently prevail in over 85% of southern California citrus soils. By extrapolation of this type of observation, Menge [14] estimated that mycorrhizal fungi must contribute an equivalent of between 112 – 556 kg phosphorus/ha. These values are similar to those calculated for other crops and soils and values upwards of 30 kg phosphorus/ha have been reported, with most around 100 kg phosphorus/ha. In phosphorus-binding soils, this phenomenon becomes paramount for efficient crop growth. Others have shown that by relating dry matter production of cassava to the available phosphorus in the soil, a critical level of 15 mg/L was obtained for mycorrhizal plants compared with 190 mg/L for nonmycorrhizal plants in sterile soil [15].

Interestingly, there is a case for dual inoculation of soils with mycorrhizal fungi and microorganisms that can help solubilize rock phosphates. For example, a strain of *Penicillium bilaji* has been shown to act in this way and can enhance growth responses of plants in calcareous soils inoculated with VA mycorrhizal fungi and receiving additions of rock phosphate [16]. Similar cases have been put forward for some of the phosphate-solubilizing bacteria. Synergism between VAM and nitrogen-fixing bacteria has also been documented. The case for dual inoculation of soils with rhizobia and VA mycorrhizal fungi has been well reviewed [17,18]. The arguments for dual inoculation become stronger as evidence accumulates for a localized nonsystemic, non-phosphorus-mediated influence of VAM on root dry weight increases and nodule activity [19] and the finding that VAM hyphae can translocate and assimilate ammonium and so aid the nitrogen nutrition of nitrogen-fixing plants [20,21].

B. Water Relations

There has been much debate about the role of VA mycorrhizae in the control of water relations of plants. Most work has been done in pots under controlled environmental conditions and has produced conflicting results on the beneficial effects of the symbiosis. Increased transpiration flux in VAM-infected plants has been reported, linked to reduced stomatal resistance and increased root hydraulic conductivity [22 – 24]. A major problem, however, has been separating the direct effects of the symbiosis from indirect effects such as better phosphorus nutrition observed in infected plants compared with nonmycorrhizal controls. Hardie [25], however, has shown that removal of the extramatrical mycelial network from roots of leek and clover caused a significant fall in transpiration flux when plants were transplanted compared with VAM-infected plants retaining the network and nonmycorrhizal plants with phosphorus added. This may indicate that the extramatrical mycelial network has an important role to play in water uptake due to the more intricate contact of the fine hyphae with the rooting medium. This may have profound effects on the survival of plants raised under a high nutrient regime and transplanted into a soil with low P availability under conditions of water stress.

Inoculation of VA mycorrhizal fungi into sterile soil has also caused conflicting reports on changes in the leaf xylem pressure potential (LXPP). Increases [26], decreases [23], and no change [27] have been recorded in LXPP of mycorrhizal plants with and without water stress when compared with nonmycorrhizal equivalents. Allen and Boosalis [28] found that stomata closed at lower LXPPs in *Glomus fasciculatum*–infected winter wheat and interpreted this as a method for increasing drought tolerance. Other workers [29] arrived at a similar conclusion for the same fungus/plant interaction undergoing cyclic water stress. Levy et al. (23) concluded that VA mycorrhizal infection enables root systems to deplete soil water more effectively even to the point of causing greater water stress in greenhouse trials with citrus and cotton. VAM-infected cotton, for example, had significantly lower LXPP leading to significantly greater water stress symptoms (higher free proline content of leaves [30] and a significantly slower recovery of turgor) compared with nonmycorrhizal plants of the same dry weight and leaf phosphorus concentration (Y. Levy, J. C. Dodd, and J. Krikun, unpublished results). A subsequent cycle of water stress did not show significant differences. It may be that VA mycorrhizae aid plants to survive temporary water stress but that this effect is more pronounced in plants of high mycorrhizal dependency such as rose [31] or *Eupatorium odoratum* [32], compared with those of low dependency such as sunflower [33], sorghum [32], or corn [34]. Recent work [35] with cassava has assessed the performance of several VA mycorrhizal fungal species under water-stressed and nonstressed conditions. Those that were ineffective under nonstressed conditions remained so when plants became stressed, while effective strains benefited the plants under both conditions. This may explain some of the conflicting results in the literature. Krikun et al. [36] reported that mycorrhizal field-grown pepper plants, fertigated with 10 mg/L and 40 mg/L, had greater stomatal conductances and lower LXPPs at midday than nonmycorrhizal plants. Mycorrhizal plants subsequently outyielded nonmycorrhizal plants at 10 mg/L P, but yields were not different at 40 mg/L P. The potential benefits of increased drought resistance of high value cash crops require further field investigations, particularly in semi-arid zones where cyclic water stress prevails. These areas hold the greatest immediate prospect for practical application of the VAM symbiosis [37] (see also Chapter 3).

C. Soil Stabilization

The extramatrical hyphal network of VA mycorrhizae can improve soil structure by binding soil particles into more stable aggregates [38,39]. This could be of vital importance in preventing large losses of tropical soils due to erosion. If particular VA mycorrhizal fungi can produce large amounts of extramatrical mycelium on a desired plant host, they could be used not only to bind the soil to the root system but to aid establishment of the host.

This is particularly relevant in a badly eroded area in the Mondomo area of Cauca, Colombia, where it is intended to establish improved pasture grasses and legumes to enable the land to be used for grazing cattle, and possibly for primary crop production with cassava. There have, however, been problems in the establishment phase of the legumes in particular. Mycorrhizal considerations have been ignored until recently but may be included in future trials.

D. Protection Against Root Diseases

Mycorrhizal infection can also protect plants from attack by soilborne root pathogens, but this aspect of mycorrhizal improvement of crop growth is reviewed elsewhere in this volume (see also Chapter 5) and will not be described in detail here. It is worth noting, however, the recent work of Thompson [40], who has shown that "long fallow disorder" of field crops is a disease resulting from suboptimal degrees of VA mycorrhizal colonization following a fallow period.

III. SUCCESSFUL FIELD RESULTS

There have been many examples of the successful use of mycorrhizal inoculants in improving the establishment or subsequent growth of forest trees or crop plants introduced into new environments. There are several reviews of the subject (e.g., 40 – 48), and there is no point in repeating the information here. Instead we will consider one or two examples of both ectomycorrhizae and VA mycorrhizae where impressive results have been obtained recently.

A. Ectomycorrhizae

The history of deliberate introduction of nonindigenous ectomycorrhizae into soils in which new tree species were being established has been a tremendous success story. The literature contains many examples where the failure to establish trees onto new sites has been completely overcome by the addition of mycorrhizal inoculum [44,45]. Here we will discuss a more recent case in which the traditional methods of mycorrhizal inoculation are now being successfully replaced by more refined technology.

In the Philippines, projected shortages of timber have led to extensive reforestation programs being undertaken since the 1960s by government or the private sector. The private sector had taken the lead in the establishment of commercial plantations using introduced fast-growing species. Many of these early efforts were failures, and survival and growth of the trees was poor either due to bad management practices, adverse environmental conditions, or the failure of the trees to form effective symbiosis. Like many tropical or semi-tropical soils, those in the reforestation areas are low in organic matter, acid-infertile, and often contain high levels of aluminum with resultant low levels of available phosphorus.

Many of the problems in establishing trees have now been overcome by mycorrhizal inoculation. The established approach to growing pine seedlings in the nursery was to collect mycorrhizal soil from under established plantations. This soil was then transported into the nursery, where it was mixed with the potting medium. Mycorrhizal soil collected in this way has the disadvantages discussed in more detail in Section IV, such as high transport costs, risks of pathogen transfer, and irregularity of inoculum quality. Nevertheless, this method has been practiced for many years. More recently, foresters in the Philippines have been planting *Eucalyptus*. Mycorrhizal inoculation had not previously been attempted with this genus in the Philippines, as most foresters or nurserymen were unaware that eucalypts are associated with mycorrhizal fungi. Since 1980, work at the National Institute of Biotechnology and Applied Microbiology (BIOTECH) has developed basidiospore inoculum techniques to the stage where a commercial product (see Section IV) has now been used successfully in a range of commercial nurseries in the Philippines for inoculation of *Pinus* and *Eucalyptus*. Sporocarps of *Pisolithus tinctorius* and a *Scleroderma* species are collected from the field and oven-dried at 27°C to a moisture content equivalent of 16 – 18% dry weight. The basidiospores are then removed and pelletized in a finely ground expanding clay. The concentration of spores in each pellet is about 9 – 10%. Pellets are added, one per pot, when one-week-old seedlings are transplanted into separate pots at the nursery stage. Significant height and diameter growth increases have been noted in seedlings of *Pinus caribaea, Eucalyptus camaldulensis*, and *E. deglupta* in the 3-month nursery phase, which were translated into highly significant growth responses in subsequent outplantings. Inoculated *E. deglupta*, for example, has shown a 64% increase in height and a 73% increase in trunk diameter after 18 months in the field when compared to uninoculated controls. Similarly, height (45%) and trunk diameter (45%) increases were noted in outplanted *Pinus caribaea* after 16 months in the field. When phosphorus dose-response curves were determined using subplots receiving fertilizers as a comparison, it was apparent that inoculation of seedlings at the nursery stage could substitute for between 67 and 86% of the inorganic fertilizer equivalent required to obtain a similar growth response (R. De La Cruz, personal communication). Field trials were conducted on commercial sites in nonsterile soils, and the results clearly indicate that the indigenous population is not able to impart these spectacular growth increases obtained over this initial 18-month period. The major problem with this form of inoculum is the variability in supplies of fresh basidiospores. Collection is obviously limited to the fruiting season, and genetic variability is introduced by collecting from geographically dispersed mycelia. Nevertheless, the encouraging results so far will ensure the adoption of this inoculum into local nursery practices. Scientists at BIOTECH are now developing alginate-bead inoculum derived from fermenter-grown mycelium of *Pisolithus tinctorius*. Significant growth increases in pot-grown *Eucalyptus* have already been attained using this approach (N. Aggangan, P. Mason, and P. Jeffries, unpublished results). Recently there has been a report [45] that *Eucalyptus dumosa* can support a brief VAM infection prior to colonization by ectomycorrhizal symbionts and that these could be important in the early establishment of the plants in phosphorus-fixing soils. The observation should also be considered in the Philippine work, possibly involving dual inoculation trials with both VA mycorrhizal fungi and *Pisolithus*.

In a temperate climate in Britain, Walker [49] has reported that inoculation of containerized sitka spruce has produced an 18% height increase in the year after outplanting, increasing to 24% by the end of the second year. This was not true, however, on all sites where trees were outplanted.

B. Vesicular-Arbuscular Mycorrhizae

Tropical Soils

Previous reviews and conference reports have highlighted the lack of published results investigating the role of VA mycorrhizae under field conditions, particularly in the tropics [42,51–54]. Marked progress has been made, however, between 1980 and 1986 in an intensive VAM research program investigating practical aspects of mycorrhizal technology in tropical crops and pastures in South America [55]. Based at CIAT (Centro Internacional Agricultura Tropicale), Colombia, the project involved the isolation and identification of indigenous VA mycorrhizal fungi from different edaphoclimatic conditions, the establishment of a large germplasm bank, greenhouse trials, field trials, on-farm tests, and training of local scientists with substantial collaboration with national scientists [56]. The success of this project has enabled large quantities (850 kg) of a high quality inoculum to be available for a subsequent large-scale field trial in the savanna ecosystem in the eastern plains of Colombia [57,131,132]. The inoculum was produced on a small farm holding in the Mondomo region of Cauca, Colombia, on a soil with similar chemical and physical properties to those of the savanna soils using a strategy designed by Sieverding [58]. This involved the establishment of many small plots (5 × 5 m) for inoculum production following soil fumigation. The plots were sown to *Brachiaria decumbens* and inoculated with various VAM fungi before being allowed to grow and mature. Subsequent inoculum was collected as the 2–20 cm soil horizon, which was mixed to improve the texture before application at a rate of 200 g/m. Spore population of the introduced VAM species remained high up to 2 years after sowing these inoculum plots (J. C. Dodd, personal observation). This simple low-cost technique can produce a high quality inoculum on-farm, given the availability of an original small quantity of a reliable starter culture such as 'MANIHOTINA' (see Section IV). In 1970, over 60% of farms in Colombia were of 5 ha or less, thus the potential for small-scale inoculum of VAM-dependent plants (cassava or forage legumes) is large given the reliance on manual labor that is usual within these family units. These acid-infertile, P-deficient soils are among nearly 1000 million ha in Latin America [56]. The aim of the trial was to determine if both effective introduced and native VA mycorrhizal fungi could be managed by precropping to the benefit of a subsequent mycorrhizally dependent crop grown under normal agronomic conditions [56,57,59,131,132].

Native savanna has been shown to have low numbers of infective propagules of VA mycorrhizal fungi compared with other soils in Colombia [55,60]. To assess the effect of different crops on the build-up of inoculum for the subsequent crop, five fodder and food crops were sown at two sites as follows: *Brachiaria dictyoneura*, *Pueraria phaseoloides* (tropical kudzu), and also a mixture of both, *Manihot esculenta* (cassava) or *Sorghum* sp. The soils of these sites differed principally in their percentage sand contents (30% and 43%, respectively), but their nutrient levels were very similar. Both soils contained less than 3 mg/L P and had a pH of around 4.8. Half the plots were inoculated with a soil mixture dominated by *Glomus manihotis* and *G. occultum*, the other half were not inoculated and contained only native endophytes. In both cases, the plots were further subdivided so that half received rock phosphate (RP) and half received triple superphosphate (SP). (In all crops, infection by VA mycorrhizae was more rapid in inoculated plots, especially in cassava, but by 14 weeks after sowing most of these differences had become insignificant as indigenous species developed within uninoculated plots.) In a survey of additional sites within the research station, there was an apparent increase in dependency on the VA mycorrhizal symbiosis by forage legumes as the sand component increased. Increased yields of inoculated *Sorghum* were observed in some of the experimental plots, but this varied between sites. Similar increases in *Sorghum* yields have been noted in the Brazilian cerrados [61].

Following the 4-month dry season (December–April), *Stylosanthes capitata* and *Vigna unguiculata* (cowpea) were sown over the same plots used in the first year. There was no significant effect on yields resulting from the previous years inoculation. All precropping treatments, however, caused a significant increase in subsequent VAM infection levels in both legumes compared with savanna controls. This was reflected in the increased yields of these crops [57,131] compared with those sown straight into native savanna. These results indicated that the benefits of inoculation occurred during the first 6 months of plant growth in these soils and were not carried over into a subsequent crop. The benefit of increasing natural VA mycorrhizal inoculum potential by precropping aided the subsequent early growth of mycorrhizally dependent legumes compared with native savanna controls. Crop rotation and management of mycorrhizal populations deserve more attention, especially in low input systems.

The counting and identification of spore populations of VA mycorrhizal fungi under precropping treatments continued throughout the trial, and counts revealed that *G. manihotis* did not persist as well as *G. occultum* in inoculated plots after the dry season. *Sorghum* specifically stimulated production of spores of the *G. occultum* component of the original inoculum, whereas cassava chiefly stimulated the *G. manihotis* isolate at both sites 6 months after sowing. Spores with the morphology in the *Entrophospora colombiana/Acaulospora mellea/Acaulospora morrowae* complex were specifically increased by tropical kudzu in the inoculated plots by the end of the trial. This build-up of VAM inoculum beneath certain crops could eventually be used to manipulate native or introduced populations in these soils for the subsequent benefit of a mycorrhizally dependent crop. This is especially relevant in that *G. manihotis* did not persist as well as *G. occultum* in these soils despite its obvious effectivity [55,62,63]. There have been other reports of hosts favoring the build-up of specific VA mycorrhizal fungi on other hosts [55,64,65]. The common belief that all VA mycorrhizal fungi are generalists and that the host plant are not selective of their symbiont is oversimplifying the ecophysiology of these fungi.

It was encouraging to note the number of field studies on various subjects outlined in the abstracts of the NACOM meeting in 1987, especially from the Far East. As Hayman [18] pointed out, most of the practical applications of VA mycorrhizae presented at the Dijon conference [66] were concerned with legumes or cassava, which are both highly mycotrophic crops. Previously, work had tended to be concentrated on soybeans, even in investigations from the tropics [58]. Other tropical forage legumes, however, such as *Stylosanthes* spp. and *Centrosema* spp., are also extremely dependent on the VAM symbiosis, even in soils which have not been presterilized [62]. The use of mycorrhizal inoculants for these crops has not been studied in as much detail in the field. There are, however, strong indications that mycorrhizal management would be of significance in improving yields. The application of benomyl to stands of *S. capitata* in a sandy soil in the savanna of eastern Colombia reduced the infection by VA mycorrhizae and caused dramatic decreases in foliar yields. In contrast, no such effect was observed on a soil with a lower sand but higher clay content due to poorer soil penetration of the fungicide (J. C. Dodd and J. L. Lenne, unpublished results). A similar application of benomyl on natural grasslands, however, indicated that infection by VA mycorrhizae is only beneficial to plants under particular environment conditions such as drought stress or at specific times in the life cycle [67]. Differences in response to inoculation depending on soil type were also shown for cowpea grown in Nigeria [68].

We have already discussed the synergism that is common when rhizobia and VA mycorrhizal fungi are inoculated together. Dual inoculation of tropical crop and forage legumes using techniques such as pelleting, which has already been successfully developed for temperate conditions [69], has great potential and must form a priority area for future investigations.

There have been several interesting studies attempted in the field to investigate natural grassland ecosystems and the roles played by VAM symbiosis. The effects of ungulates and cattle grazing on drought-stressed mycorrhizal or nonmycorrhizal grasses has been studied in a mixed grass prairie [70]. Although grazing caused severe stress to the host plants studied, no obvious beneficial effect of the VAM association could be detected when the usual growth parameters were measured. Subsequently, though, it has been found that increased photosynthetic activity occurs in the leaves of graminaceous plants under defoliation stress [71] or drought stress [72] in the field, and mycorrhizal plants must be at an advantage here. This type of study is particularly relevant to work currently occurring in the savanna ecosystem of Latin America on improving beef production by the introduction of new forage germplasm. Most of these grasses and legumes are highly mycorrhizal dependent, and increased interest in stabilizing these grass/legume associations has necessitated investigations into the ecophysiology of these plants. Such studies in nutrient recycling, grazing pressures, and drought stress have had to take into account the potential role of VA mycorrhizae in these acid-infertile soils following the impressive field results outlined previously.

Semi-Arid Soils

As stated earlier, recent reviews have detailed results of field experiments which mainly concentrated on growth increases [42,54]. Field results on more varied effects of VAM associations have been published more recently. Interesting applied work has been carried out in the semi-arid areas of southern Israel. Soil fumigation is commonplace in the intensive agricultural systems operating in this region, and initial field experiments used this procedure to evaluate the role of the indigenous VAM populations in soils with a variety of crops including peppers, onions, citrus, melon, celery, and avocados. A similar trial to that of Yost and Fox [12] was established on a loessial soil site incorporating four levels of broadcast superphosphate (0 – 3 tons/ha). Table 2 shows the stunting effect of fumigation on the yield of peppers, the reduction in growth being explained by the elimination of the VAM symbiosis. A similar result was observed for both celery and onion. In contrast, yields of melon were higher in fumigated plots, due to the eradication of root pathogens and the fact that this crop is not mycorrhizally dependent. All fumigated plants were nonmycorrhizal, whereas nonfumigated plants had >70% of their root length infected by VAM irrespective of P application. Leaf analyses showed that

TABLE 2. Yield of Bell Peppers (g/plant) in Fumigated and Nonfumigated Soil 3 Months after Transplanting

P added (tons/ha)	Fumigated	Nonfumigated
0	0	488
1	0	496
2	1	525
3	8	538

nonmycorrhizal plants were all deficient in P except melon, indicating that a mycorrhizal effect was operating, water-soluble P levels in this soil were shown to remain low even after a large addition of superphosphate [73]. Subsequent work with transplanted mycorrhizal and nonmycorrhizal peppers, and applying P as phosphoric acid at 10 and 40 ppm P, produced significantly larger fruit yields in mycorrhizal plants at the lower P level compared to the nonmycorrhizal controls [74]. The results of the investigations on the water relations of mycorrhizal and nonmycorrhizal plants in this trial were outlined in Section II.

Temperate Soils

Much more work has been done under temperate field conditions where plants are grown in less stressful conditions of high soil fertility. Nonetheless, positive effects of inoculation with mycorrhizal fungi have been reported. Buwalda and co-workers [35,74] showed that inoculation of winter and spring barley and wheat resulted in increased grain yields. The fact that substantial colonization of winter wheat root systems by indigenous VAM fungi can occur in early autumn in certain soils [75,77] could be of great benefit to a crop which has a peak of P inflow during this period [35]. The VA mycorrhizal symbiosis has also been shown to play an important role in P uptake during the reproductive phase of spring barley and during the growth of field peas in Danish soils [68,78]. Further work has also shown the role played by climatic factors, temperature and rainfall, as well as soil type and sowing date on this development of VAM infection during the cereal season and has indicated how this fact might be used to increase earlier infection or sustained infection [74,77].

The case for the use of VA mycorrhizal inoculants in higher input agrosystems remains less obvious even when improved sources of inoculum are used. A recent field trial using maize in West Germany has been described [79]. An isolate of *Glomus etunicatum* was used and inoculum was prepared in porous expanded clay as described by Dehne and Backhaus [80]; (see also Section IV). Application was by means of a combine drill adapted for experimental purposes. Two field sites were chosen, one of moderate P content (2.0 mg/g) and one of higher P content (4.7 mg/g). Soil pH values were relatively low (4.9 and 5.7, respectively). Inoculum was applied at a rate of 150 kg/ha. The growth season described (1983) corresponded with a period of long-lasting drought, but significant increases (10 – 20%) in dry weight yields of maize cobs were achieved in inoculated plots compared to noninoculated controls. Addition of 104 kg P/ha resulted in equivalent growth responses in inoculated plants at one of the sites, suggesting that inoculation would have little advantage if this amount of fertilizer was an economical alternative. On the second site, however, addition of phosphorus fertilizer resulted in growth yield decreases in the drought conditions. The relationship between soil moisture levels and phosphorus mobility in relation to VAM infection of cereals has previously been discussed. Much more extensive field experimentation is necessary to prove the usefulness of this approach in high-input agrosystems.

IV. SOURCES OF INOCULUM

There is a variety of sources of mycorrhizal inoculum. They vary from commercial suppliers to semi-commercial producers or research organizations that can supply limited amounts of fungus from experimental stocks. Marx and his colleagues [81] and Menge [14] have previously summarized some of the factors to be considered in producing inoculum of ectomycorrhizal fungi or VA mycorrhizal fungi, respectively. Inoculum can come in a range of forms: as soil, spore preparations, dried mycelium, infected roots, or more sophisticated products such as alginate beads with incorporated mycelial homogenates. Some ectomycorrhizal fungi also produce small sclerotia [82], but the usefulness of these as inocula has not been assessed in detail. The majority of commercial interest in mycorrhizal inoculants has developed over the past 20 years, mostly involving adaptation of existing techniques for larger-scale production. Relatively few patent applications have been made, most involving particular growth substrates [83 – 85], modifications of established methodology, or seed pelletization [91]. The novelty of some of these patents is questionable, and their strict application could jeopardize the future exploitation of well-established procedures already being used.

The examples discussed here mostly consist of schemes to produce mycorrhizal inoculum to sell to a customer for inoculation of his plants. This needs to be aimed at defined crop situations due to the specificity of the mycorrhizal effect. The criteria that have been recommended in the selection of the candidate isolates to choose for commercial exploitation are not discussed here, as previous reviews are available [e.g., 42,51,92].

Besides the direct marketing of inoculum, an aspect that has not been well developed is the use of mycorrhizae in industries that produce plants for resale. The plants would then be marketed as likely to have improved survival characteristics and to be superior (and possibly more "natural") than their nonmycorrhizal equivalents. Various factors would need to be considered when determining whether the production of such plants was economically viable, including the cost of initial research and development, the cost of inoculation, the size of market, the postsale performance, and, of course, the increased revenue that could be recovered from such products. A good marketing strategy would be needed but the potential for this area is large. Leaving this opportunity aside, we will now discuss some of the sources of mycorrhizal inoculants currently available or under development.

A. Ectomycorrhizae

Initially mycorrhizal fungi were introduced into virgin soils by incorporation of raw soil brought from established plantations of the particular crop in question. One of the earliest examples concerns the first trial introductions of pines in Kenya in 1902 [93]. These first seedlings, as well as those planted in subsequent attempts, failed completely until in 1910, on the advice of Kew Gardens, mycorrhizal soil inoculum was brought from a South African pine plantation and its incorporation resulted in good growth of *Pinus radiata*. A better-known example is the introduction of pines into Puerto Rico, where numerous unsuccessful attempts were made over a period of nearly 30 years until mycorrhizal soil was brought over in 1955 [44]. With hindsight this solution is now obvious to those with mycorrhizal experience but is useful as a reminder of the knowledge gap that existed and is still in evidence elsewhere.

Introduction of small quantities of soil from foreign sources can offer the cheapest form of introducing nonindigenous species to some ecosystems and is widely practiced at present for ectomycorrhizal species [44]. Raw soil inoculum has its drawbacks in being bulky and expensive to transport, and the collection and inoculation is labor intensive. In addition, the biological component is unpredictable, and there is always a risk of introducing unwanted pests or pathogens into a new environment. The mycorrhizal complement is also variable, and although the general rule is that any mycorrhizae are better than none, there is clear evidence now that some are better than others and an ill-defined inoculum can consequently be suboptimal. There is thus little scope for the commercialization of raw soil inoculum as a consistent source of mycorrhizal fungi.

It was stated above that addition of soil amendments to aid the establishment of nonindigenous trees should have been obvious to agriculturalists with a knowledge of mycorrhizae. There are still problems, though, in getting the mycorrhizal message fully integrated into agricultural practices. In Australia, for example, an ectomycorrhizal inoculation program has been implemented for pines, but it was widely assumed that eucalypt growth was adequate without added inoculants [94]. Recent results, however, from an outplanting of ectomycorrhizal seedlings have shown that dramatic increases in tree growth could be achieved following inoculation.

A purer and more consistent form of inoculum is the introduction of macerated mycelium of laboratory-grown cultures. Again there is extensive documentation of this technique for ectomycorrhizal fungi [44,81]. This approach, however, is time-consuming and requires considerable space. A certain level of expertise is also required and cultures must be kept under controlled conditions without becoming contaminated by other microorganisms. The final preparation of the inoculum from flask-grown batch cultures is also inconvenient and labor intensive. If the expertise and facilities are available for this approach, it is now more feasible to opt for batch fermentation of larger quantities of submerged mycelium under optimized conditions. This can be followed by a systematic and partially mechanized method of product extraction and packaging into an appropriate delivery medium. Providing that costs can be kept low this could be done in a large commercial operation as an alternative means of using established fermentation facilities.

The involvement of commercial interests in mycorrhizal technology is not without its problems, however, and the recent breakdown of the relationship between the agricultural and commercial priorities in the production of Mycorrhiz discussed below is an example of overzealousness by industry. There is still much to be learned about the consistent production of fungal inoculum of any sort, and a block on open discussion is counterproductive to progress regardless of market advantages. The most ambitious attempt at production of commercial mycorrhizal inoculum arose from the involvement of the Abbott Laboratories with Marx of the USDA Forest Service. Marx had been instrumental in showing the potential benefits to be derived in commercial forestry from inoculation of pines with the ectomycorrhizal fungus *Pisolithus tinctorius* (Pt). This work [81] was carried out under the auspices of the USDA by research procedures devised at the Institute for Mycorrhizal Research and Development (IMRD). Pt was chosen on the basis of its availability, ease of manipulation, wide geographic and host range, and demonstrated host effects. Abbott Laboratories became involved with the intention of producing larger amounts of saleable inoculum for use in commercial reforestation schemes. Development of the appropriate growth conditions and fermenter arrangements and the final presentation of the product was largely the responsibility of the company with advice from Marx, while nursery and field trials were done by Marx. Trial batches of inoculum were successfully produced, although changes in production conditions between different batches led to inconsistencies in product which affected the validity of the nursery results. Results of tests in 1978 and 1980 showed, however, that viable vegetative inoculum of Pt could be produced by industrial fermentation procedures in a substrate of peat-vermiculite containing nutrients. Following leaching and drying, this could then be used to form abundant Pt mycorrhizae on container-grown and bare root seedlings for practical use in forest regeneration. Results obtained were equal to those achieved using the inoculum produced at IMRD.

Extensive nursery trials across different geographical regions of the U.S. were carried out and the beneficial responses that were frequently recorded in the eastern and southern states were not always achieved in other areas of the United States. The pressures involved in getting a product quickly onto the market and near-marketing uncertainties led to a withdrawal of the collaboration and the termination of Abbott's involvement in this field. Fortunately, interest in this product was maintained through Sylvan Spawn of the Butler County Mushroom Farms, a company involved in mushroom spawn production. Batches of inoculum of *P. tinctorius* were grown in a peat-vermiculite substrate, and field results were achieved which matched those of the earlier product. The starter cultures were grown in batches of nutrient broth and then injected into plastic bags containing the solid growth medium. Care was taken to ensure that the nutrients available within the bag were not in excess but would allow the fungus to completely colonize the substrate before it became growth-limited. This eliminated the need to leach and dry and ensured that the final product would not encourage growth of other microorganisms when placed in the field. In 1987 about 3.5 million trees were inoculated operationally using Sylvans Pt inoculum. A breakaway biotech company, Mycorr Tech Incorporated (MTI), was formed by one of the researchers from Sylvan to concentrate on the Pt product. MTI assumed production this year and the company has obtained a license to use the patents and procedures for the production of Pt inoculum (Maul, personal communication). MTI produced vegetative mycorrhizal inoculum of Marx's isolate of Pt in March 1988. It is also possible for them to produce, on request, inoculum of *Laccaria laccata, Hebeloma crustuliniforme, Suillus luteus*, and *Scleroderma citrinum*. It is estimated that inoculum of Pt will only add about \$7.50 to a retail cost of around \$25–50 per thousand pine seedlings (1-0 stock). Roughly 0.7ℓ of inoculum is required per 1000 seedlings and will be added to the seed bed by means of a machine which will inoculate below eight seed drills simultaneously prior to the addition of the seeds. Marx has previously shown that it is difficult to establish Pt on pine seedlings already containing native mycorrhizae due to competition for feeder roots, and thus a fumigated seed bed is the only practical site for inoculation. Successful performance in the field demands that good initial establishment of Pt is achieved at this stage. It has been shown that survival and growth of loblolly pine 8 years after outplanting is positively correlated with initial amounts of Pt ectomycorrhizae on the seedling roots at outplanting [95]. The reason behind some of the inconsistencies in performance of Pt inoculum is only now becoming apparent after 9 years of field trials [96]. Much of the success of this inoculant fungus is apparently due to its ability to protect plants from drought stress. Furthermore, Pt is able to grow well and form mycorrhizae at relatively high temperatures compared to many of the other fungi suggested as candidates for inoculants. Field success is thus most apparent in warm climates when dry seasons are experienced. In cooler climates or in wetter seasons, indigenous fungi, especially *Thelephora terrestris*, are able to successfully colonize outplanted pines and dilute the beneficial effect that Pt can confer. In the southeastern U.S. there is a warm and dry season 2 or 3 years out of 5 [95], and it is in these cases that the benefits of inoculation are realized. In other years or in other climes, the benefits are not apparent and inoculation not worthwhile. An additional problem arose from the use of a new systemic fungicide (Bayleton) to control the Southern pine rust *Cronartium fusiforme*. This fungicide has a detrimental effect on Pt and native mycorrhizae and can reduce the efficacy of mycorrhizal inoculation. Careful documentation and presentation of these facts is necessary before a successful launch of the final product can be achieved. This stage has now been reached for Pt, and Marx has carried out field studies which include over 100 bare-root nursery tests in 38 states in the U.S., 18 container nursery tests in 9 states [81], and over 100 discrete field outplantings in 13 states. These studies have clearly demonstrated that commercial Pt inoculum can be effectively utilized in large-scale nursery operations in the eastern U.S. Although a variety of limiting factors have been repeatedly encountered, these can now be explained. The field application program has advanced considerably in recent years and is expected to accelerate in the future [97].

In addition to the example discussed above, there are other suppliers of vegetative ectomycorrhizal inoculum. Some sources are still at the developmental stage, others are able to supply inoculum now. Current markets are still small, however, and commercial acceptance awaits definitive field trials and successes. In the U.K. the only company with an ectomycorrhizal product is Biotal Limited (formerly Biotechnica Ltd.) of Cardiff. They supply a mixed inoculum of four mycorrhizal fungi, *Thelephora terrestris*, *Laccaria laccata*, *Paxillus involutus*, and *Hebeloma crustuliniforme*. The inoculum is formulated in alginate beads approximately 1 mm in diameter, each containing a total of 10 colony-forming units of the four fungi. The inoculum has a shelf life of 2 months if stored at 4°C. One ml of product contains approximately 350 beads, and the recommended application rate for each planting pit varies from 2.5 to 10 ml per tree, depending on tree size. Alternatively, beads can be added to compost prior to seeding containerized trees at a rate of 60 beads per 15 ml pot. At the current price of £60 per liter, this means about 4000 trees could be treated at around 1.5 p per tree. The recommended reduction of about 50% in fertilizer application should compensate for this additional expense. Biotal is currently involved in field trials with a number of commercial nurseries throughout the U.K. So far the results obtained from winter 1986 to spring 1987 have not shown marked improvements over nontreated plants, but this again might reflect the occurrence of good growing conditions during that particular trial season. It must be expected that growth responses will vary in intensity from season to season and definitive results will only be obtained over several successive years. A product that is being developed by the Agricultural Genetics Company, Cambridge, is apparently not unusual but is an alginate-encapsulated fermenter-grown vegetative inoculum of ectomycorrhizal fungi, in particular *Laccaria* sp. (P. Rogers, personal communication). Field trials have been initiated on Sitka spruce (*Picea sitchensis*), and again the emphasis must be to complete several seasons of the trials under a variety of conditions before a product enters the market. Precocious release of an unreliable product, as was likely to happen with Mycorrhiz, will do irrepairable harm to the market. Mycorrhizal inoculum development is a long-term undertaking with technical and stability problems that are best discussed in a wide and open forum. Realistically a company should not expect to market a saleable product until at least 4 years from completion of product development.

Elsewhere in Europe, the most progress has been made in France through the work of the Institut National de la Recherche Agronomique (INRA). At present there is no commercial ectomycorrhizal product available, but two types of preparation are under development. A conventional peat–vermiculite inoculum of *Laccaria laccata*, *Hebeloma crustuliniforme*, and *Paxillus involutus*, grown in plastic bags, has been developed by the firm Somycel. Although the product is now considered technically good, there is still a problem of developing the marketing strategy (J. L. Garbaye, personal communication). Second, INRA has collaborated with Rhône-Poulenc in the production of inoculum of *Hebeloma* species and has followed the now conventional pathway of entrapping the subsequent mycelium in alginate beads for delivery to the plant. This work was reported [98,99], and development continued in parallel with field trials organized through INRA's national network. Inoculum of *Hebeloma crustuliniforme* was grown in liquid culture for 25 days in shake flasks. It was then homogenized and immobilized in a calcium alginate gel. Mycelial viability was not affected by curing the beads or by incorporating peat or bentonite within the gel beads. At 4°C the entrapped mycelium retained its viability for over 4 months in humid conditions, but drying the beads reduced the survival time.

In Australia, development of an alginate-based inoculum is currently underway in a joint project between the University of Western Australia and CSIRO. Batch fermented starter cultures are homogenized and incorporated into hydrogel beads before being introduced into an air-lift fermenter for final production by Interbac. Aust. Pty. Ltd. The mycelium grows within the alginate bead but does not apparently grow out into the suspending nutrient medium. Discrete 2 mm diameter spheres ("Mycobeads") of encapsulated, growing mycelium are thus produced and are L-dried and stored before use. The shelf life is over 6 months at 4°C. Growth of several fungi has been initiated using this method, including *Laccaria laccata*, *Hebeloma* spp., and *Descolea maculata*. Inoculation trials are being carried out on containerized *Eucalyptus* and *Pinus*. When seeds of *Eucalyptus* were sown with the beads (10 beads per pot) 94–100% of the resultant seedlings developed mycorrhiza.

A very different strategy of product formulation has already been referred to for inoculation of tree seedlings in the Philippines. *Pisolithus tinctorius* is again the fungus used, sometimes combined with a species of *Scleroderma*, and is ideally suited to the Philippine climate. Pt is indigenous and fruit bodies are produced abundantly in established forests each season. Collection of the fruit bodies is organized on a local basis, and spores are extracted and mixed with sterilized earth before being pelletized in a conventional tablet-making machine. The pellets are easy to apply, and an application rate of one pellet placed directly below the seedling is recommended during transplanting at the nursery stage. Pellet production has been organized through the Philippine Agricultural Biotechnology Company (PABCO) in collaboration with the BIOTECH Centre of the University of the Philippines at Los Banos. They are now marketed under the trade name MYCOGROE. Some additional suppliers also produce spore-based inoculum of Pt. The International Tree Seed Company, and South Pine Inc., both from Alabama, produce spore pellets, spore-encapsulated seeds, or bulk spores. In general, though, none of these formulations have been as effective in the U.S. as the vegetative inoculum used by Marx for the establishment of large numbers of Pt ectomycorrhizae on pines [97].

The final example of the successful use of ectomycorrhizae in agriculture concerns the unusual situation where the fungal fruit body is the crop in question. Considerable progress has been made in the marketing of trees inoculated with mycelium of the truffle fungi. In western Europe truffles are one of the most expensive delicacies available (prices in 1987 averaging around £1000 per kg). *Tuber melanosporum*, the Perigord truffle, occurs naturally in a few places in Europe, but fruiting is erratic and the collection of the hypogeous sporocarps is difficult. Synthesis of tree seedlings and truffle mycelium was established in France using oak and hazel seedlings inoculated with spores, minced fruit bodies, or fragments of mycelial roots. More than 300,000 "truffle trees" were produced and sold in the 1970s [100] and in some cases have gone on to provide truffle harvests within 4 years of outplanting [42]. Attempts are also being made to establish and market trees with ectomycorrhizae of other fungi with desirable fruit bodies such as *Boletus edulis* and *Lactarius deliciosus* [101].

B. Vesicular-Arbuscular Mycorrhizae

Progress in the development of inoculum of VA mycorrhizal fungal inoculants has been severely hampered by inability to grow any of the mycobionts in axenic culture. All cultural systems used so far involve the use of dual culture of suitable host plant roots and the appropriate fungus. This should not, however, be used as an excuse for not applying current knowledge of inoculation technology in areas where a known benefit to crop production could be attained by managing the VAM symbiosis. We should not be primarily concerned with whether or not the application will be commercially profitable or more difficult in the absence of pure culture techniques. These concerns are not of relevance to a subsistence farmer in the developing world who would benefit now from the application of mycorrhizae to help solve problems of soil erosion, aid plant establishment (e.g., pasture legumes), and improve plant water relations [25]. The major problem will be the transfer of the technology onto the farm, and a considerable amount of training in the developing countries should be a priority now, initially to introduce the idea of the potential that mycorrhizae have and then to follow up with practical application. There is still an alarming ignorance of these symbioses among scientists working with P-deficient soils, and this needs to be corrected. The recent Asian Conference on Mycorrhizas (ACOM) in India in 1988 was useful in initiating exchanges of information in the East, but far more needs to be done in Africa and Latin America. There remains a problem of divorcing the need for applied and practical use of mycorrhizae using current methodology from the commercial need to develop improved techniques of inoculum production.

As with the ectomycorrhizal situation, the pioneering work on the production of VAM inoculum was carried out in the U.S. by noncommercial interests and has now been taken over by private industry. The companies involved have varied much in their approach, and marketable products are now available from small companies that have openly adapted the existing expertise and consulted widely to come up with a suitable formulation. Other examples exist where secrecy agreements have been signed, and information about the progress of interesting work has now become difficult to obtain or evaluate.

Much of the impetus for the use of VA mycorrhizal inoculants in agriculture has come from the work of Menge at the University of California in relation to citriculture. Citrus trees are obligately dependent on good VAM development for growth, and a solution was needed to alleviate symptoms of nutrient deficiency when citrus trees were reintroduced into sterilized soils. Soil sterilization prior to planting of new trees was compulsory by law, so the problem was widespread. Addition of VAM inoculum overcame the problem and Menge made supplies of VAM inoculum available to local growers from University research stocks. The operation has now been taken over by Brokaw Nursery, Inc., which has been supplying inoculum of *Glomus deserticola* since 1980. Inoculum is produced using conventional dual pot cultures [14] and consists of the growth medium containing the VAM spores and infected roots of the host plants which have been grown in sterilized, fungicide-treated containers.

This approach has also been taken up by Native Plants Inc. (NPI), located in Utah. They market VAM inoculum for use in the horticultural industry under the trade name NUTRI-LINK. Again, the product is essentially the contents of containers in which dual cultures of host plants and VAM have been grown. In this case the fungus used is *Glomus intraradices*, and the sand-like medium supplied contains both VAM spores (approximately 1000 viable spores/ml) and infected root fragments. Recommended application rates vary depending on the system of plant propagation employed, but 1 liter of Nutri-Link is recommended for mixing into 2.5 ft^3 of soil or for spreading over 20 ft^2 of nursery bed. Treatment costs per plant will thus vary but examples range from 2¢ per rooted cutting in a nursery flat up to 6¢ for some tree seedlings individually grown in tubes. A reduction of ammonium and phosphate fertilizers is recommended, and guidelines are provided for the use of compatible pesticides. NPI also sells single species material containing other VAM strains for research purposes; these include *G. deserticola*, *G. etunicatum*, *G. mosseae*, and *Gigaspora margarita*. A strain selection program is underway with the intention of isolating additional strains which perform well in acid soils, neutral/basic soils, high P soils, and/or peat-based soilless media. It is hoped to have some of these available by late 1988 (S. Larson, personal communication).

Although not directly aimed at a commercial market, an important development has been the setting up of an International Culture Collection of VA Mycorrhizal Fungi (INVAM) at the University of Florida. INVAM has been set up on an international basis and has the following objectives: (1) to preserve germplasm of VA mycorrhizal fungi, (2) to develop improved procedures for multiplying and preserving VA mycorrhizal fungi, (3) to serve as a center for personnel training on the culture and identification of VA mycorrhizal fungi, and (4) to provide mycorrhizal scientists with a reliable source of starter cultures of these fungi for research purposes. INVAM began operation in late 1985 and currently holds stocks of over 600 isolates representing 60 of the 120 described species of VA fungi (for further details see Ref. 102). Sieverding [56] discussed the need for a germplasm bank of VA mycorrhizal fungi as a basic requirement of any long-term research projects using these fungi. It is also important because its establishment would necessitate the classification of VA mycorrhizal fungi isolated from populations sampled prior to in-depth ecophysiological studies. Sieverding had already established a culture collection of over 1200 ecotypes collected mainly in Colombia during 1980–1986. This collection remains at CIAT stored as soil from pot cultures of the different isolates and probably represents the most extensive collection of tropical VA mycorrhizal fungal germplasm in the world. A slide collection of the most common 25 species within the collection is deposited at CIAT headquarters. Three hundred of the isolates have now been transferred to the collection at INVAM.

In Europe the development of inoculum sources of VA mycorrhizal fungi has taken a different route. Most emphasis has been placed on deriving new methods of culturing bulk root material infected with appropriate fungi. Initial studies were made at Rothamsted Experimental Station by a number of scientists. Standard pot culture methods were first used and modified, but more recently nutrient film techniques (NFT) have been developed and patented [89]. The construction and development of this system using *Phaseolus vulgaris* or *Zea mays* as the main plant hosts has also been described [103,104]. Inoculum of *Glomus mosseae*, *G. fasciculatum*, *G. caledonium*, and *Acaulospora laevis* was successfully produced by this approach. Bean roots grown by this technique and infected with both *Glomus fasciculatum* and *Rhizobium phaseoli* were successfully used as inoculum for red clover growing in nonsterile field soil [105]. The endophyte was introduced into soil at a rate of 6 g of fresh roots/m and became well established in the presence of indigenous mycorrhizal fungi. The effects of nutrient levels and nitrogen sources on improving the production of inoculum from NFT growth of cereals has been discussed more recently [106].

An interesting development of hydroponic systems of VA mycorrhizal inoculum production has been the incorporation of expanded clay substrates as the inoculum carrier [79,80,83,107]. Various host plants have been grown in a light expanded clay with a particle size that was either medium or large (>4 mm). Eight species of VA mycorrhizal fungi have been shown to grow well in this system and sporulate profusely within the pores of the clay substrate, which acts as an inorganic, solid carrier. Once the stock cultures have been allowed to grow up, the substrate is collected, washed, and air-dried prior to storage. Inoculum can be stored for up to 12 months without loss of infectivity. Intensive VA mycorrhizal infection is achieved in host plants by using substrate mixtures containing 0.1–5% of the inoculum [107]. The development of both NFT and expanded clay–hydroponic inoculum has now been taken over by the Agricultural Genetics Company, and information about progress is more difficult to obtain. Another recent development has been the use of aeroponic systems [108,109] for the growth of VA mycorrhizal roots. Abundant colonization and fungal sporulation has been achieved using this system, which may offer an improvement over previous hydroponic systems.

Finally, mention must be made of the potential use of a technique described for obtaining infected roots for physiological studies [110]. VA mycorrhizal infection has been achieved in root organ cultures of *Convolvulus sepium* transformed by hairy-root–inducing T-DNA from *Agrobacterium rhizogenes*. Transformed roots have been grown in bulk using a 2L fermenter, and if mycorrhizal infection could be sustained under these growth conditions, they might provide an ideal material for inoculum production.

In contrast to these more sophisticated sources of mycorrhizal inoculum, there also exist more basic products. The commercial potential of VAM has been realized in Colombia, for example, where a local company is producing pure pot culture inoculum of *Glomus manihotis* using the trade name MANIHOTINA. The same organization began earlier by selling "mycorrhizal soil" introduced onto local markets and sold for garden use. This product was raw tropical rain forest soil from the Choco area of western Colombia and contained a range of VAM species of low inoculum potential. The product has now been replaced by Manihotina (E. Sieverding, personal communication).

A novel and interesting use of VAM in a biotechnological industry was developed by Twyford Plant Laboratories (TPL) in the U.K. This company specializes in the production of cloned and tissue-cultured plantlets under strictly controlled and clean conditions. TPL has been producing cloned plantlets of date palm for several years and have now devised appropriate inoculation protocols to establish VAM on their products. Four isolates of VA mycorrhizal fungi—*Glomus mosseae*, *G. monosporum*, *G. intraradices*, and *G. aggregatum*—were obtained from soils of neutral pH in the Middle East and California which had similar characteristics to those soils into which the date palm would eventually be outplanted. In glasshouse experiments, each of the isolates significantly increased growth of the date palm plantlets after 8 months in a sandy substrate when compared with noninoculated control plants. This work will not be continued, however, as the recent organizational changes at TPL have meant an end to their mycorrhizal program.

C. Ericaceous Mycorrhizae

Although VA mycorrhizae and ectomycorrhizae make up most of the commercial interest in inoculants, some progress has also been made in the introduction of ericaceous mycorrhizae in New Zealand for inoculation of established plants or rooted cuttings of blueberry (*Vaccinium corymbosum*) to increase fruit yields. The inoculum, produced by Minitech (N.Z.) Limited and marketed under the name MYCOAID, comprises of a homogenized and dried mycelial preparation of *Pezizella ericae*. It is mixed with water and then poured over the roots of cuttings or established plants. The product has a recommended shelf-life, at 4°C, of 3 months from purchase. Initial field trials were carried out in mycorrhizal-deficient soils in 1979–1981, and significant increases (25–40%) in fruit yield were obtained. Widespread adoption of the product has not occured, however, as it appears that most of the blueberry crop in New Zealand already acquires sufficient infection by the symbiont from the nursery bed or from indigenous inoculum in the soil.

V. FUTURE TARGETS FOR APPLIED RESEARCH

The successful development and use of mycorrhizal inoculants described in the previous sections have highlighted areas where problems exist or where new technologies could be used to improve the quality of inoculum now available. More work is always needed to clarify the relationships between mycorrhizal isolates and to help investigators distinguish one from another. This is necessary for the study of the spread of inoculants into agricultural soils and natural ecosystems, as well as for the investigation of competition between different inoculant fungi or between the inoculant and the indigenous mycorrhizal flora. Much of this type of work done in the past has been hindered either by a reluctance to consider the identity of the fungi involved or a tendency to misidentify them [111]. Besides conventional taxonomic studies, which should now be facilitated through the setting up of organizations such as INVAM, there is a need for a biochemical or genetical approach.

One of the major problems facing mycorrhizal investigators is the in situ identification of the fungi involved. The problem is difficult enough in the case of ectomycorrhizae in which an experienced worker can tentatively name the symbiont from clean mycorrhizal roots on the basis of their morphological characteristics. An atlas of color photographs is now available to facilitate this procedure [112]. In VA mycorrhizal symbioses, however, it is impossible to confidently distinguish between infections of related species unless fungi with very particular characteristics are selected [113].

Progress is being made to devise biochemical or immunological probes to overcome this difficulty. Sen and Hepper [114,115] have analyzed enzyme migration patterns following polyacrylamide gel electrophoresis (PAGE) of infected roots and can identify the presence of particular species by their gel profiles. This procedure is more advanced in the case of ectomycorrhizal isolates, and Zhu and his colleagues [116], for example, have used PAGE of isozymes to study genetic varibility in 43 isolates of *Suillus tomentosus*. In the development of immunological approaches, polyclonal antibodies have been raised against various species and ELISA techniques used to identify the presence of specific fungi in infected roots [117,118]. Problems of cross-reactivity of the sera exist, though, and it may be that the more specific approach using monoclonal antibodies will be necessary for definitive results. Several laboratories, including our own, are taking this route, and some antibodies are now available. Two monoclonals against *Glomus occultum* have been produced which do not cross-react with 29 other mycorrhizal fungi tested [119]. Specificity is such that the antibodies may also differentiate strains within the same species. Dr. Millner at USDA, Beltsville, has also been successful in raising a monoclonal antibody that distinguishes a *Glomus mosseae* isolate from other species and is now testing cross-reactivity within the species complex (P. D. Millner, personal communication). These techniques will have value in the selection of the most appropriate fungi as inoculants. In those cases where successful studies of competition between different fungi on the same root systems has been accomplished, it is clear that the major mycorrhizal fungus in terms of root occupancy does not necessarily have the major effect, stimulatory or otherwise, on plant growth [113]. Thus an inoculant fungus established at the nursery stage, or placed beneath the seed in direct-drilled crops, can have determining effects on plant growth which are not necessarily related to its subsequent spread or its ability to become the main mycorrhizal symbiont as the plant matures.

The potential for genetic manipulation of mycorrhizal fungi remains unexplored. Our inability to grow VA mycorrhizal fungi in axenic culture has retarded investigations of this nature. Some progress has been made recently for ectomycorrhizal fungi, however, with the development of protoplast fusion techniques for the isolation of mutants [120] or for the characterization of plasmids of *Morchella* species as a prerequisite for vector development [121]. There is considerable scope in the future for the development of "superstrains" as inoculants in the field providing the current difficulties in releasing genetically altered organisms into the environment are overcome.

Finally, we must not forget the influence of the host genotype on the development of mycorrhizae within the root system. Different cultivars of pearl millet have been shown to become colonized to differing degrees when faced with a similar mycorrhizal challenge [122]. Genetic differences between cultivars of leafless peas have also been shown [123] to markedly affect the development of the mycorrhizal symbiosis. There is a consequent need to consider host-genotype dependent differences in the mycorrhizal symbiosis in current programs of plant breeding.

VI. CONCLUSIONS

There remains a great potential for increasing the efficiency of growth of agricultural crops by mycorrhizal inoculation and management, particularly in semi-arid and tropical zones. Although the role of mycorrhizae is principally the same in these zones as in temperate climates [124], these regions present different problems and management strategies may need alternative approaches to those adopted in developed countries. This concept is not new, of course, but there is a tendency for pessimism or frustration in certain quarters because increased crop yields have not been rapidly or universally achieved or because inoculation with mycorrhizal fungi does not have direct application in well-developed areas of agriculture. The usefulness of mycorrhizae is especially appropriate to the development of sustainable systems of agriculture [125]. We must stress again the continued need for extensive field trials over a series of seasons before definitive statements can be made about the benefits that mycorrhizal inoculation can bring. These trials should be targeted into those regions and crops where improvements in the efficiency of agriculture might be expected irrespective of the likely monetary gains. This long-term approach to devising successful strategies of mycorrhizal management means that continued funding can be problematical. The desire for short-term profits must not prevent investment into this area of research. Mycorrhizal inoculants will never be financially lucrative, but their tremendous value lies in their ability to improve the efficiency and sustainability of crop yields in areas of low soil fertility. The introduction of effective fungi into such soils must remain a priority, especially in relation to their ability to improve the efficiency of plants to use limited amounts of phosphorus fertilizers [126].

In some cases it may be more realistic to stress the improved survival capability of the mycorrhizal plants as this aspect is often more important for the third world subsistence farmer. Alternatively, the ability of some mycorrhizal fungi to help stabilize soil and reduce erosion might be just as important as an increased growth yield. This property can be rapidly and easily assessed using simple methodology in situ [127]. Soil erosion can itself adversely affect an established VA mycorrhizal symbiosis [88], and this factor must also be considered in terms of soil conservation. One way to alleviate the problem of funding mycorrhizal research is to ensure that larger-scale projects that are concerned with improving soil fertility by a variety of means also include a mycorrhizal component. This is more easily accomplished if a wider dissemination of mycorrhizal knowledge is imparted through traditional means of education. There are still articles describing the development of low-input cropping systems in which mycorrhizae are not mentioned, even though it is acknowledged that phosphorus is the most critical plant nutrient involved [128]. Technology transfer is called for in terms of ongoing collaborative projects involving reciprocal visits between partners. Far more will eventually be achieved in this way than by the short crash-course given by a "foreign expert," which can then founder after the return of the mycorrhizast to the home country. This can lead to local resentment and a lack of commitment between the partners involved.

While we have already addressed the need for more field trials on operational farms in developing countries, we would also like to highlight some other areas where more work is needed. Manipulation of indigenous or inoculated mycorrhizal populations is now clearly possible, as the work discussed in this review has shown in Colombia [55–57,131,132], and there should be a move towards integrated ways of mycorrhizal management through precropping with appropriate crop species or by manipulating the ectomycorrhizal succession that occurs on forest trees [129]. The role that mycorrhizal management could play in conservation policies is another area that has received scant attention in the past. Much of the current emphasis in global terms is to slow the destruction of natural forests in tropical zones. This is usually a result of slash-and-burn procedures necessitated by the traditional shifting agricultural practices of the local inhabitants. The indigenous mycorrhizal population beneath the virgin forests will have become highly adapted to that particular ecosystem and may not be the most suitable for the newly planted crops. Thus the plants grown on the reclaimed soils may not be mycorrhizal and frequently exhaust the available nutrients within a very short period. The farmer must then move on to a new area of forest. The soils in question are typically acid-infertile, and it would clearly be possible to design a management policy that included the introduction of appropriate mycorrhizal inoculants along with slow-release phosphate fertilizers which would ensure the maintenance of a more sustainable cropping regime. This would negate the need to constantly shift to fresh sites of cultivation and hence conserve existing resources.

In other disturbed ecosystems, the use of mycorrhizae in land rehabilitation has already been reviewed [e.g., 130]. We will not discuss this further other than reemphasizing the crucial role that mycorrhizae can play in the establishment of vegetation onto these soils. There is already considerable field work being carried out in this respect, and this will continue to be one of the expanding areas of mycorrhizal technology.

ACKNOWLEDGMENTS

We thank a large number of people for providing information about available inoculants, and especially thank Don Marx and Irene Koomen for amending parts of the manuscript.

REFERENCES

1. Plenchette, C., Fortin, J. A., and Furlan, V. Growth responses of several plant species to mycorrhizae in a soil of moderate P-fertility. I. Mycorrhizal dependency under field conditions. *Pl. Soil*, *70*:199–209, 1983.
2. Baylis, G. T. S. The magnolioid mycorrhiza and mycotrophy in root systems derived from it. In *Endomycorrhizas* (F. E. Saunders, B. Mosse, and P. B. Tinker, eds.), Academic Press, London, 1975.
3. Tinker, P. B. Role of rhizosphere microorganisms in phosphorus uptake by plants. In *The Role of Phosphorus in Agriculture* (F. E. Khanawneh, E. C. Sample, and E. J. Kemprath, eds.), American Soc. Agronomy, Madison, 1980.
4. Pacovsky, R. S. Micronutrient uptake and distribution in mycorrhizal or phosphorus-fertilized soybeans. *Pl. Soil*, *95*:379–388, 1986.
5. Dueck, T. A., Visser, P., Ernst, W. H. O., and Schat, H. Vesicular-arbuscular mycorrhizae decrease zinc toxicity to grasses growing in zinc-polluted soil. *Soil Biol. Biochem.*, *18*:331–333, 1986.
6. Dixon, R. K. and Buschena, C. A. Response of ectomycorrhizal *Pinus banksiana* and *Picea glauca* to heavy metals in soil. *Pl. Soil*, *105*:265–271, 1983.
7. Moore, P. D. Essential elements from waste. *Nature*, *333*:706, 1988.
8. Wright, D. A., Arnold, P. T., Wilms, C. E., and Kapustka, L. A. Metal pH toxicity of municipal sludge on VAM inoculum. *Proc. 7th NACOM*, 1987, p. 76.
9. Gildon, A. and Tinker, P. B. A heavy metal tolerant strain of a mycorrhizal fungus. *Trans. Br. Mycol. Soc.*, *77*:648–649, 1981.
10. Bowen, G. D., Skinner, M. F., and Bevege, D. I. Zinc uptake by mycorrhizal and uninfected roots of *Pinus radiata* and *Araucaria cunninghamii*. *Soil Biol. Biochem.*, *6*:141–144, 1974.
11. Gerdemann, J. W. Vesicular-arbuscular mycorrhizae. In *The Development and Function of Roots* (G. D. Torrey and D. T. Clarkson, eds.), Academic Press, London, 1975.
12. Yost, R. S. and Fox, R. L. Contribution of mycorrhizae to nutrition of crops growing on an oxisol. *Agron. J.*, *71*:903–908, 1979.
13. Menge, J. A. Utilization of vesicular-arbuscular mycorrhizal fungi in agriculture. *Can. J. Bot.*, *61*:1015–1024, 1983.
14. Menge, J. A., Labanauskas, C. K., Johnson, E. L. V., and Platt, R. G. Partial substitution of mycorrhizal fungi for phosphorus fertilization in the greenhouse culture of citrus. *Soil Sci. Soc. Am. Proc.*, *42*:926, 1978.

15. Howeler, R. H., Cadavid, L. F., and Burkhardt, E. Response of cassava to VA mycorrhizal inoculation and phosphorus application in greenhouse and field experiments. *Pl. Soil*, *69*:327–339, 1982.
16. Kucey, R. M. N. Increased phosphorus uptake by wheat and field beans inoculated with a phosphorus-solubilizing *Penicillium bilaji* strain and with vesicular-arbuscular mycorrhizal fungi. *Appl. Environ. Microbiol.*, *53*:2699–2703, 1987.
17. Barea, V. M. and Azcon-Aguilar, C. Mycorrhizas and their significance in nodulating nitrogen-fixing plants. *Adv. Agron.*, *36*:1–54, 1984.
18. Hayman, D. S. Mycorrhizae of nitrogen-fixing legumes. *MIRCEN Journal*, *2*:121–145, 1986.
19. Ames, R. N. and Bethlenfalvay, G. J. Localized increase in nodule activity but no competitive interaction of cowpea rhizobia due to pre-establishment of vesicular-arbuscular mycorrhiza. *New Phytol.*, *106*:207–215, 1987.
20. Ames, R. N., Reid, C. P. P., Porter, L. K., and Cambardella, C. Hyphal uptake and transport of nitrogen from two ^{15}N-labelled sources by *Glomus mosseae* a vesicular-arbuscular mycorrhizal fungus. *New Phytol.*, *95*:381–396, 1983.
21. Barea, J. M., Azcon-Aguilar, C., and Azcon, R. Vesicular-arbuscular mycorrhiza improve both symbiotic N_2 fixation and N uptake from soil as assessed with a ^{15}N technique under field conditions. *New Phytol.*, *106*:717–725, 1987.
22. Hardie, K. and Leyton, L. The influence of vesicular-arbuscular mycorrhiza on growth and water relations of red clover. I. In phosphate deficient soil. *New Phytol.*, *89*:599–608, 1981.
23. Levy, Y., Syvertsen, J. P., and Nemec, S. Effect of drought stress and vesicular-arbuscular mycorrhiza on citrus transpiration and hydraulic conductivity of roots. *New Phytol.*, *93*:61–66, 1983.
24. Nelsen, C. E. and Safir, G. R. Increased drought tolerance of mycorrhizal onion plants caused by improved phosphorus nutrition. *Planta*, *154*:407–413, 1982.
25. Hardie, K. H. The role of extraradical hyphae in water uptake by vesicular-arbuscular mycorrhizal plants. In Physiological and genetical aspects of Mycorrhizae. *Proc. 1st Eur. Symp. Mycorr.*, INRA, Paris, 1986, pp. 651–656.
26. Huang, R-S., Smith, W. K., and Yost, R. S. Influence of vesicular-arbuscular mycorrhiza on growth, water relations and leaf orientation in *Leucaena Leucocephala* (LAM) De Wit. *New Phytol.*, *99*:229–243, 1985.
27. Allen, E. B. and Allen, M. F. Water relations of xeric grasses in the field: Interactions of mycorrhizas and competition. *New Phytol.*, *104*:000, 1986.
28. Allen, M. F. and Boosalis, M. G. Effects of two species of VA mycorrhizal fungi on drought tolerance of winter wheat. *New Phytol.*, *93*:67–76,559–571, 1983.
29. Ellis, J. R., Larsen, H. J., and Boosalis, M. G. Drought resistance of wheat plants inoculated with vesicular-arbuscular mycorrhizae. *Pl. Soil*, *86*:369–378, 1985.
30. Levy, Y. Field determination of free proline accumulation and water-stress in lemon trees. *Hort. Sci.*, *15*:302–303, 1980.
31. Auge, R. M., Schekel, K. A., and Wample, R. L. Greater leaf conductance of well-watered VA mycorrhizal rose plants is not related to phosphorus nutrition. *New Phytol.*, *103*:107–116, 1986.

32. Sieverding, E. Influence of soil water regimes on VA mycorrhiza. I. Effect on plant growth, water utilization and development of mycorrhiza. *Z. Acker-u. Pflunzenbau*, *150*:400–411, 1981.
33. Koide, R. The effect of VA mycorrhizal infection and phosphorus status on sunflower hydraulic and stomatal properties. *J. Exp. Bot.*, *36*:1087–1098, 1985.
34. Daniels-Hetrick, B. A., Gerschefske-Kitt, D., and Thompson-Wilson, G. Effects of drought stress on growth response in corn, sudan grass and big bluestem to *Glomus etunicatum*. *New Phytol.*, *105*:403–410, 1987.
35. Buwalda, J. G., Stribley, D. P., and Tinker, P. B. Vesicular-arbuscular mycorrhizae of winter and spring cereals. *J. Agric. Sci. Cambridge*, *105*:649–657, 1985.
36. Krikun, J., Shimshi, D., and Haas, J. H. Mycorrhizal effects in field grown pepper. *Proc. 7th NACOM*, 1987, p. 49.
37. Haas, J. Management of VAM in irrigated crops. *Proc. 7th NACOM*, 1987, p. 44.
38. Koske, R. E. and Halvorson, W. L. Ecological studies of vesicular-arbuscular mycorrhizae in a barrier sand dune. *Can. J. Bot.*, *59*:1413–1422, 1981.
39. Nicolson, T. H. and Johnston, C. Mycorrhiza in the Graminae. III. *Glomus fasciculatus* as the endophyte of pioneer grasses in a maritime sand dune. *Trans. Br. Mycol. Soc.*, *72*:261–268, 1979.
40. Thompson, J. P. Soilless culture of vesicular-arbuscular mycorrhizae of cereals: Effects of nutrient concentration and nitrogen source. *Can. J. Bot.*, *64*:2282–2294, 1986.
41. Ferguson, J. J. (Ed.) Applications of mycorrhizal fungi to crop production. Gainesville, University of Florida Press, 1984.
42. Jeffries, P. Use of mycorrhizae in agriculture. *CRC Crit. Rev. Microbiol.*, *5*:319–357, 1987.
43. Kendrick, B. and Berch, S. Mycorrhizae: Applications in agriculture and forestry. In *Comprehensive Biotechnology*, Vol. 4 (C. W. Robinson, ed.), Pergamon Press, Oxford, 1985.
44. Mikola, P. Mycorrhizal inoculation in afforestation. *Int. Rev. For. Res.*, *3*:123–196, 1970.
45. Mikola, P. Application of mycorrhizal symbiosis in forestry practice. In *Ectomycorrhizae: Their Ecology and Physiology* (G. C. Marks and T. T. Kozlowski, eds.), Academic Press, New York, 1983, pp. 383–411.
46. Powell, C. L. and Bagyaraj, D. J. *VA Mycorrhiza*, CRC Press, Boca Raton, 1984.
47. Rhodes, L. H. The use of mycorrhizae in crop production systems. *Outlook on Agriculture*, *10*:275–281, 1980.
48. Schenck, N. C. *Methods and Principles of Mycorrhizal Research*. American Phytopathological Society, St. Paul, MN, 1982.
49. Walker, C. Sitka spruce mycorrhizas. *Proc. R. Soc. Edin.*, *93*:117–129, 1987.
50. Chilvers, G. A., Lapeyrie, F. F., and Horan, D. P. Ectomycorrhizal vs. endomycorrhizal fungi within the same root system. *New Phytol.*, *107*:441–448, 1987.
51. Abbott, L. K. and Robson, A. D. The role of vesicular-arbuscular mycorrhizal fungi in agriculture and the selection of fungi for inoculation. *Aust. J. Agric. Res.*, *33*:389–408, 1982.
52. Fitter, A. H. Functioning of vesicular-arbuscular mycorrhizas under field conditions. *New Phytol.*, *99*:257–265, 1985.

53. Gianinazzi-Pearson, V. and Diem, H. G. Endomycorrhizae in the tropics. In *Microbiology of Tropical Soils and Plant Productivity* (Y. R. Dommergues and H. G. Diem, eds.), Martinus Nijhoff/Dr. W. Junk Publishers, The Hague, 1982, pp. 209–251.
54. Powell, C. I. Field inoculation with VA mycorrhizal fungi. In *VA Mycorrhiza* CRC Press, Boca Raton, FL, 1984, pp. 205–222.
55. Howeler, R. H., Sieverding, E., and Saif, S. Practical aspects of mycorrhizal technology in some tropical crops and pastures. *Pl. Soil*, *100*:249–283, 1987.
56. Sieverding, E. Research model towards practical application of VA mycorrhizal fungi in tropical agriculture. In Physiological and genetical aspects of mycorrhizae. *Proc. 1st Eur. Symp. Mycorr.* INRA, Paris, 1986, pp. 475–478.
57. *CIAT. Annual Report 1987: Pastures program—Soil Microbiology* (In press).
58. Sieverding, E. On-farm inoculum production of VAM inoculum. *Proc. 6th NACOM*, 1985.
59. Hayman, D. S. Methods for evaluating and manipulating vesicular-arbuscular mycorrhiza. In *Microbiological Methods for Environmental Biotechnology* (J. M. Lynch and J. M. Grainger, eds.), Academic Press, London, 1984, pp. 95–117.
60. Saif, S. R. Vesicular-arbuscular mycorrhizae in tropical forage species as influenced by season, soil texture, fertilizers, host species and ectoypes. *Angew. Botanik*, *60*:125–139, 1986.
61. Miranda, J. C. C. Influencia de fungos endomicorrizicos inoculados a campo, na cultura de sorgo e soja em um solo sob cerrado. *R. bras. Ci Solo*, *6*:19–23, 1982.
62. Saif, S. R. Growth responses of tropical forage plant species to vesicular-arbuscular mycorrhizae. I. Growth, mineral uptake and mycorrhizal dependency. *Pl. Soil*, *97*:25–35, 1987.
63. Salinas, J. G., Sanz, J. I., and Sieverding, E. Importance of VA mycorrhizae for phosphorus supply to pasture plants in tropical oxisols. *Pl. Soil*, *84*:347–360, 1985.
64. Pellet, D. and Sieverding, E. Host preferential multiplication of fungal species of the endogonaceae in the field, demonstrated with weeds. In Physiological and genetical aspects of mycorrhizae. *Proc. 1st Eur. Symp. Mycorr.* INRA, Paris, 1986, pp. 555–558.
65. Daniels-Hetrick, B. A. and Bloom, J. The influence of host plant on production and colonization ability of vesicular-arbuscular mycorrhizal spores. *Mycologia*, *78*:32–36, 1986.
66. Gianinazzi-Pearson, V. and Gianinazzi, S. In Physiological and genetical aspects of mycorrhizae. *Proc. 1st Eur. Symp. Mycorr.* INRA, Paris, 1986.
67. Fitter, A. H. Effect of benomyl on leaf phosphorus concentration in alpine grasslands: A test of mycorrhizal benefit. *New Phytol.*, *103*:767–776, 1986.
68. Jakobsen, I. Effects of VA mycorrhiza on yield and harvest index of field-grown pea. *Pl. Soil*, *98*:407–415, 1987.
69. Hayman, D. S. Mycorrhizae of nitrogen-fixing legumes. *MIRCEN Journal*, *2*:121–145, 1986.
70. Wallace, L. L. Mycorrhizae in grasslands: Interactions of ungulates, fungi and drought. *New Phytol.*, *105*:619–632, 1987.
71. Wallace, L. L. and Svejcar, T. Mycorrhizal and clipping effects on *Andropogon gerardii* photosynthesis. *Amer. J. Bot.*, *74*:1138–1142, 1987.

72. Bildusas, I. J., Dixon, R. K., Pfleger, F. L., and Stewart, E. L. Growth, nutrition and gas exchange of *Bromus inermis* inoculated with *Glomus fasciculatum*. *New Phytol.*, *102*:303–311, 1986.
73. Dodd, J. C., Krikun, J., and Haas, J. Relative effectiveness of indigenous populations of vesicular-arbuscular mycorrhizal fungi from four sites in the Negev. *Isr. J. Bot.*, *32*:10–21, 1983.
74. Buwalda, J. G., Stribley, D. P., and Tinker, P. B. Effects of vesicular-arbuscular mycorrhizal infection in first, second and third cereal crops. *J. Agri. Sci. Cambridge*, *105*:631–647, 1985.
75. Dodd, J. C. and Jeffries, P. Early development of vesicular-arbuscular mycorrhizas in autumn-sown cereals. *Soil Biol. Biochem.*, *18*:149–154, 1986.
76. Dodd, J. C. and Jeffries, P. Effect of overwinter environmental conditions on vesicular-arbuscular mycorrhizal infection of autumn-sown winter cereals. *Soil Biol. Biochem.*, *21*:453–455, 1989.
77. Baltruschat, H. and Dehne, H-W. The occurrence of vesicular-arbuscular mycorrhiza in agro-ecosystems. I. Influence of nitrogen fertilization and green manure in continuous monoculture and in crop rotation on the inoculum potential of winter wheat. *Pl. Soil*, *107*:279–284, 1988.
78. Jakobsen, I. Vesicular-arbuscular mycorrhiza in field-grown crops. II. Effect of inoculation on growth and nutrient uptake in barley at two phosphorus levels in fumigated soil. *New Phytol.*, *94*:595–604, 1983.
79. Baltruschat, H. Evaluation of the suitability of expanded clay as carrier material for VA mycorrhizal spores in field inoculation of maize. *Angew. Botanik*, *61*:163–169, 1987.
80. Dehne, H-W. and Backhaus, G. F. The use of vesicular-arbuscular mycorrhizae fungi in plant production. I. Inoculum production. *Z. Pflanzentrank. Pflanzensch.*, *93*:415–424, 1986.
81. Marx, D. H., Ruehle, J. L., Kenney, D. S., Cordell, C. E., Riffle, J. W., Molina, R. J., Pawuk, W. H., Navratil, S., Tinus, R. W., and Goodwin, O. C. Commercial vegetative inoculum of *Pisolithus tinctorius* and inoculation techniques for development of ectomycorrhizae on container-grown tree seedlings. *Forest Sci.*, *28*:373–400, 1982.
82. Grenville, D. J., Piche, Y., and Peterson, R. L. Sclerotia as viable sources of mycelia for the establishment of ectomycorrhizae. *Can. J. Microbiol.*, *31*:1085–1088, 1985.
83. Baltruschat, H. and Dehne, H-W. Culture substrate for inoculation of plants with mycorrhiza fungi consisting of an adsorbent e.g. vermiculite for pumice in which mycorrhiza-infected plants have been grown. German patent 3416315, 1985.
84. Hepper, C. M. Non-biotrophic culture of vesicular-arbuscular mycorrhizal fungi in culture media containing inorganic sulphur compounds, preferably potassium sulphite. UK patent GB 2142933, 1985.
85. Wood, T., Biermann, B., and Grainger, H. Improved production of axenic vesicular-arbuscular mycorrhizal fungi using porous substrates inoculated with root inoculum. U.S. patent 755943, 1985.
86. Garrett, H. F., Cox, G. S., and Dixon, R. K. Enhancing mycorrhizal development by foliar fertilization of plants. U.S. patent 160976, 1985.

87. Giovannetti, G. Producing plants with symbiotic mycorrhiza by transfer from pre-inoculated pilot plants, useful for example for truffle production. U.S. patent 434403, 1982.
88. Habte, M., Fox, R. L., Aziz, T., and El-Swaify. Interaction of vesicular-arbuscular mycorrhizal fungi with erosion in an oxisol. *Appl. Environ. Microbiol.*, *54*:945–950, 1988.
89. Mosse, B. and Thompson, J. P. Improvements in or relating to the production of mycorrhizal fungi. UK patent GB 2043688B, 1983.
90. Stensaas, L. J. Microbiological systems for distributing phosphorus to plant roots using symbiosis with mycorrhizal fungi. U.S. patent 4589226, 1986.
91. Warner, A. Vesicular-arbuscular mycorrhizal fungus containing seed pellets consists of seed of host plant, inoculum, peat and a binder. UK patent GB 212006, 1983.
92. Trappe, J. M. Selection of fungi for ectomycorrhizal inoculation in nurseries. *Annu. Rev. Phytopathol.*, *15*:203–222, 1977.
93. Mikola, P. Application of mycorrhizal symbiosis in forestry practice. In *Ectomycorrhizae: Their Ecology and Physiology* (G. C. Marks and T. T. Kozlowski, eds.), Academic Press, New York, 1983, pp. 383–411.
94. Malajczuk, N. Ecology and management of ectomycorrhizal fungi in regenerating forest ecosystems in Australia. *Proc. 7th NACOM*, 1987, p. 54.
95. Marx, D. H., Cordell, C. E., and Clark, A. Eight-year performance of loblolly pine with *Pisolithus* ectomycorrhizae on a good quality forest site. *Southern J. For.*, *12*:275–280, 1988.
96. Marx, D. H. and Cordell, C. E. *Pisolithus* ectomycorrhizae improve 4-year performance of loblolly and slash pines in South Georgia. *Georgia For. Comm. Publ.*, No. 4, December, 1988.
97. Cordell, C. E., Marx, D. H., Maul, S. B., and Owen, J. H. Production and utilization of ectomycorrhizal fungal inoculum in the Eastern United States. *Proc. 7th NACOM*, 1987, pp. 287–289.
98. Le Tacon, F., Jung, G., Mugnier, J., Michelot, P., and Mauperin, C. Efficiency in a forest nursery of an ectomycorrhizal fungus inoculum production in a fermentor and entrapped in polymeric gels. *Can. J. Bot.*, *63*:1664–1668, 1985.
99. Mauperin, C., Mortier, F., Garbaye, J., Le Tacon, F., and Carr, G. Viability of an ectomycorrhizal inoculum produced in a liquid medium and entrapped in a calcium alginate gel. *Can. J. Bot.*, *65*:2326–2329, 1987.
100. Chevalier, G. and Frochot, H. Truffle production from artificially inoculated mycorrhizal plants: first results. *Proc. 5th NACOM*, 1980, p. 35.
101. Poitou, N., Mamoun, M., and Delmas, J. Quelques resultats obtenus concernant la mycorrhization de plantes-hotes par des champignons mycorrhiziens comestibles. *Proc. 5th NACOM*, 1980, p. 61.
102. Schenck, N. C. The international culture collection of VA mycorrhizal fungi (INVAM). *Angew. Botanik*, *61*:153–155, 1987.
103. Elmes, R. P. and Mosse, B. Vesicular-arbuscular mycorrhizal inoculum production. II. Experiments with maize (*Zea mays*) and other hosts in nutrient flow culture. *Can. J. Bot.*, *62*:1531–1536, 1984.

104. Mosse, B. and Thompson, J. P. Vesicular-arbuscular endomycorrhizal inoculum production. I. Exploratory experiments with bean (*Phaseolous vulgaris*) in nutrient flow culture. *Can. J. Bot.*, *62*:1523–1530, 1984.
105. Elmes, R. P., Happer, C. M., Hayman, D. S., and O'Shea, J. The use of vesicular-arbuscular mycorrhizal roots grown by the nutrient film technique as inoculum for field sites. *Ann. Appl. Biol.*, *104*:437–441, 1983.
106. Thompson, J. P. Soilless culture of vesicular-arbuscular mycorrhizae of cereals: effects of nutrient concentration and nitrogen source. *Can. J. Bot.*, *64*:2282–2294, 1986.
107. Dehne, H-W., Backhaus, G. F., and Baltruschat, H. Inoculation of plants with VA mycorrhizal fungi in inorganic carrier materials. *Proc. 7th NACOM*, 1986, p. 38.
108. Hung, L. L-L. and Sylvia, D. M. Production of vesicular-arbuscular mycorrhizal fungus inoculum in aeroponic culture. *Appl. Environ. Microbiol.*, *54*:353–357, 1988.
109. Sylvia, D. N. and Hubbell, D. H. Growth and sporulation of vesicular-arbuscular mycorrhizal fungi in aeroponic and membrane systems. *Symbiosis*, *1*:259–267, 1986.
110. Mugnier, J. and Mosse, B. Vesicular-arbuscular mycorrhizal infection in transformed root-inducing T-DNA roots grown axenically. *Phytopathology*, *77*:1045–1050, 1987.
111. Walker, C. Formation and dispersal of propagules of endogonaeous fungi. In *Fungal Infection of Plants* (G. F. Pegg and P. G. Ayres, eds.), Cambridge University Press, Cambridge, 1988.
112. Agerer, R. *Colour Atlas of Ectomycorrhizae*. Schwabisch Gmund, Einhorn-Verlag, Munich, 1987.
113. Lopez-Aguillon, R. and Mosse, B. Experiments on competitiveness of three endomycorrhizal fungi. *Pl. Soil*, *97*:155–170, 1987.
114. Sen, R. and Hepper, C. M. Characterization of vesicular-arbuscular mycorrhizal fungi (*Glomus* spp.) by selective enzyme staining following polyacrylamide gel electrophoresis. *Soil Biol. Biochem.*, *18*:29–34, 1986.
115. Sen, R., Hepper, C. M., Azcon-Aguilar, C., and Grace, C. Variation in certain isozymes amongst different geographical isolates of the vesicular-arbuscular mycorrhizal fungi *Glomus clarum*, *Glomus monosporum* and *Glomus mosseae*. *Soil Biol. Biochem.*, *20*:51–59, 1988.
116. Zhu, H., Higginbotham, K. O., and Dancik, B. P. Intraspecific genetic variability of isozymes in the ectomycorrhizal fungus *Suillus tomentosus*. *Can. J. Bot.*, *66*:588–594, 1988.
117. Aldwell, F. E. B., Hall, I. R., and Smith, J. M. B. Enzyme-linked immunosorbent assay (ELISA) to identify endomycorrhizal fungi. *Soil Biol. Biochem.*, *15*:377–378, 1983.
118. Aldwell, F. E. B., Hall, I. R., and Smith, J. M. B. Enzyme-linked immunosorbent assay as an aid to taxonomy of the Endogonaceae. *Trans. Brit. Mycol. Soc.*, *84*:399–402, 1985.
119. Wright, S. F., Morton, J. B., and Sworobuk, J. E. Identification of a vesicular-arbuscular mycorrhizal fungus by use of monoclonal antibodies in an enzyme-linked immunosorbent assay. *Appl. Environ. Microbiol.*, *53*:2222–2225, 1987.

120. Hebraud, M. and Fevre, M. Protoplast production and regeneration from mycorrhizal fungi and their use for isolation of mutants. *Can. J. Microbiol.*, *34*:157–161, 1988.
121. Meinharat, F. and Esser, K. The plasmids of the morels, characterization and prerequisites for vector development. *Appl. Microbiol. Biotechnol.*, *27*:276–282, 1987.
122. Krishna, K. R., Shetty, K. G., Dart, P. J., and Andrews, D. J. Genotype dependent variation in mycorrhizal colonization and response to inoculation of pearl millet. *Pl. Soil*, *86*:113–125, 1985.
123. Estaun, V., Calvet, C., and Hayman, D. S. Influence of plant genotype on mycorrhizal infection: Response of three pea cultivars. *Pl. Soil*, *103*:295–298, 1987.
124. Mikola, P. Mycorrhizae under tropical stress. *Angew. Botanik*, *61*:15–23, 1987.
125. Mosse, B. Mycorrhiza in a sustainable agriculture. *Biol. Agric. Hort.*, *3*:143–152, 1986.
126. Gianinazzi-Pearson, V. Mycorrhizae: A potential for better use of phosphate fertilizer. *Fertilizers and Agriculture*, *92*:3–12, 1986.
127. Pacioni, G. A simple apparatus for estimating soil aggregation by extramatrical mycelia. In Physiological and genetical aspects of mycorrhizae. *Proc. 1st Eur. Symp. Mycorr.*, INRA, Paris, 1986, pp. 665–666.
128. Sanchez, P. A. and Benites, J. R. Low-input cropping for acid soils of the humid tropics. *Science*, *238*:1521–1527, 1987.
129. Mason, P. A., Last, F. T., Wilson, J., Deacon, J. W., Fleming, L. V., and Fox, F. M. Fruiting and successions of ectomycorrhizal fungi. In *Fungal Infection of Plants* (G. F. Pegg and P. G. Ayres, eds.), Cambridge University Press, Cambridge, 1988.
130. Skujins, J. and Allen, M. F. Use of mycorrhizae for land rehabilitation. *MIRCEN Journal*, *2*:161–176, 1986.
131. Dodd, J. C., Arias, I., Koomen, I., and Hayman, D. S. (1990). The management of vesicular-arbuscular mycorrhizal populations in acid-infertile soils of a savanna ecosystem. I. The effect of pre-cropping and VAMF inoculation on plant growth and nutrition in the field. *Pl. Soil*, *122*:229–240.
132. Dodd, J. C., Arias, I., Koomen, I., and Hayman, D. S. (1990). The management of vesicular-arbuscular mycorrhizal populations in acid-infertile soils of a savanna ecosystem. II. The effects of inoculation and pre-crops on the native VAMF spore populations. *Pl. Soil*, *122*:241–247.

7

ECTOMYCORRHIZA

SUVERCHA and K. G. MUKERJI *University of Delhi, Delhi, India*

DILIP K. ARORA *Banaras Hindu University, Varanasi, India*

I. INTRODUCTION

Mycorrhiza forms a symbiotic association with the host plant. There are various kinds of mycorrhizae but this chapter deals only with ectomycorrhizal (ECM) fungi. Meyer [1] has reported ECM association with 3% of the phanerogams. They are more prevalent in the members of Pinaceae, Fagaceae, Betulaceae, and Salicaceae. However, few genera from other families such as *Eucalyptus*, *Tilia*, and *Arbutus* and also the members of Caesalpiniaceae and Dipterocarpaceae show the ECM association. One monocot, *Kobresia*, has also been reported to show ECM colonization. Among cryptogams, ectotrophic mycorrhiza has been described in some fern sporophytes [2,3]. Studies have shown that the fungal counterpart mostly belongs to higher basidiomycetes and ascomycetes [4]. However, a few phycomycetous fungi belonging to Endogonaceae and sterile imperfect fungi such as *Cenococcum* and many sterile unnamed mycelia belonging to basidiomycetes and ascomycetes are also involved [5,6].

The ectomycorrhizal roots possess three main features (Fig. 1), namely, (1) a fungus mantle, which is formed by fungal sheathing or colonization of the short feeder roots, (2) the Hartig net, which is an intercellular hyphal penetration between cortical cells, and (3) colonized feeder roots, showing a morphological differentiation through increased branching and elongation. Increased root/rootlet number along with dense mycelial connections occupy a very large volume of soil and thereby becomes a highly active organ of absorption. This symbiotic association benefits the host plant by increased water and nutrient uptake, especially of immobile ions like phosphates. ECM-infected plants show increased resistance to root pathogens, drought, high soil temperatures, pH, and soil toxins. On the other hand, the fungal symbiont is also benefitted. It derives carbohydrates (mostly simple sugars), vitamins, and amino acids from the autobiont. It has been observed in most cases that if the ECM fungus is detached from the autotroph, it is unable to form a fruiting body [7,8].

In this chapter we briefly review some of the ecological aspects of ECM which have practical values in forestry.

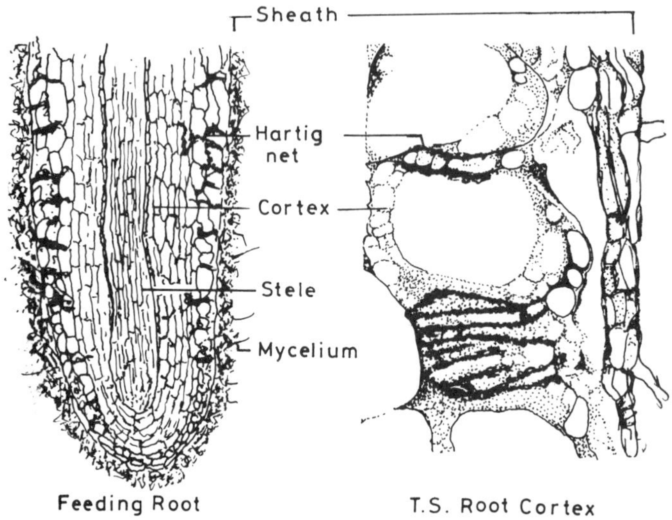

FIGURE 1. Morphology of ectomycorrhizal root.

II. ROOTS OF MYCORRHIZAL TREES

Most of the trees have a taproot system at the seedling stage. As the seedling grows towards maturity, the main axis of the root grows to extend the root system, and each axis becomes woody after developing secondary thickenings with age. To avoid full dependence on photosynthesis, the permanent axis stores carbohydrates and other compounds. The root system branches out into two components—long roots of unlimited extension and short stubby roots of restricted lifespan. In the latter, most of the ultimate laterals are of limited growth. This kind of root system is termed heterorhizic. All the apices grow and may become fully or partially colonized by a mycorrhizal fungus, those that are fully colonized develop a sheath around them, continue to grow slowly, branch dichotomously, and form a racemose root system. Those that are partially infected continue to grow actively, have a narrow diameter, and may abort or become dormant [9].

The development of mycorrhizal infection in the short roots depends on the distribution of the Hartig net in the parent long root [10]. The lateral root initials become colonized, if they pass through a region of the cortex in which the Hartig net is present. Then they rapidly become ensheathed and fully colonized. Chilvers and Gust [11] compared the rates of growth of mycorrhizal and nonmycorrhizal roots of *Eucalyptus* seedlings, and they concluded that mycorrhizal roots grow five times slower than nonmycorrhizal roots.

Interestingly, the genus *Pinus* deviates from the general trend. Its short roots are differentiated from the main axis. Their stelar construction is much simpler, and the apical meristem is restricted. Long roots of *P. resinosa* may have a dichotomous structure of mycorrhizal roots [12,13]. Irrespective of the short life of mycorrhiza, the colonization of roots is permanent. Wilcox described how these long roots might be fully ensheathed and the branch initials become infected in the cortex. The fungus grows acropetally along the new growth both superficially and between the cortical cells. In *Fagus* many long roots are fully ensheathed except for their fast-growing escaped apices.

The primary colonization occurs as soon as the seedlings are established. Chilvers [14] found that the superficial spread of the fungus is a gradual process and the spread of the hyphae and rhizomorph to form colonies depended on the efficacy of the particular soil type. The spores germinate and a weft of hyphae (sheath) is formed around the root system. Some hyphae spread inwards in the outer cortex, forming a Hartig net. The hyphal sheath bridges the intercellular Hartig net and soil system. The fungus in the cortex of long roots also—like that in the senescent mycorrhizal organs—penetrates cells when they age. Hence the total recycling of nutrients increases [9]. Unlike pathogenic infection, mycorrhizal fungi seem to prolong the life of cortical cells.

III. MODE OF INFECTION

In the seedlings of *Eucalyptus*, Chilvers and Gust [11,15] distinguished five different stages in the initiation and development of mycorrhiza:

1. A preinfection colonization of the root surface by fungal hyphae
2. Primary infection, from primary inoculum of spores or propagules in the soil
3. Penetration of the root tissue by fungal hyphae following the growth of the root
4. Proliferation of subsidiary branches of infected axis which do not escape the sheath
5. Secondary infections by hyphal strands or rhizomorphs to bring about infection of neighboring roots

These processes together bring about the colonization of the root system and the formation and spread of mycorrhizae within it.

Mycorrhizal fungi exist in soil as spores or as vegetative propagules in root fragments. Propagules of mycorrhizal fungi apparently respond to the stimulation of root exudates, and their hyphae or germ tubes grow and penetrate root epidermal cells. The colonization of host tissues progresses both internally and externally along the root surface, the external infection results in initiation of new colonization sites. In the case of ectotrophic mycorrhizal roots, morphology changes dramatically from the nonmycorrhizal state with the loss of root hairs, the development of a thick fungal mantle surrounding the epidermis, and extensive branching of the developing root. An extensive network of fungal hyphae and rhizomorphs are formed that may extend some distance out into the surrounding soil. The increased root branching and the hyphal network greatly increase the absorptive surface area. The physiological changes that accompany the development of mycorrhizae are altered permeability of membranes, which will affect the quality and quantity of root exudation, secretions, and the changes in nutrient and elemental composition of the host tissues. These changes, coupled with the chemical and physical impact of the fungal symbiont hyphae in the surrounding soil, result in a very different potential in the rhizosphere, so different that Rambelli [16] suggested the term "mycorrhizosphere" to describe the soil surrounding and influenced by mycorrhizae.

A. Penetration

Certain ectomycorrhizae have been demonstrated to produce cell wall degrading enzymes like pectinases and cellulases in culture [17]. However, attempts to demonstrate the production of these enzymes from other ECM in culture have not been successful [18].

It seems unlikely that there is a localized production of cellulase or pectinase at the hyphal tip. However, through electron microscopic studies it appears that the tips of the hyphae are wedge shaped, which is helpful in forcing their way through the cells by the hydrostatic pressure. Harley and Smith [9] hypothesized that the proteins present on the hyphal wall probably inhibit or form a complex with the host enzymes responsible for cell wall synthesis.

B. Development of Mycorrhiza

Anderson [18] has given two possible mechanisms involved in balancing fungal growth in the mycorrhiza. (a) Regulation of the flow of carbon to fungus from the plant cell may limit the extent of fungal growth. This possibility is supported by the fact that glucose supplementation of roots caused hyphae of ECM to ramify extensively and result in plant cell necrosis [19]. (b) It has been observed that a certain level of nutrients from the root cells is essential for mycorrhiza formation. Leakage of nutrients from roots was demonstrated to be higher in plants that have greater potential to become mycorrhizal. Functional mycorrhizae have been formed even on a nonhost plant upon treatment with a herbicide that increased release of nutrients from the roots. However, in general, induced phenolic responses are limited in ECM plants. It has been observed that the activities of two key enzymes in phenolic metabolism (phenylamine ammonia lyase and peroxidase) were similar in extracts from uncolonized and ectomycorrhizal roots [20]. Piche et al. [21] observed no elevated phenolic levels in their ectomycorrhizal system.

C. Host Specificity and Recognition

Interconnections between plants by mycorrhizal fungi are abundantly known in nature [22]. However, fruit bodies of many fungi appear for only brief periods and not all through the year. There are relatively few examples where fruiting bodies of one species of symbiont occur only in association with one genus of host, e.g., *Alpova diplophloeus*, which associates only with species of *Alnus*, and this has been upheld in cultural experiments. The specificity of host and ectotrophic mycorrhiza for their partners is almost always lower when tested in laboratory conditions than when examined in the field. Godbout and Fortin [23] synthesized ectomycorrhiza of *Aspen* with 29 species of fungi from 12 different genera, while from field observations they had recorded only 5 associated species. Some species of pine and birch can associate with over 40 species of fungi in laboratory experiments. Molina and Trappe [24] and Malajczuk et al. [25] demonstrated that certain fungi form complete mycorrhizae only with restricted host species.

Anderson [18] described host specificity and recognition. He explained that rhizosphere components may promote spore germination. Enhanced germination in the presence of root exudates has been observed in Melin's classical studies for ectomycorrhiza [9]. The structure of the root surface may promote mycosymbiont hyphae to contact the root and attach firmly. A search for lectin-like interactions between pine roots and surface components of an ectomycorrhizal fungus was unsuccessful [26]. Piche and Peterson [27] detected a fucose-rich polysaccharide associated with hyphae of an ectomycorrhiza only when in contact with conifer roots.

Incompatibility

Certain ectomycorrhizae display defined host specificities [9]. Mechanisms controlling incompatibility may occur at each of the stages in the life cycle of the fungus, and plant factors may play key roles in controlling some of these events. The presence of plant toxins or lack of nutrients in the root exudate may limit germination. The absence of a recognition structure at the root surface may limit adhesion. However, results of several studies indicate that incompatibility occurred after germination, attachment, and limited penetration by the fungus. These studies include the interactions of conifers or eucalyptus with ectomycorrhiza [24,25]. Induced production of phenolics may be significant. Phenols accumulate at the site of the incompatible fungus in conifers and eucalyptus [24,25].

Ectomycorrhizins

More recently, the ectomycorrhizae-specific proteins which are involved in the development of the *Pisolithus–Eucalyptus* symbiosis have been identified [28]. This association presumably involves specific expression of both fungal and plant genes. In an effort to examine for the presence of any "symbiosis-specific" genes that are expressed specifically during the ectomycorrhiza development, the polypeptide profiles of ectomycorrhiza (*Eucalyptus–Pisolithus*), uninfected roots (*E. globulus*) and free-living mycelia (*P. tinctorius*) were examined. It was found that the majority of the polypeptides observed in mycorrhiza were already synthesized in either uninfected roots or mycelia. The protein content of symbiotic tissues appeared to be made mostly from fungal polypeptides. Four polypeptides (E_1, E_2, E_3, E_4) are ectomycorrhiza-specific. Approximately 30% of the fungal polypeptides and more than 50% of the plant peptides resolved on gels are regulated during the development either by the fungal or the plant genome. Hilbert et al. [28] suggested the name "ectomycorrhizins" for this class of ectomycorrhiza-specific polypeptides, which were present in mycorrhiza but were undetectable in nonmycorrhizal roots and free-living mycelia. It seems likely that they have a role in the development of symbiosis.

IV. ASSOCIATION WITH ECOSYSTEMS

It is known that ectotrophic is more prevalent in temperate or subtemperate regions than in tropical and subtropical regions where the climatic conditions lead to restricted vegetative activity.

Tropical rain forests feature high species diversity and net primary productivity. Warmth and moisture are two environmental factors responsible for their luxuriant growth. The highly fragile nature of these forests is accompanied by highly leached and nutrient-poor soils. Besides, the living biomass is the chief storage compartment for the mineral nutrients rather than the soil [29]. In a denuded forest, nutrients stored in the living biomass are soon removed through runoff or by leaching [30]. Since tropical soils are very often deficient in phosphorus, phosphorus nutrition deficiency is one of the bottlenecks of tropical forestry. Therefore, even a slight improvement of mycorrhizal efficiency in forestry may have great practical importance. The role of ectomycorrhizal association was first experienced after introduction of exotic pines into tropical countries [5,31,32]. Inoculation of pines is probably the field of research where practical application of the results can most likely be expected. The occurrence of ectotrophic mycorrhiza in some tropical and subtropical tree families, such as Dipterocarpaceae, Caeselpiniaceae, and Myrtaceae, is very significant [33].

On the other hand, temperate forests are more stable ecosystems where the soil acts as a more important storage compartment for the mineral nutrients so that if the forest is cut over large areas, regeneration is quicker. In temperate forests, the active roots of the ectomycorrhizal trees are intensely developed in the surface and subsurface layers of the soil. Their life span may well depend upon climatic conditions or their location in the soil horizon. Under extreme winter conditions root growth may cease in the surface layers leading to considerable mortality of roots. By contrast, in less extreme areas under snow cover, root growth may continue throughout the winter, and active young mycorrhizae may be formed during the whole year. In the humic layers, ectomycorrhizae, which form the main component of the feeding system, often lie below the F layer in great quantities [1,34–36].

In a Douglas fir forest, ECM fungi form about 6% of the total tree weight in standing crops [37,38]. The fungal component of net production is large, playing an important role in nutrient cycling because it concentrates nutrients more than any other component of the ecosystem [39,40]. Richards [41] predicted that mycorrhizae might be responsible for closed nutrient cycles with minimal loss by decomposing litter directly and transporting the nutrients thereby released to the hosts in tropical rain forests. Potential involvement of tree mycorrhiza in direct nutrient cycling was demonstrated by Herrera et al. [42]. They showed that radioactive phosphorus was present in a mycosymbiont and the living roots that are connected to leaf litter labeled with the isotope. This direct nutrient cycling can only be resolved by demonstration of enzymatic capability of mycorrhizal fungi to mineralize litter. The ability of ECM to produce cellulolytic or lignolytic enzymes are conflicting. Vogt et al. [43] have shown that ECM along with conifer fine roots contribute 45% of net plant productivity in the young stand and 75% in the nature stand of a Douglas fir ecosystem.

Dighton [44] has described ecological successions of ECM-colonized roots and suggested that there is a change from a broad spectrum of mycorrhizal association in young seedlings to more specific fungal partnerships with increasing age of the host. He stated that nutrient cycling in the forest floor gives us indications of changes in the biotic, physical, chemical, and environmental conditions, which are consistent with the theoretical concepts of ecological strategies [45]. In the primary succession, changes in the resources available to the decomposer subsystem and in microclimate are likely to occur. As a stable ecosystem develops, the energy transformations to maintain homeostasis increase, while net ecosystem productivity decreases and organisms tend to move from r(reproductive)-

strategists towards k(kinn)-strategists, i.e., from quantity towards quality. As the forest ages, herbaceous ground flora is shaded out, and an increasingly greater amount of recalcitrant, tree-derived litter is included in the soil subsystem. Resource quality available to the decomposer community decreases, litter residue time increases, and litter accumulates on the forest floor. At and around canopy closure, the fungal species diversity is maximal when the forest floor litter deposition is highest and the trees reach a peak of nutrient deficiency. It was indicated that the scenario can be constructed of early mycorrhiza colonizers, r-strategists, optimizing nutrient uptake from available inorganic nutrient pools in the soil, derived from the mineralization of readily decomposable herbaceous litter. However, litter colonizers may not have the benefit of large inorganic nutrient pools as the rate of mineralization of nutrients from recalcitrant tree litter is slow. As there is a close relationship between tree growth and rate of mineralization of that litter, Dighton [44] presumed that the increasing prominence of k-strategist ECM from canopy closure onwards in a plantation forest is linked to the accumulation of tree litter. It has been observed that inorganic nutrient availability to the second rotation crop is replaced by a large reserve of organic nutrients locked up in brash, tree slumps, and roots. The successful association of later-colonizing ECM on seedlings can be established where the fungus is linked to a food base, where nutrient and carbon transfer from one plant to another through inter-connecting ECM hyphae is highly probable. A second rotation transplant may acquire mycorrhizal association (normally with older trees). This may be of some benefit in terms of nutrient and carbon acquisition from decomposing roots and stumps of the first rotation forest. By virtue of association with ECM, young plants may be able to exploit the heterogenous distribution of nutrient resource units more efficiently. A conservative estimate by Harley [46] indicates that the mycorrhizal fungus mantle on root surface of *Fagus* produced 25% of the root-mycorrhizal fungus respiration but comprised only 4% of the combined biomass. In a mycorrhizal system, the nutrient ions are protected against reabsorption and remained accessible to the host plant. Gardner and Malajczuk [47] studied the succession of ECM associated with *Eucalyptus* on rehabilitated bauxite mines in south western Australia. They examined the ECM population from a series of even age plant stands on rehabilitated sites and compared these with fungi associated with surrounding forest. Their results support the general concept proposed by Mason et al. [48] that ECM follow a successional development with tree age, and it is one of the important factors influencing these changes in fungal genera in litter buildup.

Griffiths et al. [49] studied the chemical and microbial variables in forest soils colonized with *Hysterangium setchellii* rhizomorphs. They reported that the levels of laminarinase, phosphatase, protease, and peroxidase were also significantly higher in the colonized soils, with the greatest difference being 8.6- to 62-fold higher peroxidase levels. While phosphatase and protease are integral to the breakdown of organic phosphorus, protein and laminarinase are indicative of the hydrolysis of the $\beta,1-3$ glucose polymers found in fungal cell wall material, the potential significance of elevated peroxidase levels is not as readily apparent. Among the several possibilities, they have roles in lignin degradation, detoxification of phenolics, or the suppression of competing microorganisms while cellulase levels are not significantly different when comparing mat-colonized and noncolonized samples together. On a sample-by-sample basis four of the five sample sets had significantly higher cellulase levels in the mat-colonized material. Amylase levels were significantly higher in the noncolonized soil samples. But no significant differences were found in xylanase levels. In another study the same workers found two- to threefold increase in amylase, cellulase, protease, and xylanase levels along with increase of up to sixfold for phosphatase and 100-fold for peroxidase in coniferous litter colonized by these hyphal mats. They concluded that elevated levels of microbial biomass, respiration, and certain key enzymes involved in organic matter degradation indicate that these mats may be the potential sites of enhanced organic matter decomposition, which can result in increased recovery of nitrogen and phosphorus or translocation to the host tree.

V. FACTORS AFFECTING INTENSITY OF INFECTION

The intensity of mycorrhiza formation depends on several factors. These factors, together or independently, may alter the quality of mycorrhiza formed.

A. Edaphic Factors

Early research showed that the roots of conifers were entirely mycorrhizal in raw humus soils but less so in mull soils [50]. Many workers reported similar results in *Fagus* [51,52]. They observed that in raw humus soils the fine roots are accumulated in the subsurface regions where the uninfected or unsheathed apices may be subjected to damage by drought and cold to which they are more susceptible than fully ensheathed mycorrhiza, while the mull soils encourage persistance of the apices of long roots in equable water and temperature regimes [53]. Results of experiments with seedlings have shown that in soils or sites moderately deficient in mineral nutrients, infection is more complete and intense than on those soils that are rich in mineral nutrients. Hatch [54] concluded that the nutrient status was the prime factor in determining the intensity of infection in *Pinus strobus*. A moderate deficiency of nitrogen, phosphorus, and potassium promoted mycorrhiza formation, but a great deficiency gave stunted growth and was detrimental to infections. The actual number of mycorrhizal tips per unit volume of mor soil greatly exceeds that in mull soil, although the intensity of infection was less. Indeed, in mor soil the tree puts a greater proportion of its growth effort into its root system and mycorrhiza equipment.

The moisture content of soil is important for the survival, development, and activity of different mycorrhizal fungi [5]. Worley and Hacskaylo [53] found that by varying the amount of soil moisture, it was possible to vary the fungal symbionts. Inoculation with dry forest humus resulted in development of black mycorrhiza, while the same inoculum when moist produced light-colored mycorrhiza [54,55].

B. Temperature

The optimum temperature for growth of most of the ectomycorrhizal fungi lies between 8 and 27°C [56]. When strains of *Rhizopogon luteolus*, *Suillus granulatus*, and *S. luteus* were grown on Melin-Norkrans agar for 26 days, strains of *R. luteolus* showed greatest colonization at 25°C, *S. granulatus* at 20 – 25°C, and two strains of *S. luteus* at 20°C. France and Reid [57] examined the ability of 18 species comprising 73 isolates of mycorrhizal fungi to withstand the low temperature of 10°C for 48 hours. Of these, 71 survived, although the time required for recovery after the period of low temperature by the beginning of active growth in culture varied very greatly. Moser [58] preferred to use "low-temperature" strains of *Suillus plorans* and other fungi for the inoculation of conifers (*Pinus cembra* in particular) at high level plantings in the Alps. There is much interest in the possibility that "high-temperature" fungi may be important as mycorrhizal fungi in tropical regions. More recently, Brundrett and Kendrick [59] concluded that plants with the ability to produce roots when soil temperature was low (close to 0°C) were typically nonmycorrhizal.

C. pH of Substrate

Mycorrhizal fungi show varied ranges of pH tolerance. pH affects the growth of mycorrhizal fungi in culture, but the effect is not significant. They usually grow at pH 3.0 – 7.0. The hyphal mats of *Hysterangium crassum* have higher oxalate crystal concentrations compared to uncolonized litter soil material [60]. With dissolution of phosphates in soil, oxalic acid forms insoluble precipitates with Ca, Fe, and Al [61]. This action provides the tolerance of some mycorrhizal trees to alkaline soil [62].

D. Surface Area

Since mycorrhizal plants exhibit enhanced nutrient and water absorption, it has been observed that the larger the mycorrhiza, the greater the surface area that remains in contact with soil, thereby increasing the efficiency of absorption. Ectomycorrhizae have geometric shapes that range from simple to complex, whereas the roots on which they are formed are simple cylindrical structures. Mycorrhizal trees develop more rootlets, thus increasing the surface area of absorption [63].

E. Water Stress

Ectomycorrhizal plants recover from water stress more quickly than nonmycorrhizal plants [64]. Tritiated water was transported relatively rapidly about 27 cm/hr along fungal strands from mycorrhizal roots of *Pinus sylvestris*. These strands seemed to have some tissue differentiation and a structure which might facilitate water conduction. Harley and Smith [9] have reported that nonmycorrhizal root tips of *P. strobus* and *P. taeda* become dormant after a period of severe stress and showed no renewed growth for 14 days, whereas mycorrhizal roots renewed growth within 3 days of relief of stress. Similarly, Marx and Cordell [65] suggested that root systems with abundant *Pisolithus tinctorius* are capable of extracting water and therefore essential nutrients from large soil volume during periods of severe water stress at a greater rate than other root systems. This capacity is likely to be mediated by the extensive hyphal strand network formed by *P. tinctorius* in the root zone. Spread of the fungus throughout the site shows that *P. tinctorius* can persist on a site stressed by periodic water deficits. They further reported that deficits may suppress less drought-tolerant fungi and prevent them from replacing *P. tinctorius* on root systems. This ability of the fungus has strong practical implications. They concluded that the forest trees are constantly confronted with stresses, therefore mycorrhiza formation is essential for their survival and growth. However, Diebolt and Mudge [66], working with containerized scots pine (*P. sylvestris*) inoculated with *Pisolithus tinctorius*, found that these trees could not adapt to stress conditions.

F. Pollution

Carpophores arising out of mycorrhizal roots are regarded as sensitive indicators of forest ecosystem. Schaffers and Termorshuizen [67] concluded that (a) air pollutants (O_3, SO_2, NO_x) have a strong negative effect on the occurrence of mycorrhiza (it is likely that the actual amount of mycorrhiza underground is affected to a similar extent and (b) a strong negative correlation was observed indicating a direct effect of air pollution on the mycorrhiza, possibly via soil changes. Pollution has a direct effect on tree physiology, which might be responsible for an indirect negative effect on mycorrhiza occurrence. A decrease in number of carpophores of mycorrhizal fungi of conifers was followed by an increase in saprophytes and parasites [68]. But they could not conclude whether air pollution initially affects the mycoflora and causes a decrease of tree vitality, or first affects the tree vitality and then causes a decrease of carpophores. They presumed that the cause of decrease of carpophores was due to delayed degradation of humus, which might be caused by acidification of Dutch forest soils.

Dighton and Skeffington [69] studied the effects of artificial acid precipitation on the mycorrhizae of Scots pine seedlings. They observed that SO_4^{2-} and Ca^{2+} concentration did not seem to affect mycorrhizal growth, whereas Al, Mn, and pH effected the growth of isolated mycorrhizal fungi. Mn concentration could not be toxic because it was present in a low concentration. But Al concentration and low pH (3.0 – 3.2) reduce growth. Mycorrhiza formation by *Suillus* sp. was Al sensitive even though in culture *Suillus* was found to be Al tolerant [70]. At the levels of Al tested, the highest Al level normally resulted in an increased length of time necessary for mycorrhiza formation, whereas the lowest Al level taken incited a slight reduction in length of time for mycorrhizae formation. There was a slight stimulation in colonization of *Hebeloma*, *Cenococcum*, *Pisolithus*, and *Rhizopogon* as a result of Al, except for the isolates of *Suillus*, with which a depression of colonization occurred. *Pisolithus tinctorius* outperformed all other fungi tested [70]. Chemical analysis showed a higher content of heavy metals available to plants in acidic soils. Digital line profile analysis showed that Ni and Fe were concentrated in the peripheral mycorrhizal mantle region of roots [71]. Concentration of Al, Cu, Zn, Ni, and other heavy metals may be responsible for the decreased frequencies of ectomycorrhizae in oaks on acid soils. Similar observations were made by Dixon [72]. Ectomycorrhizal colonization rates of *Suillus luteus* were significantly reduced in *Quercus* seedlings in presence of Cd, Ni, and Pb. Leaf and root heavy metal contents were lower in ectomycorrhizal plants. The size of nonmycorrhizal seedlings exposed to heavy metals was reduced as compared to mycorrhizal plants.

VI. NUTRITIONAL UPTAKE

ECM is known to enhance uptake of P, N, K, and other elements like Zn, etc. Plant roots take up phosphorus from the soil solution which is present in a very low concentration, i.e., 0.01 ppm. The phosphorus is in equilibrium with phosphate ions, adsorbed in reverse on clay particles. These ions represent a reserve of potentially available phosphorus that can go into solution when the concentration in soil solution drops owing to plant uptake. Ninety-eight percent of the phosphorus is present either in mineral form, i.e., rock phosphate, or in organic phosphate. Therefore, this large percentage of phosphorus is unavailable to plants. In phosphate-deficient soils there will be greater unsaturation of the adsorption sites and, hence, a slower diffusion rate. The rate of phosphorus uptake at the root surface exceeds that during the development of a depletion zone [73] around the roots, further reducing the number of ions available for uptake [74]. The ECM hyphae extend out of the root system and grow beyond the depletion zone. Tinker [75] classified phosphorus sources in the soil as phosphate ions present in solution or held on surfaces, mineral phosphate held in crystal lattices, and phosphorus in organic compounds. Inorganic phosphate tends to be immobile because it binds to soil colloids or is fixed as aluminium or iron phosphates. According to Naryanan and Bhattacharyya [74], the fungal hyphae absorb phosphate ions from the soil solution in the same way as the roots, and they promote dissolution of adjacent particles of mineral phosphate. The mean distance between an absorbing surface and a mineral phosphate surface could be decreased by subdividing the minerals or by having root hair on the root or by having hyphae. The main mode of increased uptake of poorly mobile ions is by fungal growth into soil, absorption of ions, and their transport back to the higher plants [76–78]. The pattern of uptake of poorly mobile ions by plants is apparently first by intensive depletion of ions from narrow zone around the roots delineated by the root hair depending on the ions and soils [76] and then with mycorrhizal infection as a result of extending hyphae in soil. Comparisons of uptake by mycorrhizal and nonmycorrhizal roots demonstrated the importance of the fungus, while a variety of experiments using whole seedlings and adult tree root systems and excised roots of beech tree provided evidence which collectively proved the significance of the sheath as a buffer or store of phosphate for the root [9]. The rate of uptake of phosphate from very dilute solutions by mycorrhizae was approximately four to six times higher than by nonmycorrhizal roots. Phosphate may be incorporated first into the metabolic pool of the sheath.

However, in beech mycorrhizae, 90% of the phosphate taken up accumulates initially in the fungal sheath and then moves slowly into the host tissue by processes which are aerobic and sensitive to metabolic inhibitors. It has been shown that the sheath is the storage organ which passes phosphate to the host in the absence of any external source. Morrison [79] demonstrated that ^{32}P passed rapidly to the shoots for a few days and then the transport ceased. In mycorrhizal plants, the isotope was transported to the shoots at a steady rate over a period of 21 days.

It has been shown by autoradiography that carbohydrates and mineral nutrients can move from one host to the other via the interconnecting hyphae [80,81]. Isotopes of ^{14}C and ^{32}P were supplied to one plant (donor plant), and their accumulation in the connected plant (receiver plant) was monitored and compared with nonmycorrhizal control plant. It was noticed that the rate of transport of ^{14}C exceeded 20 cm/hr over distances exceeding 20 cm. Read and Finlay [80] observed that the labeled carbon accumulated in the sheath of receiver roots, which is later transferred to the shoots. Prolonged shading of receiver plants led to the accumulation of significantly more labeled carbon in the shoots. In these conditions over 6% of donor carbon can be transferred to the receiver. However, phosphorus applied to the distal parts of the mycelial system moves freely through the anastomosing network of strands to all infected seedlings. Label accumulates first in the mycorrhizal roots, but then moves rapidly to the shoots. Again isotopes move readily to seedlings of a number of species interconnected by a common mycorrhizal mycelium. Similar results were also reported by Duddridge et al. [64]. Read and Finlay [80] have further concluded that the interconnections provide channels through which carbon, phosphorus, and water can be transferred along gradients of potential between the linked individual plants.

The classical theory of mycorrhizal function is a simple concept of exchange in which the fungal partner receives carbon compounds from the product of photosynthetic activity of its host while the increased level of mineral nutrient supply enhances the rate of C_6-fixation [82]. The extent to which the fungus influences carbon metabolism is not clear because of the difficulty in separating the sink effect of the fungus in the root system from the increased demand of the host for carbon compounds associated with its own enhanced metabolism and growth [82]. This is related to the possible alterations by the fungus of the hormonal balance of the host. The establishment of the fungal sheath commands the extensive input of carbon compounds at an early stage. The fungal partner is able to make limited growth on simple carbon compounds. While discussing interrelationships between carbon nutrition and morphogenesis, Lewis [83] makes a distinction between carbon metabolism of biotrophic plant disease situations which generally involve disruption of source-sink movements since the infections are located at sites of photosynthetic activity and that of mycorrhiza. Lewis has made an assessment of the evidence relating to utilization of reserve carbohydrates of plants by ECM in culture. ECM association has shown the occurrence of fungal carbohydrates such as trehalose. Lewis and Harley [84] placed agar blocks containing 10% ^{14}C-sucrose on the ends of excised mycorrhizal roots of the beech tree; this ^{14}C was later recovered from the fungal sheath as trehalose, glycogen, and mannitol. More recently Soderstrom [85] reported trehalose, mannitol, and arabitol from $^{14}CO_2$ fed mycelia of *Suillus bovinus*, *Pisolithus tinctorius*, or *Paxillus involutus* associated with seedlings of *Pinus sylvestris*, *P. contorta*, *P. radiata*, and *Betula pedula*. Proportions of these carbohydrates are different in species of ECM [85]. Norby et al. [86] found that the seedlings growing at 695 ul/l CO_2 allocated proportionately more ^{14}C-labeled photosynthete to fine roots than do the seedlings growing at ambient concentration of CO_2. This is associated with greater fine root mass and mycorrhizal density in the plants treated with high CO_2 concentrations at 34 weeks. The exudation of soluble ^{14}C-labeled compounds from roots is also greater in these plants.

It is a well-established fact that ECM improves plant vigor by enhancing the nutrient uptake. However, Harley [87] and Harley and Smith [9] have reported a growth depression in terms of dry weight and size in inoculated plants at young seedling stage because there is a competition between the partners for limited supply of carbon. But as more foliage appears, competition for the available nutrients decreases due to increased photosynthesis, and the deleterious effect of the fungus is replaced by the benefits it brings to the host nutrition.

VII. PHYTOHORMONAL INVOLVEMENT

A. Hormone Production by ECM

Hormonal involvement in the development of ectomycorrhiza is well known [10,88–90]. Ectomycorrhizal fungi are known to produce auxins, ethylene, and gibberellic acid [90,91]. But very few of the ECM fungi are reported to produce cytokinin, which induces structural modification of ECM. Most physiological and biochemical processes are apparently regulated by a delicate balance of phytohormones. Therefore, phytohormones contributed by ECM fungus can strikingly influence the mycorrhizal host [88].

Auxin

Production of auxin by ectomycorrhizal fungi was responsible for dichotomous branching and other morphological changes which occur in feeder roots of pine [88]. Many ECM liberate IAA and related indole compounds. Harley and Smith [9] pointed out that nonhormonal factors might not account for the morphological and anatomical characteristics of mycorrhizal roots. These include the genetic tendency of pine roots to dichotomize as they age even in the absence of auxin or mycorrhizal infection and the physical influence of the fungal sheath and of compact soil on root structure. It has been reported by Barnes and Naylor [92] that pine root culture will undergo dichotomous branching in response to certain vitamins and the phytohormone cytokinins using GC-MS analysis. Ek et al. [93] identified IAA in culture media of 17 out of 19 species or isolates of ECM fungi. IAA-related indole compounds and cytokinins regulate division, expansion, and differentiation of cells and synthesis of RNA and proteins in hosts and ECM [89,94–97,100]. High nitrogen and low light intensity have inhibitory effects on mycorrhiza formation. Frankenberger and Poth [98] detected and isolated IAA from *P. tinctorius*. The amount of IAA detected ranged from 4 to 5 μmol/1 by HPLC and ELISA. Another metabolite detected from GC-MS with a typical indole nucleus (m/z = 130) indicates that it could be an intermediate in auxin metabolism. Plant (Douglas fir) response was monitored after treating with *P. tinctorius* and L-tryptophan. A consistent increase in plant height and stem diameter was observed as a result of two treatments with statistical difference in dry weights of the roots and shoots. Nylund and Unestam [99] exposed seedlings of Scot pine to synthetic auxin-like IBA in concentrations of 5, 10, and 20 mg/1. They obtained enhanced translocation and root respiration of a tracer pulse in IBA-treated plants. On the other hand, the effect of IBA on photosynthesis was less clear at 20 mg/1. There was a significant 25% increase while lower concentrations resulted in both increase and decrease in different experiments. Using HPLC/MS, Rouillon et al. [100] confirmed the production of IAA and two of its degradation products by *Hebeloma hiemale*. Bruchet et al. [101] compared the in vitro activity of IAA-synthesizing enzymes between numerous species and isolates in *Hebeloma*. But no attempt was made to correlate the magnitude of auxin production by a given fungus with the intensity of its effects on host root development. Mitchell et al. [102] reported a threefold increase in IAA level in short roots of *Pinus echinata* associated with *P. tinctorius* as compared to nonmycorrhizal short roots. Mudge [90] concluded that this can be considered as first confirmation for the notion that hyperauxin condition is found in mycorrhizal symbiosis. But it is still not clear whether the elevated level of auxin in the mycorrhiza is of fungal origin, plant origin, or of both.

It has been demonstrated that most of the mycorrhizal fungi release indole compounds into media containing tryptophan. Moser [103], using a medium containing tryptophan as a nitrogen source, grew the mycelia of mycorrhiza formers, including several strains of some species and a few of fungi of other ecological groups. Most of the mycorrhiza formers produced IAA in culture except species of *Phlegmacium*, which produced indole propionic acid. Some of the other species of *Phlegmacium* produced it together with IAA, and two species of *Suillus* produced IBA and IAA. It was found that if another source of nitrogen is present in the medium, the production of indole compounds was reduced. Most of the amino acid addition was inhibitory to IAA production, but they increased mycelial growth. Moser [103] found that species and strains of ECM differed greatly in rate and quantity of auxin formation, but their activity in this regard usually exceeded that of fungi of other ecological groups. Ulrich [104] showed that *Suillus variegatus*, *S. granulatus*, and *S. luteus* could form IAA on Hagen's medium in the absence of tryptophan. But the rate of IAA formation could be enhanced by the addition of tryptophan. IAA production could be destroyed by decreasing the pH of the culture medium, or it may be subjected to enzymic destruction or oxidation in culture. Horak [105], Shemakhanova [106], and Tomaszewski and Wojceichowska [107] also reported auxin production by different species of ECM fungi in culture medium. Harley and Smith [9] concluded that most of the known ECM fungi can produce auxin in culture. Its quantities must vary both according to the rate of oxidative deamination brought about by the species concerned and according to the presence of enzymes oxidizing or destroying the indole compounds, as well as production and presence of phenols and other substances inhibiting these enzymes.

Cytokinin

In higher plants, cytokinin acts as a regulator to endogenous IAA formation, activity, and transport [108]. Cytokinin is known to activate IAA oxidase at low concentration and inhibit its activity at higher concentration. Cytokinin is also known to delay senescence in plants. Miller [109] tested the cytokinin production ability of ECM fungi in culture by growing colonies upon agar in the neighborhood of soy callus deprived of cytokinin. This method can be used for producing cytokinin in culture by many fungi, but some may produce growth inhibitors, or sometimes callus itself produces the inhibitors. Miller [109] and Crafts and Miller [110] demonstrated that *Rhizopogon roseolus*, *Suillus cothurnatus*, *Suillus punctipes*, and *Amanita rubescens* produced cytokinins, while some others may form it in traces. Some calli of pine associated with ECM did not grow significantly more than the control. They probably had produced a callus inhibitor. Ng et al. [111] confirmed that *R. luteolus* produces cytokinin in culture, as did the strains of *Suillus luteus* and *S. grevillii*. The latter was also reported to produce a callus inhibitor in its culture filterate. Ho [89] measured cytokinin–IAA ratio in extracts of fresh fruiting bodies of ECM. Ho and Trappe [112] found the cytokinin–IAA ratio to be relatively constant within species of many ECM fungi. This ratio can be used to determine the characteristic morphology of mycorrhizal and seedling growth [89]. The highest cytokinin–IAA production was obtained in isolates of *Suillus brevipes*. A significantly higher cytokinin–IAA ratio was found in one of the isolates of *Astraeus pteridis*. Gibberellin-related compounds were detected in culture media and fruiting bodies of *Boletus edulis* [88]. Harley and Smith [9] indicated that further investigations are required before all the ECM fungi attribute gibberellin-dependent activities. However, Hanley et al. [113] have reported a significant quantity of gibberellin-like compounds in the filtrates of *P. tinctorius* and *Thelephora terrestris* associated with pine seedlings, using the lettuce hypocotyl bioassay. GA treatments of seedlings of *Pinus sylvestris* did not influence fresh or dry weights or sugar content in roots and shoots. But it reduced starch in both the shoots and roots and showed a linear increase in shoot length.

Ethylene

Graham and Linderman [91] hypothesised that ethylene might be involved in mycorrhizal symbiosis. Much earlier, Lynch [114] and Lynch and Harper [115] found that ethylene production occurred in soils of low aeration containing a carbohydrate source and methionine. The hormonal activities of ethylene resemble those of IAA in many respects. Graham and Lindermann [116] tested 23 ECM species for presence of ethylene in pure culture supplied with methionine, a precursor of ethylene. They observed a considerable variation in ethylene production. Exogenous ethylene application to roots of conifer seedlings induce lateral branching and hypertrophy of cortical cells, which is comparable to mycorrhizal infection. When seedlings are inoculated with ECM fungi, it results in an increase in rhizosphere ethylene levels. In contrast to pines, Douglas fir does not show quantifiable response in the form of dichotomous branching [90]. Wilson and Field [117] observed ethylene-stimulated dichotomous branching of short roots of seedlings in *Pinus radiata*. Rupp and Mudge [118] reported a similar kind of ethylene-stimulated dichotomy and root hair inhibition in root organ cultures of *Pinus mugo* when treated with ethephon.

Exogenous treatment with auxin to mugo pine root cultures also produced ethylene. On treating the roots with silver thiosulfate (inhibitor of ethylene action), the auxin-induced branching could be inhibited. In a study on ethylene production by several ECM fungi and its effects on host root morphology, De vries [119] reported that *Laccaria laccata* and *P. tinctorius* isolate produced ethylene in amounts significantly different from the uninoculated control. He also observed stimulation of lateral root formation by ethylene, identical to its formation by *L. laccata* in Douglas fir as earlier reported by Graham and Linderman [91]. But isolates of *P. tinctorius* could not stimulate lateral root formation due to very low levels of ethylene produced. These results can be considered evidence that ethylene is involved in auxin-stimulated dichotomous branching.

B. Effect of ABA and Jasmonic Acid on ECM

Gogala [120] investigated the influence of ABA and Jasmonic acid on the growth of *Boletus bovinus* and *Suillus variegatus*. These studies are of special interest. It was found that ABA did not have any inhibitory effect on the growth of fungal mycelium. On the other hand, Jasmonic acid is a new endogenous plant growth regulator. Its methyl ester group is known as a senescence-promoting substance in plants. It is identified as a plant growth inhibitor and is present in leaves, roots, seeds, and fruits.
Gogala [120] concluded that the only plant growth substance which strongly inhibits the mycelial growth of mycorrhizal fungi even at the lowest concentration is Jasmonic acid, and, therefore, it seems to be an important factor

in the regulation of mycorrhiza. Further investigations are needed into the effect of Jasmonic acid on ECM-associated plants.

VIII. INTERACTIONS WITH MICROBES

Hiltner [121] applied the term rhizosphere to the portion of soil subjected to the effect of plant roots where significant microbial activity occurs. Since then it is acceptable that rhizosphere effect is a dynamic process triggered by root exudation and dependent on (a) host factors, e.g., age, etc. [122], (b) edaphic factors like its physical properties, fertility, and moisture-holding capacity, (c) environmental conditions, (d) cultural practices including applications of chemicals, and (e) microbial interactions in soil [123]. As indicated, Rambelli [16] suggested the term "mycorrhizosphere" for the rhizosphere around mycorrhizal roots. Linderman [123] has further elucidated the word "mycorrhizosphere" in depth. Mycorrhiza significantly changes the physiology and/or morphology of roots and the plant in general [123]. The changes in root exudates cause a new microbial equilibrium to be established. These changes presumably involve the same types of organisms as were involved in the rhizosphere before the formation of mycorrhizae, but quantitative changes occur within these types as a result of direct metabolic interactions with the mycosymbiont hyphae or spores or the indirect effects mediated by the host.

The microbial interactions in the mycorrhizosphere influence plant growth by giving microbes specific functional capabilities. These microbes may be strict or facultative anaerobes, extracellular chitinase producers, phosphate solubilizers, siderophore producers, antibiotic producers, hormone producers, pathogen suppressors, exopathogens, mycorrhiza stimulators or suppressors, etc. Studies have shown that the mycorrhiza colonization is enhanced if certain associative bacteria and fungi are present at the time of inoculation [124]. The mode of action has not been documented, but it is probably due to the production of growth-promoting substances, such as thiamin, which stimulated both the plant roots and mycorrhizal fungi. Many workers have reported interactions of soil microbes around mycorrhiza. Katznelson et al. [125] have described rhizosphere effect of mycorrhizal and nonmycorrhizal roots of yellow birch seedlings. Oswald and Ferchau [126] reported a significant difference between bacterial population of nonmycorrhizal and mycorrhizal conifer roots. Similarly, Neal et al. [127,128] also reported a difference in the microbial population of the rhizosphere associated with the mycorrhizae of Douglas fir and red alder. Linderman [129] reported that concomitant colonization and infections of roots by mycorrhizal fungi and by pathogens and other microorganisms inevitably lead to alteration of the other's activities.

A. Mechanisms of ECM Interactions with Root Diseases

Due to root diseases, the reduction in root volume disturbs nutrient and water uptake. In nutrient-poor soils, root loss may lower the yield appreciably. If a pathogen infects and destroys the feeder root prior to colonization by ectomycorrhizal fungi, ectomycorrhiza cannot form. But, if the ectomycorrhizal fungi colonize the feeder roots prior to infection of the pathogen, the roots become resistant to the attack by the pathogen [74].

ECM increases the longevity of feeder roots by preventing soil pathogens from infecting roots by various mechanisms.

Antibiosis

Zak [130] and Marx [131] have indicated that some ECM protect plant roots from pathogen damage by antibiosis. Antibodies produced should be ideally located for inhibiting effects on pathogens. These should also exert a selective effect upon rhizosphere microorganisms. Marx [131] pointed out that it is likely that a well-developed fungus sheath of ECM cannot be easily penetrated by a pathogen such as *Phytophthora cinnamomi*. It is possible that mycorrhizal infection may produce plant defense compounds against other microorganisms [132].

The penetration of the ectomycorrhizal mycosymbionts, i.e., basidiomycetes and ascomycetes, between and into the cortical cells appears to be limited by host production of phenolics and phytoalexins [24,133].

Improved Nutrition

Improved nutrition caused by mycorrhizal development can indirectly affect diseases, e.g., increased arginine content by mycorrhiza formation results in reduced chlamydospore formation of *Thielaviopsis basicola* in tobacco roots. Mycorrhizal fungi may protect absorbing roots of trees by utilizing root carbohydrates and other chemicals, thereby reducing the susceptibilitity for root pathogens [130].

Compensation

Roots lost to disease in a deficient soil will lead to reduced yield, but active mycorrhiza formation and hyphal growth in soil can compensate for such losses. Furthermore, this compensating tissue, i.e., the fungal hyphae, is not susceptible to the same diseases as the root, and this provides a nice example of the many "insurance systems" that plants have developed during evolution [122].

Mechanical Barrier

Suberization or cutinization of newly formed roots confers a degree of disease protection, but it also causes a very large decline in ion uptake by that part of the root. However, if the root has become mycorrhizal before suberization it can maintain an efficient uptake function via mycorrhizae, while enjoying the disease protection by suberization [122,134].

However, Bowen and Theodoru [135] and Garbaye and Bowen [136] have shown that some microorganisms isolated from soil or rhizosphere have a stimulating effect on ECM development. Garbaye and Bowen [137] hypothesized that these microbes are likely to be associated with the ECM fungus in the symbiotic state, i.e., isolated from the mycorrhiza itself. Such a close association suggests some mutual benefits for the ECM fungus and the surrounding microflora.

B. Interactions with Bacteria

Bacteria from fungal sporocarps of Lycoperdaceae having either broken or unbroken peridia have been isolated, but their functional importance could not be explained. Larsen et al. [138] demonstrated nitrogenase activity measured by acetylene reduction in sporocarps of certain fungi. They observed the presence of bacteria in a SEM studies. Li and Castellano [139] and Li [140] identified acetylene-reducing bacteria, *Azospirillum*, isolated from within sporocarps of ectomycorrhizal fungi associated with Douglas fir. In a follow-up study sporocarps of *Barssia oregoensis* and *Rhizopogon parksii* were examined for nitrogenase-producing bacteria. *Barssia oregoensis*, *H. setchellii*, and *R. parksii* are common ectomycorrhiza associates of Douglas fir. *L. scabrum* is commonly associated with birch.

Rambelli [16] reviewed the microbial interactions in the mycorrhizosphere. He indicated the close association of nitrogen-fixing bacteria with the ectomycorrhizal mantle of *Pinus radiata*, wherein atmospheric nitrogen fixed by the bacteria was available for use by the fungus and plant host, and the bacteria could derive nutrients from the fungal hyphae. Bowen and Theodorou [135] isolated bacteria from the mycorrhizosphere of pine and demonstrated through in vitro studies that some have deleterious and others beneficial effects on mycorrhiza. However, they could not show the actual influence of mycorrhiza on plant growth and development. Krywolap [141] observed the differences between species of mycorrhizal fungi in their activity against bacteria. Out of 26 species of ECM, 6 were active against both gram-positive and gram-negative bacteria, 11 were active against gram-negative only. But others did not show any inhibitory activity in culture media. Marx [142] identified active compounds produced by *Leucopaxillus cerealis* var. *piceina* as diatretyne nitrile and diatretyne-3 in the culture media. The amide and diatretyne-3 have only bacteriostatic activity whereas the nitrile is very potent against both bacteria and fungi. Garbaye and Bowen [137] have shown that the total number of microorganisms in 1 g of mycorrhiza was 2×10^6. Among them 75% were nonfluorescent gram-negative bacteria and 10% fluorescent pseudomonands. It was observed that the isolated microorganisms were inside the mycorrhizae. These results elucidated that a large population of microorganisms (predominantly gram-negative bacteria) lives in close association with the ectomycorrhizal fungus in its symbiotic state. They further discussed that stimulation of *Rhizopogon luteolus* by ECM at different stage of its development commonly occurs [132]. It was observed that many strains of microbes used change the growth habit of ECM; the mycelium is loose instead of being aggregated in strands as in the control. They concluded that the inside of the ECM seems to be the right place to look for interesting helpers for improving ectomycorrhizal inoculation techniques in the nursery. Malajczuk and McComb [143] compared the mycorrhizosphere and rhizosphere populations of eucalyptus seedlings grown in soils suppressive or conducive to *Phytophthora cinnamomi*. They found that microflora associated with mycorrhizae was distinctly different from that occurring on nonmycorrhizal roots, and high counts of bacteria were present throughout the mantle, both within and between symbiont cells. They found that many of the bacteria were highly antagonistic to *P. cinnamomi* and *Pythium* sp. Schisler and Linderman [144] reported that volatiles emanating from ectomycorrhizae exert a selective pressure on segments of the microflora, resulting in shifts in populations of bacteria.

C. Interactions with Fungi

The mechanisms involved in the induction of disease resistance by ectomycorrhizal fungi are poorly understood. Studies from different sources have presented contradictory results. Antibiotic production, synthesis of plant-produced antimicrobial compounds, and a physical barrier effected by ECM have been implicated in the induction of disease resistance [9]. Krupa and Fries [145] identified the volatile substances produced by a strain of *Suillus variegatus* in culture as ethanol, isobutanol, isoamyl alcohol, acetoin, and isobutyric acid. Krupa and Nylund [146] reported that 1% (v/v) concentrations of isobutanol, isoamyl alcohol, and isobutyric acid in agar media inhibited the radial growth of the mycelia of *Phytophthora cinnamomi* and *Fomes annosus*, while ethanol was stimulatory at this concentration. The production of bacteriostatic and fungistatic substances by mycorrhizal fungi has been known for a long time. Morimoto et al. [147], Levisohn [148], and Bjorkman [149] have examined mycorrhizal fungi for their antibiotic-producing ability. According to these workers, spruce becomes susceptible to *Fomes annosus* when its roots are without mycorrhiza. Similar results have also been reported by other workers where mycorrhizal fungi antagonized the pathogen [148,150–155]. However, some other mycorrhizal fungi like *Amanita* and *Pisolithus tinctorius* did not produce any antibiotic on the medium. Harley and Smith [9] concluded that chemical protection of the host against pathogens is unlikely to be a universal trait of ectomycorrhiza, but it may indeed be extremely important under appropriate ecological conditions. Duchesne et al. [156] have presented the evidence that pine root exudate stimulates the synthesis of antifungal agents by ectomycorrhizal fungus *Paxillus involutus*. Bioassay of root exudate suggested that the fungal toxic activity of pine root exudate inoculated with *P. involutus* is identical to that of exudate of *P. resinosa* seedlings inoculated with *P. involutus*. They postulated that the increase in the survival of *P. resinosa* seedlings to further infection by *Fusarium oxysporum* f. sp. *pini* is associated with an increase in the fungal toxic activity in the rhizosphere. Schisler and Linderman [144] indicated that ECM produced certain volatiles in vitro which are deleterious to the hyphal growth of several root pathogenic fungi [146]. Their results indicated that volatiles from ECM and the rhizosphere can influence the qualitative and quantitative differences in the microbial populations of mycorrhizosphere vs. nonmycorrhizosphere. Volatilies alone are not likely to be responsible for the exclusion of *Fusarium* from coniferous forest soils but presumably act in conjunction with a variety of chemical, physical, and biological factors unique to forest soils [157].

Recently Duchesne et al. [158] examined the effect of root exudate of *Pinus resinosa* on the synthesis of ethanol-soluble antifungal compounds of *Paxillus involutus*. The influence was determined by a bioassay based on microconidial germination of *Fusarium oxysporum* f. sp. *pini*. Inoculation of *P. resinosa* root exudate with both *P. involutus* and a spore suspension of *F. oxysporum* led to an 80% suppression of the sporulation of *F. oxysporum* when compared with controls lacking *P. involutus*. The increased resistance of *P. resinosa*–*P. involutus* association to *Fusarium* root rot may be the result of biosynthesis of antifungal compounds by *P. involutus*, which is stimulated by *P. resinosa* root exudate.

In another study Duchesne [159] inoculated *Pinus resinosa* seedlings with *P. involutus* following inoculation with *F. oxysporum*, which increased survival from 45.7% for nonmycorrhizal plants to 92.5%. This increase in survival was associated with a sixfold decrease in sporulation by *F. oxysporum* in the rhizosphere of seedlings inoculated with *P. involutus*. Bioassay of extracts from seedling rhizosphere, roots, and other tissues showed that the ECM caused a fivefold increase in fungitoxic activity in the rhizosphere, while the plant tissues showed no fungal toxicity.

Zak [130] and Marx [152] suggested that a growth-stimulating mycorrhiza does not necessarily have plant-protecting properties and the mycorrhizae that protect the plants from pathogens may not have a stimulating effect on the growth of the plant. This has been proved recently by Unestam et al. [160]. They have shown that *Hebeloma crustuliniforme* protected plants against *Fusarium* and *Rhizoctonia* but could not stimulate short root formation. They indicated that it is not necessary that the mechanical barrier provided by the mantle protect the root tips, because the nonmantle roots are protected even though carbohydrates may leak out into the rhizosphere and support pathogen growth.

It is possible that mycorrhiza formation may elicit production of plant defense compounds against other microorganisms. Fosters and Marks [161] found enhanced polyphenol production in the root cortex of mycorrhizal *Pinus radiata*. Similarly, Krupa and Fries [145] detected an up to eightfold increase in volatile terpenes and sesquiterpenes (fungistatic) with ECM inoculation of *Pinus sylvestis* with *Boletus variegatus*. Unestam et al. [160] did not observe any antagonistic effects in vitro, but most of the ECM placed against *Fusarium* and *Rhizoctonia*. However, they found that root extracts and exudates of ectomycorrhizal roots inhibited the pathogen growth in vitro. Extracts of *Laccaria*- and *Hebeloma*-colonized roots separated by HPLC were found to contain two nonpolar components, which were absent in non-ECM roots. The quantities of both the nonpolar compounds in the *Laccaria* roots colonized were several times larger than those in the less protected *Hebeloma*-colonized roots [160].

IX. CONCLUSIONS

It is evident that mycorrhizal studies are required for better forest management. Attempts should be made to produce inoculum on a large scale. It is essential to work out the methods of production and storage of high quality inoculum with maximum potential. Further, more concentrated efforts should be made to work out the methods of applying mycorrhizal inoculum to the seed or soil and inoculum placement in order to get maximum response.

REFERENCES

1. Meyer, F. H. Distribution of ectomycorrhizae in native and man made forests. In *Ectomycorrhizae* (G. C. Marks and T. T. Kozlowski, eds.), Academic Press, New York, 1973, pp. 79–105.
2. Cooper, K. M. A field survey of mycorrhizas in New Zealand ferns. *N.Z.J. Bot.*, *14*:169–181, 1976.
3. Iqbal, S. H., Yousaf, M., and Younus, M. A field survey of mycorrhizal associations in ferns in Pakistan. *New Phytol.*, *87*:69–79, 1981.
4. Trappe, J. M. Fungus associates of ectotrophic mycorrhizae. *Bot. Rev.*, *28*:538–606, 1962.
5. Bakshi, B. K. Mycorrhiza and its role in forestry. P. L. 480 report, Dehradun, India, 1974.
6. Warcup, J. H. A culturable *Endogone* associated with *Eucalyptus*. In *Endomycorrhizas* (F. E. Sanders, B. Mosse, and P. B. Tinker, eds.), Academic Press, London, 1975, pp. 53–63.

7. Rommell, L. G. A trenching experiment in spruce forest and its bearing on the problems of mycotrophy. *Svensk. Bot. Tidkr.*, *32*:89–99, 1938.
8. Laiho, O. *Paxillus involutus* as a mycorrhizal symbiont of forest trees. *Acta For. Fenn.*, *106*:1–65, 1970.
9. Harley, J. L., and Smith, S. E. *Mycorrhizal Symbiosis*. Academic Press, New York, 1983, p. 483.
10. Robertson, N. F. Studies on the mycorrhiza of *Pinus sylvestris*. *New Phytol.*, *53*:253–283, 1954.
11. Chilvers, G. A. and Gust, L. W. Comparisons between growth rates of mycorrhizas, uninfected roots and mycorrhizal fungus of Eucalyptus St. Johnii R.T. Bak. *New Phytol.*, *91*:453–456, 1982.
12. Wilcox, H. E. Morphological studies of the root of red pine, *Pinus resinosa*. I. Growth characteristics and branching patterns. *Am. J. Bot.*, *55*:247–254, 1968a.
13. Wilcox, H. E. Morphological studies of the roots of red pine, *Pinus resinosa*. II. Fungal colonization of mycorrhizae. *Am. J. Bot.*, *55*:687–700, 1968b.
14. Chilvers, G. A. Some distinctive types of eucalypt mycorrhiza. *Aust. J. Bot.*, *26*:49–70, 1968.
15. Chilvers, G. A. and Gust, L. W. The development of mycorrhizal populations on pot-grown seedlings of *Eucalyptus* St. Johnii R.T. Bak. *New Phytol.*, *90*:667–679, 1982.
16. Rambelli, A. The rhizosphere of mycorrhizae. In *Ectomycorrhizae* (G. L. Marks and Koslowski, eds.), Academic Press, New York, 1973, pp. 299–343.
17. Giltrap, N. J. and Lewis, D. H. Catabolite repression of the synthesis of pectin-degrading enzymes by *Suillus luteus* (L. ex fr.) S. F. Gray and *Hebeloma oculatum bruchet*. *New Phytol.*, *90*:485–493, 1982.
18. Anderson, A. J. Mycorrhizae—Host specificity and recognition. *Phytopathology*, *78*:375–378, 1988.
19. Duddridge, J. A. and Read, D. J. Modification of the host-fungus interface in mycorrhizal synthesized between *Suillus bovinus* (Fr.) O. Kunty and *Pinus sylvestris* L. *New Phytol.*, *96*:583–588, 1984.
20. Ronald, P. and Soderhall, K. Phenylamine ammonia lyase and peroxidase activity in mycorrhizal and non-mycorrhizal short roots of scots pine, *Pinus sylvestris* L. *New Phytol.*, *101*:487–494, 1985.
21. Piche, Y., Fortin, J. A., and Lafotaine, J. G. Cytoplasmic phenols and polysaccharides in ectomycorrhizal and non-mycorrhizal roots of pine. *New Phytol.*, *88*:695–703, 1981.
22. Smith, D. C. and Douglas, A. E. *The Biology of Symbiosis*. Edward Arnold, London, 1987, p. 302.
23. Godbout, C. and Fortin, J. A. Synthesized ectomycorrhizal of aspen; fungal genus level of structural characterization. *Can. J. Bot.*, *63*:252–262, 1985.
24. Molina, R. and Trappe, J. M. Patterns of ectomycorrhizal host specificity and potential among Pacific Northwest conifers and fungi. *For. Sci.*, *28*:423–458, 1982.
25. Malajczuk, N., Molina, R., and Trappe, J. M. Ectomycorrhiza formation in Eucalyptus I. Pure culture synthesis, host specificity and mycorrhizal compatibility with *Pinus radiata*. *New Phytol.*, *91*:467–482, 1982.
26. Ramstedt, M. Physiology of the ectomycorrhizal fungus *Piloderma croceum*, with special emphasis on mannitol metabolism. Ph.D. thesis, University of Uppsala, Sweden, 1985.

27. Piche, Y. and Peterson, R. L. Early events in root colonization by mycorrhizal fungi. In *Proc. 6th NACOM* (R. Molina, ed.), Oregon State University, Corvallis, OR, 1984, p. 179.
28. Hilbert, J. L., Gaudin, V., Lapeyric, F., and Martin, F. Identification of 'Ectomycorrhiza-Specific' proteins involved in the development of the *Pisolithus-Eucalyptus* symbiosis. In *Proc. 7th NACOM* (D. M. Sylvia, L. L. Hung, and J. H. Graham, eds.), Gainesville, FL, 1987, p. 251.
29. Ramakrishnan, P. S. and Swamy, P. S. Role of mycorrhiza in the management of tropical rain forest. In *Mycorrhiza Round Table Proc. of Workshop* (A. K. Varma, A. K. Oka, K. G. Mukerji, K. V. B. R. Tilak, and Janak Raj, eds.), New Delhi, India, 1987, pp. 471–487.
30. Ramakrishnan, P. S. Research on humid tropical forests. In *Regional Meeting of the National MAB Committees of Central and South Asian Countries*, New Delhi, 1985.
31. Ofosu, A. A field performance of *Pinus caribaea* inoculated with pure cultures of four mycorrhizal fungi. In *Tropical Mycorrhiza Research* (P. Mikola, ed.), Clarendon Press, Oxford, 1980, pp. 87–92.
32. Abeynayake, K. Mycorrhizal associations in tropical humid forests. In *Regional Workshop on The Ecology and Conservation of the Tropical Humid Forests of the Indomalayan Realm*, Colombo, Sri Lanka, 1987.
33. Mikola, P. (ed.) *Tropical Mycorrhiza Research*, Oxford University Press, Oxford, 1980, pp. 3–10.
34. Werlich, I. and Lyr, H. Über die mykorrhiza Ausbilding von Kiefer (*Pinus sylvestris* L.) und Buche (*Fagus sylvetica* L.) auf verschiedenen Standorten. *Arch. Forstn.*, *6*:1–23, 1957.
35. Mikola, P. and Laiho, O. Mycorrhizal relations in the raw humus layer of northern spruce forests. *Communs. Inst. For. Fenn.*, *55*:1–3, 1962.
36. Meyer, F. H. and Gottsche, D. Distribution of root tips and tender roots of beech. In *Ecological Studies: Analysis and Synthesis* (H. Ellenberg, ed.), Springer-Verlag, New York, 1971, pp. 48–52.
37. Fogel, R. Mycorrhizae and nutrient cycling in natural forest ecosystems. *New Phytol.*, *86*:199–212, 1980.
38. Fogel, R. and Hunt, G. Fungal and arboreal biomass in a western Oregon Douglas fir ecosystem: Distribution patterns and turnover. *Can. J. For. Res.*, *9*:245–256, 1979.
39. Stark, N. Nutrient cycling pathways and litter fungi. *Biosciences*, *22*:355–360, 1972.
40. Vogt, K. A. and Edmunds, R. L. Patterns of nutrient concentration in basidiocarp in Western Washington. *Can. J. Bot.*, *58*:694–698, 1980.
41. Richards, P. W. *The Tropical Rain Forest*, Cambridge University Press, Cambridge, 1952.
42. Herrera, R., Jordan, C., Klinge, H., and Mediana, E. Amazon ecosystems. Their structure and functioning with particular emphasis on nutrients. *Interciencia*, *3*:223–232, 1980.
43. Vogt, K. A., Grier, C. C., Meier, C. E., and Edmunds, R. L. Mycorrhizal role in net primary production and nutrient cycling in *Abies anabilis* (Dougl.) forbes. ecosystems in western Washington. *Ecology*, *63*:370–380, 1982.

44. Dighton, J. Ecology and management of ectomycorrhizal fungi in the U.K. In *Proc. 7th NACOM* (D. M. Sylvia, L. L. Hung, and J. H. Graham, eds.), Gainesville, FL, 1987, p. 75.
45. Dighton, J. and Mason, P. A. Mycorrhizal dynamics during forest tree development. In *Development of Higher Fungi* (D. Moore, L. A. Casselton, D. A. Wood, and J. C. Frankland, eds.), Cambridge University Press, Cambridge, 1985, pp. 119–139.
46. Harley, J. L. Fungi in ecosystems. *J. Ecol.*, *59*:653–668, 1971.
47. Gardner, J. H. and Malajczuk, N. Succession of ectomycorrhizal fungi associated with Eucalyptus on rehabilitated bauxite mines in South Western Australia. In *Proc. 6th NACOM* (R. Molina, ed.), Oregon State University, Corvallis, OR, 1984, p. 265.
48. Mason, P. A., Wilson, J., Last, F. T., and Walker, C. The concept of succession in relation to the spread of sheathing mycorrhizal fungi in inoculated tree seedlings growing in unsterile soil. *Pl. Soil*, *71*:247–256, 1987.
49. Griffiths, R. P., Caldwell, B. A., Cromack, K., Jr., Castellano, M. A., and Morite, R. Y. A study of chemical and microbial variables in forest soils colonized with *Hysterangium setchellii* rhizomorphs. In *Proc. 7th NACOM* (D. M. Sylvia, L. L. Hung, and J. H. Graham, eds.), Gainesville, FL, 1987, p. 96.
50. Melin, E. Experimentelle Untersuchungen über die Konstitution and Okologie der Mykorrhizen von *Pinus sylvestris* und *Picea abies*. *Mykol. Unters. Ber.*, *2*:72–331, 1923.
51. Hatch, A. B. and Doak, K. D. Mycorrhiza and other features of *Pinus*. *J. Arnold Arbor.*, *14*:85–99, 1933.
52. Harley, J. L. A study of the root system of the beech in woodland soils with especial reference to mycorrhizal infection. *J. Ecol.*, *28*:107–117, 1940.
53. Worley, J. F. and Hacskaylo, E. The effect of available soil moisture on the mycorrhizal association of Virginian pine. *For. Sci.*, *5*:267–268, 1959.
54. Hatch, A. B. The physical basis of mycotrophy of the genus *Pinus*. *Black Rock For. Bull.*, *6*:168, 1937.
55. Mikola, P. On the physiology and ecology of *Cenococcum graniforme*. *Communs. Inst. For. Fenn.*, *36*:1–104, 1948.
56. Theodorou, C. and Bowen, G. D. Influence of temperature on the mycorrhizal association of *Pinus radiata* D. Don. *Aust. J. Bot.*, *19*:13–20, 1971.
57. France, R. C. and Reid, C. P. P. Amino acid inhibition analysis of pine ectomycorrhizas. In *Proc. 4th NACOM*, Fort Collins, CO, 1979.
58. Moser, M. Der Einfluss liefer Temperaturen auf den Wachstum und Lebenstandigheit hoherer Pilze mit spezieller Beruchsichtigung von Mykorrhizapilzen. *Sydowia*, *12*:386–399, 1958.
59. Brundrett, M. and Kendrick, B. Mycorrhizae in a hard wood forest. In *Proc. 7th NACOM* (D. M. Sylvia, L. L. Hung, and J. H. Graham, eds.), Gainesville, FL, 1987, pp. 87–88.
60. Cromack, K., Sollins, P., Granstein, W. C., Speidel, K., Todd, A. W., Spycher, G., Ching, Y. L., and Todd, R. L. Calcium oxalate accumulation and soil weathering in mats of the hypogeous fungus *Hysterangium crassum*. *Soil Biol. Biochem.*, *11*:463–468, 1979.

61. Honston, H. W. Chelation between calcium and organic anions. *N.Z.J. Sci. Technol.*, *37*:522–537, 1956.
62. Clement, A., Garbaye, J., and Le Tacon, F. Importance des ectomycorrhizes dans la resistance au calcaire du Pin noir (*Pinus nigra* Arn. ssp. *nigricans* Host). *Oecol. Pl.*, *12*:111–131, 1977.
63. Muchovej, J. J., Muchovej, R. M. C., and Marx, D. H. Surface area of ectomycorrhizae as a measure of comparative fungal efficiency. In *Proc. 7th NACOM* (D. M. Sylvia, L. L. Hung, and J. H. Graham, eds.), Gainesville, FL, 1987, p. 213.
64. Duddridge, J. A., Malibari, A., and Read, D. J. Structure and function of mycorrhizal rhizomorphs with special reference to their role in water transport. *Nature* (London), *287*:834–836, 1980.
65. Marx, D. H. and Cordell, C. E. Ecology and management of ectomycorrhizal fungi in regenerating forests in the eastern United States. In *Proc. 7th NACOM* (D. M. Sylvia, L. L. Hung, and J. H. Graham, eds.), Gainesville, FL, 1987, p. 69.
66. Diebolt, K. S. and Mudge, K. W. Do ectomycorrhizae influence host plant response to drought? In *Proc. 7th NACOM* (D. M. Sylvia, L. L. Hung, and J. H. Graham, eds.), Gainesville, FL, 1987, p. 246.
67. Schaffers, A. P. and Termorshuizen, A. J. Vitality of *Pinus sylvestris* L. Mycorrhizal Corpophores production and air pollution. In *Proc. 7th NACOM* (D. M. Sylvia, L. L. Hung, and J. H. Graham, eds.), Gainesville, FL, 1987, p. 263.
68. Termorshuizen, A. J. and Schaffers, A. P. Mycorrhizal fungi of *Pinus sylvestris* L. related to air pollution and tree vitality. In *Proc. 7th NACOM* (D. M. Sylvia, L. L. Hung, and J. H. Graham, eds.), Gainesville, FL, 1987, p. 266.
69. Dighton, J. and Skeffington, R. A. Effects of artificial acid precipitation on the mycorrhizas of Scots pine seedlings. *New Phytol.*, *107*:191–202, 1987.
70. Muchovej, R. M. C. and Kasuya, M. C. M. Temperature or aluminium level on mycorrhizal formation in *Pinus carbaea*. In *Proc. 7th NACOM* (D. M. Sylvia, L. L. Hung, and J. H. Graham, eds.), Gainesville, FL, 1987, p. 130.
71. Wasserman, J. L., Mine, O. L., Majumdar, S. K., and Tyne, C. V. Detection of heavy metals in oak mycorrhizae of northeastern Pennsylvanian forests, using x-ray microanalysis. *Can. J. Bot.*, *65*:2622–2627, 1987.
72. Dixon, R. K. Response of ectomycorrhizal *Quercus rubra* to soil cadmium, nickel and lead. *Soil Biol. Biochem.*, *20*:555–559, 1988.
73. Nye, P. H. and Tinker, P. B. *Solute Movement in the Soil Root System*, Blackwell Scientific Pub., Oxford, 1977.
74. Narayanan, R. and Bhattacharyya, A. K. Ectomycorrhizas: Their role in tropical environment. In *Progress in Microbial Ecology* (K. G. Mukerji, V. P. Agnihotri, and R. P. Singh, eds.), Print House, Lucknow, India, 1984, pp. 609–622.
75. Tinker, P. B. Soil chemistry of phosphorus and mycorrhizal effects on plant growth. In *Endomycorrhizas* (F. E. Sanders, B. Mosse, and P. B. Tinker, eds.), Academic Press, London and New York, 1975, pp. 353–371.
76. Bowen, G. D. Mineral nutrition of mycorrhizas. In *Ectomycorrhizas* (C. Marks and T. T. Kozlowski, eds.), Academic Press, New York, 1973, pp. 151–201.

77. Bowen, G. D. Mycorrhizal roles in tropical plants and ecosystems. In *Tropical Mycorrhizal Research* (P. Mikola, ed.), Oxford University Press, Oxford, 1980, pp. 165–190.
78. Skinner, M. F. and Bowen, G. D. The penetration of soil by mycelial strands of ectomycorrhizal fungi. *Soil Biol. Biochem.*, *6*:57–81, 1974.
79. Morrison, T. M. Absorption of phosphorus from soils by mycorrhizal plants. *New Phytol.*, *61*:10–20, 1962.
80. Read, D. J. and Finlay, R. D. Mycorrhizas and inter-plants transfer of nutrients. I. Ectomycorrhizas. In *Proc. 6th NACOM* (R. Molina, ed.), Oregon State University, Corvallis, OR, 1984, p. 319.
81. Finlay, R. D. and Read, D. J. The structure and function of the vegetative mycelium of ectomycorrhizal plants. I. Translocation between plants interconnected by a common mycelium. *New Phytol.*, *103*:143–156, 1986.
82. Hadley, G. Mycorrhizas and plant growth and development. In *Beltsville Symposium on Agricultural Research (12) Biomechanisms Regulating Growth and Development* (G. L. Steffens and S. Theron, eds.), Rumsey, BARC, Maryland, Kluwer Academic Pub., Netherlands, 1987.
83. Lewis, D. H. Interrelationships between carbon nutrition and morphogenesis in mycorrhizas. In *Physiological and Genetical Aspects of Mycorrhizae, Proc. 1st ESM* (V. Gianinnazi Pearson and S. Gianinnazi, eds.), INRA, Paris, 1986, pp. 84–100.
84. Lewis, D. H. and Harley, J. L. Carbohydrate physiology of mycorrhizal roots of beech. III. Movement of sugars between host and fungus. *New Phytol.*, *64*:256–269, 1965.
85. Soderstrom, B., Finlay, R. D., and Read, D. J. The structure and function of the vegetative mycelium of ectomycorrhizal plants. IV. Qualitative analysis of carbohydrate contents of mycelium interconnecting host plants. *New Phytol.*, *109*:163–166, 1988.
86. Norby, R. J., O'Neil, E. G., Hood, W. G., and Luxmoore, R. J. Carbon allocation, root exudation and mycorrhizal colonization of *Pinus echinata* seedlings grown under CO_2 enrichment. *Tree Physiol.*, *3*:203–210, 1987.
87. Harley, J. L. *The Biology of Mycorrhiza*, 2nd ed. Leonard Hill, London, 1969, p. 334.
88. Slankis, V. Hormonal relationships in mycorrhizal development. In *Ectomycorrhiza* (G. C. Marks and T. T. Kozylowski, eds.), Academic Press, New York, 1973, pp. 231–298.
89. Ho, I. Phytohormonal control of seedling growth by ectomycorrhizal fungi. In *Proc. 7th NACOM* (D. M. Sylvia, L. L. Hung, and J. H. Graham, eds.), Gainesville, FL, 1987, p. 252.
90. Mudge, K. Hormonal involvement in the regulation and maintenance of ectomycorrhizae. In *Proc. 7th NACOM* (D. M. Sylvia, L. L. Hung, and J. H. Graham, eds.), Gainesville, FL, 1987.
91. Graham, J. H. and Linderman, R. G. Ethylene production of ectomycorrhizal fungi, *Fusarium oxysporum* f. sp. *pini*, and by aseptically synthesized ectomycorrhizae and fusarium infected Douglas-fir roots. *Can. J. Microbiol.*, *26*:1340–1347, 1980.
92. Barnes, R. L. and Naylor, A. W. In vitro culture of pine roots and the use of *Pinus serotina* roots in preliminary metabolic studies. *For. Sci.*, 5:158–168, 1959.

93. Ek, M., Ljungguist, P. O., and Stenstorm, E. Indole-3-acetic acid production by mycorrhizal fungi determined by gas chromatography–mass spectrometry. *New Phytol.*, *94*:401–407, 1983.
94. Bates, G. W. and Goldsmith, M. H. M. Rapid response of the plasmamembrane potential in oat coleoptiles to auxin and other weak acids. *Planta*, *159*:231–237, 1983.
95. Cleland, R. E. *Plant Growth Substances*, Academic Press, New York, 1982, pp. 22–31.
96. Hagen, G. A., Kleinschmidt, A., and Guilfoyle, T. Auxin regulated gene expression in intact soybean hypocotyl and excised hypocotyl sections. *Planta*, *162*:147–153, 1984.
97. Letham, D. S. and Palni, L. M. S. The biosynthesis and metabolism of cytokinins. *Ann. Rev. Plant Physiol.*, *34*:163–197, 1983.
98. Frankenberger, W. T., Jr. and Poth, M. Biosynthesis of indole-3-acetic acid by the pine ectomycorrhizal fungus *Pisolithus tinctorius*. *App. Environ. Microbiol.*, *53*:2908–2913, 1987.
99. Nylund, J. and Unestam, T. Structure and physiology of ectomycorrhizae. I. The process of mycorrhiza formation in Norway spruce *in vitro*. *New Phytol.*, *91*:63–79, 1987.
100. Rouillon, R., Guy, G., Bermillon, J., Favre-Bonvin, J., and Bruchet, G. Analysis by HPLC-mass spectrometry of the indole compounds released by the ectomycorrhizal fungus *Hebeloma hiemale* in pure culture. *Can. J. Bot.*, *64*:1893–1897, 1986.
101. Bruchet, G., Debaud, I. C., and Gay, G. Genetic variations in the physiology of *Hebeloma*. In *Proc. of the 1st Eur. Symp on Mycorrhiza* (V. Gianinazzi Pearson and S. Gianinazzi, eds.), France Inst. Natl. de la Recherche Agronomique Paris, Dijon, France, 1986, pp. 121–131.
102. Mitchell, R. J., Garrett, H. E., Core, G. S., and Atalay, A. Boron and ectomycorrhizal influence on indole-3-acetic acid levels and indole-3-acetic acid oxidase and peroxidase activities of *Pinus euchinate* Mill. roots. *Tree Physiol.*, *1*:1–8, 1986.
103. Moser, M. Beitrage zur Kentnis der Wuchstoffbezieungen in bereich ectotrophen Mykorrhizen. *Arch. Mikrobiol.*, *34*:251–264, 1959.
104. Ulrich, J. M. Auxin production by mycorrhizal fungi. *Physiol. Plant.*, *13*:429–443, 1960.
105. Horak, E. Die Bildung von IES-Derivation durch ectotrophe Mikorrhizapilze (*Phlegmacium* sp.) von *Picea abies Karsten*. *Phytopathology*, *51*:491–515, 1961.
106. Shemakhanova, N. M. *Mycotrophy of Wood Plant.* USDA and NSF, Washington, DC, Israel Program for Scientific Translations, 1967, p. 329.
107. Tomaskewski, M. and Wojciechowska, B. The role of growth regulators released by fungi in pine mycorrhizae. In *8th Int. Conf. on Plant Growth Substances*, Tokyo, 1973, pp. 217–227.
108. Hemberg, T. and Larsson, U. Interaction of kinetin and indole acetic acid in the Avene straight growth test. *Physiol. Plant.*, *26*:104–107, 1972.
109. Miller, C. O. Cytokinin production by mycorrhizal fungi. In *Mycorrhizae* (E. Hacskaylo, ed.), *Proc. 1st NACOM*, Washington, D.C., 1971.
110. Crafts, C. B. and Miller, C. D. Detection and identification of cytokinins produced by mycorrhizal fungi. *Pl. Physiol.*, *54*:586–588, 1974.

111. Ng, P. P., Cole, A. L. J., Jameson, P. E., and McWha, J. A. Cytokinin production by ectomycorrhizal fungi. *New Phytol.*, *91*:57–62, 1982.
112. Ho, I. and Trappe, J. M. Enzymes and growth substances of *Rhizopogon* sp. in relation to mycorrhizal hosts and intrageneric taxonomy. *Mycologia*, *79*:553–558, 1987.
113. Hanley, K. M. and Greene, D. W. Gibbrellin like compounds from two ectomycorrhizal fungi and the GA_3 response on Scots pine seedlings. *Hort. Sci.*, *22*:591–594, 1987.
114. Lynch, J. M. Identification of substrate and isolation of microorganisms responsible for ethylene production in the soil. *Nature* (Lond.), *240*:45–46, 1972.
115. Lynch, J. M. and Harper, S. H. T. Formation of ethylene by soil fungus. *J. Gen. Microbiol.*, *80*:187–195, 1974.
116. Graham, J. H. and Linderman, R. G. Ethylene production by ectomycorrhizal fungi on *Fusarium oxysporum* sp. *pini* in vitro and by aseptically synthesized mycorrhizal and *Fusarium* infected roots of Douglas fir. In *Proc. 4th NACOM*, Fort Collins, CO, 1979.
117. Wilson, E. R. and Field, R. J. Dichotomous branching in lateral roots of pine. The effect of 2-chloroethyephosphoric acid on seedlings of *Pinus radiata* D. Don.. *New Phytol.*, *98*:465–473, 1984.
118. Rupp, L. A. and Mudge, K. W. Ethephon and auxin induce mycorrhiza like changes in the morphology of root organ cultures of Mugo pine. *Physiol. Plant.*, *64*:316–322, 1985.
119. De vries, H. E., II, Mudge, K. W., and Lardner, J. P. Ethylene production by several ectomycorrhizal fungi and effects on host root morphology. In *Proc. 7th NACOM* (D. M. Sylvia, L. L. Hung, and J. H. Graham, eds.), Gainesville, FL, 1987, p. 245.
120. Gogala, N. Jasmonic acid and methyl jasmonate-growth inhibitors for mycorrhizal fungi. In *Proc. 7th NACOM* (D. M. Sylvia, L. L. Hung, and J. H. Graham, eds.), Gainesville, FL, 1987, p. 198.
121. Hiltner, L. Über neuer Erfahrungen und Probleme auf den Gebeit der Bodenbacteriologie und unter besonderer Beruchsichtigun der Grundungen und Brache. *Arb. dt. Land W. Ges.*, *98*:59–78, 1904.
122. Bowen, G. D. Misconceptions, concepts and approaches in rhizosphere biology. In *Contemporary Microbial Ecology* (D. C. Ellwood, J. N. Hedger, M. J. Latham, J. M. Lynch, and J. H. Slater, eds.), Academic Press, London, 1980, pp. 283–304.
123. Linderman, R. G. Mycorrhizal interactions with the rhizosphere microflora: The mycorrhizosphere effect. *Phytopathology*, *78*:366–371, 1988.
124. Davey, C. B. Nonpathogenic organisms associated with mycorrhiza. In *Mycorrhizae* (E. Hacskaylo, ed.), USDA, Beltsville, MD, Misc. Publ., 1971, pp. 114–121.
125. Katznelson, H., Rouatt, J. W., and Peterson, E. A. The rhizosphere effect of mycorrhizal and non-mycorrhizal roots of yellow birch seedlings. *Can. J. Bot.*, *40*:377–382, 1962.
126. Oswald, E. T. and Ferchau, H. A. Bacterial association of coniferous mycorrhizae. *Pl. Soil*, *28*:187–192, 1968.
127. Neal, J. L., Bollen, W. B., and Zale, B. Rhizosphere microflora associated with the mycorrhizae of Douglas-fir. *Can. J. Microbiol.*, *10*:259–265, 1964.

128. Neal, J. L., Lu, K. C., Bollen, W. B., and Trappe, J. M. A. Comparison of rhizosphere microflora associated with mycorrhizae of red alder and Douglas fir. In *Biology of Alder* (J. M. Trappe, J. F. Franklin, R. F. Tarrant, and G. M. Hansen, eds.), USDA for Serv. Pac. Northwest. for Range Exp. Stn., Corvallis, OR, 1968, p. 292.
129. Linderman, R. G. Microbial interactions in the mycorrhizosphere. In *Proc. 6th NACOM* (R. Molina, ed.), Corvallis, OR, 1984, pp. 117–120.
130. Zak, B. Role of mycorrhizae in root diseases. *Annu. Rev. Phytopath.*, *2*:377–392, 1964.
131. Marx, D. H. Mycorrhiza and feeder root diseases. In *Ectomycorrhizae-Their Ecology and Physiology* (G. C. Marks and T. T. Kozolowski, eds.), Academic Press, New York, 1973, pp. 351–382.
132. Smith, S. E. Mycorrhizal fungi. *Crit. Rev. Microbiol.*, *3*:275–313, 1974.
133. Graham, J. H. Interactions of mycorrhizal fungi with soil borne plant pathogens and other organisms: An introduction. *Phytopathology*, *78*:365–366, 1988.
134. Bowen, G. D., Bevege, D. I., and Mosse, B. Phosphate physiology of vesicular arbuscular mycorrhizae. In *Endomycorrhizae* (F. E. Sanders, B. Mosse, and P. B. Tinker, eds.), Academic Press, London and New York, 1975, pp. 241–260.
135. Bowen, G. D. and Theodorou, C. T. Interaction between bacteria and ectomycorrhizal fungi. *Soil Biol. Biochem.*, *11*:119–126, 1979.
136. Garbaye, J. and Bowen, G. D. Effect of different microflora on the success of ectomycorrhizal inoculation of *Pinus radiata*. *Can. J. For. Res.*, *17*:941–943, 1986.
137. Garbaye, J. and Bowen G. D. The effect of mycorrhizospheric mycoorganisms on ectomycorrhizal infection of *Pinus radiata*. In *Proc. 7th NACOM* (D. M. Sylvia, L. L. Hung, and J. H. Graham, eds.), Gainesville, FL, 1987, p. 196.
138. Larsen, M. J., Jurgense, M. F., Harvey, A. E., and Ward, J. C. Dinitrogen fixation associated with sporophores of *Fomentarius* and *Echinodontium tinctorium*. *Mycologia*, *70*:1217–1222, 1978.
139. Li, C. Y. and Castellano, M. A. Nitrogen fixing bacteria isolated from within sporocarp of three ectomycorrhizal fungi. In *Proc. 6th NACOM* (R. Molina, ed.), Corvallis, OR, 1984, p. 471.
140. Li, C. Y. Diazotrophic bacteria in sporocarps of ectomycorrhizal fungi *Barssia oregonensis, Hysterangium setchalli, Laccinum scabrum* and *Rhizopogon parksii*. In *Proc. 7th NACOM* (D. M. Sylvia, L. L. Hung, and J. H. Graham, eds.), Gainesville, FL, 1987, p. 205.
141. Krywolap, G. Production of antibiotics by certain mycorrhizal fungi. In *Mycorrhizae* (E. Hackskaylo, ed.), USDA Forest Service Misc., Washington, DC, 1971, pp. 210–222.
142. Marx, D. H. Ectotrophic mycorrhizae as biological deterrents to pathogenic root infections by *Phytophthora cinnamomi*. *Proc. 14 IUFRO Congress* (Munich), *24*:172–181, 1967.
143. Malajczuk, N. and McComb, A. J. The microflora of unsuberized roots of *Eucalyptus calophylla* R.Br. and *Eucalyptus marginata* Donn. ex Sm. seedlings grown in soils suppressive and conducive to *Phytophthora cinnamomi* Rands I. Rhizoplane bacteria, actinomycetes and fungi. II. Mycorrhizal roots and associated microflora. *Aust. J. Bot.*, *27*:235–272, 1979.

144. Schisler, D. A. and Linderman, R. G. The influence of volatiles purged from soil around Douglas-fir ectomycorrhizal on soil microbial populations. In *Proc. 7th NACOM* (D. M. Sylvia, L. L. Hung, and J. H. Graham, eds.), Gainesville, FL, 1987, p. 218.
145. Krupa, S. and Fries, N. Studies on ectomycorrhizae of pine. I. Production of volatile compounds. *Can. J. Bot.*, *49*:1425–1432, 1971.
146. Krupa, S. and Nylund, J. E. Studies on ectomycorrhizae of pine. III. Growth inhibition of two root pathogenic fungi by volatile organic constituents of ectomycorrhizal root systems of *Pinus sylvestris* L. *Physiol. Plant.*, *25*:1–7, 1971.
147. Morimoto, M., Iawi, M., and Lukumoto, J. Antibiotic substances from mycorrhizal fungi. I. Isolation of antibiotic strains. *Kagaku to Kagyo.*, *28*:111–116, 1954.
148. Levisohn, I. Aberrant infections of pine and spruce seedlings. *New Phytol.*, *53*:284–290, 1954.
149. Bjorkman, E. Über die Natur der Mycorrhizabildung unter besonderer Beruchsichtingung der Waldbäume und die Anwendung in der forstilichen Praxis. *Forstwiss. Zent. Bl.*, *75*:265–286, 1956.
150. Hyppel, A. Antagonistic effects of some soil fungi on *Fomes annosus* in laboratory experiments. *Stud. For. Suecica.*, *64*:1–18, 1968.
151. Hyppel, A. Effects of *Fomes annosus* on seedlings of *Picea abies* in the presence of *Boletus bovinus*. *Stud. For. Suecica.*, *66*:1–16, 1968.
152. Marx, D. H. Ectomycorrhiza as biological deterrents to pathogenic root infections. *Annu. Rev. Phytopathol.*, *10*:429–454, 1972.
153. Marx, D. H. Mycorrhizae and feeder root disease. In *Ectomycorrhizae* (G. C. Marks and T. T. Kozlowski, eds.), Academic Press, New York and London, 1973, pp. 351–382.
154. Marx, D. H. Role of ectomycorrhizae in the protection of pine from root infection by *Phythophthora cinamomi*. In *Biology and Control of Soil-Borne Plant Pathogens* (J. W. Bruehl, ed.), American Phytopathological Soc., St. Paul, MN, 1975.
155. Richard, C., Fortin, J. A., and Fortin, A. Protective effect of ectomycorrhizal fungus against the root pathogen *Mycelium radicis altovirens*. *Can. J. Bot.*, *1*:246–251, 1971.
156. Duchesne, L. C., Peterson, R. L., and Ellis, B. E. Pine root exudates stimulate antibiosis by *Paxillus involutus* against *Fusarium oxysporum*. In *Proc. 7th NACOM* (D. M. Sylvia, L. L. Hung, and J. H. Graham, eds.), Gainesville, FL, 1987, p. 193.
157. Schisler, D. A. Microbiol analysis of coniferous forest and nursery soils. Effect of ectomycorrhizal volatiles and humic rich organic substrates. Ph.D. thesis, Oregon State University, Corvallis, OR, 1986, p. 152.
158. Duchesne, L. C., Peterson, R. L., and Ellis, B. E. Pine root exudate stimulates the synthesis of antifungal compounds by the ectomycorrhizal fungus *Paxillus involutus*. *New Phytol.*, *108*:471–476, 1988.
159. Duchesne, L. C. and Peterson, R. L. Interaction between the ectomycorrhizal fungus *Paxillus involutus* and *Pinus resinosa* induces resistance to *Fusarium oxysporum*. *Can. J. Bot.*, *66*:558–562, 1988.
160. Unestam, T., Chakravarty, P., and Damm, E. Mycorrhizal protection on pine roots against pathogens. In *Proc. 7th NACOM* (D. M. Sylvia, L. L. Hung, and J. H. Graham, eds.), Gainesville, FL, 1987, p. 267.
161. Foster, R. C. and Markers, G. C. Observations on the mycorrhizas of forest trees. II. The rhizosphere of *Pinus radiata* D. Don. *Aust. J. Biol. Sci.*, *20*:915–926, 1967.

8

ECTOMYCORRHIZAL FUNGI AS EXPERIMENTAL ORGANISMS

VICTORIA BARRETT *Auburn University, Auburn, Alabama*

I. INTRODUCTION

Mycorrhizae are symbiotic associations of nonpathogenic soil fungi with living cells of plant roots. Mycorrhizae form the primary absorptive organs of most vascular plants and have been described as "metabolic harmony" between two organisms [2]. The two major types of mycorrhizae are the endomycorrhizae (vesicular-arbuscular mycorrhizae, VAM) and the ectomycorrhizae (EM), so named and distinguished by the anatomical nature of the symbiotic relationship. The VAM associations involve a limited number of species of zygomycetous fungi associated mainly with herbaceous plants. These fungi penetrate the cells of the root and are obligate symbionts [19]. In contrast, EM develop as an association with the primary roots of woody plants and involve more than a thousand species of mainly basidiomycetous and ascomycetous fungi, most with the ability to grow saprobically. The fungal partner of EM can be grown on defined, synthetic medium and thus can be experimentally manipulated apart from the plant partner [43]. The use of these fungi as experimental systems is the subject of this chapter.

Several criteria describe a microorganism for experimental use:

1. The organism should be able to be grown in the laboratory.
2. Some knowledge of the mating system and sexual behavior of the organism is important to make informative crosses and analyze the progeny.
3. The ability to induce mutations and compare the mutant strain with the wildtype contributes to understanding gene function.
4. Specific genes need to be identified with assays for their function in order to correlate phentype with genotype.
5. The field performance of different strains, including specifically altered strains, must be evaluated and compared to that of the wildtype.

The ectomycorrhizal fungi meet many of these criteria, and research is being done to apply current techniques to understanding the complex genetic behavior and to identifying traits that affect the symbiosis. In general, the ectomycorrhizal fungi hold as much promise and exhibit as many problems as other filamentous fungi. The goal of experimental manipulation of these fungi is to develop plant inocula which are easily applied, cost effective, and viable under suboptimal conditions and that positively impact plant productivity.

II. THE ECTOMYCORRHIZAL FUNGI

The fungal partner of EM is most commonly a basidiomycete, but ascomycetes and a limited number of zygomycetes and deuteromycetes also form EM [41]. Trees forming EM are generally temperate and tropical species and include many with economic value [48]. Many trees grow poorly or not at all without an appropriate fungal partner. Pines (*Pinus*) in particular are exclusively ectomycorrhizal [4]. The specificity of plant–fungus associations in EM is low, especially relative to the degree of host specificity exhibited by pathogenic fungi.

Natural selection has favored development of the symbiosis [44]. Ectomycorrhizal trees benefit from enhanced uptake of ions such as phosphorus and nitrogen [15,19], increased access to and absorption of water [7,13], protection from pathogens [38,54], and tolerance to soil toxins [12]. The fungus in turn derives most of its carbon nutrition from the plant [19].

Ectomycorrhizal roots are characterized by proliferation of lateral roots, which is due at least in part to production of growth regulators by the fungus including indole compounds (auxins) and ethylene [19]. The intimacy of the ectomycorrhizal symbiosis is revealed by the formation of the Hartig net, intercellular growth of the fungus among the cells of the cortex and epidermis [30]. The identity of the plant partner of an ectomycorrhiza is rarely in doubt, but familiarity with subtle microstructural characteristics is often required to identify the fungus (or fungi) involved [17].

III. GENETICS AND GENETIC VARIABILITY OF FUNGI

The ectomycorrhizal basidiomycetes are known to exhibit exceptional genetic behavior including hyphal fusion, nuclear migration, heterokaryosis, and parasexuality [57]. The genetic system of ectomycorrhizal fungi (and indeed most filamentous fungi) allows for immense variability both from sexual recombination and from asexual processes. Naturally occurring colonies of ectomycorrhizal fungi are probably heterokaryotic [57], which provides each colony with two (or more) functional genomes. Somatic recombination through fusion of hyphae or hypha and spore may produce additional of genetic variability allowing for ecological adaptation. Such variability and plasticity can complicate experimental results.

The potential to select improved strains of ectomycorrhizal fungi makes commercial inoculation feasible [65]. Exogenous inoculation of tree seedlings with fungi is important, especially when the seedlings are destined for planting on disturbed sites. It is likely that EM performance can be improved by classical breeding and natural selection. Ectomycorrhizal fungi grow vegetatively in laboratory culture with fruit body formation restricted to a few species in mixed culture with an appropriate host plant [10,32]. While the ability to form fruit bodies may decline with prolonged culture [10], fruiting allows production of controlled dikaryons, the result of crosses between compatible homokaryons. It should be possible to exploit some of the inherent genetic variation by making appropriate crosses [11]. Monokaryons (cells containing a single nuclear type) and dikaryons (cells containing two nuclei of opposite and compatible mating type) of *Laccaria laccata* are each capable of ectomycorrhizal formation [31].

Marx [39] reported selection of *Pisolithus tinctorius* for increased EM formation with improved plant growth by repeated isolation of the fungus from experimental seedlings. Natural selection of strains may involve optimizing the performance of fungi on seedlings growing in drought, saline, or acid soils or with pressure from particular plant pathogens. It may also be possible to identify the factors crucial for adaptation to such adverse conditions and genetically alter the fungus.

The ectomycorrhizal fungi exhibit physiological and ecological diversity within and among species, which affects the inoculation success of the fungi. There is diversity in fungal compatibility toward the host plant, varying from broad host range to restriction to a single host species. Some fungi are more beneficial to the host than are other species [65]. Trees also vary in their receptivity to a particular fungus [44].

Isolates of *L. laccata* differ in their inoculation potential and subsequent influence on plant growth [11,33]. Monokaryons of *Laccaria bicolor* with different mating types have significantly different effects on seedling growth and EM formation [32]. Variation in the shoot:root ratio, even when the extent of the EM formation was the same, suggests that different strains have the ability to influence carbohydrate allocation.

Variation is also reflected in differences of biochemical function measured by enzyme production and regulation (alkaline phosphatase, acid phosphatase, and nitrate reductase) and phytohormone production, substances which may be crucial to the symbiotic balance [22]. Gay and Debaud [16] reported that the production of indole compounds varied over an order of magnitude among strains of *Hebeloma cylindrosporum* and among species of *Hebeloma*. The levels of phytohormone are most likely regulated by a suite of genes, and even with complex regulatory mechanisms, production of indoles can be altered by crosses among polymorphic strains [16].

Some of the problems in quantifying the performance of ectomycorrhizal fungi stem from the difficulty in macroscopically distinguishing mycorrhizae of different species and reisolating the fungus. Axenic culture of the fungus is crucial to the study of the basic biology of the organism, but as a partner in a symbiosis, coculativation with the host plant is essential. Both of these growth habits can be simulated in the laboratory [43].

Variability can be managed or exploited to improve the symbiosis providing strains can be identified. Cooksey and Graham [8] demonstrated the differences among strains of phytopathogenic bacteria by genomic fingerprinting. The technique compared the electrophoretic pattern of large DNA fragments generated by restriction enzymes which recognize rare sequences in genomic DNA. These patterns reveal taxonomic relationships and can be used to identify individual strains. Similar fingerprinting of ectomycorrhizal fungi should be possible.

Taxonomic clarification of the ectomycorrhizal fungi will illustrate the basis of important differences in the physiology and ecology of the fungi. The genetics and taxonomy of the ectomycorrhizal fungi are interrelated and largely unexplored. The evolutionary relationships of organisms are largely responsible for their physical traits [57], and the genetic relationships among strains and species of organisms can be investigated by examining morphological and biochemical traits and, at a more fundamental level, the similarities of DNA sequences.

There has been recognition of the need to "improve inoculum performance" specifically by molecular approaches. The identification and study of appropriate genes, including genes involved in recognition and regulation of the mycorrhizal association, will contribute to strain improvement [2,48,57]. The criteria for evaluation of an improved inoculum must include the ability of the fungus to form mycorrhizae, stimulate plant growth, and compete with naturally occurring microbes.

The inheritance of the ability of a fungus to maintain the mycorrhizal symbiosis is complex with probable regulatory control by major genes or gene complexes [32]. *L. laccata* strain differences include varied infectivity of a compatible host [33]. Recognition of this variability is certainly important for selection of strains for use in particular environments, and yet virtually nothing is known about the genetics of any of these factors.

The intimacy of the association requires that genes regulate the growth of the fungus to prevent pathogenicity and that genes function to overcome the resistance of the plant [2]. Because ectomycorrhizae form with many plant hosts, the recognition signals must be fairly nonspecific. It has been suggested that the fungus produces molecules which affect the permeability of the plant membrane and may affect plant ATPases [2]. Research on suppressors and elicitors, from both plant and fungal tissues, is needed to determine the nature of the signals.

A promising approach to identifying the genes involved in the ectomycorrhizal symbiosis has been described by Hilbert and Martin [21]. They distinguished changes in accumulation of individual polypeptides separated in two-dimensional electrophoretic gels and compared the free-living fungus, uninfected plant roots, and established ectomycorrhizae using *Eucalyptus* and *P. tinctorius*. Ectomycorrhizal formation results in physical and biochemical changes, and it is reasonable to assume that specific gene functions in both the plant and the fungus are involved. Comparing polypeptide products is an indirect way to visualize gene expression.

The effect of ectomycorrhizal formation could be seen as changes in the relative abundance of polypeptides. Most of the changes were apparent within one week of infection and persisted in 5-week-old mycorrhizae. Nearly 740 fungal and 790 root polypeptides were differentially expressed with the appearance of 10 novel polypeptides synthesized upon colonization. The ectomycorrhiza-specific polypeptides could be a result of induced gene activity or posttranslational modification.

These results support the contention that the ectomycorrhizal association involved a complex set of genes and identify specific proteins which can be matched to specific genes. Genes involved in the establishment and maintainance of the symbiosis are of interest to study the nature of the interaction and also to be used to manipulate the relationship. Isolated polypeptides yield sequence information for synthesis of specific oligomeric nucleotide sequences to be used to probe for specific genes from an ectomycorrhizal fungus genomic library. Alternatively, antibodies made against the polypeptides can be used to probe a cDNA expression library. The positive (but admittedly time-consuming) assay for genes involved in ectomycorrhizal symbiosis would involve genetic transformation of a nonmycorrhizal strain of fungus to one with symbiotic potential.

Success with fruit-body formation and sporulation in the laboratory is encouraging, but restricted to a few species. Thus, opportunities for classical genetic analysis, involving crosses and examination of progeny, are currently limited. The alternative approach is to manipulate the genome either by induction of mutation or by DNA-mediated transformation. Both approaches can best be achieved using fungal protoplasts as the cellular material rather than intact hyphae.

IV. PROTOPLAST LIBERATION AND REGENERATION IN FUNGI

The fungal cell wall provides an effective barrier that limits access to the genome of the organism. Removal of the cell wall to liberate protoplasts provides a more malleable substrate for genetic alteration than intact hyphal cells. Protoplasts can be used for the isolation of cytoplasmic organelles [5], investigation of cell wall synthesis and synchronized cell growth [9], for cell fusions [42], and for purification of high molecular weight DNA for karyotyping and cloning. Protoplasts are also essential for induction of mutations [20,23] and genetic transformation [14], experiments designed for strain improvement.

The utility of protoplasts is limited by the ability of the cell to synthesize new cell wall material and continue growth. The minimal requirement for protoplast regeneration is possession of a nucleus. Fungal nuclei are small, often migratory through the mycelium, and may be absent from relatively large regions of hyphae [3]. Protoplasts are often formed as buds through localized weaknesses in the cell wall and frequently do not include a nucleus [9]. Only a subpopulation of protoplasts containing nuclei have potential for continued growth, and those that do not regenerate are probably the victims of uneven distribution of other cellular components.

Protoplasts have been formed from a wide variety of fungi [9,47] including several ectomycorrhizal basidiomycetes [3,20,31]. The commercially available enzyme preparation Novozyme 234 is generally effective for protoplast liberation with filamentous fungi, but the conditions of time of exposure, age of culture, and osmotic medium must be determined for each individual species. Regeneration rates range from nearly 100% for *Trichoderma reesei* [49] to 1% for some ectomycorrhizal basidiomycetes [3].

The ability to form and regenerate protoplasts is requisite for molecular genetic experiments with the ectomycorrhizal fungi. *Laccaria* and *Hebeloma* were the only genera of several evaluated that had potential for further experimentation [3,20,31].

V. MUTATION INDUCTION IN FUNGI

An approach to genetic alteration of the ectomycorrhizal fungi is the induction of mutations by exposure to chemical mutagens or ultraviolet (UV) light. Determination of the conditions for mutation induction and selection or screening will allow specific mutations to be obtained, which will eventually aid in the isolation and characterization of particular genes. Lesions in genes coding for biosynthetic functions will allow the isolation of auxotrophic strains, which can be used for the study of biosynthetic pathways or as hosts for transformation by complementation to prototrophy.

Designated traits, such as tolerance to fungistatic agents or stress factors, may be selected from cells exposed to mutagens. Protoplasts are better suited for mutation induction because these wall-less cells are often more susceptible to mutagenic agents than are hyphae. Mutagenesis is a valuable technique for development of strains deficient for single traits which can be compared to the wildtype.

Mutants of the phytopathogenic fungus *Ustilago maydis* defective in both structural and regulatory genes of the ferrichrome biosynthetic pathway have been isolated after exposure to the chemical nitrosoguanidine [34]. Auxotrophic strains and strains with increased antibiotic production have been identified after treatment of fungi with mutagens [23,26,27].

Mutants of the ectomycorrhizal fungus *H. cylindrosporum* resistant to several different fungicides were identified after exposure to UV [20]. The frequency of mutants was 10^{-6}, and these homokaryotic strains were capable of growth on concentrations of the fungicides two to five times greater than the concentration that completely inhibited growth of the wildtype heterokaryotic mycelia. The long term goal of mutation induction experiments is to better understand the structure and regulation of genes important for the exploitation or control of the fungi.

VI. TRANSFORMATION OF FILAMENTOUS FUNGI

Genetic transformation of the ascomycete *Aspergillus* provides a model for transformation of filamentous fungi [14]. The ability to transform *Aspergillus* species allows more detailed study of the genetic mechanisms and traits revealed by more classical methods [55]. Transformation provides access to the genetic system of filamentous fungi less amendable to other methods of genetic analysis. Genetic transformation offers new ways to alter the biological characteristics of an organism in order to study growth regulation, development, and metabolism [56] as well as to introduce novel genetic functions or control mechanisms.

Genetic transformation requires both uptake and expression of exogenous DNA. Fungal transformation typically involves use of regenerative protoplasts, identification of selectable marker genes, and stable integration of the transforming DNA into the chromosome of the organism. Both calcium and polyethelene glycol (PEG) are required components of the protoplast transformation mixture.

Aspergillus species have been transformed by incubation of protoplasts with DNA containing genes for utilization of novel nitrogen sources (acetamidase [55]), genes for amino acid synthesis (tryptophan synthetase [64]), and genes for drug resistance (hygromycin phosphotransferase [50]). Transformants were selected by requisite function of the introduced gene.

The use of positive selection may be limited by the genetic background of the organism. For example, the acetamidase gene may not be used in strains that have complex requirements for nitrogen-containing compounds that interfere with acetamidase selection [50]. The amino acid synthesis genes require an auxotroph incapable of endogenous amino acid synthesis. In contrast, a gene conferring resistance to a drug may be used as a marker for transformation in strains with no other requirements except sensitivity to the drug and drug resistance is generally a dominant trait.

Genes that code for resistance to several antibiotic drugs have been cloned from a number of sources. Kanamycin resistance has been used extensively as a selectable marker for plants [28], and G418 resistance has been useful for selection of yeast and *Penicillium* transformants [24,53]. Unfortunately, many commonly used antibiotics are not effective against filamentous fungi. Hygromycin B, however, is an aminoglycoside antibiotic that inhibits protein synthesis in a variety of bacteria, fungi, and higher eukaryotes. A gene conferring resistance to hygromycin B has been cloned from *Escherichia coli* [25] and has been used as a marker for transformation of several fungi (*Fulvia fulva* [46], *Glomerella cingulata* [52], *Asperigillus species* [50], *Ustilago maydis* [60]).

A number of basidiomycetous fungi have been transformed by complementation of auxotrophic mutations. Auxotrophs of both *Coprinus cinereus* and *Schizophyllum commune* were transformed to prototrophy using the homologous *trpC* genes and regulatory signals [6,45]. An adenine auxotroph of *Phanerochaete chrysosporium* was transformed with an adenine biosynthetic enzyme gene from *S. commune* [1]. Transformants of the basidiomycetous yeast *Rhodosporidium toruloides* were selected for the ability to use phenylalanine as the sole nitrogen source by virtue of the expression of the homologous phenylalanine ammonia-lyase gene [58].

Examples of transformation of basidiomycetes illustrate several general features of fungal transformation. Transforming DNA is integrated into high molecular weight DNA and stably maintained with the fungal chromosomes. There is some flexibility of expression of genes from closely related species. *Phanerochaete chrysosporium* will express the *S. commune ade* gene [1], and *F. fulva* (an ascomycete) apparently uses the *Aspergillus nidulans* promoter to express hygromycin phosphotransferase [46]. These are conservative examples of heterologous gene expression involving fungi that belong to the same taxonomic class.

The aspartyl protease of the filamentous fungus *Mucor miehei* however, was expressed in *A. nidulans* apparently from the *Mucor* promoter [18]. This is significant because *M. miehei* belongs to the class Zygomycetes, taxonomically unrelated to the class Ascomycetes, which includes *A. nidulans*. The plasticity of the fungal genome may allow for expression of genes from diverse regulatory sequences.

The hygromycin B-resistance phenotype was selected in the ectomycorrhizal fungus *L. laccata* regenerated from protoplasts exposed to a plasmid carrying the *E. coli* aminocyclitol phosphotransferase gene bracketed by the *A. nidulans* glyceraldehyde-3-phosphate dehydrogenase promoter and the transcription terminator region of the *A. nidulans* tryptophan synthetase gene (Barrett, Dixon, and Lemke, *Appl. Microbiol. Biotechnol.*, in press). The transformation resulted from the integration of one or more copies of the gene into high molecular weight DNA and expression of the drug resistance gene may be from the *Aspergillus* promoter. The ability of a basidiomycete to recognize an ascomycete promoter should be investigated further. The transformed phenotype is stable in the absence of selection, and the expression of the foreign gene does not hinder the ability to form ectomycorrhizae with the host loblolly pine.

The ability to transform *L. laccata* suggests that other ectomycorrhizal fungi can be transformed if viable protoplasts can be formed. Some species of fungi will not be suitable for these experiments because enucleate protoplasts result from removal of the cell wall [3]. Incubation of intact hyphae in cation-containing buffers [6] has provided an alternative, although inefficient, means of introduction of DNA into fungal cells may prove feasible for transforming species of ectomycorrhizal fungi not suitable for protoplast formation.

Transformation of fungi using a high-voltage electric pulse has recently been reported [51]. Electroporation is the reversible permeabilization of biological membranes by short-duration, high-amplitude electric fields [51]. Protoplasts in an osmotic buffer with the transforming DNA can be exposed to the appropriate electric field and then plated on medium to select regenerating colonies expressing the antibiotic resistance of the transforming DNA. This technique provides a method for introducing DNA into species of fungi not amendable to the traditional calcium and PEG treatment.

VII. PERSPECTIVES FOR GENETIC MANIPULATION

The ease with which fungal cells integrate foreign DNA into their chromosomes and recognize heterologous expression signals, facilitates the ability to transform these organisms. It must be remembered that many products are not the results of expression of single genes but may require a gene complex, thus it may not always be possible to transfer gene function.

The ability to genetically manipulate ectomycorrhizal fungi has enormous potential. An advantage exists for introduction of genes for root-related functions into a mycorrhizal fungus rather than into the plant genome per se, especially if it were beneficial to delimit transgenic functions to the root system in order to avoid undesirable responses in the remainder of the plant. Genetic transformation of the fungal symbiont would obviate the need to involve root-specific promoters to attain root-specific responses. There are a number of regulatory sequences responsive to external factors which may be included in transformation systems. An homolgous heat shock promoter has been used in experiments with *Ustilago maydis* [60]. Induction by anaerobiosis or heavy metal exposure may allow controlled expression of specific genes.

The obvious extension of transformation experiments is the identification of genes and regulatory sequences which positively influence the symbiosis between fungus and plant. Genes important to root development and plant survival might be identified from any source. The genes for the synthesis of IAA (auxin) from tryptophan have been cloned from *Agrobacterium tumifaciens* [28]. Stable expression of maintainance levels of phytohormones may promote root proliferation.

The intimacy of association with the root surface renders the ectomycorrhizal fungi especially well suited for targeting genetic strategies in the control of stress factors in the rhizosphere. In dry soils the root–soil interface may have very low water potential, and osmotolerance may aid in survival. Simple molecules common to plants, bacteria, and animals serve as osmolytes by aiding in the preservation of the macromolecular structure of the cell [63]. The *osm* genes of *E. coli* are responsible for the production of osmoprotectant molecules such as betaines and proline [35], which could potentially be expressed in fungi. Manipulation of the expression of the genes coding for the production of osmolyte molecules in ectomycorrhizal fungi could increase the tolerance of the associated plants to low water potential because of the exchange of material between the cells of the EM.

Genes that code for compounds repulsive to insects could be used to protect seedlings from herbivory [29]. Plasmids resident in *Bacillus thuringiensis* (*Bt*) code for an insecticidal protein(s) specific for lepidopteran pests [62]. The gene for a *Bt* toxin has been cloned in tobacco plants, principally to protect against leaf-eating larvae [59]. Similar genes, if introduced and functional in an ectomycorrhizal fungus, might provide localized protection from damage by root-associated insects without direct plant tissue involvement in synthesis of the toxin.

Rhizosphere organisms in general are ideal for use as biocontrol agents [61]. Exploitation of mycorrhizal fungi and other soil microorganisms could lead to the biological control of pathogens. Even if the mycorrhizal fungus is not directly deleterious to pathogens, the intimacy of the fungus and the root requires any biocontrol agent be part of the mycorrhizosphere. Mycorrhizal fungi indirectly affect other soil microbes by improvement in the phosphorus nutrition of the plant. Microbial activity around EM is enhanced relative to nonmycorrhizal roots and may be due to changes in root exudate and the nutritional and elemental content of the root [36]. Changes in the plant and fungus affect the rhizosphere and the overall ectomycorrhizal effect may be due to synergism with other rhizosphere organisms [36]. It is not unreasonable to assume that some contributing factors, both biological and abiotic, can be introduced into ectomycorrhizal fungi by transformation to enhance the performance of inocula especially if soils are depleted for synergistic organisms which is often the case in disturbed soils [40]. This also reminds us that modified ectomycorrhizal fungi must remain compatible with soil microbes.

The transformed phenotypes of these fungi should be stable due to the expected integration of the foreign DNA into the fungal chromosomes. This stability adds to the desirability of developing ectomycorrhizal fungi as biocontrol agents and growth-promoting inocula.

Any genetically altered organism destined for environmental release will be subject to the rules of the various regulatory agencies with jurisdiction over biotechnology products. Currently, any request for release of recombinant bacteria is reviewed on a case-by-case basis [37]. The potential for gene exchange among strains (and possible species) of fungi needs to be investigated. The ectomycorrhizal fungi could be genetically marked by transformation and used in contained facilities such as rhizotrons to assess the mobility of a strain in the rhizosphere, the propensity of exchange of genes with other fungi, as well as the competetive ability of the altered strain.

The ability to genetically alter these fungi should provide strains that enhance the mycorrhizal symbiosis and thereby positively influence the growth and yield of plants. Success of these experiments is expected provided that specific and beneficial genes that influence the ectomycorrhizal symbiosis can be identified.

REFERENCES

1. Alic, M., Kornegay, J. R., Pribnow, D., and Gold, M. H. Transformation by complementation of an adenine auxotroph of the lignin-degrading basidiomycete *Phanerochaete chrysosporium*. *Appl. Environ. Microbiol.*, *55*:406–411, 1989.
2. Anderson, A. J. Mycorrhizae – host specificity and recognition. *Phytopathol.*, *78*:375–378, 1988.

3. Barrett, V., Lemke, P. A., and Dixon, R. K. Protoplast formation from selected species of ectomycorrhizal fungi. *Appl. Microbiol. Biotechnol.*, *30*:381–387, 1989.
4. Berch, S. M., Miller, O. K., and Thiers, H. D. Evolution of mycorrhizae. In *Proc. 6th N. Am. Conf. on Mycorrhizae* (R. Molina, ed.), Forest Research Laboratory, Corvallis, 1985, pp. 189–191.
5. Billich, A., Keller, U., Kleinhauf, H., and Zocher, R. Production of protoplasts from *Fusarium scirpi* by lytic enzymes from *Streptomyces tsusimaensis*. *Appl. Microbiol. Biotechnol.*, *28*:442–444, 1988.
6. Binninger, D. M., Skrzynia, C., Pukkila, P. J., and Casselton, L. A. DNA-mediated transformation of the basidiomycete *Coprinus cinereus*. *The EMBO J.*, *6*:835–840, 1986.
7. Brownlee, C., Duddridge, J. A. Malibari, A., and Read, D. J. The structure and function of mycelial systems of ectomycorrhizal roots with special reference to their role in forming inter-plant connections and providing pathways for assimilate and water transport. *Plant Soil*, *71*:433–443, 1983.
8. Cooksey, D. A. and Graham, J. H. Genomic fingerprinting of two pathovars of phytopathogenic bacteria by rare-cutting enzymes and field inversion gel electrophoresis. *Phytopathol.*, *79*:745–750, 1989.
9. Davis, B. Factors influencing protoplast isolation. In *Fungal Protoplasts* (J. F. Peberdy and L. Ferenczy, eds.), Marcel Dekker, New York, 1985, pp. 45–71.
10. Debaud, J. C. and Gay, G. In vitro fruiting under controlled conditions of the ectomycorrhizal fungus *Hebeloma cylindrosporum* associated with *Pinus pinaster*. *New Phytol.*, *105*:429–435, 1987.
11. Debaud, J. C., Gay, G., Prevost, A., Lei, J., and Dexheimer, J. Ectomycorrhizal ability of genetically different homokaryotic and dikaryotic mycelia of *Hebeloma cylindrosporum*. *New Phytol.*, *108*:323–328, 1988.
12. Dixon, R. K. and Buschena, C. A. Response of ectomycorrhizal *Pinus banksiana* and *Picea glauca* to heavy metals in soil. *Plant Soil*, *105*:265–271, 1988.
13. Duddridge, J. A., Malibari, A., and Read, D. J. Structure and function of mycorrhizal rhizomorphs with special reference to their role in water transport. *Nature*, *287*:834–836, 1980.
14. Fincham, J. R. S. Transformation in fungi. *Microbiol. Rev.*, *53*:148–170, 1989.
15. Finlay, R. D., Ed, H., Odham, G., and Soderstrom, B. Mycelial uptake, translocation and assimilation of ammonium by *Pinus sylvestris* plants infected with four different ectomycorrhizal fungi. *New Phytol.*, *110*:59–66, 1988.
16. Gay, G. and Debaud, J. C. Genetic study on indole-3-acetic acid production by ectomycorrhizal *Hebeloma* species: Inter- and intraspecific variability in homo- and dikaryotic mycelia. *Appl. Microbiol. Biotechnol.*, *26*:141–146, 1987.
17. Godbout, G. and Fortin, J. A. Classification of ectomycorrhizae: What's new and what to do. In *Proc. 6th N. Am. Conf. on Mycorrhizae* (R. Molina, ed.), Forest Research Laboratory, Corvallis, 1985, pp. 186–188.

18. Gray, G. L., Hayenga, K., Cullen, D., Wilson, L. J., and Norton, S. Primary structure of *Mucor miehei* aspartyl protease: Evidence for a zymogen intermediate. *Gene*, *48*:41–53, 1986.
19. Harley, J. L. and Smith, S. E. *Mycorrhizal Symbiosis*. Academic Press, New York, 1983.
20. Hebraud, M. and Fevre, M. Protoplast production and regeneration from mycorrhizal fungi and their use for isolation of mutants. *Can. J. Microbiol.*, *34*:157–161, 1988.
21. Hilbert, J. L. and Martin, F. Regulation of gene expression in ectomycorrhizas I. Protein changes and the presence of ectomycorrhiza-specific polypeptides in the *Pisolithus-Eucalyptus* symbiosis. *New Phytol.*, *110*:339–346, 1988.
22. Ho, I. Enzyme activity and phytohormone production of mycorrhizal fungus, *Laccaria laccata*. *Can. J. For. Res.*, *17*:855–858, 1987.
23. Homolka, L. Vyskocil, P., and Pilat, P. Use of protoplasts in the improvement of filamentous fungi I. Mutagenization of protoplasts of *Oudemansiella mucida*. *Appl. Microbiol. Biotechnol.*, *28*:166–169, 1988.
24. Jimenez, A. and Davies, J. Expression of a transposable antibiotic resistance element in *Saccharomyces*. *Nature*, *287*:869–871, 1980.
25. Kaster K. R., Burgett, S. G., Rao, R. N., and Ingolia, T. D. Analysis of a bacterial hygromycin B resistance gene by transcriptional and translational fusions and by DNA sequencing. *Nuc. Acids Res.*, *11*:6895–6911, 1983.
26. Kim, K. S., Cho, N. Y., Pai, H. S., and Ryu, D. D. Y. Mutagenesis of *Micromonospora rosaria* by using protoplasts and mycelial fragments. *Appl. Environ. Microbiol.*, *46*:689–693, 1983.
27. Keller, U. Highly efficient mutagenesis of *Claviceps purpurea* by using protoplasts. *Appl. Environ. Microbiol.*, *46*:580–584, 1983.
28. Klee, H. J., Horsch, R. B., Hinchee, M. A., Hein, M. B., and Hoffman, N. L. The effects of overproduction of two *Agrobacterium tumefaciens* T-DNA auxin biosynthetic gene products in transgenic petunia plants. *Genes and Devel.*, *1*:86–96, 1987.
29. Klier, A. and Rapoport, G. *Bacillus* larval toxin crystal protein. *Microbiol. Sci.*, *4*:274–276, 1987.
30. Kottke, I. and Oberwinkler, F. Mycorrhiza of forest trees–structure and function. *Trees*, *1*:1–24, 1986.
31. Kropp, B. R. and Fortin, J. A. Formation and regeneration of protoplasts from the ectomycorrhizal basidiomycete *Laccaria bicolor*. *Can. J. Bot.*, *64*:1224–1226, 1986.
32. Kropp, B. R. and Fortin, J. A. The incompatibility system and relative ectomycorrhizal performance of monokaryons and reconstituted dikaryons of *Laccaria bicolor*. *Can. J. Bot.*, *66*:289–294, 1988.
33. Kropp, B. R., McAfee, B. J., and Fortin, J. A. Variable loss of ectomycorrhizal ability in monokaryotic and dikaryotic cultures of *Laccaria bicolor*. *Can. J. Bot.*, *65*:500–504, 1987.
34. Leong, S. A., Wang, J., Budde, A., Holden, D., Kinscherf, T., and Smith, T. Molecular strategies for the analysis of the interaction of *Ustilago maydis* and maize. In *Molecular Strategies for Crop Protection* (C. J. Arntzen and C. Ryan, eds.), A. R. Liss, New York, 1987, pp. 95–106.

35. Le Rudulier, D., Strom, A. R., Dandekar, A. M., Smith, L. T., and Valentine, R. C. Molecular biology of osmoregulation. *Science, 224*:1064–1068, 1984.
36. Linderman, R. G. Mycorrhizal interactions with the rhizosphere microflora: The mycorrhizosphere effect. *Phytopathol., 78*:366–370, 1988.
37. Lindow, S. E., Panopoulos, N. J., McFarland, B. L. Genetic engineering of bacteria from managed and natural habitats. *Science, 244*:1300–1307, 1989.
38. Marx, D. H. Ectomycorrhizae as biological deterrents to pathogenic root infections. *Ann. Rev. Phytopathol., 10*:429–434, 1972.
39. Marx, D. H. Ectomycorrhizal fungus inoculations: A tool for improving forestation. In *Tropical Mycorrhiza Research* (P. Mikola, ed.), Clarendon Press, Oxford, 1980, pp. 13–71.
40. McAfee, B. J. and Fortin, J. A. Comparative effects of the soil microflora on ectomycorrhizal inoculation of conifer seedlings. *New Phytol., 108*:443–449, 1988.
41. Miller, O. K. Taxonomy of ecto- and ectendomycorrhizal fungi. In *Methods and Principles of Mycorrhizal Research* (N. C. Schenck, ed.), Am. Phytopathol. Soc., St. Paul, 1982, pp. 91–101.
42. Minuth, W. and Esser, K. Intraspecific, interspecific, and intergeneric recombination in B-lactam producing fungi via protoplast fusion. *Appl. Microbiol. Biotechnol., 18*:38–46, 1983.
43. Molina, R. and Palmer, J. G. Isolation, maintenance, and pure culture manipulation of ectomycorrhizal fungi. In *Methods and Principles of Mycorrhizal Research* (N. C. Schenck, ed.), Am. Phytopathol. Soc., St. Paul, 1982, pp. 115–129.
44. Molina, R. and Trappe, J. M. Patterns of ectomycorrhizal host specificity and potential among Pacific Northwest conifers and fungi. *Forest Sci., 28*:423–458, 1982.
45. Munroz-Rivas, A. M., Specht, C. A., Drummond, B. J., Froelinger, E., Novotny, C. P., and Ullrich, R. C. Transformation of the basidiomycete, *Schizophyllum commune*. *Mol. Gen. Genet., 205*:103–106, 1986.
46. Oliver, R. P., Roberts, I. N., Harling, R., Kenyon, L., Punt, P. J., Dingemanse, M. A., and van den Hondel, C. A. M. J. J. Transformation of *Fulvia fulva*, a fungal pathogen of tomato, to hygromycin B resistance. *Curr. Genet., 12*:231–233, 1987.
47. Peberdy, J. F. *Protoplasts – Applications in Microbial Genetics*. University of Nottingham, Nottingham, 1979.
48. Perry, D. A., Molina, R., and Amaranthus, M. P. Mycorrhizae, mycorrhizospheres, and reforestation: Current knowledge and research needs. *Can. J. Forest Res., 17*:929–940, 1987.
49. Picataggio, S. K., Schamhart, D. H. J., Montenecourt, B. S., and Eveleigh, D. E. Sphaeroplast formation and regeneration in *Trichoderma reesi*. *Appl. Microbiol. Biotechnol., 17*:121–128, 1983.
50. Punt, P. J., Oliver, R. P., Dingemanse, M. A., Powels, P. H., and van den Hondel, C. A. M. J. J. Transformation of *Aspergillus* based on the hygromycin B resistance marker from *Escherichia coli*. *Gene, 56*:117–124, 1987.
51. Richey, M. G., Marck, E. T., Schardl, C. L., and Smith, D. A. Transformation of filamentous fungi with plasmid DNA by electroporation. *Phytopathol., 79*:884–887, 1989.

52. Rodriguez, R. J. and Yoder, O. C. Selectable genes for transformation of the fungal plant pathogen *Glomerella cingulata* f. sp. *phaseoli* (*Collectotrichum lindemuthianum*). *Gene, 24*:73–81, 1987.
53. Stahl, U., Leitner, E., and Esser, K. Transformation of *Penicillium chrysogenum* by a vector containing a mitochondrial origin of replication. *Appl. Microbiol. Biotechnol., 26*:237–241, 1987.
54. Sylvia, D. M. Role of *Laccaria laccata* in protection primary roots of Douglas-fir from root rot. *Plant Soil, 71*:299–302, 1983.
55. Tilburn, J., Scazzocchio, C., Taylor, G. G., Zabicky-Zissman, J. H., Mockington, R. A., and Davies, R. W. Transformation by integration in *Aspergillus nidulans*. *Gene, 26*:205–221, 1983.
56. Timberlake, W. E. and Marshall, M. A. Genetic engineering of filamentous fungi. *Science, 244*:1313–1317, 1989.
57. Trappe, J. M. and Molina, R. Taxonomy and genetics of mycorrhizal fungi: Their interactions and relevance. *Proc. 1st Eur. Symp. Mycorrhizae*, INRA, Paris, 1986.
58. Tully, M. and Gilbert, H. Transformation of *Rhodosporidium toruloides*. *Gene, 36*:235–240, 1985.
59. Vaeck, M., Reynaerts, A., Hofte, H., Jansens, S., DeBeuckeleer, M., Dean, C., Zabeau, M., VanMontagu, M., and Leemans, J. Transgenic plants protected from insect attack. *Nature* (London), *328*:33–37, 1987.
60. Wang, J., Holden, D. W., and Leong, S. H. Gene transfer system for the phytopathogenic fungus *Ustilago maydis*. *Proc. Natl. Acad. Sci. USA, 85*:865–869, 1988.
61. Weller, D. M. Biological control of soilborne plant pathogens in the rhizosphere with bacteria. *Ann. Rev. Phytopathol., 26*:379–407, 1988.
62. Whiteley, H. R. and Schnepf, H. E. The molecular biology of parasporal crystal body formation in *Bacillus thuringiensis*. *Ann. Rev. Microbiol., 40*:549–576, 1986.
63. Yancy, P. H., Clark, M. E., Hand, S. C., Bowlus, R. D., and Somero, G. N. Living with water stress: Evolution of osmolyte systems. *Science, 217*:1214–1222, 1982.
64. Yelton, M. M., Hamer, J. E., and Timberlake, W. E. Transformation of *Aspergillus nidulans* by using a *trpC* plasmid. *Proc. Natl. Acad. Sci. USA, 81*:1470–1474, 1984.
65. Zak, B. Classification of ectomycorrhizae. In *Ectomycorrhizae, Their Ecology and Physiology* (G. C. Marks and T. T. Kozlowski, eds.), Academic Press, New York, 1973, pp. 43–78.

9

ENDOPHYTIC FUNGI OF GRASSES

CHARLES W. BACON *Russell Research Center, Agricultural Research Station, United States Department of Agriculture, Athens, Georgia*

JOSE De BATTISTA *University of Georgia, Athens, Georgia*

I. INTRODUCTION

Fungi that live their entire life cycle within the aerial portions of grasses and sedges by forming a nonpathogenic, systemic and usually intercellular association are referred to as endophytic fungi or grass endophytes [1,2]. Inherent in this descriptive definition is a continuum of expressions, with each association showing varying degrees of habits and, as can be expected, some associations more intimate than others. Regardless of the degree of the association, all endophytic fungi of grasses form biotrophic associations, in which there is little or no destruction of either grass or fungus, and as will be presented later, certain members exploit the biochemical properties of the other, suggesting that in specific cases the relationship is mutualistic.

This curious symbiosis has long been recognized as De Bary [3] in 1863 demonstrated for the first time that the fungus *Epichloë typhina* (Pers. ex Fr.) Tul. formed internal hyphae within the tissues of its grass host, thus establishing the foundation for our current concept of this unique parasitic habit. More elaborate studies by Sampson [1] established the perennial nature and distribution of this fungus in several grasses. Diehl [2] established the endophytic habit for other related fungal species and in his monograph, classified them within the tribe of the family Clavicipitaceae.

Historically, forage grasses not infected with species of *Claviceps* have been implicated in the etiology of animal toxicity resembling ergotism [4–6]. It was not, however, until 1975 and succeeding years that related fungi, endophytes, were discovered to be the cause of animal toxicities [7–9]. Research on animal toxicity not only established the biosynthesis of ergot alkaloids was a feature of the class, but it also stimulated basic biological research on this neglected group of fungi for economic reasons. An important result of this research was the revelation that certain grass–fungus associations are mutualistic. In any mutualistic relationship there are advantages, often competitive, which can be exploited, and the current thinking of researchers is to use these to improve the quality of grasses, primarily turf and conservation species. Thus, a new biological control system emerges that promises to be a convenient, contained, and self-perpetuating system that, in addition to protection from environmental stresses, confers insect and disease tolerances.

In this chapter we will present a review of the literature on endophytes and related techniques that may be useful for biotechnological improvements of grasses. Information contained in earlier reviews, which emphasize toxicity [10,11], agronomic and host-parasite relations [12-14], and endophyte biology [14,15], will not be presented.

II. BIOLOGY OF ENDOPHYTIC FUNGI

A. Taxonomy

The brief morphological description used above restricts the fungi in this review to include those pyrenomycetous fungi of the tribe Balansiae (family Clavicipitaceae, class Ascomycetes) and their related anamorphs. These fungi form a primary anamorphic stromata within or around stems, leaves, or florets of grasses from which either effuse, pulvinate, or stalked perithecial stromata arise. There are, however, several species characterized by a lack of one or the other state [2]. Moreover, there are species that completely lack both states in nature, but produce the conidial state in culture [8,16].

Diehl [2] distinguished genera of the Balansiae on the basis of their anamorphs. Species with an *Ephelis* macroconidial anamorph exclusively are classified in the genus *Balansia* or *Myriogenespora*. *Epichloe* differs in that it produces only an *Acremonium* microconidial anamorph. Further, these two general can be distinguished according to the specific type of conidiation [17], although it was concluded that the structure of the conidiomata was too variable to be of taxonomic value [17,18].

As currently considered, endophytic fungi include those telemorphic species *Epichloe, Balansia, Atkinsonella*, and *Myriogenespora* and their anamorphic states *Acremonium, Ephelis*, and *Typhodium*. Another genus, *Balansiopsis*, was considered by Diehl [2], because it lacks an anamorphic state. However, Rykard et al. [17] demonstrated that the anamorphic state does exist, and it is transitional between *A. hypoxylon* and *B. epichloe*. Demonstration of the conidial state by them and following the criteria established by Diehl [2] invalidated *Balansiopsis pilulaeformis* and placed it in the *Balansia* as *B. pilulaeformis* (Berk. et Cur.) Diehl [17].

All genera are identical in the developmental pattern of the typical clavicipitaceous perithecium, as well as the centrum characters which consist of unitunicate asci in fascicles, containing eight hyaline, filiform ascospores [19-21]. Each ascospore is septate and may break up into part spores before or after its discharge from the ascus.

Conidiogenesis in the Balansiae is holoblastic for those species with the *Ephelis* state, and phialidic for those with the *Acremonium* state. The *Ephelis* conidial states are further characterized by hyaline, nonseptate, fusoid to scolecosporous conidia borne singly and sympodially at the tips of simple conidiophores located usually in apothecioid cavities [17]. The *Acremonium* conidial states are characterized by hyaline, nonseptate, ellipsoid conidia produced through collarettes on simple phialidic conidiophores. In this respect the *Acremonium* state is identical to the *Sphacelia* state of *Claviceps* [22].

The Genus *Epichloe*

This endophyte is predominantly associated with grasses, although it was reported on one species of *Juncus* [21]. It is characterized by its brightly colored, externally produced effuse ascomata on the internodes and nodes of flower stems. The flower stalk is rarely produced in infected plants, and in those cases where it is produced, a high percentage of the resulting seed is infected (C. W. Bacon, personal observations). Typically the infection produces stem elongation of the nodes [16]. The conidiomata of the *Acremonium*-state form first over the stroma as a thin white layer consisting of simple phialidic conidiophores bearing small, hyaline, nonseptate, ellipsoid conidia [17]. The type for the genus is *E. typhina*, and of the many species reported, two have been discussed recently, *E. cinerea* [20] and *E. bambusae* [23], but these two might be invalid.

The Genus *Balansia*

Members of this genus have an effuse basal hypothallus upon which the conidiomata and ascomata are formed. These fungi are associated with grasses and sedges. The other major distinction used in separating species is the location of the external hypothallus and associated conidiomata [2]. Using these criteria, Diehl [2] recognized 14 American species, although there are several other species, primarily Indian [24,25]. The ascomata of each species vary in color, but are usually black to brown, although some are yellow to green.

The Genus *Atkinsonella*

This genus was erected by Diehl [2] to include the monotypic species *Atkinsonella hypoxylon* (pk.)Tul. characterized by producing two conidial states *Ephelis* and *Typhodium*. Otherwise, this fungus closely resembles *B. pilulaeformis*, and on this basis it was suggested that there was very little justification for removing it from the genus *Balansia* [17].

The Genus *Myriogenospora*

The anamorphic state of this monotypic genus, *M. atramentosa* (Berk. & Curt. apud Berk.)Diehl is morphologically similar to the *Ephelis* state of *Balansia* species, and on the basis of its clusters of aparaphyste asci on a basal cushion, lateral paraphyses, and indistinct peridia of the perithecia allowed it to be placed in the tribe Balansiae of the Clavicipitaceae [19]. It differs from the *Balansia* species in that its mycelium does not penetrate host tissues. Infection of leaves and inflorescences occurs on unexpanded host organs from superficial mycelium located systemically, within the bud [19,26,27]. This fungus, therefore, is not typically endophytic, and our mention of it here is strictly nomenclatorial.

The Genus *Acremonium*

Section *Alba-lanosa* of *Acremonium* Morgan-Jones and Gams, primarily referred to as endophytes of fescue and perennial ryegrass, was erected to accommodate the anamorph of *Epichloe typhina* and anamorphic endophytes causing symptomless infections [28]. The latter endophytes are seed transmitted, do not produce external stromata, and are not known to possess a teleomorph. The two most frequently reported species, *A. coenophialum* Morgan-Jones & Gams and *A. lolii* Latch, Christensen & Samuels, are associated with tall fescue and perennial ryegrass, respectively, and are similar in morphology to the anamorphic state of *E. typhina*. They can be distinguished from each other only in culture by observing the anamorphic state [1,16,28,29]. However, conidiation by isolates are transitory, and cannot be relied upon in older subcultured material [30]. Also several endophytes do not produce sporulating structures in culture or otherwise and cannot be satisfactorily classified [29]. Endophytes of this genus have great potential for use in improving grasses.

Gliocladium and *Phialophora*

These are endophytes of tall fescue and perennial ryegrass that are morphologically similar to each other, differing only in pigmentation, and are probably not closely related phylogenetically to other endophytes. They have only rarely been reported [1,31], and any significance to their grass host is unknown.

B. Distribution

Grass species belonging to six subfamilies of the Gramineae [32] have been reported as hosts for endophytic fungi of the Balansiae and their relatives. In Table 1 we present a list of endophytic species that have been reported to infect grasses from different tribes and subfamilies. In addition, the table presents the total number of grass host species infected by endophytes within a subfamily and the total number of specific endophytic associations within a tribe. A total of 232 grass-fungus associations are currently known, representing 209 host grass species. If we consider only the North American species of grasses, this would indicate that only 15% of all the new world's grass species are hosts for endophytes. However, as research continues, this number will undoubtedly go higher.

The geographical distribution of infected grass species ranges from temperate zones to tropical zones. The distribution of infected material follows habitat, climatic conditions, and occurrence of grass species. Indeed, those grass species with widly different geographic distribution are infected with the same fungus at each end of their geographic extremes [2].

The concept of host specificity for endophytic fungi is far from complete. In general, a single plant is usually not infected by more than one endophyte, but a grass species can serve as a host for several different endophytic species. However, double infection of a single grass plant by different general has been reported, as in *Panicum anceps* infected by *M. atramentosa* and *B. henningsiana* [19,33]. This may reflect occupation of different in situ locations. *Balansia* species are endophytic, whereas infection by *M. atramentosa* is localized and systemic. Nevertheless, Latch et al. [31] reported the co-occurrence of two endophytic fungi, *Gliocladium*-like and *Phialophora*-like, colonizing identical tissues along with *Acremonium lolii* and *A. coenophialum* in plants of perennial ryegrass and tall fescue, respectively.

In natural grass communities, although several host species are growing sympatrically, only one species is usually infected by a particular *Balansia* taxon, suggesting a certain degree of compatibility between fungal

Table 1. Distribution of Endophytes in the Gramineae

Subfamily and tribe	Number of host species	Number of hosts recorded for each endophyte species[a]
Festucoideae	89	
Festuceae	34	*Epichloe typhina* (21), *Acremonium coenophialum* (3), *A. lolii* (1), *A. starrii* (4), *A. huerfanum* (1), *Acremonium*-like (10)
Aveneae	29	*E. typhina* (28), *Balansia epichloe* (1), *A. typhinum* (1)
Triticeae	13	*E. typhina* (13)
Melicaceae	6	*E. typhina* (6)
Stipeae	7	*A. chisosum* (1), *A. tryphinum* (1), *Acrenomium*-like (3), *Atkinsonella hypoxylon* (2), *B. epichloe* (1)
Panicoideae	72	
Paniceae	64	*B. claviceps* (7), *B. epichloe* (1), *B. henningsiana* (8), *B. obtecta* (3), *B. sclerotica* (4), *B, strangulans* (33), *Balansiopsis pilulaeformis* (3), *Ephelis japonica* (1), *Myriogenespora artramentosa* (12)
Andropogoneae	8	*B. claviceps* (1), *B. epichloe* (1), *B. henningsiana* (3), *B. sclerotica* (1), *E. typhina* (2), *M. atramentosa* (3)
Eragrostoideae	37	
Eragrosteae	9	*B. epichloe* (7), *E. typhina* (2), *M. atramentosa* (1)
Chlorideae	10	*B. epichloe* (3), *B. henningsiana* (1), *B. obtecta* (1), *B. strangulans* (5), *E. typhina* (1)
Unioleae	1	*Balansiopsis pilulaeformis* (1)
Aristideae	17	*B. aristidae* (14), *B. hemicrypta* (3)
Bambusoideae	3	
Bambuseae	3	*Balansiopsis guaduae* (3)
Oryzoideae	4	
Oryzeae	4	*B. ambiens* (1), *B. oryzae* (1), *B. pallida* (1), *E. typhina* (1), *Ephelis oryzae* (1)

TABLE 1. (Continued)

Subfamily and tribe	Number of host species	Number of hosts recorded for each endophyte species[a]
Arundinoideae	4	
Arundineae	1	*E. typhina* (1)
Danthonieae	2	*A. hypoxylon* (2), *E. typhina* (1)
Centotheceae	1	*B. epichloe* (1)

[a]Figures in parentheses represent the number of host species, within the tribe, parasitized by each fungal species.

strains and grass species [2]. Experimental infections of grasses and sedges with *Atkinsonella hypoxylon* and *B. cyperi* established that these two fungi were broadly cross-compatible within a specific host population, but *B. cyperi* was less host specific than *A. hypoxylon* [27].

Among fungal species reviewed here, *E. typhina* is the most cosmopolitan endophytic species infecting 76 grass species from all subfamilies except Bambusoideae. However, this fungus appparently occurs unevenly among the subfamilies, as 90% is associated with the Festucoideae, which are cool season grasses characterized by the C-3 photosynthetic pathway (see Table 1). Related endophytic *Acremonium* anamorphs have a more restricted host range, infecting only 25 grass species belonging to the tribes Festuceae and Stipeae of the Festucoideae (see Table 1). Due to recently renewed interest in these symptomless endophytes and their value for improving turf grass species [34], more information on host ranges has been obtained [35,36]. As more grasses are examined, additional hosts will undoubtedly be reported.

Balansia species have been found infecting 102 different grass species from all subfamilies except Bambusoideae, but these are mainly warm season grasses of the Panicoideae [64%) and Eragrostoideae (33%), subfamilies characterized by the C-4 photosynthetic pathway (see Table 1). *Balansia epichloe* is the most cosmopolitan of the *Balansia* species; its host range varies from cool season grasses of the Festucoideae to warn season grasses of the Panicoideae, Eragrostoideae, and Arundinoideae. With the exception of *B. ambiens*, *B. oryzae*, and *B. pallida*, which are associated only with Oryzeae grasses, species of *Balansia* are found infecting only panicoid or eragrostoid grasses. Some Balansieae endophytes, e.g., *B. aristidae* and *B. hemicrypta*, are restricted to the grass genus *Aristida* (see Table 1). *Balansia oryzae* is one of two species found in association with a food crop, rice [25].

The genera *Atkinsonella* and *Myriogenospora* are very restrictive in grass associations. *Atkinsonella hypoxylon* is only reported in association with two species of *Danthonia* (Arundinoideae) and two species of *Stipa* (Festucoideae), whereas *M. atramentosa* and species originally identified as *Balansiopsis* [2] are found in panicoid and eragrostoid grasses. The only endophyte reported as infecting species of the Bambusoideae is *B. pilulaeformis* (see Table 1). *Balansia cyperi* is the only *Balansia* species that is associated with a nongraminaceous host and is specific for *Cyperus*.

This is another example of host specificity, and its occurrence on this genus extends the range of endophytes.

C. Grass–Fungus Relationships

Life Cycles

Grass–endophyte relationships are discussed using different criteria, such as presence or absence of external stromata and their location on the host, and the degree and resulting nature of the symbiotic relationship between fungus and grass.

Most of the grasses infected by species of the Balansiae produce external stromata characteristic of each endophyte, including the genera *Atkinsonella, Balansia, Balansiopsis, Epichloe*, and *Myriogenespora* [1,2,19]. These fungi grow without any external indications of infection during vegetative phase of grass development, but when the grasses reach the reproductive stage, the fungi infecting them become evident by production of conidiomata and ascomata, localized externally on different plants parts according to fungal species [2,26,27]. The main effect of sporulation on the host is prevention of inflorescence development, resulting in total or partial sterility [1,2]. Two different and general mechanisms of flowering prevention have been reported [1,2,37].

The first of these is shown by *E. typhina, B. strangulans, B. obtecta*, and *B. claviceps*, which usually cause mechanical abortion of host inflorescences by surrounding the developing reproductive tiller with stromatic tissue. A second means of preventing flowering is typified by *B. epichloe* and *B. henningsiana*, where fruiting bodies are produced on the expanded leaves but the grass hosts remain sterile. A disruption of the plant hormonal balance has been suggested as a possible cause for this effect [37,38]. However, flower inhibition may be host-fungal specific, as *Eragrostis sp.* infected with *B. epichloe* produces normal seed and infected plants produce ascomata [2,7].

The generalized life cycle of endophytes reflects two basic differences in which endophytic infection and transmission occur (see Fig. 1). The basic difference between infections is the external or internal portals of entry into seed for the endophyte. Thus, endophytic fungi are categorized into seedborne direct or indirect according to their general mode of infection during the generalized life cycle [39] (Fig. 1). These two groups are similar to those proposed by White [40], but differ since they are based on morphological relationships. The earlier terminologies of Halisky et al. [41], choke and chokeless, are too restrictive, applying only to *Epichloe* and *Acremonium* endophytes and suggestive of a disease process.

Fungi belonging to the seedborne indirect group are characterized by infecting the developing florets and/or seed by either conidia and ascospores produced from stromata, and include those endophytes that produce external fruiting bodies. Examples of this latter type of infection can be found among *Balansia, Epichloe*, and *Myriogenespora* species.

In contrast, those endophytes of the seedborne direct life cycle are characterized by direct transmission of the fungus into the flower and seed from internally borne hyphae. In this regard endophytes that do not produce stromata cannot be transmitted to other grasses except to progeny through seed, i.e., these grasses are self-contained, perpetuating their own internal infection vegetively to seed. Fungi of this latter category

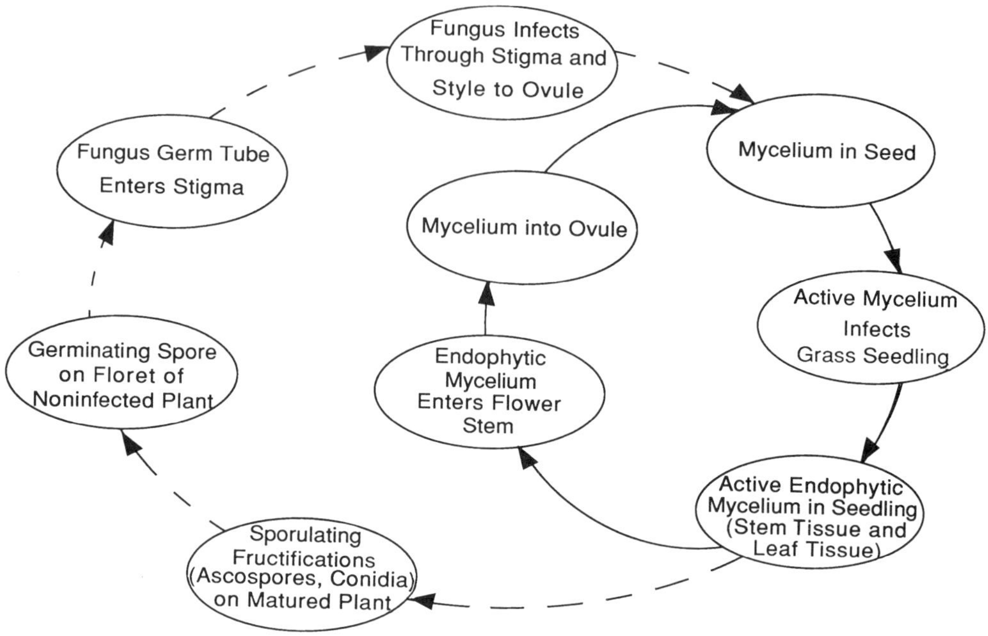

FIGURE 1. Life cycle of endophytes of grasses. Broken lines represent seedborne indirect endophytes; solid lines represent seedborne direct endophytes.

grow intercellularly but fail to produce conidia or ascospores, hence no visible reproductive structure is formed. These endophytes include, but are not restricted to, members of the genus *Acremonium* [8,26,28,31,42].

Endophytic fungi are not always clearly classifiable as either direct or indirect. For example, some grasses infected with *E. typhina* may fail to show, in a large proportion of all flowering stems produced within each tiller, external characteristics of infection, referred to as "choke disease." Such plants would therefore include tillers of both normal seed heads and choked seed heads. Thus, infection can be either direct and indirect, and both mechanisms can function in the same fungus–grass combination. Infection of seed on apparently healthy inflorescences was first reported by Sampson [1]. In *A. hypoxylon* infecting *Danthonia spicata*, although fertile stromata are produced, grass fertility is only prevented in the upper portion of the inflorescence which bears chasmogamous flowers. In the lower portion, cleistogamous flowers remain enclosed in the sheath and produce viable seed which bear hyphae, thus resulting in direct transmission [43,44].

Morphology of Infection

Grass–endophyte associations are known to vary from the very casual, as in *M. atramentosa* which lives superficially but systemically on meristematic tissue and leaves of its hosts, to the truly endophytic habit of the genera *Balansia*, *Epichloe*, and *Acremonium*, whose nonhaustoriate hyphae grow intercellularly within leaves and stems [1,2,26,45]. We feel the degrees of host–fungus interactions expressed by endophytes parallel descriptive terminology applied to lichens [56]. Therefore the term endosymbiotic may be used for those fungi with the endophytic habit, while ectosymbiotic may well describe superficially systemic fungi.

In contrast to earlier reports [2], recent investigations indicate that, in addition to *M. atramentosa* [19], *A. hypoxylon* infecting *Danthonia spicata* and *Stipa leuchotricha*, respectively, and *B. cyperi* infecting *Cyperus virens* were associated superficially on meristematic tissue of its hosts [27]. These ectosymbiotic fungi are found forming dense mats of mycelium on the surface of leaf primordia, between the bases of young leaves and, as the leaves expand, nonconnected hyphae extended onto the leaf blade. Hyphae never penetrate tissues, and no alteration of the cuticle was observed [26, 46,47]. Physical prevention of flowering is a common feature of these fungi, but the results vary for those infected hosts that do develop flowers. For example, *A. hypoxylon* is able to infect the seedlings produced from cleistogamous seed of *D. spicata* [44], but the few seed produced by *Paspalum notatum* infected with *M. atramentosa* developed into uninfected plants [26].

Despite the casualness suggested by the ectosymbiotic habit, nutrient uptake from the host does occur [46], and insect deterrence by infected plants has been shown to occur [48]. This suggests a close biochemical and ecological association, which led Rykard et al. [26] to compare *M. atramentosa* and its host with the truly endophytic associations characteristic of other members of the Balansiae.

Endosymbiotic fungi are the most widespread among the Balansiae. Endosymbionts are either seedborne direct as in *E. typhina* and most *Balansia*. These fungi perpetuate themselves by colonizing the developing seed, by external infection by spores, or by perennating in living host tissue (see Fig. 1). Hyphae of *A. coenophialum*, *A. lolii*, and *E. typhina* are found in aleurone layers of the seed, surrounding the endosperm and between the endosperm and the scutellum [49]. No hyphal penetration of the embryo by the hyphae has been detected in *Acremonium* endophytes. In contrast, embryo infection by *E. typhina* was reported in a small percentage of the infected seeds [1,16,45,50]. Seedling infection of tall fescue occurs a few days after germination. Hyphae occur in the first internode within two days after germination and in the sheath of the first leaf within three days [51]. Evidence of infection in seedlings by the tall fescue endophyte was correlated with the sheath differentiation in the first leaf [51].

The developing tillers become infected during vegetative growth of plants. Occassionally some tillers escape infection, probably because of differential growth rates of the fungus and grass tiller due to environmental factors [1,2,16,37]. Once infection is established in a tiller, hyphae can be found growing intercellularly in various leaf tissues, stems and inflorescences. Although hyphae have been found in roots of seedlings grown on agar media [52], no hyphae have been found in root tissues of soil-grown infected grasses [1,45,53]. While *Balansia* species and *E. typhina* infect both sheaths and blade (laminae) [1,2,10,26], *Acremonium* endophytes are confined to leaf sheaths. Hinton and Bacon [45] observed that tall fescue endophyte invades only areas where there is a network of intercellular spaces, probably because of a restriction imposed by the ligule. This differentiates very early in leaf development and might function as a physical barrier because of its lack of continuous intercellular spaces. After the flowering stem begins to elongate, the endophyte grows within intercellular spaces of pith tissues surrounding vascular bundles. It develops along with the flowering stem and can be found in inflorescence branches, in ovaries of florets and stamen filaments, but not in anthers, resulting in no infection of pollen grains [1,45].

Ultrastructural studies of *Acremonium*-infected tall fescue and perennial ryegrass and *E. typhina*-infected hard fescue (*Festuca longifolia* Thuil.) have shown no evidence of intracellular growth [45,53]. However, Sampson [1] observed that in the grass *Dactylis glomerata* hyphae of *E. typhina*, prior to the formation of the conidiomata, become intracellular, disrupting the middle laminae and causing tissue necrosis. Philipson and Christey [50] observed hyphae of *A. lolii* growing between host cell walls and plasmalemma when the fungus invades the embryo sac, but it was always separated from the host cytoplasm by a membrane, and host cell integrity was retained. In general, the common characteristic among different endophyte species is a nonhaustoriate intercellular growth, especially during growth of host grasses [1,2,45,53].

Hinton and Bacon [45] reported presence of a hyphal matrix of unknown origin in the interface of the host and fungal cell walls, but no host cell modification was observed. They concluded that the ultrastructural relationship of the endophyte and tall fescue appeared similar to that reported for lichens and certain mycorrhizae. Endophyte hyphae produced in culture are usually devoid of some characteristic ultrastructural components found in hyphae growing within host tissue. Crystalline inclusions of varying size are known in *A. coenophialum*, *A. lolii*, and *E. typhina*, while tubular or microtubular structures were found in the last two species [53,54]. Based on the differences between in vitro and in vivo observations, Fineran et al. [54] suggested that the crystalline and tubular inclusions might be related to the production of lolitrem neurotoxins, which have been found only in plant–fungus association and not in fungus culture [55].

Mutualism

Several criteria have been proposed to describe the type of symbiotic interaction between organisms [56,57]. All endophytes establish a permanent obligate biotrophic relationship with grasses. The degree of this interaction differs and the division line follows those relationships that fall within either the seedborne direct or indirect life cycles. In some cases necrotrophic phases can be observed, as in *E. typhina* infecting *Dactylis glomerata* [1]. In some hosts infected by *Balansia* sp., the leaf areas immediately below the ascomata turn yellow and become necrotrophic when the ascomata disintegrates (personal observation). This then suggests that yet another criterion exists for assessing the nature of these symbiotic interactions. Based on the potential fitness of the host, grass–endophyte interactions range from agonistic associations to mutualistic ones.

Differences in fitness between uninfected and endophyte-infected perennial grasses by endophytes that produce external sporulating structures, i.e., seedborne indirect, are difficult to assess. On the one hand, there is a deleterious effect resulting from the prevention of flowering and seed production, although grass sterility is not always complete. On the other hand, it has been observed that there is frequently an increase in vegetative growth of infected grasses enhancing their competitive ability and longevity [43,58]. At least in the short term, if persistence and vegetative reproduction are the dominant factors influencing the success of a grass, the relationship is essentially mutualistic. If, however, seed dissemination is a prerequisite for success because of community dynamics, the aspect of the relationship would be agonistic. Insect and animal toxicities resulting from seedborne indirect endophytes are certainly suggestive of a mutualistic symbioses, and this advantage might counterbalance those relationships suggested to be agonistic [13,15,48].

There are, however, some clear cases of agonistic interactions such as in annual grasses infected by species of *Balansia* [2] where seed production is crucial for the survival of the population, and in *M. atramentosa*, where not only seed production is prevented, but also vegetative growth in inhibited [26].

Multualistic symbioses are strongly indicated and considered common characteristic of those endophytes of the seedborne direct life cycle. In all cases these are *Acremonium*-grass associations, and the benefits to the hosts are increased growth, deterrences to insect and mammalian herbivory, and tolerance to drought [59–63]. The fungus receives nutrients, and a niche which provides protection, reproduction, and dissemination [13].

D. Mode of Infection and Dissemination

Infection and dissemination are two important aspects of each endophyte–grass system as they are the means of maintaining and perpetuating infected grasses. In both life-cycle types, the seed is considered the primary infection unit from which seedling infection takes place [2,16]. Since seeds are the infected unit, endophytes are maternally transmitted, and the seed serves as the dissemination unit.

Successful dissemination of an endophyte depends on its viability within seed. Infected seeds germinate under favorable conditions the season following their production, therefore perpetuation of endophyte infection is guaranteed. However, seeds stored for periods longer than one year are not likely to have viable endophytes [13,64]. Viability of endophyte-infected seed can be extended provided they are stored at low humidity, and at 4 to -20°C in sealed containers [13,65]. Under these conditions endophytes and seed remain viable for at least 2 years, and possibly longer.

Following germination, the fungus infects the seedlings. This apparently is a natural course of infection in perennial ryegrass and tall fescue and possibly the reason for the only successful artificial infections achieved by others [47,66]. It is not known how long the seedling is predisposed to infection, but fescue seedlings at the four- to six-leaf stage cannot be infected. The results were also negative with several endophytes used to infect other mature grass species [26,33,66].

Perennation of the infection results from the growth of a fungus into meristematic area of the grass and subsequent infection of new tillers. Again, the fungus infects young developing tissue. It has not been established that infection of matured grasses occurs naturally; this has been attempted experimentally, but with little or no success [2,26]. Observation of matured plants of tall fescue inoculated with its endophyte show that leaf sheaths can be inoculated. The endophyte grows into the tissue of the leaf, but the leaf is abscised just short of the fungus reaching the apical meristematic area (Bacon and De Battista, personal observations). Success might be obtained by inoculating the apical meristematic tissue directly, but a means of accomplishing this is unknown.

E. In Vitro Culture of Endophytes

Use of endophytes for increased pest resistances, improved appearance, and tolerance to environmental stresses requires complete understanding of the fungus in vitro relative to predicting its activity when introduced into other grasses. The fungus should maintain its original in vivo effects when transferred to either native or nonnative hosts since specific performance aspects of infected grasses may be directly related to the fungus.

Isolation

The methods used for isolating endophytes from grasses depends on whether they are seedborne direct or indirect. Endophytes that do not sporulate in vivo must be isolated from tissue of either recently collected field or greenhouse-grown grasses [30,67]. These fungi are located in the sheath portion of a grass leaf and an inner sheath should be used since it is relatively free of debris and other fungi. All plant material intended as a source for endophytes should be used immediately or refrigerated in dampened paper for a few hours.

Portions of surface sterilized plant parts are cut into smaller sections and placed in either liquid medium or on agar isolation media [30,67]. Additional isolation sources, such as seedlings [39,46], ground tissue [5,39], and seeds [67], have also been used. The time for isolating a fungus from grass depends on tissue type and its preparation, genetic makeup of the fungus and plant, and specific media [7,30,67,68,].

Endophytes characterized by the seedborne indirect life cycle may be isolated from ascospores [7,19,69]. Freshly collected mature ascomata of any species will discharge ascospores, many of which are matured and capable of forming colonies. Discharged ascospore germination may be followed under a compound microscope, an if the medium is suitable [68,69], colonies freed of bacteria and yeast will become visable within 2 to 4 weeks. The time required depends on the species. In vivo growth resembles in vitro as the tall fescue endophyte cultured on these media yielded cell wall structural properties similar to the endophyte growing in tall fescue [70]. For a more detailed account of isolating endophytes and problems associated with fresh isolates, see the recent review devoted to this aspect [39].

In Vitro Growth Requirements

Most media for endophytes have not been defined, are complex and, at best, semisynthetic [71–74]. One isolate of the tall fescue endophyte was induced to grow on defined and relatively simple media after it was maintained in the laboratory for a considerable period of time [74,75]. Species of *Balansia* have been grown on both semisynthetic and synthetic media [72,76]. Use of defined media for initial isolation of endophytes is not recommended. Once endophytes have been isolated, a defined medium may be tried, but this may select various auxotrophs [30,72]. Vitamin heterotrophy was established for six species of *Balansia* [72] and *A. ceonophialum* of tall fescue [73].

Shake culture is favored over stationary culture when it is desired to obtain a rapid growth of endophytes. Once maximum amounts of mycelium are achieved, the endophyte may be continued as stationary cultures during which several secondary products have been reported to occur [30,76].

Competence of an endophyte to produce the desired biologically active compound should be established on the initial isolate, since this capacity may not be stable in culture [30]. If the objective is to determine whether an isolate is toxic to cattle or insects, the procedure involves a two-stage fermentation method [68,77]. Addition of specific substrates or precursors should be included since variations in nitrogen source and quantity are known to influence growth rates [73], pigment production [76,78], and indole and ergot alkaloid biosynthesis [72,76,78].

The culture media and endophytes may be tested directly to determine toxicity using one of many insect feeding bioassays [59,71,79]. Negative results need not mean that an isolate is biologically inactive; it may mean that specific in vitro requirements are not met. In order to circumvent this problem, an in vivo test of isolates should always precede one in vitro. It has been established that for three species of *Balansia* qualitative production of ergot alkaloids occurred both in vivo and in vitro [68,77,80].

III. FUNGUS–GRASS AND ENDOPHYTE BIOLOGICALLY ACTIVE COMPOUNDS

The multiplicity of animal symptomatology, as well as effects on grass reproduction, suggest that in addition to nutritional relationships, there are biochemical interactions. Some of these may have some bearing on the overall survival status of the symbiosis. This interaction is probably complex and fungus and plant specific. Most of the compounds isolated from infected grasses have been tested for animal toxicities. Compounds isolated from infected grasses can be grouped according to their biological affects on the various groups of animals and their structural similarity.

A. Ergot Alkaloids

Ergot alkaloids are derivatives of lysergic acid. Several grasses infected with different species of endophytes have been shown to contain these compounds [68,80–82]. Compounds of this class are strong candidates for the various cattle toxicity symdromes reported for tall fescue [10,11,83], perennial ryegrass [84,85,77], and toxicity of the weed grass species found in pastures [10]. For detailed reviews of the animal symptomatology and specific forage toxicity, see the reviews of Bacon et al. [10], Yates [83], Siegel et al. [15], Gaynor and Rowan [86], and Studemann and Hoveland [28].

All endophytic species, primarily *Balansia*, found growing in association with major forage grass have been shown to produce ergot alkaloids [10,76]. Other fungi, such as *Atkinsonella* and *Myriogenospora*, have not been examined for this group of toxins. While most produce ergot alkaloids, there is a difference in the type of ergot alkaloids produced by each of the two groups of endophytes. Many species of *Balansia* and other members of the seedborne indirect group produce the clavine type of ergot alkaloid as the major class [68,78,87]. Species of *Acremonium* and other members of the seedborne direct group produce the peptide types of ergot alkaloids [88–90]. However, both types of alkaloids have been reported as minor constituents in each endophytic group.

Endophytes have been shown to produce ergot alkaloids in culture and in situ, indicating that these toxins are produced by the fungus [76,81,91]. Those produced by the tall fescue endophyte are thought to be translocated throughout the plant [81]. It has also been shown that soil nutritional and moisture conditions affect the biosynthesis of these compounds [63,81]. Not all isolates within a given location can produce these compounds, and each isolate varies both quantitatively and qualitatively.

B. Loline Alkaloids

Alkaloids of this class are pyrrolizidine bases, and there are several of these, none of which are found in uninfected grasses. This class has not been isolated from *Acremonium* cultures, which suggests that these alkaloids require either specific fungus and plant interaction or that in vitro culture requirements have not been met. This class has not been looked for in endophytes other than the *Acremonium*-types, primarily the endophyte of tall fescue and perennial grasses [92]. Belesky et al. [93] reported that there was not only a relationship between infection and occurrence of lolines, but that there was also direct relationship between the amount of lolines present and endophyte biomass in tall fescue. Since these alkaloids are not isolated from cultures and uninfected grasses, they may be synthesized in response to endophyte infection, suggesting that they may by phytoalexins. However, since we have no information to suggest that they are antimicrobial, one property of phytoalexins, this conclusion is unfounded.

Toxicity of the loline alkaloids varies, the N-acetyl and N-formyl forms are considered more toxic than the others [83,94]. This toxicity is considered mild since the loline alkaloids do not have the chemical structures essential for general toxicity usually characteristic of the pyrrolizidine alkaloids [95]. These two forms may reach concentrations as high as 0.8% of the grass dry weight [92]. This was one of the most studied groups of compounds from infected grass, and the chemistry and toxicity of it has recently been reviewed [94].

C. Peramine

Peramine, {3-(3'-guanidinopropyl)-2-methyl-1,2-dihydropyrrolo [1,2-a]-pyrazin-1-one}, was only recently identified as the substance responsible for feeding deterrents to the Argentine stem weevil in perennial ryegrass [96–98]. This compound has been isolated from culture filtrates of *A. lolii*, therefore it is made by the fungus [96]. Peramine has been reported from other endophytes such as *A. coenophialum* and certain isolates of *E. typhina* [99]. It has been established that other endophyte-infected grasses are toxic to insects [48], but it is not known of peramine or similar compounds are responsible for this toxicity.

D. Ryegrass Neurotoxins

Toxins of this class, commonly called the lolitrems, are very complex isoprenoid-substituted indole metabolites, similar in structure and biological activity to the tremorgenic compounds produced by species of *Penicillium* [100]. Of the four neurotoxins isolated from perennial reygrass, lolitrem B is the major one, but all are considered responsible for the ryegrass staggers syndrome [55,86,101].

To date, fungal cultures have been negative for the lolitrems, but a closely related compound, paxilline, has been isolated from cultures of *A. lolii* and from ryegrass [102]. This compound is considered the indole-isoprenoid unit of the lolitrems, and it was suggested that final conversion to the lolitrems was a consequence of ryegrass and possibly fungus metabolism [102].

E. Miscellaneous Compounds

Included here are compounds isolated from cultures or infected materials whose importance may or may not be recognized but whose isolation from infected forage imply an ecological relationship. The first of these are the simple indole derivatives related to plant growth hormones. These compounds have been isolated from cultures of species of *Balansia* and include 3-indole acetic acid, indole ethanol, and several other related compounds [38,78,103,104]. Some of these compounds have been tested for growth-promoting activity and found to be positive [38]. Information on their production and effects of these substances in situ is unavailable. It has recently been determined that the endophyte of tall fescue produces IAA in culture [75].

Of similar uncertainty is the occurrence of harmane, perlolidine, norharmane, and halostachine, all of which were isolated from tall fescue whose infection status was not reported (see Ref. 94 and references cited therein). Since perlolidine and perloline are found in both infected and uninfected grasses [94] and various triterpens and steroids are constituents of the endophyte cell wall [103,104], they are not considered part of a defense mechanism. However, the steroids ergosterol, ergosterol peroxide and one of the tetraenones have been shown to be toxic to several test systems [104].

In a series of papers [105,106] it was established that *E. typhina* associated with *Phleum pratense* produced a series of biologically active compounds some of which were considered of fungal origin and others of plant origin. The bioassay used was the inhibition of *Cladosporium phlei* to produce disease symptoms on *P. pratense*, its natural host. This is the first report of a disease-deterring mechanism from an endophyte-infected grass. The range of activity for compounds originating from the endophyte was 5 μg to 50 μg, while for those originating from the grass activity was 50 μg and several had no activity [106]. There were two additional groups of compounds that had activity, but their origins were unknown. Most of these compounds were derivatives of phenolic acid; the simplest structures originated from the fungus ascomata, and these were the most active. Involvement of phenolic acid derivatives in disease development in other endophyte-infected grasses is unknown. Preliminary work in vitro suggests that the tall fescue endophyte inhibits saprophytic growth of two pathogens [107].

IV. FUNGUS–GRASS ECOLOGICAL INTERACTIONS

A. Stress Physiology and Survival Mechanisms

The environment of endophyte-infected grasses is considered to influence the production of metabolites, secondary and primary, and hence any associated stress sparing or animal and fungus disease defense mechanisms associated with these metabolites. Such environmental influences on the accumulation of metabolites by plants and in endophyte-infected grasses are documented [63,81,86 108]. Most studies have assessed effects of soil moisture and nitrogen content and type on the accumulation of ergot alkaloids and or drought tolerance [62,63,81,86]. Improved plant growth and persistence have been established for endophyte-infected perennial ryegrass, tall fescue, and certain weed grass species. Infected grasses have more dry matter, total leaf area, tiller number, and root growth [48,55,63]. Seedling vigor and germination rates have also been shown to be increased in infected grasses [71]. Leaf roll has been suggested [63] but not well documented to serve as a survival mechanism for infected grasses. There is also an interest in studying the morphological alterations in endophyte-infected grasses, i.e., rhizomatous and prostrate, but currently the significance of this on survival is unknown. It has not been established which plant or fungus product is responsible for each specific survival mechanism, although a complex interaction is suggested.

Many other agronomically important traits of infected grasses have been studied but not under environmentally stressing conditions. Performance of infected grasses for any stress or pest tolerance should be measured under severe environmental conditions. Most studies have measured the performance of infected grasses under greenhouse conditions of constant temperature and moisture. Under such conditions beneficial effects in infected grasses, as opposed to uninfected grasses, are not exhibited. The results of one study under these conditions indicated that while photosynthetic rates were significantly different, there probably was no biological effect among the five clones tested, and stomatal resistance was increased in infected tall fescue [77]. However, this latter aspect indicates a mechanism which supports the idea that infected tall fescue is more drought tolerant [62,63]. The effect of endophyte on photosynthesis during stressful conditions are unknown. Studies of infected grasses under severe environmental conditions can establish the value of an endophyte for forage grass improvement. However, the ecological success observed in naturally infected grasses may not be transferred to nonhosts. As an alternative, it may be necessary to select naturally infected grasses and improve these for whatever the desired characteristic [3,34].

The high performance of endophyte-infected grasses under severe conditions has attracted a number of researchers to attempt to develop turf grass cultivars resistant to insects [3,34,109]. More success has been achieved by using naturally infected grasses, and increasing the desirable line, rather than by infecting either grasses not known to be hosts or uninfected grasses known to serve as hosts.

B. Herbivorous Toxicities and Pest Deterrents

A grass possessing animal and fungus disease resistances has an ecological advantage. Just how much of these processes are in operation in endophyte-infected grasses is just emerging. Information now exists which suggests that endophyte-infected grasses produce a variety of metabolities, as indicated above, which serve to deter insects, fungus pathogens, and grazing animals. Yet to be established are compounds that are defensive to root pathogens such as nematodes. These defensive mechanisms are the most promising aspects of endophyte-infected grasses. Infected grasses are ideal for biological controls of insects and diseases of grasses, and they can serve as ecological models for studying the evolution and nature of mutualistic relationships between fungi and grasses.

As indicated above, endophyte-infected grasses are known to be responsible for the poor performance of livestock consuming forage or weed grasses. The range of vertebrate hervibores known to be naturally affected by infected grasses are cattle, sheep, goats, and horses [10,11,94,101]. Some toxic grasses exhibit animal specificities, while others affect all animal species, although differently.

Invertebrate toxicities include both natural and unnatural pests of grasses. The nature of the deterrence depends on the grass species and the invertebrate species. For example, oatbird cherry aphid, *Rhopalosiphum padi* L., was not repelled by perennial ryegrass infected with *A lolii* or the *Gliocladium*-like species but was repelled by the *A. coenophialum*–infected tall fescue [79]. Three additional species of aphids were not affected by *A. coenophialum*-infected tall fescue or *A. lolii*-perennial ryegrass [59,79].

In addition to aphids, the following is a list of insect and other pests deterred by endophyte-infected grasses: chince bugs [109], leafhoppers [37], fall armyworms [110], sod webworms [3], crickets [111], corn flea beetle [112], black beetle [15], Argentine stem weevil [113], bluegrass billbug [114], flour beetle [115], and spiral stubby root nematodes [116]. Pest deterrence in perennial ryegrass is affected by the systemic infection of other fungi. Root infection of perennial ryegrass by a mycorrhizal fungus depressed the deterrence of Argentine stem weevil [117].

The only fungus disease reported as being deterred by the presence of an endophyte was *Cladosporium phlei*, which was repelled by *E. typhina* in timothy grass [105]. Some in vitro data from the endophyte of tall fescue suggest that this fungus can prevent the saprophytic growth of certain pathogenic fungi [107].

Peramine is the only insect-specific deterring compound isolated from *A. coenophialum*- and *A. lolii*-infected grasses. This compound is reported as being active at 1.3 ppm [98], but its spectrum of activity on natural or unnatural insects of grasses is unknown. Since peramine is also found in combination with other chemicals, the possibility of synergism should be considered. Synergistic effects are expected to extend the list of pests deterred and possibly increase the number of animals known to be affected by these toxins. In addition to synergism, Yates [118] indicated that perloline from tall fescue was a potentiator for toxicities from ergot alkaloids. Perloline is found in both infected and uninfected tall fescue and perennial ryegrass; its presence in other endophyte-infected grasses is unknown, but its importance in the overall ecological success of grasses may be more important than assumed.

C. Inter- and Intraspecific Grass Interactions

Since endophyte-infected plants usually occur in mixed populations with uninfected ones of the same grass species and with other grass taxa in community, it is important to consider the effect of endophyte infection on the intraspecific and interspecific competitive ability of the plants. Few reports have addressed this question, mainly due to the difficulty in manipulating grass and fungus genotypes.

Allelopathy of endophyte-infected grasses may be a contributing factor in competition. Although allelopathy has not been examined in most endophyte-infected grasses, it has been reported to occur in tall fescue [119, 120], and soil levels of loline and its derivatives are considered responsible for inhibition of plant growth [121].

Clay [43] and Kelley and Clay [122] studied the competitive ability of *Danthonia specata* genotypes uninfected of infected with *A. hypoxylon*. Eight *Danthonia* genotypes (three infected and five uninfected) and eight *Anthoxantum odoratum* genotypes were arranged in pairs in a diallele design and replanted in a natural community of these species. The design allowed them to measure interspecific and intragenotypic competitive ability. Growth, expressed as tiller number and reproductive effort, measured as inflorescence number (both adjusted by initial tiller number), were significantly larger for the infected *Danthonia* genotypes when grown in competition with *Anthoxantum* genotypes, but no significant differences were noted when plants were grown under conditions of intragenotypic competition.

Although fungal infection increased reproductive effort, there was a reduction in fecundity of host due to abortion of the chasmogamous portion of inflorescences, yielding fewer flowers per plant than uninfected plants. *Danthonia* genotypes infected with *A. hypoxylon* survived better than uninfected ones under both interspecific ($p < 0.01$) and intragenotypic ($p < 0.09$) growing conditions

In general it seems that *A. hypoxylon* infection confers an advantage to *D. spicata* plants growing in competition through enhanced survival and vegetative growth. Reduction of fecundity may not be translated in a reduction of fitness if biotic and abiotic environmental conditions favor vegetative reproduction and establishment of new individuals by seed has a low impact on the demography of the community.

However, these data are not without objections. In these experiments, grass genotype and fungal infection effects were confounded, and, taking into account the small sample size (three and five genotypes), it is difficult to conclude whether the observed differences are due to the presence of the fungus, the grass intergenotypic variation, or a combination of both.

In an unrelated experiment, Bouton et al. [123] provided some information on interspecific competitive ability of infected and uninfected tall fescue genotypes with *Acremonium coenophialum*. Six clones, two infected and four uninfected, were grown with or without competition of bermudagrass (*Cynodon dactylon*) for a period of 22 months. At the end of the experiment, number of tillers and dry matter yield were measured along with some other characters on the surviving plants. In general, infected plants produced more tillers, more dry matter, and survived better than uninfected ones. Although competition of bermudagrass depressed all characters measured, the effect of endophyte infection was yet more pronounced under this stressful condition. Again, this experiment consisted of confounding effects of fungal and grass genotypes, thus these results should be examined with the same precautions as the previous ones.

In more controlled experiments, Hill et al. [124] used *Acrenomium*-infected and uninfected versions of two tall fescue clones that were freed of infection by the use of a fungicide. Under this system fungal infection and grass genotype were no longer confounded. Each clone was grown in a modified Nedler's design that allows one to compare the competitive ability of infected and uninfected material growing pure or in mixtures under different plant densities. Tiller production was greater for infected plants than uninfected of each genotype regardless of the population density. The competitive ability was measured by relative crowding coefficient [36] based on dry matter yield. In one genotype infected material appeared to have outcompeted the uninfected at all populations densities, while in the uninfected at low densities, but at high plant densities infected material was more competitive.

Differences in response of different grass clones to endophyte infection might be due to differences in grass genotypes, endophyte genotypes, or interaction of both [125]. In previous experiments, these sources of variation were confounded in the grass genotype X infection status interaction term, due to the current impossibility of manipulating endophyte genotypes through artificial infection of adult plants.

V. CONCLUSIONS

Evolution of a relatively small group of fungi and their association with one of the largest groups of plants shows a continuum of expressions. All endophytes show similarity in classes of chemicals, host-induced expressions, and biology. There are of course differences. Some of the more primitive members show agonistic responses, but overall these associations obviously are compatible. Others are clearly mutualistic endosymbiotic, and it is these that may offer the greatest potential for technological exploitation.

Endophyte-infected grasses are self-perpetuating, usually contained within a grass, stable, and resistant to many environmental conditions. Knowledge of improved fitness in endophyte-infected grasses was an outgrowth of pasture grass toxicity studies. So far, endophytes have only a few applications as biological controls and to improve performance under stressful conditions. Future research will undoubtedly make possible other applications for this curious association of grasses with fungi.

ACKNOWLEDGMENT

We dedicate this chapter to the memory of Everett S. Luttrell, mycologist, plant pathologist, and educator.

REFERENCES

1. Sampson, K. The systemic infection of grasses by *Epichloe typhina* (Pers.) Tul. *Trans. Br. Mycol. Soc.*, *18*:30–47, 1933.
2. Diehl, W. W. *Balansia* and the Balansiae in America. U.S. Department of Agriculture, Washington, D.C. 1950.
3. De Bary, A. *Comparative Morphology and Biology of Fungi, Mycetozoa and Bacteria*. Clarendon Press, Oxford, United Kingdom, 1887.
4. Nobindro, U. Grass poisoning among cattle and goats in Assam. *Indian Vet. J.*, *10*:235–236, 1934.
5. Bailey, V. Sleepy grass and its effect on horses. *Science*, *17*:392–393, 1903.
6. Hance, H. F. On a mongolian grass producing intoxication in cattle. *J. Botany*, *14*:210–212, 1876.
7. Bacon, C. W., Porter, J. K., and Robbins, J. D. Toxicity and occurrence of *Balansia* on grasses from toxic fescue pastures. *Appl. Microbiol.*, *29*:553–556, 1975.
8. Bacon, C. W., Porter, J. K., Robbins, J. D., and Luttrell, E. S. *Epichloe typhina* from toxic tall fescue grasses. *Appl. Environ. Microbiol.*, *34*:576–581, 1977.

9. Hoveland, C. S., Haaland, R. L., King, C. C., Anthony, W. D., Clark, E. M., McGurire, J. A., Smith, L. A., Grimes, H. W., and Holliman, J. L. Association of *Epichloe typhina* fungus and steer performance on tall fescue pasture. *Agron. J.*, *72*:1064–1065, 1983.
10. Bacon, C. W., Lyons, P. C., Porter, J. K., and Robbins, J. D. Ergot toxicity from endophyte-infected grasses: A review. *Agron. J.*, *78*:106–116, 1986.
11. Stuedemann, J. A. and Hoveland, C. S. Fescue endophyte: History and impact on animal agriculture. *J. Prod. Agric.*, *1*:39–44, 1988.
12. Pottinger, R. P., Baker, G. M., and Prestidge, R. A. A review of the relationships between endophytic fungi of grasses (*Acremonium* spp.) and Argentine stem weevil. In *Proc. 4th Australasian Con. Grassland Invertebrate Ecology*, 1985, pp. 322–331.
13. Bacon, C. W. and Siegel, M. R. The endophyte of tall fescue. *J. Prod. Agric.*, *1*:45–55, 1988.
14. Clay, K. Grass endophytes. In *Microbiology of the Phyllosphere*, Vol. 1 (N. J. Fokkema and J. Van Den Heuvel, eds.). Cambridge Univeristy Press, Cambridge, 1986, pp. 188–204.
15. Siegel, M. R., Latch, G. C. M., and Johnson, M. C. Fungal endophyte of grasses. *Annu. Rev. Phytopath.*, *25*:293–315, 1987.
16. Sampson, K. The presence and absence of an endophytic fungus in *Lolium temulentum* and *L. perenne*. *Trans. Br. Myco. Soc.*, *19*:337–343, 1935.
17. Rykard, D. M., Luttrell, E. S., and Bacon, C. W. Conidiogenesis and conidiomata in the Clavicipitoideae. *Mycologia*, *76*:1095–1103, 1984.
18. Gauman, E. A. Class III. Ascomycetes. In *The Fungi*. Hafner Pub. Co., New York, 1952, pp. 199–207.
19. Luttrell, E. S. and Bacon C. W. Classification of *Myriogenespora* in the Clavicipitaceae. *Can. J. Bot.*, *55*:2090–2097, 1977.
20. Mhaskar, D. N. and Rao, V. G. Development of the ascocarp in *Epichloe cinerea* (Clavicipitaceae). *Mycologia*, *68*:994–1001, 1976.
21. Kilpatrick, R. A., Rich, A. E., and Conklin, J. G. *Juncus effuses*, a new host for *Epichloe typhina*. *Plant Dis. Rep.*, *45*:889, 1961.
22. Luttrell, E. S. Host-parasite relationships and development of the ergot sclerotium in *Claviceps purpurea*. *Can. J. Bot.*, *58*:942–958, 1979.
23. Von Gaumann, E. Mykologische Mitteilungen III. *Annales Mycologici*, *25*:167–177, 1927.
24. Singh, R. A. and Pavgi, M. S. Some hosts of *Balansia sclerotica* from Varanas, Uttar Pradesh. *Indian Phytopath.*, *25*:132–134, 1972.
25. Mohanty, N. N. Studies on udbatta disease of rice. *Indian Phytopath.*, *17*:308–316, 1964.
26. Rykard, D. M., Bacon, C. W., and Luttrell, E. S. Host relations of *Myriogenespora atramentosa* and *Balansia epichloe* (Clavicipitaceae). *Phytopath.*, *75*:950–956, 1985.
27. Leuchtmann, A. and Clay, K. *Atkinsonella hypoxylon* and *Balansia cyperi*, epiphytic members of the Balansiae. *Mycologia*, *80*:192–199, 1988.
28. Morgan-Jones, G. and Gams, W. Notes on Hyphomycetes,XLI. An endophyte of *Festuca arundinacea* and the anamorph of *Epichloe typhina*,

new taxa in one of two new sections of *Acremonium*. *Mycotaxon*, *15*:311–318, 1982.

29. White, J. F., Jr. and Morgan-Jones, G. Endophyte-host associations in forage grasses. VII. *Acremonium chisosum*, a new species isolated from *Stipa eminens* in Texas. *Mycotaxon*, *28*:179–189, 1987.
30. Bacon, C. W. Procedure for isolating the endophyte from tall fescue and screening isolates for ergot alkaloids. *Appl. Environ. Microbiol.*, *54*:2615–2618, 1988.
31. Latch, G. C. M., Christensen, M. J., and Samuels, G. J. Five endophytes of *Lolium* and *Festuca* in New Zealand. *Mycotaxon*, *20*:535–550, 1984.
32. Gould, F. W. *Grass systematics*. McGraw-Hill, New York, 1968, pp. 1–382.
33. Rykard, D. M. Comparative morphology of the conidial states and host-parasite relationship in members of Balansiae (Clavicipitaceae). Ph.D. thesis, University of Georgia, Athens, 1983, pp. 1–84.
34. Saha, D. C., Johnson-Cicalese, J. M., Halisky, P. M., and Van Heenstra, M. I. Occurrence and significance of endophytic fungi in fine fescues. *Plant Dis.*, *71*:1021–1024, 1987.
35. White, J. F., Jr. Endophyte-host associations in forage grasses. XI. A proposal concerning origin and evolution. *Mycologia*, *80*:442–446, 1988.
36. Harper, J. L. *Population Biology of Plants*. Academic Press, London, 1977.
37. Kirby, E. J. M. Host-parasite relations in the choke disease of grasses. *Trans Br. Mycol. Soc.*, *44*:493–503, 1961.
38. Porter, J. K., Bacon, C. W., Cutler, H. G., Arrendale, R. F., and Robbins, J. D. In vitro auxin production by *Balansia epichloe*. *Phytochemistry*, *24*:1429–1431, 1985.
39. Bacon, C. W. Isolation, culture and maintenance of endophytic fungi of grasses. In *The Isolation and Screening of Microorganisms from Nature*, Vol. 1 (D. P. Labeda, ed.). McGraw-Hill, New York, 1989, pp. 259–282.
40. White, J. F., Jr. Endophyte-host associations in forage grasses. XI. A proposal concerning origin and evolution. *Mycologia*, *80*:442–446, 1988.
41. Halisky, P. M., Saha, D. C., and Funk, C. R. Prevalence of non-choke-inducing endophytes in turf and forage grasses. *Phytopath.*, 75:1331, 1985 (Abstract).
42. Johnson, M. C., Siegel, M. R., and Schmidt, B. A. Serological reactivities of endophytic fungi from tall fescue and perennial ryegrass and of *Epichloe typhina*. *Plant Dis.*, *69*:200–202, 1985.
43. Clay, K. The effect of the fungus *Atkinsonella hypoxylon* (Clavicipitaceae) on the reproductive system and demography of the grass *Danthonia spicata*. *New Phytol.*, *98*:165–175, 1984.
44. Clay, K. and Jones, J. P. Transmission of *Atkinsonella hypoxylon* (clavicipitaceae) by cleistogamous seed of *Danthonia spicata* (Gramineae). *Can. J. Bot.*, *62*:2893–2895, 1984.
45. Hinton, D. M. and Bacon, C. W. The distribution and ultrastructure of the endophyte of toxic tall fescue. *Can. J. Bot.*, *63*:36–42, 1985.

46. Smith, K. T., Bacon, C. W., and Luttress, E. S. Recriprocal translocation of carbohydrates between host and fungus in Bahia grass infected with *Myriogenespora atramentosa*. *Phytopath.*, *75*:407–411, 1985.
47. Leuchtmann, A. and Clay, K. Experimental infection of host grasses and sedges with *Atkinsonella hypoxylon* and *Balansia cyperi* (Balansiae, Clavicipitaceae). *Mycologia, 80*:291–297, 1988.
48. Clay, K., Hardy, T. N., and Hammond, A. M., Jr. Fungal endophytes of grasses and their effects on an insect herbivore. *Oecologia, 66*:1–6, 1985.
49. White, J. R., Jr. and Bultman, T. L. Endophyte-host associations in forage grasses. VIII. Heterothallism in *Epichloe typhina*. *Amer. J. Bot.*, *74*:1716–1722, 1987.
50. Philipson, M. N. and Christey, M. C. The relationship of host and endophyte during flowering, seed formation, and germination of *Lolium perenne*. *N. Z. J. Bot.*, *24*:125–134, 1986.
51. Lyons, P. C. and Bacon, C. W. Growth of the tall fescue endophyte into seedlings as related to temperature and sheath differentiation. *Phytopath.*, *75*:501, 1985.
52. Welty, R. E., Azevedo, M. D., and Cook, K. L. Detecting viable *Acremonium* endophytes in leaf sheaths and meristems of tall fescue and perennial grasses. *Plant Dis.*, *70*:431–435, 1986.
53. Siegel, M. R., Jarlfors, U., Latch, G. C. M., and Johnson, M. C. Ultrastructure of *Acremonium coenophialum, Acremonium lolii*, and *Epichloe typhina* endophytes in host and nonhost *Festuca* and *Lolium* species of grasses. *Can. J. Bot.* *65*:2357–2367, 1987.
54. Fineran, B. A., Harvey, I. C., and Ingerfield, M. Unusual sheaths. *Protoplasma, 117*:17–23, 1983.
55. Gallagher, R. T., Hawkes, A. D., Steyn, P. S., and Vlleggaar, R. Tremorgenic neurotoxins from perennial ryegrass causing ryegrass staggers disorder of livestock: Structure and elucidation of lolitrem B. *J. Chem. Soc.*, *Chem Commun.*, *1984*, pp. 614–616, 1984.
56. Lewis, D. H. Symbiosis and mutualism: Crisp concepts and soggy semantics. In *The Biology of Mutualism*. (D. H. Boucher, ed.). Oxford University Press, New York, 1985, pp. 29–39.
57. Starr, M. P. A generalized scheme for classifying organismic associations. *Symp. Soc. Exper. Bot.*, *29*:1–20, 1975.
58. Bradshaw, A. D. Population differentiation in *Agrostis tenuis* Sibth. II. The incidence and significance of infection by *Epichloe typhina*. *New Phytol.*, *58*:310–315, 1988.
59. Latch, G. C. M., Christensen, M. J., and Gaynor, D. L. Aphid detection of endophyte infection in tall fescue. *N. Z. J. Agric. Res.*, *28*:129–132, 1985.
60. Prestidge, R. A., Lauren, D. R., van der Zijpp, S. G., and Di Menna, M. E. Isolation of feeding deterrents to Argentine stem weevil of endophytes of perennial ryegrass and tall fescue. *N. Z. J. Agric. Res.*, *28*:87–92, 1985.
61. Clay, K., Hardy, T. N., and Hammond, A. M., Jr. Fungal endophytes of *Cyperus* and their effects on an insect herbivore. *Amer. J. Bot.*, *72*:1284–1289, 1985.

62. Read, J. C. and Camp, B. J. The effect of the fungal endophyte *Acremonium coenophialum* in tall fescue on animal performance, toxicity, and stand maintenance. *Agron. J.*, *78*:848–850, 1986.
63. Arechaveleta, M., Bacon, C. W., Hoveland, C. S., and Radcliffe, D. E. Effect of the tall fescue endophyte on plant response to environmental stress. *Agron. J.*, *81*:1989.
64. Williams, M. J., Backman, P. A., Clark, E. M., and White, J. F., Jr. Seed treatments for control of the tall fescue endophyte *Acremonium coenophialum*. *Plant Dis.*, *68*:49–52, 1984.
65. Latch, G. C. M. Incidence of endophytes in seedlings and their control with fungicides. In *44th Proceeding New Zealand Grassland Association*, 1983, pp. 251–253.
66. Latch, G. C. M. and Christensen, M. J. Artificial infection of grasses with endophytes. *Ann. Appl. Biol.*, *107*:17–24, 1985.
67. Cole, G. T. and White, J. F., Jr. *Fungal endophytes of forage grasses.* In *Proc. Indian Acad. Sci.*, *94*:129–136, 1985.
68. Bacon, C. W., Porter, J. K., and Robbins, J. D. Ergot alkaloid biosynthesis by isolates of *Balansia epichloe* and *B. henningsiana*. *Can. J. Bot.*, *59*:2534–2538, 1981.
69. Bacon, C. W. and Hinton, D. M. Ascosporic iterative germination in *Epichloe typhina*. *Trans. Br. Mycol. Soc.*, *90*:563–569, 1988.
70. Roberts, C. A., Burton, F. E., II, and Moore, K. J. Estimation of *Acremonium coenophialum* mycelium in infected tall fescue. *Agron. J.*, *80*:737–740, 1988.
71. Clay, K. Effects of fungal endophytes on the seed and seedling biology of *Lolium perenne* and *Festuca arundinacea*. *Oecologia*, *73*:358–362, 1987.
72. Bacon, C. W. A chemically defined medium for the growth and synthesis of ergot alkaloids by the species of *Balansia*. *Mycologia*, *77*:418–423, 1985.
73. Kulkarni, R. K. and Nielsen, B. Nutritional requirements for growth of fungus endophyte of tall fescue. *Mycologia*, *78*:781–786, 1986.
74. Davis, N. D., Clark, E. M., Schrey, K. A., and Diener, U. L. In vitro growth of *Acremonium coenophialum*, an endophyte of toxic tall fescue grass. *Appl. Environ. Microbiol.*, *52*:888–891, 1986.
75. De Battista, J. P., Bacon, C. W., Severson, R., Plattner, R. D., and Bouton, J. H. Indole acetic acid production by the fungal endophyte of tall fescue. *Agron. J.*, 1990 (in press).
76. Bacon, C. W., Porter, J. K., and Robbins, J. D. Laboratory production of ergot alkaloids by species of *Balansia*. *J. Gen. Microbiol.*, *113*:119–126, 1979.
77. Belesky, D. P., Devine, O. J., Pallas, J. E., Jr., and Stringer, W. C. Photosynthetic activity of tall fescue as influenced by a fungal endophyte. *Photosynthetica*, *21*:82–7, 1987.
78. Porter, J. K., Bacon, C. W., Robbins, J. D., Himmelsbach, D. S., and Higman, H. C. Indole alkaloids from *Balansia epichloe* (Weese). *J. Agric. Food Chem.*, *25*:88–93, 1977.
79. Johnson, M. C., Dahlman, D. L., Siegel, M. R., Bush, L. P., and Latch, G. C. M. Insect feeding deterrents in endophyte-infected tall fescue. *Appl. Environ. Microbiol.*, *49*:568–571, 1985.

80. Rowan, D. D. and Shaw, G. J. Detection of ergopeptine alkaloids in endophyte-infected perennial ryegrass by tandem mass spectrometry. *N. Z. Vet. J.*, *35*:197–198, 1987.
81. Lyons, P. C., Plattner, R. D., and Bacon, C. W. Occurrence of peptide and clavine ergot alkaloids in tall fescue. *Science*, *232*:487–489, 1986.
82. Yates, S. G., Plattner, R. D., and Garner, G. B. Detection of ergopeptine alkaloids in endophyte infected, toxic Ky-31 tall fescue by mass spectrometry/mass spectrometry. *J. Agric. Food Chem.*, *33*:719–722, 1985.
83. Yates, S. G. Toxicity of tall fescue forage: A review. *Econ. Bot.*, *16*:295–303, 1962.
84. Pottinger, R. P., Barker, G. M., and Prestidge, R. A. A review of the relationships between endophytic fungi of grasses (*Acremonium* spp.) and Argentine stem weevil (*Listronotus bonarienses* Kuschel). *4th Proceedings Australasian Conference Grassland Invertebrate Ecology.* Lincoln College, Canterbury, 1985, pp. 322–331.
85. Latch, G. C. M., Christensen, M. J., and Hickson, R. E. Endophytes of annual and hybrid ryegrasses. *N. Z. J. Agric. Res.*, *31*:57–63, 1988.
86. Gaynor, D. L. and Hunt, W. F. The relationship between nitrogen supply, endophytic fungus, and Argentine stem weevil resistance in ryegrass. In *44th Proc. N. Z. Grassland Assoc.*, 1983, pp. 257–263.
87. Porter, J. K., Bacon, C. W., and Robbins, J. D. Major alkaloids of a *Claviceps* isolated from toxic bermuda grass. *Agric. Food Chem.*, *22*:838–841, 1974.
88. Porter, J. K., Bacon, C. W., Robbins, J. D., and Betowski, D. Ergot alkaloid identification in Clavicipitaceae systemic fungi of pasture grasses. *J. Agric. Food Chem.*, *29*:653–657, 1981.
89. Porter, J. K., Bacon, C. W., and Robbins, J. D. Ergosine, ergosinine, chanoclavine I from *Epichloe typhina*. *J. Agric. Food Chem.*, *27*:595–598, 1979.
90. Porter, J. K., Robbins, J. D., Bacon, C. W., and Himmelsbach, D. S. Determination of epimeric 1-(3-indoyl)propane-1,2,3,triol isolated from *Balansia epichlow*. *Lloydia*, *41*:43–49, 1978.
91. Porter, J. K., Bacon, C. W., and Robbins, J. D. Lysergic acid amide derivatives from *Balansia epichloe* and *Balansia claviceps* (*Clavicipitaceae*). *J. Nat. Prod.*, *42*:309–314, 1979.
92. Bush, L. P., Cornelius, P. C., Buckner, R. C., Varney, D. R., Chapman, R. A., Burrus, P. B., Kennedy, C. W., Jones, T. A., and Saunders, M. J. Association of N-acetyl loline and N-formyl loline with *Epichloe typhina* in tall fescue. *Crop Sci.*, *22*:941–943, 1982.
93. Belesky, D. P., Robbins, J. D., Sutedemann, J. A., Wilkinson, S. R., and Devine, O. J. Fungal endophyte infection-loline derivative alkaloid concentration of grazed tall fescue. *Agron. J.*, *79*:217–220, 1987.
94. Yates, S. G. Tall fescue toxins. In *Handbook of Naturally Occurring Food Toxicants* (M. Recheigel, ed.). CRC Press, Boca Raton, FL, 1983, pp. 249–273.
95. McLean, E. K. The toxic actions of pyrrolizidine (*Senecio*) alkaloids. *Pharm. Rev.*, *22*:429–483, 1970.

96. Rowan, D. D., Hunt, M. B., and Gaynor, D. L. Peramine, a novel insect feeding deterrent from ryegrass infected with the endophyte *Acremonium loliae*. *J. Chem. Soc., Chem. Commun.*, *1986*:935–936, 1986.
97. Prestidge, R. A. and Gallagher, R. T. Lolitrem B-A stem weevil toxin isolated from *Acremonium*-infected ryegrass. In *Proceedings 38th N. Z. Weed Pest Control Conference*. 1985, pp. 38–40.
98. Rowan, D. D. and Gaynor, D. L. Isolation of feeding deterrents against Argentine stem weevil from ryegrass infected with the endophyte. *J. Chem. Ecol.*, *12*:647–658, 1986.
99. Siegel, M. R., Latch, G. C. M., Bush, L. P., Fannin, N. F., Rowan, D. D., Tapper, B. A., Bacon, C. W., and Johnson, M. C. Alkaloids and insecticidal activity of grasses infected with fungal endophytes. *Appl. Environ. Microbiol.*, 1990 (in press).
100. Wilson, B. J., Wilson, C. H., and Hayes, A. W. Tremogenic toxin from *Penicillium cyclopium* grown on food materials. *Nature*, *220*:77–78, 1968.
101. Mortimer, P. H. and di Menna, M. Ryegrass staggers: Further substantiation of a *Lolium* endophyte aetiology and the discovery of weevil resistance of ryegrass pastures infected with *Lolium* endophyte. In *44th Proc. N. Z. Grassland Assoc.*, 1983, pp. 240–243.
102. Christopher, W. M. and Mantle, P. G. Paxilline biosynthesis by *Acremonium loliae*; a step toward defining the origin of lolitrem neurotoxins. *Phytopath.*, *26*:969–971, 1987.
103. Porter, J. K., Bacon, C. W., Robbins, J. D., and Higman, H. C. A field indicator in plants associated with ergot-type toxicities in cattle. *J. Agric. Food Chem.*, *23*:771–775, 1975.
104. Davis, N. D., Cole, R. J., Dorner, J. W., Weete, J. D., and Backman, P. A. Steroid metabolites of *Acremonium coenophialum*, an endophyte of tall fescue. *J. Agric. Food Chem.*, *34*:105–108, 1986.
105. Koshino, H., Tereda, S., Yoshihara, T., Sakamura, S., Shimanuki, T., Sato, T., and Tajimi, A. Three phenolic acid derivatives from stromata of *Epichloe typhina* on *Phleum pratense*. *Phytochemistry*, *27*:1333–1338 1988.
106. Yshihara, T., Togiya, S., Koshino, H., Sakamura, S., Shimanuki, T., Sato, T., and Tajimi, A. Three fungitoxic sesquiterpenes from stromata of *Epichloe typhina*. *Tetrahedron Lett.*, *26*:5551–5554, 1985.
107. White, J. R., Jr. and Cole, G. T. Endophyte-host associations in forage grasses. III. In vitro inhibition of fungi by *Acremonium coenophialum*. *Mycologia*, *77*:487–489, 1985.
108. Belesky, D. P., Stuedemann, J. A., and Wilkinson, S. R. Ergopeptine alkaloids in grazed tall fescue. *Agron. J.*, *80*:209–212, 1988.
109. Funk C. R., Halisky, P. M., Ahmad, S., and Hurley, R. H. How endophytes modify turfgrass performance and response to insect pests in turfgrass breeding and evaluation trials. In *Proceeding 5th International Research Conference, Avignon* (F. Lemaire, ed). INRA, Versailles, 1985, pp. 137–145.
110. Hardy, T. N., Clay, K., and Hammond, A. M., Jr. Leaf age and related factors affecting endophyte-mediated resistance to fall armyworm

(Lepidoptera: Noctuidae) in tall fescue. *Environ. Entomol.*, *15*:1083–1089, 1986.
111. Asay, K. H., Minnick, T. R., Garner, G. B., and Harmon, B. W. Use of crickets in bioassay of forage quality in tall fescue. *Crop Sci.*, *15*:585–588, 1975.
112. Kirfman, G. W., Brandenburg, R. L., and Garner, G. B. Relationship between insect abundance and endophyte infestation level in tall fescue in Missouri. *J. Kans. Entomol. Soc.*, *59*:552–554, 1986.
113. Prestidge, R. A., Pottinger, R. P., and Barker, G. M. An association of Lolium endophyte with ryegrass resistance to Argentine stem weevil. In *35th Proceedings New Zealand Weed Pest Control Conference*, 1982, pp. 119–122.
114. Ahmad, S., Johnson-Cicalese, J. M., Dickson, W. K., and Funk, C. R. Endophyte-enhanced resistance in perennial ryegrass to the bluegrass bill-bug *Sphenophorus parvalus*. *Entomol. Exp. Appl.*, *41*:3–10, 1986.
115. Clay, K. Fungal endophytes of grasses: A defensive mutualism between plants and fungi. *Ecology*, *69*:10–16, 1988.
116. Pedersen, J. F., Rodriguez-Kabana, R., and Shelby, R. A. Ryegrass cultivars and endophyte in tall fescue affect nematodes in grass and succeeding soybean. *Agron. J.*, *80*:811–814, 1988.
117. Barker, G. M. Mycorrhizal infection influences *Acremonium*-induced resistance to Argentine stem weevil in ryegrass. *Proc. 40th N. Z. Weed and Pest Control Conf.*, 1988, pp. 199–203.
118. Yates, S. G., Fenster, J. C., Bartelt, R. J., and Powell, R. G. Toxicity assay of tall fescue extracts, fractions and alkaloids using the large milkweed bug. *Proc. Tall Fescue Mtg., Southern Region Information Exchange Group 37*, Nov. 17–18, Memphis, TN, 1987.
119. Peters, E. J. Toxicity of tall fescue to rape and birdsfoot trefoil seeds and seedlings. *Crop Sci.*, *8*:650–653, 1968.
120. Walters, D. T. and Gilmore, A. R. Alleopathic effects of fescue on growth of sweetgum. *J. Chem. Ecol.*, *2*:469–479.
121. Petroski, R. J., Dornbos, D. L., and Powell, R. G. Germination and growth inhibition of annual ryegrass (*Lolium multiflorum* L.) and Alfalfa (*Medicago sativa* L.) by loline alkaloids and synthetic N-acyl loline derivatives. *J. Agric. Food Chem.*, (in press).
122. Kelley, S. E. and Clay, K. Interspecific competitive interactions and the maintenance of genotypic variation within two perennial grasses. *Evol.*, *41*:92–103, 1987.
123. Bouton, J. H., Whitehead, F. C., and De Battista, J. P. Tall fescue rhizome production as influenced by bermudagrass competition and cutting frequencies. *Agron. J.*, *81*:220–223, 1989.
124. Hill, N. S., Belesky, D. P., and Stringer, W. C. Relative competitiveness of tall fescue (*Festuca arundinacea* Schreb.) as influenced by endophyte (*Acremonium coenophialum* Morgan-Jones and Gams). *Crop Sci.* (Submitted)
125. Hill, N. S., Stringer, W. C., Rottinghaus, G. E., Belesky, D. P., Parrott, W. A., and Pope, D. A. Growth, morphological, and chemical components responses of tall fescue to *Acremonium coenophialum*. *Crop Sci.*, *30*:156–161, 1990.

Part II

BIOLOGICAL CONTROL OF PHYTOPATHOGENS AND WEEDS

10

MOLECULAR BIOLOGY IN CONTROL OF FUNGAL PATHOGENS

RALPH BAKER *Colorado State University, Fort Collins, Colorado*

I. INTRODUCTION

If the narrow definition of molecular biology, which involves only the molecules associated with DNA and RNA is adopted, there is scarcely a reason for including this chapter on control of fungal plant pathogens. The new technology is of such recent origin that no direct link between genetic engineering and fungal pathogen control is yet established. Fortunately, however, enzymes, lectins, antibiotics, and other things produced by living cells are molecules. Therefore, the task of this chapter is to construct a bridge between the basic and the applied, although the platform may have many gaps between the shores.

II. MOLECULAR BIOLOGY IN DIAGNOSTICS

Lankow et al. [1] contributed a fine review of the status of immunoassays applied to crop management systems. The rapid diagnosis of fungal diseases of plants by such methods, however, has lagged behind applications related to viruses due to the complex nature of fungi. Monoclonal antibody technology [2] may provide diagnostic reagents useful for these complex pathogens. Rapid diagnoses, presymptomatic detection of infection, and quantification, according to Lankow et al. [1], can lead to more intelligent crop management decisions related to crop rotation, cultivar selection and choice of control treatment.

Monoclonal antibodies were tested for detection of differences between two *Tilletia* spp. [3] and *Fusarium* spp. [4] with limited success. However, a double antibody enzyme-linked immunosorbent assay (ELISA) was developed for the detection of *Pythium* sp. in turf grass [5]. Kits for diagnosis of this and other turf grass diseases (dollar spot, brown patch) as well as *Rhizoctonia solani* (AG 1-5), *Sclerotinia* spp., and *Phytophthora* spp. are available commercially (Ari-Diagnostics, Cinnaminson, NJ 09077). The assays may detect a number of species or be specific for only one. Sensitivity is usually on the order of 5–10 μg protein per ml.

Necogen Corporation (Lansing, MI 48912) specializes in development of ELISA techniques for detecting mycotoxins under the trade name Agri-Screen^R. Rapid immunoassays are available for aflotoxin B_1 and M_1, zearalenone T_2 toxin and DON (vomitoxin).

III. MOLECULAR BIOLOGY AND BIOLOGICAL CONTROL

Biological control involves the utilization of living agents to suppress plant pathogens [6]. The methods for accomplishing control of fungal diseases are many and diverse [7]; however, only recently have there been concerted efforts to explain activities of such agents in terms of their molecular biology. The first phase of such investigations was concerned with a better definition and analysis of mechanisms [6]. Once these are understood, creative strategies for enhancement can be initiated [8].

A. Antibiosis

When the first awakening of interest in biocontrol occurred, scientists jumped to the conclusion that production of antibiotics by antagonists afforded a logical explanation for mechanisms. A period of disillusionment followed, illustrated by the statement of Pramer [9] that there was "a lack of convincing evidence to support the thesis that antibiotics produced under natural conditions are capable of exerting a significant . . . effect." Antibiosis in the petri dish did not correlate with the ability of such potential antagonists to suppress disease. Basic research, summarized by Baker [10], provided a basis for an understanding of this failure to find a role for antibiotics in biocontrol. Antibiotic production requires substantial substrates, whereas the lifestyle of soil microorganisms is characterized by starvation with brief periods of activity in the saprophytic and/or parasitic condition. Once produced, antibiotics may be bound and inactivated on clay particles, and there is even evidence that they can be substrate sources for some voracious saprophytes! These factors, therefore, lead to the obvious conclusion that antibiotics function within an ecosystem in a defensive manner to maintain possession of a relatively rich substrate. Therefore, the challenging strategy in biocontrol involves manipulation of such antagonists so that they not only produce antibiotics for their own survival, but also interfere with pathogenic activity in the infection court by reducing inoculum potential (e.g., inducing death or inhibiting germination of dormant structures) or by interfering with saprophytic increase.

Especially in recent years, there has been a revival of interest in the role of antibiotics in biocontrol. Aenko and Hering [11] established that *Gliocladium virens*, which produces viridin [12] and gliotoxin [13], was responsible for death of sclerotia of *R. solani* on potato tubers. Howell [14] eliminated the mycoparasitic attributes of the same antagonist by mutation and demonstrated that it still induced biocontrol of Rhizoctonia damping-off in cotton seedlings. This suggested that antibiotics were involved in biocontrol. Lifshitz et al. [15] eliminated all possible mechanisms except antibiosis in the control of Pythium damping-off of peas by seed treatment with *Trichoderma harzianum*. In all these examples, there was substantial substrate in the form of dead seed or tuber coat to promote antibiotic production and relative isolation from avaricious colloids. Antibiotic-producing agents located at such microsites, therefore, have potential in biocontrol.

An example of the use of biotechnical aspects of molecular biology to identify the role of antibiotics in biocontrol is developing in studies of soil suppressive to take-all [16]. Developing evidence suggests that floureres-cent pseudomonads suppress *Gaeumannomyces graminis* var. *tritici* (Ggt) by a number of mechanisms including the potential for production of a phenazine antibiotic and by iron (Fe) competition. Transposons (Tn) are used in these studies that carry genes unrelated to transposition. These confer important properties on the host bacterium. Among these is the deletion of bacterial genes adjacent to the site of insertion. Tn5 pseudomonad mutants defective in phenazine production were not inhibitory to Ggt [17]. Moreover, when antibiotic production was restored in the mutants by complementation with a cosmid from the parent antibiotic-producing strain DNA library, inhibition of Ggt in vitro and suppressiveness was observed. The substrate for antibiotic production by the bacteria presumably might occur in the lesions on roots associated with take-all and, thus, limit their size [18].

Such studies could measure the quantitative impact of antibiotic production in biological control systems; however, conclusion are not yet possible. M H. Ryder (at the 5th International Congress of Plant Pathology in Japan) reported that *Pseudomonas fluorescens* (strain 2-79 isolated in the United States) did, indeed, produce a phenazine antibiotic inhibitory to the take-all fungus in vitro, but it was inhibitory only below pH 7. Since laboratory and field assays for biocontrol are done in soils above pH 7, this presents a dilemma that cannot be resolved at this time.

Antibiotics also may be involved in the biocontrol of preemergence damping-off (induced by *Pythium ultimum*) by seed treatment with *Pseudomonas fluorescens*. The bacterium inhibited growth of *P. ultimum* on a rich medium such as potato dextrose agar, and several antifungal substances were responsible for this inhibition [19]. By use of a broad host range cosmid vector, Gutterson et al. [20] constructed a library and used it to complement mutants deficient in fungal inhibition for identification of corresponding genes. A minimum of three genes in three distinct genomic regions were required for inhibition by antibiotics; however, two cosmids from these loci mapped by transposon insertion mutagenesis contained at least two genes each. This suggests that at least five genes function as determinants of fungal inhibition.

Gutterson and associates also advanced the laudable objective of determining how antibiotic biosynthesis is regulated under the environmental conditions present in a spermosphere or rhizosphere. When *P. fluorescens* (HV 37a) was grown on potato-extract agar, one antibiotic was synthesized only in the presence of glucose, whereas two others were synthesized only in its absence [19]. This regulation occurred at the transcriptional level and involved *afuA* and *afuB* gene products in the regulation of transcription of a *afuE* operon. There also was evidence that there was a rapid increase in the *afuE* operon transcription when cell growth was in the log phase, suggesting the presence of a metabolic switch to allow expression. Such studies are notable examples of how mechanisms elaborated at the gene-regulating level may be useful in the ultimate enhancement of efficiencies of biocontrol agents in new infection courts. Indeed, Gutterson et al.

[21] recently improved production of antibiotics produced by *P. fluorescens* inhibitory to certain fungal phytopathogens by use of recombinant DNA technology.

B. Competition

Microorganisms compete for mineral and organic nutrients to proliferate and survive in soil [22]. For successful germination in and penetration of the infection court, many fungal plant pathogens require exogenous sources of nutrients [23]. In other cases, propagules may be large enough to have endogenous reserves adequate for successful infection [24] or may be associated with previously infected host materials (like wheat straw) that can supply substrates to increase inoculum potential [16]. In all these cases, there is opportunity to exploit competition by biocontrol agents so that pathogens are deprived of elements essential for the successful completion of the infection process or life cycle.

Elements potentially essential for germination and penetration of host tissue by soilborne pathogens requiring exogenous nutrients are carbon (C), nitrogen (N) [25], and Fe [23]. Although there could be opportunities for studies of competition for C and N on the molecular level, attention has centered on Fe relationships associated with iron-acquiring compounds called siderophores. All groups of microorganisms examined so far produce such compounds, and a voluminous literature exists [e.g., 26,27].

Members of the *Pseudomonas fluorescens-P. putida* groups of rhizosphere-colinizing bacteria produce siderophores of the catechol-hydroximate type [26,28,29]. Since such siderophores have stability or iron-binding constants many log units above those of the hydroximide type produced by fungal pathogens, fluorescent pseudomonads have proved useful in biocontrol [24]. As is true for all biocontrol agents, however, a knowledge of the optimum environmental components necessary for efficient exploitation of antagonism is critical [8]. Therefore, it is essential that the iron available to the host, fluorescent pseudomonads, and/or fungal pathogens (in the form of Fe^{3+}) be at low levels in the rhizosphere. Siderophores are produced only when Fe^{3+} is in short supply [27], and this ensures that there is also fierce competition for this element [30]. Low levels of Fe^{3+} are characteristic of alkaline soils [31], and this provides at least a partial explanation for the suppressiveness to Fusarium wilt pathogens in the Salinas Valley in California where soil pH may be 7.2 – 8.2 [32]. In addition, when the ferrated form of ethylenediaminedi-o-hydroxyphenyl-acetic acid (FeEDDHA) is added to the system, the host is able to utilize the iron, the ligand then combines selectively to Fe^{3+} in the soil pool, and fungal pathogens are unable to acquire the element from the FeEDDHA because of the lower stability constant of their hydroximate siderophores [30]. Thus, by proper management of the biology of the system and use of appropriate chelating agents, additive or even synergistic biocontrol can be achieved [33].

An interesting aspect of developing theory related to siderophores is that certain organisms possess the capacity to take up and utilize iron from siderophores produced by other microbes but do not secrete siderophores themselves. This phenomenon was reported by Marugg et al. [34] for both Tn5 mutants and wild types of fluorescent pseudomonads. The ability of a given fluorescent pseudomonad to utilize the siderophores of another is dependent upon the possession of an outer membrane reception protein for that pseudomonad's ferric siderophore [35]. This led Buyer et al. [28] to the hypothesis that the most beneficial strains used in biocontrol would produce siderophores that are not used by deleterious strains in the rhizosphere but could use iron transport systems produced by the deleterious strains [26]. This suggests that research directions in genetic engineering could be directed toward appropriate manipulations of outer membrane receptors to provide increased efficiencies in iron competition among beneficial microorganisms. Again, there is wide genotypic diversity among the fluorescent pseudomonads (revealed by Southern hybridization analysis) so that strain selection and/or genetic engineering should provide ample opportunity for development of superior iron-acquiring biocontrol agents [36].

By construction of a genomic library, Loper et al. [37] detected at least four distinct genes or gene clusters involved in the biosynthesis of fluorescent pigment in *P. syringae* pv. *syringae*, which is associated with siderophores [26]. Fluorescent-minus mutants (Flu^-), however, could not be restored to Flu^+ by the four recombinant plasmids, suggesting the additional genes or gene clusters were involved in fluorescent pigment biosynthesis by this microorganism. By use of such techniques in genetic engineering, however, Loper [38] later produced Flu^- strains of *P. fluorescens* and compared their biocontrol activity against colonization of cotton seeds by *P. ultimum* with the parental Flu^+ strain. Flu^- strains did not protect against damping-off, whereas Flu^+ strains did. Even so, the Flu^+ strain influenced preemergence damping-off more than could be accounted for by reduction in initial colonization by *P. ultimum*, suggesting that siderophore production is not completely responsible for biocontrol activity. Similar conclusions related to antagonistic activity in vitro of wild types of *Pseudomonas* spp. and their siderophore-negative mutants were reported by Weger et al. [39].

C. Induced Resistance

Some of the most intriguing phenomena observed in biological control are associated with the effects on resistance of a host to pathogens that are induced by other living agents capable of colonizing plant tissue [40]. Examples include reduction in disease severity of stem rots and vacular wilts by precolonization of host tissue with related nonpathogenic strains of fusaria [41–43], the numerous systems employed by Kúc and associates [e.g., 44,45], and even an interkingdom interaction between spider mites with Verticillium wilt of cotton [46].

Basic theory associated with such phenomena has been advanced by Kúc [44,45]. All hosts contain the genetic potential for resistance to plant pathogens. Plants vary, however in the speed and magnitude with which this potential is expressed and in the quantitative activity of the resultant gene products.

The mechanism involved in these systems are complex, but theoretical foundations for understanding the phenomena on a molecular and physiological basis have been advanced. These include recognition phenomena [47], protection with lectins [48], enhanced peroxidase activity and lignification [44], and/or production of phytoalexins [45].

Although induced resistance has been noticed by researchers for decades [8,9], none of the systems has been applied in agriculture. Perhaps the interest of commercial developers has been diverted to other biocontrol products which have more direct effect on pathogens. Certainly the magnitude of control by strategies involving induced resistance equal those of other systems including application of chemicals. Once enhancement [8] of such agents occurs and delivery systems are developed, the potential for practical uses may be realized.

D. Hypovirulence

Abnormal "white" isolates of *Endothia parasitica* frequently were associated with nonkilling chestnut blight cankers in a number of geographical areas such as Italy and North America. These isolates had reduced virulence (hypovirulence) and induced less canker development when co-inoculated with virulent isolates or when inoculated into a developed lesion. This history, causal agents responsible for the abnormality of white isolates, epidemiology, as well as implications in biological control of chestnut blight were comprehensively and elegantly reviewed recently by Griffin [49]. A summary of his review in relation to the molecular biology of the phenomenon is presented here. (Griffin lists 135 references, and the reader is referred to his review for details.)

The cytoplasmic determinants associated with hypovirulence in *E. parasitica* are double-stranded RNA (dsRNA). Type strains, identified by disk gel electrophoresis, vary in different parts of the world and molecular weights range from 0.4 to 6.0×10^6 daltons. The transfer of these cytoplasmic agents from white isolates to virulent strains by hyphal anastomosis induces hypovirulence in the previously virulent isolates. Even so, the correlations with casuality of dsRNA is not complete. In some American isolates, dsRNA has not been detected. Alternatively, dsRNA was detected in one virulent strain of *E. parasitica*.

Obviously, the possibilities for use of hypovirulence in control of chestnut blight has intrigued investigators. Hypovirulent strains usually only produce pycnidia; however, conidia produced by these can be disseminated by rain, insects, mites, birds, and mammals. Therefore, there is opportunity for spread of cytoplasmic avirulence mechanisms found in nature. Apparently, this was observed in Italy, Switzerland, and the United States and is responsible in part or whole for chestnut survival. In Italy, control induced by hypovirulence is widespread, and this appears also to be true in Michigan, suggesting that biological control is more efficient when hypovirulent strains are present in an area outside the natural range of the trees. In contrast to the situation in Michigan, resistance appears to be an important interacting factor in biological control for trees in their natural habitat in the Appalachians. Management of tree regrowth to allow natural ecological succession of hypovirulence may be beneficial in such situations.

Attempts to establish hypovirulence by introducing cytoplasmic determinants into chestnut forests has met with some success in France and the United States; however, such treatments to date failed to control blight completely. The abundant virulent inoculum present in nature, the slow dissemination of hypovirulent inoculum, its low sporulation capability, and the delicate balance between virulence and reduced virulence among the biocontrol agents are factors that limit efficiency in control. Conversion to hypovirulence, of course, is determined by hyphal anastomoses. There are numerous vegetative-compatability (V-C) groups within the species *E. parasitica*. Therefore, mixtures of cytoplasmic hypovirulent V-C strains introduced into areas where they do not occur is also beneficial.

E. Mycoparasitism

There is a vast array of fungi that parasitize other fungi. Many of these are necrotrophic [50] or destructive hyperparasites and, predictably, are useful as biocontrol agents. Others are biotrophic (balanced) parasites and may be less useful. However, Adams and Ayers [51] described an interesting system involving *Sporidesmium sclerotivorum* in which the mycoparasite itself is biotrophic but secondary colonization of host fungi may confer necrotrophic attributes. Again, basic studies on biotrophs that elucidate the mechanisms associated with determinants of host specificity of a mycoparasite and resistance of hosts [52] could amplify research directions with necrotrophs.

The ecology and physiology of necrotrophic hyperparasitism recently was reviewed by Baker [53]. Chet [54] and Elad and Misaghi [55] reviewed the complex processes involved with *Trichoderma* spp. in mycoparasitism.

Recently the physiology and molecular biology leading to internal colonization of fungal hosts by mycoparasites has received attention. Predictably, some of the events e.g., recognition phenomena, appear to be similar to those demonstrated in plant parasitic interactions. In others, the processes differ in the enzymes involved because there may be wall components in fungi that are different from those found in plant cells.

In dual culture, coiling around the hyphae of a fungal host may occur. The phenomenon is not due to thigmotropism [56] but to chemotropism. Agglutinins appear to be likely candidates for the mechanism [54].

When a host is approached, the mycoparasite may form hook-like structures [54, 57–60], which are probably appressoria. These structures may aid in penetration of the host hyphal cells just as they do in higher plant–parasitic relationships.

Another parallel in mycoparasitism in comparison with those found in higher plant relationships is recognition phenomena. A hemagglutinin was isolated by Elad et al. [61] from hyphae of *T. harzianum* and the culture filtrate of *R. solani*. This agglutinated blood type O erythrocytes. Agglutination was dependent on Ca^{2+} and Mg^{2+}. Galactose, a component of the cell walls of *Trichoderma* spp., counteracted agglutination. This suggests that a lectin present in the hyphae of *R. solani* binds to galactose residue on the cells walls of *Trichoderma* spp.

Agglutinin was demonstrated also in extracts and culture filtrates of *Sclerotium rolfsii* [62]. In this case sugar specificity, hemagglutination activity, and stability of the agglutinin was quite different from that involved with *R. solani* [61]. Activity depended on the cations Mn^{2+} and Ca^{2+} and was inhibited by glucose and mannose. The agglutinin is apparently a glycoprotein with lectin characteristics. Barak et al. [63] observed binding of a L-fructose–specific lectin to *T. harzianum*. Thus, L-frucosyl residues may serve as receptors to the agglutinin of *R. solani*. In summary, recognition of the pathogen by its antagonists apprently leads to the mycoparasitic process.

Intracellular fibrillar material was observed in host hyphae shortly after contact with a fungal host was observed [57,58]. This also was observed by Hoch [64] and, apparently, is a mucilagenous polysaccharide originating from either of the interacting fungi.

Some mycoparasites breach host cell wall barriers by enzymatic action. Digested zones on host hyphae underneath coils and holes at penetration sites were observed when hyphae of *T. harzianum* were removed from host hyphae [e.g., 54,58]. Fluorescent tests and wheat germ agglutinin in coiling zones indicated that chitin fibrils are exposed in the cell walls of the host as a result of extracellular β-1, 3-glucanase excreted by a mycoparasite [57,58,65,66].

Mycoparasites may penetrate fungal cell walls mechanically [67] or by enzymatic activity [68]. The cell walls of oomycetes are composed of β-glucans cellulose and <1.5% chitin. Basidiomycetes and ascomycetes contain relatively more chitin, as well as β-glucans, but no cellulose. Mycoparasites may produce hydrolytic enzymes appropriate to the substrate – β-1, 3-glucanase and chitinase, for example, being produced by *Pythium nunn* in liquid cultures containing cell walls of two pathogenic fungi belonging to the class Basidiomycetes and cellulase produced in the presence of cell walls of two pathogens belonging to the oomycetes but not chitinase [65]. In contrast, oomycetous hosts of *P. nunn* do not produce cell wall–degrading enzymes.

Factors related to host specificity are just beginning to be understood. Recognition could be a key factor [63]. Qualitative or quantitative variability in hydrolytic enzymes that are available for penetration of host cell wall hyphae determined host range and/or virulence of *T. harzianum* and *P. nunn* [65,68].

As recently as a decade ago, doubts were expressed as to whether hyperparasitism was a factor in biocontrol of plant diseases or even significant in population suppression of other soil microorganisms [69]. A recent review [53], however. suggests that mycoparasitism may be used to advantage in a number of plant disease systems if associated with appropriate management of the ecological parameters. Although the impact of the new technology on improvement of efficiency in biocontrol is yet to be realized, a basic understanding of the molecular biology of hyperparasitism (summarized above) should provide exciting opportunities for exploitation.

IV. MONITORING AGENTS RELEASED INTO THE ENVIRONMENT

Introduction of recombinant microorganisms into the environment to control diseases as well as to enhance crop growth, for waste treatment, pollution control, and mineral control has attracted considerable public and regulatory attention. One important factor in decisions to release these engineered strains into ecosystems is monitoring their subsequent environmental distribution. A sensitive, selective marker system was developed for tracing fluorescent pseudomonads in nature subsequent to release [70].

Wild types of *Pseudomonas* spp. do not express the o-nitrophenylgalactoside (ONPG) phenotype and do not utilize lactose as a sole carbon source. Broad host range plasmids containing *Escherichia coli lac* operon genes *lac* Z and *lac* Y were constructed and introduced into *P. fluorescens*. This enabled such transformants to grow on minimal lactose media. When these are released into the environment, this marker system enables detection of *Lac*$^+$ transformants at a sensitivity of <10 cfu/g of soil. On solid nutrient medium containing the chromogenic substrate X-Gal (5-chloro-4-bromo-3-indyl-β-D-galactopyranoside), production of β-galactosidase from either *lac* Z or *lac* ZY plasmids gives rise to blue-green colonies by *Pseudomonas* in distinct contrast to yellow-white colonies of nontransformants. Therefore, detection of transformants is easily accomplished and a number of biocontrol systems utilizing fluorescent pseudomonads are now being labeled in this way for experimental purposes.

V. CONCLUSIONS

There are things constantly dying, bleeding, and groaning in the universe. The task of a plant pathogist is to control such phenomena so that the things that die are the voracious pathogens rather than their hosts. In this context, the task of researchers in molecular biology is to enhance the microbiological *persona grata* to the detriment of our dearest enemy pathogens. Therefore, enhancement becomes the objective – to so manipulate molecules in biotechnology that efficiencies are improved. It is comforting to know that powerful manipulations may be accomplished by traditional as well as advanced methods for insertion of genetic material into biocontrol agents [e.g., 71]. For example, prolonged and repeated exposure fungicides increased tolerance of *T. harzianum* to chlorothalonil, iprodione, procymidone, and vinclozolin [72]. Mutant resistance of *Trichoderma* spp. to fungicides of the MBC group (e.g., benomyl) was also obtained by exposure to ultraviolet light. In these cases, some isolates had enhanced ability to suppress selected soilborne plant pathogens. Again, fungi are typically not rhizosphere competent. Conventional treatment with a mutagenic agent induced rhizosphere competence in members of the genus *Trichoderma* [73, 74]. Such mutations apparently involved substantial increases in molecular activity of cellulase systems that utilized substrates in the mucigel of the root [75,76]. The result was not only protection in the spermosphere from a seed treatment, but additional beneficial activity in the developing rhizosphere [77].

"Nature is armed to the teeth" (Bond). The challenge is to manipulate her in beneficial ways at a molecular level.

REFERENCES

1. Landow, R. K., Grothaus, G. D., and Miller, S. A. Immunoassays for crop management systems and agricultural chemistry. In *Biotechnology in Agricultural Chemistry* (H. M. LeBaron, R. O. Mumma, R. C. Honeycutt, and J. H. Duesing, eds.). Amer. Chem. Soc., Washington, D. C., 1987. pp. 227–252.
2. Halk, E. L. and De Boer, S. H. Monoclonal antibodies in plant-disease research. *Annu. Rev. Phytopathol., 23*:321–350, 1985.
3. Banowetz, G. M., Trione, E. J., and Kryger, B. B. Immunological comparisons of teliospores of two wheat bunt fungi, *Tilletia* species, using monoclonal antibodies and antisera. Mycologia, 76:51–52, 1984.
4. Iannelli, D., Capparelli, R., Marziano, F., Scala, F., and Noviello, C. Production of hybridomas secreting monoclonal antibodies to the genus *Fusarium*. *Mycotaxon, 17*:523–532, 1983.
5. Miller, S. A., Grothaus, G. D., Petersen, F. P., and Papa, S. L. Detection of Pythium blight in turfgrass using a monoclonal antibody-based diagnostic test. *Phytopathology, 76*:1057 (Abstr.), 1986.
6. Baker, R. Biological control of plant pathogens: Definitions. In *Biological Control in Agricultural IPM Systems* (M. A. Hoy and D. C. Herzog, eds.), Academic Press, Orlando, 1985, pp. 25–39.

7. Cook, R. J. and Baker, K. F. *The Nature and Practice of Biological Control of Plant Pathogens.* Amer. Phytopathol. Soc., St. Paul, 1983.
8. Baker, R. and Scher, F. M. Enhancing the activity of biological control agents. In *Innovative Approaches to Plant Disease Control* (I. Chet, ed.), Wiley, New York, 1987, pp. 1-17.
9. Baker, K. F. and Snyder, W. C. *Ecology of Soil-Borne Plant Pathogens.* University of California Press, Berkeley, 1965, pp. 16 – 17.
10. Baker, R. Mechanisms of biological control of soil-borne pathogens. *Annu. Rev. Phytopathol.*, *6*:263 – 294, 1968.
11. Aleko, M. O. and Hering, T. F. The mechanisms associated with the antagonistic relationship between *Corticium solani* and *Gliocladium virens.* *Trans. Br. Mycol. Soc.*, *55*:173 – 179, 1970.
12. Bryan, P. W. and McGowan, J. C. Viridin: A highly fungistatic substance produced by *Trichoderma viride.* *Nature (London)*, *151*:144 – 145, 1945.
13. Weindling, R. and Nelson, O. H. The isolation of a toxic substance from the culture filtrate of *Trichoderma.* *Phytopathology*, *26*:1068 – 1070, 1936.
14. Howell, C. R. Relevance of mycoparasitism in the biological control of *Rhizoctonia solani* by *Gliocladium virens.* *Phytopathology*, 77:992 – 994, 1987.
15. Lifshitz, R., Windham, M. T., and Baker, R. Mechanism of biological control of preemergence damping-off of pea by seed treatment with *Trichoderma* spp. *Phytopathology*, *76*:720 – 725, 1986.
16. Cook, R. J. Biological control of plant pathogens, with special reference to the take-all fungus in suppressive soils. *Plant Prot. Bull.*, *27*:179 – 198, 1985.
17. Thomashow, L. S. and Weller, D. M. Role of a phenazine antibiotic in disease suppression by *Pseudomonas fluorescences.* *Phytopathology*, 77:1724 (Abstr.), 1987.
18. Cook, R. J., Wilkinson, H. T., and Alldredge, R. Evidence that microorganisms in suppressive soil associated with wheat take-all decline do not limit the number of lesions produced by *Gaeumannomyces graminis* var. *tritici.* *Phytopathology*, *76*:342 – 345, 1985.
19. James, D. W. and Gutterson, N. I. Multiple antibiotics produced by *Pseudomonas fluorescens* HV37a and their differential regulation by glucose. *Applied and Environ. Microbiol.*, *52*:1183 – 1189, 1986.
20. Gutterson, N. I., Layton, T. J., Ziegle, J. S., and Warren, G. J. Molecular cloning of genetic determinants for inhibition of fungal growth by a fluorescent pseudomonad. *J. Bacteriol.*, *165*:696 – 703, 1986.
21. Gutterson, N., Ziegle, J., Warren, G., and Layton, T. Genetic determinant of catabolite induction of antibiotic biosynthesis in *Pseudomonas fluorescens* HV37a. *J. Bacteriol.*, *170*:380 – 385, 1988.
22. Clark, F. E. The concept of competition in microbial ecology. In *Ecology of Soil-Borne Pathogens* (K. F. Baker and W. C. Snyder, eds.), University of California Press, Berkeley, 1963, pp. 339 – 345.
23. Baker, R. Ecology of the fungus, *Fusarium*: Competition. In *Fusarium: Diseases, Biology, and Taxonomy* (P. E. Nelson, T. A. Toussoun, and R. J. Cook, eds.), Pennsylvania State University Press, University Park, 1981, pp. 245 – 249.

24. Baker, R., Elad, Y., and Smith, B. Physical, biological and host factors in iron competition in soils. In *Iron, Siderophores, and Plant Diseases* (T. R. Swinburne, ed.), Plenum, New York, 1986, pp. 77–84.
25. Griffin, G. J. Physiology of conidium and chlamydospore germination in *Fusarium*. In *Fusarium: Disease, Biology, and Taxonomy* (P. E. Nelson, T. A. Toussoun, and R. J. Cook, eds.), Pennsylvania State University Press, University Park, 1981, pp. 245–249.
26. Leong, J. Siderophores: Their biochemistry and possible role in the biocontrol of plant pathogens. *Annu. Rev. Phytopathol.*, *24*:187–209, 1986.
27. Swinburne, T. R. (ed.) *Iron, Siderophores, and Plant Diseases.* Plenum, New York, 1986.
28. Buyer, J. S., Wright, J. M., and Leong, J. Structure of pseudobactin A214, a siderophore from a bean-deleterious *Pseudomonas*. *Biochemistry*, *25*:5492–5499, 1986.
29. Teintze, M., Hassain, M. B., Baines, C. L., Leong, J., and van der Helm, D. Structure of ferric pseudobactin, a siderophore from a plant growth promoting *Pseudomonas*. *Biochemistry*, *20*:6446–6457, 1980.
30. Scher, F. M. and Baker, R. Effect of *Pseudomonas putida* on a synthetic iron chelator on induction of soil suppressiveness to Fusarium wilt pathogens. *Phytopathology*, *72*:1567–1573, 1982.
31. Lindsey, W. L. Role of chelation in micronutrient availability. In *The Plant Root and Its Environment* (R. W. Carlson, ed.), University of Virginia Press, Charlottesville, 1984, pp. 507–524.
32. Scher, F. M. and Baker, R. Mechanism of biological control in a *Fusarium*-suppressive soil. *Phytopathology*, *70*:412–417, 1980.
33. Scher, F. M., Dupler, M., and Baker, R. Effect of synthetic iron chelates on population densitites of Fusarium oxysporum and the biological control agent *Pseudomonas putida*. *Can. J. Microbiol.*, *30*:1271–1275, 1984.
34. Marugg, J. D., Spanje van, M., Hockstra, W. P. M., Schippers, B., and Weisbeck, P. J. Isolation and analysis of genes involved in siderophore-biosynthesis in the plant growth-stimulating *Pseudomonas putida* strain WCS 358. *J. Bacteriol.*, *164*:563–570, 1985.
35. Magazin, M., Moores, J. C., and Leong, J. Cloning of the gene coding for the outer membrane receptor protein for ferric pseudobactin, a siderophore from a plant growth-promoting *Pseudomonas* strain. *J. Biol. Chem.*, *261*:795–799, 1986.
36. Lawson, E. C., Jonsson, C. B., and Hemming, B. C. Genotypic diversity of fluorescent pseudomonads as revealed by southern hybridization analysis with siderophore-related gene probes. In *Iron, Siderophores, and Plant Diseases* T. R. Swinburne, ed.), Plenum, New York, 1986, pp. 77–84.
37. Loper, J. E., Orser, C. S., Panopoulos, N. J., and Schroth, M. N. Genetic analysis of fluorescent pigment production in *Pseudomonas syringae* pv. *syringe*. *J. Gen. Microbiol.*, *130*:1507–1515, 1984.
38. Loper, J. E. Role of fluorescent siderophore production in biological control of *Pythium ultimum* by a *Pseudomonas fluorescens* strain. *Phytopathology*, *78*:166–172, 1988.
39. Weger de, L. A., Boxtel van, R., Burg van der, B., Gruters, R. A., Geels, F. P., Schippers, B., and Lugtenberg, B. Siderophores and

outer membrane proteins of antagonistic plant-growth-stimulating root colonizing *Pseudomonas* spp. *J. Bacteriol.*, *165*:585–594, 1986.

40. Smedegaard-Peterson, V. The limiting effect of disease resistance on yield. *Annu. Rev. Phytopathol.*, *23*:475–490, 1985.
41. Baker, R., Hanchey, P., and Dottarar, S. D. Protection of carnation against Fusarium stem rot by fungi. *Phytopathology*, *68*:1495–1501, 1975.
42. Paulitz, T. C., Park, C. S., and Baker, R. Biological control of Fusarium wilt of cucumber with nonpathogenic isolates of *Fusarium oxysporum*. *Can. J. Microbiol.*, *33*:349–353, 1987.
43. Wymore, L. A. and Baker, R. Factors affecting cross-protection in control of Fusarium wilt of tomato. *Plant Dis.*, *66*:908–910, 1982.
44. Kúc, J. Induced immunity to plant disease. *Bioscience*, *32*:854–860, 1982.
45. Kúc, J. Induced systemic resistance to plant disease and phytointerferons – are they compatible? *Fitopatol. Braz.*, *10*:9–33, 1985.
46. Karban, R., Adamchak, R., and Schnathorst, W. C. Induced resistance and interspecific competition between spider mites and a vaccular wilt fungus. *Science*, *235*:678–679, 1987.
47. Daly, J. M. The role of recognition in plant disease. *Annu. Rev. Phytopathol.*, *22*:273–307, 1984.
48. Rudger, H. On the physiological role of plant lectins. *Bioscience*, *34*:95–99, 1984.
49. Griffin, G. J. Chestnut blight and its control. *Horticultural Rev.*, *8*:291–336, 1986.
50. Barnett, H. L. and Binder, F. L. The fungal host-parasite relationship. *Annu. Rev. Phytopath.*, *11*:273–292, 1973.
51. Adams, P. B. and Ayers, W. A. Histological and physiological aspects of infection of sclerotia of two *Sclerotinia* species by two mycoparasites. *Phytopathology*, *73*:1072–1076, 1983.
52. Manocha, M. S. Host specificity and mechanism of resistance in a mycoparasitic system. *Physiolo. Plant Pathol.*, *18*:257–265, 1981.
53. Baker, R. Mycoparasitism: Ecology and physiology. *Can. J. Plant Pathol.*, *9*:370–379, 1987.
54. Chet, I. *Trichoderm* – application, mode of action, and potential as a biocontrol agent of soilborne plant pathogenic fungi. In *Innovative Approaches to Plant Disease Control* (I. Chet, ed.), Wiley, New York, 1987, pp. 137–160.
55. Elad, Y. and Misaghi, I. J. Biochemical aspects of plant microbe and microbe-microbe interactions in soil. In *Recent Advances in Phytochemistry. Vol. 19. Chemically Mediated Ineractions Between Plants and Other Organisms* (G. A. Cooper-Driver and T. Swain, eds.), Plenum, New York, 1985, pp. 21–46.
56. Dennis, C. and Webster, J. Antagonistic properties of species-groups of *Trichoderma*. III. Hyphal interaction. *Trans. Br. Mycol. Soc.*, *57*:363–369, 1971.
57. Elad, Y., Barak, R., Chet, I., and Henis, I. Ultrastructural studies of the interaction between *Trichoderma* spp. *Phytopath. Z.*, *107*:168–175, 1983.
58. Elad, Y.. Chet, I., Boyle, P., and Henis, Y. Parasitism of *Trichoderma* spp. on Rhizoctonia and *Sclerotium rolfsii* – Scanning electron

microscopy and fluorescence microscopy. *Phytopathology*, *73*:85–88, 1983.
59. Lifshitz, R., Dupler, M., Elad, Y., and Baker, R. Hyphal interactions between *Pythium nunn* and several soil fungi. *Can. J. Microbiol.*, *30*:1482–1487, 1987.
60. Tee, J. C. *Gliocladium virens*, a destructive mycoparasite of *Sclerotinia sclerotiorum*. *Phytopathology*, *70*:670–674, 1980.
61. Elad, Y., Barak, R., and Chet, I. Possible role of lectins in mycoparasitism. *J. Bacteriol.*, *154*:1431–1435, 1983.
62. Barak, R., Elad, Y., Mirelam, D., and Chet, I. Lectins: A possible basis for specific recognition in the interaction of *Trichoderma* and *Sclerotium rolfsii*. *Phytopathology*, *75*:458–462, 1985.
63. Barak, R., Elad, Y., and Chet, I. The properties of L-fucose binding agglutinin associated with the cell wall of *Rhizoctonia solani*. *Arch. Microbiol.*, *44*:346–349, 1986.
64. Hoch, H. C. Mycoparasite relationships – IV. *Stephanoma phaeospora* parasitic on a species of *Fusarium*. *Mycologia*, *70*:370–379, 1978.
65. Elad, Y., Lifshitz, R., and Baker, R. Enzymatic activity of the mycoparasite *Pythium nunn* during interaction with host and non-host fungi. *Physiol. Plant Pathol.*, *27*:131–143, 1985.
66. Elad, Y., Sadooski, Z., and Chet, I. Detection of mycoparasitisms by infrared photography. *Microb. Ecol.*, *9*:185–187, 1983.
67. Sutherland, E. D., Baker, K. K., and Lockwood, J. L. Ultrastructure of *Phytophthora megasperma* f. sp. *glycinea* oospores parasitized by *Actinoplanes missouriensis* and *Humicola fuscoatra*. *Trans. Br. Mycol. Soc.*, *82*:726–729, 1984.
68. Elad, Y., Chet, I., and Henis, Y. Degradation of plant pathogenic fungi by *Trichoderma harzianum*. *Can. J. Microbiol.*, *28*:719–725, 1982.
69. Cooke, R. *The Biology of Symbiotic Fungi*. John Wiley and Sons, London, 1977.
70. Drahos, D. J., Hemming, B. C., and McPherson, S. M. Tracing recombinant organisms in the environment: β-Glactosidase as a selective non-antibiotic marker for fluorescent pseudomonads. *Biotechnology*, *4*:439–444, 1986.
71. Barry, G. F. Permanent insertion of foreign genes into the chromosome of soil bacteria. *Biotechnology*, *4*:446–449, 1986.
72. Papavizas, G. C. Genetic manipulation to improve the effectiveness of biocontrol fungi for plant disease control. In *Innovative Approaches to Plant Disease Control* (I. Chet, ed.), Wiley, New York, 1987, pp. 193–212.
73. Ahmad, J. S. and Baker, R. Rhizosphere competence in *Trichoderma harzianum*. *Phytopathology*, *77*:182–189, 1987.
74. Ahmad, J. S. and Baker, R. Rhizosphere-competence of benomyl tolerant mutants of *Trichoderma* spp. *Can. J. Microbiol.*, *34*:694–696, 1988.
75. Ahmad, J. S. and Baker, R. Competitive saprophytic ability and cellulolytic activity of rhizosphere-competent mutants of *Trichoderma harzianum*. *Phytopathology*, *77*:358–362, 1987.
76. Ahmad, J. and Baker, R. Growth of rhizosphere-competent mutants of *Trichoderm aharzianum* on carbon substrates. *Can. J. Microbiol.*, *34*:807–814, 1988.
77. Ahmad, J. S. and Baker, R. Implications of rhizosphere-competence of *Trichoderm aharzianum*. *Can. J. Microbiol.*, *34*:229–234, 1988.

11

PHYSIOLOGY AND BIOCHEMISTRY OF BIOTROPHIC MYCOPARASITISM

MOHAN S. MANOCHA *Brock University, St. Catharines, Ontario, Canada*

I. INTRODUCTION

Review articles and books aimed at research on the plant host–fungal pathogen interactions to elucidate the mechanism of host specificity and resistance have inundated the literature in plant pathology. A number of models have been suggested to explain the mechanism of specificity in plant host–pathogen systems [1–4]. There are very few comparable studies with the mycoparasites. In an article in *Annual Review of Phytopathology*, Barnett and Binder [5] discussed the interrelationship between a number of fungal host–parasite combinations, with emphasis on morphology and special nutritional requirements of the biotrophic mycoparasites. Since this 1973 review, research on the cellular and molecular aspects of biotrophic mycoparasitism has pointed to some similarities with the biotrophic haustorial pathogen–plant host systems such as rusts and powdery mildews [6]. Studies with mycoparasitic systems have contributed to our knowledge on various aspects of parasitism, host specificity, and resistance. This chapter is devoted to recent findings in the field of mycoparasitism. These findings are used to illustrate a new model based on the induced imbalance in the growth regulatory metabolism of the host, which determines compatibility and incompatibility in the biotrophic mycoparasitic system. The pivotal role of proteinases in such regulatory mechanisms is discussed.

II. MYCOPARASITISM: DEFINITION AND DISTRIBUTION

Mycoparasitism is generally defined as parasitism of a fungus (host) by another fungus (mycoparasite). The term hyperparasitism is used in a restricted sense, implying fungus parasitic on other fungal parasites or pathogens. Based on their mode of parasitism, mycoparasites are separated into two major groups, i.e., necrotrophs and biotrophs [5]. The necrotrophic or destructive mycoparasites kill the host cells slightly before or after the invasion and absorb nutrients from the dying or dead host cells. Necrotrophic mycoparasites, like the necrotrophic fungal parasites of higher plants, tend to be more aggressive, have a broad host range extending to wide taxonomic groups, and are relatively unspecialized in their mode of parasitism. The antagonistic activity of necrotrophic mycoparasites is attributed to the production of antibiotic, toxin, or hydrolytic enzymes in such proportions as to cause the death and destruction of their host. Recent upsurge in research efforts to exploit the potential of necrotrophic mycoparasites for biological control of plant pathogens is evidenced by several review articles [7–9] and books [10,11].

The biotrophic or balanced mycoparasites, on the other hand, are difficult to define. Biotrophs are usually regarded as those organisms that are able to obtain nutrients from living cells and are recognized by the fact that invaded tissues do not die. This type of parasitic relationship is physiologically balanced, and the parasite seems to be highly adapted to this mode of life. Biotrophic mycoparasites, unlike the necrotrophs, tend to have a more restricted host range and produce specialized structures to absorb nutrition from their host. Barnett and Binder [5] recognized three distinct types of biotrophic mycoparasites, which differ in their morphology, physiology, and mode of nutrition. These are as follows: (a) Intracellular or the internal mycoparasites, which develop within the cells of other fungi, are found in the Chytridiales, Blastocladiales, Plasmodiophorales, and Lagenidiales [12,13]. (b) The contact mycoparasites do not penetrate the host cell, but develop specialized absorptive branches or cells (buffer cells) which contact the host hypha. The contact cells are delineated by a septum in the germ tube. Eventually the interface between host and contact cell is dissolved, giving rise to a pore through which increased nutrient uptake by the mycoparasite is accomplished [14–16]. Contact mycoparasites and their hosts belong to Deuteromycetes, and in some cases the hosts are Ascomycetes. (c) The haustorial mycoparasites penetrate the host cells and produce distinct haustoria within the host cell to absorb nutrition [17–19]. The haustorial mycoparasites belong to the morphological group merosporangiferous Mucorales. The principal genera *Piptocephalis* and *Syncephalis* belong to Zoophagales, and *Dispira, Dimargaris* and *Tieghemiomyces* belong to order Dimargaritales of class Zygomycetes. *Syncephalis* represents an intermediate type, which has both biotrophic and necrotrophic characteristics. The initial stages of its infection are similar to biotrophic haustorial mycoparasites, but once inside the host cell, the penetration peg does not differentiate into a haustorium. Rather, it continues apical extension and develops branched hyphae to colonize the host cytoplasm. The host cytoplasm degenerates and the mycoparasite satisfies its nutritional requirements by absorption directly from the dying cytoplasm. Growth and sporulation of the host is severely inhibited.

Mycoparasitism is a phenomenon of common occurrence among all groups of fungi from chytrids to higher basidiomycetes parasitizing aquatic fungi, saprophytic soil fungi, mushrooms, and parasites of plants and animals. The reader is referred to Lumsden's article on ecology of mycoparasitism for details on the population dynamics, distribution, and ecological importance of mycoparasitism in nature [20]. For mycoparasitism within zygomycetes see Ref. 21.

III. SPORE GERMINATION AND AXENIC GROWTH

Internal mycoparasites represented by chytrids and other aquatic phycomycetes produce zoospores. When a zoospore of the parasite comes in contact with the host hypha, the movement of the zoospore slows down and it becomes amoeboid. The amoeboid movement of the zoospore can occur for several minutes prior to encystment. Sykes and Porter [13] reported that zoospores of *Catenaria allomycis* do not accumulate rapidly around the branches of its host, *Allomyces arbuscula*. The host does not produce a chemical stimulus, at least not as strong as reported for *Rozella allomycis* by Held [22]. These mycoparasites have not been grown in axenic cultures and little is known about their nutritional requirements.

The conidia of contact mycoparasites do not germinate in distilled water or water agar, but require a complex natural product such as yeast extract, malt extract, or mycelial extract [5]. Germinating conidia of *Calcarisporium parasiticum* and *Gonatobotrys simplex* apparently secrete an attractant to induce a tropic response of their host hyphae. According to Barnett and Binder [5], this unusual ability to cause a tropism of the host is a survival mechanism because the parasite at this stage does not produce long germ tubes. In contrast, conidia of *Gonatobotryum fuscum*, *G. highlei*, and *Stephanoma phaeospora* produce long germ tubes of germination, and the host hyphae do not show a positive tropism. Interestingly, all five contact mycoparasites have in common a requirement for a nutrient present in hot water extracts of the host mycelium. This nutrient is referred to in literature as mycotrophein, and its presence is demonstrated in all host fungi and in many nonhost Ascomycetes and Deuteromycetes [23]. Other contact, biotrophic mycoparasites, *Acladium tenellum* and *Melanospora zamie* do not require mycotrophein for their axenic growth. Both *A. tenellum* and *M. zamie* are deficient for thiamine and biotin but not mycotrophein. All contact mycoparasites that have been tested can grow axenically on a variety of carbon and nitrogen sources supplemented with either mycotrophein or thiamine and biotin [16,24]. Hwang et al. [25] have partially purified the growth factor mycotrophein using a combination of silicic acid and high performance liquid chromatography. The partially purified compound promoted axenic growth of *Stephanoma phaeospora*, *Gonatobotrys simplex*, *Calcarisporium parasiticum*, and *Gonatorrhodiella highlei*. Electron impact mass spectral measurements on the purified mycotrophein tentatively identified the presence of a mixture of tetra- and pentaethylene glycol mono-(nonylphenyl) ethers – nonionic surfactants. However, similar commercial products did not show any growth-promoting activity.

Spores of haustorial mycoparasites, like the conidia of the contact mycoparasites, do not swell or germinate in water or on water agar. Spore germination requires one or more nutrients, which may be furnished by several complex natural products, by excretions of a nearby living host fungus, or by filtrates of host culture [5]. Mycotrophein, unlike its role in contact mycoparasites, does not sustain the continuous growth of the haustorial mycoparasites. Similarly, various nutrients acting as stimulants of germination do not support the continuous growth of these mycoparasites. In an attempt to optimize germination conditions for *Piptocephalis virginiana* spores in the absence of the host, Balasubramanian and Manocha [26] found that besides the addition of host extract into nutrient broth, shake cultures promoted considerably better germination than still cultures. Probably the shake cultures prevent aggregation of spores, thus relieving them from overcrowding, a condition generally not encountered in the natural environment. The cylindrical spores of *P. virginiana* swell to several times their volume and become globose. Each spore is capable of producing one to four germ tubes, although two germ tubes are more common under normal conditions. The germ tubes branch at random and continue to advance until the endogenous reserves of essential nutrient(s) become exhausted, resulting in autolysis or the germ tubes make contact with the potential host [27]. Jeffries and Young [28] have described details of spore structure and germination observed by using light and electron microscopy in *P. unispora* and in other haustorial mycoparasites which belong to order *Dimargaritales* [29].

Germ tubes of biotrophic haustorial mycoparasites show strong positive tropism toward living host cells [17,30,31] with an accelerated rate of growth [32]. The directed growth of *P. fimbriata* germ tubes is considered by Evans and Cooke [32] to be due to a diffusible, nonvolatile, high molecular weight substance, probably proteinaceous, produced by the host *Mortierella vinacea*. Germ tubes of the mycoparasite also grow towards the wall fragments of hyphae of *M. vinacea*, but not towards the intact hyphae or wall fragments of *Circinella mucoroides*. The latter fungus does not produce the growth-directing factors, but it is still a susceptible host to *P. fimbriata* [32]. These findings suggest that the growth-directing factors occur in the cell wall of the host fungus and are of no consequence in parasitism. Nevertheless, Jeffries [21] believes directed growth response to be of paramount importance in initiation of the parasitic attack. In our laboratory we observed the positive tropism of *P. virginiana* germ tubes towards the hosts and nonhosts alike. There was no marked difference in the percent germination of parasite spores in the presence of the host or nonhost species. The frequency of apparent contacts between parasite germ tubes and host or nonhost hyphae was also the same [27,33]. There could be some other explanation besides chemotropism to account for the observed directed growth of the mycoparasites. Fimbriae, reported on the surfaces of many fungi, may direct growth towards other fungi nonspecifically. Fimbriae range in size from 2 to 20 μm in length, 7 nm in width, and are composed of one protein with molecular weight 74,000. They may provide the submicroscopic first physical contact between the two organisms before any biochemical interaction occurs [34].

Several haustorial mycoparasites, such as *Dimargaris*, *Dispira*, and *Tieghemiomyces*, have been grown successfully in axenic cultures, and their nutritional requirements have been determined. These mycoparasites prefer glycerol over glucose as a carbon source and a high concentration of casein hydrolysate as the nitrogen source [35,36]. Their amino acid requirements are different. For example, *Tieghemiomyces parasiticus* made good axenic growth on a glycerol medium with L-cystine, L-valine, and L-leucine, whereas *Dispira cornuta* utilized L-alanine, L-aspartic, and L-glutamic acid [37]. The haustorial mycoparasites also show deficiencies for thiamine and biotin [35]. Nonetheless, *Piptocephalis* species have not been grown in axenic culture despite numerous attempts using a number of nutritional and cultural conditions. Spores of *P. virginiana* make limited axenic growth on various media and produce thin, slender hyphae with a few sporophores before they cease to grow [38]. The sporophores are thin, dwarf, branched once or unbranched, and terminate in one or two heads of spores. Each head produced 6 – 12 spore chains at its tip with 4 – 6 spores in each chain. The spores are cylindrical and appear normal, with their average dimensions similar to those of the parasitic culture [39]. These spores either do not germinate or fail to produce a hypha on germination. Similar observations are reported for *P. unispora* by Jeffries and Young [31].

IV. ALTERATION IN MEMBRANE PERMEABILITY AND LIPID METABOLISM

Alterations in cell permeability as a characteristic of host response in a variety of plant host–fungal parasite systems have been extensively investigated. To date a number of review articles have been published dealing mainly with the effect of pathogen-produced toxins and enzymes on the permeability of plant cells and tissue [40–42]. These articles dealt with necrotrophic parasites, which kill their host cells and utilize the nutrients that are released. Except for a few studies, including the one by Hoppe and Heitefuss [43] on *Phaseolus vulgaris* infected with *Uromyces phaseoli*, alterations of cell permeability in plant host–fungal pathogen systems involving biotrophic parasites have been neglected.

A study with a biotrophic haustorial mycoparasite has shown an altered cell permeability in infected host as measured by leakage of electrolytes, amino acids, and sugars [44]. The leakage was more pronounced during early stages of parasitic infection. The efflux of ions, sugars, and amino acids was not as large as that in rust-infected bean leaves as observed by Hoppe and Heitefuss [43] during later stages of infection, when uredosori were well developed and the sporulating postules had reached the maximum size. The high loss of cell constituents could be due to fungal mycelium and spores. However, Hoppe and Heitefuss [43] did not observe any further leakage during sporulation. They suggested that loss of cellular material from the infected leaves is due mainly to higher permeability of the host membranes whose function as diffusion barriers is decreased by infection. Differences between the two studies could be due to the very nature of the two host–parasite systems. In the biotrophic mycoparasitic system, there is little or no apparent inhibition of host growth by *P. virginiana*. The rust fungus, on the other hand, is a biotrophic pathogen which eventually causes host cells to senesce and die. Furthermore, tissue damage in host plants as a result of excision or production of inhibitors and phytoalexins [45,46] may account for unspecific leakage. In the mycoparasitic system, such damage does not occur since intact mycelia are used and no toxin production is reported either by the host or the mycoparasite [44].

Membrane permeability plays an important role in the nutrient supply available to the parasite, and hence in determining the success of the parasite and susceptibility of the host. There is evidence that a nutrient required by the mycoparasite is normally held within the cells of most hosts and that very little escapes into the substrate before autolysis. The contact cell of the mycoparasite must therefore function in some way to increase the permeability of the host membrane to that nutrient [5]. The fact that the invading parasite reduces the diffusion barrier property of the host membrane, which in turn is determined by the membrane lipids, stimulated interest in the lipid metabolism of infected host.

Besides a few isolated studies, most of the data on the alterations of lipid metabolism in plant hosts infected by biotrophic fungal pathogens comes mainly from two laboratories: those of Hoppe and Heitefuss [47–50] in Göttingen and Lösel and Lewis [51] in Sheffield. Both the groups used rust fungi, and their investigations emphasized the detection of changes in (a) total lipid levels, (b) individual phospholipid classes, (c) glycolipids and sterols, and (d) fatty acid composition, especially of the phospholipids. These studies are not without inherent problems expected of the biotrophic pathogens on plant hosts. It is difficult to attribute changes in lipid concentration to either changes in host lipid metabolism or to direct contribution by the biotrophic pathogens. Different experimental designs used by the two groups further complicated the interpretation of the data (for details see Ref. 52). However, both groups obtained some parallel results such as (a) the increase in total lipids, (b) decrease in chloroplast lipids, (c) increase in phosphotidylserine and phosphatidic acid, and (d) higher unsaturation of phospholipids following the rust infection.

A similar study using a biotrophic mycoparasite has extended our knowledge of the lipid metabolism in host–parasite interaction and its role in parasitic physiology [39]. This study demonstrated that the mycoparasite *P. virginiana* produces long, erect, branching sporophores when growing on its host, *Choanephora cucurbitarum*. The spores from this parasitic culture contain two terminal lipid bodies. On germination on malt-yeast extract and in the absence of their host, these spores produce thin, flexuous, unbranched sporophores bearing a few spores. The axenic culture spores lack the two terminal lipid bodies present in the parasitic culture spores. Corresponding to this, the lipid content of the parasitic spores was 205 mg/g dry weight, compared with 95 mg/g dry weight in the axenic culture spores. Thin-layer and gas-liquid chromatographic analyses of spore extracts showed that the major qualitative differences were the absence of (a) γ-linolenic acid, a polyunsaturated fatty acid, and (b) diphosphotidylglycerol from the axenic culture spores. The proportions of other major fatty acids, such as palmitic, palmitoleic, stearic, oleic, and linolenic acids, and lipid classes, such as phosphatidylcholine, phosphatidylethanolamine, phosphatidylglycerol, digalactosyl diglyceride, monogalactosyl diglyceride, phosphatidic acid, phosphatidylinositol, and phosphatidylserine, were otherwise similar in spores from both parasitic and axenic culture [39,53]. Earlier, Manocha and Lee [19] reported frequent presence of sphaerosome-like bodies of variable sizes in the germ tubes and hyphae of the mycoparasite and the host, respectively. These bodies tested positive for lipids. The young haustorium contains many sphaerosome-like bodies, while only few are observed in the mature haustorium. These bodies are either consumed during growth of the parasite or have migrated to the developing sporophores where they become the main constituent of the spore. In another mycoparasitic system, *Catenaria allomycis* on *Allomyces arbuscula*, Sykes and Porter [13] made comparable observations. Lipid droplets developed in the host in response to infection, and soon after the parasite thallus could be observed associated with accumulating lipid droplets in the host cytoplasm. Later, as sporangium development continued, the isthmus had lost most of the cytoplasmic contents that were previously present, apparently by the contents being drawn into the developing sporangia. A large amount of presumed lipid material could be seen in the developing sporangia of the parasite.

Alterations in lipid metabolism of the host *Choanephora cucurbitarum* infected by *P. virginiana* parallel those observed by Lösel and Lewis [51] in the rust (*Puccinia poarum*)-infected leaves of *Tussilago farfara*. In both cases there was an increase of total lipid in the host during progress of the infection, reaching a maximum when sporulation of the parasite commenced, thereafter decreasing steadily as spores with lipid reserves were dispersed. This increase in total lipid of the infected fungal host was due mainly to increased concentration of phospholipids, but in the rust-infected host plant a much greater increase in neutral lipids also occurred. Hoppe and Heitefuss [47] reported greater incorporation of ^{32}P label by rust-infected tissue than by control leaves. They also observed a marked increase in specific activity in phosphatidylserine and phosphatidic acid following infection. Another point of significance in their work was the observation of higher levels of unsaturated fatty acids in phosphatidylethanolamine and phosphatidycholine from the rust-infected bean leaves and fungal uredospores [48,49]. Further similarity between plant host–parasite and fungal host–mycoparasite systems is shown by the work reported from our laboratory [53] demonstrating increased levels of γ-linolenic acid, a polyunsaturated fatty acid, in diphosphatidylglycerol, phosphatidylethanolamine, and phosphotidylcholine from the infected fungal host. As a result of the infection process, no new lipid class was observed from any of the host–parasite systems investigated.

The absence of γ-linolenic acid in lipids from the axenic culture of *P. virginiana* and its presence in the parasite's spores when grown on its host indicate some kind of block in the synthesis of polyunsaturated fatty acid by the parasite. As mentioned earlier, *Piptocephalis* is a biotrophic mycoparasite with its host range limited to mucoraceous fungi. These host fungi are characterized by the presence of (a) chitosan and chitin in their cell wall and (b) γ-linolenic acid in their lipids [54]. The amount of γ-linolenic acid found in the mucoraceous host fungus, *C. cucurbitarum*, is influenced by cultural conditions such as temperature, light, age of the culture, and C/N ratio of the growth media [55]. These conditions also affect the degree and rate of development of the mycoparasite *P. virginiana*. Manocha and Deven [55] showed a direct correlation between the levels of γ-linolenic acid present in *C. cucurbitarum* during the early stages of growtn (24 hr) and the degree of parasitism of *P. virginiana*. The amount of γ-linolenic acid present in the host mycelium was found to be unrelated to either dry weight of the mycelium or total lipid contents. Unequivocal evidence in support of direct correlation between levels of γ-linolenic acid and degree of parasitism was provided through growth temperature manipulations of two slightly psychrophilic fungi, *Thamnidium elegans* and *Helicostylum elegans*, which are also mucoraceous hosts to the mycoparasite *P. virginiana* [53]. *Thamnidium elegans* is highly susceptible at 25°C and slightly susceptible at 20°C. On the other hand, *Helicostylum elegans* is highly susceptible at 20°C and slightly susceptible at 25°C. Both host fungi were grown at different temperatures and the following three parameters were examined: (a) degree of unsaturation of lipids, (b) level of γ-linolenic acid, and (c) degree of parasitism. The observed relationship between γ-linolenic acid and parasitism was found to be exclusive since neither the degree of unsaturation nor other polyunsaturated fatty acid or its isomer could replace γ-linolenic acid in its support of parasitism. Caution must be exercised in interpreting these results. The presence of γ-linolenic acid in a mycelium does not necessarily mean it is a susceptible

TABLE 1. Fatty Acid Composition of Total Lipid Extracted from Axenic Cultures of *Piptocephalis virginiana* and Some of Its Host and Nonhost Species Grown on Malt Yeast Extract at 23°C

Organism	Fatty acid % composition								Degree of unsaturation (Δ/mole)	Degree of parasitism	Host reaction
	14:0	16:0	16:1	18:0	18:1	18:2	γ18:3	Others			
Piptocephalis virginiana	4.5	37.0	8.1	6.9	17.9	24.5	ND	—	0.75	N.A.	N.A.
Choanephora cucurbitarum	0.6	19.7	2.4	6.8	21.1	21.8	26.4	—	1.46	+++	Susceptible
Mortierella pusilla	2.5	12.0	4.7	6.9	36.3	20.3	17.3		1.36	+++	Susceptible
Thamnidjum elegans	2.4	10.6	4.1	6.7	35.2	17.3	17.2	6.5[a]	1.39	+++	Susceptible
Mycotypha microspora	4.5	22.0	9.1	5.0	22.7	19.6	12.1	—	1.23	++	Susceptible
Helicostylum elegans	4.2	12.8	3.4	7.9	26.4	20.1	12.1	12.8[b]	1.09	+	Susceptible
Phascolomyces articulosus	2.4	11.2	12.7	10.9	23.3	17.4	18.2	4.0[b]	1.26	-	Resistant
Mortierella candelabrum	2.1	12.5	6.0	8.8	33.1	24.4	13.1	—	1.27	-	No Reaction
Linderina pennispora	3.8	11.6	19.2	4.0	23.0	25.6	ND	11.4[c]	0.93	-	No Reaction
Chaetomium cochlioides	0.8	11.5	2.3	6.5	17.4	44.6	ND	12.4[d]	1.46	-	No Reaction

ND = not detected; N. A. = not applicable.

[a]Others include octadecatetraenoic acid (3.5%).

[b]Others include long-chain fatty acids > C22:0.

[c]Others include C17:0.

[d]Others include α-18:3.

host, as sometimes erroneously interpreted. The parasite must gain entry into the potential host and establish a nutritional relationship. Some nonmucoraceous fungi have been reported to contain γ-linolenic acid, e.g., *Blastocladiella emersonii* and *Chaetomium cochliodes*, but *P. virginiana* fails to parasitize these fungi. On the other hand, mucoraceous fungi such as *Phascolomyces articulosus*, a resistant host, and *Mortierella candelabrum*, a nonhost, also possess γ-linolenic acid, but the parasite does not establish a nutritional relationship with the former and shows no interaction whatsoever with the latter (Table 1). Details of mycoparasite interaction with a resistant host and nonhost are discussed later in this chapter.

The role of γ-linolenic acid in mucoraceous fungi is unknown, but the role of long chain polyunsaturated fatty acids in membrane permeability is well documented in the literature [56]. Since γ-linolenic acid is the major component of phospholipids, such as diphosphatidylglycerol, phosphatidylcholine and phosphatidylethanolamine, it is not unreasonable to assume that γ-linolenic acid plays an important role in permeability and in maintaining the integrity and fluidity of the membranes. Phospholipids are also intimately involved in the functional organization of a number of complex enzyme systems in the membranes, especially of mitochondria. In yeast, diphosphatidylglycerol is exclusively associated with mitochondria and the amount present is closely correlated to the state of mitochondrial membrane development [57]. The respiratory deficient mitochondria obtained from anaerobically grown yeast has less diphosphatidylglycerol and more phosphatidylinositol as well as alterations in the other phospholipids [58]. In *Candida sloofi*, lack of diphosphatidylglycerol is correlated to its respiratory deficiency [59]. Absence of diphosphatidylglycerol from the axenic culture of the mycoparasite *Piptocephalis* might explain the impaired growth of the parasite in the absence of its host. A defective membrane would lead to great difficulties for a free-living organism but might be a useful adaptation to parasitism. It seems worthwhile to point out here that the addition of either γ-linolenic acid or diphosphatidylglycerol or both did not support the continuous growth of the mycoparasite in axenic culture.

V. CELL–CELL INTERACTION IN MYCOPARASITISM

Host cell surfaces provide the first point of cell–cell interaction between host and mycoparasite. It could be the important center of focus where the attachment of the mycoparasite and development of the appressorium occurs. Attachment can be divided into three components encompassing initial recognition followed by the directed (?) growth of the mycoparasite germ tube towards the host and binding to the host surface. Virtually nothing is known about the nature of initial recognition stimulus. Very little is known about the directed growth (as mentioned before) and the binding process. The latter is probably dependent upon the presence of specific binding sites on the host surface. Nonetheless, attachment of the mycoparasite to the host cell surface is essential for the exchange of signal molecules and for the development of the appressorium. The mycoparasite may establish numerous contacts with the nonhost, but it fails to attach to the surface of the nonhost and produce an appressorium [27]. To my knowledge, there is no record of any study illustrating the process of attachment in internal and contact biotrophic mycoparasites. Most of the information (described below) comes from studies with haustorial biotrophs and necrotrophs.

Analyses of cell surfaces of fungi acting as host and nonhost to the mycoparasite *P. virginiana* revealed no major differences among them. Analytical techniques used included microchemical methods, electron microscopy of thin sections and surface replicas, infrared spectrophotometry, and x-ray diffraction [54,60,61]. Cell wall composition of host and nonhost species used in the above study was typical of mucoraceous fungi with chitin and chitosan as the major polysaccharides [62]. Chitin and chitosan contents in cell walls of hosts and nonhost species were as much as 41–46% of the total dry weight. However, the relative proportion of the two polysaccharides was slightly different in the nonhost, *Mortierella candelabrum*, as compared to that in the host species. There was more chitin than chitosan in the cell wall of the nonhost. The amount of neutral sugars also varied, but the qualitative profiles obtained by gas-liquid chromatography were similar. The neutral sugars detected were fucose, mannose, galactose, and glucose. Besides carbohydrates, the cell walls of these fungi also contained lipids and proteins in varying amounts. Such quantitative differences in cell wall components can also occur in cultures of different ages [63] and different stages in life cycle [62]. Electron microscopic studies have shown that the susceptible hosts of the mycoparasite *P. virginiana* possess a single layer of cell wall with a multinet arrangement of microfibrils [54,61]. The resistant host *Phascolomyces articulosus* possesses two layers of cell wall differing in electron opacity as well as in microfibrillar arrangement [54]. The parallel arrangement of microfibrils in *P. articulosus* was also observed in the nonhost, *M. candelabrum*. But, unlike the nonhost, the cell wall of the resistant host is contacted and penetrated by the mycoparasite. It is obvious that no functional significance can be attached to differences in gross chemical composition and cytological features to explain the attachment of the mycoparasite to host and not to the nonhost cell surface.

A direct binding assay developed in our laboratory provides a convenient method to quantify attachment of the parasite germ tubes to the isolated cell wall fragments of the host and nonhost. This procedure consists of artificially inoculating the host or nonhost cell wall fragments previously fixed on a glass slide with germinating spores of the mycoparasite. After 2 hours of incubation, the slides are washed in a steady stream of distilled water, stained with cotton blue, and examined for number of attachments (for details of experimental protocol and controls, see Ref. 64). The germ tubes of the mycoparasite attach irreversibly to the cell wall fragments of the susceptible as well as resistant hosts and form distinct appressoria at the points of contact similar to those observed on intact hyphal walls. This suggests that mycoparasite attachment and appressorium formation does not require a viable host. Germ tubes previously inactivated by exposure to 65°C for 15 min or to 3% glutaraldehyde for 30 min did not attach to the host cell wall fragments, thus indicating a requirement for vital activity of the germ tubes to seek out specific binding sites on the host cell surface. Coiling of the mycoparasite around the host hypha, as observed in number of cases, could be a result of such a search for binding sites. Inhibition of attachment in the presence of competitive sugars, chitobiose and chitotriose, or by treatment of cell wall fragment with alkali or acid suggests that the attachment of the mycoparasite, *P. virginiana*, involves certain glycoproteins present at the host surface. This hypothesis is supported by the results obtained with a fluorescent lectin binding assay, using 10 different lectins with varied fine structure of their binding sites for specific sugars. All species tested, both hosts and nonhosts, showed a positive binding reaction with fluorescent lectins specific for N-acetylglucosamine oligomers [33,64]. The tip of the mycoparasite germ tube also bound positively with wheat germ agglutinin (WGA). These germ tubes (previously treated with WGA) did not attach to the host cell surface. This could also be due to the inhibitory effect of WGA on parasite germ tubes as reported for other fungi [2,65]. The sugar residues, e.g., N-acetylglucosamine, glucose, and mannose, observed at the cell surface of the mycoparasite *P. virginiana* were similar to those observed at the surface of the host species. In addition to N-acetylglucosamine, glucose, and mannose, the cell surfaces of the nonhost fungi also show the presence of other sugar residues such as fucose, galactose, or galactosamine. Could the presence of other sugars, i.e., fucose or galactose, at the surface of nonhost fungi interfere with the attachment of the mycoparasite? This question remains unanswered.

Protein and glycoprotein profiles obtained by SDS-PAGE from alkali extracts of isolated cell wall fragments of host and nonhost species show quantitative and qualitative differences after staining with Coomassie brilliant blue (R250) and periodic acid-Schiff's reagent [33,64]. These differences are well marked in band patterns of glycoproteins. Extracts of host and nonhost species possess four common bands, but two bands of high molecular weight (68 and 66 Kd) are observed only in the extract of the host cell wall. It is likely that glycoproteins represented by these bands play a role in the attachment process.

The role of agglutinin–carbohydrate binding in mycoparasitic system involving a necrotroph has been investigated by Barak et al. [65a]. These authors isolated and partially purified an agglutinin from the hyphal extracts as well as culture filtrate of plant pathogen, *Sclerotium rolfsii*. Agglutinating activity was specifically inhibited by D-glucose, D-mannose, and several of their derivatives. SDS-PAGE showed two protein bands after staining with Coomassie brilliant blue, corresponding to MW 60 and 55 Kd. Barak et al. [65a] have demonstrated a positive correlation between the ability of different isolates of the mycoparasite *Trichoderma* spp. to attach *S. rolfsii* and the agglutination of conidia of *Trichoderma* by *S. rolfsii* agglutinin. They have suggested a role of this agglutinin in specific recognition in fungus–fungus interactions. Chet's group [66] has also reported that *Rhizoctonia solani* produces an L-fucose–specific agglutinin, which agglutinates erythrocytes type O but not A or B and cells of *Escherichia coli* B. The agglutination was inhibited by L-fucose, L-galactose, and their derivatives. A correlation was demonstrated between inhibition by α-methyl L-fucoside and the prevention of *Trichoderma* coiling around its host. In a completely different system involving a zygomycetous fungus, *Conidiobolus lamprauges*, Ishikawa et al. [67,68] demonstrated the role of chitin binding hemagglutinin in the infection of insects, especially during the initial adheasion of *C. lamprauges* to the insect cuticle. The chitin binding hemagglutinin isolated from the cell wall and culture filtrate of *C. lamprauges* was strongly inhibited by N-acetylglucosamine oligomers.

By using gold-labeled lectins, Bonfante-Fasolo et al. [69] detected Con A-binding sites on extracellular material commonly radiating from the wall of the infective strain of *Pezizella ericae*, which has a strong tendency to form ericoid mycorrhiza. The Con A-binding sites were particularly abundant when the infective strain was in contact with the host but were uncommon on the surface of the noninfective strain, whether the latter was in contact with the host or not. According to these authors, Con A-binding sites, i.e., mannose or glucose residues as oligosaccharides or as part of glycoproteins, may play an important role in establishing the first contact between fungus and plant. Whether the extracellular fibrillar material at the surface of *P. ericae* has any resemblance to commonly occurring fimbriae on fungal cell surfaces is yet to be determined.

Among biological systems, plant host-bacterial pathogen or symbiont are probably the best characterized. It has become evident that plant cell wall polysaccharides and lectins can have important biological regulatory functions and can specifically induce gene expression, which affects the development of the plant-microbe interaction (for details see Ref. 70). The nature of the signal molecule involved in the induction of host defense responses or in establishment of bacterial pathogen or symbiont is still unclear. On the other hand, the mechanism of recognition and attachment in plant host-fungal pathogen is not well understood. Kojima and Uritani [71] reported the ability of sweet potato tuber extracts to preferentially agglutinate spores of nonpathogenic strains of *Ceratocystes fimbriata*, but the glycoprotein-like factor isolated from sweet potato did not exhibit specificity. Young and Kauss [72] investigated the agglutination of mycelial cell wall fragments and spores of *Colletotrichum lindemuthianum* by seed and hypocotyl extracts from three resistant and susceptible cultivars of the host plant *Phaseolus vulgaris* and by seed extracts from two nonhosts, *Vigna radiata* and *Ricinus communis*. All extracts and some commercial proteins were capable of causing agglutination, and no correlation between agglutination and host-pathogen specificity was observed. These authors concluded that the agglutination by plant extracts result primarily from nonspecific interactions of proteins with the fungal cell wall rather than from interactions involving hapten-specific lectins. Our results also showed that the crude cell wall extract did not exhibit the same degree of specificity as did the intact cell walls of the host [33]. It appears that harsh extraction treatments result in loss of specificity by producing serious structural alterations, especially of the glycoproteins. Precise structural and spatial arrangement of agglutinin at the host cell surface may be responsible for host-specific interaction. Evidence to support the above statement is provided by the work of Tamai et al. [73]. These workers observed that antisera prepared to intact yeast cells or to isolated cell walls gave the strongest immunochemical reaction with the surface mannan extracted with pronase and considerably less to mannan extracted by autoclaving cell walls. Besides extraction methods, the loss of specificity in the crude extract may also result from the presence of agglutinin-like proteins such as carbohydrate-binding enzymes and toxins [74].

VI. DIFFERENTIATION OF APPRESSORIUM

Appressoria are described as structures having the capacity to adhere to the host surface and the ability to effect penetration. Emmett and Parbery [75] wrote a comprehensive review on cytology, morphology and biochemistry of appressoria produced by plant pathogenic fungi. More recently, Staples and Hoch [76] have reviewed events leading to signalling and initiation of appressorium development in rust fungi. The uredospore germlings of rust fungi respond to various chemical and physical stimuli to form appressoria. Some advances have been made in understanding the processes involved in site recognition and in cell response to differentiate appressorium in biotrophic plant pathogens.

In contrast, the information on appressorium differentiation in biotrophic mycoparasites is nonexistent. Jeffries and Young [77] described the morphology of appressorium development in *Piptocephalis unispora*. Within 2 hr of establishing contact with the host hyphal surface or cell wall fragments, the elongating germ tubes of *P. virginiana* form swollen structures, appressoria, at their tips. Whereas in plant pathogens appressoria form readily on hosts (including resistant hosts) as well as on nonhost plants [75,78], appressoria are rarely formed by the germ tubes of the mycoparasite *P. virginiana*, which fail to attach to the cell surface of nonhosts [33,64]. Treatment of mycoparasite germ tubes with pronase results in loss of their ability to attach to the host cell wall fragments and form appressoria. Epstein et al. [79] also reported that treatment of bean rust uredospore germlings with pronase prevents the thigmodifferentiation response, i.e., differentiation of appressoria. The simplest explanation offered for the pronase effect is that surface proteins, possibly glycoproteins, are required for binding the germ tubes to the inductive surfaces. Attachment seems to be a prerequisite for the differentiation of an appressorium. The mycoparasite *P. virginiana* forms appressoria with equal ease on the surfaces of the susceptible as well as the resistant hosts. Appressorium development can occur on isolated cell wall fragments and does not require a physiologically active or living host cell. It appears that the mycoparasite attachment as a result of recognition of complementary macromolecules at the cell surface and the subsequent development of appressorium may not impart specificity, but it does help in determining its host range.

VII. HOST-PARASITE INTERFACE AND NUTRITIONAL RELATION

Attachment of the mycoparasite to its host surface precedes the establishment of nutritional relationships. Biotrophic mycoparasites have evolved a range of unique host-parasite interfaces to obtain nutrition from their hosts. Some of the structures of host-parasite interfaces in mycoparasites resemble those of the plant pathogens which have been reviewed by Bracker and Littlefield [80]. Biotrophic relationships occurring during mycoparasitism are dealt with herein, and an attempt has been made to avoid repetition of information already published.

A. Contact Mycoparasites

In contact mycoparasites the mode of parasitism involves the development of specialized branches which establish contact with the host hypha. The specialized branches may partially or completely surround the host hypha or touch end-to-end with a short branch of the host [5]. A septum develops within the hypha near the contact zone to form a distinct buffer cell. The presence of these specialized absorptive structures or buffer cells is considered as evidence of parasitism, however, no attempt has been made to examine the transfer of nutrients. As shown by light and electron microscopic studies, these mycoparasites presumably obtain their nutrition through the contact cells and do not penetrate the host cell. Hoch [14,15, 81] has provided ultrastructural details of the contact zone between the mycoparasitic fungi *Gonatobotrys simplex*, *Stephanoma phaeospora*, and *Calcariosporium parasiticum* on their respective hosts *Alternaria tenius*, *Fusarium* sp., and *Physalospora obtusa*. The interfacial wall between the buffer cell and the host hypha dissolves, and cytoplasmic continuity between the two partners is achieved. In *Calcarisporium–Physalospora* interaction, the dissolution of intervening wall between the parasite and host forms a large pore (0.2 – 1.0 μm in diameter) at their interface. The size of the pore is large enough to allow exchange of organelles. Hoch [15] reported the transfer of nuclei of the mycoparasite into the host hypha. The contact cell of the mycoparasite is always occluded after development of the mature host-parasite interface, thus the transport of organelles does not occur further than this region [15]. In the other two mycoparasites, *Gonatobotrys* and *Stephanoma*, the dissolution of intervening walls between partners results in development of several fine channels (24 – 31 nm in diameter), resembling the plasmodesmata of higher plants [14], at their interface. The small size of their interconnecting channels (plasmodesmata) prevents the exchange of organelles but obviously permits cytosolic movements and diffusion of soluble macromolecules. For example, spherical virus-like particles (44 – 47 nm in diameter) were observed in every *Stephanoma* nucleus examined but were never observed in the ground cytoplasm of *Stephanoma* or in any part of the host cells [81].

These mycoparasites are capable of obtaining all nutrients necessary for their growth and sporulation from washed living mycelia of their hosts which has been separated from the culture medium [16]. The required nutrients are held within the host mycelium and are not excreted into the culture medium. The contact mycoparasites generally cause no apparent harm to their hosts except for a reduction in rate of growth [5].

B. Internal Mycoparasites

Internal mycoparasites represented by chytrids penetrate and grow within the host and are completely surrounded by host cytoplasm. Sykes and Porter [13] have examined in detail the fine structure of the host–parasite interface during infection of *Allomyces arbuscula* by an obligate parasite *Catenaria allomycis*. The encysted zoospore produces a germ tube which attaches to the host hypha by an appressorium. Attachment to the host is strong at this stage, and the cysts are not easily dislodged. An infection tube arises from the appressorium and penetrates the host wall. The mechanism of penetration is believed to be both enzymatic and mechanical. The whole process of attachment and penetration is completed within 1 to 1.5 hr. A double-layered papilla is produced by the host in response to invasion by the parasite. The host plasma membrane and the papilla are ruptured by the continued growth of the infection tube, resulting in direct contact of the mycoparasite with the host cytoplasm. The thallus released from the cyst into the cytoplasm is surrounded by a wall throughout its intracellular development, although there is a change in its form from the thin cyst and infection tube wall to the thicker endobiotic thallus wall after penetration. In contrast, when *Rozella* of *Rozellopsis* infect their hosts, the invading protoplast (bound only by plasma membrane) is released into the host cytoplasm [82]. The hyphal wall of the host surrounds its own cytoplasm as well as the protoplast of the mycoparasite. During early stages of development these internal mycoparasites appear to cause little or no harm to the host. Later, they undergo rapid growth with a corresponding degenration of host cytoplasm. The mycoparasite forms a surrounding wall and produces zoospores. Nothing is known about the nutritional requirement of these mycoparasites.

C. Haustorial Mycoparasites

Attachment to host surface and development of appressorium appears to be a common feature among the filamentous haustorial mycoparasites. However, the process of penetration, development of the haustorium and host-parasite interface, and establishment of nutritional relationship varies with different host parasite combinations.

Piptocephalis species are the most extensively investigated haustorial mycoparasites. Morphological and cytological aspects of the infection process, i.e., development of infection peg, haustorium and host-parasite interface, have been examined using light and electron (transmission and scanning) microscopes [17–19,31,83,84]. These studies support the commonly held view that penetration by mycoparasites is accomplished by both enzymatic and mechanical means. Enzymatic penetration of the host by necrotrophic mycoparasites is well known. Necrotrophs produce and release into their environment high levels of hydrolytic enzymes to facilitate the digestion of host cell walls [85,86]. However, the role of hydrolytic enzymes in penetration of biotrophic mycoparasites is open to question. Recent studies show that not only many necrotrophs, but also some biotrophs produce cell wall degrading enzymes [87], although the levels of enzymes are low and are generally not released into their external environments. Balasubramanian and Manocha [26] investigated the cell wall degrading enzymes produced by the mycoparasite *P. virginiana* in culture filtrate as well as in cell extract. Chitinase and chitosanase activity was found to be higher in cell extracts than in the culture filtrates. The enzyme activity was stimulated by the addition of isolated host cell wall fragments in the nutrient medium. Presumably these enzymes are under strict regulatory control and the mycoparasite must deploy them in a specific manner during the penetration of a living host so as not to cause extensive cell wall damage. The penetration of the host by the mycoparasite is believed to be partially facilitated by the localized release of enzymes which degrade the major cell wall polymers of the host.

Ultrastructural studies support the involvement of both enzyme and mechanical pressure in the penetration of host cell wall by *Piptocephalis* species [18,19,83]. In a compatible interaction the infection peg of the mycoparasite pushes inward the thin, single-layered host cell wall, which is eventually breached, and the mycoparasite penetrates the host cell. On the other hand, in an incompatible interaction the outer layer of the two-layered cell wall of the resistant host, *Phascolomyces articulosus*, is enzymatically eroded by the mycoparasite. There is no evidence of the dissolution or rupture of the inner layer during early stages of infection. However, papilla formation as a response to mycoparasite is evident [27,84]. Evans et al. [83] also observed differences in penetration by an aggressive mycoparasite, *Piptocephalis fimbriata*, of thin-walled M-phase cells and thick-walled Y-phase cells of *Mycotypha microspora*. The thick-walled Y-phase cells were eroded enzymatically at the point of contact with the appressorium of the mycoparasite. The M-phase cells were penetrated by a narrow infection peg arising from the appressorium. There was no sign of erosion of the host cell wall during infection, but some invagination or depression was noted. Differences in penetration of susceptible and resistant hosts are elucidated further by scanning electron microscopy. Manocha and Golesorkhi [27] observed, using the scanning electron microscope, depressions at points of penetration on the surface of the susceptible host which mirrored the inward push by the infection peg. Penetration of the resistant host was marked with a distinct annular zone "halo" around the penetration peg indicating apparent dissolution of the host cell wall at that point.

The sequence of events leading to the maturation of the haustorium and to the development of fungal host – mycoparasite interface have been investigated in a number of combinations [17 – 19,83,88]. A young haustorium of *Piptocephalis virginiana* is produced 16 – 18 hr after inoculation by mixing spore suspensions of both host and mycoparasite on the surface of sterilized solid medium. At this early stage of development the haustorial wall is closely appressed to the extrahaustorial membrane. A thin translucent zone becomes apparent between the haustorium and the extrahaustorial membrane at 20 – 22 hr after incoulation [19]. An extensive network of smooth-surface endoplasmic reticulum and increased vesicular activity of the host cytoplasm is observed in the vicinity of the haustorium. The extrahaustorial matrix increases in volume. This appears to be a normal process during maturation of haustoria in susceptible hosts. In a mature haustorium, 32 – 36 hr after inoculation, the extrahaustorial matrix is well developed and appears to be composed of material with the same electron opacity as that of the host cell wall [19]. Vesicles containing material similar in density to the extrahaustorial matrix are observed near the extrahaustorial membrane at the host-parasite interface. The role, origin, and composition of extrahaustorial matrix have been a matter of speculation. During later stages of infection, 48 hr or more after infection, the host cell organelles and membranes undergo degeneration.

The extrahaustorial matrix is a common occurrence at the interface of haustorial biotrophs and their hosts. However, the information on its origin and composition is scant. The origin of extrahaustorial matrix is attributed to the activity of the parasite, the host, or both. Its exact composition remains uncertain. Use of biotrophic haustorial mycoparasites has provided some insight into this problem. High resolution autoradiographic studies have shown that [^{3}H]-N-acetylglucosamine ([^{3}H]-GlcNAc), a precursor of chitin, accumulates preferentially in the cell wall of the growing hypha of the host *Choanephora cucurbitarum* and in the extrahaustorial matrix (sheath) around the haustorium of the mycoparasite *Piptocephalis virginiana* 24 and 30 hr after inoculation [89]. Further confirmation of incorporation of [^{3}H]-GlcNAc in the chitin fraction of the host cell wall and the extrahaustorial matrix was obtained by use of polyoxin D, a competitive inhibitor of the enzyme chitin synthase. Polyoxin D inhibited the label incorporation in the host cell wall as well as the deposition of electron dense material in the extrahaustorial matrix which is observed in untreated samples 32 hr after inoculation [90]. These results provide good experimental evidence that the extrahaustorial matrix is composed, at least in part, of host cell wall material. The presence of polysaccharides, cellulose, β-1, 3 glucan, and protein has been indicated in the extrahaustorial matrix of a biotrophic plant pathogen [91]. So far, there is no experimental evidence to indicate the parasite's contribution to the extrahaustorial matrix.

That the extrahaustorial matrix may function by preventing transport of nutrients or molecules from host to parasite and vice versa was investigated in plant host-fungal pathogen systems by using ultrastructural and micro-autoradiographic techniques. Electron microscopy of interfaces between stem rust and susceptible and resistant wheat varieties and wheat with temperature sensitive gene *Sr6* has shown that the extrahaustorial matrix develops earlier in the resistant than in the susceptible host [92, 93]. These results were later confirmed by Coffey [94], who investigated the fine structure of interface between the rust fungus *Melampsora lini* and flax with major resistance *k* gene. High resolution autoradiographic studies have shown that there is more incorporation of [^{3}H]-leucine in the haustorium of *Puccinia graminis* produced in the susceptible than in the resistant host [95]. The incorporation of label declined in the susceptible host with the age of the infection and closely coincided with the development of the extrahaustorial matrix. Mendgen [95] also observed reduced uptake of nutrients by bean rust haustoria in a resistant host. A more recent report has associated induced resistance in wheat and barley to powdery mildew with the haustorial efficiency impaired by the enlargement of extrahaustorial matrix [96]. These studies do indicate that the extrahaustorial matrix deposited by the host cell may render the haustoria nonfunctional by acting as a barrier to nutrient uptake in partially resistant hosts.

VIII. MECHANISM OF HOST DEFENSE

Research work on fungal host defense reaction to biotrophic mycoparasites is almost negligible. The information obtained so far does not indicate any involvement of toxins or inhibitors produced by the mycoparasite or by the fungal host in response to the attack by the mycoparasite. It is hard to believe that the fungal hosts do not resist the advances of their parasites. Probably their lack of economic importance has not attracted the attention of workers in the field of fungal host resistance. Furthermore, the genetics and biochemistry of fungal host and their mycoparasites are poorly understood. In contrast, research on plant disease resistance provides us with numerous books and reviews every year. Experimental approaches used to compare biochemical responses in resistant and susceptible plants, including genetics of plants and pathogens, environmentally altered resistance, induced resistance, and susceptibility, have not been tried using mycoparasitic systems.

There is virtually no information in literature on the structural or biochemical responses of fungal hosts to contact mycoparasites. There is little work on the structural aspect of host resistance in the internal mycoparasites. Papilla formation, as a host response to challenge by the internal mycoparasite, has been reported in most of the cases. Held [12] considered the presence of papilla as a requirement for successful penetration in the *Allomyces–Rozella* system. Aist and Israel [97] reported that the failure of some cysts of *Olpidium* species to penetrate host root hairs is unrelated to papilla formation. In the *Allomyces–Catenaria* system Sykes and Porter [13] observed that a papilla is always associated with the infection and that it is produced by the host in an attempt to stop the invading parasite.

Evidence that papillae play a role in host resistance comes from studies with haustorial mycoparasites. *Piptocephalis* species develop compatible relationships with *Choanephora cucurbitarum* and *Cokeromyces recurvatus* and incompatible relationships with *Phascolomyces articulosus*. In contrast, *Dimargaris cristalligena* is a vigorous parasite of both *C. recurvatus* and *Ph. articulosus* [88]. In *Piptocephalis–Phascolomyces* interaction, host resistance is attributed to the early development of papilla, even before the penetration of the host [84]. In *Dimargaris–Phascolomyces* interaction, the success of the mycoparasite is due to delayed host response; i.e., papillae do not develop until the later stages of infection and the haustorial development is not prevented [88]. Manocha and Graham [98] studied the incorporation of [^{3}H]-GlcNAc at infection sites 18, 24, and 36 hr after inoculation of susceptible and resistant hosts with the mycoparasite *P. virginiana*. The degree of incorporation varied as a function of host-parasite compatibility and duration of the incubation period. Infection sites on the resistant host, *Phascolomyces articulosus*, showed incorporation of label after all three incubation periods, although labeling was extensive at 24 and 36 hr. This pattern of label incorporation coincided with papilla formation (18 hr) and early development of well-differentiated sheath around the haustorium, if the latter was formed. In contrast, negligible levels of label accumulated at infection sites 18 hr after inoculation in susceptible host, *C. cucurbitarum*. Although incorporation increased significantly by 24 hr, levels of grain density comparable to those observed in the resistant host were not apparent until 36 hr following inoculation. This coincides with the time when a sheath appears in *C. cucurbitarum*. In other susceptible hosts, such as *Cokeromyces recurvatus* and *Mycotypha microspora*, papillae and cell wall thickening are missing or poorly developed. Thus it seems papillae are an important component of resistance, and they probably contain host cell wall material.

Three possible explanations for the suppression of chitin synthesis at the infection sites in *Choanephora cucurbitarum* at 18 hr after inoculation are (a) the inability of the host to respond to the mycoparasite by failing to mobilize chitin synthase to the infection site, (b) its inability to activate its chitin synthase by either lack of activator or presence of an inhibitor, and (c) differential activation of chitin synthase and chitinase by host proteinases. These possibilities were investigated by examining the incorporation of [^{3}H]-GlcNAc in susceptible and resistant hyphae after wounding by sonication. Label incorporation at the penetration and wounding sites was inhibited by polyoxin D. Cycloheximide treatment greatly increased the label in the subapical region and reduced it at the hyphal tip in both hosts. However, this treatment did not prevent localized incorporation at the wounding or the penetration sites on the resistant host. These results indicated that suppression of localized incorporation of [^{3}H]-GlcNAc at penetration sites in the susceptible host was not due to the lack of mobility of chitin synthase to the penetration sites or its activating factor. However, this work [99] did not rule out the possibility of differential activation of chitinase and chitin synthase in the two hosts. Differences in the level of activation of two zymogens, chitin synthase and chitinase, as determined by the specificity of the proteinases could explain the outcome of the interaction.

It is obvious from the results of above mentioned investigations that resistance in *Phascolomyces articulosus* to the mycoparasite, *P. virginiana*, is nonspecific, whereas susceptibility in *Choanephora cucurbitarum* is a specific resonse.

Properties of the enzyme chitin synthase, which is responsible for the incorporation of [^{3}H]-GlcNAc into chitin, were examined using extracts from young (24 hr) hyphae of *Choanephora cucurbitarum* and *Phascolomyces articulosus* [100]. The enzyme activity was mainly associated with the mixed membrane fraction and displayed basically the same features as the corresponding enzymes reported from other mucoraceous fungi [101–103]. However, the two preparations (from *C. cucurbitarum* and *Ph. articulosus*) differed from each other in their response to various proteinases. Enzyme preparation from the resistant host, *Ph. ariticulosus*, was activated by all three proteinases used in this study, whereas the *C. cucurbitarum* preparation was activated by acid proteinase, slightly activated by trypsin over a narrow concentration range, and was inhibited by neutral proteinase.

Acid and neutral proteinases produced by the two hosts, *C. cucurbitarum* and *Ph. articulosus*, were isolated and characterized and were found to be remarkably different in their properties [104–106]. The acid proteinase from *C. cucurbitarum* is of the serine type [6], whereas the acid proteinase obtained from *Ph. articulosus* is of the cysteine type [105]. *Phascolomyces articulosus* is unique among the mucoraceous fungi in being the only known resistant host of the mycoparasite *P. virginiana*. In all other known cases of host-parasite interactions, the mucoraceous fungi are either compatible hosts with a varying degree of susceptibility or nonhosts showing no interaction whatsoever with the mycoparasite. *Phascolomyces* is also unique in producing a cysteine type of proteinase, whereas all the other known studies of proteinases in mucoraceous fungi have reported serine types [107]. Neutral proteinases obtained from these two hosts are also different in their molecular weight, sensitivity to various inhibitors and heavy metals, and substrate specificity [106].

The role of proteinases in host-specific reaction to a mycoparasite is suggested by the following work [108]. Effects of proteinases extracted from the susceptible and resistant host were tested on the activities of chitinase and chitin synthase of the same host. Both chitinase and chitin synthase are membrane-bound, zymogenic, and activated by partial proteolysis. Host proteinases of acid and neutral type stimulated chitinase activity, but only the acid proteinase stimulated chitin synthase activity of the two hosts. Neutral proteinase of the susceptible host, *C. cucurbitarum*, inhibited its chitin synthase, whereas that of the resistant host, *Ph. articulosus*, stimulated its chitin synthase activity. The mycoparasite *Piptocephalis virginiana* also produces a neutral proteinse, but this proteinase enhanced chitinase and suppressed chitin synthase activity of both the hosts alike. It seems that localized incorporation of [^{3}H]-GlcNAc at infection sites on the resistant host and its absence on the susceptible host could be due to the differential activation of chitinase and chitin synthase in the two hosts. Activation of chitinase in excess of chitin synthase may explain the lack of [^{3}H]-GlcNAc accumulation in the susceptible host. Such chitinase has been shown to be much more effective against nascent chitin than on preformed chitin in *Mucor rouxii* [109]. Similar results were obatined for yeast endochitinase by Correa et al. [110].

Based on the above information, it is tempting to speculate that in vivo activation of the two zymogens, chitinase and chitin synthase, will be highly specific because of the specificity of membrane-bound proteinases. However, there is no experimental proof of in vivo activation by and specificity of proteinases. On the contrary, studies on the mutants of *Saccharomyces cerevisiae* lacking proteinase B have shown that the cell-free extract of these mutants did not affect chitin synthase activation or chitin synthesis [111,112]. However, the mutants are able to grow normally and do not show any aberrant budding behavior. Previously, proteinase B was thought to be involved in the activation of chitin synthase zymogen to allow septum formation during sporulation [113]. The importance of a genetic approach to solve problems concerning the relationship between proteinase activity, proteinase function, and selectivity in vivo has been emphasized. As a result, proteinase mutants have helped uncover new multiple forms of proteinases. According to Wolf [114], it seems likely that unknown proteinases responsible for more specific proteolysis exist in *S. cerevisiae*. A similar situation could also account for the differential activation of chitinase and chitin synthase by specific proteinases in the susceptible and resistant hosts of the mycoparasite.

IX. CONCLUSION

Studies on the physiology and biochemistry of biotrophic mycoparasitism have shown that the system can serve as a convenient model for investigations on the mechanism of host specificity and recognition. Use of mycoparasites has added to the existing bank of information on (a) the role of lipid physiology and biotrophism, (b) the mechanism of attachment of parasite to the host cell through the mediation of surface glycoproteins, (c) the development and host cell wall-like composition of extrahaustorial matrix, and (d) the role of host enzymes necessary for normal metabolism in host response to parasite.

Most of the results on the physiology and biochemistry of biotrophic mycoparasites have been obtained using only one mycoparasite, i.e., *Piptocephalis virginiana*. Whether these observations are applicable to other host-mycoparasite combinations remains to be seen. Other mycoparasites such as *Piptocephalis xenophila* with host range outside the mucoraceous fungi and *Dimargaris crystalligena* with ability to establish a compatible relationship with *Phascolomyces articulosus*, the only known resistant host of *P. virginiana*, have not been studied extensively. The work on *P. xenophila* and *D. crystalligena* has been limited to investigations on their morphology and fine structure of their haustoria. There is a need for comparative study of these mycoparasites. Use of *P. xenophila* may provide insight to the extended host range of this mycoparasite, and the use of *D. crystalligena* may explain the basis of host resistance in *Phascolomyces articulosus*.

The major drawback of using the mycoparasitic system to elucidate certain aspects of host-parasite interactions is the lack of information on the genetics of these mycoparasites as well as their hosts. This limits the validity of any model offered to elucidate the mechanism of host specificity in mycoparasitic systems. Until such studies were undertaken the proposed model suggesting the involvement of proteinases to induce imbalance in the growth regulatory mechanism of the host will remain untested. The importance of using genetic approaches to solve these problems cannot be overemphasized. For example, the isolation of neutral proteinase minus mutants of susceptible host, *Choanephora cucurbitarum*, will resolve the question of differential activation of chitinase and chitin synthase at the penetration point. Furthermore, the question of whether mycoparasites produce elicitors recognized by their hosts remains uncertain. Similarly, the nature of the host product, termed suppressor, which supposedly impedes elicitor detection in compatible interaction, is not known.

Efforts need to be expanded to obtain further information, especially on the cell–cell interaction in fungi and on the regulation of fungal host metabolism for compatible and incompatible interaction. Recent advancements in the chemistry of proteinases, their catalytic specificity, cellular localization, and their involvement in regulatory processes of the cell may help understand the role of these enzymes in host-parasite relations. Alternatively, other regulatory mechanisms such as phosphorylation-dephosphorylation of key enzymes of the host (in the present case chitinase and chitin synthase) could also provide some insight into the mechanism of host response to parasite. Future work along these lines will unravel some of the mysteries of the host-parasite interaction in mycoparasitic systems.

ACKNOWLEDGMENTS

Financial support from the Natural Sciences and Engineering Research Council for investigation on mycoparasitism is thankfully acknowledged. Thanks are due to my former and present graduate students and research associates for their contributions to this project. I am also thankful to Mrs. M. Ferracuti for assisting in the preparation of this manuscript.

REFERENCES

1. Albersheim, P. and Anderson-Prouty, A. J. Carbohydrates, proteins, cell surfaces and biochemistry of pathogenesis. *Annu. Rev. Plant Physiol.*, *26*:31–52, 1975.
2. Callow, J. A. Recognition, resistance, and role of plant lectins in host-parasite interactions. *Adv. Bot. Res.*, *4*:1–49, 1977.
3. Ellingboe, A. H. Inheritance of specificity: The gene-for-gene hypothesis. In *Recognition and Specificity in Plant Host-Parasite Interactions* (J. M. Daly and I. Uritani, eds.), Jpn. Sci. Soc. Press, Tokyo, 1979, pp. 3–17.
4. Keen, N. T. Phytoalexins — Progress in relation of their accumulation in gene-for-gene interactions. In *Plant Infection. The Physiological and Biochemical Basis* (Y. Asada, W. R. Bushnell, S. Ouchi, and C. P. Vance, eds.), Jpn. Sci. Soc. Press, Tokyo, 1982, pp. 281–300.

5. Barnett, H. L. and Binder, F. L. The fungal host–parasite relationship. *Annu. Rev. Phytopathol.*, *11*:279–292, 1973.
6. Manocha, M. S. Cellular and molecular aspects of fungal host-mycoparasite interaction. *Zeitschrift für Pflanzenkranheiten und Pflanzenschutz.*, *94*:431–444, 1987.
7. Blakeman, J. P. and Fokkema, N. J. Potential for biological control of plant diseases on the phylloplane. *Annu. Rev. Phytopathol.*, *20*:167–192, 1982.
8. Papavizas, G. C. *Trichoderma* and *Gliocladium*: Biology, ecology, and potential for biocontrol. *Annu. Rev. Phytopathol.*, *23*:23–54, 1985.
9. Papavizas, G. C. and Lumsden, R. D. Biological control of soilborne fungal propagules. *Annu. Rev. Phytopathol.*, *18*:389–413, 1980.
10. Cook, R. J. and Baker, K. F. *The Nature and Practice of Biological Control of Plant Pathogens.* Am. Phytopathol. Soc., St. Paul, MN, 1988.
11. Windels, C. E. and Lindow, S. E. (eds.) *Biological Control of the Phylloplane.* Am. Phytopathol. Soc., St. Paul, MN, 1985.
12. Held, A. A. Host-parasite relations between *Allomyces* and *Rozella.* Parasite penetration depends on growth response of host cell-wall. *Arch. Microbiol.*, *82*:128–139, 1972.
13. Sykes, E. E. and Porter, D. Infection and development of the obligate parasite *Catenaria allomycis* on *Allomyces arbuscula. Mycologia*, *72*:288–300, 1980.
14. Hoch, H. C. Mycoparasitic relationships: *Gonatobotrys simplex* parasitic on *Alternaria tenuis. Phytopathology*, *67*:309–314, 1977.
15. Hoch, H. C. Mycoparasitic relationships. III. Parasitism of *Physalospora obtusa* by *Calcarisporium parasiticum. Can. J. Bot.*, *55*:198–207, 1977.
16. Jordan, E. G. and Barnett, H. L. Nutrition and parasitism of *Melanospora zamiae. Mycologia*, *70*:300–312, 1978.
17. Armentrout, V. N. and Wilson, C. L. Haustoria-host interaction during mycoparasitism of *Mycotypha microspora* by *Piptocephalis virginiana. Phytopathology*, *59*:897–905, 1969.
18. Jeffries, P. and Young, T. W. K. Ultrastructure of infection of *Cokeromyces recurvatus* by *Piptocephalis unispora* (Mucorales). *Arch. Microbiol.*, *109*:277–288, 1976.
19. Manocha, M. S. and Lee, K. Y. Host-parasite relations in a mycoparasite. *I*. Fine structure of host, parasite and their interface. *Can. J. Bot.*, *49*:1677–1681, 1971.
20. Lumsden, R. D. Ecology of mycoparasitism. In *The Fungal Community and Its Role in the Ecosystem* (D. T. Wicklow and G. W. Caroll, eds.), Marcel Dekker, New York, 1980, pp. 295–318.
21. Jeffries, P. Mycoparasitism within the Zygomycetes. *Bot. J. Linnean Soc.*, *91*:135–150, 1985.
22. Held, A. A. Attraction and attachment of zoospores of the parasitic chytrid *Rozella allomycis* in response to host-dependent factors. *Arch. Microbiol.*, *95*:97–114, 1974.
23. Whaley, J. W. and Barnett, H. L. Parasitism and nutrition of *Gonatobotrys simplex. Mycologia*, *55*:199–210, 1963.
24. Kuykendall, W. R., Hindal, D. E., and Barnett, H. L. Nutrition and parasitism of the contact mycoparasite *Acladium tenellum. Mycologia.*, *75*:656–665, 1983.

25. Hwang, K., Stelzig, D. A., Barnett, H. L., Roller, P. P., and Kelsey, M. I. Partial purification of the growth factor mycotrophein. *Mycologia, 77*:109–113, 1985.
26. Balasubramanian, R. and Manocha, M. S. Proteinase, chitinase, and chitosanase activities in germinating spores of *Piptocephalis virginiana. Mycologia, 78*:157–163, 1986.
27. Manocha, M. S. and Golesorkhi, R. Host-parasite relations in a mycoparasite. VII. Light and scanning electron microscopy of interactions of *Piptocephalis virginiana* with host and nonhost species. *Mycologia, 73*:976–987, 1981.
28. Jeffries, P. and Young, T. W. K. Physiology and fine structure of sporangiospore germination in *Piptocephalis unispora* prior to infection. *Arch. Microbiol., 107*:99–107, 1976.
29. Jeffries, P. and Young, T. W. K. Sporangiospore structure and germination in Dimargaritales. *Trans. Br. Mycol. Soc., 83*:223–232, 1984.
30. Berry, C. R. and Barnett, H. L. Mode of parasitism and host range of *Piptocephalis virginiana. Mycologia, 49*:374–386, 1957.
31. Jeffries, P. and Young, T. W. K. Mycoparasitism by *Piptocephalis unispora:* Host range and reaction with *Phascolomyces articulosus. Can. J. Bot., 56*:2449–2459, 1978.
32. Evans, G. H. and Cooke, R. C. Studies on mucoralean mycoparasites. III. Diffusible factors from *Mortierella vinacea* Dixon-Steart that direct germ tube growth in *Piptocephalis fimbriata* Richardson and Leadbeater. *New Phytol., 91*:245–253, 1982.
33. Manocha, M. S., Balasubramanian, R., and Enskat, S. Attachment of a mycoparasite with host but not with nonhost *Mortierella* species. In NATO ASI Series Vol. 41. *Biology and Molecular Biology of Plant-Pathogen Interactions* (J. Bailey, ed.), Springer-Verlag, Berlin, 1986, pp. 59–69.
34. Gardiner, R. B. Fungal fimbriae: Their structure and distribution. Ph.D. dissertation. University of Western Ontario, London, Canada, 1985.
35. Barnett, H. L. Nutritional requirements of axenic growth of some haustorial mycoparasites. *Mycologia, 62*:750–760, 1970.
36. Binder, F. L. Differences in uptake of glucose and glycerol by *Tieghemiomyces parasiticus in axenic culture. Can. J. Bot., 52*:2571–2574, 1974.
37. Barker, S. M. and Barnett, H. L. Nitrogen and vitamin requirements for axenic growth of the haustorial mycoparasite *Dispira cornuta. Mycologia, 65*:21–27, 1973.
38. Berry, C. R. Factors affecting parasitism of *Piptocephalis virginiana* on other Mucorales. *Mycologia, 51*:824–832, 1959.
39. Manocha, M. S. Host-parasite relations in a mycoparasite. III. Morphological and biochemical differences in the parasitic- and axenic-culture spores of *Piptocephalis virginiana. Mycologia, 67*:382–391, 1975.
40. Page, O. T. Effect of phytotoxins on the permeability of cell membranes. In *Phytotoxins in Plant Diseases* (R. K. S. Wood, A. Ballio, and A. Graniti, eds.), Academic Press, New York, 1972, pp. 211–225.
41. Wheeler, H. Disease alterations in permeability and membranes. In *Plant Disease: An Advanced Treatise*, Vol. 3 (J. G. Horsfall and E. B. Cowling, eds.), Academic Press, New York, 1978, pp. 327–347.

42. Wood, R. K. S. The killing of plant cells by soft rot parasites. In *Phytotoxins in Plant Diseases* (R. K. S. Wood, A. Ballio, and A. Graniti, eds.), Academic Press, New York, 1972, pp. 273–288.
43. Hoppe, H. H. and Heitefuss, R. Permeability and membrane lipid metabolism of *Phaseolus vulgaris* infected with *Uromyces phaseoli*. I. Changes in efflux of cell constituents. *Physiol. Plant Pathol.*, *4*:5–9, 1974.
44. Manocha, M. S. and Maharaj, R. Host-parasite relations in a mycoparasite. VI. Leakage of electrolytes, amino acids, and sugars after infection with *Piptocephalis virginiana*. *Can. J. Bot.*, *59*:1271–1276.
45. Elnaghy, M. R. and Heitefuss, R. Permeability changes and production of antifungal compounds in *Phaseolus vulgaris* infected with *Uromyces phaseoli*. I. Role of spore germination self inhibitor. *Physiol. Plant Pathol.*, *8*:253–267, 1976.
46. Elnaghy, M. R. and Heitefuss, R. Permeability changes and production of antifungal compounds in *Phaseolus vulgaris* infected with *Ulromyces phaseoli*. II. Role of phytoalexins. *Physiol. Plant Pathol.*, *8*:269–278, 1976.
47. Hoppe, H. H. and Heitefuss, R. Permeability and membrane lipid metabolism of *Phaseolus vulgaris* infected with *Uromyces phaseoli*. II. Changes in lipid concentration and ^{32}P incorporation into phospholipids. *Physiol. Plant Pathol.*, *4*:11–23, 1974.
48. Hoppe, H. H. and Heitefuss, R. Permeability and membrane lipid metabolism of *Phaseolus vulgaris* infected with *Uromyces phaseoli*. III. Change in relative concentration of lipid bound fatty acids and phospholipase activity. *Physiol. Plant Pathol.*, *4*:25–35, 1974.
49. Hoppe, H. H. and Heitefuss, R. Permeability and membrane lipid metabolism in *Phaseolus vulgaris* infected with *Uromyces phaseoli*. IV. Phospholipids and phospholipid fatty acids in healthy and rust infected leaves resistant and susceptible to *Uromyces phaseoli*. *Physiol. Plant Pathol.*, *5*:263–271, 1975.
50. Hoppe, H. H. and Heitefuss, R. Permeability and membrane lipid metabolism in *Phaseolus vulgaris* infected with *Uromyces phaseoli*. V. Sterols in healthy and rust-infected leaves, resistant and susceptible to *Uromyces phaseoli*. *Physiol. Plant Pathol.*, 5:273–281, 1975.
51. Lösel, D. M. and Lewis, D. H. Lipid metabolism in leaves of *Tussilago farfara* during infection by *Puccinia poarum*. *New Phytol.*, *73*:1157–1169, 1974.
52. Brennan, P. J. and Lösel, D. M. Fungal lipid physiology. *Adv. Microbial. Physiol.*, *17*:47–179, 1978.
53. Manocha, M. S. Role of lipids in host-parasite relations of a mycoparasite. *Physiol. Plant Pathol.*, *17*:319–328, 1980.
54. Manocha, M. S. Host specificity and mechanism of resistance in a mycoparasitic system. *Physiol. Plant Pathol.*, *18*:257–265, 1981.
55. Manocha, M. S. and Deven, J. M. Host-parasite relations in a mycoparasite. IV. A correlation between the levels of γ-linolenic acid and parasitism of *Piptocephalis virginiana*. *Mycologia*, *67*:1148–1157, 1975.
56. Chapman, D. Fluidity and phase transition of cell membranes. In *Biomembranes*, 7 (H. Eisenberg, E. Katchalski-Katzir, and L. A. Mason, eds.), Plenum, New York, 1975, pp. 1–9.

57. Jakovcic, S., Getz, G. S. Rabinowitz, M., Jacob, H., and Swift, H. Cardiolipin content of wild type and mutant yeasts in relation to mitochondrial function and development. *J. Cell Biol.*, *48*:490 – 502, 1973.
58. Paultauf, F. and Schatz, G. Promitochondria and anaerobically grown yeast. II. Lipid composition. *Biochem.*, *8*:335 – 339, 1969.
59. Arthur, H. and Watson, K. Thermal adaptation in yeast: Growth temperature, membrane lipid and cytochrome composition of psychrophilic, mesophilic, and thermophilic yeasts. *J. Bacteriol.*, *128*:56–68, 1976.
60. Letourneau, D. R., Deven, J. M., and Manocha, M. S. Structure and composition of the cell wall of *Choanephora cucurbitarum*. *Can. J. Microbiol.*, *22*:486 – 494, 1976.
61. Manocha, M. S. Cell surface characteristics of *Mortierella* species and their interaction with a mycoparasite. *Can. J. Microbiol.*, *30*:290 – 298, 1984.
62. Bartnicki-Garcia, S. Cell wall chemistry, morphogenesis, and taxonomy of fungi. *Annu. Rev. Microbiol.*, *22*:87 – 108, 1968.
63. Manocha, M. S. and Campbell, C. D. Host-parasite relations in a mycoparasite. VIII. Age-related host resistance in *Choanephora cucurbitarum*. *Mycologia*, *75*:588 – 596, 1983.
64. Manocha, M. S. Specificity of mycoparasite attachment to the host cell surface. *Can. J. Bot.*, *63*:772 – 778, 1985.
65. Mirelman, D., Galun, E., Sharon, M., and Lotan, R. Inhibition of fungal growth by wheat germ agglutinin. *Nature* (London), *256*:414 – 416, 1975.
65a. Barak, R., Elad, Y., Mirelman, D., and Chet, I. Lectins: A possible basis for specific recognition in the interaction of *Trichoderma* and *Sclerotium rolfsii*. *Phytopathology*, *75*:458 – 462, 1985.
66. Barak, R., Elad, Y., and Chet, I. The properties of L-fucose binding agglutinin associated with the cell wall of *Rhizoctonia solani*. *Arch. Microbiol.*, *144*:346 – 349, 1986.
67. Ishikawa, F., Oishi, K., and Aids, K. Purification and some properites of chitin-binding hemagglutinin from *Conidiobolus lamprauges*. *Agric. Biol. Chem.*, *45*:577 – 574, 1981.
68. Ishikawa, F., Oishi, K., and Aida, K. Chitin-binding hemagglutinin associated with cell wall of *Conidiobolus lamprauges*. *Agric. Biol. Chem.*, *47*:587 – 592, 1983.
69. Bonfante-Fasolo, P., Perotto, S., Testa, B., and Faccio, A. Ultrastructural localization of cell surface sugar residues in ericoid mycorrhizal fungi by gold-labeled lectins. *Protoplasma*, *139*:25 – 35, 1987.
70. Halverson, L. J. and Stacey, G. Signal exchange in plant-microbe interactions. *Microbiol. Rev.*, *50*:193 – 225, 1986.
71. Kojima, M. and Uritani, I. The possible involvement of a spore agglutinating factor(s) in various plants in establishing host specificity by various strains of black rot fungus, *Ceratocystis fimbriata*. *Plant Cell Physiol.*, *8*:97 – 111, 1974.
72. Young, D. H. and Kauss, H. Agglutination of mycelial cell wall fragments and spores of *Colletotrichum lindemuthianum* by plant extracts, and by various proteins. *Physiol. Plant Pathol.*, *20*:285 – 297, 1982.
73. Tamai, Y., Nakashima, T., Takakuwa, M., and Misaki, A. Antigenicities of several cell wall mannan preparations of cell envelope preparations from Baker's yeast. *Agric. Biol. Chem.*, *44*:49 – 53, 1980.
74. Barondes, S. H. Lectins: Their multiple endogenous cellular functions. *Annu. Rev. Biochem.*, *50*:207 – 232, 1981.

75. Emmett, R. W. and Parbery, D. G. Appressoria. *Annu. Rev. Phytopathol.*, *13*:147–167, 1975.
76. Staples, R. C. and Hoch, H. C. Infection structures: Form and Function. *Expt. Mycology, 11*:163–169, 1987.
77. Jeffries, P. and Young, T. W. K. Scanning electron and light microscopy of appressorium development in *Piptocephalis unispora* (Mucorales). *Arch. Microbiol.*, *103*:293–296, 1975.
78. Bushnell, W. R. Hypersensitivity in rusts and powdery mildews. In *Plant Infection: The Physiological and Biochemical Basis* (Y. Asada, W. R. Bushnell, S. Ouchi, and C. P. Vance, eds.), Jpn. Sci. Soc. Press, Tokyo, 1982, pp. 97–116.
79. Epstein, L., Laccetti, L., and Staples, R. C. Cell-substratum adhesive protein involved in surface contact responses of the bean rust fungus. *Physiol. Mole. Plant Pathol.*, *30*:373–388, 1987.
80. Bracker, C. E. and Littlefield, L. J. Structural concepts of host-parasite interfaces. In *Fungal Pathogenecity and the Plant Response* (J. W. Bryde and C. V. Cutting, eds.), Academic Press, London, 1973, pp. 159–313.
81. Hoch, H. C. Mycoparasitic relationships. IV. *Stephanoma phaeospora* parasitic on a species of *Fusarium*. *Mycologia, 70*:370–379, 1978.
82. Held, A. A. *Rozella* and *Rozellopsis*: Naked endoparasitic fungi which dress-up as their hosts. *Bot. Rev.*, *47*:451–515, 1981.
83. Evans, G. H., Cooke, R. C., and Parker, M. L. Penetration of yeast-phase cells of *Mycotypha microspora* by *Piptocephalis fimbriata*. *Trans. Br. Mycol. Soc.*, *82*:540–546, 1984.
84. Manocha, M. S. and Golesorkhi, R. Host-parasite relations in a mycoparasite. V. Electron microscopy of *Piptocephalis virginiana* infection in compatible and incompatible hosts. *Mycologia, 71*:565–576, 1979.
85. Bateman, D. Plant cell wall hydrolysis by pathogens. In *Biochemical Aspects of Plant-Parasite Interactions* (J. Friend and D. R. Threlfall, eds.), Academic Press, New York, 1976, pp. 79–103.
86. Elad, Y., Chet, I., and Henis, Y. Degradation of plant pathogenic fungi by *Trichoderma harzianum*. *Can. J. Microbiol.*, *28*:719–725, 1982.
87. Cooper, R. M. The mechanism and significance of enzymic degradation of host cell walls by parasites. In *Biochemical Plant Pathology* (J. A. Callow ed.), John Wiley, New York, 1983, pp. 101–135.
88. Cuthbert, A., Jeffries, P., and Young, T. W. K. Ultrastructure of the infection of *Phascolomyces articulosus* by *Dimargaris cristalligena*. *Trans. Br. Mycol. Soc.*, *90*:383–389, 1988.
89. Manocha, M. S. and Lee, K. Y. Host-parasite relations in a mycoparasite. II. Incorporation of tritiated N-acetylglucosamine into *Choanephora cucurbitarum* infected with *Piptocephalis virginiana*. *Can. J. Bot.*, *50*:35–37, 1972.
90. Manocha, M. S. and Letourneau, D. R. Structure and composition of host-parasite interface in a mycoparasitic system. *Physiol. Plant Pathol.*, *12*:141–150, 1978.
91. Coffey, M. D. Cytochemical specialization at the haustorial interface of a biotrophic fungal parasite, *Albugo candida*. *Can. J. Bot.*, *61*:2004–2014, 1983.

92. Manocha, M. S. Autoradiography and fine structure of host-parasite interface in temperature-sensitive combinations of wheat stem rust. *Phytopathol. Z.*, *82*:207–215, 1975.
93. Shaw, M. and Manocha, M. S. The physiology of host-parasite relations. XV. Fine structure in rust-infected wheat leaves. *Can. J. Bot.*, *43*:1285–1292, 1965.
94. Coffey, M. D. Flax rust resistance involving the *K* gene: An ultrastructural survey. *Can. J. Bot.*, *54*:1443–1457, 1976.
95. Mendgen, K. Nutrient uptake in rust fungi. *Phytopathology*, *71*:983–989, 1981.
96. Stenzel, K., Steiner, U., and Schonbeck, F. Effect of induced resistance on the efficiency of powdery mildew haustoria in wheat and barley. *Physiol. Plant Pathol.*, *27*:357–367, 1985.
97. Aist, J. R. and Israel, H. W. Papilla formation: Timing and significance of papilla formation during host penetration by *Olpidium brassicae*. *Phytopathology*, *67*:187–194, 1977.
98. Manocha, M. S. and Graham, L. L. Host cell wall synthesis and its role in resistance to a mycoparasite. *Physiol. Plant Pathol.*, *20*:157–164, 1982.
99. Manocha, M. S. and McCullough, C. M. Suppression of host cell wall synthesis at penetration sites in a compatible interaction with a mycoparasite. *Can. J. Bot.*, *63*:967–973, 1985.
100. Manocha, M. S. and Begum, A. Properties of chitin synthase from mucoraceous hosts of a mycoparasite. *Can. J. Microbiol.*, *31*:6–12, 1985.
101. Jan, Y. N. Properties and cellular localization of chitin synthetase in *Phycomyces blakesleeanus*. *J. Biol. Chem.*, *249*:1973–1979, 1974.
102. Peberdy, J. F. and Moore, P. M. Chitin synthetase in *Mortierella vinacea*, properties, cellular localization, and synthesis in growing cultures. *J. Gen. Microbiol.*, *90*:228–236, 1975.
103. Ruiz-Herrara, J., Lopez-Romero, E., and Bartnicki-Garcia, S. Properties of chitin synthetase in isolated chitosomes from yeast cells of *Mucor rouxii*. *J. Biol. Chem.*, *252*:3338–3343, 1977.
104. Balasubramanian, R. and Manocha, M. S. Purification and properties of an acid proteinase from *Choanephora cucurbitarum*. *Can. J. Microbiol.*, *32*:151–155, 1986.
105. Balasubramanian, R. and Manocha, M. S. Cysteine proteinase of *Phascolomyces articulosus*: Purification and properties. *Can. J. Bot.*, *64*:2441–1445, 1986.
106. Balasubramanian, R. and Manocha, M. S. Neutral proteinases from two mucoraceous fungi: A comparative study. *Can. J. Microbiol.*, *33*:281–285, 1987.
107. North, M. J. Comparative biochemistry of the proteinases of eukaryotic microorganisms. *Microbiol. Rev.*, *46*:308–340, 1982.
108. Manocha, M. S. and Balasubramanian. R. In vitro regulation of chitinase and chitin synthase activity of two mucoraceous host of a mycoparasite. *Can. J. Microbiol.*, *34*:1116–1121, 1988.
109. Lopez-Romero, E., Ruiz-Herrera, J., and Bartnicki-Garcia, S. The inhibitory protein of chitin synthase from *Mucor rouxii* is a chitinase. *Biochim. Biophys. Acta*, *702*:233–236, 1982.
110. Correa, J. U., Elango, N., Polachek, I., and Cabib, E. Endochitinase, a mannan-associated enzyme from *Saccharomyces cerevisae*. *J. Biol. Chem.*, *257*:1392–1397, 1982.

111. Wolf, D. H. and Ehmann, C. Studies on a proteinase B mutant of yeast. *Eur. J. Biochem.*, *98*:375–384, 1979.
112. Zubenko, G. S., Mitchell, A. P., and Jones, E. W. Septum formation, cell division, and sporulation in mutants of yeast deficient in proteinase B. *Proc. Natl. Acad Sci. USA 76*:2395–2399, 1979.
113. Ulane, R. E. and Cabib, E. The activating system of chitin synthetase from *Saccharomyces cerevisiae*. *J. Biol. Chem.*, *251*:3367–3374, 1976.
114. Wolf, D. H. Proteinase action in vitro versus proteinase function in vivo: Mutants shed light on intracellular proteolysis in yeast. *Trends in Biochemical Sciences*, *7*:35–37, 1982.

12

BIOLOGICAL CONTROL OF POSTHARVEST FRUIT DISEASES

WOJCIECH JANISIEWICZ *Agricultural Research Service, United States Department of Agriculture, Kearneysville, West Virginia*

I. INTRODUCTION

Limitations of disease control in plants by conventional methods (e.g., chemical control or breeding for resistance) have become apparent in recent years. A number of once effective chemicals are now ineffective due to resistance by pathogens [1,2]. Other chemicals have been banned because of recently recognized hazards to humans and the environment [3]. Major postharvest pathogens of fruit and vegetables have developed resistance to a number of fungicides [4–7]. Most fruit and vegetable storage diseases are caused by fewer than 30 pathogen species, while only a limited number of chemicals are registered for their control. There are few adequate fungicide replacements when pathogens develop resistance.

Biological control as an alternative to chemical control of plant pathogens has generated much interest in recent years [8–13]. However, despite this interest, few biocontrol products are commercially available [14]. Factors contributing to this situation are the inconsistent results in the field, the lack of systematic search for biocontrol agents, and frequent abandonment of projects after initial failures. Biocontrol agents generally reduce losses caused by disease to acceptable levels, though rarely do they eliminate disease entirely. Often biocontrol of disease is less spectacular than chemical control and less appealing to potential users.

During the last few decades most of the attention in biocontrol research has been focused on control of root pathogens, because of wide dissatisfaction with chemical control. Research on plant tops is a less attractive area of research because of more severe and fluctuating conditions, which adversely affect many biocontrol agents, resulting in inconsistent control. However, the ecosystem of above-ground plant parts is less buffered and less stable than that existing in the soil, which may allow better establishment of introduced foreign microorganisms. Generally, less than 1% of the above-ground plant surface is colonized by microbes. Leaf colonization occurs mainly in clusters along leaf veins and anticlinal walls [15]. Thus, biocontrol on above-ground parts of plants has great potential, providing that environmental conditions can be controlled to the antagonists' advantage. The environment in fruit storage buildings, for example, could be modified to enhance growth and survival of antagonists, which could be applied directly to the fruit.

This review focuses on the development of microbial treatment for control of postharvest diseases of fruits. I intend to present the current status of this subject, to discuss approaches and tactics, and to separate fact from speculation, which in the past frequently raised expectations for success to a greater level than was eventually realized.

II. DEVELOPMENT OF POSTHARVEST FRUIT DISEASES

To develop effective biocontrol of postharvest fruit diseases, one must understand disease etiology, particularly time and mechanisms of fruit infection and factors affecting disease severity. I will present some key points in the development of different postharvest diseases of fruit. A number of excellent review papers and books have been written on this subject [4,16–21].

It is important to realize that mature fruit, even after harvest, is still a living plant part which is relatively resistant to most microorganisms [22]. Rots around the lenticels of apples, for example, are almost never noticed at the time of harvest, but can be frequently seen in fruit in storage. As fruit ripens during storage, it may become increasingly more susceptible to various pathogens [22]. Some postharvest pathogens infect when fruit are still attached to the plant [19,21,23–25]. With others, infection is initiated through injuries made at harvest (e.g., cut stems) or through mechanical wounds [18–21].

Pathogenic fungal spores may germinate on fruit surfaces and form appressoria [4]. Hyphae grow from the bottom of an appressorium into the area between the cuticle and outer walls of the epidermal cells. At this point, resistance of fruit tissue prevents further fungal growth. Pathogens can elicit fungitoxic compounds, which may play a significant role in their latent behaviors as in the cases of latency of *Nectria galligena* on apples [26] or *Colletotrichum musae* on bananas [27]. As fruit reaches a certain stage of maturity, the resistance level decreases and the pathogen resumes activity and invades fleshy tissue.

A number of pathogens can infect senescent flower parts early in the growing season and exist in a dormant state until fruit ripens [19,23], when they resume growth and invade fruit. Infections also occur through lenticels. For example, spores of *Gloeosporium album* and *G. perennans* pass through lenticels of apples at certain stages of development during summer and become quiescent after limited growth [28,29]. After a few months of storage, when fruit reaches an advanced stage of maturity, the fungus resumes growth, invades surrounding tissue, and causes characteristic lesions around lenticels.

Some of the most devastating postharvest pathogens enter through mechanical and physiological (cold, heat, oxygen deficiency, etc.) injuries created during and after harvest. Certain injuries resulting from detachment of fruit from plants are unavoidable and may serve as the infection court for some pathogens. Pathogens causing such diseases as stem rot of pears, mango, papaya, and avacado, pedicel rot of pineapple, or crown rot of banana enter through this type of wound [4].

Mechanical injuries resulting from harvest and postharvest handling expose susceptible fleshy tissue to various wound-invading pathogens whose spores are always abundant in the surrounding atmosphere or in storage treatment solutions [6]. These wounds also create a vacant niche for colonization by saprophytes. Sometimes a skin break is not necessary for disease development. *Penicillium expansum* frequently infests apple lenticels, and a simple bruising of the fruit might initiate disease development [17].

After harvest, fruit are frequently subjected to water for various reasons, e.g., hydrocooling peaches, applying antioxidants for protection of apples and pears from superficial scald, removing latex from cut stems of mangoes and bananas, or treatment of fruit with water formulations of antimicrobial chemicals. The solutions are used again many times, and the pathogen inoculum brought in on fruits and containers continually increases [6].

Control of postharvest fruit diseases can be achieved using different strategies. For some diseases, control might start in the field where spraying at bloom or a few days or weeks before harvest would substantially reduce the incidence of postharvest disease. Applying fungicides just before harvest to protect fruit injured during harvest is generally not recommended except when substantial injuries are expected (as in the case of mechanical harvesting).

Sanitation, retention of fruit resistance through inhibition of the ripening process, and maintenance of appropriate storage conditions are frequently applied pathogen control measures which are well developed and understood. More challenging problems, however, involve preventing infection by protecting fruit surface injuries made during harvest and eradicating established infections (latent and quiescent) and inoculum present in lenticels and stomates.

Chemicals are usually applied just after harvest, although in some cases the treatment could be more effective if delayed 10 hours to a few days [30] depending on pathogen and type of fruit. Presumably, this period allows pathogen spores to germinate at the wound site and become more susceptible to subsequently applied fungicides.

Eradication of quiescent infections can be accomplished by systemic fungicide application [17]. Heat treatment alone or in combination with fungicides has also been shown to be an effective control measure in some cases [17]. High temperatures weaken the pathogen and might also stimulate resistance in fruit [31]. Fungicides are applied to harvested fruit by vapors, in solutions, or suspended in water or waxes at various stages of operations (e.g., cleaning, brushing, or waxing) before storage [4,5,7].

III. MICROFLORA OF FRUITS

Fruits are rich in sugars and other nutrients, resulting in a surface favorable to the growth of various microorganisms. Since yeasts appear to be the major component of flora of fruit surfaces, most studies have focused on yeast development. Because of their importance in vinification and postharvest handling, the microflora of grapes [33–36] and apples [32,37–42] have been the most extensively studied. A few reports on other fruit have also appeared [43–51].

The surfaces of fruits are covered with substances which are either produced by the plant, such as sugars, amino acids, and waxes, or deposited there from the environment. Examples of the latter include soil particles, pollen grains, microorganisms, insects and their exudates, such as honeydew. Microbial development on the surface can be influenced by these factors as well as the morphology of the fruit surface [52,53]. Water is a critical factor in microbial development and substances on fruit surfaces play important roles in determining the amount of available moisture and its solute content. The fruit microflora is not confined to the surface and can occur in the center of the fruit, such as the apple core [41]. However, it does not penetrate into flesh unless the tissue is damaged. Damaged tissue not already infected by a pathogen can be colonized rapidly by various microorganisms, particularly yeasts. On Grenadier apples, the populations of acetic-acid and lactic-acid bacteria declined with the rise of yeast populations during development [52]. The inhibition of bacterial growth by yeasts has been ascribed to the secretion of unsaturated fatty acids [54]. On the other hand, promotion of bacterial growth may be due to release of nicotinic acid and thiamine or the reduction of osmotic pressure by the utilization of sugars [52].

Bowen and Beech [37] raised the possibility that some yeasts might be associated with specific cultivars. However, geographic distribution appears to play a major role in the kinds of yeasts colonizing fruit surfaces. They isolated yeasts from bitter-sweet, sharp, and bitter-sharp cultivars of apple and compared them with yeasts from the cultivar Sweet Adolf. *Candida pulcherima* accounted for only 8% of the total yeasts isolated from sweet cultivars, but 80% from other cultivars.

Apple yeast flora may change rapidly after harvest depending on conditions. The yeast *Klockeria apiculata* displaced the predominant *C. pulcherima* when apples were stored in sacks [38]. If the apples were left on the ground, *C. pulcherima* was displaced by *Candida* spp., *Torulopsis* spp., and *Rhodoturula* spp.

In general, the surface microflora of soft fruits such as raspberry, blackberry, loganberry, gooseberry, blackcurrant, blueberry, and strawberry seem to be similar, and specific microorganisms are not associated with specific fruit. However, Dennis [45,55] reported that *Colletotrichum gloeosporioides* and *Didymella applanata* were common on raspberries, whereas *Alterniaria* spp. and *Epicoccum purpurascens* were associated with blackberries.

The antagonistic effects of saprophytic microflora on postharvest fruit pathogens can be demonstrated by the occurrence of what has been termed iatrogenic diseases [56–59]. Extensive use of benomyl to control *Botrytis cinerea* on strawberry resulted in a higher incidence of rots induced by benomyl-resistant *Rhizopus* spp. [60,61]. This is thought to result from the action of benomyl against microorganisms antagonistic to *Rhizopus* spp., since control treatments without fungicides had significantly less rhizopus rot. Other examples of iatrogenic disease include increase in *Alternaria citri* after the use of benomyl to control *Penicillium* spp. [62] and *C. gloeosporioides* [63] on citrus and an increase in *Phytophtora syringae* [64] and *Alternaria tenuis* [65] on apples and pears, respectively, after application of benzimidazols to control *Gloeosporium* spp. and *P. expansum*.

The fruit surface supports the growth of a variety of interacting microorganisms. This suggests that a form of biological control occurs on fruits in nature and that some of these organisms may be potential biocontrol agents for fruit pathogens.

IV. APPROACHES AND TACTICS

The control of postharvest diseases of fruit by means of biocontrol agents raises some important questions. What is the acceptable level of control? Should the effectiveness of postharvest biocontrol be measured by the same criteria used in chemical control? Should the same standards be applied to postharvest control as are used to judge biocontrol of field diseases? Harvested fruit is a product of a season's investment, and substantial losses can result from a small proportion of damaged fruit. The criteria of what constitutes damage may be as important as the damage itself, if only perfect fruit are considered acceptable fruit. These observations suggest that besides economics, consumer preference will play a key role in determining the acceptability of biocontrol.

Different biocontrol strategies are needed for different host-pathogen interactions. In the case of floral infections [19,23,25], for example, *B. cinerea* rot of strawberry [66 – 69] and grape [70 – 72], or dry-eye rot of apple [73 – 76], control starts in the field at bloom. The disease cycle must be broken by precolonization of senescent parts of flowers with antagonists that prevent pathogen infections. Biocontrol of quiescent infections, such as anthracnose caused by *Colletotrichum* spp. on a variety of fruits or *Gloeosporium* spp. and *Nectria galligena* on apples, should also be started in the field similar to chemical control [77,78].

Fruit protection by antagonists having a direct adverse effect on the pathogen would be the most effective measure in these cases. These are examples of where protection of fruits would be most effective if initiated prior to harvest.

For the majority of diseases, however, biocontrol should begin shortly after harvest and before placing the fruit into storage. A primary strategy appears to be the prevention of wound infections. This could be accomplished by precolonization of wounds by antagonists capable of excluding pathogens by competition for nutrients and space, inhibiting pathogen spore germination, growth, and sporulation, or stimulating plants to initiate their own mechanisms of disease resistance. Dip or drench operations are constant sources of inocula, and it is important to realize that fresh pathogen inocula are constantly introduced during these operations [6].

The frequency and rate of application of biocontrol agents depends on the pathogen to be controlled or host to be protected. If the disease cycle can be broken and there is no danger of further infection, only one application may be required. However, repeated applications may be needed if subsequent infections are likely. The frequency of application will also depend on the growth and survivability of the control agent after application. Fruit also becomes more susceptible to infections with ripening and additional applications of antagonists at higher rates may be needed on fruit during storage. The latter practice represents a costly and difficult task. Effectiveness and duration of biocontrol may also depend on the type of propagule used. For example, vegetative bacterial cells or germinated fungal spores are generally more effective than bacterial spores or chlamydospores at the time of application. However, the bacterial spores and chlamydospores can be more effective at the latter stage because of their greater ability to survive. Therefore, the duration of biocontrol may be extended by using mixtures of propagules.

Application of biocontrol agents will depend on the task to be accomplished. For field application, antagonists are sprayed on the plant to runoff. Postharvest application could be accomplished by drenching or dipping fruit in a suspension of antagonists. Pressure infiltration of fruit with antagonist suspensions could be used to affect pathogens located under the cuticle or in lenticels. Antagonists could also be applied with waxes on sorting lines.

Biocontrol agents could be applied alone or in combination with diluted fungicides effective against the same pathogen. A number of fungicide-resistant biocontrol agents have been obtained through various genetic manipulations without losing their effectiveness [59]. Heat treatment of fruits followed by application of antagonists is another possibility. The benefit of integrated control results in the delay of development of the pathogen by diluted fungicides or heat treatment until biocontrol becomes effective. Blending with full-strength fungicides is needed to control diseases other than those targeted by biocontrol agents. This approach was effective in biocontrol of *Monilinia fructicola* with *Bacillus subtilis* on peaches where dicloran was used for the control of *Rhizopus* spp., another important postharvest pathogen [79].

V. CHARACTERISTICS OF BIOCONTROL AGENTS

A number of factors must be considered in the selection of biocontrol agents to be used in the control of postharvest fruit diseases. The type of organism to be used is of primary importance. Postharvest fruit pathogens are mainly necrotrophs and depend on exogenous nutrients for germination and initiation of the pathogenic process. Antagonists which can reduce the nutrient base rapidly could stop this infection process. Fruit wounds are a moist, nutrient-rich environment which provide an excellent food source for various microorganisms. Bacteria and yeasts, in contrast to filamentous fungi, can utilize available nutrients, increase in numbers, and colonize surfaces in a rapid fashion. Rapid colonization is an important feature since fruit pathogens must be inhibited early in the infection process.

Antibiotic-producing microorganisms are also potential candidates for biocontrol agents. Energy needed for production of antibiotics is available in the fruit. The fruit surface has a small buffering capacity and is readily accessible. Biocontrol agents able to elicit resistance responses in the fruit at the wound site could also be valuable, particularly on freshly harvested fruit, where responses to elicitors are strong and relatively rapid. Hyperparasitism is a slow process and requires prior presence of pathogens to increase hyperparasitic populations. Thus, this type of antagonist could be too slow to inhibit aggressive fruit pathogens.

Survivability of the antagonist is a major factor in determining its usefulness. Antagonists must survive and be effective after exposure to postharvest treatments and storage conditions. Ability to colonize wound surfaces is particularly important. Dessication, a major cause of death of microorganisms applied to plant surfaces, could be controlled in storages to the antagonists' advantage.

Relative humidity (RH) of 98% and higher maintained during fruit storage is conducive to the antagonists' survival. Some organisms, notably yeasts, can grow and survive without free water if a high RH atmosphere exists. Low temperatures may allow good antagonist survival, but may decrease its effectiveness. An example is the biocontrol of botrytis rot on strawberries using *Trichoderma* spp. in the field. The failure to control botrytis rot at low temperatures was attributed to the lack of the antagonist growth and subsequent colonization of plant surfaces [68].

Antagonists must be compatible with chemicals used for fruit treatment (e.g., anti-scald diphenylamine) and for handling in storage, and with fungicides used to control other diseases. Biological agents should not be adversely affected by the storage environment particularly under controlled atmosphere (CA) conditions with reduced oxygen and elevated carbon dioxide and nitrogen levels. The adverse effects of CA on some postharvest pathogens is well known [17]. Finally, the antagonist cannot be a health hazard since it is expected that it will be ingested with the fruit.

VI. SELECTING BIOCONTROL AGENTS

A number of steps must be taken if the search for biocontrol agents is to lead to the development of effective and environmentally sound commercial products. To illustrate, an example of the development of biological control of postharvest diseases of pome fruits will be used where the most important steps that have been taken and those planned for the future will be described.

The first and most crucial step, neglected by many investigators, is selecting a search site for potential antagonists. This determines the kind of organisms to be found and used in further studies. In my program, we chose fruits and leaves of apple and pear as search sites for isolates. No consideration was given to using any known antagonists or other culture collections until antagonists from our isolations were found. Culture collections were then compared with isolated antagonists.

Isolations were made throughout a growing season. However, special emphasis was placed on organisms isolated closer to harvest. Isolations were also made from apples in storage for a few months because this site had been subjected to conditions under which a potential biocontrol agent will have to thrive.

Washing fruit and leaves in phosphate buffer with a short period of mild sonication recovered the most microorganisms. Simple isolation media which allowed the growth of a large variety of organisms were selected. The media included nutrient yeast dextrose agar (NYDA), NYDA + 10% apple juice, and malt extract agar, which is particularly suitable for yeast isolation. After colonies were visible on plates, one set of plates was seeded with pathogen spores to detect those organisms that inhibited growth. However, except for the selection of a few organisms with the largest zones of inhibition, little emphasis was placed on this selection process because it has been well documented that inhibition tests in vitro rarely coincide with effective biocontrol in vivo [80]. The second set of plates was used to randomly isolate organisms having different colony characteristics. At this stage, the majority of filamentous fungi were rejected unless they possessed yeast-like growth.

The next step was to intensively screen potential antagonists on fruit for inhibition of rot development under laboratory conditions. The potential antagonists were applied to wounded fruit at high concentrations (3 loops per 5 ml of water). Three fruit were used for each isolate because of the large number of isolates screened. Isolates giving obvious protection were retained for secondary screening. More elaborate tests were possible using fewer isolates.

The minimum effective concentration of the selected antagonists was determined. The compatibility of antagonists with postharvest treatments such as antioxidants, flotation salts (for pears), and surfactants was determined. For the compatible antagonists, we attempted to determine the mode of action and the pathogen life cycle stage they affected. Compatibility among antagonists was tested on fruit treated with a mixture of antagonists effective against different pathogens by challenging them with a spore mixture of the respective pathogens. The ability of antagonists to colonize fruit was tested and attempts were made to enhance their activity. These different tests confirmed the usefulness of the antagonists.

Before going to large, expensive tests, small trials should be conducted under commercial conditions. Fruit should be treated as in regular commercial operations except for deliberately wounding fruit and replacing fungicides with biocontrol agents. After storage in regular or controlled atmosphere, the fruit should be evaluated for biocontrol effectiveness. If successful, work to develop the formulation of biocontrol agents should begin. At this point, estimates of treatment cost and toxicological and safety data should be obtained.

The final step will be a pilot test conducted on a large scale in a commercial storage facility. This test will be expensive and should be done only when the chances for success are high. In this test, effect of storage practices, antagonist dose rates, and formulations on biocontrol effectiveness are determined.

VII. ENHANCING ACTIVITY OF BIOCONTROL AGENTS

Efficacy and fitness of biocontrol agents can be improved by changing not only the chemical and physical environment of fruit, but the antagonist itself. Nutrients are frequently a limiting factor in growth of epiphytic organisms. Application of nutrients to various plants stimulate epiphytic microflora [81]. It is conceivable that selected nutrients applied to fruit may favor antagonist growth. Epiphytic microflora were affected by changes in plant exudates after application of fertilizers [82,83] or systemic fungicides [84]. Similar effects on fruit microflora can be expected and warrants further exploration. Composition of fruit microflora might be under genetic control of plants as with cultivar-specific microflora discussed previously. Thus, plants could be subject to genetic manipulations to favor organisms antagonistic to postharvest pathogens. Moisture, temperature, and gas composition can be easily controlled in a postharvest environment to accommodate antagonists. However, conditions favoring antagonists must be weighed against requirements for maximum physiological storage life of fruit.

Antagonist effectiveness can be greatly affected by growth conditions (e.g., media, aeration, temperature), age of the culture, and type of propagules used. Optimal conditions for all of these components must be determined for each antagonist. Low temperatures used in storage can have a detrimental effect on antagonist performance. Tronsmo and Ystaas [74] obtained biocontrol of dry eye rot on apples at low temperatures by selecting isolate of *Trichoderma harzianum* able to grow under these conditions.

Improving antagonists through genetic manipulation will be limited in the near future to bacterial antagonists since genetic systems in fungi are poorly characterized [85,86]. Bacteria are among the best antagonists of postharvest fruit diseases. Their effectiveness could be increased by incorporation of genes allowing better utilization of compounds of fruit origin or applied compounds favoring antagonists, better tolerance to pesticides, and better colonization of the fruit surfaces under unfavorable conditions. Genetic engineering could also be used to delete pathogenic genes of antagonists which are pathogenic to other crops or to engineer the antagonist from the fruit pathogen to control the wild type organisms.

VIII. FUNGI AS POTENTIAL BIOCONTROL AGENTS

A number of fungi have been shown to have antagonistic activity against various fruit pathogens, however, *Trichoderma* spp. have been most intensely investigated because they have a broad spectrum of activity against various soilborne and above-ground pathogens (Table 1). Different species of this fungus antagonize fungal plant pathogens by producing volatile and nonvolatile antibiotics, by competing for nutrients and space, and by mycoparasitism [87–90]. Thus, it is not surprising that *Trichoderma* spp. have also been used to control fruit diseases such as gray-mold on grapes and strawberries and dry-eye rot of apples. Control in these cases relies mainly on preventing infection of floral parts by *B. cinerea*, from which fruit infection can occur later in the season. While control of this stage of disease development is critical, the antagonist can also repress other stages of pathogen development. Application of the antagonist begins at bloom time and might continue until near harvest. The antagonists are generally applied as a conidial suspension often supplemented with nutrients such as carboxyl-methyl-cellulose, yeast extract, and sucrose, which improve conidial germination and allow them to adhere better to plant surfaces.

Little is known about the nature of antagonism on flowers and fruit, however, it has been hypothesized that nonvolatile antibiotics and hyphae interaction play a major role [72]. As already mentioned, some *Trichoderma* spp. do not grow and control fruit diseases below certain temperatures, thus cold-tolerant species have been selected which provide good control at lower temperatures [74,75]. Fungicide-tolerant *Trichoderma* spp. have been obtained [90,91], which raises the possibility of their use in integrated controls with fungicides in alternating applications and as mixtures.

There are many indications that yeasts can be excellent candidates for biocontrol of fruit diseases. Since the discovery of yeasts, fruits have been recognized as their perfect habitat. The high sugar content and low pH of fruit, particularly at wound sites, allow yeasts to flourish while competing with bacteria and filamentous fungi. Yeasts have a more favorable surface-to-volume ratio than do filamentous fungi, which allows them to absorb nutrients and colonize fruit surfaces more rapidly.

Yeasts can tolerate dry conditions up to 0.62 a_W whereas bacteria generally do not grow below 0.75 a_W. This might be the result of large quantities of extracellular polysaccharides produced by yeasts. A number of yeasts can grow at subzero temperatures on frozen or partly processed fruits, thus it is conceivable that some of the yeasts used in biocontrol could also be active at the lowest storage temperature (-1°C).

A few yeasts with antagonistic activity against pathogens of pome and citrus fruit have been isolated. They provided rot control at low concentrations sufficient to consider them commercial prospects. In no case was

TABLE 1. Potential Fungal Biocontrol Agents for Control of Fruit Pathogens

Biocontrol agent	Pathogen/Disease controlled	Ref.
Acremonium breve	Gray-mold of apples caused by *Botrytis cinerea*	100
Debaryomyces hansenii	Blue-mold and green-mold of grapefruit and oranges caused by *Penicillium italicum* and *P. digitatum*, respectively. Sour-rot of grapefruit and lemons caused by *Geotrichum candidum*	120
	Gray-mold of apples caused by *B. cinerea*	103
Athelia bombacina	Apple scab caused by *Venturia inaequalis*	97
Chaetomium globosum	Apple scab caused by *V. inaequalis*	94–96
Cladosporium herbarum *Pullularia pullulans*	Gray-mold of strawberries caused by *B. cinerea*	66,67
Myrothecium roridum *M. verrcaria*	Green-mold of oranges caused by *P. digitatum*	122,123
Penicillium fumiculosum (nonpathogenic strain)	Interfruitlet corking, leathery pocket, and fruitlet core rot of pineapple caused by *P. fumiculosum*	125
Trichoderma aureoriridi *T. harzianum* *Penicillium* spp. *Aspergillus* spp.	Antracnose of banana caused by *Colletotrichum musae*	126
T. harzianum *T. viride*	Gray-mold of strawberries caused by *B. cinerea*	68,69,76
T. harzianum	Gray-mold of grapes caused by *B. cinerea*	70–72
T. harzianum *T. pseudokoningii*	Dry-eye rot of apple caused by *B. cinerea*	73–75

the mode of action fully determined, but nutrient competition and stimulation of fruit resistance seem to be the most likely possibilities. Unlike most of the filamentous fungi, yeast antagonists would probably be used exclusively to prevent pathogen infection at the wound site where they can grow best and out-compete pathogens.

IX. CONTROL OF THE DISEASES

Different approaches to the microbial control of postharvest diseases of fruits have been tried in recent years [92,93]. Significant progress has been made, which is reflected by the increasing number of investigations reporting successful control. To present this, it is helpful to group various investigations by the kinds of fruit on which the research was conducted.

A. Pome Fruits

One of the most serious diseases of apples in some parts of the world is dry eye rot caused by *B. cinerea* [25]. The fungus infects during bloom through sepals where it can expand to the fruit and cause rot. The development of rot ceases at 0.3 to 3 cm below the sepals and the rot becomes dry [24]. The rot area is always surrounded by a layer of lignin. If infected fruit is stored, the rot can expand as fruit matures and other fruit may be infected, resulting in serious economic loss.

Tronsmo and Raa [73] reported a reduction in incidence of dry eye rot by *Trichoderma pseudokoningii* following artificial inoculation with *B. cinerea*, however, similar treatments failed to control natural infections. This failure was attributed to a lack of tolerance of *T. pseudokoningii* to low temperatures occurring during flowering. When another antagonist, *T. harzianum*, tolerant to low temperatures [74,75], was used with a food base, significant reduction in fruit rot was obtained under field conditions.

Another disease where field infection of fruit may lead to development of disease symptoms in storage is apple scab, caused by *Venturia inaequalis* [21]. This disease causes damage mainly in the field; however, late-season infection of fruit may occur during wet years. The infection may not be visible at harvest, but can develop into small spots in storage. Andrew et al. [94] conducted an extensive screening of microorganisms isolated from apple leaves for antagonistic activity toward *V. inaequalis*. *Chaetomium globosum* was rated as the most effective antagonist after in vitro and in vivo tests. The antagonist inhibited *V. inaequalis* mainly through production of the antibiotic chaetomin [95]. Ascospore application of some of the strains of *C. globosum* to apple seedlings approached a 90% reduction in apple scab severity under controlled conditions. However, at best only 25% disease reduction was obtained in field trials [96]. This loss in effectiveness probably resulted from degradation of the antibiotic under field conditions. The antagonist did not colonize leaf surfaces well; therefore, control probably was a result of antibiotics released from applied spores and not a new synthesis by the antagonist on leaves. In another study, the basidiomycete antagonist *Athelia bombacina* applied to *V. inaequalis* infected apple leaves resulted in decreased production of ascospores under laboratory and field conditions [97]. The antagonist probably acts through competition for nutrients and antibiosis.

TABLE 2. Lesion Development (mm) on Golden Delicious Apples After Protection with Different Concentrations of White Yeast (strain F-43-31) and Inoculation with Various Concentrations of Spore Suspension of *Penicillium expansum* and Reinoculation after 7 Days

Yeast concentration (cells/ml)	Pathogen spore concentration				
	10^7	10^6	10^5	10^4	10^3
Initial inoculation only					
1×10^6	23.0	23.3	23.0	13.3	0.0
2.5×10^6	18.0	21.7	23.3	3.3	4.0
5×10^6	26.0	21.7	15.3	0.0	0.0
1×10^7	6.7	8.7	3.0	0.0	0.0
6×10^7	9.3	2.0	2.6	0.0	0.0
1×10^8	0.0	0.0	0.0	0.0	0.0
0-CK-P[a]	26.3	29.0	24.0	21.3	24.0
0-CK-W[b]	0.0	0.0	0.0	0.0	0.0
d'[c]	14.4	8.7	6.9	5.6	5.5
1 week after reinoculation					
1×10^6	44.0	40.0	49.0	32.0	0.0
2.5×10^6	36.0	40.0	58.7	8.0	12.7
5×10^6	56.7	48.3	41.7	0.0	0.0
1×10^7	13.7	22.3	5.0	0.0	0.0
6×10^7	23.3	8.7	13.3	0.0	0.0
1×10^8	0.0	0.0	0.0	12.3	0.0
0-CK-P[a]	41.7	56.3	46.7	36.7	40.7
0-CK-W-P[d]	10.0	0.0	0.0	0.0	0.0
d'[c]	27.7	23.0	22.4	20.6	17.3

[a]Apple inoculated with *P. expansum* as a control.

[b]Apple treated with sterile water only as a control.

[c]Means within a column of the antagonist-treated apples are compared with the control treated with *P. expansum* according to Dunnett's Procedure ($p = 0.05$).

[d]Apple water control treated with *P. expansum* only during reinoculation procedure.

Source: Ref. 98.

Studies with postharvest application of biocontrol agents to control the two most important postharvest pathogens of pome fruits, *P. expansum* (incitant of blue-mold) and *B. cinerea* (incitant of gray-mold), were conducted using antagonists isolated from apple fruit and leaves [98]. *Pseudomonas syringae* pv. *lachrymans* and white yeast F-43-31 gave excellent protection against *P. expansum* on wounded apples (Table 2). There was a quantitative relationship between concentration of the pathogen used for fruit inoculation and the amount of antagonist needed for disease control. Apples inoculated with the highest concentration of the pathogen (10^7 conidia/ml) were completely protected from disease development if sufficiently high concentrations of the antagonists were used. The protection was ongoing since reinoculation of apple wounds with the pathogen 7–10 days later did not produce new lesions. In similar experiments, control of gray-mold of apples was achieved using *Acremonium breve* [99]. Spores of both pathogens occur in commercial drench tanks at the same time so that drenched fruit can be easily contaminated with both of them. Thus, biocontrol agents should be effective against these pathogens concurrently.

In tests where *P.s.* pv. *lachrymans* and *A. breve* were combined, control of both diseases was obtained (Fig. 1) [99]. As in the earlier tests, pathogen's and antagonist's concentrations were important in effective biocontrol of fruit. Both antagonists were good colonizers and survived on apple wounds very well over an extended period of time [100]. *Acremonium breve* was not adversely affected by exposure to the antioxidant diphenylamine [DPA], which is commonly used in storages for preventing superficial scald, a physiological disorder. However, a decline in the viability of *P.s.* pv. *lachrymans* was observed. This negative effect of DPA was largely eliminated when the surfactant Ortho X-77 was added to the solution. Both antagonists protected apples against the respective pathogens after exposure to these postharvest treatments.

Control of gray-mold and reduction in blue-mold was achieved on Golden Delicious apples and Bosc pears protected with *Pseudomonas cepacia* isolated from apple leaves [101,102]. This bacterium produces a strong antifungal compound in artificial media. The compound was isolated from the bacterium and from growth medium and was identified as pyrrolnitrin. It inhibited growth of both fungi in vitro at concentrations below 1 mg/L. Complete control of gray-mold and blue-mold on apples and pears was obtained with this compound. Therefore, the organism or antifungal compound itself could be considered for commercial disease control for various fruits. Blue-mold and gray-mold of pears were also controlled with *P.s.* pv. *lachrymans*. This antagonist was effective at 24 and 2°C over 5- and 30-day periods, respectively (Table 3, Fig. 2).

Debaromyces hansenii, a yeast effective in controlling postharvest decay on citrus, reduced the size of lesions induced by *B. cinerea* by 63% in apples. Other isolates of this yeast from culture collections reduced the lesion size 12–37% [103].

B. Stone Fruits

Reports related to biological control of *Monilinia* spp. describe in vitro interactions [104,105], antagonism on fruit [79,106–109], and twigs [110,111] in the orchard. Jenkins [106,107] isolated *Bacillus* sp. from conidia of *M. fructicola* obtained from the tree canopy and soil. This bacterium was frequently associated with lysis of the conidia and germ tubes. Significant

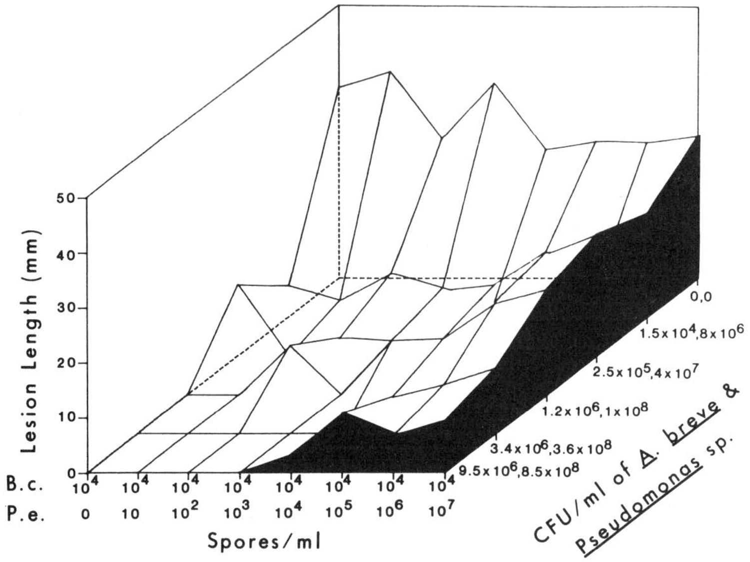

FIGURE 1. Effect of the antagonist mixture of *Acremonium breve* and *Pseudomonas syringae* pv. *lachrymans* (isolate L-22-64) on lesion development resulting from inoculation of wounded Golden Delicious apples with mixtures of spores of *Botrytis cinerea* (B.c.) and *Penicillium expansum* (P.e.). CFU = colony-forming units. (From Ref. 99.)

reduction in brown rot development was observed on wounded peach fruit treated with this bacterium and incubated under optimal conditions for disease development.

Pusey and Wilson [108] used antibiotic producing *B. subtilis* (B3) to control *M. fructicola* on stone fruit after harvest. The treatment was very effective on peaches and apricots and less effective on nectarines and plums. Similar control was achieved when culture filtrates were used. Important for the control of this disease is inhibiting spore germination or early germ tube development. Once infection takes place it progresses regardless of the treatment. The antibiotic produced by the antagonist was described as iturin [112,113]. The bacterium is compatible with commercial fruit waxes and the fungicide dicloran, used for control of rhizopus rot (another important postharvest disease of peach) [79]. Pilot tests for commercialization of this biocontrol agent gave variable results. It appears that efficacy of this antagonist might be improved by developing better fermentation systems and formulations.

Wilson et al. [114] obtained partial control of peach fruit rot caused by *Rhizopus stolnifer* with *E. cloacae*. Ripe fruit were more susceptible to rhizopus rot, but biocontrol was not affected by fruit maturity. Nutrient competition was suggested as a major factor in the interaction between these two organisms [115].

Utkhede and Sholberg [116] tested a number of *B. subtilis* isolates obtained from various sources for antagonistic activity aginst *Alternaria alternata* (incitant of alternaria rot) and *M. fructicola* (incitant of brown

TABLE 3. Control of Gray-Mold on Wounded Red Bartlett Pears Treated with a Cell Suspension of *Pseudomonas syringae* pv. *lachrymans* (L-59-66) and Conidia of *Botrytis cinerea*[a]

	Storage conditions			
	24°C — 5 days		2°C — 30 days	
Antagonist concentration (CFU[b]/ml)	Lesion[c] diameter (mm)	Wounds infected (%)	Lesion diameter (mm)	Wounds infected (%)
Control	37.0	100.0	30.1	85.7
1.5×10^8	3.9	22.3	6.5	29.0
2.2×10^8	0.3	2.0	1.8	12.0
3.7×10^8	0	0	1.0	6.7

[a]Concentration of *B. cinerea* conidia was 1×10^4/ml.

[b]CFU = colony forming units.

[c]Average of three replicates, 15 fruit each, two wounds per fruit.

rot) on cherries after harvest. Most of the isolates significantly reduce the incidence of both diseases, however, they were generally more effective in preventing brown rot than alternaria rot. This could have resulted from a different type of inoculum used, macerated mycelium of *A. alternata* and conidia of *M. fructicola*.

C. Soft Fruits

"Soft fruits," or small fruits, are terms used on the basis of culinary characteristics and include strawberries, raspberries, currants, gooseberries, blueberries, and blackberries. Susceptibility to fungal infection varies considerably among fruits and within cultivars. Major diseases on these fruits are caused by *B. cinerea* (gray-mold), *Mucor piriformis*, and *Rhizopus* spp. [117].

Botrytis cinerea can infect through wounds of mature fruit or through flower parts and remain latent until fruit ripens [117]. Thus, healthy fruit after harvest is obtained by control of the pathogen in the field early in the season [60,61]. Bhatt and Vaughan demonstrated antagonistic activity of *Cladosporium herbarum, Dendrophoma obscurans*, and *Pullularia pullulans* against mycelial growth of *B. cinerea* in vitro [66,67]. They obtained a significant reduction in fruit rot (up to 42% in the case of *C. herbarum* treatments) under greenhouse conditions applying antagonists to flowers at petal fall and then at 7–10-day intervals. However, field trials were ineffective.

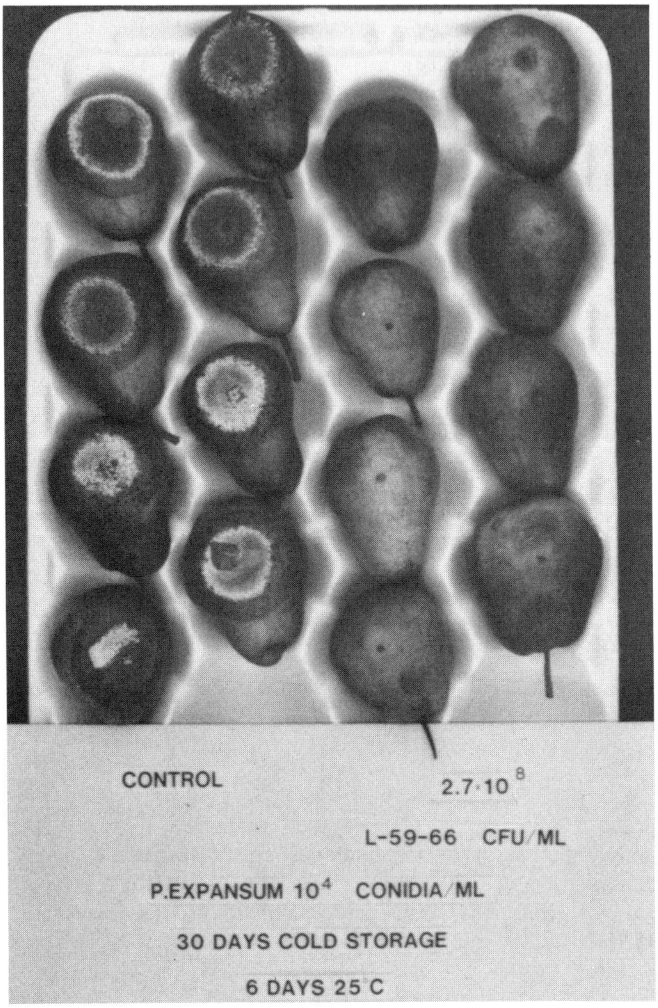

FIGURE 2. Lesion development on wounded Red Bartlett pears dipped for 2 min in a suspension of the antagonist *Pseudomonas syringae* pv. *lachrymans* (isolate L-59-66) and conidia of *Penicillium expansum* (1×10^4 conidia/ml) stored for 30 days at 2°C and then for 6 days at 25°C.

Tronsmo and Dennis [68,69] sprayed strawberry plants with different *Trichoderma* spp. at 14-day intervals from early flowering to within 14 days of harvest. *Trichoderma viride* or *T. harzianum* treatments reduced *B. cinerea* infection to the level obtained with the fungicide dichlofuanid. The antagonists were recovered from harvested fruit. In a later study, they [76] showed that *T. harzianum* sprayed on plants survived for 10 months (including winter months). The protection of strawberries in large-scale experiments was variable, and further work is needed before this biocontrol measure can be tested on a commercial scale.

In grapes, as in strawberries, the length of time between harvest and consumption is largely limited by rot development on fruit. *Botrytis cinerea* is a major field and postharvest pathogen of grape, although other organisms may also contribute to postharvest spoilage [19]. The primary postharvest rots are initiated in the field. Infection occurs through flowers and, under favorable conditions late in the season, directly through uninjured skin of the berry, or through wounds and cracks in the skin.

Biocontrol of gray-mold disease of grapes has been successful in France [70–72]. Control begins at bloom when the saprophytic phase of the disease cycle is broken. Colonization of floral parts by *T. harzianum* significantly reduced subsequent colonization by *B. cinerea.* Three additional sprays with *T. harzianum*, one at the bunching stage of grapes, one at the onset of ripening, and another 3 weeks preharvest, significantly reduced incidence of gray-mold. The antagonist also reduces the development of *B. cinerea* on grapes by hyperparasitism and reduces inoculum for the next season by limiting sclerotia formation on wood. Currently, formulations of the antagonist are produced by French industry. Modern control methods of *B. cinerea* in vineyards are based on models that consider interactions between the pathogen, plant, and climate. It has been suggested that another factor, the antagonist, be included in the model.

D. Subtropical and Tropical Fruits

In vitro antagonism between antibiotic producing *B. subtilis* and citrus pathogens was first conducted by Gutter and Littauer [118] in dual cultures on potato dextrose agar (PDA). The antibiotic produced by the bacterium was fungistatic to most pathogens, but it appears to be fungicidal to *Penicillum digitatum* and a few other fungi. The investigators suggested using this principle for control of citrus fruit rots. Effective control methods at that time caused this finding to go unnoticed.

More recently, Singh and Deverall [119] initiated a study on biological control of *Alternaria citri, Geotrichum candidum*, and *P. digitatum.* This study was initiated after finding an increasingly higher incidence of *A. citri* rots after benomyl treatment, which probably resulted from suppression of antagonistic fungi on fruit treated with benomyl [62]. Neither *A. citri* nor *G. candidum* are susceptible to benomyl. The screening for potential antagonists was designed to isolate microorganisms with greatest antibiotic activity in vitro. *Bacillus subtilis* produced the strongest inhibition of growth of all three pathogens on PDA. Protection against all three pathogens was achieved on fruit dipped in the bacterial suspension and inoculated with the pathogen 48 hours later. However, the 48-hour delay in inoculation may not reflect adequately the efficacy of control that might occur under natural conditions.

Chalutz and Wilson [120] used the antagonistic yeast *Debaromyces hansenii* to control *P. digitatum* on grapefruit, *P. italicum* on grapefruit and oranges, and *G. cadidum* on grapefruit and lemon. The best protection was obtained on grapefruit. *Penicillum digitatum* and *P. italicum* infections were reduced by 94% after 6 days of incubation. The antagonist was isolated from oranges and does not produce antibiotics in culture. *Aureobasidium pullulans, D. hansenii,* and *Pseudomonas syringae* also controlled *P. italicum* and *P. digitatum* on grapefruit and lemon [121].

Appel et al. [122] obtained more than 87% control of *P. digitatum* on lemons by dipping fruit that had been inoculated with the pathogen 18 hours earlier in a conidial suspension (1×10^8 conidia/ml) of *Myrothecium roridum* or *M. verrcaria*. The control was effective at all temperatures tested (12 – 24°C). Mutant strains of *M. roridum* reduced the incidence of green-mold (*P. digitatum*) 93%; however, wild types failed to provide any control on oranges [123].

Biocontrol of diseases of tropical fruit has received little attention, although potential for success might be even greater than with other fruit. Most tropical fruit probably support much higher microbial populations than do temperate zone fruit, similar to the microbial colonization of leaf surfaces [124]. Higher temperatures used for storage of tropical fruit may increase the pool of potential antagonists by including organisms performing well only at those higher temperatures.

The biological control of pineapple fruit diseases caused by *Penicillium fumiculosum*, though it relates more to field control, is worth mentioning here [125]. The pathogen is associated with interfruitlet corking, leathery pocket, and fruitlet core rot of pineapple. Application of spores from nonpathogenic strains (red-pigmented reverse on PDA medium) of this fungus to the hearts of the plants 1 – 4 weeks after chemical forcing reduced severity of all three diseases and increased fruit size. The mode of action of antagonistic strains was not elucidated, but stimulation of phytoalexin production and nutrient competition were suggested as possibilities.

Kanapathipillai [126] tested microorganisms isolated from the surface of banana fruit for control of anthracnose caused by *C. musae*. The disease was controlled with *Trichoderma aureoriridi*, *T. harzianum*, *Penicillium* spp., and *Aspergillus* spp. Culture filtrates of these fungi greatly reduced spore germination and germ tube growth of *C. musae* and controlled anthracnose when applied to the fruit.

Spalding [127] attempted to control anthracnose (incited by *Colletotrichum gloeosporioides* and stem-end rot (incited by *Diplodia netalensis* or *Phomopsis citri*) with *B. subtilis* (B3) on mango fruit after harvest. The antagonist strongly inhibited all three pathogens in vitro. However, dipping or pressure infiltration of infected mango fruit with the antagonist suspension did not control the diseases. Pressure infiltration was used to facilitate penetration of the antagonist beneath the cuticle or into stem tissue where the pathogens reside. Quiescent infections on mango are also difficult to control with fungicides, and heat treatment is frequently used to eradicate the pathogens.

X. CONCLUSIONS

To meet the demand for alternatives to chemical control of postharvest fruit diseases,biological control emerges as a possible option.

Although great progress has been made in recent years, we are still in the pioneering stage of this technology. We know that the biocontrol can work on fruits, but we still do not know its limitations or how well it will fit into current management practices and storage technology. There are indications that a number of antagonists are effective on harvest mature, but not on ripe fruit. The relationship between the physiological stage of fruit and effectiveness of the antagonist must be thoroughly investigated. More emphasis should be placed on biocontrol of minor (weak) pathogens. This can provide valuable information which may be difficult to detect with aggressive nectrotrophic pathogens. There is considerable lack of knowledge of fruit ecosystems. Partial characterization of fruit microflora has been reported in a few cases and successions of organisms were observed, but little is known of what governs this ecosystem. More needs to be known about the chemical environment of both undamaged and damaged fruit surfaces and its effect on epiphytic microflora. This information will be useful not only in selecting, but also in the manipulating of the antagonist to enhance effectiveness. Mechanisms of the antagonist need to be clearly understood so that selection of efficient biocontrol agents can be based on a rational approach.

The integration of antagonists of fruit pathogens with chemical and heat treatment seems to have great potential but remains largely unexplored. The economic feasibility of postharvest biocontrol has to be determined. This will differ greatly from costs calculated for field crop applications because of the high value of stored crops. The effect of biocontrol agents on humans must be critically examined. Interdisciplinary efforts will be required to realize these goals and to bring postharvest biocontrol to commercialization.

ACKNOWLEDGMENTS

Appreciation is expressed to B. D. Horton and W. Conway for their critical reviews of the manuscript, and to L. Yourman for editorial comments.

REFERENCES

1. Ogawa, J. M., Gilpatrick, J. D., and Chiarappa, L. Review of plant pathogens tolerant to fungicides and bacteria. *FAO Plant Prot. Bull.*, *25*:97–111, 1977.
2. Delp, C. J. Coping with resistance to plant disease control agents. *Plant Dis.*, *64*:652–657, 1980.
3. Anonymous. *Regulating Pesticides in Food.* National Academy Press, Washington, D.C., 1987, p. 272.
4. Eckert, J. W. Postharvest diseases of fresh fruits and vegetables — etilogy and control. In *Postharvest Biology and Handling of Fruits and Vegetables* (N. F. Haard and D. K. Salunkhe, eds.), AVI, Westport, CT, 1981, pp. 81–117.
5. Eckert, J. W. and Ogawa, J. M. The chemical control of postharvest diseases: Subtropical and tropical fruits. *Annu. Rev. Phytopathol.*, *23*:421–456, 1985.
6. Spotts, R. A. and Cervantes, L. A. Populations pathogenicity, and benomyl resistance of *Botrytis* spp., *Penicillium* spp., and *Mucor piriformis* in packinghouses. *Plant Dis.*, *70*:106–108, 1986.
7. Eckert, J. W. The chemical control of postharvest diseases: Deciduous fruits, berries, vegetables and root/tuber crops. *Annu. Rev. Phytopathol.*, *26*:433–469, 1988.
8. Baker, K. F. and Cook, R. J. *Biological Control of Plant Pathogens.* W. H. Freeman, San Francisco, 1974, p. 433.
9. Cook, R. J. and Baker, K. F. *The Nature and Practice of Biological Control of Plant Pathogens.* American Phytopathol. Soc., St. Paul, MN, 1983, p. 539.

10. Blakeman, J. P. and Fokkema, N. J. Potential for biological control of plant diseases on the phylloplane. *Annu. Rev. Phytopathol.*, *20*:167–192, 1982.
11. Schroth, M. N. and Hancock, J. G. Selected topics in biological control. *Annu. Rev. Microbiol.*, *35*:453–476, 1981.
12. Vidaver, A. K. Biological control of plant pathogens with prokaryotes. In *Phytopathogenic Prokaryotes* (M. S. Mount and G. H. Lacy, eds.), Academic Press, New York, 1982, pp. 387–397.
13. Windels, C. E. and Lindow, S. E. (eds.) *Biological Control on the Phylloplane*. Amer. Phytopathol. Soc., St. Paul, MN, 1985, p. 169.
14. Upadhyay, R. S. and Rai, B. Biocontrol agents of plant pathogens: Their use and practical constrains. In *Biocontrol of Plant Diseases*, Vol. 1 (K. G. Mukerji and K. L. Garg, eds.), CRC Press, Boca Raton, FL, 1988, pp. 153–165.
15. Collins, M. A. Colonization of leaves by phylloplane saprophytes and their interactions in this environment. In *Microbiology of Aerial Plant Surfaces* (C. H. Dickinson and T. F. Preece, eds.), Academic Press, New York, 1976, pp. 401–418.
16. Dennis, C. (ed.) *Postharvest Pathology of Fruits and Vegetables*. Academic Press, New York, 1983, p. 264.
17. Eckert, J. W. Pathological diseases of fresh fruits and vegetables. In *Postharvest Biology and Biotechnology* (H. G. Hultin and N. Milner, eds.), Food and Nut. Press., Westport, CT, 1978, pp. 161–209.
18. Harvey, J. M., Smith, W. L., and Kaufman, J. *Market Diseases of Stone Fruits: Cherries, Peaches, Nectarines, Apricots, and Plums*. U.S. Gov. Printing Office, Washington, D.C., 1972.
19. Harvey, J. M. and Pentzer, W. T. *Market Diseases of Grapes and Other Small Fruits*. Agric. Handbk. No. 189. U.S. Gov. Printing Office, Washington, D.C., 1960.
20. Smoot, J. J., Houck, L. G., and Johnson, H. B. *Market Diseases of Citrus and Other Subtropical Fruits*. Agric. Handbk. No. 398. U.S. Gov. Printing Office, Washington, D.C., 1971.
21. Pierson, C. F., Ceponis, M. J., and McColloch, L. P. *Market Diseases of Apples, Pears, and Quinces*. Agric. Handbk: No. 376. U.S. Gov. Printing Office, Washington, D.C., 1971.
22. Eckert, J. W. and Ratnayake, M. Host-pathogen interaction in postharvest diseases. In *Postharvest Biology and Handling of Fruit and Vegetables* (N. F. Haard and D. K. Salunkhe, eds.), AVI, Westport, CT, 1981, pp. 81–117.
23. Powelson, R. L. Initiation of strawberry fruit rot caused by *Botrytis cinerea*. *Phytopathology*, *50*:491–499, 1960.
24. Tronsmo, A. and Raa, J. Morphological/cytological description of dry eye rot in apple fruit caused by *Botrytis cinerea* Pers. *Acta Agric. Scandinavica*, *28*:218–220, 1978.
25. Tronsmo, A. and Raa, J. Life cycle of the dry eye rot pathogen *Botrytis cinerea* Pers. on apple. *Phytopathol. Z.*, *89*:203–207, 1977.
26. Swinburne, T. R. and Brown, A. E. The biosynthesis of benzoic acid in Bramley's seedling apples infected by *Nectria galligena* Bres. *Physiol. Plant Pathol.*, *6*:259–271, 1975.
27. Brown, A. E. and Swinburne, T. R. The resistance of immature banana fruits to anthracnose (*Colletotrichum musae*). *Phytopathol. Z.*, *99*:70–79, 1980.

28. Edney, K. L., Fiedler, J. C., Mann, G., and North, C. J. Prestorage heat treatment for reduction of lenticel rotting of apples. *Intern. Inst. Refrig. Paris Bull. Annexe.*, *1*:243–249, 1966.
29. Edney, K. L. An investigation of persistant infection of stored apples by *Gleosporium* spp. *Ann. Appl. Biol.*, *82*:355–360, 1976.
30. Eckert, J. W. and Kolbezen, M. J. Influence of formulation and application method on the effectiveness of benzimidazole fungicides for controlling postharvest diseases of citrus fruit. *Neth. J. Plant Pathol.*, *83*:343–352, 1977.
31. Spotts, R. A. and Chen, P. M. Prestorage heat treatment for control of decay of pear fruit. *Phytopathology*, *77*:1578–1582, 1987.
32. Davenport, R. R. Distribution of yeasts and yeast-like organisms from aerial surfaces of developing apples and grapes. In *Microbiology of Aerial Plant Surfaces* (C. H. Dickinson and T. F. Preece, eds.), Academic Press, New York, 1976, pp. 325–359.
33. Diwakar Rama Wadia, K., Manoharachary, C., and Janaki, C. H. Fruit surface mycoflora of *Vitis vinifera* L. and *Capsicum annuum* L. in relation to their fruit rot diseases. *Proc. Indian Nat. Sci. Acad.*, *49*:371–376, 1983.
34. Guerzoni, E. and Marchetti, R. Analysis of yeast flora associated with grape sour rot and of the chemical disease markers. *Appl. Environ. Microbiol.*, *53*:571–576, 1987.
35. LeRoux, G., Eschenbruch, R., and DeBruin, S. I. The microbiology of South African winemaking. VIII. The microflora of healthy and *Botrytis cinerea* infected grapes. *Phytophylactica*, *5*:51–54, 1973.
36. Singh, J. P. and Kainsa, R. L. Microbial flora of grapes in relation to storage and spoilage. *Indian Phytopathol.*, *36*:72–76, 1983.
37. Bowen, J. F. and Beech, F. W. The distribution of yeasts on cider apples. *J. Appl. Bact.*, *27*:333–341, 1964.
38. Bowen, J. F. and Beech, F. W. Yeast flora of cider factories. *J. Appl. Bact.*, *30*:475–483, 1967.
39. Clark, D. S., Wallace, R. H., and David, J. J. Yeasts occurring on apples and in apple cider. *Can. J. Microbiol.*, *1*:145–149, 1954.
40. Clark, D. S. and Wallace, R. H. *Candida malicola*, nov. sp., isolated from apples. *Can. J. Microbiol.*, *1*:275–276, 1954.
41. Marshall, C. R. and Walkley, V. T. Some aspects of microbiology applied to commercial apple juice production. I. Distribution of microorganisms on the fruit. *Food Res.*, *16*:448–456, 1951.
42. Williams, A. J., Wallace, R. H., and Clark, D. S. Changes in the yeast population on Quebec apples during ripening. *Can. J. Microbiol.*, *2*:645–648, 1956.
43. Buhagiar, R. W. M. *Torulopsis bacarum, Torulopsis pustula* and *Torulopsis multis-gemmis* sp. nov., three new yeasts from soft fruit. *J. Gen. Microbiol.*, *86*:1–11, 1975.
44. Buhagiar, R. W. M. and Barnett, J. A. The yeasts of strawberries. *J. Appl. Bact.*, *34*:727–739, 1971.
45. Dennis, C. The microflora of the surface of soft fruits. In *Microbiology of Aerial Plant Surfaces* (C. H. Dickinson and T. F. Preece, eds.), Academic Press, New York, 1976, pp. 419–432.
46. Stollarova, V. The presence of yeasts and yeast-like microorganisms on cherry fruits (*Cerasus avium (L.) moench*). *Biologia*, *37*:1115–1120, 1982.

47. Stollarova, V. Share of the species *Hanseniaspora uvarum* in natural populations of yeasts. *Biologica, 39*:313–319, 1984.
48. Badr El-Din, S. M. S., Moawad, H., Salem, S. H., Khater, T., and Iskandar, M. Yeasts in the phyllosphere of field-grown plants in Egypt. *Zentralbl. Mikrobiol., 141*:488–492, 1986.
49. Kamra, N. and Maden, M. Ecology of yeasts associated with fleshy fruits. *Microbios Lett., 34*:79–85, 1987.
50. Starmer, W. T., Lachance, M., and Phaff, H. J. A comparison of yeast communities found in necrotic tissue of cladodes and fruits of *Opuntia stricta* on islands in the Caribbean Sea and where introduced into Australia. *Microb. Ecol., 14*:179–192, 1987.
51. Sasaki, Y. and Yoshida, T. Distribution and classification studies on the wild yeasts or budding fungi on the fresh fruits in Hokkaido. *Journ. Facul. Agr. Hokkaido Univ., Sapporo, 51*:194–220, 1959.
52. Last, F. T. and Price, D. Yeasts associated with living plants and their environs. In *The Yeasts*, Vol. 1 (A. H. Rose and J. S. Harrison, eds.), Academic Press, London, 1969, pp. 183–218.
53. Do Carmo-Sousa, L. Distribution of yeasts in nature. In *The Yeasts*, Vol. 1 (A. H. Rose and J. S. Harrison, eds.), Academic Press, London, 1969, pp. 79–105.
54. Uroma, E. and Virtanen, O. E. Antibiotics of yeasts. *Chem. Abstr., 43*:2277, 1949.
55. Dennis, C. and Mountford, J. The fungal flora of soft fruits in relation to storage and spoilage. *Ann. Appl. Biol., 79*:141–147, 1975.
56. Griffiths, E. Iatrogenic plant diseases. *Annu. Rev. Phytopathol., 19*:69–82, 1981.
57. Hislop, E. C. and Cox, T. W. Effects of captan on the non-parasitic microflora of apple leaves. *Trans. Br. Mycol. Soc., 52*:223–235, 1969.
58. Hislop, E. C. Some effects of fungicides and other agrochemicals on the microbiology of the aerial surfaces of plants. In *Microbiology of Aerial Plant Surfaces* (C. H. Dickinson and T. F. Preece, eds.), Academic Press, New York, 1976, pp. 41–74.
59. Vyas, S. C. *Nontarget Effects of Agricultural Fungicides.* CRC Press, Boca Raton, FL, 1988, p. 258.
60. Dennis, C. Effect of pre-harvest fungicides on the spoilage of soft fruit after harvest. *Ann. Appl. Biol., 81*:227–234, 1975.
61. Jordan, V. W. L. The effects of prophylactic spray programmes on the control of pre- and post-harvest diseases of strawberry. *Plant Path., 22*:67–70, 1973.
62. Singh, V. Control of *Alternaria* rot in citrus fruit. *Australas. Plant Pathol., 9*:12–13, 1980.
63. Laville, E. Evolution des pourritures d'entreposage des argrumes avec l'utilisation de nouveaux fongicides de traitement apres recolte. *Fruits., 26*:33–35, 1971.
64. Upstone, M. Evaluation of chemicals for control of Phytophtora fruit rot in stored apples. *Proc. Br. Crop Prot. Conf. Pest and Diseases, 1*:203–205, 1977.
65. Spalding, D. H. Postharvest use of benomyl and thiabendazole to control blue-mold rot development in pears. *Plant Dis. Reptr., 54*:655–657, 1970.

66. Bhatt, D. D. and Vaughan, E. K. Preliminary investigations on biological control of gray-mold (*Botrytis cinerea*) of strawberries. *Plant Dis. Reptr.*, *46*:342–345, 1962.
67. Bhatt, D. D. and Vaughan, E. K. Inter-relationships among fungi associated with strawberries in Oregon. *Phytopathology*, *53*:217–220, 1963.
68. Tronsmo, A. and Dennis, C. The use of *Trichoderma* species to control strawberry fruit rots. *Neth. J. Plant Pathol.*, *83*:449–455, 1977.
69. Tronsmo, A. and Dennis, C. Effect of temperature on antagonistic properties of *Trichoderma* species. *Trans. Br. Mycol. Soc.*, *71*:469–474, 1978.
70. Dubos, B., Bulit, J., Bugaret, Y., and Verdu, D. Possibilities d'utilisation du *Trichoderma viride* Pers. comme moyen biologique de lutte contre la Pourriture grise (*Botrytis cinerea* Pers.) et l'Excoriose (*Phomo-* (*Phomopsis viticola*) Sacc. de la vigne. *C. R. Acad. Agric.*, 1159–1168, 1978.
71. Dubos, B. and Bulit, J. Filamentous fungi as biocontrol agents on aerial plant surface. In *Microbial Ecology of the Phylloplane* (J. P. Blakeman,ed.), Academic Press, London, 1981, pp. 353–367.
72. Dubos, B. Biocontrol of *Botrytis cinerea* on grapevines by antagonistic strain of *Trichoderma harzianum.* In *Current Prespectives in Microbial Ecology* (M. J. Klub and C. A. Reddy, eds.), American Society for Microbiology, Washington, D.C., 1984, pp. 370–373.
73. Tronsmo, A. and Raa, J. Antagonistic action of *Trichoderma pseudokoningii* against the apple pathogen *Botrytis cinerea. Phytopathol. Z.*, *89*:216–220, 1977.
74. Tronsmo, A. and Ystaas, J. Biological control of *Botrytis cinerea* on apple. *Plant Dis.*, *64*:1009, 1980.
75. Tronsmo, A. *Trichoderma harzianum* used as a biocontrol agent against *Botrytis cinerea* on apple. *Les Colloques de l'INRA*, *18*:109–113, 1983.
76. Tronsmo, A. *Trichoderma* used as a biocontrol agent against *Botrytis cinerea* rots on strawberry and apple. *Scientific Reports of the Agricultural University of Norway*, *65*:1–22, 1986.
77. Burchill, R. T. and Edney, K. L. The control of *Gleosporium album* rot of stored apples by orchard sprays which reduce sporulation of wood infection. *Ann. Appl. Biol.*, *72*:249–255, 1963.
78. Corke, A. T. K. and Sneh, B. Antisporulant activity of chemicals toward fungi causing cankers on apple branches. *Ann. Appl. Biol.*, *91*:325–330, 1979.
79. Pusey, P. L., Wilson, C. L., Hotchkiss, M. W., and Franklin, J. D. Compatibility of *Bacillus subtilis* for postharvest control of peach brown rot with commercial fruit waxes, dicloran, and cold-storage conditions. *Plant Dis.*, *70*:587–590, 1986.
80. Andrews, J. H. Strategies for selecting antagonistic microorganisms from the phylloplane. In *Biological Control on the Phylloplane* (C. E. Windels and S. E. Lindow, eds.), The American Phytopathological Society, St. Paul, MN, 1985, pp. 31–44.
81. Morris, C. E. and Rouse, D. I. Role of nutrients in regulating epiphytic bacterial populations. In *Biological Control on the Phylloplane* (C. E. Windels and S. E. Lindow, eds.), The American Phytopathological Society, St. Paul, MN, 1985, pp. 63–82.

82. Weissman, G. S. Effect of ammonium and nitrate nutrition on protein level and exudate composition. *Plant Physiol.*, *39*:947–951, 1964.
83. Sol, H. H. Influence of different N sources on 1) the sugars and amino acids leached from leaves, and 2) the susceptibility of *Vicia faba* to attack by *Botrytis fabae*. *Meded. Fac. Landbouwwet. Rijksuniv. Gent.*, *32*:768–775, 1967.
84. Fokkema, N. J. and de Nooij, M. P. The effect of fungicides on microbial balance in the phyllosphere. *EPPO Bull.*, *11*:303–310, 1981.
85. Napoli, C. and Staskawicz, B. Molecular genetics of biological control agents of plant pathogens: Status and prospects. In *Biological Control in Agriculture IMP Systems* (M. A. Hoy and D. C. Herzog, eds.), Academic Press, New York, 1985, pp. 455–463.
86. Panopoulos, N. J. Tactics and feasibility of genetic engineering of biocontrol agents. In *Microbiology of the Phyllosphere* (N. J. Fokkema and J. Van Den Heuvel, eds.), Cambridge University Press, New York, 1986, pp. 312–332.
87. Dennis, C. and Webster, J. Antagonistic properties of species-groups of *Trichoderma*. I. Production of non-volatile antibiotics. *Trans. Br. Mycol. Soc.*, *57*:25–39, 1971.
88. Dennis, C. and Webster, J. Antagonistic properties of species-groups of *Trichoderma*. II. Production of volatile antibiotics. *Trans. Br. Mycol. Soc.*, *57*:41–48, 1971.
89. Dennis, C. and Webster, J. Antagonistic properties of species-groups of *Trichoderma*. III. Hyphal interaction. *Trans. Br. Mycol. Soc.*, *57*:363–369, 1971.
90. Papavizas, G. C. *Trichoderma* and *Gliocladium*: Biology, ecology, and potential for biocontrol. *Annu. Rev. Phytopathol.*, *23*:23–54, 1985.
91. Papavizas, G. C., Lewis, J. A., and Abd-El-Moity, T. H. Evaluation of new biotypes of *Trichoderma harzianum* for tolerance to benomyl and enhanced biocontrol capabilities. *Phytopathology*, *72*:126–132, 1982.
92. Janisiewicz, W. J. Biological control of diseases of fruits. In *Biocontrol of Plant Diseases*, Vol. 2 (K. G. Mukerji and K. L. Garg, eds.), CRC Press, Boca Raton, FL, 1988, pp. 153–165.
93. Wilson, C. L. and Pusey, P. L. Potential for biological control of postharvest plant diseases. *Plant Dis.*, *69*:375–378, 1985.
94. Andrews, J. H., Berbee, F. M., and Nordheim, E. V. Microbial antagonism to the imperfect stage of the apple scab pathogen *Venturia inaequalis*. *Phytopathology*, *73*:228–234, 1983.
95. Cullen, D. and Andrews, J. H. Evidence for the role of antibiosis in the antagonism of *Chaetomium globosum* to the apple scab pathogen, *Venturia inaequalis*. *Can. J. Bot.*, *62*:1819–1823, 1984.
96. Cullen, D., Berbee, F. M., and Andrews, J. H. *Chaetomium globosum* antagonizes the apple scab pathogen, *Venturia inaequalis*, under field conditions. *Can. J. Bot.*, *62*:1814–1818, 1984.
97. Heye, C. C. and Andrews, J. H. Antagonism of *Athelia bombacina* and *Chaetomium globosum* to the apple scab pathogen, *Venturia inaequalis*. *Phytopathology*, *73*:650–654, 1983.
98. Janisiewicz, W. J. Postharvest biological control of blue-mold on apples. *Phytopathology*, *77*:481–485, 1987.

99. Janisiewicz, W. J. Biocontrol of postharvest diseases of apples with antagonist mixtures. *Phytopathology, 78*:194–198, 1988.
100. Janisiewicz, W. J. Population dynamics of two biological control agents on apple and their compatibility with postharvest treatments. *Phytopathology, 77*:723(Abstr.), 1987.
101. Janisiewicz, W. J. and Roitman, J. Postharvest mucor rot control on apples with *Pseudomonas cepacia. Phytopathology, 77*:1776(Abstr.), 1987.
102. Janisiewicz, W. J. and Roitman, J. Biological control of blue-mold and gray-mold on apple and pear with *Pseudomonas cepacia. Phytopathology, 78*:1697–1700, 1988.
103. Wisniewski, M., Wilson, C., and Chalutz, E. Variability in biocontrol of fruit rots among isolates of the yeast *Debaryomyces hansenii. Phytopathology, 78*:1592 (Abstr.), 1988.
104. Ale-Agha, N., Dubos, B., Grosclaude, G., and Ricard, J. L. Antagonism between nongerminated spores of *Trichoderma viride* and *Botrytis cinerea, Monilia laxa, M. fructigena* and *Phomopsis viticola. Plant Dis. Reptr., 58*:915–917, 1974.
105. Melgarejo, P. and Sagasta, E. M. Destructive morphological changes induced in *Monilinia laxa* by *Penicillium purpurogenum. Trans. Br. Mycol. Soc., 86*:343–346, 1986.
106. Jenkins, P. T. The longevity of conidia of *Sclerotinia fructicola* (Wint.) Rehm under field conditions. *Aust. J. Biol. Sci., 21*:937–945, 1968.
107. Jenkins, P. T. A method to determine the frequency of airborne bacteria antagonistic to *Sclerotinia fructicola. Aust. J. Exp. Agric. Anim. Husb., 8*:434–435, 1968.
108. Pusey, P. L. and Wilson, C. L. Postharvest biological control of stone fruit brown rot by *Bacillus subtilis. Plant Dis., 68*:753–756, 1984.
109. Pusey, P. L., Hotchkiss, M. W., Dulmage, H. T., Baumgardner, R. A., Zehr, E. I., Reilly, C. C., and Wilson, C. L. Pilot tests for commercial production and application of *Bacillus subtilis* (B3) for postharvest control of peach brown rot. *Plant Dis., 72*:622–626, 1988.
110. Melgarejo, P., Carrillo, R., and M.-Sagasta, E. Mycoflora of peach twigs and flowers and its possible significance in biological control of *Monilinia laxa. Trans. Br. Mycol. Soc., 85*:313–317, 1985.
111. Melgarejo, P., Carrillo, R., and M.-Sagasta, E. Potential for biological control of *Monilinia laxa* in peach twigs. *Crop Protect.,* 5:422–426, 1986.
112. McKeen, C. D., Reilly, C. C., and Pusey, P. L. Production and partial characterization of antifungal substances antagonistic to *Monilinia fructicola* from *Bacillus subtilis. Phytopathology, 76*:136–139, 1986.
113. Gueldner, R. C., Reilly, C. C., Pusey, P. L., Costello, C. E., Arrendale, R. F., Cox, R. H., Himmelsbach, D. S., Crumley, F. G., and Cutler, H. G. Isolation and identification of iturine as antifungal peptides in biological control of peach brown rot with *Bacillus subtilis. J. Agric. Food Chem., 36*:366–370, 1988.
114. Wilson, C. L., Franklin, J. D., and Pusey, P. L. Biological control of Rhizopus rot of peach with *Enterobacter cloacae. Phytopathology,* 77:303–305, 1987.

115. Wisniewski, M., Wilson, C. L., and Hershberger, W. Characterization of inhibition of Rhizopus stolnifer germination and growth by *Enterobacter cloacae*. *Can. J. Bot.*, *67*:2317–2323, 1989.
116. Utkhede, R. S. and Sholberg, P. L. In vitro inhibition of plant pathogens by *Bacillus subtilis* and *Enterobacter aerogenes* and in vivo control of two postharvest cherry diseases. *Can. J. Microbiol.*, *32*:963–967, 1986.
117. Dennis, C. Soft fruits. In *Postharvest Pathology of Fruits and Vegetables* (C. Dennis, ed.), Academic Press, New York, 1983, pp. 23–42.
118. Gutter, Y. and Littauer, F. Antagonistic action of *Bacillus subtilis* against citrus fruit pathogens. *Bull. Res. Council of Israel*, *3*:192–196, 1953.
119. Singh, V. and Deverall, B. J. *Bacillus subtilis* as a control agent against fungal pathogens of citrus fruit. *Trans. Br. Mycol. Soc.*, *83*:487–490, 1984.
120. Chalutz, E. and Wilson, C. L. Postharvest biocontrol of green- and blue-mold and sour rot of citrus fruit by *Debaryomyces hansenii*. *Plant Dis.*, *74*:134–137, 1989.
121. Wilson, C. L. and Chalutz, E. Postharvest biological control of Penicillium rots of citrus with antagonistic yeasts and bacteria. *Scientia Horticulturae*, *40*:105–112, 1989.
122. Appel, D. J., Gees, R., and Coffey, M. D. Biological control of the postharvest pathogen *Penicillium digitatum* on Eureka lemons. *Phytopathology*, *78*:1593 (Abstr.), 1988.
123. Gees, R., Droby, S., and Coffey, M. D. Postharvest disease control of green-mold on oranges by a mutant strain of *Myrothecium roridum*. *Phytopathology*, *78*:1593–1594 (Abstr.), 1988.
124. Ruinen, J. The phyllosphere. I. An ecologically neglected milieu. *Plant and Soil*, *15*:81–109, 1961.
125. Lim, T. K. and Rohrbach, K. G. Role of *Penicillium funiculosum* strains in the development of pineapple fruit diseases. *Phytopathology*, *70*:663–665, 1980.
126. Kanapathipillai, V. S. Nonchemical control of *Colletotrichum musae*, the anthracnose pathogen of bananas. *Proc. 5th Int. Congr. of Plant Pathol.*, Kyoto, Japan, 1988, p. 422.
127. Spalding, D. H. Evaluation of various treatments for control of postharvest decay of Florida mangos. *Proc. Fla. State Hort. Soc.*, *99*:97–99, 1986.

13

CURRENT LIMITS TO BIOLOGICAL CONTROL OF FUNGAL PHYTOPATHOGENS

Eric B. Nelson *Cornell University, Ithaca, New York*

I. INTRODUCTION

Biological control of plant pathogens has been the subject of intense excitement and discovery as well as scientific and political controversy over the past decade. In just the past 5–10 years, biological control research in plant pathology has been one of the most rapidly expanding areas within this discipline as evidenced by the number of recent books [1–8] and reviews [9–22] devoted to the subject. However, 15 years ago, Baker and Cook [1] proposed a number of circumstances and misconceptions, stemming primarily from the attitudes of growers and consumers, that had limited the development of biological controls up to that time. These included such things as familiarity with chemical controls, ease of use and fast curative action of chemical controls, amounts of money being spent on fungicide development, etc. They also indicated the general neglect of in-depth studies by plant pathologists beyond the initial curiosity of the biological control phenomenon. Several years later, Schroth and Hancock [20] outlined additional scientific and regulatory constraints to the development of biological controls that included such things as lack of funding for basic research in biological control, improper and contrived experimentation, lack of quantitative data, ignorance of system complexity, rigid government regulations, etc. Although outstanding advances have been made in all aspects of biological control since 1974, many of the constraints that were brought to our attention as many as 15 years ago remain unaddressed today. For example, we are still faced with grower misconceptions, reluctance of growers to adopt alternative disease control practices, ignorance among researchers of the complexities of biological control, rigid government regulations, and so forth.

Since the 1920s, when interest in biological control first arose [4], there have been only three commercial biological controls targeted for plant diseases put into the marketplace in the United States. Two of those, QUANTUM-4000 (Gustafson Chemical Co., Dallas, TX) and DAGGER G (Ecogen, Inc., Langehorne, PA) are targeted for fungal pathogens, while the third, GALLTROL-A (AgBioChem, Inc., Orinda, CA), is effective against *Agrobacterium tumefaciens*, the causal agent of crown gall on stone fruits.

There are only a few commercial products available in Europe and the Middle East.

My purpose in this review is not to cover all of the technical advances in biological control of fungal plant pathogens in the past 60 years or even in the past 10 years; these have been covered adequately in many of the reviews and books cited above. But rather, I wish to highlight some of the more recent advances made in this rapidly developing field and to emphasize current directions and needs of biological control research, implementation, and public relations in light of the expectations that scientists, producers, consumers, and the general public have of biological control in agriculture.

In this review, I hope to address some of the reasons for the slow progress in the development of biological controls for fungal pathogens as well as some of the current advances that, in all likelihood, will propel biological control development forward. I will also consider biological controls in a more narrow sense and discuss only those examples that are mediated by microorganisms or microbial genes and gene products, for this is where much of the intense research is currently being focused.

II. DIRECTIONS AND EXPECTATIONS OF BIOLOGICAL CONTROL

During the early years of biological control following some of the initial and exciting discoveries, it was thought that biological control of plant pathogens might be the end to fungicide use. However, this was quickly met with pessimism as control failures eventually outnumbered successes [1]. As the science of biological control moved from the laboratory to the field, we came to realize that many of our in vitro successes were not always realized in natural systems and that biological controls, in many cases, may never completely eliminate chemical controls in crop production. However, through continued evaluation in agricultural situations over the past decade, it has become evident that biological controls have a very important place in commercial agriculture in providing levels of disease control that, in many cases, facilitate reduced applications of fungicides and, in a few cases, eliminate the need for fungicide applications altogether. In addition, we recognize that biological controls realistically offer important alternatives to plant health management. For example, aside from reducing fungicide inputs, biological control is a potentially important tool in managing fungicide resistance among pathogen populations. This is becoming more of a problem with many of the newer systemic fungicides on the market today. In addition, the success of sustainable agriculture is largely dependent on the integration of biological and other nonchemical means of control into disease management strategies. Recent developments in Integrated Pest Management (IPM) systems are a direct result of the awareness of the importance of biological controls in holistic approaches to plant health management. Most importantly, however, biological control offers an environmentally safe alternative to traditional chemical control programs.

A. The "Climate" for Biological Control

The Need for Reductions in Fungicide Use

As we become increasingly aware of the damage we are inflicting upon ourselves and our planet by pouring noxious gases into the atmosphere, dumping toxic and infective wastes into our oceans, and contaminating our food and water supplies with industrial pollutants and pesticides, more emphasis is being placed on environmental protection. More than ever, the general public is expressing increasing concern over unnecessary exposure to pesticides and other environmental pollutants. In response to this outcry, alternative disease control strategies are now in much more demand to reduce or eliminate the need for environmentally hazardous pesticides. As a result, more pressure is being placed not only on state and federal agencies to institute legislation to help regulate the indiscriminate use of pesticides, but also on producers of food and fiber crops, lawn care companies, landscapers, producers of other nonfood crops, and even homeowners. All are being asked to use pesticides more judiciously than in the past. There could therefore be no better time than the present to encourage the development of biological controls. This has been evident in the tendency of state and federal agencies as well as private companies to increase funding in biological control research and also the interest of grower organizations to fund small-scale biological control research programs.

Economic Considerations of Chemicals Versus Biologicals

Currently, biological controls, along with effective disease detection, diagnosis and sound cultural management techniques, offer the best alternative to reductions in fungicide use. Despite this, surprisingly little consideration has been given to the economics of biological controls as compared with conventional chemical controls. Reichelderfer [24] has addressed this subject with insect pests, yet comparisons for fungal diseases have not been reported. Due to the lack of biological fungicides currently on the market and the absence of biological control systems in use in commercial agriculture, it is difficult to say at this time whether biological controls will prove to be more economical than chemical disease control approaches in the long term. Nevertheless, in anticipation of an economically successful product, several companies in the United States that have traditionally marketed agricultural chemicals are now in the process of actively pursuing research and development of biological controls for a number of different diseases on a number of different commodities.

Impact of Biological Controls

Although much of the current focus of biological control development in the United States is on food crops, the greatest chances of success as well as the largest impact on fungicide use will, in all likelihood, be with nonfood crops such as ornamentals and turfgrasses, since producers and managers of these commodities constitute one of the largest groups of fungicide consumers in the United States and possibly the world. For example, in a 1978 study in New York State [25], it was estimated that approximately 81,000 acres of turfgrass in that state were on golf courses with a total maintenance cost of nearly $43 million. It has been estimated that approximately 10% of that $43 million is spent on fungicide applications alone (G. Ferrentino, personal communication). That represents over $4 million spent on fungicide applications every year on golf courses. It is apparent, therefore, that any reductions in fungicide use with this commodity will have dramatic economic as well as environmental and social impacts.

In addition to high fungicide use crops, biological controls should also be targeted toward those diseases for which effective fungicides are either not available or not totally effective. The recent development of the biological fungicide Dagger G in the United States for control of cotton seedling diseases caused by *Pythium* spp. and *Rhizoctonia solani* is an example of a product designed to fill the gap in control for which chemical fungicides were not totally effective. Trends in the chemical fungicide industry indicate that one of the areas receiving the greatest attention with product and label expansion is specialty crops. These are generally high-value commodities that occupy considerably less acreage than many agronomic or common horticultural crops and have disease problems for which few or no fungicides are currently labeled. Through the targeting of new control strategies for high-value, high-fungicide-use crops, biological controls may more easily and quickly find their way into commercial agriculture.

The Potential for Antagonist Manipulation

Recent developments in molecular biology have tremendously increased our abilities to answer questions about the function of antagonists. As a result of studies on antagonist biology, we are now gaining a better understanding of how antagonists can be manipulated to get the most out of them in the tasks they are being asked to perform. For example, biocontrol technology is developed such that we now have the potential to introduce and establish antagonists on specific plant parts or in specific ecosystems, the techniques to identify genes conferring biological control activity, and the ability to understand gene regulation. Moreover, we have the technology to utilize the genes or gene products of antagonists to more precisely manipulate biological control expression. Undoubtedly, advances in antagonist molecular biology have been one of the principal reasons biological control of fungal pathogens has become more of a reality today than just a few years ago. Future directions in biological control research will inevitably take more of a molecular approach in answering not only biochemical and genetic questions but ecological and epidemiological questions as well.

In summary, the "climate" for the development of biological controls in commercial agriculture is more promising than even before, despite many of the concerns surrounding this rapidly developing technology. Many of the tools and concepts as well as the attitudes of producers and consumers are sufficiently developed to hopefully see the rapid introduction and adoption of biological controls into diverse agricultural production systems and used as a common tool to plant pest management.

III. CURRENT STATE OF BIOLOGICAL CONTROL RESEARCH

A. Systems and Tools Available

The systems that have been most studied in relation to the biological control of fungal plant pathogens have been those utilizing resident antagonists such as suppressive soils and composts [7,16,26] as well as those involving antagonist introductions via seed treatments [e.g., 27–29], cutting inoculations [4,30], colonized organic substrates [26,31–33], trunk and stem injections [34], soil sprays [35], foliar sprays [36], alginate pellets [37], and fluid drilling systems [38]. The approaches and strategies that have been selected in the discovery and evaluation of effective biological control agents have been to select those individual strains that are effective in reducing inoculum, particularly in soils, so that subsequent plant infection is reduced or to select those that are effective in preventing plant infection, either by protecting the plant surface directly or by inducing plant defenses such that infection does not occur [4]. Much of the attention has been placed on the interactions of antagonists with pathogens with little attention given to the antagonist–host interaction. However, much more emphasis is now being placed on this area. Particularly now with the advent of recombinant DNA technology, we no longer must consider systems involving whole microbial cells but may now address systems involved with the expression of antagonist genes in the host plant [39]. It is becoming clear that the host can play a major role in influencing the associations and activities of allied microorganisms, including those with biological control potential [40–43].

Many of the diseases that have been commonly considered as candidates for biological control studies are listed in Table 1 and Table 2. It is apparent from studies of soilborne diseases that the number of pathogens and antagonists studied is restricted to relatively few genera. This has been fortunate in that we now have an enormous amount of information available regarding the biocontrol potential of these genera. As a result, much of the information needed to implement biological controls using these systems has, in many instances, already been obtained. However, there has also been a fair amount of redundancy in research efforts.

Current research trends indicate that fewer biological control systems are being investigated, but those receiving attention are being studied in much greater detail than in the past. The situation is somewhat different with foliar diseases. Although the numbers of researchers in this field is limited, diverse pathogen and antagonist genera have been studied. As a result of this diversity accompanied by a limited research effort, progress in biocontrol of fungal foliar diseases has been slow. Additionally, many of the currently available fungicides are quite effective, thus reducing the impetus for a search for alternatives. Therefore, a large body of the information needed to implement foliar biological controls using these systems has generally not been generated. However, recent advances with biological frost protectants [44,45] may help move foliar disease biocontrol forward.

B. Target Sites in Pathogen Disease Cycles

Without an understanding of the processes involved in pathogenesis among pathogens in various pathosystems, a meaningful approach to biological control cannot be effectively conceived or implemented. Knowledge of these processes is vital to our understanding of how, when, and where biological control activity must be expressed. This will also influence not only what agents we use, but how we apply and manipulate them in a crop management system. As our knowledge of the biological control of soilborne fungal diseases develops, it is apparent that we know very little about the processes in soil by with host–pathogen interactions are initiated. Our knowledge of propagule dormancy, the factors triggering their activation and germination, and the preinfection behavior of fungal pathogens in response to and in association with the host plants is all poorly understood. However, more studies are currently being directed toward this important aspect of the host-plant–antagonist interaction.

Dormant Propagules

In general, only those biological control systems involving hyperparasites have been directed toward the eradication of dormant fungal

TABLE 1. Commonly Studied Antagonist–Soilborne Fungal Pathogen Systems

Diseases	Causal agents	Antagonists evaluated	References
Damping-off	*Pythium, Rhizoctonia*	*Pseudomonas, Bacillus, Enterobacter, Pythium, Trichoderma, Gliocladium, Penicillium, Laetisaria,*	27–29, 31, 32, 67, 83, 87, 88, 112, 120, 125, 134, 136–138, 140, 146
Root and crown rots	*Pythium, Thielaviopsis, Rhizoctonia, Sclerotium, Phytophthora, Fusarium, Sclerotinia, Gaeumannomyces*	*Pseudomonas, Sporidesmium, Bacillus, Trichoderma, Coniothryium, Gliocladium, Verticillium,*	30, 33, 46, 48, 52, 53, 55, 57, 58, 60, 62, 79, 80, 82, 85, 133, 135, 142–145
Vascular wilts	*Fusarium, Verticillium*	*Bacillus, Talaromyces, Trichoderma, Pseudomonas*	34, 35, 75, 77, 139, 141

TABLE 2. Commonly Studied Antagonist–Foliar Fungal Pathogen Systems

Diseases	Causal agents	Antagonists evaluated	References
Leaf spots and blights	*Cercospora, Venturia, Alternaria*	*Pseudomonas, Bacillus, Chaetomium, Athelia, Alternaria*	8, 36, 106, 147
Rusts	*Poria, Cronartium, Uromyces, Puccinia Melampsora*	*Scytalidium, Tuberculina, Verticillium, Eudarluca, Darluca, Pseudomonas*	12, 165
Powdery mildews	*Sphaerotheca, Podosphaera*	*Ampelomyces*	12
Stem blights/Cankers	*Nectria*	*Trichoderma, Bacillus*	12, 148
Fruit rots	*Botrytis, Monilinia, Penicillium, Mucor, Rhizopus*	*Trichoderma, Enterobacter, Bacillus, Coniothryium, Gliocladium, Pseudomonas*	51, 149, 150–152, 154
Tree decays	*Heterobasidion, Cryphonectria, Ophiostoma, Phellinus, Chondrostereum*	*Peniophora, Trichoderma, Fusarium, Pseudomonas, Bacillus, Cryphonectria*	153

propagules, primarily sclerotia and oospores. If analogies were to be made with chemical or physical controls, this approach would be similar to fumigation or solar heating, which is directed toward reducing preplant soil populations of fungal pathogens. Sclerotia of *Sclerotinia* and *Botrytis* spp. are parasitized by species of *Coniothyrium* [46, 48–51], *Sporidesmium* [52–54], and *Gliocladium* [55–57], where reductions in soil sclerotial populations are directly related to reduced disease severity. *Gliocladium* was shown to parasitize sclerotia of *R. solani* [58]. Likewise, Kuter et al. [59] recovered a number of isolates of *Gliocladium virens* and *Trichoderma* spp. from decaying sclerotia of *Rhizoctonia solani* recovered from hardwood bark composts. Similarly, *Trichoderma harzianum* was found to parasitize sclerotia of *Sclerotium rolfsii* [60–62] in culture and in natural soil. To my knowledge, bacterial parasites of fungal sclerotia have not been described.

In addition to hyperparasites of fungal sclerotia, those attacking other survival propagules have been described. Sneh et al. [63] found oospores of *Phytophthora, Pythium*, and *Aphanomyces euteiches* recovered from natural soils to be parasitized by a number of different microorganisms including chytrids, hyphomycetes, actinomycetes, and bacteria. Likewise, Pratt [166] observed the frequent recovery of various organisms believed to be hyperparasites from germinating oospores of *Sclerospora sorghi.* Hoch and Abawi [64] demonstrated the mycoparasitism of oospores of *Pythium ultimum* by *Fusarium merismoides.*

Much of the work cited above was done nearly 10 years ago and has been receiving less and less attention. It is unlikely that much applied research effort will continue with sclerotial or oospore parasites since the establishment and maintenance of hyperparasite populations in soil for biological control is, in most cases, not an economical venture at this time. The one exception might be systems with *Sporidesmium sclerotivorum* [52]. The primary constraint to this approach is that in order to place hyperparasites in contact with target propagules, mass inoculations of soil are required. Currently this is not a practical agricultural management practice. The most effective means of introducing antagonists will be to place them on the plant surface at the site where they need to be active. This has been accomplished largely by the use of seed treatments.

Propagule Germination and Ectotrophic Colonization

Dormant propagules of soilborne as well as many foliar fungal pathogens require the proper host stimulus in order for germination to occur and a compatible host–pathogen interaction to be initiated. For soilborne fungal propagules, seed and root exudates provide the stimulus for such germination. Interference with exudate stimulants is a promising approach by which to interfere with pathogen activity and achieve biological control. A few studies have been directed toward this important step in pathogenesis. Although Ahmad and Baker [65] observed reductions in sporangium germination of *P. ultimum* in the presence of *Trichoderma*-treated seeds as compared with untreated seeds, the mechanism by which this phenomenon occurred was not elucidated. It is believed to occur through competition for exudate stimulants.

Pathogen propagules germinate quickly in response to exudates released from germinating seeds and developing roots [66–68]. Yet, once germination occurs, pathogens are at an extremely susceptible stage for biological control. This may be one reason biological seed treatments have been so effective in suppressing seed and root rots. It has been reported in several studies that seed treatments with biological control agents can also reduce the build-up of pathogen biomass on plant surfaces prior to infection [27,67,69]. Deacon [15] has recently pointed out that all of the commercially successful examples of biological control systems are targeted for those early stages of pathogen development (i.e., preinfection stages of pathogen development).

C. Antagonist Function

There has been considerable interest in the past few years in identifying the mechanisms by which bacterial and fungal antagonists suppress fungal plant pathogens. In the search for mechanisms of biological control, it has become apparent that biological control activity is not a simple process mediated by a single microbial metabolite. The capacity of an organism to simply produce a fungicidal compound in and of itself is not always sufficient to explain biological control activity and cannot be considered "the mechanism." Rather, biological control activity involves a series of events not unlike those involved in plant pathogenesis or in the establishment of symbiotic associations with each of these events mediated by a specific biologically active compound under genetic or enzymatic regulation. Current research in this area has focused on establishing the involvement of certain metabolites in the biological control process, particularly antibiotics, siderophores, and enzymes.

Major microbial metabolites involved in disease suppression by biological control agents include antibiotics, siderophores, NH_3, enzymes, and hydrogen cyanide. The role of antibiotics in biological control processes has been reviewed recently [70] and will therefore not be covered here. Additionally, siderophores are another important microbial metabolite with important roles in biological control processes. Their biochemistry and role in biological control have also been reviewed recently [71] and only recent developments will be covered. Siderophores have been implicated in the biocontrol processes by which fluorescent pseudomonads suppress plant infection by *Fusarium* spp. [72–77] and *Gaeumannomyces graminis* var. *tritici* [74,78] but until recently had not been demonstrated to play a role in the suppression of other fungal diseases. Loper [69] employed Tn5 mutants of a siderophore-producing strain of *Pseudomonas fluorescens* to demonstrate that these compounds were important in the suppression of Pythium damping-off of cotton. However, siderophore production could not account for all of the suppression observed. This difference could also not be explained on the basis of differential population levels between mutants and parental strains in the spermosphere. Becker and Cook [79] demonstrated that chemically generated mutants of fluorescent pseudomonads deficient in siderophore production could no longer protect wheat seedlings from Pythium root rot and damping-off. With some of the strains used in that study, antibiotics were also produced.

It is unlikely that siderophores act alone in the biocontrol process. For example, with *G. graminis* var. *tritici*, antibiotics produced by fluorescent pseudomonads, in addition to siderophores, have been shown to play a role in biological control [80–82]. However, from genetic data as well as extractions from soil, antibiotics are the major Pseudomonad metabolite involved in pathogen suppression in this system [82]. Likewise, antibiotics also play a role in the suppression of *Pythium* spp. by fluorescent pseudomonads [83,84].

Siderophores, along with hydrogen cyanide, have also been implicated in the suppression of *Thielaviopsis basicola* by fluorescent pseudomonads [85]. Mutagenized or naturally occurring HCN^- strains of *P. fluorescens* are less inhibitory to *T. basicola* than HCN^+ wild-type strains [86]. Complementation of HCN^- mutants or wild-type strains with cloned HCN^+ genes restores biological control activity [86].

Other mechanisms by which biological control agents inhibit fungal pathogens are becoming evident in some studies. For example, in studies with the biological control agent *Enterobacter cloacae*, it has been found that certain strains are particularly effective inhibitors of Pythium seed infection of cucumber, cotton, and rye [27,40,87,88]. Seeds of these plants release low levels of carbohydrates during initial stages of germination. However, these same strains of *E. cloacae* are not effective seed protectants when coated onto snapbean, lima bean, corn, and soybean seeds, which release high levels of sugars during germination [40]. Possible reasons for this differential behavior have become apparent from in vitro studies of the interaction between *E. cloacae* and *P. ultimum*.

In paired cultures of *E. cloacae* and *P. ultimum*, untreated bacterial cells adhere firmly to hyphal cell wells resulting in over 50% reductions in fungal growth [40]. However, pretreating cell suspension of *E. cloacae* with either D-form monosaccharides and disaccharides, selected amino sugars, or β-linked glucosides prevents cells from attaching to hyphae and inhibiting fungal growth. On the other hand, pretreating bacterial cells with L-form or methylated sugars or α-linked glucosides does not interfere with the ability of *E. cloacae* to attach to hyphae and inhibit the growth of *P. ultimum*. It appears from these results that bacterial attachment to hyphal cell walls involves the binding of a bacterial cell receptor to specific sugar residues, possibly glucosides, associated with the fungal cell wall and that this phenomenon is intimately linked to fungal growth inhibition and biological control activity. To lend support to this hypothesis, addition of attachment-blocking sugars to *E. cloacae*-treated cucumber seeds or pretreatment of bacterial cells with these sugars prior to seed coating renders *E. cloacae* ineffective as a biological seed protectant [40].

Although the elimination of biological control activity in the presence of specific sugars has been attributed to the capacity of certain sugars to interfere with the ability of *E. cloacae* to bind to hyphae of *Pythium* spp. by blocking bacterial surface receptors involved in hyphal attachment [40], other mechanisms of inhibition may also be operative. For example, the same sugars that block bacterial attachment to hyphae also interfere with biological control activity of *E. cloacae* by impairing the ability of these strains to deaminate proteins and amino acids and liberate volatile ammonia which is toxic to *P. ultimum* [88]. It is clear that the mechanism of biological control in this system is quite complex and may involve an array of bacterial metabolites and cell–cell interactions.

In other recent studies, it has been found that *Talaromyces flavus*, an effective biological control agent against *Verticillium dahliae* [35], produces a metabolite with antibiotic properties that is active in inhibiting growth and microsclerotial germination of the pathogen [89]. It has been discovered that the active compound has enzymatic properties [70] and has recently been identified as glucose oxidase [90]. Although glucose or glucose oxidase alone is not inhibitory to *V. dahliae*, the reaction product, hydrogen peroxide, is quite toxic to microsclerotial germination [70].

D. Establishment on Plant Parts

It is evident that if biological control agents are not able to colonize sites where pathogen entry into plants occurs, the ability of strains to control that target pathogen is greatly diminished. Thus, much work during the past 5 years has been directed toward improving the colonization of plant parts, particularly plant roots, by introduced biological control agents.

Below-Ground Parts

Much of the success of biological control agents targeted for root-infecting pathogens can be attributed to their ability to colonize plant roots. This is opposed to agents targeted for seed-rotting and damping-off pathogens where root colonization is not likely to be an important character. Much of the current work with root colonization by biological control agents has been reviewed recently [22], and only the highlights and more recent developments will be covered here. Much of our recent advances in predicting the colonization potential and survivability of introduced strains have come with a better understanding of the mechanisms underlying root colonization and rhizosphere competence [91]. It has recently been discovered that colonization of plant roots by fluorescent pseudomonads can be initiated when bacterial cells bind to exudate agglutinins present in the rhizosphere and on the root surface [92–94]. These agglutinins have greater affinities for saprophytic strains than pathogenic strains [92]. Mutants of *Pseudomonas putida* strains that lack the ability to bind to root agglutinins are incapable of extensive colonization of bean and cucumber roots [42,95]. A direct relationship has also been observed between the ability of strains to colonize roots and their ability to suppress Fusarium wilt of cucumber [42]. Chao et al. [96] reported similar findings in that the ability of rhizosphere bacteria to colonize roots, when introduced as seed treatments, was directly related to their ability to bind to agglutinins present in pea root washings. Although the pseudomonad agglutinins are apparently a glycoprotein complex, the nature of the other bacterial agglutinins are unknown.

Other recent studies have focused on the bacterial receptors potentially responsible for binding of cells of *Pseudomonas, Klebsiella*, and *Enterobacter* to root agglutinins [97–101]. Receptors in these strains are apparently proteinaceous and located on fimbriae. Direct relationships have been shown between the degree of fimbriation and root-colonizing ability of introduced strains [97]. Future studies are likely to be directed toward understanding the nature of this receptor complex and how it is genetically regulated.

Other studies on factors possibly contributing to root-colonization capacity of introduced biological control agents have failed to reveal additional characters important in root colonization. Bacteria motility and chemotaxis to seed exudates are apparently not related to successful root colonization [102]. However, there are contradictory results in the literature [103]. Similarly, the role of osmotolerance among fluorescent pseudomonads [22] and cellulase production among *Trichoderma* strains [91,104] as an aid to root colonization is currently under debate. Interestingly, gram-positive bacteria, when applied as seed treatments, do not effectively colonize roots of developing seedlings [105]. The reasons for these differences are not understood.

Above-Ground Parts

Little research is available on colonization of above-ground plant parts by introduced biological control agents. Much of our failure with foliar biocontrol systems may be a result of our lack of understanding of antagonist establishment and proliferation on above-ground plant parts. Blakeman and Fokkema [12] have indicated that an antagonist threshold of 10^4-10^5 cells per cm^2 of leaf area is necessary before biological control will be expressed. In a few systems that have been studied, the mortality of introduced antagonists on leaves is quite high [23,106]. However, in field experiments with ice-nucleating *P. syringae* strains, populations of introduced antagonists fluctuated from 10^5 to 10^7 cells per gram fresh weight of leaves and flowers [45]. Their spatial distribution on the leaf or flower surface was not determined.

Much of the information currently available on above-ground colonization by microorganisms has been developed from studies of foliar plant pathogens [107]. From studies of the distribution of *Pseudomonas syringae* pathovars following inoculation of cucumber terminal buds, it has been observed that bacterial cells become distributed uniformly on the youngest leaves, erratically on intermediate-aged leaves, and sparsely on the oldest leaves after 3 days of incubation at 24°C [108]. Most of the cells on intermediate and oldest leaves were distributed along leaf veins. Perhaps these types of studies will be useful in future studies of antagonist population dynamics on foliar plant surfaces.

Seed treatments provide an attractive means of introducing antagonists for the colonization of above-ground plant parts by antagonists. However, no published reports of the feasibility of this technology have been noted. Recently work in my laboratory (Mason and Nelson, unpublished) has revealed that populations of the antagonist *Enterobacter cloacae* can reach high populations on above-ground parts of cotton seedlings grown in sterile soil systems (Table 3) and can survive for extended periods of time. Further studies in the field and with appropriate disease pressure should reveal the significance of these populations to biological control of foliar plant pathogens.

E. Strain Improvement

Various approaches have been taken to improve the activity of biological control agents either to directly increase their suppressive abilities or to increase their rhizosphere competence. These approaches have been summarized recently [109]. Harman et al. [28] were able to improve the performance of *Trichoderma hamatum* when applied to seeds by incorporating specific food bases into seed treatments. Nelson et al. [110] used the same strategy to increase the biological control activity of *Trichoderma harzianum* and *T. koningii* strains against *P. ultimum*. Activity was increased by as much as 48% by incorporating specific food bases such as organic acids, polysaccharides, and polyhydroxy alcohols into seed coatings. Likewise, Harman and Taylor [29] recently reported dramatic increases in the efficacy of *Trichoderma* and *Enterobacter* seed treatments by priming treated seeds in a solid matrix and adjusting seed surface pH to levels favorable for activity. In solid matrix priming, seeds are treated with the appropriate biological control agent and placed in a finely ground lignite or coal substance that is moistened sufficiently to allow seed to imbibe water but to not allow radicle emergence. Once seeds are planted in soil, seedlings

TABLE 3. Above-Ground and Below-Ground Distribution of Populations of *Enterobacter cloacae* (E6) on Developing Cotton Seedlings after Application as a Seed Treatment[a]

Distance from seed (cm)[b]		Log CFU/g dry wt of cm root segments at various weeks after planting				
		1	2	3	4	5
Above ground	PL[c]	—	—	—	4.8e	1.6l
	CT[d]	8.0a	8.0a	7.8b	4.8e	3.9gh
	8	—[e]	—	—	—	2.7jk
	7	—	—	3.2h	4.1e	2.7jk
	6	—	—	3.4h	3.1f	2.6jk
	5	—	—	3.8gh	4.0e	2.7jk
	4	—	4.5f	3.8fg	4.4e	2.5k
	3	7.0d	5.3e	3.9f	4.6e	3.1i
	2	7.5b	5.5d	3.7g	4.7e	3.2i
Below ground	1	6.9e	5.2e	5.5e	5.8d	3.7h
	-1	5.9f	7.4b	8.2a	6.4cd	6.8a
	-2	7.3c	7.4b	8.3a	8.0a	6.5b
	-3	7.5b	5.3de	7.9b	6.7bc	5.3d
	-4	5.9f	3.8g	7.9b	7.3b	5.8c
	-5	—	6.2c	7.4c	6.1cd	4.6e
	-6	—	—	6.1d	6.1cd	6.6ab
	-7	—	—	—	—	4.3f
	-8	—	—	—	—	4.0g

[a]Initial populations of *E. cloacae* on seed was 3.9×10^9 CFU/seed; seed planted 1 cm below soil surface and incubated at 25°C; plants were watered from below.

[b]Numbers preceded with a (-) sign indicate centimeter sections of root taken from below the seed; other numbers indicate centimeter section of hypocotyl taken from above the seed.

[c]PL = first primary leaf.

[d]CT = cotyledon (seed coat remained attached in most cases).

[e](—) = plant hadn't developed to that height.

Numbers in each column followed by the same letter are not significantly (p = 0.05) different according to Duncan's Multiple Range Test.

Source: Mason and Nelson, unpublished

emerge more rapidly and are more tolerant of *Pythium* infection. The level of control provided by solid matrix-primed *Trichoderma*-treated seeds is substantially better than a conventional fungicide seed treatment.

Further enhancement of fungal biological control agents applied to seeds has been achieved through nonspecific genetic manipulations [11]. Whereas traditional genetic approaches for manipulating biocontrol fungi have involved UV and chemical mutagenesis [104,112], sexual and parasexual hybridization [113], and protoplast fusion [114], the most promising strategies involve gene transfer using fungal transformation [115]. However, biocontrol genes in many instances are not known and have not been cloned. Likewise, transformation systems for many of our known filamentous biocontrol fungi have not yet been developed. I suspect that this will be an area of intense research in the next few years.

Genetic approaches for enhancing the performance of bacterial biocontrol agents have also not yet received wide attention. However, a classic example of how this approach can be used is with the development of delta–endotoxin–producing *Pseudomonas* strains for control of lepidopteran insects [116,117]. Since toxigenic strains of *Bacillus thuringiensis*, the source of delta-endotoxin, are poor root colonizers, the toxin is produced only on above-ground plant parts where it is susceptible to biodegradation and inactivation [118]. To increase the efficacy of the toxin, target the application, and minimize the need for repeated applications, toxin-producing capabilities were introduced into *Pseudomonas* strains, which are ideally adapted to root colonization and survival in soils [22]. By placing the genes for delta-endotoxin production in pseudomonads, more prolonged and efficacious protection of root systems can be achieved. As for fungal systems, however, the genes involved in biological control activity in currently described bacterial systems are just now beginning to be identified.

F. Variability

Variability in performance of biological control agents as compared with chemical fungicides is well known among those involved in biological control research. This constitutes one of the greatest obstacles in biological control development. The reasons for this variability are many and may involve unfavorable soil edaphic factors [119], inability to compete for iron [120], inability to compete with indigenous microflora [121], inability to proliferate on and in plant tissues and in the rhizosphere [65], or inability to express biological control activity [40,122]. Our success in reducing variability will come only from an understanding of the ecology of antagonistic strains and the mechanisms by which biocontrol activity is expressed and regulated. Much of the impetus for understanding the mode of action of antagonists is to develop technologies to overcome some of the variability inherent in biocontrol systems.

IV. CURRENT LIMITS TO OUR UNDERSTANDING OF BIOLOGICAL CONTROL

A. Mechanisms of Biological Control

With most of the known biological control systems being studied today, we still know very little about the genes, gene products, and regulatory systems involved in the expression of biological control activity. In many instances, we have identified only one or two characters of any particular agent that are involved in biological control. To date, our most advanced knowledge of biological control processes is with *Pseudomonas* systems [86] and with *Trichoderma* systems [17,87]. The process, as we understand it, by which *Trichoderma* strains inhibit several soilborne plant pathogens is outlined in Figure 1.

Our knowledge of mechanisms will ultimately allow us to manipulate antagonist behavior more effectively as well as to reduce much of the variability inherent in biological control systems. For example, knowledge that *Enterobacter cloacae* is an effective biological control agent against *Pythium* spp. in part because of its sugar-reversible attachment to hyphae of *P. ultimum* helps to explain why these strains are effective in suppressing Pythium seed rot and preemergence damping off only when coated onto low-sugar-exudation seeds such as cucumber or rye [40]. Had this information not been generated, evaluations of biological control activity on various plant species would have appeared simply as extreme and unknown

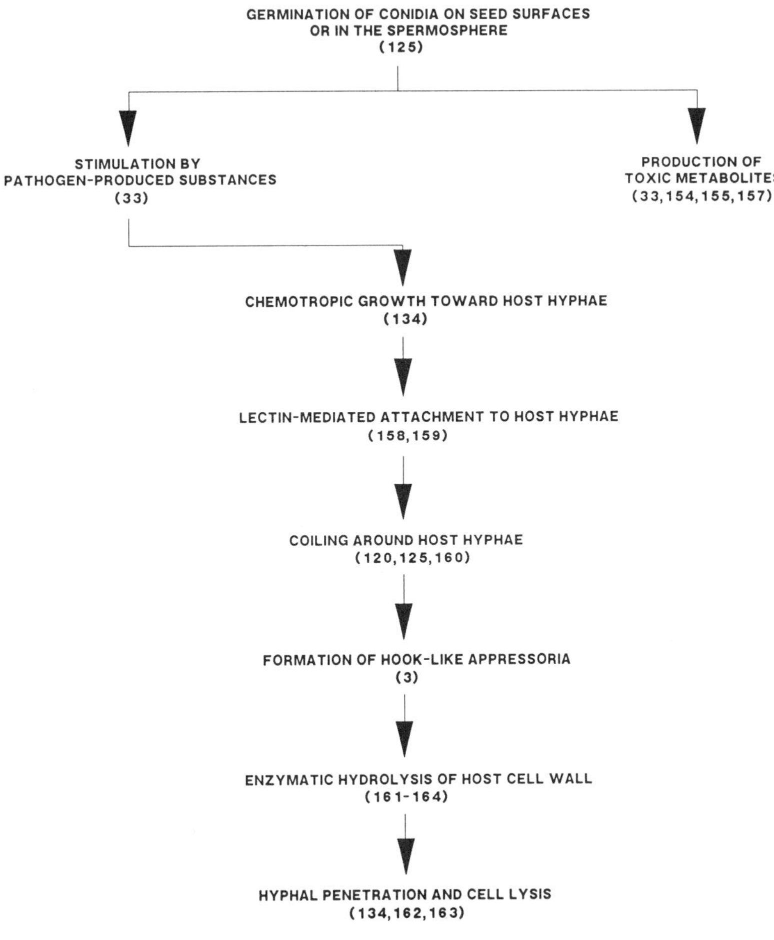

FIGURE 1. Proposed mechanisms by which *Trichoderma* spp. inhibit *Rhizoctonia solani*, *Pythium* spp., and *Sclerotium rolfsii*.

variability in activity making predictions of biocontrol performance impossible. Likewise, knowledge that siderophores are produced primarily in alkaline, iron-limiting soils helps to explain why fluorescent pseudomonads are more effective antagonists in these and not acidic soils [74]. Had this not been known, performance in various soils would have been considered simply as extreme variability of unknown origin. Similarly, predictions of action of antagonists have provided and will continue to provide us with important biological and biochemical information that should greatly increase our chances of success in developing effective commercial biological control agents.

B. Antagonist Ecology

With the exception of a limited number of studies on root colonization and population dynamics in soil, we know very little about the ecology of introduced biological control agents. Little is known of the interactions of biological control agents with other plant and soil microorganisms [120,123], their interactions with different plant genotypes [43], and the environmental factors influencing survival [124] and biological control expression [27, 40,125]. This is particularly true for the biological control of foliar fungal diseases but applies to soilborne diseases as well. A much increased research effort is needed in these areas before we will be able to reliably maintain and manipulate biological control agents.

In the past we have taken an idealistic approach to the discovery and study of effective antagonists in expecting that the best antagonists will be those that control multiple target pathogens and will survive and be effective for extended periods of time. Spurr and Knudsen [21] have emphasized the futility in searching for such ideal biological control agents. A more logical and reasonable approach is to target antagonists for specific pathogens under specific situations and allow them to survive long enough to get the job done but not long enough to cause environmental concern or to increase the chances of mutations developing that could make the management of introduced strains much more difficult. Additionally, biological controls may actually have the greatest impact on those pathogens that may be more restricted in their host range or their tissue preferences since targeting agents for these pathogens should be more feasible. In general, our mind set has been to expect everything from one antagonist instead of targeting antagonists for specific functions. Over the years it has become increasingly difficult to find or develop strains with extremely broad range activity [4]. An alternative widely discussed among scientists for many years is to combine strains with specific activity in order to achieve the desired broad spectrum control. Yet rarely have studies of this type been conducted, and rarely have the efficacies of individual antagonists been compared with those of mixtures. Until microbial diversity and community structure in soils and in association with plants is understood, the research focus will be to understand the nature of specific antagonist activity so that integration of various and diverse agents into disease management strategies might be more practical.

C. Preinfection Pathogen Behavior

In order to successfully develop effective and sustainable biological control strategies, a more thorough understanding of the host–pathogen interaction is necessary. Particularly with seed- or root-infecting pathogens, chemical control measures are generally not effective once plant penetration occurs. Therefore, most conventional disease control strategies, aside from the use of host plant resistance and some systemic fungicides, are designed to protect seeds or developing roots from infection by soilborne pathogens by interfering with the prepenetration activity of these microorganisms. Despite the importance of the prepenetration phase of pathogenesis, little is known concerning the spermosphere and/or rhizosphere behavior of soilborne pathogens prior to plant infection. Our knowledge of soilborne pathogen behavior is based largely upon indirect and correlative observations of population fluctuations and plant symptom expression. Under these experimental conditions, specific pathogen behavior can only be inferred. Likewise, differences in the activity of soilborne pathogens internal and external to the plant can rarely be discerned. Our ability to predict and ultimately manage the interaction of soilborne pathogens with seeds or roots will be determined by the extent to which changes in the rhizosphere and spermosphere environment as well as propagule behavior in relation to those changes can be defined and predicted.

D. Integration with Other Agricultural Practices

Little is known about how other production practices unique to a given commodity will influence the performance of introduced antagonists. Certainly the questions of how fertilizers, pesticides, tillage, organic amendments, etc. influence pathogen activity and disease development have been addressed. Many of the benefits derived from these practices directly result from influences on resident antagonists. However, the direct relationship these practices have on antagonist activity and disease suppression have been studied only in a few cases. For example, in studies of the effects of nitrogen fertilization on the incidence of take-all disease of wheat, it was found that applications of ammonium fertilizers increased the populations of pseudomonads antagonistic to *Gaeumannomyces graminis* var. *tritici* on wheat roots [126–128]. Conversely, Hubbard et al. [120] found that pseudomonads antagonistic to the biocontrol agent *Trichoderma hamatum* were responsible for its failure to control Pythium damping-off of pea in low iron soils. By increasing the available iron in those soil, pseudomonads were no longer inhibitory to *T. hamatum* and it effectively protected seedlings from damping-off. More studies of this nature are needed in order to gain information on ways to maintain biological control activity and reduce variability in ways compatible with crop production systems.

Although small-scale field and laboratory or greenhouse trials can generate some of this information, larger-scale field trials on growers' premises using their equipment may provide more meaningful information concerning compatibility with other production practices. As more and more biocontrols find their way into the hands of producers, many of the questions concerning feasibility of use and crop production compatibility may come to light. It is essential in the future that grower experiences with biological controls be documented so that formulation, delivery, and maintenance of high activity can be approached in a more meaningful way.

E. Delivery Systems

Stack et al. [129] have recently reviewed some of the traditional strategies used for delivering biological control agents to production systems. Others were mentioned earlier in this review. Of those particularly targeted for soilborne plant pathogens, seed treatments and in-furrow treatments appear to be the most promising. However, as biological control technology continues to evolve, other delivery systems will become more feasible, particularly as our knowledge of genes conferring biological control becomes more developed. No longer do we need to consider delivery systems involving various preparations of microorganisms. With the development of efficient plant transformation systems [130], we are now able to consider the delivery of biological control genes into plant genomes to generate pathogen suppressive plants. Other technologies currently developed for the control of root-feeding insects are those involving the insertion of biological control genes into xylem-colonizing bacteria (i.e., *Clavibacter xyli*) to achieve systemic protection instead of limited site protection of plant parts (P. Carlson, Crop Genetics International, personal communication). This technology could also be developed for the control of root-infecting and vacular wilt pathogens.

Our understanding of delivery systems is still in its infancy. Yet, as our understanding of biological control mechanisms grows, our conceptions of new delivery systems designed to promote and maintain maximum biological control expression will also have to evolve.

F. "Natural" Biological Control Systems

As mentioned above, the trend among biocontrol researchers and practitioners is toward introducing both native and exotic, laboratory-manipulated antagonists into agricultural production systems. However, we have not taken full advantage of natural biological control systems provided through the use of suppressive soils and composted organic wastes. For years suppressive soils have provided significant levels of disease control for Fusarium wilt diseases of vegetables, Phytophthora root rots of Eucalyptus, take-all disease of wheat, and many others [7]. Yet we have not exploited the full potential of disease control inherent with these systems. Undoubtedly, suppressive soils are more widespread in agroecosystems than we currently recognize. Future research on the identification and exploitation of suppressive soils should be encouraged.

Similarly, organic wastes are becoming more and more of a problem for many enterprises and communities to dispose of. Composting has proven to be a safe, clean means of converting unwanted wastes into agriculturally useful disease-suppressive products [26]. For example, composted hardwood bark has been used successfully for years in Ohio nurseries for the control of root diseases of container-grown woody ornamentals [131]. The potential for use of these and other composts in the ornamentals industry is particularly exciting.

V. CURRENT LIMITS TO IMPLEMENTATION

A. Commercialization

Scher and Castagno [132] outlined several criteria that must be met in order for the successful development of a commercially viable biofungicide. In part, the product should be:

1. Needed by someone in the marketplace
2. Technically feasible

3. Economically feasible
4. Competitively attractive
5. Acceptable to environmentalists and regulatory agencies
6. Compatible with the company's activities and interests

Currently, limits to the commercialization of biological controls imposed by criteria 1, 2, and 6 have apparently been overcome in many instances. Likewise, the economic feasibility of biofungicide development seems quite attractive. For example, the development of microbial fungicides in the United States is estimated to take approximately 2 – 3 years at a cost of less than $500,000, while chemical fungicides are estimated at approximately 7 – 10 years at a cost exceeding $20 million (T. Currier, personal communication). Current costs of applying one of the more recent biofungicides, Dagger G, is estimated at approximately $9.50 per acre (T. Currier, personal communication). If these figures can be used as a general standard with which to base future product economics, biologicals will be extremely attractive assuming the product is used by a large portion of the target audience. Because biological control products are relatively new to the marketplace, it is not yet clear, particularly in the United States, whether they will compete well with chemical fungicides and be acceptable to environmetalists and regulatory agencies. Although it is encouraging that more and more biofungicides are being put into the marketplace, suggesting that at least some of the concerns of environmentalists and regulatory agencies are waning, time will tell whether biologicals turn out to be effective enough to either replace or augment traditional fungicides. It is critical that some of these initial products consistently perform as well as or outperform conventional fungicides if the future of biological fungicides is to be bright.

B. Grower Acceptance and Public Awareness

Biological controls will likely gain acceptance slowly among growers unless biologicals are shown to provide economic incentives as well as demonstrations of their ease of use. Growers have become so accustomed to chemical fungicides that they expect to be able to use biological controls as they would conventional fungicides at a comparable cost. In addition, biological controls, in contrast with some chemical controls, rarely result in dramatic plant recovery following application — an attribute considered desirous by many growers. It is encouraging that many users of conventional fungicides are aware of biological controls, yet their adoption of simple and readily available integrated approaches to plant disease management is slow. Educational efforts are needed to inform growers and the general public of the virtues and limitations of biological control approaches. This is perhaps one of our biggest challenges in biological control implementation today, since acceptance by growers and the general public will ultimately determine the success or failure of biological controls in commercial agriculture.

VI. CONCLUSIONS

There are many limitations to the development and implementation of biological controls for fungal plant pathogens. Many of our limitations today are the same as those posed 10 – 15 years ago. However, the science of biological control has made extraordinary technical advances in the last few years, overcoming many of these developmental hurdles. Despite these technical advances, and most likely as a result of them, public concern over "new technologies" associated with biological control implementation have reached an all-time high. As our technology develops, public education will be an ever-increasing challenge.

Today biological control is on the verge of a new era of discovery and commercialization. One must believe that the benefits of biological controls, once realized, will overcome any limitations currently impeding development and ultimately change the way in which disease control is approached.

REFERENCES

1. Baker, K. F. and Cook, R. J. *Biological Control of Plant Pathogens.* Am. Phytopathol. Soc., St. Paul, 1974.
2. Burges, H. D. (Ed.) *Microbial Control of Pests and Plant Diseases.* Academic Press, New York, 1981.
3. Chet, I. *Innovative Approaches to Plant Disease Control.* Wiley, New York, 1987.
4. Cook, R. J. and Baker, K. F. *The Nature and Practice of Biological Control of Plant Pathogens.* Am. Phytopathol. Soc., St. Paul, 1983.
5. Mukerji, K. G. and Garg, K. L. (Eds.) *Biocontrol of Plant Diseases,* Vols. 1 and 2. CRC Press, Boca Raton, FL, 1988.
6. Papavizas, G. C. (ed.) *Biological Control in Crop Production.* Allanheld Osmun, London, 1981.
7. Schneider, R. W. (ed.) *Suppressive Soils and Plant Diseases.* Am. Phytopathol. Soc., St. Paul, 1982.
8. Windels, C. E. and Lindow, S. E. (eds.) *Biological Control on the Phylloplane.* American Phytopathological Society, St. Paul, 1985.
9. Baker, K. F. The future of biological and cultural control of plant disease. In *Challenging Problems in Plant Health* (T. Kommedahl and P. H. Williams, eds.), Am. Phytopathol. Soc., St. Paul, 1983, pp. 422–430.
10. Baker, K. F. Evolving concepts of biological control of plant pathogens. *Annu. Rev. Phytopathol., 25*:67–85, 1987.
11. Baker, R. Biological control of plant pathogens: Definitions. In *Biological Control in Agricultural IPM Systems* (M. A. Hoy and D. C. Herzog, eds.), Academic Press, Orlando, 1985, pp. 25–39.
12. Blakeman, J. P. and Fokkema, N. J. Potential for biological control of plant diseases on the phylloplane. *Annu. Rev. Phytopathol., 20*:167–192, 1982.
13. Brown, M. E. Seed and root bacterization. *Annu. Rev. Phytopathol., 12*:181–197, 1974.
14. Burr, T. J. and Caesar, A. Beneficial plant bacteria. *CRC Crit. Rev. Plant Sci., 2*:1–20, 1984.
15. Deacon, J. W. Biocontrol of soil-borne plant pathogens with introduced inocula. *Phil. Trans. R. Soc. Lond. B, 318*:249–264, 1988.
16. Hornby, D. Suppressive soils. *Annu. Rev. Phytopathol., 23*:23–54, 1985.
17. Papavizas, G. C. Trichoderma and Gliocladium: Biology, ecology, and potential for biocontrol. *Annu. Rev. Phytopathol., 23*:23–54, 1985.

18. Papavizas, G. C. and Lumsden, R. D. Biological control of soilborne fungal propagules. *Annu. Rev. Phytopathol., 18*:389–413, 1980.
19. Schroth, M. N. and Hancock, J. G. Disease-suppressive soil and root-colonizing bacteria. *Science, 216*:1376–1381, 1982.
20. Schroth, M. N. and Hancock, J. G. Selected topic in biological control. *Annu. Rev. Microbiol., 35*:453–476, 1981.
21. Spurr, H. W., Jr., and Knudsen, G. R. Biological control of leaf diseases with bacteria. In *Biological Control on the Phylloplane* (C. E. Windels and S. E. Lindow, eds.), American Phytopathological Soc., St. Paul, 1985, pp. 45–62.
22. Weller, D. M. Biological control of soilborne plant pathogens in the rhizosphere with bacteria. *Annu. Rev. Phytopathol., 26*:379–407, 1988.
23. Knudsen, G. R. and Spurr, H. W., Jr. Management of bacterial populations for foliar disease biocontrol. In *Biocontrol of Plant Diseases*, Vol. I (K. G. Mukerji and K. L. Garg, eds.), CRC Press Inc., Boca Raton, 1988, pp. 83–92.
24. Reichelderfer, K. H. Economic feasibility of biological control of crop pests. In *Biological Control in Crop Production* (G. C. Papavizas, ed.), Allenheld, Totowa, NJ, 1981, pp. 403–417.
25. Gruttadaurio, J., Hardy, E. E., and Lieberman, A. S. Turfgrass in New York State: A billion dollar agricultural production and service industry. Cornell University Department of Floriculture and Ornamental Horticulture Publication, Ithaca, 1978.
26. Hoitink, H. A. J. and Fahy, P. C. Basis for the control of soilborne plant pathogens with composts. *Annu. Rev. Phytopathol., 24*:93–114, 1986.
27. Nelson, E. B. Biological control of Pythium seed rot and pre-emergence damping-off of cotton with *Enterobacter cloacae* and *Erwinia herbicola* applied as seed treatments. *Plant Dis., 72*:140–142, 1988.
28. Harman, G. E., Chet, I., and Baker, R. Factors affecting *Trichoderma hamatum* applied to seeds as a biocontrol agent. *Phytopathology, 71*:569–572, 1981.
29. Harman, G. E. and Taylor, A. G. Improved seedling performance by integration of biological control agents at favorable pH levels with solid matrix priming. *Phytopathology, 78*:520–525, 1988.
30. Baker, R., Hanchey, P., and Dottarar, S. D. Protection of carnation against Fusarium stem rot by fungi. *Phytopathology, 68*:1495–1501, 1978.
31. Hadar, Y., Chet, I., and Henis, Y. Biological control of *Rhizoctonia solani* damping-off with wheat bran culture of *Trichoderma harzianum*. *Phytopathology, 69*:64–68, 1979.
32. Martin, S. B., Hoch, H. C., and Abawi, G. S. Population dynamics of *Laetisaria arvalis* and low temperature *Pythium* spp. in untreated and pasteurized beet field soils. *Phytopathology, 73*:1445–1449, 1983.
33. Sivan, A., Elad, Y., and Chet, I. Biological control effects of a new isolate of *Trichoderma harzianum* on *Pythium aphanidermatum*. *Phytopathology, 74*:498–501, 1984.
34. Hall, T. J., Schreiber, L. R., and Leben, C. Effects of xylem-colonizing *Bacillus* spp. on Verticillium wilt in maples. *Plant Dis., 70*:521–524, 1986.

35. Marois, J. J., Johnston, S. A., Dunn, M. T., and Papavizas, G. C. Biological control of Verticillium wilt of eggplant in the field. *Plant Dis.*, *66*:1166–1168, 1982.
36. Cullen, D., Berbee, F. M., and Andrews, J. H. *Chaetomium globosum* antagonizes the apple scab pathogen, *Venturia inaequalis*, under field conditions. *Can. J. Bot.*, *62*:1814–1818, 1984.
37. Fravel, D. R., Marois, J. J., Lumsden, R. D., and Connick, W. J., Jr. Encapsilation of potential biocontrol agents in an alginate-clay matrix. *Phytopathology*, *75*:774–777, 1985.
38. Conway, K. E. Use of fluid-drilling gels to deliver biological control agents to soil. *Plant Dis.*, *70*:835–839, 1986.
39. Willmitzer, L. Genetic engineering for plant disease resistance. In *Innovative Approaches to Plant Disease Control* (I. Chet ed.), Wiley-Interscience, New York, 1987, pp. 353–364.
40. Nelson, E. B., Chao, W. L., Norton, J. M., Nash, G. T., and Harman, G. E. Attachment of *Enterobacter cloacae* to hyphae of *Pythium ultimum*: Possible role in the biological control of Pythium pre-emergence damping-off. *Phytopathology*, *76*:327–335, 1986.
41. Stenhagen, G., Alborn, J., and Lundborg, T. Variation of root and microflora exudates in genotypes of barley. In *Allelochemicals: Role in Agriculture and Forestry* (G. R. Waller, ed.), American Chemical Society Symposium Series 330, American Chemical Society, Washington, 1987, pp. 76–88.
42. Tari, P. H. and Anderson, A. J. Fusarium wilt suppression and agglutinability of *Pseudomonas putida*. *Appl. Environ. Microbiol.*, *54*:2037–2041, 1988.
43. Weller, D. M. Effects of wheat genotype on root colonization by a take-all suppressive strain of *Pseudomonas fluorescens*. *Phytopathology*, *76*:1059, 1986.
44. Lindow, S. E. The role of bacterial ice nucleation in frost injury to plants. *Annu. Rev. Phytopathol.*, *21*:363–384, 1983.
45. Lindow, S. E. Competetive exclusion of epiphytic bacteria by ice *Pseudomonas syringae* mutants. *Appl. Environ. Microbiol.*, *53*:2520–2527, 1987.
46. Ahmed, A. H. M. and Tribe, H. T. Biological control of white rot of onion (*Sclerotium cepivorum*) by *Coniothyrium minitans*. *Plant Pathol.*, *26*:75–78, 1977.
47. Chet, I. *Trichoderma*-Application, mode of action, and potential as a biocontrol agent of soil-borne plant pathogenic fungi. In *Innovative Approaches to Plant Disease Control* (I. Chet, ed.), Wiley-Interscience, New York, 1987, pp. 137–160.
48. Huang, H. C. Importance of *Coniothyrium minitans* in survival of sclerotia of *Sclerotinia sclerotivorum* in wilted sunflower. *Can. J. Bot.*, *55*:289–295, 1977.
49. Huang, H. C. and Hoes, J. A. Penetration and infection of *Sclerotinia sclerotiorum* by *Coniothyrium minitans*. *Can. J. Bot.*, *56*:2243–2246, 1976.
50. Trutman, P., Keane, P. J., and Merriman, P. R. Reduction of sclerotial inoculum of *Sclerotinia sclerotiorum* with *Coniothyrium minitans*. *Soil Biol. Biochem.*, *12*:461–465, 1980.
51. Turner, G. J. and Tribe, H. T. On *Coniothyrium minitans* and its parasitism of *Sclerotinia* species. *Trans. Br. Mycol. Soc.*, *66*:97–105, 1976.

52. Adams, P. B. Comparisons of isolates of *Sporidesmium sclerotivorum* in vitro and in soil for potential as active agents in microbial pesticides. *Phytopathology, 77*:575–578, 1987.
53. Adams, P. B. and Ayers, W. A. Factors affecting parasitic activity of *Sporidesmium sclerotivorum* on sclerotia of *Sclerotinia minor. Phytopathology, 70*:366–368, 1980.
54. Ayers, W. A. and Adams, P. B. Mycoparasitism of sclerotia of *Sclerotinia* and *Sclerotium* species by *Sporidesmium sclerotivorum. Can. J. Microbiol., 25*:17–23, 1979.
55. Huang, H. C. Control of Sclerotinia wilt of sunflower by hyperparasites. *Can. J. Plant Pathol., 2*:26–32, 1980.
56. Moody, A. R. and Gindrat, D. Biological control of cucumber black root rot by *Gliocladium roseum. Phytopathology, 67*:1159–1162, 1977.
57. Tu, J. C. *Gliocladium virens*, a destructive mycoparasite of *Sclerotinia sclerotivorum. Phytopathology, 70*:670–674, 1980.
58. Tu, J. C. and Vaartaja, O. The effect of the hyperparasite (*Gliocladium virens*) on *Rhizoctonia solani* and on Rhizoctonia root rot of white beans. *Can. J. Bot., 59*:22–27, 1981.
59. Kuter, G. A., Nelson, E. B., Hoitink, H. A. J., and Madden, L. V. Fungal populations in container media amended with composted hardwood bark suppressive and conducive to Rhizoctonia damping-off. *Phytopathology, 73*:1450–1456, 1983.
60. Henis, Y., Adams, P. B., Lewis, J. A., and Papavizas, G. C. Penetration of sclerotia of *Sclerotium rolfsii* by *Trichoderma* spp. *Phytopathology, 73*:1043–1046, 1983.
61. Henis, Y. and Papavizas, G. C. Factors affecting germinability and susceptibility to attack of sclerotia of *Sclerotium rolfsii* by *Trichoderma harzianum* in field soil. *Phytopathology, 73*:1469–1474, 1983.
62. Elad, Y., Barak, R., and Chet, I. Parasitism of *Sclerotium rolfsii* sclerotia by *Trichoderma harzianum. Soil Biol. Biochem., 16*:381–386, 1984.
63. Sneh, B., Humble, S. J., and Lockwood, J. L. Parasitism of oospores of *Phytophthora megasperma* var. *sojae, P. cactorum, Pythium* sp., and *Aphanomyces euteiches* in soil by oomycetes, chytridiomycetes, hyphomycetes, actinomycetes, and bacteria. *Phytopathology, 67*:622–628, 1977.
64. Hoch, H. C. and Abawi, G. S. Mycoparasitism of oospores of *Pythium ultimum* by *Fusarium merismoides. Mycologia, 71*:621–625, 1979.
65. Ahmad, J. S. and Baker, R. Implications of rhizosphere competence of *Trichoderma harzianum. Can. J. Microbiol., 34*:229–234, 1988.
66. Nelson, E. B. Rapid germination of sporangia of *Pythium* species in response to volatiles from germinating seeds. *Phytopathology, 77*:1108–1112, 1987.
67. Lifshitz, R., Windham, M. T., and Baker, R. Mechanism of biological control of preemergence damping-off of pea by seed treatment with *Trichoderma* spp. *Phytopathology, 76*:720–725, 1986.
68. Stanghellini, M. E. and Hancock, J. G. Radial extent of the bean spermosphere and its relation to the behavior of *Pythium ultimum. Phytopathology, 61*:165–168, 1971.
69. Loper, J. E. Role of fluorescent siderophore production in biological control of *Pythium ultimum* by a *Pseudomonas fluorescens* strain. *Phytopathology, 78*:166–172, 1988.

70. Fravel, D. R. Role of antibiosis in biocontrol of plant diseases. *Annu. Rev. Phytopathol.*, *26*:75–91, 1988.
71. Leong, J. Siderophores: Their biochemistry and possible role in the biocontrol of plant pathogens. *Annu. Rev. Phytopathol.*, *24*:187–209, 1986.
72. Elad, Y. and Baker, R. Influence of trace amounts of cations and siderophore-producing pseudomonads on chlamydospore germination of *Fusarium oxysporum*. *Phytopathology*, *75*:1047–1052, 1985a.
73. Elad, Y. and Baker, R. The role of competition for iron and carbon in suppression of chlamydospore germination of *Fusarium* spp. by *Pseudomonas* spp. *Phytopathology*, *75*:1053–1059, 1985b.
74. Kloepper, J. W., Leong, J., Teintze, M., and Schroth, M. N. *Pseudomonas* siderophores: A mechanism explaining disease-suppressive soils. *Curr. Microbiol.*, *4*:317–320, 1980.
75. Scher, F. M. and Baker, R. Effect of *Pseudomonas putida* and a synthetic iron chelator on induction of soil suppressiveness to Fusarium wilt pathogens. *Phytopathology*, *72*:1567–1573, 1982.
76. Simeoni, L. A., Lindsay, W. L., and Baker, R. Critical iron level associated with biological control of Fusarium wilt. *Phytopathology*, *77*:1057–1061, 1987.
77. Sneh, B., Dupler, M., Elad, Y., and Baker, R. Chlamydospore germination of *Fusarium oxysporum* f. sp. *cucumerinum* as affected by fluorescent and lytic bacteria from Fusarium-suppressive soil. *Phytopathology*, *74*:1115–1124, 1984.
78. Weller, D. M., Howie, W. J., and Cook, R. J. Relationship between in vitro inhibition of *Gaeumannomyces graminis* var. *tritici* and suppression of take-all by fluorescent pseudomonads. *Phytopathology*, *78*:1094–1100, 1988.
79. Becker, J. O. and Cook, R. J. Role of siderophores in suppression of *Pythium* species and production of increased-growth response of wheat by fluorescent pseudomonads. *Phytopathology*, *78*:778–782, 1988.
80. Brisbane, P. G. and Rovira, A. D. Mechanism of inhibition of *Gaeumannomyces graminis* var. *tritici* by fluorescent pseudomonads. *Plant Pathology*, *37*:104–111, 1988.
81. Poplawsky, A. R., Peng, Y. F., and Ellingboe, A. H. Genetics of antibiosis in bacterial strains suppressive to take-all. *Phytopathology*, *78*:426–432, 1988.
82. Thomashow, L. S. and Weller, D. M. Role of phenazine antibiotic from *Pseudomonas fluorescens* in biological control of *Gaeumannomyces graminis* var. *tritici*. *J. Bacteriol.*, *170*:3499–3508, 1988.
83. Howell, C. R. and Stipanovic, R. D. Suppression of *Pythium ultimum*-induced damping-off of cotton seedlings by *Pseudomonas fluorescens* and its antibiotic, pyoluteorin. *Phytopathology*, *70*:712–715, 1980.
84. James, D. W., Jr., and Gutterson, N. I. Multiple antibiotics produced by *Pseudomonas fluorescens* HV37 and their differential regulation by glucose. *Appl. Environ. Microbiol.*, *52*:1183–1189, 1986.
85. Ahl, P., Voisard, C., and Defago, G. Iron bound-siderophores, cyanic acid, and antibiotics involved in suppression of *Thielaviopsis basicola* by a *Pseudomonas fluorescens* strain. *J. Phytopathol.*, *116*:121–134, 1986.

86. Defago, G. and Haas, D. Pseudomonads as antagonists of soilborne plant pathogens: Modes of action and genetic analysis. *Soil Biochemistry, 6*:(in press), 1990.
87. Hadar, Y., Harman, G. E., Taylor, A., and Norton, J. M. Effects of pre-germination of pea and cucumber seeds and of seed treatment with *Enterobacter cloacae* on rots caused by *Pythium* spp. *Phytopathology, 73*:1322–1325, 1983.
88. Howell, C. R., Beier, R. C., and Stipanovic, R. D. Production of ammonia by *Enterobacter cloacae* and its possible role in the biological control of Pythium pre-emergence damping-off by the bacterium. *Phytopathology, 78*:1075–1078, 1988.
89. Fravel, D. R., Kim, K. K., and Papavizas, G. C. Viability of microsclerotia of *Verticillium dahliae* reduced by a metabolite produced by *Talaromyces flavus*. *Phytopathology, 77*:616–619, 1987.
90. Kim, K. K., Fravel, D. R., and Papavizas, G. C. Identification of a metabolite produced by *Talaromyces flavus* as glucose oxidase and its role in the biocontrol of *Verticillium dahliae*. *Phytopathology, 78*:488–492, 1988.
91. Ahmad, J. S. and Baker, R. Competetive saprophytic ability and cellulolytic activity of rhizosphere-competent mutants of *Trichoderma harzianum*. *Phytopathology, 77*:358–362, 1987.
92. Anderson, A. J. Isolation from root and shoot surfaces of agglutinins that show specificity for saprophytic pseudomonads. *Can. J. Bot., 61*:3438–3443, 1983.
93. Anderson, A. J. and Jaslavich, C. Agglutination of pseudomonad cells by plant products. *Physiol. Plant Pathol., 15*:149–159, 1979.
94. Jaslavich, C. and Anderson, A. J. Isolation from legume tissues of an agglutinin of saprophytic pseudomonads. *Can. J. Bot., 59*:264–271, 1981.
95. Anderson, A. J., Habidazdegah-Tari, P., and Tepper, C. S. Genetic studies on the role of an agglutinin in root colonization by *Pseudomonas putida*. *Appl. Environ. Microbiol., 54*:375–380, 1988.
96. Chao, W. L., Li, R. D., and Chang, W. T. Effect of root agglutinin on microbial activities in the rhizosphere. *Appl. Environ. Microbiol., 54*:1838–1841, 1988.
97. Vesper, S. J. Production of pili (fimbriae) by *Pseudomonas fluorescens* and correlation with attachment to corn roots. *Appl. Environ. Microbiol., 53*:1397–1403, 1987.
98. Haahtela, K. and Korhonen, T. K. In vitro adhesion of N_2-fixing enteric bacteria to roots of grasses and cereals. *Appl. Environ. Microbiol., 49*:1186–1190, 1985.
99. Haahtela, K., Tarkka, E., and Korhonen, T. K. Type 1 fimbria-mediated adhesion of enteric bacteria to grass roots. *Appl. Environ. Microbiol., 49*:1182–1185, 1985.
100. Korhonen, T. K., Nurmiaho-Lassila, E.-L., Laakso, T., and Haahtela, K. Adhesion of fimbriated nitrogen-fixing enteric bacteria to roots of grasses and cereals. *Pl. Soil, 90*:59–69, 1986.
101. Korhonen, T. K., Tarkka, E., Ranta, H., and Haahtela, K. Type 3 fimbriae of *Klebsiella* sp.: Molecular characterization and role in bacterial adhesion to plant roots. *J. Bacteriol. 155*:860–865, 1983.
102. Scher, F. M., Kloepper, J. W., Singleton, C., Zaleska, I., and Laliberte, M. Colonization of soybean roots by *Pseudomonas* and

Serratia species: Relationship to bacterial motility, chemotaxis, and generation time. *Phytopathology*, *78*:1055–1059, 1988.
103. De Weger, L. A., van der Vlugt, C. I. M., Wijfjes, M., Baker, P. A. H., Schippers, B., and Lugtenberg, B. Flagella of a plant-growth-stimulating *Pseudomonas fluorescens* strain are required for colonization of potato roots. *J. Bacteriol.*, *169*:2769–2773, 1987.
104. Ahmad, J. S. and Baker, R. Rhizosphere competence of *Trichoderma harzianum*. *Phytopathology*, *77*:182–189, 1987.
105. Scher, R. M., Ziegle, J. S., and Kloepper, J. W. A method for assessing the root-colonizing capacity of bacteria on maize. *Can. J. Microbiol.*, *30*:151–157, 1984.
106. Boudreau, M. A. and Andrews, J. H. Factors influencing antagonism of *Chaetomium globosum* to *Venturia inaequalis:* A case study in failed biocontrol. *Phytopathology*, *77*:1470–1475, 1987.
107. Hirano, S. S. and Upper, C. D. Ecology and epidemiology of foliar bacterial plant pathogens. *Annu. Rev. Phytopathol.*, *21*:243–269, 1983.
108. Leben, C. Relative humidity and the survival of epiphytic bacteria with buds and leaves of cucumber plants. *Phytopathology*, *78*:179–185, 1988.
109. Baker, R. and Scher, F. M. Enhancing the activity of biological control agents. In *Innovative Approaches to Plant Disease Control* (I. Chet, ed.), Wiley-Interscience, New York, 1987, pp. 1–17.
110. Nelson, E. B., Harman, G. E., and Nash, G. T. Enhancement of *Trichoderma*-induced biological control of Pythium seed rot and pre-emergence damping-off of peas. *Soil Biol. Biochem.*, *20*:145–150, 1988.
111. Papavizas, G. C. Genetic manipulation to improve the effectiveness of biocontrol fungi for plant disease control. In *Innovative Approaches to Plant Disease Control* (I. Chet, ed.), Wiley-Interscience, New York, 1987, pp. 193–212.
112. Howell, C. R. Relevance of mycoparasitism in the biological control of *Rhizoctonia solani* by *Gliocladium virens*. *Phytopathology*, *77*:992–994, 1987.
113. Katan, T., Dunn, M. T., and Papavizas, G. C. Genetics of fungicide resistance in *Talaromyces flavus*. *Can. J. Micriobiol.*, *30*:1079–1087, 1984.
114. Peberdy, J. F. Fungal protoplasts: Isolation, reversion, and fusion. *Annu. Rev. Microbiol.*, *33*:21–29, 1979.
115. Turgeon, B. G., Garber, R. C., and Yoder, O. C. Development of a fungal transformation system based on selection of sequences with promoter activity. *Mol. Cell Biol.*, *7*:3297–3305, 1987.
116. Obukowicz, M. G., Perlak, F. J., Kusano-Kretzmer, K., Mayer, E. J., Bolten, S. L., and Watrud, L. S. Tn5-mediated integration of the delta-endotoxin gene from *Bacillus thuringiensis* into the chromosome of root-colonizing pseudomonads. *J. Bacteriol.*, *168*:982–989, 1986.
117. Obukowicz, M. G., Perlak, F. J., Kusano-Kretzmer, K., Mayer, E. J., and Watrud, L. S. Integration of the delta-endotoxin gene of *Bacillus thuringiensis* into the chromosome of root-colonizing strains of pseudomonads using Tn5. *Gene*, *45*:327–331, 1986.

118. West, A. W. Fate of the insecticidal, proteinaceous parasporal crystal of *Bacillus thuringiensis* in soil. *Soil Biol. Biochem.*, *16*:357–360, 1984.
119. Kommedahl, T., Windels, C. E., Sarbini, G., and Wiley, H. B. Variability in performance of biological and fungicidal seed treatments in corn, peas, and soybeans. *Protection Ecology*, *3*:55–61, 1981.
120. Hubbard, J. P., Harman, G. E., and Hadar, Y. Effect of soilborne *Pseudomonas* spp. on the biological control agent, *Trichoderma hamatum*, on pea seeds. *Phytopathology*, *73*:655–659, 1983.
121. Chao, W. L., Nelson, E. B., Harman, G. E., and Hoch, H. C. Colonization of the rhizosphere by biological control agents applied to seeds. *Phytopathology*, *76*:60–65, 1986.
122. Nelson, E. B., Kuter, G. A., and Hoitink, H. A. J. Effects of fungal antagonists and compost age on suppression of Rhizoctonia damping-off in container media amended with composted hardwood bark. *Phytopathology*, *73*:1457–1462, 1983.
123. Hubbard, J. P., Harman, G. E., and Eckenrode, C. J. Interaction of a biological control agent, *Chaetomium globosum*, with seed coat microflora. *Can. J. Microbiol.*, *28*:431–437, 1982.
124. Fravel, D. R. and Marois, J. J. Edaphic parameters associated with the establishment of the biocontrol agent, *Talaromyces flavus*. *Phytopathology*, *76*:643–646, 1986.
125. Harman, G. E., Chet, I., and Baker, R. *Trichoderma hamatum* effects on seed and seedling disease induced in radish and pea by *Pythium* spp. or *Rhizoctonia solani*. *Phytopathology*, *70*:1167–1172, 1980.
126. Smiley, R. W. Antagonists of *Gaeumannomyces graminis* from the rhizoplane of wheat in soils fertilized with ammonium- or nitrate-nitrogen. *Soil Biol. Biochem.*, *10*:169–174, 1978a.
127. Smiley, R. W. Colonization of wheat roots by *Gaeumannomyces graminis* inhibited by specific soils, microorganisms, and ammonium-nitrogen. *Soil Biol. Biochem.*, *10*:175–179, 1978b.
128. Halsey, M. and Powelson, R. The influence of NH_4Cl with KCl on antagonism between fluorescent pseudomonads and *Gaeumannomyces graminis* var. *tritici* on the wheat rhizoplane. *Phytopathology*, *71*:105 (Abstr.), 1981.
129. Stack, J. P., Kenerley, C. M., and Pettit, R. E. Application of biological control agents. In *Biocontrol of Plant Diseases*, Vol. II, (K. G. Mukerji and K. L. Garg, eds.), CRC Press, Inc., Boca Raton, FL, 1988, pp. 43–54.
130. Schreier, P. H., Kuntz, M., Lipphardt, S., Lorz, H., Baker, B., Simons, A., de Bruijn, F., Schell, J., Bohnert, H., Reiss, B., and Wasmann, C. C. New developments in plant transformation technology: Its application to cellular organelles, cereals, and dicotyledonous crop plants. In *Biotechnology in Plant Science, Relevance to Agriculture in the Eighties* (M. Zaitlin, P. Day, and A. Hollaender, eds.), Academic Press, Orlando, 1985, pp. 237–246.
131. Hoitink, H. A. J. Composted bark, a lightweight growth medium with fungicidal properties. *Plant Dis.*, *64*:142–147, 1980.
132. Scher, F. M. and Castagno, J. R. Biocontrol: A view from industry. *Can. J. Plant Pathol.*, *8*:222–224, 1986.

133. Aldrich, J. and Baker, R. Biological control of *Fusarium roseum* f. sp. *dianthi* by *Bacillus subtilis*. *Plant Dis. Rptr.*, *54*:446–448, 1970.
134. Chet, I., Harman, G. E., and Baker, R. *Trichoderma hamatum*: Its hyphal interactions with *Rhizoctonia solani* and *Pythium ultimum*. *Microb. Ecol.*, *7*:29–38, 1981.
135. Jager, G. and Velvis, H. Biological control of *Rhizoctonia solani* on potatoes with antagonists. 5. The effectiveness of three isolates of *Verticillium biguttatum*. *Neth. J. Pl. Path.*, *92*:231–238, 1986.
136. Kaiser, W. J. and Hannan, R. M. Biological control of seed rot and preemergence damping-off of chickpea with *Penicillium oxalicum*. *Plant Dis.*, *68*:806–811, 1984.
137. Martin, F. N. and Hancock, J. G. The use of *Pythium oligandrum* for biological control of preemergence damping-off caused by *P. ultimum*. *Phytopathology*, *77*:1013–1020, 1987.
138. Paulitz, T. C. and Baker, R. Biological control of Pythium damping-off of cucumbers with *Pythium nunn*: Population dynamics and disease suppression. *Phytopathology*, *77*:335–340, 1987.
139. Scher, R. M. and Baker, R. Mechanism of biological control in a Fusarium-suppressive soil. *Phytopathology*, *70*:412–417, 1980.
140. Singh, P. J. and Mehrotra, R. S. Biological control of *Rhizoctonia bataticola* on gram by coating seed with *Bacillus* and *Streptomyces* spp. and their influence on plant growth. *Pl. Soil*, *56*:475–483, 1980.
141. Sivan, A. and Chet, I. Biological control of *Fusarium* spp. in cotton, wheat, and muskmelon by *Trichoderma harzianum*. *J. Phytopathol.*, *116*:39–47, 1986.
142. Sivan. A., Ucko, O., and Chet, I. Biological control of Fusarium crown rot of tomato by *Trichoderma harzianum* under field conditions. *Plant Dis.*, *71*:587–592, 1987.
143. Stutz, E. W., Defago, G., and Kern, H. Naturally occurring fluorescent pseudomonads involved in suppression of black root rot of tobacco. *Phytopathology*, *76*:181–185, 1986.
144. Utkhede, R. S. Antagonism of isolates of *Bacillus subtilis* to *Phytophthora cactorum*. *Can.J. Bot.*, *62*:1032–1035, 1984.
145. Weller, D. M. and Cook, R. J. Increased growth of wheat by seed treatments with fluorescent pseudomonads, and implications for *Pythium* control. *Can. J. Plant Pathol.*, *8*:328–334, 1986.
146. Windels, C. E. and Kommedahl, T. Factors affecting *Penicillium oxalicum* as a seed protectant against seedling blight of pea. *Phytopathology*, *68*:1656–1661, 1978.
147. Austin, B., Dickinson, C. H., and Goodfellow, M. Antagonistic interactions of phylloplane bacteria with *Dreschlera dictioides* (Drechsler) Shoemaker. *Can. J. Microbiol.*, *23*:710–715, 1977.
148. Swineburne, T. R. Microflora of apple leaf scars in relation to infection by *Nectria galligena*. *Trans. Br. Mycol. Soc.*, *60*:389–403, 1973.
149. Tronsmo, A. and Ystaas, J. Biological control of *Botrytis cinerea* on apple. *Plant Dis.*, *64*:1009, 1980.
150. Pusey, P. L. and Wilson, C. L. Postharvest biological control of stone fruit brown rot by *Bacillus subtilis*. *Plant Dis.*, *68*:753–756, 1984.

151. Janisiewicz, W. J. Post-harvest biological control of blue mold on apples. *Phytopathology, 77*:481–485, 1987.
152. Tronsmo, A. and Dennis, C. The use of *Trichoderma* species to control strawberry fruit rots. *Neth. J. Pl. Path., 83*:449–455, 1977.
153. Mercer, P. C. Biological control of decay fungi in wood. In *Biocontrol of Plant Diseases*, Vol. I (K. G. Mukerji and K. L. Gard, eds.), CRC Press, Inc., Boca Raton, FL, 1988, pp. 177–198.
154. Wilson, C. L., Franklin, J. D., and Pusey, P. L. Biological control of Rhizopus soft rot of peach with *Enterobacter cloacae*. *Phytopathology, 77*:303–305, 1987.
155. Dennis, C. and Webster, J. Antagonistic properties of species-groups of *Trichoderma*. I. Production of non-volatile antibiotics. *Trans. Br. Mycol. Soc., 57*:25–39, 1971.
156. Dennis, C. and Webster, J. Antagonistic properties of species-groups of *Trichoderma*. II. Production of volatile antibiotics. *Trans. Br. Mycol. Soc., 57*:41–48, 1971.
157. Claydon, N., Allan, M., Hanson, J. R., and Avent, A. G. Antifungal alkyl pyrones of *Trichoderma harzianum*. *Trans. Br. Mycol. Soc., 88*:503–513, 1987.
158. Elad, Y., Barak, R., and Chet, I. Possible role of lectins in mycoparasitism. *J. Bacteriol., 154*:1431–1435, 1983a.
159. Barak, R., Elad, Y., Mirelman, D., and Chet, I. Lectins: A possible basis for specific recognition in *Trichoderma-Sclerotium rolfsii* interactions. *Phytopathology, 75*:458–462, 1985.
160. Elad, Y., Chet, I., Boyle, P., and Henis, Y. Parasitism of *Trichoderma* spp. on *Rhizoctonia solani* and *Sclerotium rolfsii*. Scanning electron microscopy and fluorescence microscopy. *Phytopathology, 73*:85–88, 1983
161. Elad, Y., Chet, I., and Henis, Y. Degradation of plant pathogenic fungi by *Trichoderma harzianum*. *Can. J. Microbiol., 28*:719–725, 1982.
162. Elad, Y., Barak, R., Chet, I., and Henis, Y. Ultrastructural studies of the interaction between *Trichoderma* spp. and plant pathogenic fungi. *Phytopath. Z., 107*:168–175, 1983.
163. Elad, Y., Sadovski, Z., and Chet, I. Detection of mycoparasitism by infrared photomicrography. *Microb. Ecol., 9*:185–187, 1983.
164. Lewis, J. A. and Papavizas, G. C. Permeability changes in the hyphae of *Rhizoctonia solani* induced by germling preparations of *Trichoderma* and *Gliocladium*. *Phytopathology, 77*:699–703, 1987.
165. Krantz, J. Hyperparasitism of biotrophic fungi. In *Microbial Ecology of the Phylloplane* (J. P. Blakeman, ed.), Academic Press, London, 1981, pp. 327–352.
166. Pratt, R. G. Germination of oospores of *Sclerospora sorghi* in the presence of growing roots of host and non-host plants. *Phytopathology, 68*:1606–1613, 1978.

14

BIOLOGICAL CONTROL OF WEEDS WITH FUNGI

C. GERALD VAN DYKE *North Carolina State University, Raleigh, North Carolina*

I. INTRODUCTION

Weeds are, and apparently have always been, a menace to mankind. A simple definition of weeds is that they are plants growing where they are not wanted. A more accepted definition is that weeds are plants that are competitive, pernicious, and persistent. Like the plant pathologist who desires to learn more about plant pathogens and their interaction with plants and the entomologist who studies insect–plant interactions, the weed scientist has as an ultimate goal to learn as much as possible about weeds and how they can be controlled. As these disciplines have become more sophisticated, there has been a desire to understand better the parameters involved in the management of pests in order to view the control process from a long-range rather than a short-range view. Of course there is still a need to control pests as quickly and efficiently as possible, and this work must continue. However, because of constraints on resources and concern for environmental imbalance caused by some control methods, it is imperative to view pest control solutions in an integrated manner that will provide solutions not just for the present, but for the future as well. Those involved in integrated solutions should strive to provide safer, less expensive, and more efficient means of controlling pests. Biocontrol is a strategy that fits, in part, these parameters. Biocontrol organisms must be carefully and aggressively considered as potentials for weed management.

II. PRINCIPLES OF BIOCONTROL

A. Historical Perspectives

All plants are known to have some pests; some plants are being limited by certain pests. Fungi are the most prevalent of the organisms that cause plant diseases, and therefore are also the most common disease-causing agents of weeds. As far as we know, fungi have always been associated with plants. Person and Christ [1] have stated that little is known about the origin of fungal–plant interactions, and such origins can not be empirically tested. Fossil plants have fungi that often so closely resemble present-day fungi that it is impossible to distinguish them. Therefore, we begin with what we know about fungal–host interactions with no help from fossil records. However, considerable research has been done on plant–fungal relationships. By bringing together what is known about fungi, their interrelationships with plants, and the biology of weeds, it is possible to have good working models for research projects in biocontrol of weeds with fungi.

In 1893, Halsted [2] published a listing of fungi new to certain weeds. As early as 1916, Cockayne [3] recognized that a rust was checking the spread of a thistle in California. Fullaway [4] described a biocontrol project, conducted between 1943 and 1949; fungal spore suspension was inoculated on and effectively controlled certain cactus species that were causing problems to ranchers in Hawaii. Other researchers had reported the potential use of biocontrol fungi before 1970; a few examples of diverse host–pathogen interactions are cited [5–8].

Although plant pathology had its modern-day beginnings in the mid-1800s, and although the examples above clearly show that successes had already been achieved early in this century, the concept of biocontrol of weeds with plant pathogens did not become a serious topic of conversation until the 1970s. The first review of the topic was by Wilson in 1969 [9]. This was followed, in 1972, by a review of biocontrol of aquatic weeds by Zettler and Freeman [10]. Freeman et al. [11] eloquently summarized the void in this subject: "The only surprise is that it took so long for plant pathologists to discover a potential that lay dormant, for the most part on their doorsteps since the birth of the science."

It has been noted in several reviews that entomologists had been studying and using insects for biocontrol of weeds some 50 years earlier. It is almost ironic to think that because an entomologist, Professor H. S. Smith, first used the expression "biological control," the term was reserved for use in entomology. It is no coincidence then that entomologists were, for many years, in the forefront in biocontrol research. Plant pathologists who study biocontrol of weeds were encouraged in 1982 when a United States Department of Agriculture (USDA) appropriation for biocontrol of weeds was divided equally between entomologists and plant pathologists. The plant pathologists were part of Regional Research Project S-136 "Biological Control of Weeds with Plant Pathogens."* As a result of that funding and industry cooperation, a 2-year regional study of the feasibility of controlling sicklepod (*Cassia obtusifolia*) with the fungal pathogen *Alternaria cassiae* was conducted and subsequently published [12].

B. Modern Developments

The first International Symposium on Biocontrol of Weeds was held in Delmont, Switzerland, in 1969. The next two were held in 1971 in Rome, Italy, and 1973 in Montpellier, France. In the same year, 1973, the Weed Science Society of America formed an ad hoc committee on biological control (now a permanent committee). That group published the results of a search to identify biocontrol efforts in North America [13]. Only five weeds were listed as having potential for control with plant pathogens.

The fourth International Symposium on Biocontrol of Weeds was held in Gainesville, Florida, in 1976, the same year that the American Phytopathological Society formed an all-inclusive committee on biocontrol. The first contributed paper sessions were organized that year. Later, in 1978, as

---

*Revised in 1989, now S-234, "Discovery and Development of Plant Pathogens for Biological Control of Weeds."

TABLE 1. Chairmen and Administrative Advisors of Southern Regional Research Project S-136, "Biological Control of Weeds with Plant Pathogens" (officially began functioning in October 1978)[a]

Year	Chairman	Affiliation	Discipline
1978–79	George E. Templeton	University of Arkansas, Fayetteville	Plant pathologist
1980	Roy J. Smith, Jr.	USDA, Stuttgart, Arkansas	Weed scientist
1981	H. Lynn Walker	USDA, Stoneville, Mississippi (present address, Louisiana Tech, Ruston)	Plant pathologist
1982	David O. Tebeest	University of Arkansas, Fayetteville	Plant pathologist
1983	R. Charudattan	University of Florida, Gainesville	Plant pathologist
1984	A. Douglas Worsham	North Carolina State University, Raleigh	Weed scientist
1985	C. Gerald Van Dyke	North Carolina State University, Raleigh	Plant pathologist
1986	William H. Ridings (deceased prior to taking office)	Clemson University, South Carolina	Plant pathologist
1986	P. C. Quimby, Jr.	USDA, Stoneville, Mississippi (Present address, USDA, Bozeman, Montana)	Weed scientist
1987	C. Douglas Boyette	USDA, Stoneville, Mississippi	Plant pathologist
1988	William L. Bruckart	USDA, ARS, Fort Detrick, Maryland	Plant pathologist
1989	Gregory J. Weidemann	University of Arkansas, Fayetteville	Plant pathologist
Advisor USDA—Cooperative States Research Service			
1978–89	John F. Fulkerson	Washington, D.C.	Plant pathologist
Administrative Advisers Agr. Res. Service			
1978–82	Curtis R. Jackson	Georgia Exp. Stat., Experiment, Georgia	Plant pathologist
1983–85	Gale A. Buchanan	Alabama Agr. Exp. Station, Auburn Univ.	Weed scientist
1986	David H. Teem	Alabama Agr. Exp. Station, Auburn Univ.	Weed scientist

[a]Now S-134, see note on p. 358.

part of that committee, Charudattan [14] compiled and edited a directory of "Biological Control Projects in Plant Pathology"; this directory included 24 projects relating to control of weeds.

Southern Regional Research Project S-136, titled "Biological Control of Weeds with Fungi," revised in 1983 as "Biological Control of Weeds with Plant Pathogens," was officially organized in October 1978.* The technical committee of this project is composed of university and USDA scientists. Persons who have served as chairmen of the technical committee, and their respective discipline and affiliation, are listed in Table 1. After S-136 formed in 1978, the technical committee sponsored an International Symposium in September of 1980 in Little Rock, Arkansas. At this symposium, the basic principles of plant pathology were reviewed by Holcomb [15], weed science was reviewed by Barrett [16], and genetics of fungal pathogens was reviewed by Leonard [17]. The proceedings were published as a book in 1982 [18].

In May 1980, the Beltsville Symposium (V) in Agricultural Research was titled "Biological Control in Crop Production." In July of the same year, the fifth International Symposium on Biocontrol of Weeds was held in Brisbane, Australia. These International Symposia previously had been dominated by entomologists, with the exception that plant pathologists had helped organize the fourth Symposium held in Gainesville, Florida. It was becoming obvious that plant pathogenic fungi were potentially excellent candidates for future exploration in biocontrol of weeds. Even entomologists were asking directly and indirectly why more plant pathogens had not been examined [19]. The next year, 1981, "DeVine" was registered as the first mycoherbicide for commercial use. The following year the fungal pathogen *Colletotrichum* was registered as "Collego" (for further details of "Collego" and "DeVine" see below). In the 1983 minutes of the board of directors for the Weed Science Society of America (WSSA), it was established that the Biological Control of Weeds Committee would have as its main objective to "keep WSSA members informed of the objectives and progress of biological weed control." At the 1985 board meeting a draft of a monograph on Biological Control of Weeds was accepted, and support was given for an initiative to establish a slide set for teaching purposes.

In February 1983, a National Interdisciplinary Biological Control Conference was held in Las Vegas, Nevada. Biocontrol of weeds with plant pathogens was a significant part of that conference [20]; plant pathogens and plant pathologists were "coming of age" and are now having more input and influence on decisions relating to biocontrol.

The sixth International Symposium on Biocontrol of Weeds was held in Vancouver, Canada, in 1984, the seventh in 1988 in Rome, Italy, and the eighth is scheduled for 1992 in Christchurch, New Zealand.

III. STRATEGIES FOR BIOCONTROL

A. Classical Biocontrol

The classical strategy is usually employed for controlling weeds found in undisturbed areas, rangelands, or waterways. These are environments that are not as frequently influenced by humans as are agricultural lands. Obligate, parasitic fungi, such as rusts or deuteromycetes that propagate

*S-136 is now S-234, see note p. 358.

and disseminate easily, are best suited for biocontrol in these habitats. These fungi must be capable of developing infections that build to epiphytotic levels to prove efficacious. The weeds are usually exotics and lack their natural fungal pathogens. When their pathogens are introduced, after careful quarantine to determine host specificity and safety, the weeds can sometimes be controlled as they were in their native habitats. There have been several successful uses of this strategy in different parts of the world (see below).

Obligate Parasites

Blackberry Rust: Two wild blackberry species of European origin, *Rubus constrictus* and *R. ulmifolius*, are important weeds in Chile. A rust fungus, *Phragmidium violaceum*, isolated from *Rubus* sp. in Europe was found to be pathogenic on wild blackberries, but not on other members of the Rosaceae. This rust fungus was released in Chile where it caused widespread damage to wild blackberry plants [21]. Hasan [22] reported that similar efforts were being undertaken in France to study the potential of controlling wild blackberries with *P. violaceum*.

Skeletonweed Rust: Skeletonweed (*Chondrilla juncea*) originated in the Mediterranean regions of the Near East and southern Russia. *Chondrilla juncea* has since spread to the United States and to Australia. A natural parasite of skeletonweed is *Puccinia chondrillina*. This fungus is a macrocyclic, monoecious rust completing its life cycle on *Chondrilla*; certain isolates have been unable to attack other composites or closely related relatives [23]. There are, however, various ecotypes of *C. juncea*, and virulence is variable in *P. chondrillina* isolates. Mediterranean rust strains which had shown reduced growth of skeletonweed in European tests were introduced into Australia in 1971. The results there were dramatic with reductions of skeletonweed populations in less than a year [24,25]. These population levels subsequently have remained stable [26]. Hasan [27] reported this to be the first introduction of a rust for weed control in Australia. From 1960 to 1970, skeletonweed infestations increased in the United States more than 4000-fold from its first discovery in 1960. The weed was well adapted to the wheat-fallow cultural practices of the Pacific Northwest. Chemical control was cost prohibitive and environmentally undesirable [28]. In light of the successful treatment with *P. chondrillina* in Australia, a cooperative program between the states of California, Oregon, Washington, and Idaho was established for release in the United States of the rust in the late 1970s. Field inoculations, which included hand-dusting, powered dusting, and water suspensions, and aerial application by helicopter and fixed-winged aircraft all resulted in establishment of infection in these states. The pathogen also survived the winter, and "secondary rust pustules were found up to 3.4 km from inoculation foci within one year of inoculation" [29]. Lee has summarized the studies in the United States [30].

Other Examples: Another rust, *Uromyces rumicis*, was found to be highly pathogenic on curly dock (*Rumex crispus*) and other *Rumex* species [31]. The rust appeared to be mostly host specific and therefore not a threat to cultivated plants. This rust affects *Rumex* infestation in Europe under natural conditions and has been considered for release for controlling *Rumex* in the United States [32]. At the time of Frank's report [32], there was concern about alternate hosts for this rust. Subsequently, a study of the host range of *U. rumicis* by Schubiger et al. [33] indicated that the rust was host specific and should be safe for use on *Rumex*. This rust is not listed in rust indices of the United States.

Hasan [34] reported results of field experiments involving a rust (*Puccinia xanthii*) of *Xanthium* species. This rust is widespread in southern Europe and Asia Minor [22] and is well known in the United States; it has been extensively surveyed in North Carolina [35]. Hasan [22] also reported that this rust caused "widespread and damaging infections on *X. strumarium* in Queensland (Australia)" shortly after its unintentional introduction. Heavy infections of the rust can cause extensive damage to leaves and stems of *Xanthium* species, and under test conditions infections have been shown to reduce dry weight and seed viability [22]. This rust is microcyclic (producing only telio- and basidiospores) and is observed only in mid-to-late-summer on *X. strumarium* in the United States. This suggests that high temperatures may be one of the requirements for rust infection. As *Xanthium* leaves mature, the teliospores become dormant and therefore are not easily used for purposes of inoculation; for these reasons further investigations of the potential use of the fungus in the United States have been deferred (Van Dyke, unpublished data).

In 1983, Phatak et al. [36] described rust epiphytotics of *Puccinia canaliculata* which occurred during 1978–1981 on dense stands of yellow nutsedge (*Cyperus rotundus*) in Georgia. Dehydration of yellow nutsedge roots and tubers and reduction of flowering in infected plants indicated that this rust could be an effective agent for control of this important weed pest. This research group and others have continued to study the possible integration of this biocontrol into the nutsedge management systems.

Dr. William H. Ridings of Clemson University was studying the use of *Puccinia heterospora* and its effect on spurred anoda *Anoda cristata*. This study ceased as a result of his untimely death in 1986. Ridings et al. [37] demonstrated in field tests that when teliospores of *P. heterospora* were applied to spurred anoda, the resulting disease caused reductions in weed dry weight. They also showed that natural spread of the fungus had occurred during the test.

Powdery mildews have been investigated for their efficacy on weeds. Hasan [22] summarized the investigations of two powdery mildews, *Erysiphe cichoracearum* and *Leveillula taurica* f. sp. *chondrillae*, that caused appreciable damage on *Chondrilla* in the Mediterranean region. In one test, seed production was reduced as a result of mildew infections.

Gardner and Hodges [38] have shown that the rust *Kuehneola uredinis*, long known to occur in Hawaii, causes heavy infection and damage to a noxious exotic blackberry (*Rubus penetrans*) but only mildly and inconsequentially attacks valued endemic *Rubus* spp.

The rust fungus *Puccinia expansa* has been shown to cause loss of dry weight and reduction of flowers when infection occurs on ragwort (*Senecio jacobaea*) according to Alber et al. [39]. They suggested that ragwort control by insects might be enhanced in combination with rust inoculations.

Bruchart and Dowler [38] presented an evaluation of the use of exotic rust fungi for classical biocontrol. Many other examples could be discussed. A few are cited [41–43].

Nonobligate Parasites

In the early 1970s in Hawaii, an introduced weed Hamakua pamakani (*Ageratina riparia* = *Eupatorium riparium*) was identified as one of the most serious weeds in certain range areas. A fungus originally listed as *Cercosporella ageratinae* was found in Jamaica on *A. riparia* and brought to Hawaii under careful quarantine. More than 25 plant families were inoculated with the fungus, but were immune. In 1975–1976 fungal spores were released into selected weed sites in Hawaii, and epiphytotics of disease on the weed caused considerable weed demise in areas where climatic factors favored the pathogen. The disease progressed to an extent that most Hamakua pamakani was killed or unable to compete with other plants. This fungus was able to produce spores on diseased plants; these spores spread and infected other plants of this weed to effect widespread control. The factors which made this effort a success were the ability of the fungus to produce large quantities of spores on abundantly available plant material and climatic conditions that were ideal for disease development. Fortunately, the fungal pathogen was host specific so there was little danger of disease on other plants. Trujillo reported on the history of this biocontrol example and updated the results at the sixth International Symposium [44] by stating that "In general, more than 50,000 ha of pasture land have been rehabilitated to their full potential by this pathogen, while no evidence of host resistance nor the presence of mutants pathogenic to other plant species has been encountered in the 8 years since its release." Trujillo [45] now suspects that the spores isolated from infected Hamakua are those of the anamorph stage of the smut *Entyloma compositarum*, however, the teleomorph stage is still uncertain.

B. Mycoherbicide Strategy

The mycoherbicide or innudative strategy involves release of massive quantities of fungal inoculum on weeds at strategic stages in their life cycle, preferably when temperature and moisture are optimum for disease development. Many mycoherbicides are being developed for use on weed seedlings in which the aim is to kill the weed or stunt it so that competition with infested crop plants is reduced or made negligible.

Host range and efficacy are two of the major concerns in the development of mycoherbicides. Most of the obligate fungi, particularly the rusts, used in classical biocontrol have very specific hosts or narrow host ranges. This is generally true also of the deuteromycetes, although their host ranges are usually broader. Therefore, it is necessary to screen carefully plant families related to the weed being tested, as well as representative members of other plant families [46]. Since mycoherbicidal fungi are usually native or endemic pathogens, generally there are fewer risks related to host specificity. This subject has been reviewed by Charudattan [47], Mortensen [48], and Watson [49]. Another difficult evaluation is that of efficacy. Mortensen [48] has addressed some of the problems associated with efficacy and host range relative to laboratory and field screening of pathogens. Many different methods are used for rating efficacy. Efficacy is sometimes considered only in terms of level of kill of the weed by the pathogen as reviewed by Charudattan [50]. It would be desirable to know also, when the weed is not killed, to what extent pathogen infection reduces the competitiveness of the weed with a crop plant.

Success Stories

Collego and DeVine: Two mycoherbicides have been registered since the early 1980s and have been used successfully in subsequent years. In 1981, scientists in Florida [51] and from Abbott Laboratories [52], North Chicago, Illinois, developed and registered with the United States Environmental Protection Agency (EPA) a formulation of the fungal pathogen *Phytophthora palmivora*. This fungus infects and kills seedling and adult plants of strangler vine *Morrenia odorata*, a serious weed problem in Florida citrus groves. This mycoherbicide is marketed under the trade name DeVine. *Phytophthora palmivora* is an oomycetous fungus, found abundantly in Florida soils.

The second mycoherbicide to be registered was Collego in 1982. The fungal pathogen *Colletotrichum gloeosporioides* f. sp. *aeschynomene* was found to control selectively northern jointvetch *Aeschynomene virginica* in rice and soybeans in Arkansas, Louisiana, and Mississippi [53]. Scientists at the University of Arkansas, USDA Rice Research Station, Stuttgart, AR, and Upjohn Company [54], of Kalamazoo, Michigan, cooperated in this development. This mycoherbicide presently is marketed by Ecogen, Inc., Langhorne, Pennsylvania. These are monumental examples of mycoherbicide success. The formulation and application procedures for these two mycoherbicides have been detailed by TeBeest and Templeton [55] and by Charudattan [56].

Scientists and others involved in the discovery, testing, registration, formulation, and other developmental phases of commercialization of these products had many hurdles to overcome. Templeton [57–61] documents the long task that brought Collego to market. This product represented the test case for many of the mycoherbicides being developed today. Many regulatory guidelines were rewritten to accommodate mycoherbicides during this period. Fermentation technology was deployed to produce and formulate inoculum; one of the most significant accomplishments was the production of the first dried formulation. Churchill [62] has reviewed that development. Host range and toxicological testing were redefined for mycoherbicides; Templeton [61] and Charudattan [47] have reviewed the testing levels, referred to as tiers, involved in these procedures.

These products were, and are, successful because of the persistent efforts of scientists and industry personnel. In addition, the environmental settings for both of these mycoherbicides are unique. The rice and soybean areas where Collego is used are irrigated systems and provide an ideal situation for fungal infection and disease development. Also, since many *Colletotrichum* species form spores in submerged liquid fermentation, they were, and still are, ideal candidates for mycoherbicide development. In the DeVine story, another unique environmental situation was favorable to mycoherbicide development. Since the weed, strangler vine, is a problem in citrus groves, the product use is confined to those counties in Florida where citrus is grown. This represents a relatively large but confined area for application. In this case the fungus is much less host specific, but application has natural geographic boundaries; it is also restricted by soil application and therefore there is little secondary spread by wind. It is interesting to note that DeVine had been so persistent in the soils of citrus groves after application that this carryover has nearly eliminated the need for reapplication of the product. This is good news for the grower, but not profitable to the manufacturer.

Formulations Near Registration

One of the most promising new mycoherbicides, being developed by the Mycogen Corporation, San Diego, California, is CASST. The fungal pathogen, *Alternaria cassiae*, was discovered by Walker [63] in Mississippi and has shown great promise for control of sicklepod (*Cassia obtusifolia*). Many field trials, including a multistate test by scientists involved in Regional Research Project S-136 [12] showed that controls of 80% or better could be achieved when spore formulations were applied to seedlings in the cotyledon to first leaf stage of growth. This fungal deuteromycete, like many others, does not produce spores in submerged fermentation. Therefore, a new method of spore production was necessary and was developed by Walker [64]. This innovative technique has proved to be effective for many deuteromycetes. However, there are still problems associated with large-scale mass production of spores suitable for marketed mycoherbicide products.

Another promising mycoherbicide, BioMal, has been tested extensively in Canada. PhilomBios of Saskatoon, in cooperation with scientists at the Regina Research Station of Agriculture Canada, developed a fungal pathogen *Colletotrichum gloeosporioides* f. sp. *malvae*, which effectively controls round-leaved mallow (*Malva pusilla*). Mortensen has recently reviewed results of several years of field testing [65].

Other Potential Mycoherbicides

Other endemic fungal pathogens that have shown various levels of success include *Colletotrichum malvarum* for control of prickly sida (*Sida spinosa*) in cotton and soybean [58], a leaf spot pathogen *Cercospora rodmanii* for suppression of waterhyacinth in Florida [66,67], a blight-inducing pathogen *Alternaria macrospora* for control of spurred anoda in cotton [68], *Chondrostereum pupureum* for control of a noxious tree species *Prunus serotina* in newly planted forests in the Netherlands [69], *Colletotrichum truncatum* for control of Florida beggarweed *Desmodium tortuosum* [70], *Bipolaris* species on grass weeds [71,72], *Fusarium solani* f. sp. *cucurbitae* for control of Texas gourd *Cucurbita texana* [73], possible control of Bathurst burr (*Xanthium spinosum*) with *Colletotrichum orbiculare* [74], and many others not previously mentioned [75–84].

IV. INTEGRATION OF BIOCONTROL AGENTS

Integrated pest management (IPM) is a relatively new discipline, although the idea is not new. Pest management that integrates the best systems available and also considers natural factors already at work is management that is both environmentally sound and economical. Both classical and mycoherbicide biocontrol strategies are being tested and, in some cases, are in use and are integrated with other management systems. Charudattan [56] has reviewed integration of microbial agents with pesticides, with other microbials, and with arthropods.

A. Economic Considerations

One of the first economic concerns has been whether existing application methods and equipment can be used for delivering microbials [85,86].

Quimby and Walker [86] indicated that it is possible to integrate certain mycoherbicides with other pesticides in tank-mix combinations.

Mycoherbicides

Smith [87] has summarized several years of successful integration of Collego into rice and soybean weed control systems. This work has provided a benchmark for future efforts to integrate mycoherbicides with other control systems. Other preliminary tests to augment mycoherbicides have been reported. Boyette et al. [88] attempted integration of two different fungi for control of two different weeds in this same rice system. Scheepens [89] has shown that combinations of the chemical herbicide atrazine with the endemic fungal pathogen *Curvularia lunata* had a positive effect on the control of barnyardgrass *Echinochloa crus-galli*. Bannon et al. [90] observed positive synergistic effects when they combined the mycoherbicide *Alternaria cassiae* (CASST) with a plant growth regulator.

Charudattan [91] studied integrated control of waterhyacinth *Eichhornia crassipes* with a pathogen, insects, and herbicides in an aquatic system and proposed ways of effectively using *C. rodmanii* with chemical and insect agents.

Classical Systems

Classical biocontrol systems, in which rust fungi have been used, have also been integrated into existing management systems. Phatak et al. [92] showed a synergistic effect when the rust fungus *Puccinia canaliculata* and paraquat were used for control of yellow and purple nutsedge (*Cyperus rotundus*).

V. FUTURE PROSPECTS

A. Exploration and Integration

Exploration research is a recent adventure; there are certainly many biocontrol agents awaiting discovery and development. Winder and Van Dyke [93] reported a rapid method for screening weed seeds to find new fungal candidates. As certain chemical herbicides are restricted or banned, there will be many opportunities to use biocontrols.

Challenges in biocontrol research include continued research on the integration of presently used organisms with existing management systems and new research on ways to integrate promising candidate fungal agents. As new information is gathered on compatibility of biocontrols and chemicals, and as these agents change or are purposely modified, additional possibilities in the area of weed control may be realized. Applications of combined treatments of mycoherbicides and chemical herbicides have been studied extensively in only a few systems.

B. Weed Biology and Economic Thresholds

There is still much to learn about weed biology and economic thresholds. Barrett [16] reviewed the status of weed biology at the 1980 International Symposium. This information will enhance the development of biocontrols. The idea that the only "good" weed is a "dead" weed may not be valid in modern, new-technology, weed-control strategies. Many biocontrols reduce weed vigor by stunting plant growth, cutting seed yields, reducing photosynthetic area of leaves, etc.

C. Manipulation of Fungal Pathogens

Selection and genetic manipulation of fungal pathogens have been studied for a few biocontrols, but much potential work is awaiting financial resources and new methodology. Charudattan [56] discussed selection of fungal strains for fungicide resistance. Potentially all of the fungi studied as biocontrols could be selected for more virulence, more host specificity, better growth and sporulation, more environmentally optimal infection and disease development, increased ability for large-scale economic production, more stable shelf life of mycoherbicidal products, etc. Since there is concern about release of genetically manipulated organisms into the environment, selection of naturally occurring strains is presently the desirable strategy; much additional research needs to be done in this area. However, with recent successes in genetic manipulation of fungi [94], the potential for genetically altered organisms for biocontrol purposes seems more realistic.

D. Other Areas with Research Potential

Other areas of research that have been undertaken and that need additional attention include microcomputer measurements of pathogen injury to the weed, as studied by Lindow and Anderson [95], host and pathogen histological interactions with light microscopy as done by TeBeest et al. [96], and light and scanning electron microscopy, by Van Dyke and Trigiano [97]. Van Dyke [98] has reviewed some of the microscopic factors that should be considered. Walker [68] has considered granular formulation of certain mycoherbicides as an effective means of application, and Walker and Connick [99] have experimented with sodium alginate to help realize this goal. There are examples of studies related to environmental parameters of disease development, such as light [100] and temperature and moisture [101,102]. Bannon and Walker [103] have explored the influence of nonionic surfactants and crop oils. Quimby and Fulgham [104] have demonstrated that the problem of long dew periods, which certain fungi need in order to infect, can be overcome or bypassed with an invert emulsion. Their group has also developed new formulation nozzles for boosting the efficacy of mycoherbicide fungi [105].

E. Fungal Toxin Development

Many of the fungal pathogens of weeds are among the known genera that produce toxins, i.e., *Alternaria, Helminthosporium* (*Bipolaris, Drechslera, Exserohilum*), *Fusarium*, etc. Some of these toxins are host specific, others are not. Chemical diversity and activity are also recognized in these metabolites. Once again the potential for fungal toxin research has been known for many years; Wright reviewed fungal toxins in 1968 [106]. The potential of these toxins (fungal metabolites) as precursors for new herbicide chemistry has been suggested recently [107] and is being investigated in other labs also [108, R. Charudattan, personal communication]. Several chemical and pharmaceutical companies have been interested in researching fungal metabolites [109].

Quimby et al. [110] have characterized a fungal metabolite from two Alternaria species having different hosts. *Alternaria crassa*, a pathogen which has been evaluated as a potential mycoherbicide of jimsonweed (*Datura stramonium*), and *A. helianthi*, a pathogen of sunflower and cocklebur (*Xanthium strumarium*), contain radicinin. The crude extract of this metabolite delayed germination of seeds of hemp sesbania *Sesbania exaltata*. Quimby et al. [110] also observed reduced plant shoot growth and a synergistic effect between radicinin and the herbicide imazaquin.

In our study of *Bipolaris* species, the metabolites have aided in and suggested implications for use in taxonomic separations at the species level [unpublished data].

VI. CONCLUSIONS

The future of biocontrol with fungi and/or other plant pathogens appears promising. It will obviously depend on many factors. There must be both scientific interest in pursuing this area of research and funding to carry it out. There is a need for industry to be more involved and for those now involved to maintain their interest. There must be more interdisciplinary interaction among plant pathologists, weed scientists, industry, fermentation technologists, entomologists, and others. If present concerns about chemical pesticides continue, and if documented deleterious pesticide effects on the biosphere continue to become realities, then it appears that the need for biocontrols will be obvious. This does not and will not assure that interdisciplinary activity will increase, nor does it mean that funding, which is urgently needed, will be available for more research in biocontrol of weeds with fungi. Hopefully, however, it will prompt many more scientists to be cognizant of biocontrols as solutions to pesticide concerns and help others see the need to encourage such efforts. The potential for biocontrol of weeds with fungi looks too promising for its efforts to falter at this time. Its future possibilities seem to be unlimited. With biotechnological efforts just "coming of age," there is a vast number of possible research directions available for those who will venture there. The multitude of different parameters involved in a biocontrol system, i.e., fungal biotypes, adjuvants, fungal toxins, fungal-chemical synergism, to name but a few, have not been adequately explored for potential enhancement of systems that have already been tested.

A. Contributions by Weed Scientists

Weed scientists must provide the following information about weeds (as discussed by Smith [87]): weed biology (including geographic distribution and origin), species relationships, extent of genetic variations, emergence patterns, seed sources and requirements for seed germination and seedling growth, weed-crop interference relationships, economic thresholds, environmental conditions during periods of mycoherbicide application, and interactions of pest and crop management strategies that might influence mycoherbicide use.

B. Contributions by Plant Pathologists

Plant pathologists must provide the following information about pathogenic fungi (as discussed by Smith [87]): characterization of host–parasite relationships on the weed and on closely related species, competitive levels of parasitism among the various fungal strains, sporulation capacity in artificial culture, host specificity, virulence, spore durability and genetic stability, lab procedures for fungal growth and sporulation, the environmental optima for infection and disease development, host range, the synergistic or inhibitory effects of pesticides, and key constraints that must be avoided in field applications.

C. Contributions by Industry or Government

Industry or government personnel must provide the following information relating to production and marketing of mycoherbicides (as discussed by Smith [87]): fermentation labs must determine whether spore production can be increased to economical, large-scale production with a viable, stable product that has season-long shelf life, and provide fresh and dry-formulated product for pilot field tests.

ACKNOWLEDGMENTS

I wish to thank Drs. W. S. Chilton, A. D. Worsham, R. L. Beckmann, and E. D. Seneca for helpful suggestions during the manuscript preparation. I am indebted to members of S-136, especially Drs. George E. Templeton, R. Charudattan, and H. Lynn Walker for stimulating my interest in biocontrol and for constructive comments concerning this manuscript.

Paper No. 11978 of the Journal Series of the North Carolina Agricultural Research Service, Raleigh, NC 27695-7601.

REFERENCES

1. Person, C. and Christ, B. Population genetics and evolution of host-parasite interactions. In *Challenging Problems in Plant Health* (T. Kommedahl and P. H. Williams, eds.), The Amer. Phytopath. Soc., St. Paul, MN, 1983, pp. 379–386.
2. Halsted, B. D. Some new weed fungi. *Bull Torrey Bot. Cl.*, *20*:250–252, 1893.
3. Cockayne, A. H. California thistle rust as a check on the spread of California thistle. *International Rev. of the Sci. and Prac. of Agri.*, *7*:451, 1916.
4. Fullaway, D. T. Biological control of cactus in Hawaii. *J. of Econ. Ent.*, *47*:696–700, 1954.
5. Dodd, A. P. The biological control of prickly pear in Australia. *Monographiae Biologicae*, *8*:565–567, 1959.
6. Dodd, A. P. Biological control of *Eupatorium adenophorum* in Queensland. *Aust. J. Sci.*, *23*:356–365, 1961.
7. Wicker, E. F. and Shaw, C. G. Fungal parasites of dwarf mistletoes. *Mycologia*, *60*:372–383, 1968.

8. French, D. W. and Schroeder, D. B. The oak wilt fungus. *Ceratocystis fagacearum* (Bretz) Hunt. as a selective silvicide. *Forest Science*, *15*:198–203, 1969.
9. Wilson, C. L. Use of plant pathogens in weed control. *Annu. Rev. Phytopathol.*, *7*:411–434, 1969.
10. Zettler, F. W. and Freeman, T. E. Plant pathogens as biocontrols of aquatic weeds. *Annu. Rev. of Phytopathol.*, *10*:445–470, 1972.
11. Freeman, T. E., Charudattan, R., and Conway, K. E. Status of the use of plant pathogens in the biological control of weeds. In *Proc. IV Int. Symp. Biol. Cont. Weeds*, Gainesville, Florida, 1976 (T. E. Freeman, ed.), University of Florida, 1978, pp. 201–206.
12. Charudattan, R., Walker, H. L., Boyette, C. D., Ridings, W. H., TeBeest, D. O., Van Dyke, C. G., and Worsham, A. D. Evaluation of *Alternaria cassiae* as a mycoherbicide for sicklepod (*Cassia obtusifolia*) in regional field tests. *South. Coop. Ser. Bull. 317*, 1986, 19 pp.
13. Goeden, R. D., Andres, L. A., Freeman, T. E., Harris, P., Pienkowski, R. L., and Walker, C. R. Present status of projects on the biological control of weeds with insects and plant pathogens in the United States and Canada. *Weed Sci.*, *22*:490–495, 1974.
14. Charudattan, R. A directory of biological control projects in plant pathology. Mics. pub. from Dept. of Plant Path., University of Florida, Gainesville, 1978.
15. Holcomb, G. E. Constraints on disease development. In *Biological Control of Weeds With Plant Pathogens* (R. Charudattan and H. L. Walker, eds.), John Wiley and Sons, New York, 1982, pp. 61–72.
16. Barrett, S. C. H. Genetic variation in weeds. In *Biological Control of Weeds With Plant Pathogens* (R. Charudattan and H. L. Walker, eds.), John Wiley and Sons, New York, 1982, pp. 73–98.
17. Leonard, K. J. The benefits and potential hazards of genetic heterogeneity in plant pathogens. In *Biological Control of Weeds With Plant Pathogens* (R. Charudattan and H. L. Walker, eds.), John Wiley and Sons, New York, 1982, pp. 99–112.
18. Charudattan, R. and Walker, H. L. *Biological Control of Weeds With Plant Pathogens*. John Wiley and Sons, New York, 1982, p. 293.
19. Harris, P. Current approaches to biological control of weeds. *Tech. Comm., Commonwealth Inst. of Biol. Con.*, *4*:67–76, 1971.
20. Charudattan, R. Biological control of weeds with plant pathogens. In *Proceedings of The National Interdisciplinary Biological Control Conference* (S. L. Battenfield, ed.), Las Vegas, NV, 1983, pp. 23–25.
21. Oehrens, E. Biological control of the blackberry through the introduction of rust, *Phragmidium violaceum*, in Chile. *F.A.O. Plant Protect. Bull.*, *25*:26–28, 1977.
22. Hasan, S. Plant pathogens and biological control of weeds. *Rev. of Plant Path.*, *59*:349–356, 1980.
23. Hasan, S. and Jenkins, P. T. The effect of some climatic factors on the infectivity of the skeleton weed rust, *Puccinia chondrilla*. *Plant Dis. Rept.*, *56*:858–860, 1972.
24. Cullen, J. M., Kable, P. F., and Catt, M. Epidemic spread of a rust imported for biological control. *Nature*, *244*:462–464, 1973.
25. Burdon, J. J., Groves, R. H., and Cullen, J. M. The impact of biological control on the distribution and abundance of *Chondrilla juncea* in southeastern Australia. *J. Appl. Ecol.*, *18*:957–966, 1981.

26. Hasan, S. and Cullen, J. M. Studies in the Mediterranean region on biological control of Australian weeds. *Proc. EWRS III Symp. Weed Problems Medit. Area*, Oeiras, Portugal, 1984, pp. 365–371.
27. Hasan, S. First introduction of a rust fungus in Australia for the biological control of skeleton weed in Australia. *Ann. Appl. Biol.*, *99*:119–124, 1974.
28. Ehrensing, D. T. and Isaacson, D. L. Biological control of skeleton-weed: Basic considerations for initiating a program in Oregon with a list of references. *Oregon Dept. Agric., Weed Control Bull. 2*, 1977.
29. Emge, R. G., Milching, J. S., and Kingsolver, C. H. Epidemiology of *Puccinia chondrillina*, a rust pathogen for the biological control of rush skeleton weed in the United States. *Phytopathology*, *71*:839–843, 1981.
30. Lee, G. A. Integrated control of rush skeletonweed (*Chondrilla juncea*) in the Western U.S. *Weed Sci. (Suppl. 1)*, *34*:26–28, 1986.
31. Inman, R. E. A primary evaluation of *Rumex* rust as a biological control agent of curly dock. *Phytopathology*, *61*:102–107, 1971.
32. Frank, P. Report on the status of the rust *Uromyces rumicis* introduced into quarantine as a control agent for curly dock *Rumex crispus*. Proc. II Int. Symp. Biol. Contr. Weeds, 1971, Rome, Italy (P. H. Dunn, ed.), *Misc., Commonw. Inst. Biol. Contr. Publ.*, *6*:121–125, 1973.
33. Schubiger, F. X., Defago, G., Sedlar, L., and Kern, H. Host range of the haplontic phase of *Uromyces rumicis*. In *Proc. VI Int. Symp. Biol. Contr. Weeds*, 1984, Vancouver, Canada (E. S. Delfosse, ed.), Agric. Can., 1985, pp. 653–659.
34. Hasan, S. *Xanthium* rust as a possible biological control agent of Bathurst and noogoora burrs in Australia. *Misc. Pub., Com. Instit. of Biol. Contr.*, *8*:137–140, 1974.
35. Van Dyke, C. G. and Spurr, H. W., Jr. Potential use of *Puccinia xanthii* (rust fungus) for biocontrol of cocklebur (*Xanthium strumarium*). In *Biological Control of Weeds with Plant Pathogens* (R. Charudattan and H. L. Walker, eds.), John Wiley and Sons, New York, 1982, pp. 244–245.
36. Phatak, S. C., Summer, D. R., Wells, H. D., Bell, D. K., and Glaze, N. C. Biological control of yellow nutsedge with the indigenous rust fungus *Puccinia canaliculata*. *Science*, *219*:1446–1447, 1983.
37. Ridings, W. H., Hoyer, P. H., and Schimmel, L. E. Biological control of spurred anoda [*Anoda cristata* (L.) Schlecht] using teliospores of *Puccinia heterospora*. *Weed Sci. Soc. of Amer.*, *26*:54–55, 1986 (abstract).
38. Gardner, D. E. and Hodges, C. S. Leaf rust caused by *Kuehneola uredinis* on native and nonnative *Rubus* species in Hawaii. *Plant Dis.*, *67*:962–963, 1983.
39. Alber, G., Defago, G., Sedlar, G., and Kern, H. Damage to *Senecio jacobaea* by the rust fungus *Puccinia expansa*. In *Proc. VI Int. Symp. Biol. Contr. Weeds*, 19–25 August 1984, Vancouver, Canada (E. S. Delfosse, ed.), Agric. Can., 1985, pp. 587–592.
40. Bruckart, W. L. and Dowler, W. M. Evaluation of exotic rust fungi in the United States for classical biological control of weeds. *Weed Sci. (Suppl. 1)*, *34*:11–14, 1986.

41. Watson, A. K. and Clement, M. Evaluation of rust fungi as biological control agents of weedy *Centaurea* in North America. *Weed Sci.* (*Suppl. 1*), *34*:7–10, 1986.
42. Dyer, W. E., Turner, S. K., Fay, P. K., Sharp, E. L., and Sands, D. C. Control of Canada thistle by rust, *Puccinia obtegens*. In *Biological Control of Weeds with Plant Pathogens* (R. Charudattan and H. L. Walker, eds.), John Wiley and Sons, New York, 1982, pp. 243–244, 1982.
43. Defago, G., Kern, H., and Sedlar, L. Potential control of weedy spurge by the rust *Uromyces scutellatus*. *Weed Sci.*, *33*:857–860, 1985.
44. Trujillo, E. E. Biological control of hamakua pa-makani with *Cercosporella* sp. in Hawaii. In *Proc. VI Int. Symp. Biol. Contr. Weeds*, 1984, Vancouver, Canada (E. S. Delfosse, ed.), Agric. Can., 1985, pp. 661–671.
45. Trujillo, E. E., Aragki, M., and Shoemaker, R. A. Infection, disease development, and axenic culture of *Entyloma compositarum*, the cause of Hamakua pamakani blight in Hawaii. *Plant Dis.*, *72*:355–357, 1988.
46. Wapshere, A. J. A strategy for evaluating the safety of organisms for biological weed control. *Ann. Biol.*, *77*:201–211, 1974.
47. Charudattan, R. Regulation of microbial weed control agents. In *Biological Control of Weeds with Plant Pathogens* (R. Charudattan and H. L. Walker, eds.), John Wiley and Sons, New York, 1982, pp. 175–188.
48. Mortensen, K. A proposal for a standardized scale of attack and its application to biocontrol agents of weeds in laboratory screening tests. In *Proc. VI Int. Symp. Biol. Contr. Weeds*, 1984, Vancouver, Canada (E. S. Delfosse, ed.), Agric. Can., 1985, pp. 643–650.
49. Watson, A. K. Host specificity of plant pathogens in biological weed control. In *Proc. VI Int. Symp. Biol. Contr. Weeds*, 1984, Vancouver, Canada (E. S. Delfosse, ed.), Agric. Can., 1985, pp. 577–586.
50. Charudattan, R. Assessment of efficacy of mycoherbicide candidates. In *Proc. VII Int. Symp. Biol. Contrl. Weeds*, 1888, Rome, Italy (E. S. Delfosse, ed.), 1990.
51. Ridings, W. H. Biological control of stranglervine in citrus—a researcher's view. *Weed Sci.* (*Suppl. 1*), *34*:31–32, 1985.
52. Kenny, D. S. DeVine—the way it was developed—an industrialist's view. *Weed Sci.* (*Suppl. 1*), *34*:15–16, 1986.
53. Templeton, G. E. Mycoherbicide research at the University of Arkansas—past, present, and future. *Weed Sci.* (*Suppl. 1*), *34*:35–37, 1986.
54. Bowers, R. C. Commercialization of Collego—an industrialist's view. *Weed Sci.* (*Suppl. 1*), *34*:24–25, 1986.
55. TeBeest, D. O. and Templeton, G. E. Mycoherbicides: Progress in the biological control of weeds. *Plant Dis.*, *69*:6–10, 1985.
56. Charudattan, R. The use of natural and genetically altered strains of pathogens for weed control. In *Biological Control in Agricultural IPM Systems* (M. A. Hoy and D. C. Herzog, eds.), Academic Press, Inc., New York, 1985, pp. 347–372.
57. Templeton, G. E. EPA issues first experimental use permit for biological control fungus. *Rice J.*, *78*:29–30, 1975.

58. Templeton, G. E., TeBeest, D. O., and Smith, R. J., Jr. Development of an endemic fungal pathogen as a mycoherbicide for biocontrol of northern jointvetch in rice. In *Proc. IV Inter. Sym. on the Biol. Con. of Weeds*, Gainesville, FL, 1976 (T. E. Freeman, ed.), University of Florida, 1978, pp. 214–216.
59. Templeton, G. E., TeBeest, D. O., and Smith, R. J., Jr. Biological control of weeds with mycoherbicides. *Annu. Rev. Phytopathol.*, *17*:301–310, 1979.
60. Templeton, G. E. Biological herbicides: discovery, development, deployment. *Weed Sci.*, *30*:430–433, 1982.
61. Templeton, G. E. Mycoherbicides—achievement, developments and prospects. *Proc. 8th Australian Weeds Conf.*, 1987, Weed Sci. Soc. New South Wales, Sydney, Australia, 1987, pp. 489–497.
62. Churchill, B. W. Mass production of microorganisms for biological control. In *Biological Control of Weeds With Plant Pathogens* (R. Charudattan and H. L. Walker, eds.), John Wiley and Sons, New York, 1982, pp. 139–156.
63. Walker, H. L. Seedling blight of sicklepod caused by *Alternaria cassiae*. *Plant Dis.*, *66*:426–428, 1982.
64. Walker, H. L. *Alternaria macrospora* as a potential biocontrol agent for spurred anoda: Production of spores for field studies. *U.S.D.A. Science & Educ. Admin. Advances in Agri. Tec., Southern Series, No. 12*, 1980, p. 5.
65. Mortensen, K. The potential of an endemic fungus, *Colletotrichum gloeosporioides*, for biological control of round-leaved mallow (*Malva pusilla*) and velvetleaf (*Abutilon theophrasti*). *Weed Sci.*, *36*:473–478, 1988.
66. Charudattan, R., Linda, S. B., Kluepfel, M., and Osman, Y. A. Biocontrol efficacy of *Cercospora rodmanii* on waterhyacinth. *Phytopathology*, *75*:1263–1269, 1985.
67. Conway, K. E. Evaluation of *Cercospora rodmanii* as a biological control of waterhyacinth. *Phytopathology*, *66*:914–917, 1976.
68. Walker, H. L. Granular formulation of *Alternaria macrospora* for control of spurred anoda (*Anoda cristata*). *Weed Sci.*, *29*:342–345, 1981.
69. Scheepens, P. C. and Hoogerbrugge, A. Control of *Prunus serotina* in forests with the endemic fungus *Chondrostereum purpureum*. In *Proc. VII Int. Symp. Biol. Contr. Weeds*, 1988, Rome, Italy (E. S. Delfosse, ed.), 1990.
70. Cardina, J., Littrell, R. H., and Hanlin, R. T. Anthracnose of Florida beggarweed (*Desmodium tortuosum*) caused by *Colletotrichum truncatum*. *Weed Sci.*, *36*:329–334, 1988.
71. Chiang, M-Y, Van Dyke, C. G., and Leonard, K. J. Evaluation of endemic foliar fungi for potential biological control of johnsongrass [*Sorghum halepense* (L.) Pers.]: screening and host-range tests. *Plant Dis.*, *73*:459–464, 1989.
72. Winder, R. S. and Van Dyke, C. G. The pathogenicity, virulence, and biocontrol potential of two *Bipolaris* species on johnsongrass (*Sorghum halepense*). *Weed Sci.*, *38*:89–94, 1990.
73. Weidemann, G. J. Effects of nutritional amendments on conidial production of *Fusarium solani* f. sp. *cucurbitae* on sodium alginate granules and on control of Texas gourd. *Plant Dis.*, *72*:757–759, 1988.

74. Auld, B. A., McRae, C. F., and Say, M. S. Possible control of *Xanthium spinosum* by a fungus. *Agr., Ecosystems and Environ.*, *21*:219–223, 1988.
75. Blaney, C. L. and Van Dyke, C. G. Fungal pathogens with potential for biocontrol of yellow nutsedge (*Cyperus esculentus*). *J. Elisha Mit. Sci. Soc.*, *103*:71–76, 1988.
76. Cartwright, R. D. and Templeton, G. E. Biological limitations of *Protomyces gravidus* as a mycoherbicide for giant ragweed, *Ambrosia trifida*. *Plant Dis.*, *72*:580–582, 1988.
77. Crawley, D. K., Walker, H. L., Riley, J. A. Interaction of *Alternaria macrospora* and *Fusarium lateritium* on spurred anoda. *Plant Dis.*, *69*:977–979, 1985.
78. Gudauskas, R. T., Teem, D. H., and Morgan-Jones, G. Anthracnose of *Cassia occidentalis* caused by *Colletotrichum dematium* F. *truncata*. *Plant Dis. Reptr.*, *61*:468–470, 1977.
79. Littlefield, L. J. Biocontrol of leafy spurge with pathogenic fungi. In *Proc. VI Int. Symp. Biol. Contr. Weeds*, 1984, Vancouver, Canada (E. S. Delfosse, ed.), Agric. Can., 1982, p. 633.
80. McRae, C. F., Ridings, H. I., and Auld, B. A. Anthracnose of *Xanthium spinosum*—quantitiative disease assessment and analysis. *Australasian Plant Path.*, *17*:11–13, 1988.
81. Ormeno-Nunez, J., Reeleder, R. D., and Watson, A. K. A foliar disease of field bindweed (*Convolvulus arvensis*) caused by *Phomopsis convolvulus*. *Plant Disease*, *72*:338–342, 1988.
82. Trujillo, E. E. and Obrero, F. P. Cephalosporium wilt of *Cassia surattensis* in Hawaii. In *Proc. IV Int. Symp. Biol. Contr. Weeds*, 1976, Gainesville, FL (T. E. Freeman, ed.), University of Florida, 1978, pp. 217–220.
83. Walker, H. L. *Fusarium lateritium*: a pathogen of spurred anoda (*Anoda cristata*), prickly sida (*Sida spinosa*), and velvetleaf (*Abutilon theophrasti*). *Weed Sci.*, *29*:629–631, 1981.
84. Wilson, S. and Hall, R. L. Potential of *Pyrenophora avenae* for biological control of wild oats, *Avena fatua*. *Australian Weeds Conf.*, *8*:105–108, 1987.
85. Auld, B. A., Menz, K. M., and Tisdell, C. A. *Weed Control Economics*. Academic Press, New York, 1987.
86. Quimby, P. C., Jr. and Walker, H. L. Pathogens as mechanisms for integrated weed management. *Weed Sci.* (*Suppl.*), *30*:30–34, 1982.
87. Smith, R. J., Jr. Biological control of northern jointvetch (*Aeschynomene virginica*) in rice (*Oryza sativa*) and soybeans (*Glycine max*)—A researcher's view. *Weed Sci.* (*Suppl. 1*), *34*:17–23, 1986.
88. Boyette, C. D., Templeton, G. E., and Smith, R. J., Jr. Control of winged waterprimrose (*Jussiaea decurrens*) and northern jointvetch (*Aeschynomene virginica*) with fungal pathogens. *Weed Sci.*, *27*:497–501, 1979.
89. Scheepens, P. C. Joint action of *Cochliobolus lunatus* and atrazine on *Echinochloa crus-galli* (L.) Beauv. *Weed Res.*, *24*:43–47, 1987.
90. Bannon, J. S., Hudson, R. A., Stowell, L., and Glatzhofer, J. Combinations of herbicides/plant growth regulators with CASST™ herbicide (*Alternaria cassiae*). *Proc. South. Weed Sci. Soc.*, *41*:268, 1988.

91. Charudattan, R. Integrated control of waterhyacinth (*Eichhornia crassipes*) with a pathogen, insects, and herbicides. *Weed Sci.*, *34*:26–30, 1986.
92. Phatak, S. C., Callaway, M. B., and Vavrina, C. S. Biological control and its integration in weed management systems for the purple and yellow nutsedge (*Cyperus rotundus* and *C. esculentus*). *Weed Technology*, *1*:84–91, 1987.
93. Winder, R. S. and Van Dyke, C. G. A rapid screening technique for detecting potential biocontrol agents for weeds. *J. Elisha Mit. Sci. Soc.*, *102*:1–6, 1986.
94. Yoder, O. C. Use of pathogen-produced toxins in genetic engineering of plants and pathogens. In *Genetic Engineering of Plants: An Agricultural Perspective* (T. Kosuge, C. P. Meredith, and A. Hollaender, eds.), Plenum Press, New York, 1983, pp. 335–353.
95. Lindow, S. E. and Andersen, G. L. Microcomputer measurements of pathogen injury to weeds. *Weed Sci. (Suppl. 1)*, *34*:38–42, 1986.
96. TeBeest, D. O., Templeton, G. E., and Smith, R. J., Jr. Histopathology of *Colletotrichum gloeosporioides* f. sp. *aeschynomene* in northern jointvetch. *Phytopathology*, *68*:1271–1275, 1978.
97. Van Dyke, C. G. and Trigiano, R. N. Light and scanning electron microscopy of the interaction of the biocontrol fungus *Alternaria cassiae* with sicklepod (*Cassia obtusifolia*). *Can. J. Plant Path.*, *9*:230–235, 1987.
98. Van Dyke, C. G. Factors in the infection process of fungal pathogens for biocontrol of weeds. In *Proc. VII Int. Symp. Biol. Contr. Weeds*, 6–11 March 1988, Rome, Italy, (E. S. Delfosse, ed.), 1990, pp. 559–563.
99. Walker, H. L. and Connick, W. J., Jr. Sodium alginate for production and formulation of mycoherbicides. *Weed Sci.*, *33*:333–338, 1983.
100. Bannon, J. S. and Hudson, R. A. The effect of application timing and lighting intensity on efficacy of CASST™ (*Alternaria cassiae*) on sicklepod (*Cassia obtusifolia* L.). *Weed Sci. Soc. Am.*, *29*:51, 1988.
101. McRae, C. F. and Auld, B. A. The influence of environmental factors on anthracnose of *Xanthium spinosum*. *Phytopathology*, *78*:1182–1186, 1988.
102. TeBeest, D. O., Templeton, G. E., and Smith, R. J., Jr. Temperature and moisture requirements for development of anthracnose on northern jointvetch. *Phytopathology*, *68*:389–393, 1978.
103. Bannon, J. S. and Walker, H. L. The influence of non-ionic surfactants and non-phytotoxic crop oils on control of sicklepod by *Alternaria cassiae*. *Proc. South. Weed Sci. Soc.*, *40*:27, 1987.
104. Quimby, P. C., Jr. and Fulgham, F. E. Formulating and applying a weed pathogen to bypass dew. 7th Pesticide Formu. and Appl. Syst. Symp. 1986, Phoenix, Arizona. *ASTM*, *7*:25–26, 1986.
105. Quimby, P. C., Jr., Fulgham, F. E., Boyette, C. D., and Hoagland, R. E. New formulation/nozzles boost efficacy of pathogens for weed control. *Weed Sci. Soc. Am.*, *29*:52, 1988.
106. Wright, D. E. Toxins produced by fungi. *Annu. Rev. Microbiol.*, *22*:269–282, 1968.

107. Sugawara, F., Strobel, G., Strange, R. N., Siedow, J. N., Van Duyne, G. D., and Clardy, J. Phytotoxins from the pathogenic fungi *Drechslera maydis* and *Drechslera sorghicola*. *Proc. Natl. Acad. Sci.*, *84*:3081–3084, 1987.
108. Pena-Rodriguez, L. M., Armingeon, N. A., and Chilton, W. S. Toxins from weed pathogens, I. Phytotoxins from a *Bipolaris* pathogen of johnson grass. *J. Nat. Prod.*, *51*:821–828, 1988.
109. Fischer, H. and Bellus, D. Phytotoxicants from microorganisms and related compounds. *Pestic. Sci.*, *14*:334–346, 1983.
110. Quimby, P. C., Jr., Schmidt, J. H., Boyette, C. D., and Hoagland, R. E. *Alternaria crassa*: A possible new source of radicinin. *Proc. South. Weed Sci. Soc.*, *41*:265, 1988.

Part III

FUNGAL DEGRADATION PROCESSES

15

DECOMPOSITION OF CELLULOSE BY FUNGI

PAUL MARKHAM and MICHAEL J. BAZIN *King's College, University of London, London, England*

I. INTRODUCTION

Cellulose is the most abundant biological polymer on the earth. It is impossible to be exact, but a reasonably conservative estimate is that at any moment there are approximately 1400 billion tonnes of cellulose in existence. Each year about 28 billion tonnes of cellulose are formed by photosynthetic plants, accounting for approximately 6% of the total of about 742.5 billion tonnes of atmospheric carbon dioxide fixed by land and sea plants annually [1–3]. This rate of cellulose production is balanced by the annual loss of an approximately equal quantity of cellulose by decomposition. Microorganisms are responsible for more than 90% of all cellulose decomposition, animals mediating virtually all of the rest [3].

Thus, the approximate stability in the mass of cellulose in existence masks a dynamic system in which cellulose is continually being synthesized and degraded by a multiplicity of processes mediated by species representing all the major groups of living organisms. The approximate steady-state quantity of cellulose on earth is at least 50 times the amount newly synthesized annually. This is because most cellulose is contained in long-lived plants, particularly trees, or in slowly degraded material, especially lignocellulose; additionally it may be deposited in environments where it is only slowly degraded due to factors such as anaerobiosis or physical inaccessibility.

Cellulose production is divided almost equally between that synthesized by land plants (about 15 billion tonnes per year), and that synthesized by marine algae (about 12.5 billion tonnes per year) [3]. Compared to this level of productivity, the quantity of cellulose synthesized by other organisms is vanishingly small, but in the context of this review it is worth noting that some fungi synthesize cellulose. Cellulose production by fungi is almost entirely restricted to the Hyphochytridiomycete and Oomycete classes of the Eumycota (true fungi) according to the Ainsworth classification system [4] and to the slime moulds (Myxomycetes) [5]. These three groups contain only about 1,100 of approximately 78,000 known species of fungi [6]. In all of these species cellulose occurs as a major wall component [5], just as it does in the major producer species of plants and algae.

In contrast to their negligible role in cellulose production, the fungi are the major group of organisms responsible for cellulose decomposition, possibly accounting for as much as 80% of global cellulolysis. Bacteria are responsible for the majority of the remaining activity, and protozoal and animal cellulose decomposition contribute only a few percent of total decomposition, mostly by physical and chemical processes rather than enzymic degradation. Plants do have cellulose-degrading enzymes, but their activity is restricted to the internal processes of cellulose turnover as it occurs in the construction and remodeling of cellular structures, primarily cell walls [7]. Plant cellulases, therefore, do not play a role in global cellulose decomposition except to a negligible extent resulting from chance release of such cellulolytic enzymes from damaged tissue.

There is virtually no instance of cellulose decomposition in the natural environment in which fungi do not take part. However, it is also important to remember that there is probably no type of cellulolysis in nature that is carried out by fungi alone. Fungi may initiate cellulolysis, for example, when they are the primary colonizers of bulk lignocellulose such as a tree, but they are always rapidly joined by other organisms, notably bacteria. Cellulolysis is therefore always a consortium activity in nature, in which fungi generally have the major role but for which other organisms, especially bacteria, may be essential. Why this is so may be explained by the nature of naturally occurring cellulose. With very few exceptions cellulose does not occur as a pure polymer in isolation, but is always associated with, and usually surrounded by, other more complex polymers. Most common and most intractable of these is lignin, which is associated with cellulose in almost all land plants except liverworts and mosses [8]. Hemicelluloses are also major associated polymers in plant walls. In algae, lignin is absent and in the marine habitat they constitute a vast supply of nonlignified cellulose. However, the cellulose is still associated with other complex polymers, particularly xylans, which surround and protect the cellulose from decomposition, much as lignin does in land plants. Thus, while cellulose itself is of relatively simple composition and polymeric structure, it occurs naturally protected by complex polymers, and this may explain the need for interactions between many types of organism in its decomposition. Many fungi alone are capable of degrading all such complex associated polymers as well as cellulose itself, but other organisms can assist in the decomposition process, probably by utilizing some of the breakdown products which might prove toxic or cause catabolite inhibition of the degradative enzymes [9–11].

The preeminence of the fungi in the processes of cellulose decomposition is not just a consequence of the efficiency and diversity of their cellulolytic enzyme systems. This is undoubtedly an important factor, but the bacteria as a group have almost certainly an equally efficient and arguably even more diverse battery of cellulolytic enzymes at their disposal than do the fungi. It is the growth form and habit of the majority of the fungi which confers their superiority over unicellular microorganisms. The mycelial growth habit of the filamentous fungi enables them to rapidly cover much greater surface areas of relevant substrates than can the unicellular bacteria. Probably most important of all, however, is the ability of a mycelial fungus to penetrate and to ramify through a substrate and to mobilize and transport the products of polymer degradation over considerable distances through the mycelial network. Thus, rapid colonization of cellulosic substrates and efficient removal of the products of cellulose decomposition are the great advantages possessed by filamentous fungi which set them apart from all other organisms as the prime decomposers of cellulosic materials.

II. NATURE AND DISTRIBUTION OF CELLULOSE

A. Molecular Structure of Cellulose

Cellulose has a very simple basic molecular structure, being just a linear polymer of glucose units joined together by β-1,4 linkages with the loss of one molecule of water for each glucose molecule incorporated. The length and therefore molecular weight of the polymer chain is indeterminate, being highly variable and dependent upon the organism from which the cellulose is obtained and on the age and metabolic state of the organism at the time of isolation. Individual cellulose molecules can contain up to about 18,000 glucose moieties as has been reported for cell wall cellulose from algae of the genus *Valonia* [12], which corresponds to a chain length of about 12.5 μm and a molecular weight in excess of 2.9×10^6 daltons. This size of molecule is significantly greater than that determined for cellulose from the cotton plant *Gossypium herbaceum* of about 14,000 glucose units per molecule. The size of cellulose in cotton is believed to be typical of the size of cellulose polymers in cell walls of higher plants in general [12]. An exception to this appears to be trees, because cellulose obtained from wood is often reported to contain only about 2,000 glucose units per molecule [13]. However, it is believed that this figure is artificially low due to the greater severity of the extraction procedures needed to remove cellulose from woody tissue as opposed to green plants, such as cotton, from which extraction is relatively easy [12]. Remarkably, the range of polymer size for the cellulose molecules of plant and algal cell walls seems to be very narrow, so that the cellulose molecules of secondary walls of higher plant cells and the walls of mature algal cells are believed to be of a single polymer length. Indeed, even separate wild isolates of *Valonia* from widely differing geographical locations gave the same figure for size of cell wall cellulose. Only in the young walls of algae and the first-formed primary walls of plant cells are there significantly shorter cellulose polymers, and in cellulose from these sources there is also a spread of sizes corresponding to many different size classes covering a continuous range up to the maximum size typical of the mature walls. Bacterial cellulose, however, is quite different in the diversity of cellulose polymer size and the maximum recorded length. Marx-Figini [12], using the same method as for her determination of the size of cotton and *Valonia* cellulose, found that *Acetobacter xylinum* pellicle cellulose never exceeded an average content of 3500 glucose units per molecule, which would be about 2.5 μm in length (a molecular weight of about 5.7×10^5 daltons), although a separate study has reported estimates of up to 7200 glucose units per molecule of cellulose obtained from the same organism [14]. Such disparities are not surprising in view of the technical difficulties associated with size determinations, differences in methods employed, different extraction procedures, and even different strains of organism. It is clear, however, that bacterial cellulose is made up of individual polymers which are considerably smaller than those that occur in plant and algal cell walls. This difference may be related to the fact that the bacterial cell, at about 2 μm in length, is considerably smaller than a typical plant or algal cell, which would be of the order of 20 – 300 μm long, or even several centimeters long in the case of some algae, including *Valonia* [15]. The low value for polymer length here cannot be artificially caused by severity of extraction procedures because bacterial cellulose is extracellular and therefore easily accessible, and it is also virtually pure and so requires only mild treatment. It is also interesting to note that there is apparently no single size class, but a continuous range of polymer lengths starting from an average size of about 2000 glucose units per molecule and increasing linearly with the age of the bacterial culture to reach the maximum average size of about 3500 glucose units per molecule at stationary phase [12].

B. Organization of Cellulose Fibers

Cellulose molecules do not occur as single polymeric chains in nature, but are always found aggregated together into larger organizational units, most commonly some type of fiber. The basic unit of association appears to be the microfibril, in which cellulose molecules associate laterally to form a highly crystalline fibril approximately 3 – 5 nm in width. These microfibrils may then themselves associate laterally in either two or three dimensions, to form much larger aggregates, which are observed by electron microscopy to be ribbons as produced by members of the bacterial genus *Acetobacter* [16] or larger fibers, which is the predominant type of organization found in the walls of fungi, algae, and higher plants.

C. Distribution of Cellulose in Fungi and Other Organisms

As the most abundant biopolymer on earth, it might be expected that cellulose would be found to be widely distributed among living organisms. However, the distribution of cellulose is in fact very restricted. The only major group of organisms in which cellulose occurs widely is the plants. It is absent from animal and only occurs in a small number of microorganisms. Among the bacteria, the occurrence of cellulose appears to be restricted to the extracellular pellicle formed by a few species of the genus *Acetobacter* [16]. While restricted to a small number of fungal species, where it is present, cellulose forms an integral and essential part of the hyphal structure. It is a major wall polymer of the oomycetes and hyphochytridiomycetes [5], about 596 species [6] belonging to the subdivision Mastigomycotina [4]. Cellulose also occurs sporadically in the walls of a few members of other fungal subdivisions, notably 37 species of the genus *Ceratocystis* and a few species of the closely related genus *Europhium* [17]. In those fungi that contain cellulose it is found, as in the plants, as a major wall polymer. Indeed this similarity between wall composition of the oomycetes and hyphochytridiomycetes and some groups of the algae, coupled with many similarities in shape, organization, and metabolism between these groups, has led to the suggestion that the oomycetes and hyphochytridiomycetes may not be true fungi at all. It is believed that they may be much more closely related to the algae, having descended from an algal ancestor that had lost its chloroplasts [18 – 20]. Thus, it may be that cellulose is not a true fungal product or constituent but merely occurs in species which superficially appear to be fungi but may be more closely related to algae. Indeed, if this is so, it would eliminate the only significant occurrence of cellulose production outside the plant kingdom. The highly restricted incidence of cellulose production by bacteria could be considered either a chance reestablishment of cellulose synthesis among the vast diversity of bacterial metabolism or, more plausibly, the restricted persistence of cellulose production among bacteria descended from the same prokaryotic cellulose-producing ancestor, which had been a progenitor of the photosynthetic eukaryotes. This suggestion seems more plausible when one considers the fact that *Acetobacter* species produce an extracellular mesh of cellulose, which could readily be viewed as the progenitor of the cellulose-based cell wall of the plants.

III. CELLULOLYTIC ENZYMES

The term "cellulase" is still often used to indicate enzyme-mediated breakdown of the cellulose polymer to low molecular weight soluble products, notably glucose. This can give the false impression that there is a single enzyme with the capacity to carry out this process. However, it has been known for many years that this is not the case and that a complex of enzymes are required for the total breakdown of polymeric cellulose to its monomeric constituent sugar. Over three-quarters of a century ago, Pringsheim [21] proposed that cellulose hydrolysis required two enzyme activities, one to release dissaccharide units from the cellulose polymer, and one to split the disaccharide into glucose units. In the 1950s Reese et al. [22] modified this basic hypothesis by proposing the existence of two enzyme activities, which they designated Cx and C_1. Their hypothesis took note of the fact that many organisms can hydrolyze amorphous or soluble derivatives of cellulose, but far fewer species can degrade the highly crystalline forms of cellulose which occur most commonly in nature. So they suggested that all organisms that have any cellulolytic activity possess Cx type activity, the ability to hydrolyze soluble cellulose derivatives, but that C_1 activity, the ability to disrupt and degrade crystalline cellulose, is only produced by true cellulose-degrading species capable of decomposing native crystalline cellulose. Intensive biochemical studies during the ensuing decades have demonstrated that this hypothesis, while a useful starting point, was also much too simplistic a view. Such investigations have now provided a level of understanding of the enzymology of cellulose degradation that has enabled highly effective cell-free systems for the hydrolysis of highly crystalline cellulose to be established and many of the central features of the process to be elucidated. It is clear from this present understanding that the term "cellulase system" is probably the best for the battery of enzyme activities now known to be required for the complete hydrolysis of crystalline cellulose to glucose by fungi [23].

A. Types of Activities of Cellulolytic Enzymes

It is now clear that there are three major types of enzyme required for the hydrolysis of crystalline cellulose to glucose. There are endoglucanases (EC 3.2.1.4), exoglucanases or cellobiohydrolases (EC 3.2.1.91), and β-glucosidases (EC 3.2.1.21), which are sometimes called cellobiases [24].

These three types of enzyme activity are present in most, if not all, true cellulolytic organisms as a multiplicity of individual enzymes. Therefore, there are from two to several different enzyme molecules for each of the three major functions, each individual enzyme having slightly or significantly differing activities from others of the same general type and often of quite different molecular structure [23-25].

At least in general terms, the role of each of these major types of enzyme activity in the breakdown of natural cellulose is as follows. The often highly crystalline and therefore largely inaccessible structure of native cellulose must initially be opened up. The mechanism by which this is achieved is still unclear, it may be that the binding of cellulolytic enzymes themselves is sufficient to bring about such a disruption of crystallinity, but it has been suggested that an enzymically mediated oxidation of cellulose is the first stage in decomposition [23]. Certainly appropriate enzymes have been found to be associated with cellulolysis in a large number of fungi. Such an oxidation would disrupt the packing of the cellulose molecules and make them accessible to cleavage by endoglucanases. This could clearly be true for the majority of cellulolytic fungi which are aerobic, but the anaerobic rumen fungi, at least, must operate a different mechanism, and it is by no means clear that such an initial oxidation step is of general occurrence. In general both endoglucanases and cellobiohydrolases appear to be inactive against crystalline cellulose in isolation, although a few exceptions to this have been reported. However, in combination endoglucanases and cellobiohydrolases act synergistically to degrade crystalline cellulose. The endoglucanases cleave internal β-1,4 links opening up the molecule to the action of the cellobiohydrolases, which strictly are considered to cleave disaccharide units from the free ends of the polymeric chain exposed by the endoglucanases. However, many of the enzymes of this type appear to have more general exoglucanase activity, and a few even seem to exhibit endoglucanase activity, such as CBH-I and CBH-II of *Trichoderma reesei* [23]. There is still considerable uncertainty in the characterization of the activities of the enzymic components of the cellulase system. This is partly because of the multiplicity of enzymes of each of the three general types produced by any one species, each individual molecule of which may have slightly differing specificities, but also because of difficulties with, and therefore uncertainties about, the purity of many of the isolated enzymes. Some activities reported for a particular type of enzyme might simply be due to trace contamination with other types of enzymes. The role of the β-glucosidases appears to rather more straightforward. They act upon the cellobiose liberated by the cellobiohydrolases and cleave it to the monosaccaride end product of cellulose decomposition, glucose [23-25].

In addition to the three major enzyme types there are often a number of ancillary enzyme activities present which are not clearly essential to the process of cellulose hydrolysis, but may assist decomposition by carrying out useful conversions, for example, oxidation, or the removal of end products which may cause catabolite repression [23,25]. Such enzymes include cellobioside phosphorylase (EC 2.4.1.20), cellobiose oxidase/dehydrogenase (EC 1.1.99.18), glucose oxidase, and cellobiose:quinone oxidoreductase (EC 1.1.5.1), which is a flavoprotein, believed to be a link between cellulolysis and lignolysis as carried out by the white-rot fungi such as *Sporotrichum pulverulentum* [26].

It is important to note that in most naturally occurring sources of cellulose, it will be necessary for a cellulolytic organism to either grow in the presence of another organism that produces, or produce itself, enzymes capable of degrading more complex polymers, particularly lignin. This is because natural cellulose is often surrounded by complex polymers such as lignin, xylans, and hemicelluloses, and the cellulose is simply inaccessible to an organism until such complex polymers are, at least partially, degraded.

There are several commonly occurring features of cellulolytic enzymes that are worth noting. First, very many such enzymes are glycoproteins, this is not unusual in fungal systems, but in bacteria other enzymic glycoproteins are rare [24]. The reason that cellulolytic enzymes are usually glycoproteins may be explained by the fact that many such enzymes have been shown to have two domains connected by a hinge region. One of the domains is the catalytic site, and the other is the cellulose-binding region, and it is this latter domain that is highly glycosylated. Thus, the glycosylation may perform an essential function in substrate recognition and binding by these enzymes [23,24].

The activity of many cellulolytic enzymes is promoted by the presence of Ca^{2+} ions and thiol-donating molecules but subject to catabolite repression by the presence of significant concentrations of low molecular weight products of cellulolysis such as cellobiose and glucose [24]. Many, but not all, cellulase complexes are induced by the presence of a suitable substrate. This is as might be expected, but it is not initially clear how such a macromolecular substrate could bring about induction, given that the inducer molecule must presumably pass across the cell wall and plasma membrane to exert its effect. The solution to this apparent paradox appears to be that even those species in which the cellulase complex must be induced have a very low level constitutive production of cellulolytic enzymes. This has been demonstrated to be true in *Trichoderma reesei* and *Myrothecium verrucaria*, for example [27,28]. Thus, when a suitable substrate is encountered by these enzymes, they release small quantities of low molecular weight breakdown products from cellulose, which then cross the plasma membrane and induce the large-scale production of the cellulase complex enzymes [28].

B. Fungal Cellulase Systems

There is a considerable body of information available about fungal cellulase systems. One genus in particular, *Trichoderma*, has proved to be especially useful because of its ability to produce and export large quantities of enzymic protein. Over 50% of the entire protein production of the organism may be exported as extracellular enzymes and on the appropriate medium much of this can be in the form of enzymes of the cellulose complex. Of all fungi, the deuteromycete *T. reesei* has been the most extensively used both for investigations of cellulolysis and for commercial production of cellulolytic enzymes [27]. Two other *Trichoderma* species, *T. viride* and *T. koningii*, have also been widely investigated [25]. However, due to the fact that the cellulolytic enzymes of *Trichoderma* spp. have a number of unsatisfactory properties, the cellulolytic enzymes of an ever-increasing range of other fungi are being investigated. The white-rot basidiomycete *Phanaerochaete chrysosporium* (Syn. *Sporotrichum pulverulentum*) is an increasingly important organism [23], largely due to its considerable capacity for lignin degradation in addition to cellulose decomposition [29]. Such a capacity is usually extremely important in the utilization of lignocellulose because the cellulose is generally protected from decomposition by lignin polymers. More recently the ascomycete genus *Chaetomium* has been extensively studied, in particular *C. cellulolyticum*, *C. globosum*, and *C. thermophile* [30], the latter species being notable for producing enzymes with desirable thermotolerant properties. Other thermophilic fungi that produce cellulolytic enzymes have also been studied, in particular *Talaromyces emersonii* [23,25,31]. Among the deuteromycetes, *Penicillium funiculosum* and *Fusarium solani* probably have the best characterized cellulase complexes after those of *Trichoderma* [23,25]. There a few species of fungi, such as *Agaricus bisporus*, whose cellulase systems have been studied for very specific reasons [32]. *Agaricus* mushrooms are by far the most valuable of the current products of lignocellulose decomposition, thus *Agaricus* cellulolytic enzymes currently constitute the most commercially significant fungal cellulase system. A number of cellulolytic representatives of the mastigomycetes have been well studied, including several species of *Pythium* [33,34]. But the cellulase system of *Neocallimastix frontalis* has probably been given the most attention [9,23], because this chytrid is one of the highly specialized fungal inhabitants of the rumen, where it is involved in the anaerobic degradation of cellulose.

The ability to produce a cellulase system capable of decomposing crystalline cellulose is clearly widespread among the true fungi. It is interesting to note that cellulolytic ability seems to be less common in the mastigomycetes, the group which contains most of the species with cellulose in their walls, than among species that do not contain cellulose. It appears, therefore, that the capacity to degrade cellulose has arisen as an adaptation of the fungi to the predominant type of carbon-containing substrate in their environment and not merely as an extension of their own wall biochemistry. Fungal species with cellulose in their walls would of course produce cellulolytic enzymes to open cellulose polymers and allow remodeling during growth [35], especially for the initiation of branching. However, extracellular cellulolytic enzymes produced for substrate utilization appear to have no relationship to such internal enzymes. Among the zygomycetes, only a very few species such as *Mortierella ramanniana* exhibit a cellulolytic capability [3].

Fungal cellulolytic enzymes in general have been shown to be subject to induction and catabolite repression by a wide range of molecules. Many studies of such regulation of fungal cellulase systems have been carried out [26,28,36–40]. It is difficult to generalize about the response of fungal cellulolytic enzymes to inducer and repressor molecules because this appears to vary considerably from species to species and for different types of enzyme activity within the cellulase system of any given species. Additionally, some molecules can act as both inducers and catabolite repressors of a single enzyme activity depending on the concentration at which they are present in the culture medium. This is especially true of cellobiose in most species, and glucose is often a strong catabolite repressor but, perhaps surprisingly, can also be a weak inducer of some cellulolytic enzyme activities. A good example is the case of the extracellular endocellulase of *Agaricus bisporus* [40]. This enzyme is induced by high concentrations of cellobiose and subject to catabolite repression at low cellobiose concentrations. Conversely, high glucose concentrations strongly inhibit enzyme activity, whereas at low glucose concentrations the enzyme is weakly induced [40]. As would be expected, a wide range of cellulose substrates induce fungal cellulolytic enzymes, but there are also some strong inducers of cellulolytic enzymes which are less obvious. One interesting case is the strong induction of

T. viride cellulase system by sophorose [38,41], a sugar which does not induce the extracellular endocellulase of *Agaricus bisporus* [40]. For the *T. viride* cellulase system, sophorose is another molecule which has a dual effect, being a potent inducer at low concentrations but rapidly exerting catabolite repression at concentrations above 0.2% [41]. Many small molecules, especially sugars, have been shown to be strong catabolite repressors of fungal cellulase systems, including mannose, galactose, arabinose, xylose, fucose, and glucuronic acid [37]. High molecular weight polymers other than celluloses are sometimes found to be moderate or weak inducers of cellulase systems, for example, glucomannan has been found to induce the carboxymethyl cellulase activity of *Sporotrichum thermophile* [36]. A few noncellulose polymers also inhibit cellulolytic activity in some species, for example, for example, xylans repress cellulolysis in some white-rot fungi [37]. Indeed, even celluloses themselves at high concentrations can be responsible for a reduction in cellulolytic activity. Many fungal cellulase systems are found to be at their optimum activity levels when the cellulose on which the organism is growing is at low concentration, often about 1%, but as low as 0.05–0.1% for the *A. bisporus* endocellulase [40].

Fungi capable of decomposing crystalline cellulose produce a cellulase system consisting of several types of enzyme molecule and activity. Two or more individual enzymes of such complexes usually work synergistically together such that the result of their combined activity is a greater degree of cellulose decomposition than the sum of the individual enzyme activities [23–25,42]. What is not clear is to what extent there is any physical association between the different enzyme molecules and between the enzymes and the fungal surface. This is an important and interesting question if we consider the equivalent cellulase complexes produced by bacteria, where there is not just a synergy in activity, but a tight physical association between the enzymes of the cellulase complex, forming a very specific aggregate which can be visualized under the electron microscope as a globular particle about 18 nm in diameter [43]. This bacterial cellulase complex has been most extensively studied in the anaerobic bacterium *Clostridium thermocellum* and has been termed a "cellulosome" by Lamed and associates [43, 44]. It contains at least one protein believed to be responsible for organizing the particle structure in addition to several proteins with cellulolytic activity [43,44]. Cellulosomes have been observed to aggregate together to form polycellulosome clusters, which are approximately hemispherical protruberances about 60 nm in diameter, bound to and covering much of the bacterial cell surface. The cellulosomes themselves seem to be positioned on the exterior of these protruberences [44]. When the bacterial cells are in close proximity to cellulose, the polycellulosomes undergo a structural change, which results in fibrous connections being made between the protruberance and the cellulose fiber to which individual cellulosomes become appressed. Indeed, after prolonged cellulolysis the bacterial cells may break their connections to the cellulose, leaving presumably active cellulosomes behind on the cellulose surface [43]. This type of physical association of all the components of the cellulase complex is believed to be of general occurrence throughout the cellulolytic bacteria [23,43].

There is some evidence for a comparable aggregate harboring the cellulolytic enzymes in a few species of fungi, but probably largely based on the example of *Trichoderma* spp., it has generally been considered that cellulose-degrading fungi simply produce and export large quantities of extracellular cellulolytic enzymes which diffuse into the surrounding substrate. In many instances, brown-rot fungi, for example, where a physical barrier such as lignin is interposed between fungus and cellulose, such freely diffusing extracellular enzymes would be required for cellulolysis to occur at adequate levels. However, there are some indications that many fungi may have adopted some kind of intermediate strategy between these two extremes. It is increasingly believed that at least some fungal cellulolytic enzymes are either bound to the hyphal wall or held in close association with the fungal hypha, for example, in a surrounding polysaccharide matrix or sheath. This latter explanation has been suggested for the case of white and soft-rot fungi, which erode cellulose within woody tissue only to a limited distance from their hyphae [45]. Some claims have also been made of direct binding of cellulolytic enzymes to hyphal walls, the β-glucosidase of *Sporotrichum pulverulentum*, for example [26], and the β-glucosidase of *T. reesei* has been clearly demonstrated to be bound to the hyphal wall [46]. Certainly it makes sense that the enzyme that releases glucose from cellobiose should be fixed to the hyphal surface. In this location it would have access to freely diffusing molecules such as cellobiose generated by the activities of endoglucanases and cellobiohydrolases, but it would also release the end product of cellulolysis, glucose, either very close to the hypha or directly into the hypha, so minimizing any danger of loss to opportunistic competitor organisms. But as Jackson and Talburt [46] have pointed out, β-glucosidase activity may be involved in other functions of the mycelium such as conidiogenesis, for which such a hyphal surface location of this enzyme activity might be essential. The difficulty many experimenters have experienced in separating individual enzyme activities from others in fungal cellulase systems has led to speculation that the existence of fungal cellulase complexes in vivo might be far more common than was once thought [23]. There are few reports indicating the existence of highly organized multienzyme aggregates containing the cellulolytic systems of fungi. Probably the best example is that of *Trichoderma reesei*, from which has been isolated a complex which has cellulase, β-glucosidase, and xylanase activity and contains at least six different proteins [47]. Interestingly, remnants of hyphal wall material were found associated with the isolated complex, implying a possible binding of the aggregate to the hyphal surface. There is evidence of a similar type of cellulase complex produced by the anaerobic rumen fungus *Neocallimastix frontalis*, which includes endoglucanase and β-glucosidase activities and at least on additional factor believed to be a cellobiohydrolase [25]. Such observations are important not just to a fundamental understanding of the molecular biology of fungal cellulolysis, but also in respect to the production and harvesting of such enzymes. Potentially valuable enzymes may have been overlooked or thought to be only produced in very low concentrations if they were indeed wallbound or in aggregates from which they are difficult to separate. Indeed, the isolation in significant quantity of such complexes could be important in obtaining highly efficient cell-free cellulose-degrading systems of fungal origin.

IV. FUNGAL CELLULOLYSIS IN THE NATURAL ENVIRONMENT

The enormous quantity of cellulose on earth means that it is ubiquitous throughout the natural environment. In virtually every type of habitat in which cellulose decomposition occurs, fungi are involved, usually as the major agent of cellulolysis, but occasionally as a minor contributor only. Plant lignocellulosic materials are even carried high into the atmosphere, one of the most inhospitable of environments, and the attendant microbial population is certain to carry on a low level of cellulolysis, as a transient phenomenon, even there. In the ocean, a few cellulolytic fungi have been recorded at depths of about one kilometer [48].

Of the many and varied natural environments in which fungal cellulolysis occurs, a few, such as forest and woodland habitats or cut timbers, have been well studied and an overall picture of the process of cellulose decomposition has begun to emerge. From other environments very little information has yet been obtained, often due to technical difficulties associated with such studies, as in the case of the oceanic environment or the animal rumen.

In the majority of natural substrates, cellulose is physically protected from decomposition. In terrestrial environments the greatest degree of physical protection is provided by lignin, although hemicelluloses also contribute to a much smaller extent. In the marine environment cellulose in wood is protected in the same way, but a vast quantity of cellulose occurs in algae which contain no lignin. In this source of cellulose some degree of physical protection from decomposition is provided by the surrounding polymers, particularly xylans. However, these are far less refractory than lignin and leave the cellulose open to decomposition by a much wider range of organisms than can degrade lignified cellulose [8]. Because fungi, particularly basidiomycetes, are supremely well adapted to the degradation of lignocellulose, they have a much greater dominance in the decomposition of terrestrial sources of cellulose than in algal sources, on which substrate they face far greater competition.

The other major factor that often diminishes the extent to which cellulose is decomposed is a relative deficiency of nitrogen. Natural cellulosic materials often have C:N ratios of 100:1 or greater, and added nitrogen is required for the majority of such cellulose to be decomposed. Positive effects of added nitrogen on cellulose degradation in natural substrates have been shown as compared to a generally neutral effect on the degradation of other polymers [49]. This is apparently because most fungi exhibit maximal cellulolytic activity at C:N ratios significantly lower than that encountered in natural substrates. While a few white-rot basidiomycetes are able to produce diffusible extracellular cellulolytic enzymes at C:N ratios as high as 2000:1, most cellulolytic fungi have a considerably reduced capacity to degrade cellulose at C:N ratios above 200:1 [50]. In a specific study on this phenomenon, Chang-Ho and Yee [51] demonstrated that the optimum C:N ratio for *Volvariella volvacea* and *Coprinus cinereus* is in the range 24–36, and *Agaricus bisporus* has a similar requirement for low C:N ratios, hence the addition of nitrogenous material to straw compost for commercial mushroom production [32,52].

A. Wood-Rotting Fungi

This topic is discussed in detail elsewhere in this volume (see Chapter 18), and is therefore only considered briefly here.

Decomposition of Standing and Fallen Natural Wood

The decomposition of wood by microorganisms is an extremely complex process, to which fungi are by far the major contributors. However, it is very rare to find circumstances in which bacteria are not also associated with the degradative process; possible exceptions are undamaged wood, where a highly pathogenic species such as *Heterobasidion annosum, Armillaria mellea*, or *Ganoderma lucidum* is the primary colonizer [53], or deep into heartwood where the relative anaerobicity and the sheer penetrative capacity of fungal hyphae might leave the fungus unchallenged by competitors, at least for a time.

The complexity of the process of wood decay is exemplified by the fact that six major classes of fungi have been identified as being involved. These are known as "primary molds" or sugar fungi [54], "stainers," "soft-rots," "white-rots," "brown-rots," and "secondary moulds" [45,55]. Brown-rot basidiomycetes degrade large quantities of cellulose but either do not degrade lignin at all or decompose very little of it. Because cellulose is white, this leaves the wood looking brown due to the residual lignin. The white-rots, on the other hand, decompose both cellulose and lignin in significant quantities and the removal of considerable amounts of lignin leaves the wood looking white. The soft-rots are a group of mainly ascomycete and deuteromycete fungi very like the brown-rot basidiomycetes, in that they degrade cellulose but leave lignin largely unaltered. However, the distinction between the two groups is not merely a taxonomic one, because they do demonstrate quite distinct differences in the way that they remove cellulose from wood. Soft-rot fungi enter the plant cell wall from the lumen of the cell, by penetrating the innermost layer of the mature cell wall (known as the S_3 layer), which they do not degrade to any significant extent because it contains large amounts of lignin. On reaching the thick inner wall layer (the S_2 layer), which is mostly composed of cellulose, the invading hyphae branch, often dichotomously at the tip to form a so-called T-branch [55] and colonize and degrade this cellulose-rich layer. Such soft-rot fungi form characteristic cavities within the S_2 layer, which are elongated parallel to the cellulose fibers of the wall and with very smooth edges and tapered ends. This characteristic shape is believed to be due partly to the architecture of the cell wall itself and partly to indicate a close association between the cellulolytic enzymes produced by the fungi and their hyphae. Certainly the cellulolytic enzymes do not appear to diffuse very far from the producing hyphae into the sourrounding tissue, which suggests that such enzymes may either be bound to the hyphal surface or held within its sphere of influence, perhaps by a surrounding mucilage [45]. The brown-rot basidiomycetes also attack the cell wall from the lumen of the cell, but they do not penetrate through the S_3 layer, remaining appressed to that layer but within the cell lumen. Their decomposition of cellulose takes place at some distance from the hyphae, mediated by a diffusable cellulolytic system, involving either enzymes, or a chemically catalyzed degradation, or both. There is certainly considerable evidence of nonenzymic cellulose-decomposing process mediated by hydrogen peroxide and Fe^{2+} ions [45], but it is not clear if this is the only means by which brown-rot fungi degrade cellulose or merely a process ancillary to an enzymic degradation system. There is a clear parallel in the enzymic degradation of lignin by fungi, which involves hydrogen peroxide and manganese ions [29]. Such a spatially separated cellulose degradation system allows brown-rot fungi to mobilize large quantities of cellulose from the S_2 and S_1 wall layers, whilst leaving the S_3 layer lignin largely intact, thus creating the brown, fragile tissue remnants characteristic of the activity of such species.

The white-rot basidiomycetes also initially remain in the cell lumen and attack the cell wall directly. However, because of their ability to decompose lignin as well as cellulose, they have a direct structural impact on the S_3 layer itself, and large areas of wall material are digested immediately around the hyphae. The only exception to this is that the wall material on which the hypha is directly resting is preferentially protected, presumably as a result of exclusion of the lytic enzymes from that area due to very tight adhesion by the hyphae. This leaves a very characteristic arrangement of the white-rot hyphae mounted on a ridge surrounded by troughs within the wall material from which polymeric material has been digested. Such areas of degradation rapidly merge together to create very extensive regions of wall degradation, and such fungi can very quickly break through the S_3 layer to attack the S_2 layer and the rest of the wall causing very considerable damage [55].

The so-called secondary molds appear to be opportunistic organisms, mainly ascomycetes and deuteromycetes, with considerable cellulolytic capacities such as *Trichoderma viride* and *Gliocladium roseum*, which utilize cellulose that has been made accessible as a result of the activities of lignin-degrading basidiomycetes [55].

Decay of Seasoned Timbers

Seasoned timers often differ in one vital respect from natural wood in being considerably drier. This single fact means that properly handled timber is virtually immune from attack by wood-degrading fungi. Invasion by damaging fungi only becomes possible if regions of the timber are allowed to remain damp, due to incorrect drying or stacking in badly aerated piles, or if dried timber subsequently becomes damp again due to being brought into contact with a source of moisture such as wet plaster or the soil. It is, however, also true that complete saturation of timber with water, causing virtually anaerobic conditions, can also give considerable protection from fungal decay. Indeed complete immersion of timber is one method that has been used for storage. This arrangement presumably reduces the likelihood of attack by physically separating the wood from potential invasive fungi as well as by the creation of essentially anaerobic conditions, although bacterial decay will generally occur in this circumstance. Timber may also differ from natural wood as a result of any chemical treatments to which it has been subjected, usually involving impregnation of the tissue with various chemicals toxic or inhibitory to fungi. Such treatment will alter significantly the range of species that can attack the wood, rarely excluding all species but usually preventing attack by the great majority of the normal wood-rotting community. In contrast, one feature of timber that increases its susceptibility to fungal attack is that it is physically altered by being cut, exposing virgin wood to potential fungal colonization in appropriately damp conditions.

Although of decreasing importance now, dry-rot, caused by *Serpula lacrimans*, remains a significant type of fungal decomposition of timber in buildings. Despite its common name, it can only initially invade wet wood. After such establishment occurs, it is then able to use its cellulolytic capacity to mobilize nutrient reserves from the woody tissue for growth and to create a large inoculum potential, which permits it to invade drier regions of timber. Nevertheless, adequately dried wood will not support the growth of even this fungus, and difficulties associated with its removal from buildings in which it has become established are largely due to the problems inherent in drying timbers adequately in such environments, where the other fabric of the building harbors too much moisture. It was once thought that the organism could release large quantities of water from the cellulose in the timber via the metabolic processes of cellulolysis; indeed the organism does exude droplets of water in appropriate circumstances. However, it is now clear that while some water is created as a result of cellulolysis, it is no more than that produced by most other cellulolytic fungi [56]. *Serpula lacrimans* is a typical brown-rot fungus, decomposing cellulose extensively in the timber, but leaving the lignin virtually unchanged. This becomes evident because extensively degraded timber shrinks and cracks to give the cubical appearance so characteristic of wood subjected to dry rot [57].

Much more common than dry-rot is wet-rot, which can be caused by several species, but *Coniophora putanea* is by far the most common [57]. This organism alone may be responsible for up to 90% of timber decay in buildings [57]. It is a brown-rot basidiomycete and therefore degrades primarily cellulose and hemicelluloses, leaving lignin virtually unaltered. It is this single organism, therefore, that is responsible for the vast majority of cellulolysis that occurs in building timbers.

B. Leaf Litter Decomposers

Leaf litter is a convenient term for the organic material that accumulates on the forest and woodland floor. The bulk of this material (~ 60 – 70%) does originate from leaves, but some of it consists of other debris of plant origin including branches (12 – 15% or more), bark (1 – 14%), and fruits (1 – 17%) [58]. On average, about 2 tons per hectare (1 – 20 tons per hectare [59]) of litter is deposited per year on the boreal forest floor [60]. This litter is supplemented by organic material of microbial and animal origin due to the microorganisms which grow upon it and the microfauna which not only die within it, but more significantly, eat it and then redeposit it slightly degraded, significantly comminuted, and supplemented especially with nitrogen [58,60].

In soils of neutral or alkaline pH, there is a large earthworm population, and these animals rapidly comminute and incorporate leaf litter into what is known as a mull soil. Such soils have little or no accumulated litter lying on their surface. It is on mor soils where litter accumulates to a considerable depth. This is in part because such soils have an acid pH, which is apparently not suitable for earthworms. It is in the litter of such mor soils that the largest population of litter-decomposing fungi occurs [54, 60]. The most commonly encountered leaf-litter decomposers are basidiomycetes belonging to the genera *Mycena*, *Marasmius*, *Collybia*, and *Clitocybe*, their fruit bodies being common on the woodland floor [60]. A few ascomycete species are also commonly found as leaf litter decomposers, for example, *Spathularia flavida* and *Cudonia confusa* [61]. The litter-decomposing basidiomycetes show many of the same polymer-degrading capacities as wood-rotting fungi, notably strong cellulolytic and lignolytic ability, and most of them are clearly analogous to the white-rots [60]. Nevertheless, there is a quite distinct group of litter decomposers, which do not generally colonize large sources of woody tissue. *Mycena galopus*, for example, grows on leaves and twigs and is strongly cellulolytic and lignolytic but does not colonize tree stumps. Some, like *Marasmius androsaceus*, which normally grows only on pine needles, are restricted to one type of material only [60,61].

Because of the large quantity of litter that is deposited annually on the forest floor and the fact that the decomposition half-life of such material is commonly about 2 years, a considerable thickness of litter accumulates above mor soils [60]. This becomes stratified, and three layers are generally recognized within it: intact leaves on top (L layer); broken up but recognizable leaf material below that (F layer); and dark powdery organic material at the bottom (H layer), below which there is an indistinct transition into the soil proper. This is a general scheme only, and there is considerable variation from one woodland to another depending on the species of trees present and the precise nature and half-life of the litter they produce.

It is in the F layer that the bulk of the basidiomycete mycelium and degradative activity is found. Here cellulose is decomposed, but it is common to find that lignin is removed much faster from the leaf material [60]. In an in vitro experiment, for example, *Collybia peronata* was found to reduce the lignin content of oak leaves by about half in 6 months while hardly decomposing any of the cellulose present [60]. This is perhaps an extreme example — most other species decompose a small but significant amount of the cellulose content of leaves along with the lignin — but it illustrates a general trend. In the H layer, however, it is deuteromycete fungi such as species of *Trichoderma* and *Penicillium* that predominate [60], and their degradative activity is likely to be responsible for far more significant levels of cellulose decomposition in that layer and in the soil (see Chapter 18).

C. Decomposition of Cellulose in Soil

The activities of leaf-litter decomposers dramatically reduce the quantities of the major macromolecules in plant material entering the soil. Nevertheless, even such a small residual proportion of the original enormous quantity of cellulose contained by plant material is a significant component of the organic content of the soil, and many soil fungi, particularly species of *Trichoderma* and *Penicillium*, are highly active cellulose decomposers. A very useful method of isolating and identifying cellulolytic fungi in the soil is the one pioneered and used extensively by Tribe [64,65] of burying pieces of cellophane. This method has the dual advantage of providing a conveniently discrete and recoverable source of cellulose which is transparent so that microscopic examination and identification of colonizing fungi can be carried out. In such experiments the major cellulolytic fungi found colonizing the cellophane were members of the genera *Humicola, Botryotrichum, Chaetomium*, and *Rhizoctonia* together with a number of chytrids. Such experiments have further served to demonstrate that some noncellulolytic species of soil fungi may still derive significant carbon nutrition from soil cellulosic substrates by growing in the presence of a strongly cellulolytic species. For example, the noncellulolytic *Pythium oligandrum* was found to be capable of growing on cellophane as sole carbon source if it was grown in the presence of the strongly cellulolytic *Botryotrichum piluliferum*, whereas it failed to grow in the absence of this species [65]. Such "parasitism" of the products of cellulolysis carried out by other fungi appears to be a common phenomenon among noncellulolytic species [10,66].

In addition to such soil saprophytes, however, many plant pathogenic fungi and at least some, but probably many, mycorrhizal fungi are also significant cellulose decomposers. Many plant pathogens appear to utilize lignocellulosic material in the soil as a means of survival in this habitat in the absence of a host plant and also probably as a means of building up an adequate inoculum potential for subsequent attack upon the host, as in the case of *Rhizoctonia solani* [62]. It is not clear to what extent mycorrhizal fungi utilize soil cellulosic material or how important such an ability is to them or their symbiosis with plants. Typical mycorrhizal basidiomycetes such as *Amanita citrina* and *Suillus variegatus* have been found to produce small quantitites of cellulolytic enzymes, which might indicate some utilization of cellulosic material in soil [60]. The orchidaceous mycorrhizal fungus *Rhizoctonia repens* has been shown to be capable of utilizing cellulosic material in competition with other soil microorganisms and, in a separate experiment, to be able to maintain a mycorrhizal association supporting growth of orchid seedlings, while growing on cellulose [63].

D. Cellulose Decomposition by Aquatic Fungi

There are three distinct bulk water environments in which aquatic fungi occur and where significant cellulose decomposition by fungi takes place. By far the largest is, of course, the sea, which harbors a small but quite distinct population of marine fungi. Freshwater environments may be divided into two on the basis of mobility of the water, which is critical in terms of aeration. Still waters, such as lakes and ponds, become rapidly anaerobic with depth, and it is on the anaerobic bottom of such environments that cellulosic material accumulates and most decomposition takes place. Thus, the species involvement and nature of cellulose decomposition here is quite distinct from that which occurs in rapidly flowing streams and rivers, which are largely aerobic and within which plant material is often carried along by the flow rather than being deposited on the bed.

Cellulolysis by Marine Fungi

It is probably due to a combination of its high ionic strength and the relative paucity of nutrients that the sea offers a highly restrictive environment for fungi. Compared to the total of approximately 63,600 species of nonlichenized true fungi known in terrestrial environments [6], the number of fungi isolated from the marine environment up to 1979 was only 488 species [67]. To a certain extent this figure must represent a degree of ignorance of the marine fungal population due largely to technical difficulties associated with isolating and growing marine fungi. Indeed, just 7 years after the compilation of that total, Kohlemeyer [68] reported that a further 32 species of ascomycetes (approximately an additional 10%) had been identified. However, it is nevertheless apparent that the diversity of marine fungi is far less than that of the land fungi. Of all the major fungal groups, it is the basidiomycetes that seem least able to cope with the marine environment. Only 7 species of marine basidiomycetes were recorded in 1979 [67], compared to about 16,000 terrestrial species [6]. Thus, basidiomycetes constitute just under 1.5% of marine fungi as opposed to about 25% of land fungi. The group showing the major gain is the lower fungi (mastigomycetes and zygomycetes), which with 100 known marine species [67] account for about 20.5% of the marine fungi compared to just 3% (1935 species) of terrestrial species [6]. This is not entirely surprising when it is noted that this group includes the mastigomycetes, which have motile zoospores, clearly a specific adaptation to an aqueous environment. In terms of cellulose decomposition, however, it is the extreme rarity of marine basidiomycetes that is of major interest. This is because in the terrestrial environment the basidiomycete fungi play the most important role, at least quantitatively, in the decomposition of cellulose, mainly because wood-rotting basidiomycetes are essential to the rapid liberation of the largest single source of cellulose. However it is worth noting that one major reason the basidiomycetes are so successful in degrading cellulose in wood is because many species can also degrade lignin and so remove the major protective polymer surrounding the cellulose. This may go some way toward explaining why there are very few marine basidiomycetes, because there is virtually no lignified cellulose produced in the marine environment. Of course lignocellulose occurs in the sea, but it is due to terrestrial material being accidentally or deliberately introduced, largely as wood, and is therefore present as small discrete items and in very small quantity relative to the volume of the oceans. The major truly marine source of cellulose is seaweed, which does not contain lignin. This is a very substantial source of cellulose often overlooked in considerations of cellulose production and decomposition, but it is of considerable interest specifically because it constitutes the only major source of lignin-free cellulose. Quantitatively this type of cellulose far exceeds the amount of lignified cellulose in the oceans and is much more widely distributed. About one third of all marine fungi are associated with algae and are found together with bacteria and protozoa growing on rotting seaweed washed up on the shore [67]. It is believed that fungi are probably less important in the decomposition of cellulose in most seaweeds than bacteria, with the exception of the Phaeophyta (brown algae). This is because most types of seaweed are relatively thin and delicate, whereas the brown algae are much more resistant and on such material the penetrative capacity of fungi provides them with considerable competitive advantage over bacteria [67].

Information about enzymes produced by marine fungi is sparse, and it is difficult to be sure of how widespread the ability to decompose cellulose is among obligate marine species, especially because many isolations are made from woody material, which would be expected to favor cellulolytic species. Certainly the few species of marine basidiomycetes seem to be capable of degrading cellulose. The wood-inhabiting species *Nia vibrissa*, *Digitatispora marina*, and *Halocyphina villosa* have all been shown to have the characteristics of white-rot fungi, including the ability to degrade crystalline cellulose as well as lignin [69,70]. No obligate marine brown-rot basidiomycetes have yet been isolated, possibly because their characteristic of degrading cellulose at a distance using a diffusible chemical and/or enzymic system is simply not practicable in a bulk water environment [71]. Many marine ascomycetes also show strong cellulolytic activity. All 37 isolates of the pyrenomycete genus *Lulworthia* tested for cellulolytic capacity proved positive, although all such strains were obtained from woody substrates [72]. One of the most extensive surveys of enzymic activity in marine fungi was reported by Molitoris and Schaumann [73], who collected a wide range of fungal species from marine environments. These included both obligate marine species and also isolates from marine habitats, such as estuaries and brakish waters, which could harbor incidental or opportunistic inhabitants. Crucially, however, they obtained isolates from a wide range of ecological groups including wood inhabitants, leaf-litter degraders, soil or sediment inhabitants, plant root–associated species, parasitic species, and species free-floating in seawater. Of 35 obligate marine species of fungi, they found cellulolytic capability in representatives of all the ecological types listed above. There was, however, variation between the ecological groups in the proportion of species showing cellulolytic activity, corresponding in general to the extent to which each group might be expected to encounter and utilize cellulose as a significant carbon source. Thus, a high percentage of cellulolytic species were found among the leaf-litter degraders, where 11 out of 18 species tested (61%) showed cellulolytic activity, followed closely by 10 out of 20 wood-inhabiting species (50%). The lowest percentage of species showing cellulolytic activity were found among the free-floating isolates, 2 out of 11 (18%); and the parasitic species, 2 out of 10 (20%). An intermediate proportion, 8 out of 18 (44%), of soil and sediment inhabitants showed cellulolytic activity. Based on this survey it seems reasonable to believe that the ability to decompose cellulose is widespread among marine fungi, both in terms of taxonomic groups and in the diversity of habitats occupied by cellulolytic fungi.

Cellulose Decomposition in Streams and Rivers

Just as there are very few marine species of fungi in comparison to those of the terrestrial community, there are also relatively low numbers of fungal species present in freshwater environments. Basidiomycetes are again very poorly represented, and the two major groups of freshwater fungi are the saprolegniales (water molds) and the aquatic hyphomycetes, although a significant number of ascomycetes and other mastigomycetes do also occur in freshwater habitats. It is the hyphomycetes that are the major decomposers of cellulose in freshwater. The saprolegniales are apparently entirely noncellulolytic [58]. Cellulolysis is carried out by a few Mastigomycetes, notably a few species of *Pythium* [33,74], although the majority of *Pythium* species are either noncellulolytic or only capable of weak cellulose decomposition [75]. Similarly a few chytrids are capable of cellulolysis [58]. The ascomycete species isolated from freshwater environments appear in general to be capable of considerable cellulose decomposition [75], although because such isolations are frequently made from cellulosic material, such as leaves and wood, this conclusion could be misleading. However, lignocellulosic substrates appear to be by far the most significant habitat of such fungi in freshwater, and it is therefore likely that cellulolytic capacity is a truly general capability of freshwater ascomycetes. It is not surprising that the vast majority of freshwater fungi should be strongly cellulolytic because the overwhelming majority of carbon sources available in this environment are provided by lignocellulose, predominantly from leaves but also a significant quantity from wood [75]. At least 1 kg of leaves per year can be deposited into each meter of the length of a stream passing through a wooded valley [58], far exceeding any other supply of carbon.

Cellulose Decomposition in Stagnant Freshwater

In contrast to the turbulent waters of streams and rivers, where a considerable proportion of the cellulosic debris is carried along in the body of the water in relatively well-aerated conditions, the bulk of the cellulosic material is found as a dense accumulation on the bottom of stagnant waters. Here the environment is virtually anaerobic, and any cellulolysis that does occur is slow. It is believed that most, possibly all, cellulose degradation is carried out by bacteria. However, this may simply reflect ignorance of what actually occurs in this ecological niche. It is clear from the example of the animal rumen that fungi can be involved in cellulolysis in anaerobic environments. Certainly if such material is dredged from the bottom of stagnant waters and incubated in appropriate conditions characteristic fungal species, mostly of the hyphomycete genera *Helicoon* and *Helicodendron*, can be identified and are found to be consistent inhabitants of such substrates

[76]. It is probable, however, that despite some degree of decomposition, the bulk of the cellulosic material reaching the bed of stagnant waters simply becomes incorporated into the soil, largely undegraded.

E. Cellulolysis by Rumen Fungi

A major but discrete and highly selective natural site of cellulolysis is the digestive tract of ruminant animals. The animals themselves are dependent upon the products of cellulolysis, yet are unable to decompose cellulose and rely entirely on the cellulolytic activities provided by microbial inhabitants of the rumen. The rumen is a very anaerobic environment, which supports a mixed population of bacteria, fungi, and protozoa. The presence and activity of fungi in the rumen was only established relatively recently, due to the fact that the species present are chytridiomycetes, which produce motile zoospores. It is the zoospore stage of the life cycle of these fungi that is most readily detected in samples taken from the rumen, and for many years such cells were considered to be protozoa. This confusion was probably at least partly due to the fact that the rumen-inhabiting chytridiomycetes are unusual in having multiflagellated zoospores bearing about 8–17 flagella [77] compared to the uniflagellate zoospores, which are typical of other chytridiomycetes. Zoospores from the rumen were therefore confused with ciliate protozoa. The genus *Neocallimastix* was created for such anaerobic rumen-inhabiting chytrids, and subsequently other genera have also been identified, including *Sphaeromonas* and *Piromonas*. Zoospores of these fungi rapidly invade fresh plant material entering the rumen [9], and several such fungal species have now been demonstrated to produce extracellular enzymes with cellulolytic action including endo 1,4-β-glucanase and cellobiohydrolase activities [78]. It is therefore clear that such fungi are directly involved in cellulolysis in the rumen, but it is not currently known how important or extensive the contribution of fungi is to the total cellulose decomposition in this environment. It is apparent that the degradation of plant material in the rumen is carried out by a complex consortium consisting of many of the microorganisms present, and there is some evidence that there is true synergy between their various activities rather than merely competition for the substrate. For example, the cellulolytic activity of *Neocallimastix frontalis* is increased when it is grown in the presence of methanogenic rumen-inhabiting bacteria as compared to growth in the absence of other organisms [11]. Such beneficial interactions between species may not be just synergism between different types of cellulolytic enzyme activity, but could be due to noncellulolytic processes which create circumstances in which cellulolysis may proceed more efficiently or quickly. For example, at least some of the cellulolytic enzymes produced by rumen fungi are, as found in many other environments, subject to catabolite and end-product repression [78,79]. Some of the noncellulolytic rumen microorganisms may facilitate celluloysis by utilizing the end products of cellulose decomposition and so minimize end-product inhibition of cellulolytic enzymes. This may well explain the interaction between *N. frontalis* and the methanogenic bacteria.

F. Fungal Decomposition of Cellulose in Herbivore Dung

Even though significant decomposition of cellulose takes place in the guts of herbivores, particularly ruminants, considerable quantities of cellulose still remain in material discarded as dung. Even in runinants, only about

30 – 60% of lignocellulosic material eaten by the animal is degraded [9]. As in most other environments, the cellulose in herbivore dung is degraded by a consortium of many different microbes and microfauna, but the fungi are the most apparent and are responsible for the largest proportion of the cellulolysis. True cellulose-decomposing ascomycetes such as *Sordaria fimicola* and species of *Podospora* and *Ascobolus* are often the first species to produce macroscopic fruit bodies on such a substrate generally followed by the appearance of the basidiocarps of the cellulose and lignin degrading basidiomycetes, of which members of the genus *Coprinus* are the most common. It is now clear that most of these species begin to grow on this substrate very soon after it becomes available, and what was once thought to be a classic succession depends rather more on the time required by each species to produce its fruit bodies than on any sequence of colonization by the different organisms [54]. This substrate is therefore colonized by many species at about the same time. Being considerably enriched in nitrogen compared to the original plant material it contains, there is little or no nitrogen limitation of cellulolysis. Herbivore dung is thus the site of intense competition between cellulolytic fungi and bacteria and a habitat for noncellulolytic microbes able to utilize low molecular weight compounds in fresh dung or released by cellulolysis [54].

G. Cellulolysis in the Guts of Nonruminant Animals

Ruminant animals are by far the most significant site of animal-based decomposition of cellulose, but cellulolysis occurs to some extent in the guts of all plant-eating animals, even man. Most, possibly all, enzymic decomposition of crystalline cellulose in animal guts is due to the activity of enzymes produced by microorganisms, either as residents of the alimentary tract or as colonizers of cellulosic material subsequently ingested by animals. There are a considerable number of nonrunimant members of the animal kingdom in whose guts cellulolysis has been found to occur at significant levels. These organisms are mainly crustacea, molluscs, and insects, but also include significantly the earthworm, *Lumbricus terrestris* [58]. Such organisms can play a very critical role in cellulose breakdown, particularly in wood, leaf litter, and in the soil. A major part of their importance is merely due to their chewing and otherwise disrupting the integrity of lignocellulosic material, rendering it more accessible to enzymic breakdown, then returning it to the environment in the nitrogen-enriched surroundings of fecal pellets. However, there is also clear evidence of cellulolytic enzyme activity in the guts of these organisms. It is unlikely that all such activity is of truly animal origin, and the available evidence suggests that the ability to decompose crystalline cellulose is in fact entirely of microbial origin. Some insects have been reported to produce enzymes capable of degrading soluble cellulose, but even for this type of cellulolysis, the majority of the activity is thought to be due to enzymes of microbial origin [80].

Two of the best studied examples of insects which utilize fungal enzymes to decompose cellulosic material within their guts are the fungus-growing higher termites and the Attine ants [80]. It has been demonstrated that the capacity to decompose crystalline cellulose within the gut of the termite *Macrotermes natalensis* is entirely due to the activity of enzymes derived from the basidiomycete fungus, a species of the genus *Termitomyces*, which it grows on a mass of chewed plant fragments and subsequently eats [81].

It is not, however, clear whether the enzymes are derived directly from the fungal biomass eaten or whether they are acquired together with ingested cellulosic material into which they have been released by the fungus. The fungus grown by the Attine ants produces a similar complement of cellulolytic enzymes, and cellulolysis in the guts of these ants is believed to occur as a result of the activity of fungal enzymes in a very similar manner to that demonstrated for the termites [80,81].

Among wood-boring animals, those that obtain any nutrition from the polymeric constituents of the wood they penetrate often harbor cellulolytic microorganisms in their guts. There is considerable variation in the extent to which bacteria or fungi are believed to be important in this respect. For example, in the marine environment it has been reported that in the guts of teredinids (shipworms), which are wood-boring molluscs, fungal cellulolysis is of little importance, and that mediated by bacteria is of major significance, whereas, fungi are considered to be very significant in the wood-based nutrition of gribbles, which are wood-boring crustacea. In this case once again, it is not clear to what extent the cellulolytic fungi are indigenous to the animal gut or whether they are merely colonizers of the woody tissue eaten by the animal, which then utilizes the products of normal fungal decomposition of the cellulose in the wood [82].

With few exceptions the whole topic of cellulolysis in nonruminant animal guts remains rather confused, and available information is still fragmentary [83]. Suggestions about the relative importance of bacteria as opposed to fungal cellulolysis and whether such activity is indigenous to the gut or merely carried in on already colonized tissue probably merely reflect the lack of information available rather than the true situation in most species.

H. Cellulolysis by Plant Pathogenic Fungi

Even though cellulose is the major macromolecular constituent of plants, there are several means by which a plant pathogen may invade a host without needing to produce cellulose degrading enzymes. Penetration into plant tissue may be achieved quite simply by entry through the plant's natural openings, such as stomata, or by utilizing a route through noncellulosic areas, for example through the cuticle and between cells in the non–cellulose-containing pectin-rich intercellular regions. Alternatively, passage through the cellulose-rich cell walls might be achieved by inducing the plant to activate its own cellulolytic enzymes as part of an autolytic response, a common feature of some plant defense mechanisms following invasion by a pathogen. Clearly, then, a plant pathogenic fungus would not need to produce cellulolytic enzymes to invade a host plant, however, many and probably most such species do produce cellulose-degrading enzymes [7,84–86], often including all the enzyme activities necessary for the decomposition of crystalline cellulose [84]. What is not clear, however, it whether such production of cellulolytic enzymes is essential to the success of the pathogen in the initial invasion of the plant, or merely an incidental reflection of the need for the organism to produce suitable enzymes to mobilize the nutrients available to it, of which cellulose would form a major part. In at least some species, such as *Fusarium oxysporum*, it has been suggested that there may be a direct link between cellulolytic activity and pathogenesis [86]. However, while it is clear that cellulose decomposition is important and probably essential to the continued growth of a plant pathogen in its host, it is generally considered that cellulolytic activity is rarely, if ever, important for invasion of the host plant by fungal pathogens. The major reason for this view is that cellulolytic activity generally only becomes detectable some time after invasion has been achieved, whereas pectolytic enzyme activity is detected very early in the attack upon the host plant, and it is these enzymes together with the less clearly defined "cutinases" that are thought to be among the most important for invasion of the plant host [87].

V. LABORATORY INVESTIGATIONS OF FUNGAL CELLULOLYSIS

A. Approaches to Observation of Cellulolysis In Vitro

Laboratory-based studies of cellulolysis require the provision of a cellulose substrate, which is not always a straightforward matter. This fact is illustrated by the great diversity of experimental cellulose and cellulose-derived substrates available and which have been used for investigations of cellulolysis. The most frequently used substrates are listed in Tables 1 and 2, together with a brief description of the nature and major experimental usefulness of each. Such a large number of potential substrates stems partly from the fact that there is a wide range of natural cellulose-containing materials, most of which can be employed in some form in laboratory experiments. However, most of the available substrates are cellulose modified to some extent or the modified and unmodified breakdown products from cellulose decomposition. This reflects the need to provide experimental systems capable of modeling the various states in which cellulose is known to occur in natural substrates, notably amorphous; highly crystalline; a mixture of amorphous and crystalline; or partially degraded cellulose in which a variety of breakdown products are available. A fundamental dichotomy between types of substrate other than whether they are natural or modified, is whether they are soluble or insoluble. This is important because many apparently cellulolytic organisms are capable of decomposing soluble cellulose but incapable of decomposing insoluble cellulose, whereas organisms capable of decomposing insoluble cellulose are always capable of degrading soluble substrates. As discussed earlier, this distinction is diagnostic of the capacity of an organism to produce a key type of cellulolytic enzyme capable of attacking highly crystalline cellulose and opening up its structure for attack by other types of enzyme. Because cellulolysis is known to be achieved as a result of the combined activity of a number of different types of enzyme, many of which act on cellulose molecules at different stages of decomposition, the various breakdown products must be available as pure substrates to permit investigations of the nature and specificity of individually purified types of enzyme. Finally, some of the substrates are very specifically modified to provide an experimentally convenient means of detecting the result of enzymic hydrolysis, such as the release of a fluorescent or colored product as in the case with methylumbelliferyl-oligosaccharides or p-nitrophenyl-oligosaccharides.

Once an appropriate substrate for experimental studies of cellulolysis has been selected, the other key requirement is a means of adequately quantifying the cellulolytic activity that ensues. This can be relatively straightforward in cell-free systems or in experiments utilizing microorganisms grown in simple defined media. In both of these cases, relatively straightforward biochemical assays for products of cellulolysis, such as

TABLE 1. Soluble Substrates Used in Experimental Investigations of Cellulolysis

Substrate	Nature of substrate	Experimental usage
Carboxymethyl cellulose (CMC)	Chemically modified cellulose with various proportions of carboxy methyl substituents	Most commonly used form of relatively high molecular weight soluble cellulose
Hydroxyethyl cellulose (HEC)	A variant of CMC with varying degrees of substitution with hydroxy ethyl residues	Alternative to CMC
Methylumbelliferyl-β-D-cellodextrins (MUC)	Various β-1,4-linked glucose polymers with fluorochrome substituents added chemically	Detection of cellulolytic activity of fluorimetric methods
p-Nitrophenyl-β-D-cellodextrins	Various β-1,4-linked glucose polymers with chromophore substituents added chemically	Detection of cellulolytic activity using direct spectrophotometric methods
β-Glucans	Low molecular weight glucose oligosaccharides containing types of linkage additional to, or instead of, β-1,4 links	Investigation of capacity of cellulolytic enzymes or organisms to hydrolyze glycosidic linkages in addition to β-1,4 links
Cellodextrins	Various β-1,4-linked glucose polymers containing four or more monosaccharides	Investigations of the later stages of cellulose decomposition
Cellotriose	β-1,4-linked glucose trisaccharide	Minimum size substrate for cellobiohydrolase activity
Cellobiose	β-1,4-linked glucose disaccharide	Investigations of end-product inhibition and of β-glucosidase activity

TABLE 2. Insoluble Substrates Used in Experimental Investigations of Cellulolysis

Substrate	Nature of substrate	Experimental usage
Wood	Natural bulk substrate of cellulose embedded in a matrix of complex polymers	Decomposition of bulk and complex substrate in a quasi-natural environment
Lignocellulosic waste, including wheat straw; sugar cane bagasse; sawdust; etc.	Natural waste materials consisting of cellulose associated with complex polymers, especially lignin. Usually partially fragmented and degraded	Investigations of the decomposition of natural waste. Fragments are convenient for introducing semi-natural substrate into solid and liquid growth media
Valonia cellulose	Cellulose purified from algal cell walls	Studies of decomposition of crystalline cellulose
Microcrystalline bacterial cellulose	Only naturally occurring pure cellulose free from associated polymers	Especially useful for electron microscopic studies of cellulolysis and as the only source of pure native cellulose
Cotton	Purified natural plant cellulose, highly crystalline	Convenient source of relatively pure, highly crystalline cellulose
Rayon	Fabric made of chemically modified, purified plant cellulose	Convenient form of cellulose for colonization studies when submerged in soil or water
Cellophane	Transparent sheet of purified, reconstituted plant cellulose	Allows direct microscopic examination of colonizing organisms and extent and nature of degradation
Filter paper	Purified and reconstituted plant cellulose	Convenient solid source of cellulose in petri dish culture. Used particularly to measure weight loss due to cellulolysis
Avicel Solka floc Whatman CF or CC	Highly crystalline purified cellulose (commercial products)	Studies of decomposition of crystalline cellulose

TABLE 2. (Continued)

Substrate	Nature of substrate	Experimental usage
Phosphoric acid swollen cellulose (Walseth cellulose)	Purified cellulose chemically treated to expand the fibers and render molecules of cellulose accessible to enzymes	Studies of organisms and isolated enzymes unable to degrade crystalline cellulose
Amorphous cellulose	Purified cellulose processed to reduce crystallinity	General class of substrate of which Walseth cellulose is a specific example
β-Glucans	High molecular weight glucose polymers, often branched, containing types of β-linkage additional to, or instead of, β-1,4 links	Investigation of ability of cellulolytic enzymes and organisms to hydrolyze glycosidic linkages in addition to β-1,4 links

glucose or cellobiose, or direct spectrophotometric measurement of concentrations of fluorescent or colored substituents released from appropriate substrates can be utilized. Somewhat more technically complex and always less precise is the frequently utilized viscometric methods for determining the degree to which highly polymeric cellulose has been cleaved by the action of cellulolytic enzymes or organisms. In less well-defined systems, however, one of the better means of quantifying cellulose decomposition is by determining weight loss of a convenient substrate, such as pieces of rayon, cotton, cellophane, or filter paper, all of which can be incorporated either directly into natural environments or introduced into ill-defined artificial cultures. Probably the most widely used of such procedures is the determination of filter paper weight loss, doubtless largely due to the convenience of this substrate being available ready-cut in convenient sizes, easily weighed, and in appropriate shape for petri dish cultures. Garrett [54,88] has developed probably the best system for quantification of cellulose decomposition by fungi based on filter paper weight loss. It involves not simply the determination of absolute weight loss per unit of fungal biomass, but relates that activity to hyphal extension rate in a parameter he has called the cellulolysis adequacy index (CAI). The actual basis of the index is arbitrary in that it is calculated by dividing percentage weight loss of filter paper after 7 weeks incubation at 22.5°C by the colony radial extension rate expressed as mm/day of the same strain grown at the same temperature on potato dextrose agar (PDA). The reason for this is that it was originally developed for use with soil-inhabiting fungi, and these are standard growth conditions for such organisms. However, the CAI has proved to be of general application to a wide range of fungi and provides a convenient means of comparing many different species [34]. CAI values range from 0 up to about 5 [34,54]. Species considered weakly cellulolytic, such as members of the genus *Pythium*, have values significantly below one, whereas typically strongly cellulolytic soil fungi, such as *Botryotrichum piluliferum* (syn. *Chaetomium piluliferum*), have CAI values approaching 5 [34]. The reason such a parameter is of particular value is that relating cellulolytic capacity to some measure of growth rate provides an indication of the capacity of a fungus to release low molecular weight carbon compounds from cellulose in relation to the demand it must have for such molecules to support its rate of growth. Many species of *Pythium* appear to have moderate cellulolytic capacities if only their ability to cause weight loss of filter paper is assessed, but because such fungi have relatively fast extension rates, it becomes clear that the contribution cellulolysis can make to their overall demand for carbon compounds to support a high growth rate is small. This is clearly shown by calculation of the CAI [34]. A slowly extending fungus with approximately the same ability to cause weight loss from filter paper would nevertheless derive a much greater proportion of its carbon nutrition from its cellulolytic activity, and this would be demonstrated by a high CAI value for such an organism. Deacon [34] has suggested that the general concept on which CAI calculation is based, relating cellulolytic ability to some measure of carbon nutrient demand, could be extended to the calculation of several similar but possibly even more generally applicable indices. The maximum specific growth rate (μ_{max}) of each organism is clearly the only adequately absolute parameter to which cellulolytic ability could be satisfactorily related. However, for many filamentous fungi a value for this parameter is still difficult to determine, and it would generally not reflect the growth rate in the natural environment, where the specific growth rate is likely to be significantly below maximum. Because such an index is primarily useful for indicating relative competitive ability of different species on natural cellulosic substrates or as an indicator of whether a given species is more likely to be saprophytic or plant pathogenic, colony radial extension rate on a standard medium may provide the most convenient estimate of growth rate despite its limitations in that respect [89]. On a limited number of filamentous fungi, Deacon [34] has demonstrated a correlation between CAI and degree of pathogenicity against cereal roots. The lower the CAI value of a strain, the greater its pathogenicity, whereas strains with high values of the CAI were likely to be saprophytic.

B. Mathematical Models of Cellulose Degradation

Several attempts have been made to formulate mathematical descriptions of cellulolysis. The function of these mathematical models has been either to obtain a quantitative description of cellulose degradation, thereby facilitating the optimization of the process, or to serve as a vehicle for hypotheses from which experimentally testable predictions can be made. In the latter case the goal of such investigations has been a better understanding of how the system works. Rather arbitrarily, we may distinguish two types of models: those that consider cellulose decomposition in a well-mixed system and those in which the process takes place in a heterogeneous environment such as a column of soil. The former class of models has been associated mostly with biotechnology and the utilization of cellulose for sugar production, while the latter have applications in ecology.

Holtzapple and collaborators [90] developed a kinetic model of cellulolysis in a homogeneous system from first principles and compared it and models suggested by other workers with results from both batch and continuous culture experiments. The reactions scheme for their HCH-1 model, as they called it, is shown in Figure 1. Here, free enzyme (E^f) combines reversibly with free sites on cellulose (G^f_x) to produce adsorbed enzyme (E^a). In this reaction, δ is the adsorption constant. The product of the reaction combines with free cellulose (G_x) to form enzyme–substrate complex (EG_x). Here η is a complexing constant. This complex yields soluble products (G_s) and free enzyme at rate k (a kinetic constant). Product inhibition is represented by G_s binding with free enzyme, adsorbed enzyme, and enzyme–substrate complex in the manner indicated. For simplicity, the product binding constant (β) is assumed to be identical for all enzyme species. The resulting mass balance equation is:

$$\frac{dG_s}{dt} = \frac{k\ G_x\ E\left(\dfrac{1}{1 + \beta G_s}\right)}{\alpha + \theta G_x + \varepsilon E}$$

where k and α are lumped constants, β is a product-binding constant, G_s, G_x, and E are concentrations of the substances previously defined, ε is total enzyme concentration, and θ is the fraction of total cellulose sites which are free, and is defined by

$$\theta = \frac{G_x - \alpha - \varepsilon E + \sqrt{(G_x - \alpha - \varepsilon E)^2 + 4\alpha G_x}}{2G_x}$$

The authors compared this model with four others with assumptions and balance equations as follows:

Michaelis-Menten kinetics with competitive product inhibition [91]:

$$\frac{dG_s}{dt} = \frac{k\ G_x\ E}{\alpha(1 + \beta G_s) + G_x}$$

Michaelis-Menten kinetics with noncompetitive product inhibition [92]:

$$\frac{dG_s}{dt} = \frac{k\ G_x\ E\ (1 + \beta G_s)}{\alpha + \beta G_s}$$

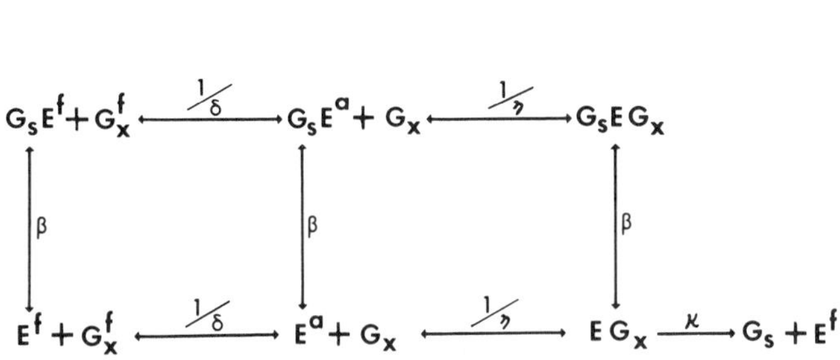

FIGURE 1. Reaction mechanism on which the HCH-1 model of enzymic cellulose hydrolysis of Holtapple and associates is based. (From Ref. 90.)

Enzyme attachment to the cellulose surface via a Langmuir adsorption equilibrium and competitive product inhibition [93]:

$$\frac{dG_s}{dt} = \frac{k\,G_x\,E}{\alpha(1 + \beta G_s) + G_x + \varepsilon E}$$

The HCH-1 model with the simplifying assumption that $\theta = 0$:

$$\frac{dG_s}{dt} = \frac{k\,G_x\,E\left(\frac{1}{1 + \beta G_s}\right)}{\alpha + G_x + \varepsilon E}$$

On the basis of their experimental data, Holtzapple and collaborators [90] considered that all except that HCH-1 model could be invalidated.

Mandenius and associates [94] investigated kinetic models of enzymic cellulose degradation using a two-phase aqueous system. Because the enzymes involved in cellulolysis are strongly inhibited by cellobiose and glucose, separation of reactants and products by such a system has the potential of significantly enhancing productivity. The authors tested the HCH-1 model and two further models. One of these is based on the model of Gong and associates [95], which takes the form:

$$\frac{dG}{dt} = \frac{c}{\left(1 + \frac{G}{K_{i1}}\right)c + \left(1 + \frac{G}{K_{i2}}\right)K}$$

where G and C are the concentrations of glucose and cellobiose, respectively, K_{i1} and K_{i2} are inhibition constants, and K is the Michaelis constant. The basis of this equation is that cellobiase (β-glucosidase) is noncompetitively inhibited by glucose. The third model, based on the work of Wald and associates [96], assumes that there are different kinetics for the hydrolysis of crystalline and amorphous cellulose. For concentrations of these two substrates at G_{2c} and G_{2a} respectively, the kinetic equations employed were:

$$\frac{dG_{2c}}{dt} = k_4 E_{ac}$$

$$\frac{dG_{2a}}{dt} = k_5 E_{aa}\left(\frac{1}{1 + K_{i4}\,G_2}\right)$$

where k_4 and k_5 are rate constants, E_{ac} and E_{aa} are enzyme concentrations, G_2 is the cellobiose concentration, and K_{i4} is the cellobiose inhibition constant. The authors found that all three models were applicable to their experimental system with relatively minor differences in their fit to the data.

A considerably more complex model of cellulose hydrolysis than those referred to above was proposed by Peitersen and Ross [97]. This considered growth and enzyme production by *Trichoderma* and consisted of 12 ordinary differential equations containing 38 parameters. It was based on

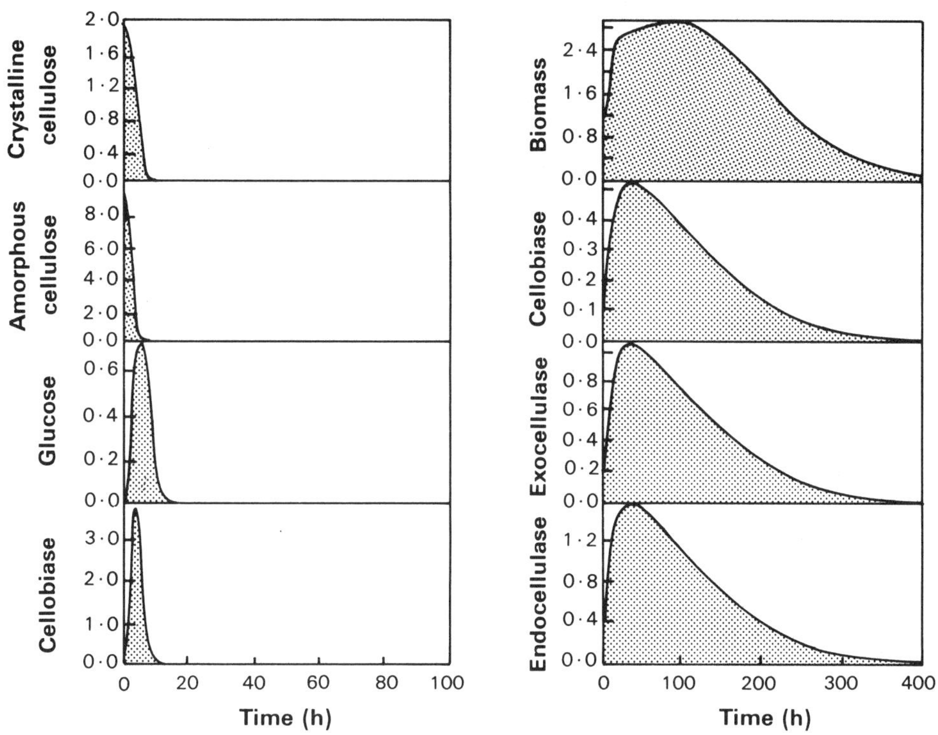

FIGURE 2. Simulation of cellulose digestion in a column reactor based on the model of Theodorou and associates. (Reproduced from Ref. 100.)

the following assumptions: crystalline cellulose is converted to amorphous cellulose before it is enzymatically degraded, cellobiose and glucose result from the enzymic action of "cellulase" and cellobiase, the kinetics of the system are pH dependent, Michaelis-Menten kinetics operate, all reactions are subject to product inhibition, and the specific growth rate of *Trichoderma* can be described by a Monod [98] function. Similar assumptions were adopted by Theodorou and collaborators [99,100]. However, these authors investigated cellulose digestion by *T. reesei* in a column reactor analogous to a soil column [101]. Thus, their work was ecologically directed rather than having a biotechnological emphasis. Their experimental system consisted of a glass tube packed with small glass beads interspersed with cellulose and inoculated with *T. reesei* spores. The fungus was allowed to germinate and grow for a short period of time and then carbon-free nutrient solution was passed down the column and the effluent analyzed for the products of cellulolysis. The system was modeled along the lines of Peitersen and Ross [97] except that the column was considered in terms of a series of well-mixed compartments, each compartment (except for the first and last) receiving the effluent from the one above and discharging into the one below. Numerical solution of the large set of

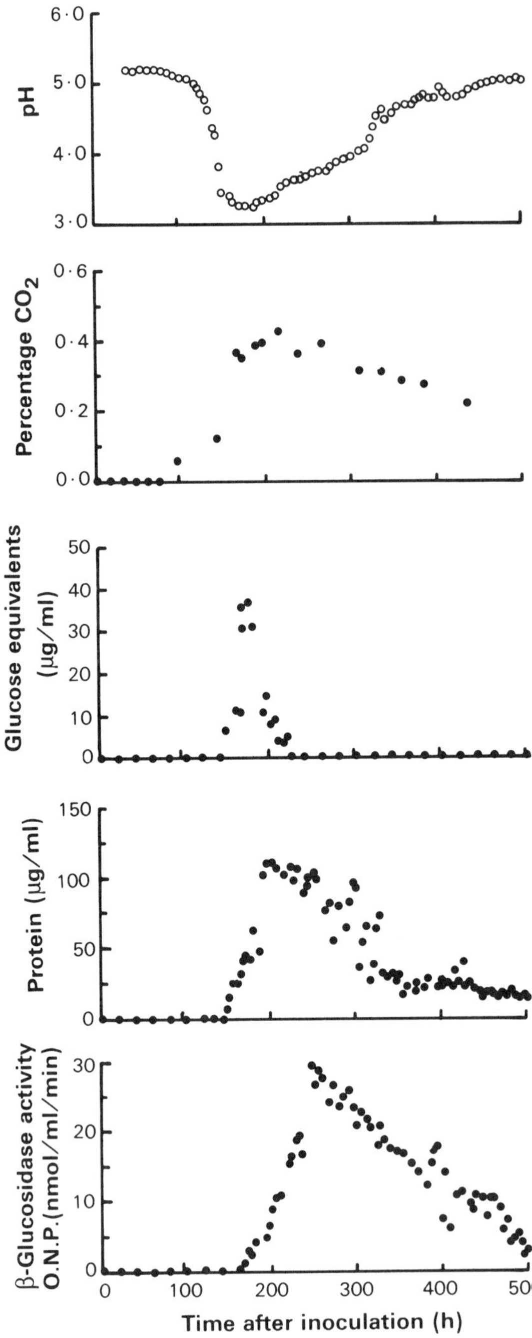

FIGURE 3. The characteristic properties of the effluent from a continuous-flow column containing *Trichoderma reesei*. The column initially contained 0.5 g cellulose and was supplied with glucose-free medium for the first 500 hr after inoculation. (Reproduced from Ref. 100.)

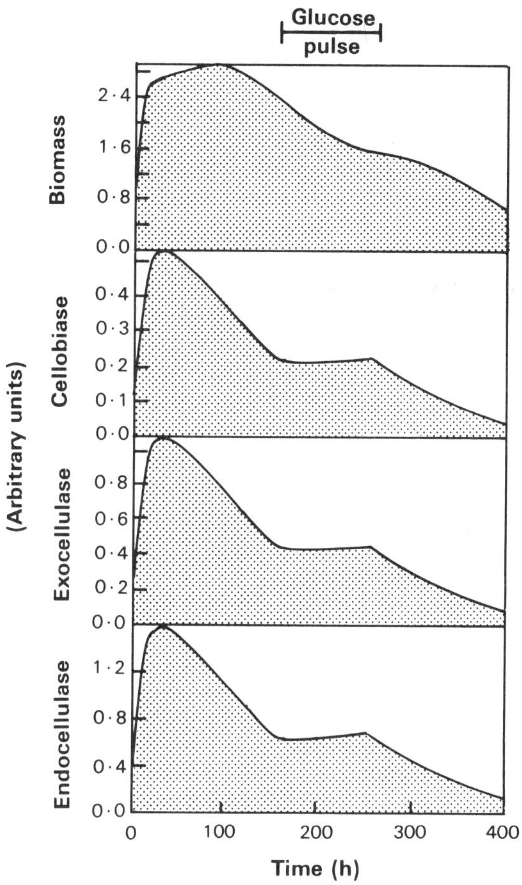

FIGURE 4. Simulated effect of pulsing glucose down a continuous flow column based on the model of Theodorou and associates. (Reproduced from Ref. 100.)

nonlinear ordinary differential equations that resulted simulated the dynamic behavior of the system. It was found that when the system was close to steady state or was changing only slowly, the experimental results matched the predictions of the model, at least qualitatively. Sudden changes in the system, such as that resulting from a pulse of glucose in the input medium, caused significant differences between predicted and experimental results. Examples of the results obtained by Theodorou and associates are shown in Figures 2 to 5.

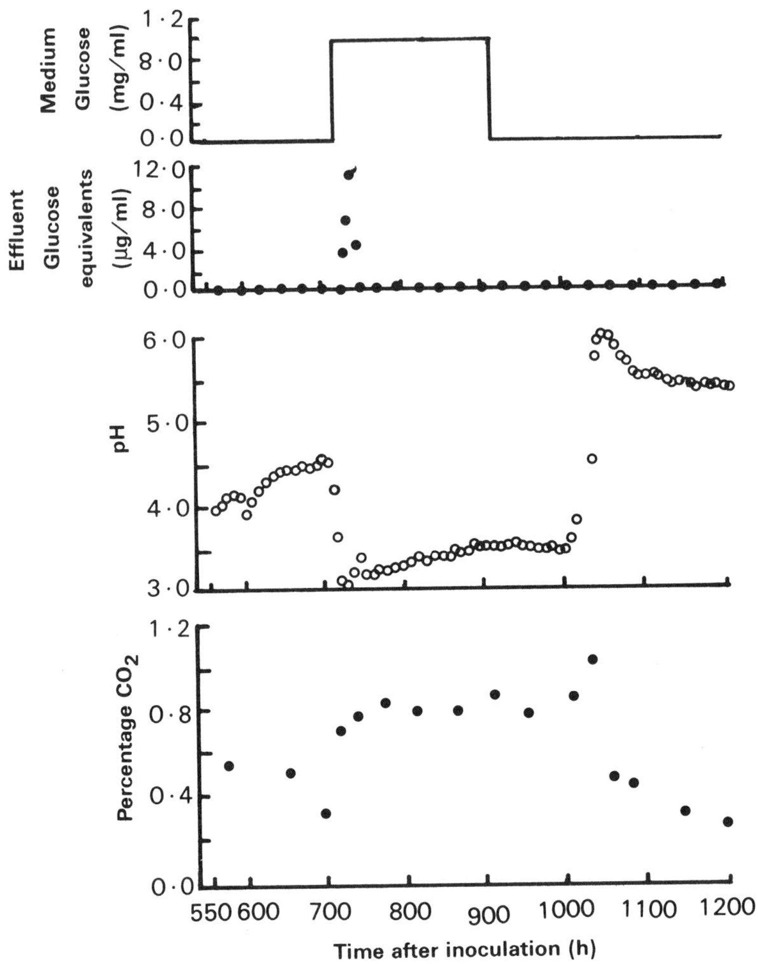

FIGURE 5. Effect of pulsing glucose into an experimental continuous-flow column containing cellulose and *Trichoderma reesei*. (Reproduced from Ref. 100.)

VI. APPLICATIONS OF FUNGAL CELLULOLYSIS

The commercial exploitation of fungal cellulolysis is, at present, very limited. However, in view of the annual accumulation of huge quantities of lignocellulosic waste, which is not only freely available, but often has to be disposed of, the potential for economically viable processes which exploit fungal degradation of such material must be enormous.

A. Cultivated Mushrooms

Mushroom production is currently the only application of fungal lignocellulosic decomposition commercially viable on a significant scale. Just one species, *Agaricus bisporus*, accounts for about 70% of worldwide production

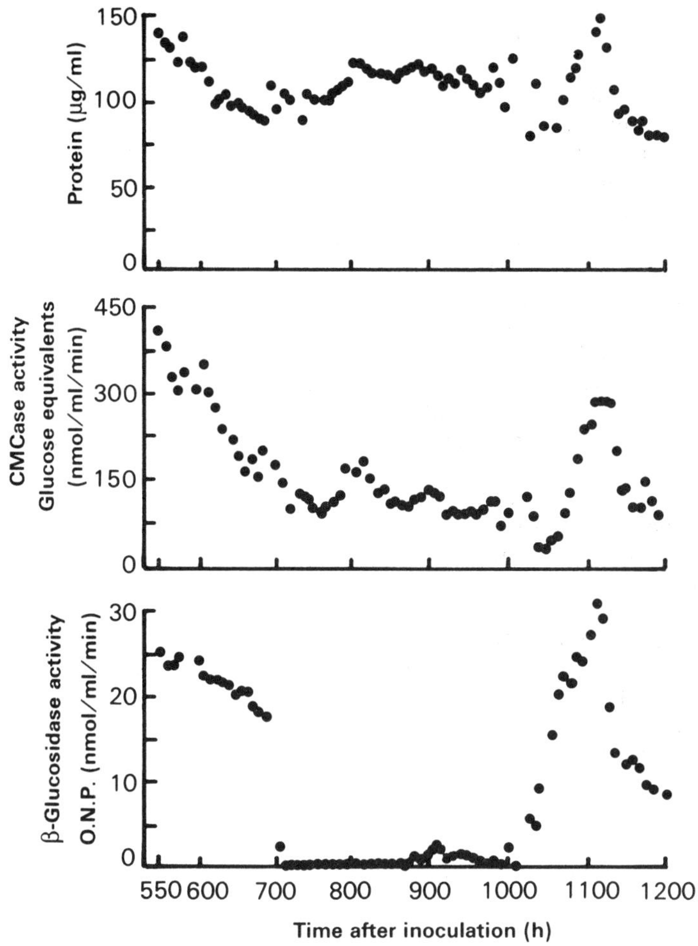

FIGURE 5. (Continued)

corresponding to an annual yield of over one million tonnes of mushrooms [32]. Three other species account for all but a tiny proportion of the remainder, *Lentinus edodes*, the "shiitake mushroom," contributes about 14% of worldwide production [102], while the "padi-straw mushroom" *Volvariella volvacea* provides approximately 7% of the annual global yield of cultivated mushrooms [52]. The "oyster mushroom" *Pleurotus ostreatus* accounts for a much smaller but increasing proportion of the worldwide production. Due to the preeminence of *A. bisporus*, mushroom production depends largely on the degradation of lignocellulosic composts based on wheat straw. The padi-straw mushroom is usually grown on rice straw composts, although it can utilize a wide range of waste plant materials as substrates [51]. The shiitake and oyster mushrooms are routinely produced on wood-based substrates, traditionally on intact logs, but more recently also on artificial substrates based on sawdust [102]. There is, therefore, a significant use of woody lignocellulose in mushroom production, a contribution that seems to be increasing.

In the case of *A. bisporus* fruit body production, the commercial process depends upon the replacement of the normal dominant species of the fungal succession on this substrate, usually a member of the genus *Coprinus*, especially *C. cinereus*, with the desired organism. This is achieved by steam pasteurization of the compost followed by heavy inoculation with *A. bisporus*, carried out when the composting has progressed beyond the thermogenic stage. Pasteurization hardly affects the thermophilic microbial species present in the compost but severely reduces the numbers of the potential competitor mesophiles such as *C. cinereus*. In contrast to *V. volvacea*, which can establish itself effectively on unpasteurized rice straw [52], *A. bisporus* does not appear able to compete very successfully against the natural microbial population of wheat straw compost. However, with such a shift in the balance of advantage provided by the extremely high inoculum potential and the severe reduction of possible competitors, *A. bisporus* becomes the dominant fungal species. It is quite clear that mobilization of the cellulose content of the compost is of considerable importance in the production of fruit-bodies by *A. bisporus*. Indirectly this is apparent from the significant depletion of cellulose in the compost during the phase of mushroom production. However, because many other organisms that could be responsible for cellulose decomposition are also present and active during this phase, more direct evidence is necessary. This has been provided by demonstration of the production by *A. bisporus* of extracellular cellulolytic enzymes and of a direct correlation between a dramatic rise in the activity of such enzymes from a low basal level during mycelial growth to a very high level at the onset of fruiting [103]. Recently it has been shown that peaks of measurable activity of *A. bisporus* endocellulase in compost culture coincide with flushes of mushroom production. This was monitored over a period of about 40 days during which five such flushes occurred [104]. A direct relationship between these two activities was demonstrated by manipulation of fruit-body harvesting. Whether mushrooms were picked continuously at the button stage or the harvesting was delayed for several days after the mushrooms were in optimum commercial condition, there was found to be a very tight correlation between the level of endocellulase activity in the compost and the mushroom biomass harvested [104]. It is, therefore, clear that the products of cellulose decomposition must contribute a significant proportion of the material converted to mushroom biomass. If this is so, then conservation of the cellulose content within the compost is an important consideration for commercial mushroom production. There is a potential drain of this component from the substrate as a result of the activity of fungi and bacteria that precede the basidiomycete in the earlier stages of composting. In particular, during the thermogenic and subsequent cooling phases there seems to be considerable loss of cellulose content due to the activity of several species, including the ascomycete fungus *Chaetomium thermophile*, in both wheat and rice straw composts. Several other strongly cellulolytic fungal species occur during composting, including the deuteromycetes *Humicola insolens* and *Thermomyces lanuginosus* in wheat straw compost and *Aspergillus fumigatus* and *Scytalidium thermophilum* in rice straw composts [52]. Such species also generally appear to be more strongly cellulolytic than *V. volvacea* or *A. bisporus* [52]. Any means of decreasing such cellulose decomposition would clearly be desirable and several approaches have been attempted, including the addition to the compost of low molecular weight sugars to bring about catabolite repression of cellulolytic enzymes.

B. Transformation of Cellulosic Wastes

There are several plausible options for the commercial transformation of lignocellulosic wastes either to useful products or simply as a means of disposal. All of these possibilities remain at present, however, largely inviable as commercial operations or are only practicable on a small scale. This is largely due to the relatively low concentration of product obtainable from a large bulk of waste material and the problems inherent in separating the desired product from the substrate. Both of these difficulties are at least partly caused by the complex composition of lignocellulosic waste, including its content of a range of high molecular weight polymers, some of them very resistant to breakdown, and also the multiplicity of breakdown products which results from any kind of chemical or biological degradation process. Nevertheless, many processes have already been attempted on a modest scale, and the large-scale commercial viability of most of them merely awaits a better understanding of the biochemistry of degradation of lignocellulosic polymers and the microbial enzyme systems which mediate such processes. Dominating thinking about commercial processes based on lignocellulosic waste for many years have been the two concepts of producing either foodstuff or fuel from such material. Both are realistic possibilities and have been exploited to a limited extent.

With the exception of mushroom production, the potential for food production from lignocellulosic waste appears to be restricted, at least at present, to the production of animal feeds, usually by employing a microbial fermentation process which renders the waste more digestible, either by releasing lower molecular weight molecules from the polymeric constituents of the waste or by conversion of the lignocellulosic biomass into more nutritious microbial biomass.

The concept of fuel production from lignocellulosic waste revolves around the idea of producing ethanol either directly from the substrate or by producing sugars, particularly glucose, which can subsequently be readily converted to ethanol. Many microorganisms can convert polymeric constituents of lignocellulose directly to ethanol, especially bacteria [27], but some filamentous fungi such as a species of *Monilia* also have this ability [105]. The idea of producing desirable chemicals has been extended to the more generalized case of utilizing lignocellulosic waste as the substrate for the production of a wide range of chemicals by microorganisms grown on such material, although this is at present a possibility rather than a commercial proposition.

A specialized application of degradation of lignocellulosic material, particularly by fungi, which has direct relevance to a large-scale commercial operation, is the production of pulp for paper making. This process in particular, however, is dependent on lignin-degrading ability, notably of the white-rot fungi, and significant loss of cellulose is considered to be disadvantageous. Thus, in this type of treatment of lignocellulosic waste some means of reducing or preventing cellulolysis by fungi is required. Two approaches seem to hold some promise. One is the addition of low molecular weight sugars, particularly glucose, to the substrate to reduce the activity of cellulolytic enzymes as a result of catabolite repression.

This is, however, expensive and due to difficulties of dispersal within the solid substrate could be problematic on a large scale. The more promising option is the use of cellulase-deficient mutants of white-rot fungi, and a number of these have been isolated including strains of *Sporotrichum pulvurelentum* (syn. *Phanaerochaete chrysosporium*) and *Bjerkandera adusta* [8,106].

There are, of course, some instances in which the important requirement is simply for disposal of lignocellulosic waste, wether or not such a process results in useful products. This is particularly important in the waste waters from industrial processes which utilize lignocellulosic materials, such as papermaking and various types of woodprocessing. The waste water flowing out from such processes can contain large amounts of lignocellulosic material. Potential problems for pollution of nearby bulk water include the toxic nature of some of the molecules present or more significantly the potential disturbance of the natural population of flora, fauna, and microorganisms as a result of the intense microbial growth which will develop on such waste materials. Often the basidiomycete fungi are not present as part of the microbial population which develops in such a situation, and one potential approach to the treatment of such waste material is the introduction of a white-rot basidiomycete. One process utilizing the fungus *Sporotrichum pulverulentum* has been developed for treating such waste water and has the added advantage of producing an animal feedstuff of value to both ruminant and nonruminant animals as a by-product [107, 108]. It is likely that one of the most profitable ways forward for the commercial utilization of lignocellulosic waste, at least in the short term, will be the development of processes such as this, which combine two or more aspects of lignocellulose transformation.

C. Commercial Production of Cellulases

On a very small scale there is a current market for cellulolytic enzymes, and a prime source of such enzymes are the fungi, particularly *Aspergillus niger* and species of the genus *Trichoderma* [109]. Several companies produce and market such enzymes but at present supply them largely for experimental and small-scale uses notably in the food industry for extraction of products such as green tea and soya protein, or to aid digestibility of baby and geriatric foods [109,24]. However, there must be considerable commercial potential for such enzyme production if their cost can be adequately reduced to permit large-scale use. This is, of course, something of a circular problem, because with higher demand and a bigger market value, the price of the product would inevitably come down. Thus, a breakthrough is required which will generate demand for large-scale use. Such higher volume use of cellulolytic enzymes in industrial processes is likely to come with the identification and production of enzymes that can achieve a modification of natural lignocellulose with significantly enhanced properties, but which cannot be achieved by traditional methods such as ball milling or steam explosion. An advance of this type may not come simply from selection of producer organisms and types of enzymes but may require the application of modern technologies such as genetic and protein engineering to design and produce an enzyme with appropriate properties [110] as discussed further below.

Trichoderma reesei is the organism most widely used for the commercial production of cellulases [27], due to the relative ease of its cultivation and its ability to export large quantities of enzymic protein into the growth medium. Such enzyme production utilizes a two-stage process in which biomass production is carried out under one set of growth conditions (35°C, pH 4) and a shift to a second set of culture conditions (27 – 28°C, pH 3) leads to enzyme production and export [27]. More recently, several workers have directed their research towards the possible production of enzymes with desirable properties in addition to the ability to decompose crystalline cellulose. A number of thermophilic fungal species have been developed for cellulolytic enzyme production in the hope that such enzymes will be able to function at higher temperatures than those of *T. reesei*. Such high-temperature enzymes would than be able to mediate faster decomposition of cellulosic materials, which would be desirable because slow rates of degradation of lignocelluloses has been one of the factors preventing many potentially commercial processes of this type from becoming viable. The thermophilic species *Talaromyces emersonii* produces a cellulase complex capable of degrading crystalline cellulose to glucose, and this is one of a new range of organisms in which the production of cellulolytic enzymes is being investigated [31]. A highly thermostable endo-1,4-β-glucanase has recently been isolated from the thermophilic fungus *Myceliophthora thermophila* with many desirable properties compared to an equivalent enzyme from *T. reesei*. Not only was the specific activity of the *M. thermophila* enzyme 2.4 times higher at 40°C, but of particular interest is the fact that it is about 300 times more thermostable at 65°C, with a half-life of 170 hr, than the *T. reesei* enzyme with a half-life of 0.4 – 0.7 hr [111]. Enzymes with such properties could be used in processes which require substrate pasteurization or are run at a high temperature to increase the speed of substrate transformation and reduce the risk of contamination [111].

D. Future Potential

With the exception of mushroom production, the whole range of commercial applications of cellulose decomposition by fungi still remain as future potential despite the fact that such processes have been suggested and investigated over several decades. The problem appears to be that while virtually all of the options are practicable, none of them has yet proved susceptible to scale-up on a commercial basis. The global fall in the oil price appears to have severely curtailed the commercial viability of fuel production from lignocellulosic waste, which at one time seemed to be the best prospect for a truly large-scale commercial process. Nevertheless, progress in our understanding of the biochemistry of cellulose decomposition and especially in the genetics and molecular biology of the cellulase complexes has been so dramatic, particularly in the last decade, that a significant advance in the application of fungal cellulolysis is now inevitable and probably imminent. Gottschalk [2] has made the interesting point that far from adding to the increasingly acute problem of the so-called "greenhouse effect" resulting from elevated atmospheric CO_2 concentration, viable commercial processes dependent on lignocellulose degradation could actually be one potential answer. This is because while degradation of lignocellulose would lead to the breakdown of polymers into small molecules more readily converted to CO_2, the need for substrate supply to sufficiently large-scale processes of this type would make it economically viable to plant and maintain increased areas of long-lived lignocellulosic biomass, particularly trees, which would serve to prevent fixed CO_2 from being rapidly returned to the atmosphere.

Potential applications of fungal cellulose decomposition may thus not depend on financial considerations alone, but also on the environmental benefits they may provide.

We have not yet referred specifically to the molecular genetics of fungal cellulase systems, in spite of the fact that there have been major advances in this field in recent years. Several genes coding for fungal cellulolytic enzymes have now been cloned, notably genes from *Trichoderma reesei* [24,109]. We have not previously considered this area of significant progress simply because while considerable information has been provided by such studies about the base sequence and organization of such genes and also corroboration of amino acid sequences of the isolated enzymic proteins, such work has not directly advanced our understanding of the mechanisms of cellulose decomposition by fungi. The area in which such molecular genetic work is of potentially enormous value is in the promise it holds for applications of fungal cellulolysis. It is the prospect of introducing cellulolytic capability into organisms that do not normally possess that capacity which is extremely exciting, and in the near future it is the yeast *Saccharomyces cerevisiae* which is the most promising vehicle for such development. Already a strain carrying the gene coding for the endoglucanase I from *Trichoderma reesei* has been constructed, and pilot scale tests have demonstrated its capacity to improve the filtration and clarity of beer due to its newly acquired ability to remove β-glucans [24]. The prospect of providing yeast with cellulolytic enzyme activities capable of transforming cellulose to sugars, which it could then ferment to ethanol, is another potentially very valuable application of genetic technology [109].

VII. CONCLUSION

Decomposition of cellulose is arguably the single most important process carried out by the fungi. Certainly if measured in terms of the contribution of fungal cellulolysis to the global carbon cycle and the quantity of fungal biomass which is derived directly from cellulose decomposition, this must be true. Thus, understanding the mechanism of fungal cellulolysis is of vital importance to our understanding of the fungi as a whole. However, the unraveling of the complexities of fungal cellulolysis goes beyond a merely esoteric interest in the fungi themselves. Because of the ubiquity and shear mass of cellulose present on the earth, an understanding of global cellulolysis, in which the fungi play such a major part, is likely to be vital to the maintenance of the natural ecosystems of the planet in which man is, of necessity, certain to take an increasingly direct role. Coupled to this is the need to exploit the vast reservoirs of carbon and energy available in the form of lignocellulosic waste, both to remove any potential pollutant effects of such material and also to utilize such resources either simply to the benefit of the human race or as a commercial operation or both.

Understanding of fungal cellulolysis has progressed enormously in the 40 years since the pioneering work of Reese and his colleagues [22]. This is exemplified by the fact that it is now possible to isolate from several fungi the constituents of an in vitro system capable of decomposing crystalline cellulose to glucose. Thus, at least in isolated preparations from specific fungi, the biochemistry of fungal cellulose decomposition is well understood. In contrast, in the natural environment details of the processes by which fungi decompose cellulosic substrates are, with few exceptions, either sparse or still virtually unknown. Nevertheless, we are now at the stage at which the techniques of molecular genetics and protein engineering will very soon permit the cell-free cellulolytic systems derived from fungi to be manipulated almost at will to achieve any type of transformation of cellulosic material that is required. It is not clear how quickly such small-scale laboratory-based techniques and processes can be made practicable on a large-scale and on a commercial basis, but the sheer capacity of microorganisms for proliferation is certain to make this rapidly feasible.

ACKNOWLEDGMENT

We are grateful to Dr. Lynne Boddy for reading some sections of the manuscript and for her very helpful comments and criticisms.

REFERENCES

1. Smith, J. E. *Biotechnology*. Edward Arnold, London, 1981.
2. Gottschalk, G. Cellulose degradation and the carbon cycle. In *Biochemistry and Genetics of Cellulose Degradation* (J.-P. Aubert, P. Beguin, and J. Millet, eds.), Academic Press, London, 1988, pp. 3–8.
3. Grant, W. D. and Long, P. E. *Environmental Microbiology*. Blackie, Glasgow, 1981.
4. Ainsworth, G. C. Introduction and keys to higher taxa. In *The Fungi: An Advanced Treatise*, Vol. IV B (G. C. Ainsworth, F. K. Sparrow and A. S. Sussman, eds.), Academic Press, New York, 1973.
5. Bartnicki-Garcia, S. Cell wall chemistry, morphogenesis, and taxonomy of fungi. *Annu. Rev. Microbiol.*, *22*:87–108, 1968.
6. Hawksworth, D. L., Sutton, R. C., and Ainsworth, G. C. *Ainsworth and Bisby's Dictionary of the Fungi*, 7th edition. Commonwealth Mycological Institute, Kew, 1983.
7. Verma, D. P. S., Kumar, V., and Maclachlan, G. A. β-Glucanases in higher plants: Localization, potential functions and regulation. In *Cellulose and Other Natural Polymer Systems* (R. M. Brown, ed.), Plenum Press, New York, 1982, pp. 459–488.
8. Kirk, T. K. Degradation and conversion of lignocelluloses. In *The Filamentous Fungi*, Vol. IV *Fungal Technology* (J. E. Smith, D. R. Berry, and B. Kristiansen, eds.), Edward Arnold, London, 1983, pp. 266–295.
9. Deacon, J. W. Decomposition of filter paper cellulose by termophilic fungi acting singly, in combination, and in sequence. *Trans. Br. Mycol. Soc.*, *85*:663–669, 1985.
10. Orpin, C. G. Genetic approaches to the improvement of lignocellulose degradation in the rumen. In *Biochemistry and Genetics of Cellulose Degradation* (J. P. Aubert, P. Beguin, and J. Millet, eds.), Academic Press, London, 1988, pp. 171–179.
11. Wood, T. M., Wilson, C. A., McCrae, S. I., and Joblin, K. N. A highly active extracellular cellulase from the anaerobic rumen fungus *Neocallimastix frontalis*. *FEMS Microbiol. Letters*, *34*:37–40, 1986.

12. Marx-Figini, M. The control of molecular weight and molecular-weight distribution in the biogenesis of cellulose. In *Cellulose and Other Natural Polymer Systems* (P. M. Brown, ed.), Plenum Press, New York, 1982, pp. 243–271.
13. Burns, R. G. Extracellular enzyme-substrate interactions in soil. In *Microbes in Their Natural Environments* (J. H. Slater, R. Whittenbury, and J. W. T. Wimpenny, eds.), Cambridge University Press, Cambridge, 1983, pp. 249–298.
14. Ring, G. J. F. A study of the polymerization kinetics of bacterial cellulose through gel-permeation chromatography. In *Cellulose and Other Natural Polymer Systems* (R. M. Brown, ed.), Plenum Press, New York, 1982, pp. 299–325.
15. Hall, J. L., Flowers, T. J., and Roberts, R. M. *Plant Cell Structure and Metabolism.* Longman, London, 1974.
16. Haigler, C. H. and Benziman, M. Biogenesis of cellulose I microfibrils occurs by cell-directed self-assembly in *Acetobacter xylinum.* In *Cellulose and Other Natural Polymer Systems* (R. M. Brown, ed.), Plenum Press, New York, 1982, pp. 273–297.
17. Jewell, T. R. A qualitative study of the cellulose distribution in *Ceratocystis* and *Europhium*. *Mycologia, 66*:139–146, 1974.
18. Beakes, G. W. Oomycete phylogeny: Ultrastructural perspectives. In *Evolutionary Biology of the Fungi* (A. D. M. Rayner, C. M. Brasier, and D. Moore, eds.), Cambridge University Press, Cambridge, 1987, pp. 405–421.
19. Cavalier-Smith, T. The origin of fungi and pseudofungi. In *Evolutionary Biology of the Fungi* (A. D. M. Rayner, C. M. Brasier, and D. Moore, eds.), Cambridge University Press, Cambridge, 1987, pp. 339–353.
20. Cavalier-Smith, T. The kingdom Chromista: Origin and systematics. In *Progress in Phycological Research*, Vol. 4 (F. E. Round and D. J. Chapman, eds.), Biopress, Bristol, 1987, pp. 309–347.
21. Pringsheim, H. Über den fermentativen Abbau der Zellulose. *Z. Physiol. Chem.*, *78*:266–291, 1912.
22. Reese, E. T., Siu, R. G. H., and Levinson, H. S. The biological degradation of soluble cellulose derivatives and its relationship to the mechanism of cellulose hydrolysis. *J. Bact.*, *59*:485–497, 1950.
23. Coughlan, M. P. and Ljungdahl, L. G. Comparative biochemistry of fungal and bacterial cellulolytic enzyme systems. In *Biochemistry and Genetics of Cellulose Degradation* (J.-P. Aubert, P. Beguin, and J. Millet, eds.), Academic Press London, 1988, pp. 11–30.
24. Knowles, J., Lehtovaara, P., and Teeri, T. Cellulase families and their genes. *Trends in Biotech.*, *5*:255–261, 1987.
25. Wood, T. M., McCrae, S. I., Wilson, C. A., Bhat, K. M., and Gow, L. A. Aerobic and anaerobic fungal cellulases, with special reference to their mode of attack on crystalline cellulose. In *Biochemistry and Genetics of Cellulose Degradation* (J. -P. Aubert, P. Beguin, and J. Millet, eds.), Academic Press, London, 1988, pp. 31–52.
26. Eriksson, K. E. Enzyme mechanisms involved in cellulose hydrolysis by the rot fungus *Sporotrichum pulverulentum*. *Biotech. Bioeng.*, *20*:317–332, 1978.

27. Tsao, G. T. and Chiang, L-C. Cellulose and hemicellulose technology. In *The Filamentous Fungi*, Vol. IV *Fungal Technology* (J. E. Smith, D. R. Berry, and B. Kristiansen, eds.), Edward Arnold, London, 1983, pp. 295–326.
28. Hulme, M. A. and Stranks, D. W. Induction and the regulation of production of cellulase by fungi. *Nature, 226*:469–470, 1970.
29. Kirk, T. K. Biochemistry of lignin degradation by *Phanaerochaete chrysosporium*. In *Biochemistry and Genetics of Cellulose Degradation* (J. -P. Aubert, P. Beguin, and J. Millet, eds.), Academic Press, London, 1988, pp. 315–332.
30. Fahnrich, P. and Irrgang, K. Cellulase and protein production by *Chaetomium cellulolyticum* strains grown on cellulosic substrates. *Biotech. Lett.*, 3:201–206, 1981.
31. McHale, A. P. Cellulase production by *Talaromyces emersonii* CBS 814.70 co-immobilised with cellulose in calcium alginate. *Biotech. Lett.*, 5:361–364, 1988.
32. Wood, D. A., Claydon, N., Dudley, K. J., Stephens, S. K. and Allan, M. Cellulase production in the life cycle of the cultivated mushroom, *Agaricus bisporus*. In *Biochemistry and Genetics of Cellulose Degradation* (J. -P. Aubert, P. Beguin, and J. Millet, eds.), Academic Press, London, 1988, pp. 53–70.
33. Park, D. A cellulolytic pythiaceous fungus. *Trans. Brit. Mycol. Soc.*, *65*:249–257, 1975.
34. Deacon, J. W. Cellulose decomposition by *Pythium* and its relevance to substrate-groups of fungi. *Trans. Brit. Mycol. Soc.*, *72*:469–477, 1979.
35. Fèvre, M. Glucanases, glucan synthases and wall growth in *Saprolegnia monoica*. In *Fungal Walls and Hyphal Growth* (J. H. Burnett and A. P. J. Trinci, eds.), Cambridge University Press, Cambridge, 1979, pp. 225–263.
36. Canevascini, G., Coudray, M.-R., Rey, J.-P., Southgate, R. J. G., and Meier, H. Induction and catabolite repression of cellulase synthesis in the termophilic fungus *Sporotrichum thermophile*. *J. Gen. Microbiol.*, *110*:291–303, 1979.
37. Sutherland, J. B. Regulation of cellulolytic activity in the white-rot fungus *Ischnoderma resinosum*. *Mycologia*, *78*:52–55, 1986.
38. Mandels, M., Parrish, F. W., and Reese, E. T. Sophorose as an inducer of cellulase in *Trichoderma viride*. *J. Bact.*, *83*:400–408, 1962.
39. Nisizawa, T., Suzuki, H., and Nisizawa, K. Catabolite repression of cellulase formation in *Trichoderma viride*. *J. Biochem.*, *71*:999–1007, 1972.
40. Manning, K. and Wood, D. A. Production and regulation of extracellular endocellulase by *Agaricus bisporus*. *J. Gen. Microbiol.*, *129*: 1839–1847, 1983.
41. Nisizawa, T., Suzuki, H., Nakayama, M., and Nisizawa, K. Inductive formation of cellulase by sophorose in *Trichoderma viride*. *J. Biochem.*, *70*:375–385, 1971.
42. Beldman, G. Voragen, A. G. J., Rombouts, F. M., and Pilnik, W. Synergism in cellulose hydrolysis by endoglucanases and exoglucanases purified from *Trichoderma viride*. *Biotech. Bioeng.*, *31*:173–178, 1988.

43. Lamed, R. and Bayer, E. A. The cellulosome concept: Exocellular/ extracellular enzyme reactor centers for efficient binding and cellulolysis. In *Biochemistry and Genetics of Cellulose Degradation* (J. -P. Aubert, P. Beguin, and J. Millet, eds.), Academic Press, London, 1988, pp. 101–116.
44. Lamed, R., Setter, E., and Bayer, E. A. Characterization of cellulose-binding, cellulase-containing complex in *Clostridium thermocellum*. *J. Bact.*, *156*:827–836, 1983.
45. Montgomery, R. A. P. The role of polysaccharidase enzymes in the decay of wood by basidiomycetes. In *Decomposer Basidiomycetes: Their Biology and Ecology* (J. C. Frankland, J. M. Hedger, and M. J. Swift, eds.), Cambridge University Press, Cambridge, 1982, pp. 51–65.
46. Jackson, M. A. and Talburt, D. E. Mechanism for β-glucosidase release into cellulose-grown *Trichoderma reesei* culture supernatants. *Exp. Mycol.*, *12*:203–216, 1988.
47. Sprey, B. and Lambert, C. Titration curves of cellulases from *Trichoderma reesei*:Demonstration of a cellulase-xylanase-β-glucosidase-containing complex. *FEMS Micro. Letts.*, *18*:217–222, 1983.
48. Ahearn, D. G. and Crow, S. A. Fungi and hydrocarbons in the marine environment. In *The Biology of Marine Fungi* (S. T. Moss, ed.), Cambridge University Press, Cambridge, 1986, pp. 11–18.
49. Fog, K. The effect of added nitrogen on the rate of decomposition of organic matter. *Biol. Rev.*, *63*:433–462, 1988.
50. Griffin, D. M. *Ecology of Soil Fungi*. Chapman and Hall, London, 1972.
51. Chang-Ho, Y. and Yee, N. T. Comparative study of the physiology of *Volvariella volvacea* and *Coprinus cinereus*. *Trans. Br. Mycol. Soc.*, *68*:167–172, 1977.
52. Hedger, J. N. and Basuki, T. The role of basidiomycetes in composts: A model system for decomposition studies. In *Decomposer Basidiomycetes: Their Biology and Ecology* (J. C. Frankland, J. N. Hedger, and M. J. Swift, eds.), Cambridge University Press, Cambridge, 1982, pp. 262–305.
53. Mercer, P. C. Basidiomycete decay of standing trees. In *Decomposer Basidiomycetes: Their Biology and Ecology* (J. C. Frankland, J. N. Hedger, and M. J. Swift, eds.), Cambridge University Press, Cambridge, 1982, pp. 143–160.
54. Garrett, S. D. *Soil Fungi and Soil Fertility*, 2nd ed. Pergamon Press, Oxford, 1981.
55. Levy, J. F. The place of basidiomycetes in the decay of wood in contact with the ground. In *Decomposer Basidiomycetes: Their Biology and Ecology* (J. C. Frankland, J. N. Hedger, and M. J. Swift, eds.), Cambridge University Press, Cambridge, 1982, pp. 161–178.
56. Jennings, D. H. The movement of *Sperpula lacrimans* from substrate to substrate over nutritionally inert surfaces. In *Decomposer Basidiomycetes: Their Biology and Ecology* (J. C. Frankland, J. N. Hedger, and M. J. Swift, eds.), Cambridge University Press, Cambridge, 1982, pp. 91–108.
57. Dickinson, D. J. The decay of commercial timbers. In *Decomposer Basidiomycetes: Their Biology and Ecology* (J. C. Frankland,

J. N. Hedger, and M. J. Swift, eds.), Cambridge University Press, Cambridge, 1982, pp. 179–190.
58. Hudson, H. J. *Fungal Saprophytism* 2nd ed. Edward Arnold, London, 1980.
59. Swift, M. J. Basidiomycete components of forest ecosystems. In *Decomposer Basidiomycetes: Their Biology and Ecology* (J. C. Frankland, J. N. Hedger, and M. J. Swift, eds.), Cambridge University Press, Cambridge, 1982, pp. 307–337.
60. Hering, T. F. Decomposing activity of basidiomycetes in forest litter. In *Decomposer Basidiomycetes: Their Biology and Ecology* (J. C. Frankland, J. N. Hedger, and M. J. Swift, eds.), Cambridge University Press, Cambridge, 1982, pp. 212–225.
61. Hintikka, V. The colonization of litter and wood by basidiomycetes in Finnish forests. In *Decomposer Basidiomycetes: The Biology and Ecology* (J. C. Frankland, J. N. Hedger, and M. J. Swift, eds.), Cambridge University Press, Cambridge, 1982, pp. 227–239.
62. Chung, Y. R., Hoitink, H. A. J., Dick, W. A., and Herr, L. J. Effect of organic matter decomposition level and cellulose amendment on the inoculum potential of *Rhizoctonia solani* in hardwood bark media. *Phytopathology, 78*:836–840, 1988.
63. Smith, S. E. Physiology and ecology of orchid mycorrhizal fungi with reference to seedling nutrition. *New Phytol., 65*:488–499, 1966.
64. Tribe, H. T. Aspects of decomposition of cellulose in Canadian soils. I. Observations with the microscope. *Can. J. Microbiol., 6*:309–316, 1960.
65. Tribe, H. T. Interactions of soil fungi on cellulose film. *Trans. Brit. Mycol. Soc., 49*:457–466, 1966.
66. Harper, S. H. T. and Lynch, J. M. Colonization and decomposition of straw by fungi. *Trans. Br. Mycol. Soc., 85*:655–661, 1985.
67. Kohlemeyer, J. and Kohlemeyer E. *Marine Mycology*. Academic Press, New York, 1979.
68. Kohlemeyer, J. Taxonomic studies of the marine ascomycotina. In *The Biology of Marine Fungi* (S. T. Moss, ed.), Cambridge University Press, Cambridge, 1986, pp. 199–210.
69. Leightley, L. E. and Eaton, R. A. *Nia vibrissa* — a marine white-rot fungus. *Trans. Brit. Mycol. Soc., 73*:35–40, 1979.
70. Mouzouras, R. Patterns of timber decay caused by marine fungi. In *The Biology of Marine Fungi* S. T. Moss, ed.), Cambridge University Press, Cambridge, 1986, pp. 341–353.
71. Jones, E. B. G. Decomposition by basidiomycetes in aquatic environments. In *Decomposer Basidiomycetes: The Biology and Ecology* (J. C. Frankland, J. N. Hedger, and M. J. Swift, eds.), Cambridge University Press, Cambridge, 1982, pp. 191–212.
72. Schaumann, K., Mulach, W., and Molitoris, H. P. Comparative studies on growth and exoenzyme production of different *Lulworthia* isolates. In *The Biology of Marine Fungi* (S. T. Moss, ed.), Cambridge University Press, Cambridge, 1986, pp. 49–60.
73. Molitoris, H. P. and Schaumann, K. Physiology of marine fungi: A screening programme for growth and enzyme production. In *The Biology of Marine Fungi* (S. T. Moss, ed.), Cambridge University Press, Cambridge, 1986, pp. 35–47.

74. Park, D. A two-year study of numbers of cellulolytic *Pythium* in river water. *Trans. Br. Mycol. Soc.*, *74*:253–258, 1980.
75. Zare-Maivan, H. and Shearer, C. A. Extracellular enzyme production and cell wall degradation by freshwater lignicolous fungi. *Mycologia*, *80*:365–375, 1988.
76. Webster, J. H. *Introduction to the Fungi* 2nd ed. Cambridge University Press, Cambridge, 1980.
77. Lowe, S. E., Griffith, G. G., Milne, A., Theodorou, M. K., and Trinci, A. P. J. The life cycle and growth kinetics of an anaerobic rumen fungus. *J. Gen. Microbiol.*, *133*:1815–1827, 1987.
78. Mountfort, D. O. and Asher, R. A. Role of catabolite regulatory mechanisms in control of carbohydrate utilization by the rumen anaerobic fungus *Neocallimastix frontalis*. *Appl. Environ. Microbiol.*, *46*:1331–1338, 1983.
79. Mountfort, D. O. and Asher, R. A. Production and regulation of cellulase by two strains of the rumen anaerobic fungus *Neocallimastix frontalis*. *Appl. Environ. Microbiol.*, *49*:1314–1322, 1985.
80. Martin, M. M. Biochemical implications of insect mycophagy. *Biol. Rev.*, *54*:1–21, 1979.
81. Martin, M. M. and Martin, J. S. The distribution and origins of the cellulolytic enzymes of the higher termite, *Macrotermes natalensis*. *Physiol. Zool.*, *52*:11–21, 1979.
82. Eaton, R. A. Preservation of wood in the sea. In *The Biology for Marine Fungi* (S. T. Moss, ed.), Cambridge University Press, Cambridge, 1986, pp. 355–365.
83. Cruden, D. L. and Markovetz, A. J. Microbial ecology of the cockroach gut. *Annu. Rev. Microbiol.*, *41*:617–643, 1987.
84. Olutiola, P. O. Cellulase enzymes in culture filtrates of *Ceratocystis paradoxa*. *Mycologia*, *68*:1083–1092, 1976.
85. Whitney, P., Chapman, J. M. and Heale, J. P. Carboxymethylcellulase production by *Verticillium albo-atrum*. *J. Gen. Microbiol.*, *56*:215–225, 1969.
86. Husain, A. and Dimond, A. E. Role of cellulolytic enzymes in pathogenesis by *Fusarium oxysporum* f. sp. *lycopersici*. *Phytopathology*, *50*:329–331, 1960.
87. Hudson, H. J. *Fungal Biology*. Edward Arnold, London, 1986.
88. Garrett, S. D. Cellulose decomposing ability of some cereal foot-rot fungi in relation to their saprophytic survival. *Trans. Br. Mycol. Soc.*, *49*:57–68, 1966.
89. Trinci, A. P. J. Influence of the width of the peripheral growth zone on the radial growth rate of fungal colonies on solid media. *J. Gen. Microbiol.*, *67*:325–344, 1971.
90. Holtzapple, N. T., Caram, H. S., and Humphrey, A. E. The HCH-1 model of enzymic cellulose hydrolysis. *Biotech. Bioeng.*, *26*:775–780, 1984.
91. Ghose, T. K. and Das, K. A simplified kinetic approach to cellulose-cellulase system. In *Advances in Biochemical Engineering*, Vol. 1 (T. K.Ghose and A. Fiechter, eds.), Springer-Verlag, New York, 1971, pp. 55–76.
92. Howell, J. A. and Stuck, J. D. Kinetics of solka floc cellulose hydrolysis by *Trichoderma viride* cellulase. *Biotech. Bioeng.*, *17*:873–893, 1975.

93. Huang, A. A. Kinetic studies on cellulose-cellulase system. *Biotech. Bioeng.*, *17*:1421–1433, 1975.
94. Mandenius, C. F., Nilsson, B., Persson, I., and Tjerneld, F. Kinetic models of enzmic cellulose degradation in aqueous two-phase systems. *Biotech. Bioeng.*, *31*:203–207, 1988.
95. Gong, C., Ladisch, M. R., and Tsao, G. T. Cellobiase from *Trichoderma viride*:Purification, properties, kinetics and mechanism. *Biotech. Bioeng.*, *19*:959–981, 1977.
96. Wald, S., Wilke, C. R., and Blanch, H. W. Kinetics of the enzymatic hydrolysis of cellulose. *Biotech. Bioeng.*, *26*:221–230, 1984.
97. Peitersen, N. and Ross, E. W. Mathematical model for enzymatic hydrolysis and fermentation of cellulose by *Trichoderma*. *Biotech. Bioeng.*, *21*:997–1017, 1979.
98. Monod, J. *Recherches sur la Croissance des Culture Bacteriennes*. Hermann, Paris, 1942.
99. Theodorou, M. K., Bazin, M. J., and Trinci, A. P. J. Growth of *Trichoderma reesei* and the production of cellulolytic enzymes in a continuous flow column. *Arch. Microbiol.*, *130*:372–380, 1981.
100. Theodorou, M. K., Bazin, M. J., and Trinci, A. P. J. The dynamics of cellulose decomposition. In *Microbiological Methods for Environmental Biotechnology* (J. M. Grainger and J. M. Lynch, eds.), Academic Press, London, 1984.
101. Theodorou, M. K., Bazin, M. J., and Trinci, A. P. J. Cellulose degradation in a structured ecosystem which is analogous to soil. *Trans. Br. Mycol. Soc.*, *75*:451–454, 1980.
102. Diehle, D. A. and Royse, D. J. Shiitake cultivation on sawdust: Evaluation of selected genotypes for biological efficiency and mushroom size. *Mycologia*, *78*:929–933, 1986.
103. Wood, D. A. and Goodenough, P. W. Fruiting of *Agaricus bisporus*. Changes in extracellular enzyme activities during growth and fruiting. *Arch. Microbiol.*, *114*:161–165, 1977.
104. Claydon, M., Allan, M., and Wood, D. A. Fruit body biomass regulated production of extracellular endocellulase during periodic fruiting by *Agaricus bisporus*. *Trans. Br. Mycol. Soc.*, *90*:85–90, 1987.
105. Gong, C. S., Maun, C. M., and Tsao, G. T. Direct fermentation of cellulose to ethanol by a cellulolytic filamentous fungus, *Monilia* sp. *Biotech. Lett.*, *3*:77–82, 1981.
106. Eriksson, K. E. Microbial delignification. Basics, potentials and applications. In *Biochemistry and Genetics of Cellulose Degradation* (J. -P. Aubert, P. Beguin, and J. Millet, eds.), Academic Press, London, 1988, pp. 285–302.
107. Ek, M. and Eriksson, K. E. Utilization of the white-rot fungus *Sporotrichum pulverulentum* for water purification and protein production on mixed lignocellulosic wastewaters. *Biotech. Bioeng.*, *22*:2273–2284, 1980.
108. Thomke, S., Rundgren, M., and Eriksson, S. Nutritional evaluation of the white-rot fungus *Sporotrichum pulverulentum* as a feedstuff to rats, pigs, and sheep. *Biotech. Bioeng.*, *22*:2285–2303, 1980.
109. Montenecourt, B. S. and Eveleigh, D. E. Fungal carbohydrases: Amylases and cellulases. In *Gene Manipulations in Fungi* (J. W.

Bennett and L. L. Lasure, eds.), Academic Press, Orlando, 1985, pp. 491–512.

110. Klysov, A. A. Cellulases of the third generation. In *Biochemistry and Genetics of Cellulose Degradation* (J. -P. Aubert, P. Beguin, and J. Millet, eds.), Academic Press, London, 1988, pp. 87–99.

111. Klyosov, A. A. and Ermolova, O. V. A thermostable endo-1,4-β-glucanase from *Myceliophthora thermophila*. *Biotech. Lett.*, 5:351–354, 1988.

16

FUNGAL DEGRADATION OF LIGNIN

JOHN A. BUSWELL* *Paisley College of Technology, Paisley, Renfrewshire, Scotland*

I. INTRODUCTION

The past decade has witnessed an exceptional upsurge in research activity aimed at a better understanding of the microbiological and biochemical processes involved in lignin biodegradation. This has happened, in part, through a general recognition of both the need and potential for new technologies which can use as feedstock the huge quantities of lignocellulose generated annually within the biosphere. Added impetus has come from energy cost considerations in other industrial processes based on lignocellulose, for example, in paper pulp manufacture, where lignin is an undesirable element and must be separated from cellulose and hemicellulose constituents. In the field of agriculture, too, development of processes for the efficient removal of the lignin component from waste lignocellulosic residues would provide a major source of forage for ruminants. Furthermore, since under natural conditions much of the available lignin is converted to humus, clarification of the interactions which occur between ligninolytic microorganisms in the soil environment are intrinsic to our knowledge of soil structure and fertility.

Several fundamental developments have punctuated this latest advance in our comprehension of ligninolytic systems. Highly sensitive and reliable assay procedures using radiolabeled lignins and lignin substructures underscored a rapid identification in the late 1970s of the physiological conditions affecting ligninolysis. In 1983, the logical extension of this knowledge culminated in the discovery of extracellular H_2O_2-dependent lignin-degrading enzymes, lignin peroxidases, in the culture fluid of the white-rot fungus *Phanerochaete chrysosporium*. Earlier insights into the chemistry of biodegradation gained from the characterization of low molecular weight components following fungal decay of hard- and soft-wood lignins have since been confirmed using lignin model compounds and purified peroxidases. Most recently, emphasis has centered on elucidating the mechanism(s) underlying the various reactions which are catalyzed by ligninases, and in extending our perception of ligninolytic systems to include a larger number of fungal types. Genetic and molecular biological research directed at engineering ligninolytic microorganisms for enhanced expression is also well established.

**Present affiliation*: The Chinese University of Hong Kong, Shatin, Hong Kong.

This review attempts to summarize current knowledge of the subject with special emphasis on more recent developments. The reader is also directed to other recent surveys [1-6] and two books [7,8] which provide both general coverage and more detailed treatment of specific aspects.

II. LIGNIN STRUCTURE AND DISTRIBUTION

Lignin is a heterogeneous, three-dimensional polymer composed of oxyphenylpropanoid units connected by several different C-C and C-O-C linkages [9,10], and together with hemicellulose and cellulose it is a major structural component of woody tissues. It is formed through random coupling of radical species which arise from peroxidase-mediated dehydrogenation of the three monomeric precursors, p-coumaryl, coniferyl, and sinapyl alcohols (Fig. 1). Softwood lignin contains mostly coniferyl-type structures (guaiacyl lignin), whereas the lignin of hardwoods (guaiacyl-syringyl-lignin) is composed of roughly equal amounts of coniferyl and sinapyl units with a smaller quantity of p-hydroxyphenylpropane residues. Grass lignin appears to be a mixture of all three precursors, with relatively high amounts of p-hydroxyphenylpropane moieties, and also contains ester-linked ferulic and p-coumaric acids [11]. Figure 2 shows the structural model of softwood lignin proposed by Sakakibara [12]. Lignin is concentrated in the middle lamella region, but the major portion is located within the secondary wall where it is mixed and occasionally covalently bonded to hemicellulose forming a matrix interspersed with the cellulose fibrils (Fig. 3).

Woody vascular tissue is generally comprised of 20-35% lignin, and, in addition to conveying rigidity, the relative resistance of the polymer to biological degradation almost certainly serves to protect cell walls from microbial attack. Since lignin in plant tissue is intimately associated with the cellulose and hemicellulose constituents, it represents a key factor in industrial processes designed to upgrade lignocellulose, e.g. production of paper pulp, pretreatment of lignocellulose residues prior to enzymic saccharification, or to increase digestibility for ruminant feedstock purposes.

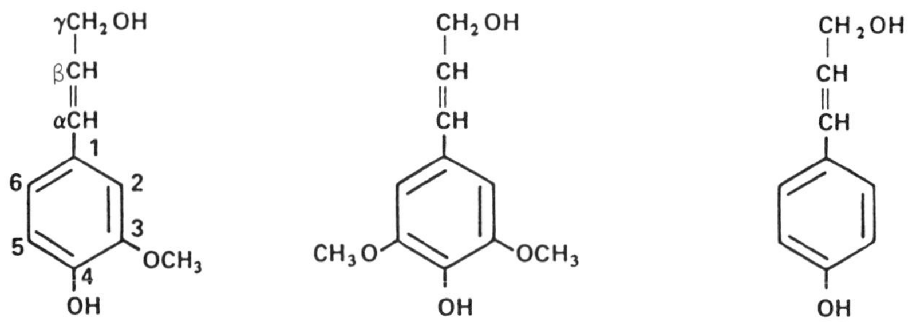

(1) Coniferyl alcohol　(2) Sinapyl alcohol　(3) *p*-Coumaryl alcohol

FIGURE 1. Cinnamyl alcohol precursors of lignin and nomenclature of carbons.

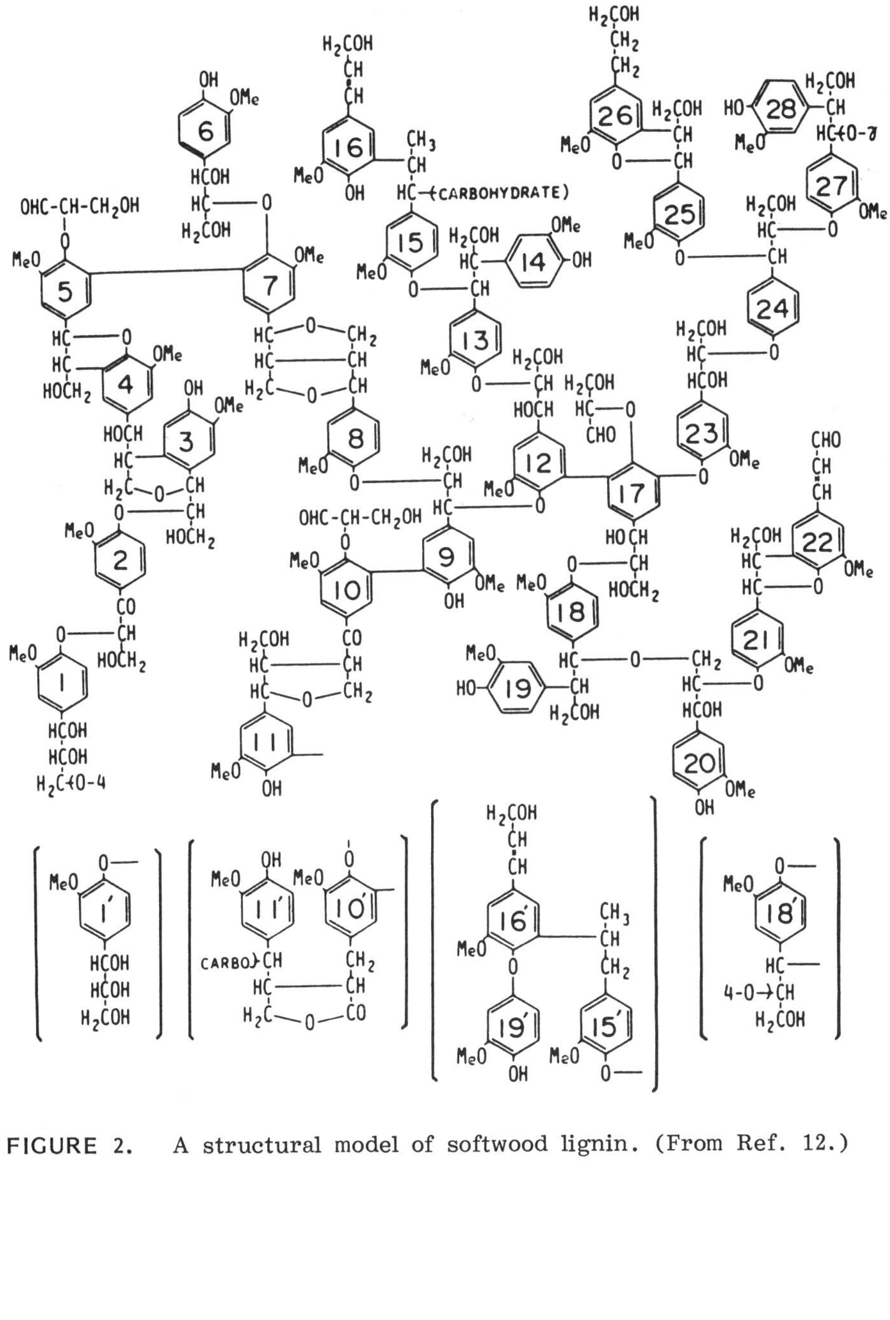

FIGURE 2. A structural model of softwood lignin. (From Ref. 12.)

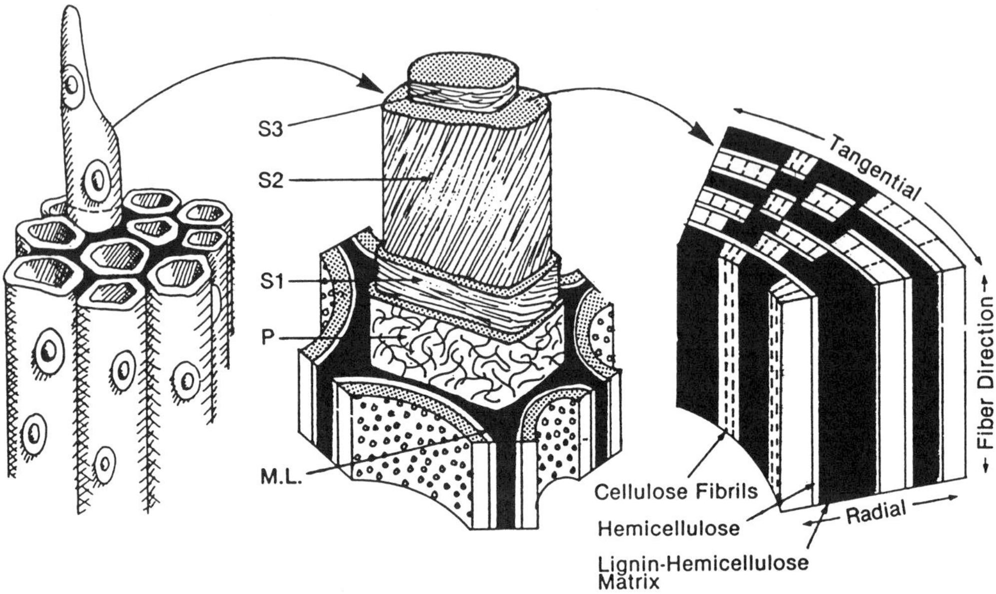

FIGURE 3. Molecular architecture of woody tissues: (a) bundle of contiguous wood cells, (b) wall layers in cut-away view of single cell, and (c) section of secondary wall illustrating the relationship of hemicellulose and lignin to the cellulose fibrils. Cell wall layers are P (primary); and S1, S2 and S3 (secondary). The middle lamella (M.L.) separates the cells. (From Ref. 345.)

III. METHODOLOGY

A. Use of Polymeric and Dimeric Substrates in Lignin Biodegradation Studies

Prior to the mid-1970s, lignin biodegradation research was greatly hampered by the difficulty in obtaining pure, unmodified substrate material and the absence of sensitive and accurate assays. Even low yields of the more representative lignin preparations (e.g., milled wood lignin [MWL]) required careful and time-consuming extraction procedures [7]. Thus, the development of ^{14}C-labeled synthetic lignins (dehydropolymers or DHPs) signified a major advance [13,14]. Synthetic lignins, labeled uniformly or in specific carbon atoms (i.e., aromatic nucleus, propyl side-chain, methoxyl group), are produced from correspondingly labeled coniferyl, sinapyl, and/or p-coumaryl alcohols using horseradish peroxidase and H_2O_2 [13,14]. ^{14}C-labeled natural lignins have also been prepared by feeding labeled precursors (e.g., L-phenylalanine, ferulic acid) to lignifying plants [13,15]. Measurement of the $^{14}CO_2$ released from these substrates during fungal attack provides a sensitive and now routine assay for overall biodegradation but does not take into account partial decomposition to water-soluble intermediates [16,17]. Macromolecular changes which occur in lignin following microbial attack can be detected using ^{13}C-enriched DHP and ^{13}C-NMR spectroscopy [18]. Radiolabeled lignins remain relatively

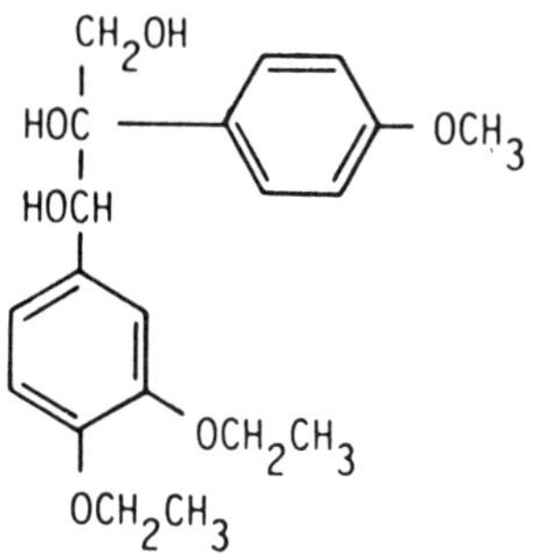

(a)

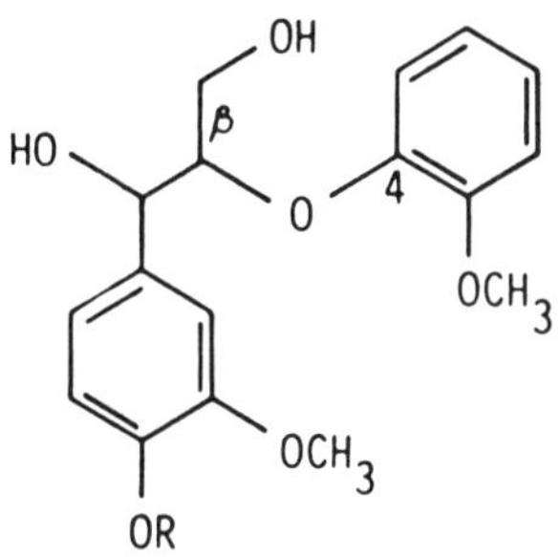

(b)

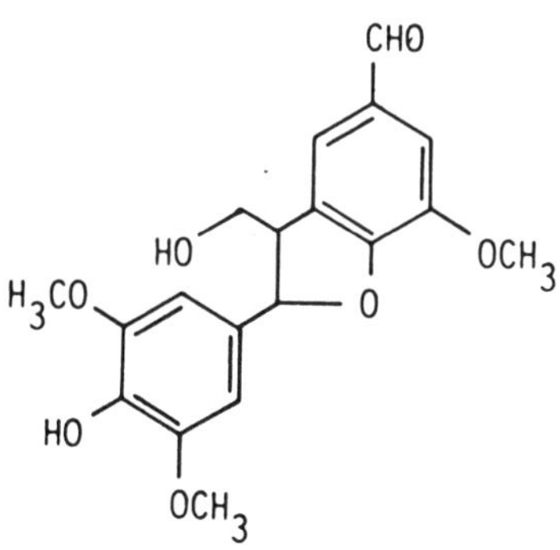

(c)

FIGURE 4. Representative dimeric lignin model compounds: (a) β-1 diarylpropane substructure; (b) β-O-4 aryl ether substructure; (c) phenylcoumaran substructure.

expensive, and cheaper alternatives such as polyguaiacol [19], a polymeric lignin model, and commercially available polymeric dyes have proved useful especially for qualitative detection of ligninolytic activity and in the selection of fungal mutants [20 – 22].

Studies adopting dimeric lignin model compounds (dilignols) have proved invaluable in earlier attempts to identify key chemical reactions contributing to lignin depolymerization [23 – 29] and, more recently, in establishing the catalytic functions of lignin-degrading enzymes [30 – 36]. Dilignols of the type shown in Figure 4 represent the β-1 diarylpropane, β-O-4 aryl ether, and phenylcoumaran substructures which together constitute over 60% of the total intermonomeric linkages in lignin [7]. The validity of using dimeric substructures in lignin biodegradation research is clearly evident following recognition that cultural and physiological parameters affecting ligninolysis are equally consequential for the metabolism of the dilignols [25]. Both dimeric and monomeric [37,38] lignin models continue to be important for elucidating both the mechanism(s) underlying the manifold reactions catalyzed by ligninolytic enzymes and the degradation pathways implicated in polymer decomposition.

B. Electron Microscopy

Scanning and transmission electron microscopy (TEM) combined with histochemical techniques have revealed a great deal about the micromorphological and ultrastructural features associated with in situ lignin transformation by wood-decaying fungi [39 – 43]. Erosion patterns at various stages of decay can be determined using lignin-specific fixatives such as potassium permanganate or osmium tetroxide [44 – 46], and microscopical observations have been confirmed by chemical analyses of decayed tissue and by X-ray microanalysis of brominated wood [47]. Nyhlen and Nilsson [48] used TEM and UV microscopy to assess the progression of decay zones in different cell wall layers during the delignification of pine caused by *Phlebia radiata* and four other white-rot fungi. Enzyme probes and electron microscopy are also being adopted in cytochemical studies of wood degradation to determine the in vivo distribution and role of ligninolytic enzymes [49 – 51].

IV. LIGNIN-DEGRADING FUNGI

Research adopting the more sophisticated detection procedures now available have greatly extended the list of microorganisms reported to attack the lignin polymer. While prokaryotes in particular actinomycetes [52,53] are capable of transforming lignin to varying degrees, there is general recognition that specific types of wood-decaying fungi constitute the major decomposers. Depending upon the form of decay, these wood-rotting fungi are normally classified as soft-rot, brown-rot, and white-rot species. Fungi that are not members of a recognized decay group also decompose lignin.

A. Soft-Rot Fungi

Varying degrees of lignin transformation are attributed to an assortment of ascomycetes and Fungi Imperfecti which cause soft-rot decay of wood. Beech wood lignin has a lower methoxyl content and is more soluble following decay by *Chaetomium globosum* [54,55]. Six soft-rot species (*Graphium* sp., *Monodictys* sp., *Paecilomyces* sp., *Papulospora* sp., *Thielavia terrestris* and *Allescheria* sp.) removed lignin from alder and poplar wood, although more slowly than polysaccharide [56]. With pine, however, *T. terrestris* and *Paecilomyces* sp. depleted lignin more rapidly in proportion to carbohydrate components, although total weight losses were lower compared to hardwoods [56].

Haider and Trojanowski [13] observed only limited release of $^{14}CO_2$ (up to 7.8% of total ^{14}C after 5 days) from variously labeled synthetic lignins and ^{14}C-maize by *Preussia fleischakii, Stachybotrys chartarum,* and two strains of *Chaetomium piluliferum*. Conversion rates were low but comparable to levels attained by two white-rot fungi tested under the same conditions. In a subsequent investigation [57], 20–30% $^{14}CO_2$ was released from ^{14}C-labeled corn-stalk lignins by *C. piluliferum* after 7 weeks. However, only negligible amounts of $^{14}CO_2$ were released from ^{14}C-ring DHP after 10 days by *C. globosum, Phialophora mutabilis,* and *Petriellidium boydii* under conditions which promote maximal lignin decomposition by *P. chrysosporium* [58]. Further study of soft-rot fungi is clearly needed if the true extent of their lignin-degrading capacity is to be established.

B. Brown-Rot Fungi

Although brown-rot fungi preferentially degrade the cellulose and hemicellulose of wood tissue [59], they also cause limited changes in the lignin component which can be oxidized in part to CO_2. Colonization is usually confined to the less-lignified S_2 and S_1 layers of the secondary wall, and the more highly lignified primary wall and middle lamella regions are generally resistant to attack [60]. Fungal hyphae advance longitudinally through the lumina of the wood cells and appear unable to bore through the fibers [60]. Under laboratory conditions, brown-rot fungi will attack hardwood, softwood, and grass lignins. Enoki et al. [61] examined the capacity of 12 brown-rot species to degrade lignin in Japanese cedar and Japanese beech wood and reported losses ranging from 0.5 to 54% after 12 weeks of decay. Antai and Crawford [62] found that *Poria placenta* decomposed grass lignin more effectively than either hardwood (maple) or softwood (spruce) lignin.

Chemical analyses of brown-rotted wood reveal a large decrease in the number of methoxyl groups in the modified lignin, up to 35% in the case of spruce wood decayed by *Lenzites trabea* (= *Gloeophyllum trabeum*) [63,64]. This reduction in methoxyl content is accompanied by an increase in phenolic hydroxyl groups due in part to direct hydroxylation of the aromatic ring [65] as well as demethylation reactions [64]. Oxidation of the propyl side chain is a further feature of brown-rot attack resulting in the formation of α-carbonyl and conjugated carboxyl groups [65]. Fukuda and Haraguchi [66] described similar changes in the functional groups of MWL isolated from spruce wood following decay by the brown-rot fungus *Tyromyces palustris*. However, compared to white-rot fungi, brown-rot species appear less able to cleave aromatic rings in lignin or, when ring-opening does occur, to further degrade the lignin fragments [1,67].

Several studies have established that certain brown-rot fungi oxidize radiolabeled lignins to $^{14}CO_2$ although with widely varying abilities and again with a much reduced capacity compared to white-rot fungi. Kirk et al. [14] found that cultures of *G. trabeum* and *Poria cocos* released

$^{14}CO_2$ from ^{14}C-methoxyl DHP more extensively than from side-chain or ring carbons. This pattern was confirmed by Haider and Trojanowski [57], who reported the conversion of 29.2, 15.2, and 10.1% of available label to $^{14}CO_2$ from ^{14}C-(methoxy)-, $^{14}CO_2$-(side-chain)-, and ^{14}C-(ring)-labeled DHP, respectively, by *G. trabeum* after 15 days. Eight other brown-rot fungi exhibited conversion rates which ranged between 0.4 and 19.8% $^{14}CO_2$ from ^{14}C-(methoxyl)-DHP, 0.5–13.0% $^{14}CO_2$ from $^{14}C_2$-side chain, and 0.3–8.7% $^{14}CO_2$ from ^{14}C-(ring)-DHP (Table 1). Evolution of $^{14}CO_2$ from radiolabeled synthetic lignins was generally higher than from plant material (corn stalks) labeled in the lignin component [57]. Ander et al. [58] reported that *G. trabeum, Fomitopsis pinicola,* and *Poria placenta* released less than 2% $^{14}CO_2$ from both methoxyl- and ring-labeled DHP in 9–15 days, whereas another study failed to detect any DHP degradation by brown-rot species [68]. The generally poor activity of brown-rot fungi towards synthetic lignins is also reflected in their ability to degrade dimeric lignin model compounds [69]. However, relatively little is known about the optimal conditions for lignin degradation by brown-rot fungi and low conversion rates reported among these species may, to some extent, be due to unfavorable culture conditions.

C. White-Rot Fungi

Capacity for complete mineralization of lignin by isolated microorganisms appears to be restricted to the white-rot fungi and to the taxonomically related litter-decomposing basidiomycetes, although it might be surmised that total decomposition under natural conditions occurs through the cooperative action of microbial consortia. Several hundred species of white-rot fungi are now recognized, which together form a heterogeneous collection of mainly basidiomycetes and a few species of ascomycetes. Most attention has focused on one white-rot species, *Phanerochaete chrysosporium*, which is both an efficient lignin degrader [69a] and amenable to laboratory experimentation. Substantial depletion of lignin is also reported during white-rot decay of hardwoods by ascomycetes (e.g., *Xylaria, Libertella*, and *Hypoxylon* spp.) and apparently involves preferential attack on the syringyl moieties [1]. Individual features of lignin degradation by white-rot fungi, in particular *P. chrysosporium*, will be described in more detail in later sections.

D. Ligninolytic Activity in Other Fungi

Several species of *Cyathus*, a gasteromycete, delignify wheat straw and kenaf [70,71], and some species exhibit a degree of specificity for the lignin component of both straw and hardwood [71].

Although soil-inhabiting *Fusarium* spp. are adept at degrading phenolic lignin substructures [72], they display a restricted capacity to decompose lignin. A strain of *Fusarium solani* released 4–5% of the total label in ^{14}C-labeled synthetic lignins as $^{14}CO_2$ after 30 days [73]. Sutherland et al. [74] examined 18 *Fusarium* isolates and found that 2.2–4.3% of C^{14}-spruce lignin was converted to $^{14}CO_2$ and 3.9–8.4% to ^{14}C-labeled water-soluble products after 60 days. Some decomposition of both hardwood and

TABLE 1. $^{14}CO_2$ Release from $O^{14}CH_3$-, $^{14}C'_2$-(Side Chain), or ^{14}C-Ring-Labeled DHP or Corn Stalk Lignin During Incubation with Several Brown Rot Fungi

	$^{14}CO_2$ released								
	DHP lignin			Corn stalk lignin					
				$O^{14}CH_3$		$^{14}C_2$		^{14}C ring	
Brown rot fungus	$O^{14}CH_3$	$^{14}C'_2$	^{14}C ring	20 days	35 days	20 days	35 days	20 days	35 days
Lenzites trabea	29.2	15.2	10.1	8.1	12.9	2.4	5.3	2.0	4.2
L. abientina	19.8	13.0	8.7	11.9	16.1	2.6	6.3	2.3	4.7
L. saepiaria	9.7	9.3	5.3	1.9	5.6	1.9	3.3	1.7	3.3
Poria contigua	11.2	11.5	6.2	4.1	12.9	2.5	5.2	1.2	2.9
P. vaporia	4.1	2.3	1.3	2.8	7.0	4.7	6.9	2.1	4.4
P. monticola	1.8	0.5	0.6	—	—	—	—	—	—
Merulius lacrimans	2.3	1.6	0.3	—	—	—	—	—	—
M. silvester	0.8	1.2	0.7	—	—	—	—	—	—
M. himantoides	0.4	0.6	0.4	—	—	—	—	—	—

Values not given are not determined.
Values indicate $^{14}CO_2$ release as a % of the applied activity after 15 days for DHP, and 20 and 35 days for corn stalk lignins, respectively.
Source: Ref. 57

softwood lignin was reported among 12 marine fungi [75]. Of the total ^{14}C-lignin in maple, 1.9–4.6% was converted to $^{14}CO_2$ and 2.6–7.9% to water-soluble material. Conversion rates for ^{14}C-spruce lignin ranged between 2.2–4.8% ($^{14}CO_2$) and 3.5–8.4% (soluble ^{14}C-label). Degradation of ^{14}C-labeled DHP and kraft lignin by *Aspergillus* species [76,76a], delignification of wheat straw by *Aspergillus japonicus* [77], and slow mineralization of ^{14}C-labeled synthetic lignins and corn lignin by mycorrhizal fungi (*Cenococcum, Amanita, Tricholoma,* and *Rhizopogon* [78]) has also been described.

V. LIGNIN DEGRADATION BY WHITE-ROT FUNGI

A. Decay Patterns in Woody Tissue

Under appropriate conditions, white-rot fungi degrade all the major constituents of woody tissue [79,80]. Species vary considerably in the relative rates at which they attack the lignin and polysaccharide components [43,81,82], but there is no evidence so far to indicate that different white-rot fungi use fundamentally different mechanisms to degrade lignin.

Two general decay patterns can be distinguished: (a) "simultaneous degradation," involving a coincident depletion of all cell wall components, and (b) a "selective decay," characterized by selective removal of lignin and hemicellulose without extensive depletion of cellulose which remains behind in the form of microfibrils [83]. Some white-rot species may switch from selective to simultaneous decay, but the reverse has not been observed [84,85].

Simultaneous degradation is normally distinguished by the formation of holes and erosion troughs in the region immediately surrounding the fungal hyphae. Decomposition of wall components occurs initially in the cell wall layers adjacent to the cell lumen and extends progressively towards the middle lamella, causing a gradual thinning of the cell walls [86,87]. In selective decay, delignification of secondary walls and the middle lamella occurs without substantial attack on the cellulose within the S_1 and S_2 layers and is apparent at considerable distances from the colonizing fungal hyphae [88–90]. A particular form of selective decomposition is white pocket-rot decay where delignification is confined to small, localized spindle-shaped zones surrounded by apparently sound wood or areas of simultaneous degradation [88–90]. Histological examination of these white pockets reveal the presence of mainly cellulosic material. Such localized delignification may result from accumulation of polymerization products during decomposition as in the case of *Xylobolus frustulatus* [91]. Deposits of manganese oxides and raised pH levels are associated with some selectively delignified areas [92].

Selective lignin removal under both natural and controlled laboratory conditions is now widely reported among white-rot fungi (Table 2). Kawase [93] found evidence of preferential lignin decomposition among a collection of naturally decayed wood samples colonized by white-rot fungi. *Cryptoderma yamanoi, Fomes ulmarius* (= *Rigidoporous ulmarius*), and *Polyporus resinosus* (= *Ishnoderma resinosum*) fungi identified in decayed samples by Kawase [93] along with *Polyporus berkeleyi* and *Polyporus frondosus* all removed lignin faster than carbohydrate from both aspen and

TABLE 2. Some White-Rot Fungi that Selectively Remove Lignin from Woody Tissue

Fungus	Refs.
Bjerkandera adusta (Polyporus adustus)	83
Cerrena unicolor (Daedalea unicolor)	83
Cryptoderma yamanoi	81, 93
Dichomitus squalens (Polyporus anceps)	83, 88
Ganoderma applanatum	42, 83
Ganoderma lobatum	83
Ganoderma oregonese	83
Ganoderma tsugae	83, 84
Hapalopilus croceus (Polyporus croceus)	83
Heterobasidion annosum (Fomes annosus)	83
Hirschioporus paragamenus	90
Inonotus dryadeus	83
Inonotus dryophilus (Polyporus dryophilus)	83, 89
Inonotus glomeratus	106
Inonotus ludovicianus	83
Inonotus rheades	83
Inonotus texanus	83
Inonotus tomentosus	83
Ischnoderma resinosum (Polyporus resinosus)	81, 83, 90, 93
Laurilia sulcata (Stereum sulcatum)	83
Peniophora sp. *(Phanerochaete P-B1)*	97
Perenniporia medulla-panis (Poria medulla-panis)	83, 90, 95, 106
Perenniporia subacida (Poria subacida)	83
Phellinus sp.	83, 95
Phellinus fastuosus	83
Phellinus ferreus	106
Phellinus ferruginosus	106
Phellinus ignarius	106
Phellinus nigrolimitatus (Fomes nigrolimitatus)	83
Phellinus pini	83, 88
Phellinus viticola (P. isabellinus)	82, 83
Phlebia radiata	82

TABLE 2. (Continued)

Fungus	Refs.
Phlebia tremellosus (Merulius tremellosus)	82, 96
Pleurotus ostreatus	81, 82, 94
Polyporus berkeleyi	81, 93
Polyporus frondosus	81, 93
Polyporus giganteus	81
Pycnoporus cinnabarinus	82
Rigidosporus ulmarius (Fomes ulmarius)	81, 82, 93
Xylobolus frustulatus	83, 90, 91
Xylobolus subpileatus	83

birch wood wafers [81]. *Fomes ulmarius* also selectively attacked the lignin component of pine and Sitka spruce wood [81]. At low weight loss (3.4%) and in the presence of malt extract, *Pycnoporus cinnabarinus* degraded up to 12.5% of the lignin in pinewood blocks with no detectable loss of cellulose or mannan [82]. Chemical, micromorphological, and ultrastructural techniques were used to identify 26 white-rot fungi which selectively removed lignin from various coniferous and hardwood tree species [83]. Other white-rot fungi which selectively attack the lignin component under certain conditions include *Pleurotus ostreatus* [94], *Poria medulla-panis* [95], *Phellinus* spp. [95], and *Phlebia (Merulius) tremellosus* [82,96].

Some fungi (e.g., *Ishnoderma resinosum*) facilitate both selective and nonselective decay [83], and observed fungal preference for the lignin component may depend on several factors including the wood species [59,83], the type of wood cell [83,89], stage of decay [81,90], nutrient availability [83], prevailing environmental conditions [83], and the work-up procedures adopted prior to analysis [83]. *Polyporus giganteus* depleted lignin at a faster rate than carbohydrates from aspen wood but favored carbohydrates from birchwood [81]. Conversely, *Pleurotus ostreatus* selectively removed lignin from birch but not aspen [81]. While some white pocket-rot fungi can selectively delignify most types of woody cell, selective decay is often restricted to certain cell types. *Dichomitus squalens* and *Phellinus pini* [88] preferentially colonized and delignified earlywood cells or latewood tracheids respectively of white pine (*Pinus monticola* Dougl.) and western larch (*Larix occidentalis*). Decay patterns in oak produced by *Inonotus dryophilus* demonstrated a preference for earlywood fibers and axial parenchyma cells but not medullary rays or latewood [89]. Interestingly, delignification did not occur in axial parenchyma cells adjacent to ray parenchyma cells which had undergone nonselective decay. However, as the fungus degraded axial parenchyma cells at an increasing distance from ray parenchyma, selective lignin degradation was again evident. Since high nitrogen levels suppress ligninolytic activity, this effect may be due to the relatively high nitrogen content of ray parenchyma cells [89]. Individual fungi also vary in selectivity towards the lignin component according to the stage of decay. Kirk and Moore [81] reported that, in general, selective lignin removal occurred during the first few percentages of weight loss and did not proceed until only carbohydrate remained. Similarly, the cellulose that remained in delignified pockets of aspen wood blocks after decay by *Xylobolus frustulatus* was subsequently utilized by the same fungus [91]. Nutrient availability greatly influenced the respective rates at which a *Phanerochaete* sp. (strain P-B1) attacked the lignin and carbohydrate moieties of birchwood [97]. Grown in the presence of malt extract or asparagine, the fungus primarily attacked the lignin component, whereas an inorganic nitrogen source such as ammonium nitrate resulted in more rapid degradation of cellulose [97]. Since many "selective" fungi cause delignification only in these isolated areas or pockets, assaying the simultaneously rotted wood with delignified wood may mask the selective delignification capacity [83].

B. Influence of Physiological Parameters

Lignin degradation in *P. chrysosporium* proceeds only in the post-primary growth phase [98] or under conditions where secondary metabolism (idiophase) is expressed simultaneously with restricted primary growth [99]. Several nutritional and cultural parameters are now described which strongly influence the manifestation and activity of the ligninolytic system and the appearance of other idiophasic features which are, or might be, directly associated with ligninolysis in this fungus.

Oxygen Tension

The largely oxidative nature of white-rot-mediated lignin decomposition [100] is reflected in the marked stimulatory effect of molecular oxygen on the ligninolytic systems of many white-rot fungi [101-104]. Kirk et al. [101] detected little lignin degradation (^{14}C-DHP $\rightarrow$ $^{14}CO_2$) in cultures of *P. chrysosporium* maintained under 5% oxygen but found that the rate and extent of polymer decomposition increased dramatically with increasing oxygen partial pressures. Similar effects were reported for *Coriolus versicolor* [101]. Release of $^{14}CO_2$ and ^{14}C-labeled water-soluble products from ^{14}C-labeled lignin in aspen wood by *C. versicolor*, *Polyporus brumalis*, *Merulius tremellosus*, *Pyncoporus cinnabarinus*, *Lentinus edodes*, *Grifola frondosa*, *Gloeoporus dichrous*, and four strains of *P. chrysosporium* also proceeded at a faster rate in an atmosphere of O_2 than in air [103]. However, while the extent and selectivity of lignin decomposition by *P. chrysosporium* increased significantly from 0.2 to 1.0 atm. O_2, pressures above 1.0 atm. had no further enhancing effect [104]. Increased oxygen concentrations had little effect on the rates of aspen lignin degradation by *P. ostreatus* and *Bondarzewia berkeleyi* [103]. In a separate study [105], elevated O_2 levels enhanced the rate and extent of $^{14}CO_2$ evolution from ^{14}C-lignin in poplar wood meal by several white-rot fungi including *P. ostreatus* although the degree of stimulation varied considerably according to the fungal species. Increased O_2 partial pressures also increased the rate of lignin loss from sweetgum sapwood caused by heartrot and saprot fungi [106]. Yang et al. [107] observed that cultures of

P. chrysosporium reduced lignin levels in thermomechanical pulps when maintained under 100% O_2 instead of air. Degradation of wheat straw lignin (^{14}C-lignin → $^{14}CO_2$) by white-rot fungi also proceeded faster in an O_2 atmosphere [108] or when oxygen transfer into fungal cultures was improved [109]. Furthermore, the amount of water-soluble lignin released from straw by fungal treatment was initially higher in oxygenated cultures [108].

Molecular oxygen increases both the titer of the ligninolytic complex of *P. chrysosporium* and the rate of lignin oxidation by the system after its formation [102]. Thus, an increase in the O_2 partial pressure of cultures from 0.2 to 0.8 atm. shortly before the onset of secondary metabolism increases by 5-fold the titer of the ensuing lignin-oxidizing system [102]. Likewise, the rate of lignin oxidation (^{14}C-DHP → $^{14}CO_2$) by fungal cultures maintained under 0.1 atm. O_2 is only 28% of the oxidation rate attained at 0.4 atm. O_2. More recently, Faison and Kirk [110] detected higher activities of a ligninase (lignin peroxidase) [30,31], thought to play a key role in lignin depolymerization, in cultures of *P. chrysosporium* grown under oxygen compared to cultures under air atmosphere. Activity was highest in cultures held under air during primary growth and then shifted to an oxygen atmosphere.

Enhanced production of H_2O_2, essential for lignin peroxidase activity, may partly explain this stimulatory effect of O_2 [111]. Peroxide levels in cultures of *P. chrysosporium* flushed with 100% O_2 are much higher than in air-grown cultures [111]. Elevated O_2 concentrations also increase de novo synthesis of the secondary metabolite veratryl alcohol [112] (see Section V.B). In turn, this may lead to higher levels of lignin peroxidase activity, either through direct induction [113,114] or as a result of protection against H_2O_2-mediated inactivation [115].

Carbon Cosubstrate Requirement

Although native lignin is potentially capable of providing the energy necessary to sustain fungal growth, more recent data reveal that, in order for polymer decomposition to proceed, white-rot fungi require a more easily assimilable carbon cosubstrate [116,117]. Earlier reports [118] that white-rot fungi could use lignin as sole source of carbon have yet to be confirmed.

Ligninolysis (^{14}C-DHP → $^{14}CO_2$) in *P. chrysosporium* is also triggered by restricting exogenous carbon [119]. Furthermore, under nitrogen-sufficient conditions, the nature of the carbon source can influence both ligninase activity [99] and the extent of lignin degradation [117]. Buswell et al. [99] reported that lignin peroxidase activities were considerably higher in glycerol-supplemented cultures of *P. chrysosporium* INA-12 than in glucose-grown ones and related this elevated activity to the relatively poor rate of fungal growth on glycerol. Other strains of *P. chrysosporium* have since been observed to exhibit the same response (P. Roch, J. A. Buswell, E. Odier, and R. B. Cain, unpublished results). Higher lignin peroxidase activities found in agitated carbon-limited cultures of *P. chrysosporium* BKM-F-1767 compared to nitrogen-limited cultures were correlated with changes in extracellular protein profiles involving preferential synthesis of certain isoenzymes [120]. The nature of the carbon source may also influence overall ligninase activity by affecting the rate of hydrogen peroxide production, leading to different rates of H_2O_2-mediated inactivation [115]. Recently, higher levels of $^{14}CO_2$ release from ^{14}C-kraft lignin by agitated cultures of *C. versicolor* were equated with increased crystallinity of cellobiose and cellulose preparations adopted as cosubstrates [76].

The effects of carbohydrate supplementation on the degradation of lignin in woody tissues are less well-defined. Reid [121] reported that $^{14}CO_2$ release from ^{14}C-labeled lignin in aspen wood by *P. chrysosporium* PRC 2750 was stimulated by increasing concentrations of cellulose. Cellulose supplementation had no effect on lignin degradation (^{14}C-poplar lignin $\rightarrow$ $^{14}CO_2$) by *Sporotrichum pulverulentum* (= *P. chrysosporium* P127-1) and *Phanerochaete sordida*, whereas glucose caused a slight repression [105]. However, addition of cellulose or glucose to the poplar wood medium stimulated $^{14}CO_2$ release by *Phlebia radiata* but repressed lignin oxidation by *Polyporus versicolor* and *Pleurotus ostreatus*. Bone and Levonen-Munoz [122] reported that glucose, xylose, or mannose enhanced lignin decomposition during solid-state fermentation of oat straw. In another study, lignin degradation (^{14}C-DHP $\rightarrow$ $^{14}CO_2$) by *Lentinus edodes* grown on oak wood medium was stimulated by fructose, xylose, or glucose supplementation even though cultures had not exhausted all available carbohydrates [123].

Since lignin is apparently unable to serve as a growth substrate, the cosubstrate is presumably the source of the energy needed for biosynthesis of individual components of the ligninolytic system (e.g., ligninolytic enzymes, hydrogen peroxide, veratryl alcohol). The cosubstrate requirement may also reflect the necessity for reducing power to counteract the *in vitro* polymerizing effects of ligninases [124].

Nutrient Nitrogen Limitation

Limiting nutrient nitrogen also triggers ligninolytic activity (^{14}C lignin $\rightarrow$ $^{14}CO_2$) in *P. chrysosporium* [125] (Fig. 5) along with de novo synthesis of the secondary metabolite veratryl alcohol [98]. Nitrogen limitation is also accompanied by the appearance of new hyphal outgrowths [126], biosynthesis of extracellular polysaccharide [127], and production of lignin peroxidase [110]. Addition of excess limiting nutrient will restore primary growth and suppress both lignin peroxidase activity and polymer decomposition [110,128]. Of several amino acids added to fungal cultures, L-glutamate, L-glutamine, and histidine most strongly repressed $^{14}CO_2$ evolution from ^{14}C-DHP [129], while different nitrogen supplements had varying effects on the biodegradation of aspen wood lignin [130]. Glutamate and histidine also inhibit mycelial binding of synthetic lignin [131]. Excess glutamate and glutamine completely suppressed lignin peroxidase production by another strain of *P. chrysosporium*, strain INA-12, even though high enzyme activities were detectable in fungal cultures supplemented with the same concentrations of nitrogen supplied as asparagine and/or NH_4NO_3 [99].

Currently, little is known about the control mechanisms that regulate secondary metabolism and lignin degradation although studies with nitrogen-limited cultures of *P. chrysosporium* have shown that high intracellular

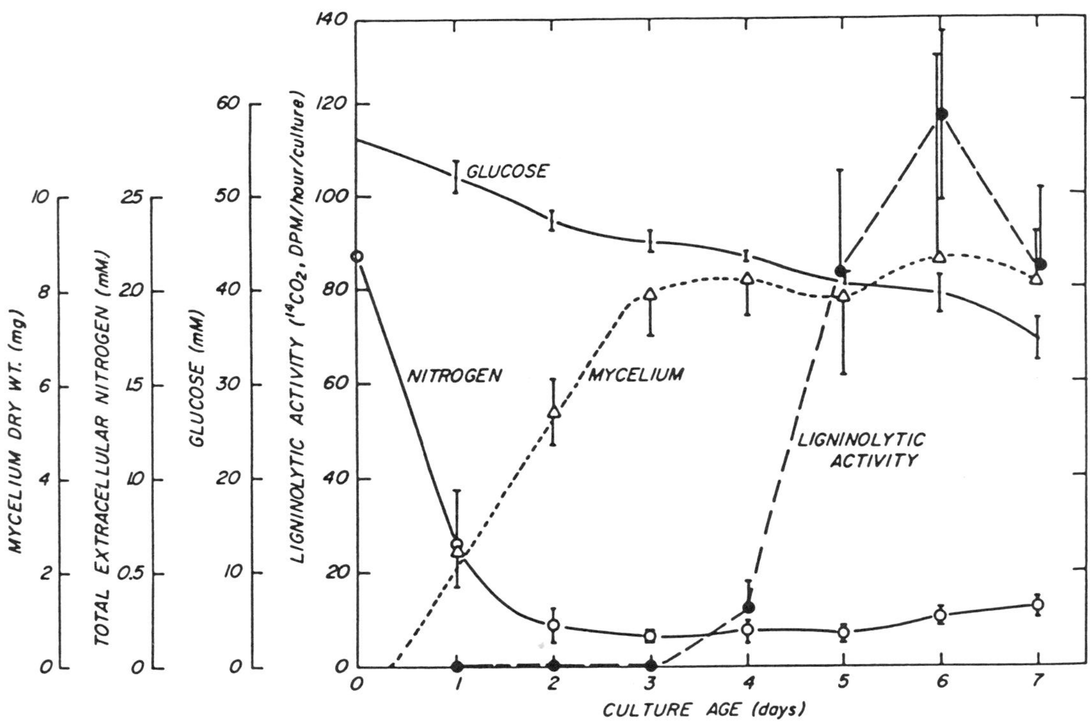

FIGURE 5. The development of ligninolytic activity in *Phanerochaete chrysosporium* in response to limitation of nitrogen. Ligninolytic activity is based on $^{14}CO_2$ evolved during 6 hr under 100% O_2 and 39°C after addition of 5×10^4 dpm of ring-labeled ^{14}C-DHP per culture: activity values are mean ± 1 SD for six replicate cultures. All other values are mean ± 1 SD for three replicate cultures. (From Ref. 98.)

levels of glutamate/glutamine are associated with nitrogen-based suppression of ligninolysis [129]. Addition of glutamate and NH_4^+ to ligninolytic cultures of *P. chrysosporium* provokes sharp increases in pool glutamate levels [128] and represses both $^{14}CO_2$ evolution from ^{14}C-DHP [128] and production of lignin peroxidases [110]. Carbon dioxide evolution remains suppressed, however, even when glutamate levels returned to values found in ligninolytic mycelia [128]. A separate study, which compared the activities of the ammonia-assimilating enzymes glutamine synthetase, NAD-, and NADP-glutamate dehydrogenase in extracts of ligninolytic and nonligninolytic mycelia, also concluded that the nitrogen response was linked to some aspect of glutamate/glutamine metabolism [132]. Onset of lignin degradation in nitrogen-limited cultures is also preceded by large increases (up to 10-fold) in intracellular cyclic adenosine monophosphate (cAMP), which are rapidly reversed on addition of glutamate [133].

While the suppressive effect of nitrogen on lignin degradation by *P. chrysosporium* and other white-rot fungi [105,133a] is wholly compatible with the low nitrogen content of wood [134], it is clear that not all ligninolytic fungi exhibit the same response [135,136]. *Lentinus edodes*, *Pleurotus ostreatus* [135], and an unidentified basidiomycete isolated from cattle dung [136] all released similar amounts of $^{14}CO_2$ from ^{14}C-lignin under both "high" and "low" nitrogen conditions. Production of lignin peroxidase was also observed in agitated flask cultures of *Phlebia radiata* containing high (20 mM-N) nutrient nitrogen although activities were lower than with low (2.0 mM-N) nitrogen concentrations [137]. A UV-derived mutant (MA-4) of *P. chrysosporium* wild-type strain ME-446, selected using rhemazole brilliant blue -R dye as an indicator of ligninolytic activity, produces lignin peroxidases and Mn(II)-dependent peroxidases (see Section V.D) under conditions of excess nitrogen supplied as ammonium tartrate [22]. HPLC and isoelectric focusing revealed that the lignin peroxidases of the mutant strain were distinct from those of the wild type.

Culture Agitation

Culture agitation was originally found to strongly inhibit lignin oxidation (^{14}C-DHP $\rightarrow$ $^{14}CO_2$) [125], veratryl alcohol biosynthesis [112], degradation of dilignols [25], and lignin peroxidase production [110] by *P. chrysosporium*. Pellet formation normally occurs in shake culture and may be partly responsible for these effects by restricting oxygen availability [101]. However, there are now several reports of lignin oxidation to CO_2 [138,139] and lignin peroxidase activity [139–141] in agitated cultures of wild-type and mutant strains of *P. chrysosporium*. Reid et al. [138] showed that agitated cultures, growing as single large pellets, released $^{14}CO_2$ from ^{14}C-DHP and degraded lignin in aspen wood to CO_2 and water-soluble products. Leisola et al. [120] found that lignin peroxidase activity in mildly agitated carbon-limited or nitrogen-limited cultures was dependent upon the concentration of added veratryl alcohol, the size and concentration of fungal pellets, and the oxygen partial pressure. Jäger et al. [139] reported lignin oxidation and lignin peroxidase production by *P. chrysosporium* growing as small mycelial pellets in agitated submerged cultures supplemented with several nonionic surfactants. Of several additives examined, Tween 20, Tween 80 (also used by Reid et al. [138] to prepare spore inocula in their experiments), and CHAPS (3-[3-colamidopropyl]-dimethylammonio 1-propanesulphonate) were most effective. Tween 80 was essential for ligninase production in agitated submerged cultures of *Phlebia radiata* [37], and enzyme activity in agitated cultures of *P. chrysosporium* INA-12 was markedly enhanced and reached maximum values earlier when Tween 80 and/or oleic acid was added to the culture medium [142]. Agitated cultures containing Tween 80 or emulsified oleic acid also converted ^{14}C-DHP to $^{14}CO_2$ at rates proportional to lignin peroxidase activity [142]. Supplementation of growth medium with emulsified vegetable oils, tall oil fatty acids (a byproduct of the pulping industry), and phospholipid sources also increased lignin peroxidase activity in cultures of this fungal strain [143].

Considerable research effort is now being directed towards the development of fermentation systems for the large-scale production of ligninolytic enzymes for use in commercial processes based on lignocellulose. Kirk et al. [144] reported increased ligninase titers using a mutant strain, *P. chrysosporium* SC26, and a bench-scale disc fermenter. Linko et al. [145] produced lignin peroxidase in a horizontal column reactor by immobilizing spores of *P. chrysosporium* in beads of agarose and agar gel. Secretion of high levels of ligninases by *P. chrysosporium* was also achieved using stirred tank fermenters equipped with a spiral of silicon tubing on which the fungus grew in the form of mycelial mats [146]. Paszczynski et al. [147] described a system for the semi-continuous production of lignin peroxidase and Mn(II) peroxidase in which the fungal mycelium was attached to the roughened interior surface of a slowly rotating 20 l glass carboy. Semicontinuous lignin peroxidase production has also been attained using *P. chrysosporium* immobilized in cubes of polyurethane foam to minimize shear effects [148]. Foam-immobilized fungus could be stored at 4°C for a period of at least 2 months and still be reactivated to the former enzyme producing state within 48 hours [148]. Nylon was found superior to polyurethane foam for lignin peroxidase production by immobilized *P. chrysosporium* in repeated batch shake cultures [149] and in a bioreactor [150]. Recently, Hatakka et al. [137] have described the production of ligninases by *P. radiata* in a laboratory reactor containing a plastic support matrix to which the fungal mycelium was loosely bound.

Other Factors Affecting Lignin Biodegradation

Influence of Other Nutrients: An additional factor known to be associated with enhanced lignin peroxidase activity is microelement concentration. Kirk et al. [144], reported a 1.7-fold increase in enzyme activity following incorporation of a 6-fold excess of a trace metal solution (containing Mn, Mg, F, Co, Ca, Zn, Cu, Mo, and Al) into the culture medium of *P. chrysosporium* BKM-7-1767 and attributed this effect to the Cu or Mn components. Increased iron, manganese, and copper all enhance lignin peroxidase activity in *P. chrysosporium* INA-12 (D. McEwan and J. A. Buswell, unpublished results) but with only marginal effects on mycelium yields. Certain inorganic elements, including Fe, Ca, and Mo, also stimulated ligninolytic activity (^{14}C-DHP $\rightarrow$ $^{14}CO_2$) in *P. chrysosporium* (ATCC 35540), whereas Mn but not Ca improved ligninolysis in *Lentinus edodes* [123].

Under certain conditions, sulphur limitation also affects the onset of lignin decomposition, and a possible interaction between sulphur and O_2 has been reported [119].

Effect of pH: The pH of the culture medium has a marked effect on lignin degradation (^{14}C-DHP → $^{14}CO_2$) by *P. chrysosporium* and the rate of $^{14}CO_2$ release drops sharply when the initial pH is set outside the range 3.5–5.5 [101]. However, while the optimum for overall degradation is around pH 4.5, activity of lignin peroxidase is maximum between pH 2–3 depending on the assay procedure [2,151].

Biosynthesis and Role of Hydrogen Peroxide

An earlier suggestion that active oxygen species might be important in lignin biodegradation [152] led several research groups to consider the direct participation of nonenzymic oxidation involving superoxide radical ($O_2^{\cdot -}$), hydroxyl radical ($^{\cdot}$OH), singlet oxygen ($^{1}O_2$), or H_2O_2. Data relating to ($^{\cdot}$OH) and ($O_2^{\cdot -}$) remains inconclusive [3], but there is unequivocal evidence to show that H_2O_2 plays an essential role in polymer decomposition.

Koenigs [153,154] detected H_2O_2 in cultures of several wood-decomposing fungi and suggested that it may be associated with lignin degradation. Forney et al. [155] correlated increased H_2O_2 generation in cell extracts of *P. chrysosporium* with the appearance of ligninolysis and subsequently proposed that H_2O_2 synthesis occurred inside peroxisome-like structures located within the "periplasmic" space [156]. Highley and Murmanis [157] also described H_2O_2 production in several white-rot species. In *C. versicolor*, peroxide generation took place predominantly in the periplasmic space as well as in the cytoplasm and hyphal sheath [157]. Both intra- and extracellular H_2O_2 production was reported in the white-rot fungus *Panus tigrinus* [158].

Faison and Kirk [111] measured extracellular H_2O_2 in culture fluids of *P. chrysosporium* and found that levels increased sharply during the period following cessation of primary growth and the onset of ligninolysis. Peroxide levels were higher in cultures maintained under oxygen rather than air [111], reflecting the observed enhancing effect of oxygen on lignin degradation. Presence of a lignin substrate also stimulated H_2O_2 production [110,159]. Addition of catalase to ligninolytic cultures strongly suppressed $^{14}CO_2$ evolution from labeled DHP [111]. Subsequently, H_2O_2-dependent ligninases have been described [30,31].

Several enzymes have been implicated as possible origins of H_2O_2 for lignin biodegradation by *P. chrysosporium*. Reddy and coworkers cite glucose 1-oxidase (EC 1.1.3.4) as the primary source of H_2O_2 in ligninolytic cultures of the fungus [160]. Like ligninolysis, appearance of glucose oxidase coincides with the onset of secondary metabolism, is triggered in response to nitrogen or carbon starvation, and is repressed by NH_4^+ and glutamate [160]. A protein catalyzing glucose-dependent H_2O_2 production is present in extracts of ligninolytic mycelia but not in extracts of nonligninolytic cultures [160]. Furthermore, glucose oxidase-negative mutants exhibit little or no capacity for lignin biodegradation (2-[^{14}C]-synthetic lignin → $^{14}CO_2$) while retaining the ability to synthesize 25–30% of wild-type levels of veratryl alcohol [161,162]. Revertants showed both increased glucose oxidase activity and lignin-degrading capacity compared to the wild type [161,162]. However, it is not clear if ligninases are produced by these mutants.

Eriksson et al. [163] have purified a H_2O_2-generating glucose 2-oxidase from extracts of glucose starved *P. chrysosporium*. This enzyme is less specific than glucose 1-oxidase and will oxidize xylose at significant rates (37% compared to glucose). Glucose 2-oxidase has also been detected in other wood-decaying fungi [164].

Extracellular Mn(II)-dependent peroxidases from ligninolytic cultures of *P. chrysosporium* can also act as an oxidase, producing H_2O_2 by oxidation of reduced glutathione, dithiothreitol, NADPH, and dihydroxymaleic acid [165,165a]. Kuwuhara et al. [166] reported that intact cultures of *P. chrysosporium* produced extracellular NADH and NADPH with highest levels present when the fungus was grown under low nitrogen conditions.

Other H_2O_2-generating enzymes detected in *P. chrysosporium* include fatty acid acyl-CoA oxidase [167], cellobiose: quinone oxidoreductase [168], and methanol oxidase [169,170]. Recently, Kersten and Kirk [171] described an extracellular glyoxal oxidase which is expressed during secondary metabolism and utilizes simple aldehyde, α-hydroxycarbonyl-, or α-dicarbonyl compounds. Two substrates, glyoxal and methylglyoxal, were identified in the culture fluid as secondary metabolites.

Biosynthesis of Extracellular Polysaccharide

Cultures of *P. chrysosporium* subjected to nitrogen starvation elaborate an extracellular/cell bound slime composed of a mixture of polysaccharides [127]. Chemical and enzymic analysis have shown the polysaccharide to contain 92% glucose and 8% arabinose and to consist of a backbone of β-D-(1→3)-linked residues, approximately every second residue substituted at 0-6, together with internal (1→6)-linked D-glucose residues in a ratio of 4.4 to 1 [172]. A separate study identified the major component as a highly branched β-D-(1→3)-glucan with single (1→6) β-D-glucopyranosyl groups linked to the backbone [173]. A similar extracellular β-1,3-glucan is produced by the white-rot fungus *Stereum sanguinolentum* especially when grown on cellobiose [174].

Although polysaccharide biosynthesis occurs simultaneously with the appearance of ligninolytic activity [127], there is no direct evidence so far to implicate the glucan in lignin degradation. Indeed, excess polysaccharide formation has an inhibitory effect on the degradation of Kraft lignin by the fungus [127]. Mucopolysaccharide sheaths are found associated with hyphae which are actively degrading woody tissue and may help to preserve a microenvironment which is optimal for components of the lignin-degrading system [175]. The glucan might also function indirectly by regulating extracellular glucose concentrations [172] and as a reserve energy source for the fungus [173]. High glucose levels appear to repress the activity of certain sugar-oxidizing enzymes which may provide the H_2O_2 necessary for ligninolysis [172]. Bes et al. [172] have described a glucan-hydrolyzing enzyme in culture filtrates of *P. chrysosporium* and suggest that the enzyme is induced when glucose levels drop below the threshold necessary for adequate H_2O_2 provision. However, addition of the hydrolyzing enzyme to ligninolytic cultures suppresses $^{14}CO_2$ release from ^{14}C-DHP [172].

Synthesis of Veratryl Alcohol and Its Roles in Lignin Biodegradation

Veratryl alcohol is a secondary metabolite synthesized de novo by *P. chrysosporium* [176] via a pathway involving phenylalanine, 3,4-dimethoxycinnamyl alcohol, and veratryl glycerol [112]. Kawai et al. [177] have also reported de novo production of this metabolite by *C. versicolor*. Veratryl alcohol "induces" [110,113,114] and protects [115,178] components of the ligninolytic system of *P. chrysosporium*, and there is evidence supporting a role as a natural mediator in the depolymerization process [179,180]. More recently, Leisola et al. [181] have proposed that veratryl alcohol enhances lignin degradation through the creation of activated oxygen species. Furthermore, the oxidation of veratryl alcohol to veratrylaldehyde serves as a convenient assay of lignin peroxidase activity [151].

Veratryl alcohol production by *P. chrysosporium* and the appearance of ligninolysis in this fungus share several common features. Both appear at the onset of idiophase triggered by either nitrogen or carbon limitation [98,119], are enhanced by elevated oxygen concentrations [112], and are suppressed by culture agitation [112] and by various nitrogenous compounds such as glutamate and ammonium salts [129]. Gold et al. [182] have also described mutant and revertant strains of *P. chrysosporium* in which lignin decomposition and veratryl alcohol biosynthesis were linked, although this relationship apparently does not exist in other mutants [182a].

It is now clear that veratryl alcohol regulates the activity of certain components of the ligninolytic system in *P. chrysosporium* and other white-rot fungi. Earlier studies had shown lignin degradation to be inducible and that maximal rates of decomposition in cultures of *P. chrysosporium* were proportional to the amount of added lignin above a required minimum concentration [21]. Veratryl alcohol ranked among the most effective of several lignin substructure models, which elicited the same response and also evoked elevated levels of lignin peroxidase and enhanced H_2O_2 production [110]. Lignin peroxidase production in agitated and agitated/immobilized cultures of *P. chrysosporium* is also dependent on veratryl alcohol supplementation of culture media [139,146,148,149]. More recently, veratryl alcohol at concentrations above 0.2 mM, guaiacol, and three dilignols markedly increased lignin peroxidase yields in agitated cultures of *Phlebia radiata* [137]. Similarly, production by several other white-rot fungi (*C. versicolor*, *M. tremellosus*, *Pleurotus ostreatus*, *Chrysosporium pruinosum*, *Polyporus versicolor*) of extracellular peroxidases resembling the lignin peroxidase of *P. chrysosporium* is also stimulated by veratryl alcohol [183,184]. However, increases in ligninase mRNA or in translation of the lignin peroxidase protein have yet to be directly attributed to veratryl alcohol [2].

Lignin peroxidases of *P. chrysosporium* are unstable in the presence of high concentrations of exogenous hydrogen peroxide and addition of veratryl alcohol to buffered mixtures of enzyme and peroxide prevents inactivation [178]. Many peroxidases may be converted by excessive amounts of hydrogen peroxide into a less catalytically active form [184a]. It is now clear that even physiological levels of peroxide generated in ligninolytic cultures of *P. chrysosporium* may be sufficiently high to cause inhibition [115]. Interestingly, while three lignin peroxide isoenzymes identified in culture supernatants were all subject to peroxide-mediated inactivation, only one isoenzyme peak could be protected by veratryl alcohol [115].

The proposal [179,180] that veratryl alcohol mediates in electron transfer processes will be discussed in Section V.D.

C. Chemistry of White-Rot Lignin Degradation

Much valuable information about the general chemistry of lignin degradation by white-rot fungi has been gleaned from structural comparisons of polymeric lignin isolated from both sound and decayed wood [4,185–191] and from the isolation and characterization of low molecular weight degradation products [4,189–193]. Taken together with data obtained from complementary studies using dimeric lignin models, this knowledge has provided a detailed, if not yet complete, picture of the various chemical reactions operative in polymer breakdown [4].

Sophisticated chemical and spectroscopic analysis of extracted lignins from decayed spruce (*Pinus glauca*), birch (*Betula papyrifera*), and beech (*Fagus crenata*) wood reveal the biodegradation process to be largely oxidative in nature [4,185–193] (Table 3). However, it is becoming increasingly apparent that reductive steps may also participate in lignin degradation by white-rot fungi [4,191]. A lower oxygen content and higher hydrogen content compared to sound MWL was found in a sample of birch lignin decayed by *P. chrysosporium*, and a number of reduced low molecular weight compounds have been identified as lignin-derived degradation products in methanol extracts of both spruce and birch wood decayed by the same fungus [191]. Furthermore, there are reports also describing the reduction of the lignin-derived compounds vanillic acid and vanillin to vanillyl alcohol [194] and the production of quinone-reducing enzymes by *P. chrysosporium* [168,195].

A second major feature of spruce and birch lignins decayed by white-rot species is the marked reduction in methoxyl content [185,186,196] (Table 3). Rapid demethoxylation of synthetic lignin (DHP) and spruce and birch MWL occurs in ligninolytic cultures of *P. chrysosporium* with up to 25.5% of the total label in $O^{14}CH_3$-DHP released as $^{14}CH_3OH$ within 22 hr [197]. Concentrated *P. chrysosporium* culture filtrates also cleave methoxyl groups from polyguaiacol [198]. Formation of several low molecular weight degradation products identified by Chen and coworkers can only be accounted for by 3-0-demeth(ox)ylation of guaiacyl units and by 3-0- and 5-0-demeth(ox)ylation of syringyl units of the corresponding 0-methyl structures [4]. However, it is not yet known if demeth(ox)ylation occurs before or after release of the 0-methyl substructures from the polymer [4].

Available data clearly indicate that lignin decomposition involves the oxidative fission of aromatic rings which still form part of the polymeric structure [18,187,190–193,199]. Earlier analyses of decayed wood demonstrated a reduced methoxyl content without a concomitant increase in methoxyl-deficient aromatic structures and α, β-unsaturated carboxyl groups not derived from the propyl side chain [187]. Characterization of degraded lignin and ^{13}C-DHP with ^{13}C-NMR has provided further evidence of ring-opening reactions [18,190,191], and identification of fungal degradation products from spruce and birch lignin suggest that aromatic nuclei in most major lignin substructures are cleaved [191,193,196,199].

Recognition of numerous low molecular weight C6-C1 products and of 4-0-alkylated vanillic acid and vanillyl alcohol structures and 4-0-alkylated syringic acid and syringyl alcohol derivatives in spruce and birch lignin, respectively, following decay by *P. chrysosporium* have firmly established Cα-Cβ cleavage of the propyl side chain as a major depolymerization reaction [190,191]. Cleavage of the Cα-Cβ bond has also been demonstrated

TABLE 3. Elemental Analyses and Methoxyl Content of Fungus-Degraded Lignins and Milled Wood Lignins from Spruce and Birch

Source	C(%)	H(%)	O(%)	OCH_3(%)	Ref.
Spruce					
MWL	62.85	6.08	31.07	15.11	186
MWL	62.81	5.88	31.31	15.24	185
Lignin decayed by					
P. anceps	58.71	4.99	36.30	11.21	186
Coriolus versicolor	57.97	4.70	37.23	11.33	186
Poria subacida	58.54	5.23	36.22	11.75	185
Birch					
MWL	57.47	6.22	36.31	20.52	196
Lignin decayed by					
Phanerochaete chrysosporium					
M-HA-1[a]	55.29	5.87	38.84	12.50	196
M-HA-2	49.09	4.81	46.10	7.52	196
M-HA-3	63.74	6.67	29.59	11.31	196

[a]M-HA-indicates the high molecular weight acidic fraction of the methanol extract of decayed wood.
Source: Ref. 4.

in ligninolytic cultures of *P. chrysosporium* using β-O-4 and β-1 lignin model compounds and is catalyzed by extracellular lignin peroxidase isolated from the fungus [30] (see Section V.D).

Also known to occur during fungal degradation of lignin is the oxidation of the α-hydroxyl group to an α-carbonyl moeity [185,187]. Depolymerization of spruce lignins by ligninolytic cultures of *P. chrysosporium* is enhanced following selective chemical Cα-oxidation [200]. Paradoxically, Cα-oxidation of dimeric β-1 and β-O-4 substructure models strongly retards Cα-Cβ cleavage by these cultures [26,28].

Other reactions implicated in lignin biodegradation include hydroxylation reactions [185,192,201] and reductive cleavage of the aryl ether bond [4].

While degradation of both hardwood and softwood lignins may involve similar general reactions, guaiacyl/syringyl lignins appear to be more susceptible to fungal attack. ^{14}C-Syringyl DHP and birch MWL were more readily degraded by *P. chrysosporium* than ^{14}C-guaiacyl DHP and spruce MWL [202]. A more recent study showed that white-rotting ascomycetes failed to degrade guaiacyl lignin and preferentially attacked the syringyl units of angiosperm lignin (T. Nilsson, J. R. Obst, and T. K. Kirk, unpublished results; see also Ref. 1). Although topochemical distribution of the two types of lignin within woody tissue may be partly responsible [191, 203], this decomposition pattern may also reflect the increased level of condensation (*ortho-coupling*) between guaiacyl propane units resulting in structures which are more resistant to attack than other linkages in lignin. Indeed, DHP (a guaiacyl lignin) was enriched in condensed ring moieties (e.g., byphenyl and aryl-alkyl C-C bonded structures) after degradation by *P. chrysosporium* [204]. However, Tai et al. [191] also reported that

P. chrysosporium cleaved aromatic rings of uncondensed syringyl propane units in birch lignin in preference to aromatic nuclei of uncondensed guaiacyl units. This greater susceptibility of the syringyl propane structure to fungal attack is now readily explained in terms of our current understanding of the mechanism of ligninase catalysis [1] (see Section V.D).

D. Involvement of Enzymes in Lignin Biodegradation

Lignin Peroxidase (ligninase)

Physicochemical Characteristics: Lignin peroxidase, an extracellular H_2O_2-dependent enzyme which is considered to participate directly in lignin biodegradation, was discovered in 1983 in ligninolytic cultures of *P. chrysosporium* [30,31]. Similar enzymes have since been reported in other white-rot fungi including *C. (Trametes) versicolor* [183,184,205], *P. ostreatus* [183], *Bjerkandera adusta* [183], *P. radiata* [137,206], *L. edodes* [207], *Chrysosporium pruinosum* [183], *P. tigrinus* [207a], and *Merulius tremellosus* [184,205].

Physical and spectroscopic characterization have shown that the ligninase from *P. chrysosporium* is a glycosylated hemoprotein with a molecular weight of 41–42 Kd and contains one mole of protohaem IX per mole of enzyme [30,31,140,147,151,208]. It exhibits the structural and kinetic properties of a true peroxidase [209,210]. Purified enzyme has a typical hemoprotein absorption spectrum with both Soret and visible absorption bands [151,208], which undergoes a red shift on addition of H_2O_2 [211]. Ferri-ligninase forms complexes with both cyanide and azide [211] and ferro-ligninase with CO, NO, and O_2 [211]. ESR and Raman resonance spectroscopy confirm that the native enzyme contains a high spin ferric iron which is pentacoordinate at room temperature [210] but hexacoordinate at low temperatures [211]. Histidine is reported to be the fifth ligand in the pentacoordinate haem [210,211]. A convenient assay for the enzyme is based on the increase in absorbance at 310 nm due to the oxidation of veratryl alcohol to veratraldehyde [2,151].

Better separation techniques using FPLC have shown that culture fluids of *P. chrysosporium* contain up to 15 different ligninase (veratryl alcohol-oxidizing) isoenzymes [115,120,139,144,147,212–215], one of which has been crystallized [216]. Those studied in detail are all glycoproteins with similar molecular weights (39–43 Kd) [147,213–215], and antibody cross-reactivity and peptide patterns produced on protease digestion reveal a considerable degree of homology [144,213]. However, the isoenzymes differ in isoelectric points, exhibit different specific activities, and vary slightly in Km values for veratryl alcohol oxidation and for H_2O_2 reduction [2,120, 139,144,147,212–215]. Three ligninase isoenzymes have also been described in *P. radiata* [206]. Multiple isoenzymes may arise from different structural genes and posttranslational modification [1,2]. Culture conditions [120,144, 212–215] and age [147] are known to influence isoenzyme patterns.

Catalytic Properties: Our current understanding of the manifold H_2O_2-dependent oxidations catalyzed by the lignin peroxidase stems from research carried out in several laboratories using a broad range of model substructures. Details of the various oxidations have been described and reviewed elsewhere [1,30–37,217], and only the major reaction categories will be summarized here.

1. Cα-Cβ cleavage: lignin peroxidase cleaves the Cα-Cβ bond of the propane side chain in nonphenolic β-O-4 aryl ether and β-1-diarylpropane models [30–33,140,217–221] (Fig. 6). Cα-Cβ cleavage is also evident in polymeric lignin following white-rot decay, occurs in intact ligninolytic cultures of *P. chrysosporium* [23,25,26,28,32,212,223] and *C. versicolor* [224], and was detected during the ligninase-mediated degradation of methylated spruce lignin [30]. Since these C-C and C-O-C bonds account for about 55% of the total intermonomeric linkages in lignin [7], Cα-Cβ cleavage is likely to lead to extensive depolymerization.

2. Cα-oxidation: β-O-4 and β-1 substructures are oxidized at Cα to form the corresponding ketone [30,31,140,151,217,218,221]. Veratryl alcohol is oxidized to veratraldehyde in an analogous reaction [151,214]. Cα-oxidation of lignin models is also observed in fungal cultures [26,200] and Cα-carbonyl groups are a feature of polymeric lignin partially degraded by white-rot fungi [4,190]. Interestingly, Cα-oxidation strongly retards the metabolism of β-1 and β-O-4 models by ligninolytic cultures of *P. chrysosporium* [26,28,200], but chemically induced Cα-oxidation in lignin enhances depolymerization in direct proportion to the content of α-carbonyl groups [200].

3. Aromatic ring-cleavage: lignin peroxidase catalyzes the cleavage of the B ring of β-O-4 substructure models to form α,β-cyclic carbonate, formate, and methyl oxalate esters of arylglycerol [32,34–36,225]. These compounds were previously recognized as ring-cleavage products by Higuchi and associates in earlier studies using β-O-4 models and intact cultures of *P. chrysosporium* [24,29,226]. The immediate ring-cleavage product is a methylmuconate derivative which arises via oxygenative cleavage between C_3–C_4 [225,227]. Demeth(ox)ylation is not a prerequisite for ring fission by the ligninase [227a]. More recently, cyclic carbonate and formate derivatives were identified as intermediates in the degradation of a synthetic lignin by lignin peroxidase [36]. Formation of aromatic ring-cleavage products is also reported in intact cultures of *C. versicolor* [228] and *Coriolus hirsutus* [36].

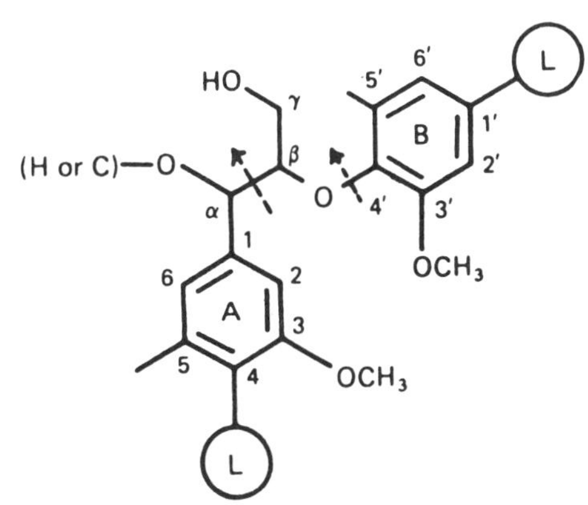

FIGURE 6. Initial cleavage sites in the degradation of the arylglycerol β-aryl ether substructure of lignin by lignin peroxidase. L = lignin polymer. Additional linkages between units, or additional methoxyl groups, can occur at C-5.

Lignin peroxidase further catalyzes the ring cleavage of veratryl alcohol to yield γ- and δ-lactones [38,229,230].

4. Aryl dealkoxylation: methanol and p-benzoquinone were identified as products of lignin peroxidase-catalyzed oxidation of 1,4-dimethoxybenzene [37]. Several other methoxybenzenes and 1,4-diethoxybenzene also serve as substrates fro the enzyme [37]. Formation of small amounts of methanol from β-O-4 dimeric models [217] suggest that lignin peroxidase is at least partly responsible for the reduction in methoxy content associated with lignin biodegradation by *P. chrysosporium*.

5. Cleavage of the arylglycerol-β-aryl ether bond: lignin peroxidase catalyzes the $O\text{-}C_{(4)}$ cleavage of the arylglycerol-β-aryl ether bond [31, 140,217–220] (Fig. 6) confirming earlier speculation as to a possible origin of arylglycerol and phenol products arising during the degradation of β-O-4 dimers by intact cultures of *P. chrysosporium* [23,223,231–233]. Isotope-labeling experiments show unequivocally that phenylglycerol products are formed by cleavage of the β-O-4 bond between the ethereal oxygen and the 4-position carbon of the B ring and not via $C_{(\beta)}\text{-}O$ cleavage [24,234,235].

Other reaction catalyzed by lignin peroxidase include:

6. Migration of phenoxy moieties from Cβ to Cγ [217,219,236], a reaction previously reported in whole cultures of *P. chrysosporium* [237].

7. Hydroxylation of benzylic methylene groups [151,214].

8. Dihydroxylation at the Cα-Cβ olefinic bond in styryl structures [140,151].

9. Phenol oxidation and polymerization of lignin [124,151].

10. Degradation of certain polyaromatic and dibenzodioxins [178,238,239].

Mechanism of Ligninase Catalysis: Intense research undertaken in several laboratories has established that ligninase exhibits the redox cycle of a true peroxidase (Fig. 7) and that substrate free radicals are intermediates in enzyme catalysis [1,2,33,37,179,181,209–211,217,218,221,238–244]. The initial step is the reaction between the native ferric state of the ligninase and H_2O_2 to yield a two-electron oxidized form of the enzyme, Compound I. This two-electron oxy-ferryl state then generates unstable radical cations from alkoxylated aromatic substrates in two consecutive single-electron transfer steps, whereby Compound I is converted to Compound II (the one-electron oxidized form of ligninase) and Compound II back to the ferric ground state. The aryl cation radicals subsequently undergo a variety of nonenzymic interactions which underly the catalytic diversity exhibited by ligninase. Substrate reactivity with ligninase and the subsequent reactions of the radical cation will depend on the nature and pattern of aromatic ring substitution. The nonspecific nature of lignin oxidation via radical cation formation has led Kirk and Farrell [1] to refer to lignin biodegradation as an example of enzymatic combustion, a process which they suggest may function in the microbial conversion of other complex molecules in the biosphere.

Schoemaker et al. [179] have proposed that veratryl alcohol functions as a natural mediator in lignin biodegradation and suggest that the first step in the process is the generation of veratryl alcohol radical cations by the Compound I form of ligninase. These radical cations then remove a single electron from a methoxylated aromatic ring of lignin, initiating bond cleavage which in turn leaves a radical in the degradation products. Oxygen is then thought to bind to these radicals, giving rise to the production

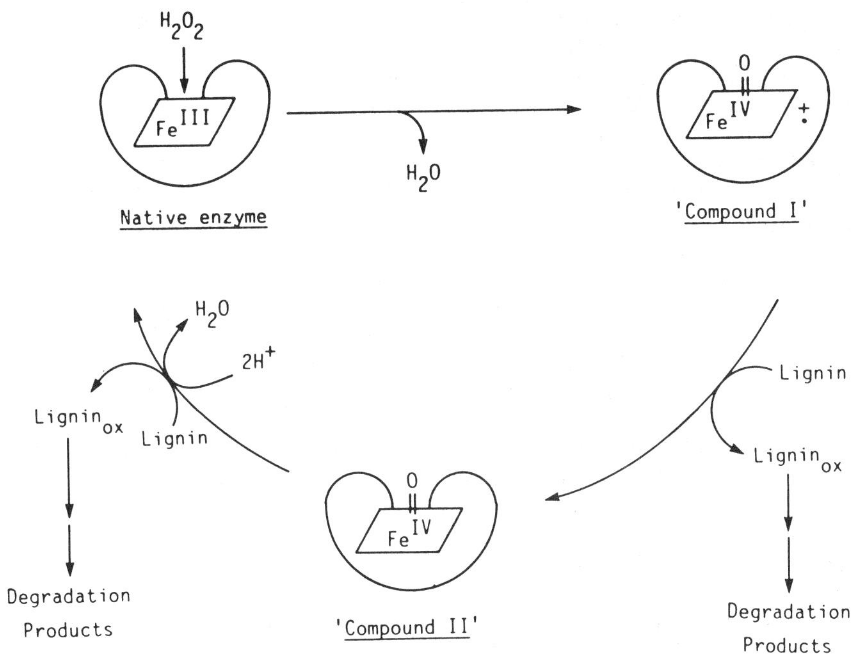

FIGURE 7. Catalytic cycle of *Phanerochaete chrysosporium* lignin peroxidase. (From Ref. 345.)

of active oxygen species capable of promoting extensive auto-oxidation of lignin at sites remote from the fungal hyphae [181].

The theory, recently refined by Harvey et al. [245], provides an attractive explanation for the enzyme-catalyzed decomposition of a three-dimensional insoluble polymer which is not readily accessible to the active center of the enzyme. Harvey et al. [180] provided indirect evidence for a mediator role when observing that the monomethoxylated compounds anisyl alcohol and mandelic acid, normally poor substrates for lignin peroxidase, were readily oxidized in the presence of catalytic amounts of veratryl alcohol or 1,4-dimethoxybenzene. Kirk and Farrell [1], however, suggest that veratryl alcohol may be simply protecting the ligninase against inactivation by the anisyl derivatives. However, the mediator concept is given additional support by a further examination of lignin peroxidase–catalyzed oxidation of veratryl alcohol [2] and recent reports that an oxidation product of veratryl alcohol participates in both lignin peroxidase-catalyzed oxidation of benzopyrene [178] and polymerization of MWL [124]. Furthermore, active oxygen species have been implicated in the oxidation of veratryl alcohol by lignin peroxidase [38].

Manganese Peroxidase

Kuwuhara et al. [208] reported a second extracellular H_2O_2-dependent enzyme in ligninolytic cultures of *P. chrysosporium*. Described as a Mn(II)-dependent, lactate-activated peroxidase, the enzyme oxidizes phenol red, o-dianisidine and a polymeric dye (Poly R), and also catalyzes the

Mn(II)-independent decarboxylation of vanillic acid. Three H_2O_2-dependent, Mn^{2+}-activated enzyme activities were cited by Huynh and Crawford [246]. An aromatic methyl ether demeth(ox)ylase converted 2-methoxy-3-phenylbenzoic acid to 2-hydroxy-3-phenylbenzoic acid and is probably the same enzyme found earlier to demeth(ox)ylate polyguaiacol [198]. A methyl ester esterase converted methyl-2-hydroxy-3-phenylbenzoate to 2-hydroxy-3-phenylbenzoic acid. The third enzyme (originally called vanillylacetone peroxidase) catalyzes the H_2O_2-dependent oxidation of several phenols and an aromatic amine and generates H_2O_2 during the H_2O_2-independent oxidation of glutathione, dithiothreitol, NADPH, and dihydroxymaleic acid [147,165,246]. This latter enzyme is very similar to the Mn(II)-dependent peroxidase of Kuwuhara et al. [208] but does not oxidize phenol red or require lactate [147,165]. Although extracellular, most enzyme activity appeared to be closely associated with the outer mycelial surface [147].

Subsequent purification and characterization revealed both enzymes to have a molecular weight between 45 and 47 Kd and to contain a single protoheme IX with a high-spin [147,247], predominantly pentacoordinate, ferric iron and with histidine coordinated as the fifth ligand [249,250]. Manganese peroxidase oxidizes Mn(II) to Mn(III), which in turn can oxidize a variety of organic substrates [147,247,248]. A Mn-peroxidase has also been purified from *C. versicolor* [251]. The enzyme described by Paszczynski et al. [147] contained 17% neutral carbohydrate. Up to six Mn-peroxidases have now been separated from ligninolytic cultures of *P. chrysosporium* [1,165,212,213], although antibody cross-reactivity reveals a very high degree of homology [213].

Gold and coworkers have recently undertaken a more detailed spectral characterization of the Mn-peroxidase from *P. chrysosporium*. Their data indicate that the enzyme undergoes a redox cycle similar to that found for lignin peroxidase. Thus, the native enzyme is first converted by H_2O_2 to the two-electron oxidized form Compound I, which is then reduced back to the ground state in two single electron steps via the one-electron oxidized form of the enzyme, Compound II. However, while both Mn(II) and various phenols are able to reduce Compound I to Compound II, only Mn(II) can reduce Compound II back to the native enzyme, thereby accounting for the manganese dependence of the enzyme.

Nonphenolic compounds are not substrates for Mn-peroxidases [252], which may function in lignin biodegradation as phenol-oxidizing enzymes and possibly by generating H_2O_2 [1,147,165,247,248,252].

Other Phenol-Oxidizing Enzymes

Other phenol-oxidizing enzymes (laccase [EC. 1.10.3.2] and peroxidase [EC. 1. 11. 1.7]), produced by most white-rot fungi and by other fungal species, have many times been linked with lignin biodegradation [e.g., 5, 80,82,253–255]. Phenoloxidase-deficient mutants of *P. chrysosporium* also exhibit impaired lignin-degrading ability [182,182a,256].

Laccase and peroxidase catalyze the one-electron oxidation of phenols to phenoxy radicals, which in turn undergo a variety of nonenzymic interactions leading to $C\alpha$-oxidation, alkyl-phenyl cleavage, and demethoxylation [5,32,72,257,258]. Such reactions may be important in the degradation of phenolic units in lignin and of phenolic intermediates released during polymer decomposition [5,32]. However, polymerization reactions predominate in in vitro experiments using purified laccases and peroxidases [e.g., 259 – 263]. In vivo, polymerization may be limited by the intervention of reducing systems such as quinone-reducing enzymes [168,195].

Quinone-Oxidoreductases

It is now recognized that reductive steps are also important in lignin biodegradation by white-rot fungi [4,191] (see Section V.D). Currently, little is known of the mechanism(s) involved although two enzymes have been identified which may play an indirect role by preventing oxidative polymerization of radical species and quinones generated during polymer decomposition.

Cellobiose:quinone oxidoreductase (CBQase) is produced by several white-rot fungi [168,264] and Fungi Imperfecti [265]. The enzyme catalyzes the oxidation of cellobiose to cellobiono-δ-lactone using either an *ortho* or *para*-quinone or a phenoxyl radical as electron acceptor which is subsequently reduced to the corresponding phenol [168]. Polymerization of kraft lignin and lignosulphonates did not occur in cultures of *P. chrysosporium* when a CBQase system was active [266,267]. However, lignin degradation still proceeds when CBQase is catabolite repressed [5] or lacking due to mutation [116]. Furthermore, CBQase did not prevent oxidative coupling of phenols or polymerization of lignin by peroxidase [268].

An intracellular NAD(P)H:quinone oxidoreductase in *P. chrysosporium* is also thought to prevent quinone accumulation and polymerization and thereby facilitate catabolism of low molecular weight phenols [195].

E. Genetics and Molecular Biology

Genetic techniques are assuming increasing importance in lignin biodegradation research [1,3], and a range of mutants of *P. chrysosporium*, prepared by UV [116,256,271], and γ-irradiation [182a] of conidia are now available (Table 4). Mutation studies with this fungus have been aided by the development of methodology for inducing colonial growth [269] and for replica plating [269], and through identification of the physiological conditions required for fruit-body formation [270]. A variety of auxotrophic marker strains have been isolated and characterized [271 – 273] and subsequently used in genetic recombination experiments [274 – 276]. Preparation, fusion, and regeneration of protoplasts from both wild-type and auxotrophic strains has also been accomplished [277].

Cellulase-deficient (Cel^-) and phenoloxidase positive (Pox^+) mutants of white-rot fungi have been prepared by UV irradiation of conidia and improved by cross-breeding in attempts to enhance selective lignin degradation for use in biopulping processes [116,275,278,279]. Kirk and coworkers [144,280] described a mutant of *P. chrysosporium*, SC26, which produces high lignin peroxidase activities in a rotating disc bioreactor originally designed for decolorizing pulp bleach effluents (see Section VI.B). Several phenoloxidase-deficient mutants have also been selected in order to examine the relationships between lignin biodegradation, phenoloxidase production, and a variety of other secondary metabolic functions [182,182a,256].

A more thorough understanding of the biochemistry of lignin biodegradation has led to the adoption of recombinant DNA technology for enhanced expression of ligninases in genetically engineered organisms. Methods for the rapid preparation of DNA from *P. chrysosporium* have been developed [281,282] and gene libraries created for the isolation of genes and their

TABLE 4. Some Mutants of *Phanerochaete chrysosporium*

Organism (ATCC number)	Strain or mutation	Phenotype[a]	Ref.
32629	44	$C^-;X^+;POx^-$	256
32629	44-2	$C^-;X^-;POx^+$	279
32629	63-2	$C^-;X^-;POx^+$	279
32629	31	$C^+;L^+$	275
32629	3113	$C^-;L^+$	275
32629	85118	$C^-;L^+$	275
24725	SC26	L-enhanced	144
34571	*leu* 3,1,2	Leucine auxotroph	272
34571	*ade* 1,2	Adenine auxotroph	272
34571	*rib* 1	Riboflavin auxotroph	272
34571	LMT 320	Riboflavin auxotroph	182a
34571	*nic* 1	Nicotinamide auxotroph	272
34571	*arg* 1,3,4	Arginine auxotroph	272
34571	*his* 1,2	Histidine auxotroph	272
34571	LMT 31	Histidine auxotroph	182a
34571	*met* 1	Methionine auxotroph	272
34571	*leu* 1	Leucine auxotroph	272
34571	LMT 30	Leucine auxotroph	182a
34571	LMT 26	POx^-; L-enhanced	182a

[a]C = cellulase; X = xylan degradation; POx = phenol oxidase activity; L = lignin degradation; + = positive.
Source: Ref. 1.

products [283]. Clones carrying genes expressed specifically in the secondary metabolic phase have been identified using complementary DNA (cDNA) probes derived from translatable mRNA isolated from nonligninolytic and ligninolytic mycelium of *P. chrysosporium* [283,284]. Zhang et al. [285] used synthetic oligonucleotide probes to identify cDNA clones for ligninase. Tien and Tu [286] reported the cloning and sequencing of a cDNA for a ligninase from *P. chrysosporium* and showed that expression of the enzyme is regulated at the mRNA level. More recently, Farrell and associates [1,287] have cloned the ligninase isoenzyme H8 [144] of *P. chrysosporium*. Recombinant ligninase exhibited specific activities comparable to the fungal enzyme when catalyzing aromatic demethylation, Cα-oxidation, and Cα-Cβ cleavage reactions. Recombinant enzyme also catalyzed substrate-dependent H_2O_2 uptake in the presence of kraft and milled wood lignin which resulted in Cα-oxidation within the polymer [287].

VI. POTENTIAL APPLICATIONS OF LIGNINOLYTIC FUNGI AND THEIR LIGNINASES

A. Biopulping and Biobleaching

Considerable impetus for lignin biodegradation research originates from energy cost considerations within the pulp and paper industry. Earlier efforts at improving the efficiency of traditional pulping methods by biological pretreatment of wood chips with cellulase-deficient mutants of *P. chrysosporium* and *P. radiata* to minimize attack on the cellulose component were only partially successful [288–290]. Poorer strength and optical properties of the final pulp offset the benefits of any reduction in refining energy requirements [288,289]. Partial delignification of coarse thermomechanical pulp by *P. chrysosporium* prior to refining also achieved energy savings but again decreased the strength of the final pulp [288]. Some retention of pulp quality is possible using glucose-impregnated wood chips [289] or by supplementing coarse pulp with glucose before fungal treatment [291]. More recently, Johnsrud and Eriksson [292] used improved cellulase-deficient mutants of *P. chrysosporium* to treat birch chips prior to kraft pulping, O_2-alkali cooking, or chemimechanical pulping, but better pulp quality and savings in alkali were still outweighed by the cost of the pretreatment. However, Cel^- mutants may be better suited for the delignification of wheat straw or bagasse [293].

Other studies have examined the feasibility of using white-rot fungi to remove residual lignin from kraft pulp (biobleaching) as an alternative to chlorine-based chemical bleaching [290,294–297].

Kirk and Chang [298] have discussed the technical difficulties inherent in commercial-scale biomechanical pulping or biobleaching processes using ligninolytic fungi, which include the need for careful control of environmental conditions on a large scale, long pretreatment times, and total selectivity for the lignin component. Enzymic delignification using cheaply produced ligninases may be more easily integrated into existing processes and therefore more viable on an industrial scale. Potential advantages include lower energy and chemical requirements leading to reduced effluent production and greater selectivity yielding a higher quality paper product [215].

B. Waste Effluent Treatment and Removal of Environmental Pollutants

White-rot fungi and their lignin-transforming enzymes may be extremely valuable in the treatment of pulp and paper mill effluents [215,290,296] and indeed for the general removal of persistent environmental pollutants [178,239,242,299]. Pulp mill and bleach plant effluents are highly colored due to polymeric lignin degradation products and contain chlorinated aromatics which are highly toxic and mutagenic [290,300,301]. Effective color removal from bleach plant effluents has been described using several white-rot fungi including *Tintoporia borbonica* [302], *P. chrysosporium* [303–305], and *C. versicolor* [306–308]. A continuous large-scale process for effluent decolorization, the MyCoR process (Mycelial Colour Reduction), uses *P. chrysosporium* immobilized in a rotating biological contactor [309–311]. Glucose, corn syrup, or residual cellulose in the pulp mill sludge can provide the energy necessary for fungal decolorization [305]. The process removes both high molecular weight chlorolignins and low molecular weight chlorinated aromatic compounds from the effluent [312]. *C. versicolor*, immobilized in beads of calcium alginate and supplemented with sucrose, reduced the color of kraft mill wastes by 80% after 3 days [307]. Paice and Jurasek [313] reported an initial rapid reduction in the color of bleach plant effluent using peroxidase and H_2O_2 emphasizing once again the potential of lignin-transforming enzymes. Decolorization of kraft black liquor by white-rot fungi including *P. chrysosporium* and *C. versicolor* [315] has also been studied.

It is now clear that a wide range of hazardous and recalcitrant environmental pollutants including polychlorinated biphenyls [299,316], DDT [1,1-bis (4-chlorophenyl)-2,2,2-trichloroethane] [299,317], polychlorinated dibenzo(p)dioxins [242,318], lindane (1,2,3,4,5,6-hexachloro-cyclohexane) [299], and benzo(a)pyrene [178,239,299] are degraded by the ligninolytic system of *P. chrysosporium*. Chloranline/lignin conjugates [319] and crystal violet [320], a triphenylmethane dye found in soil and river sediments as a consequence of improper waste treatment, are also transformed.

C. Delignification of Lignocellulose for Feedstock and Saccharification

Polysaccharides in the huge quantities of lignocellulosic residues generated by the agricultural and forest industries represent a potential source of dietary energy for ruminants and a feedstock for industrial fermentations. However, before enzymic hydrolysis of the polysaccharides can proceed, they must be separated from the lignin component [321,322]. Therefore, attention has been focused on fungal systems which increase the digestibility of wood [e.g., 96,323], wheat straw [e.g., 324–327], and other plant wastes [328] and on identifying the physical and chemical parameters which control delignification in solid-state fermentation [e.g., 96,329,330]. Other studies have examined the pretreatment of wheat straw by white-rot fungi for enzymic saccharification of cellulose [e.g., 331–333] and the conversion of wood [334], bark [335], and lignocellulosic wastes [e.g., 336] into crude fungal protein. Current economics of straw and wood fermentation are not particularly favorable but might be improved by linking upgrading of the lignocellulose with the production of valuable edible fungi [337–339]. *Lentinus edodes*, the shiitake mushroom, has been cultivated successfully on wood chips, corn cobs, and cereal straw, and fungal growth enhances the digestibility of the lignocellulosic substrate [337].

D. Bioconversion of Polymeric Lignin Byproducts

Another potential application of lignin-transforming enzymes is in the creation of high value chemicals from the large amounts of kraft lignin and lignosulfonates formed as byproducts of the kraft and sulfite pulping processes [340]. Most of these byproduct lignins are burned as fuel and to allow recovery of inorganic pulping chemicals [298], but some are used, for example, as emulsifiers and dispersants in the oil industry or in the production of adhesives [341,342]. Low molecular weight chemicals such as vanillin and dimethyl sulfoxide are also prepared on an industrial scale from lignosulfonates and kraft lignin, respectively, but the market for these compounds is relatively small [343]. Other chemical conversions are of limited value in being costly, nonspecific and in requiring extensive purification of the end-products [340].

Two types of biological transformation of byproduct lignins have been envisaged [298]. One type would retain the polymeric structure, but selective modification would create additional reactive groups. Thus, selective demethylation of lignins by brown-rot fungi might provide a feedstock for the manufacture of high value polymers [298,344]. Alternatively, the polymer would undergo controlled biodegradation and be converted into desirable low molecular weight products [67,298]. However, the latter approach in particular remains a remote possibility in view of the heterogeneous structure of lignin and the nonspecific and nonstereoselective nature of lignin-transforming enzymes.

VII. CONCLUSIONS

Recent progress in our understanding of lignin biodegradation has significantly improved the prospects for major improvements in existing processes and for the development of new technologies, based on lignocellulose. Despite this rapid advance, however, much remains to be accomplished, not only in relation to the ligninolytic complex of *P. chrysosporium*, but in characterizing lignin-transforming systems of other fungi. Commercial exploitation of ligninases will require the cheap, large-scale production and stabilization of these enzymes and alternative strategies for achieving these aims are already being explored. However, any approach involving gene cloning for enhanced expression of ligninase proteins in alternative hosts will inevitably require a deeper knowledge of the genetical and biochemical control systems regulating enzyme synthesis and secretion. Furthermore, observed in vitro polymerization of lignin by lignin peroxidases suggest that other enzymes, as yet unidentified, are essential for complete mineralization of the polymer. Indeed, data already exist to indicate that normal functioning of the ligninolytic system may be interdependent on the biochemical networks responsible for the degradation of cellulose and hemicellulose. There is a need to explain how the various ligninases already characterized function in catalyzing lignin degradation in situ. Other authors have recognized, too, that further study of these novel enzymes offers much to basic science, for example, in perhaps extending our understanding of oxygen activation. These and other questions facing scientists in this field ensure a challenging future for lignin biodegradation research.

REFERENCES

1. Kirk, T. K. and Farrell, R. L. Enzymatic "combustion": The microbial degradation of lignin. *Annu. Rev. Microbiol.*, *41*:465–505, 1987.
2. Tien, M. Properties of ligninase from *Phanerochaete chrysosporium* and their possible applications. *CRC Crit. Rev. Microbiol.*, *15*:141–168, 1987.
3. Buswell, J. A. and Odier, E. Lignin biodegradation. *CRC Crit. Rev. Biotechnol.*, *6*:1–60, 1987.
4. Chen, C.-L. and Chang, H.-m. Chemistry of lignin biodegradation. In *Biosynthesis and Biodegradation of Wood Components* (T. Higuchi, ed.), Academic Press, San Diego, 1985, pp. 535–556.

5. Kirk, T. K. and Shimada, M. Lignin biodegradation: The microorganisms involved and the physiology and biochemistry of degradation by white-rot fungi. In *Biosynthesis and Biodegradation of Wood Components* (T. Higuchi, ed.), Academic Press, San Diego, 1985, pp. 579–605.
6. Leisola, M. S. A. and Fiechter, A. New trends in lignin biodegradation. *Adv. Biotechnol. Processes.*, *5*:59–89, 1985.
7. Crawford, R. L. *Lignin Biodegradation and Transformation.* Wiley-Interscience, New York, 1981.
8. Higuchi, T. Chang, H.-m., and Kirk, T. K. (eds.), *Recent Advances in Lignin Biodegradation Research*, Uni Publishing Co., Tokyo, 1983.
9. Sarkanen, K. V. and Hergert, H. L. Classification and distribution. In *Lignins: Occurrence, Formation, Structure and Reactions* (K. V. Sarkanen and C. H. Ludwig, eds.), Wiley-Interscience, New York, 1971, pp. 43–94.
10. Adler, E. Lignin chemistry—past, present and future. *Wood Sci. Technol.*, *11*:169–218, 1977.
11. Higuchi, T. Shimada, M., Nakatsubo, F., and Tanahashi, M. Differences in biosynthesis of guaiacyl and syringyl lignins in woods. *Wood Sci. Technol.*, *11*:153–167, 1977.
12. Sakakibara, A. A structural model of softwood lignin. *Wood Sci. Technol.*, *14*:89–100, 1980.
13. Haider, K. and Trojanowski, J. Decomposition of specifically ^{14}C-labelled phenols and dehydropolymers of coniferyl alcohol as models for lignin degradation by soft and white rot fungi. *Arch. Microbiol.*, *105*:33–41, 1975.
14. Kirk, T. K., Connors, W. J., Bleam, R. D., Hackett, W. F., and Zeikus, J. G. Preparation and microbial decomposition of synthetic (^{14}C) lignins. *Proc. Natl. Acad. Sci. U.S.A.*, *72*:2515–2519, 1975.
15. Crawford, D. L. Crawford, R. L., and Pometto A. L., III. Preparation of specifically labelled ^{14}C-(lignin) and ^{14}C-(cellulose) lignocellulose and their decomposition by the microflora of soil. *Appl. Environ. Microbiol.*, *33*:1247–1251, 1977.
16. Leisola, M., Ulmer, D., Haltmeier, T., and Fiechter, A. Rapid solubilization and depolymerization of purified kraft lignin by thin layers of *Phanerochaete chrysosporium*. *Appl. Microbiol. Biotechnol.*, *17*:117–120, 1983.
17. Reid, I. D., Abrams, G. D., and Pepper, J. M. Water soluble products from the degradation of aspen lignin by *Phanerochaete chrysosporium*. *Can. J. Bot.*, *60*:2357–2364, 1985.
18. Ellwardt, P.-CHR., Haider, K., and Ernst, L. Investigations of microbial degradation of lignin by ^{13}C NMR spectroscopy of specifically ^{13}C enriched DHP-lignin from coniferyl alcohol. *Holzforschung.*, *35*:103–109, 1981.
19. Crawford, R. L., Robinson, L. E., and Foster, R. D. Polyguaiacol: A useful model polymer for lignin biodegradation research. *Appl. Environ. Microbiol.*, *41*:1112–1116, 1981.
20. Glenn, J. K. and Gold, M. H. Decolourization of several polymeric dyes by the lignin degrading basidiomycete *Phanerochaete chrysosporium*. *Appl. Environ. Microbiol.*, *45*:1741–1747, 1983.
21. Ulmer, D. C., Leisola, M. S. A., and Fiechter, A. Possible induction of the ligninolytic system of *Phanerochaete chrysosporium*. *J. Biotechnol.*, *1*:13–24, 1984.

22. Kuwuhara, M. and Asada, Y. Production of ligninases, peroxidases and alcohol oxidase by *Phanerochaete chrysosporium* and its mutant. *Proc. Int. Symp. Lignin Enzymic and Microbial Degradation*, Paris, 1987, pp. 171–176.
23. Umezawa, T., Nakatsubo, F., and Higuchi, T. Degradation pathway of arylglycerol-β-arylethers by *Phanerochaete chrysosporium. Agric. Biol. Chem.*, *47*:2677–2681, 1983.
24. Umezawa, T. and Higuchi, T. Aromatic ring cleavage in the degradation of β-O-4 lignin substructure by *Phanerochaete chrysosporium. FEBS Lett.*, *182*:257–259, 1985.
25. Weinstein, D. A., Krisnangkura, K., Mayfield, M. B., and Gold, M. H. Metabolism of radiolabelled β-guaiacyl ether-linked lignin dimeric compounds by *Phanerochaete chrysosporium. Appl. Environ. Microbiol.*, *39*:535–540, 1980.
26. Enoki, A. and Gold, M. H. Degradation of the diarylpropane lignin model compound 1-(3',4'-diethoxyphenyl)-1,3-dihydroxy-2-(4"-methoxyphenyl)-propane and derivatives by the basidiomycete *Phanerochaete chrysosporium. Arch. Microbiol.*, *132*:123–130, 1982.
27. Kirk, T. K. What have studies with model compounds revealed about the fungal degradation of lignin? *Proc. Int. Symp. Wood Pulping Chem., Tsukuba.*, *3*:7–12, 1983.
28. Kirk, T. K. and Nakatsubo, F. Chemical mechanism of an important cleavage reaction in the fungal degradation of lignin. *Biochem. Biophys. Acta.*, *756*:376–384, 1983.
29. Higuchi, T. Degradative pathways of lignin model compounds. In *Biosynthesis and Biodegradation of Wood Components* (T. Higuchi, ed.), Academic Press, San Diego, 1985, pp. 557–578.
30. Tien, M. and Kirk, T. K. Lignin-degrading enzyme from the hymenomycete *Phanerochaete chrysosporium* Burds. *Science*, *221*:661–663, 1983.
31. Glenn, J. K., Morgan, M. A., Mayfield, M. B., Kuwuhara, M., and Gold, M. H. An extracellular H_2O_2-requiring enzyme preparation involved in lignin degradation by white rot basidiomycete *Phanerochaete chrysosporium. Biochem. Biophys. Res. Commun.*, *114*:1077–1083, 1983.
32. Higuchi, T. Catabolic pathways and role of ligninases for the degradation of lignin substructure models by white-rot fungi. *Wood. Res.*, *73*:58–81, 1986.
33. Hammel, K. E., Tien, M., Kalyanaraman, B., and Kirk, T. K. Mechanism of oxidative Cα-Cβ cleavage of a lignin model dimer by *Phanerochaete chrysosporium* ligninase: Stoichiometry and involvement of free radicals. *J. Biol. Chem.*, *260*:8348–8353, 1985.
34. Umezawa, T., Shimada, M., Higuchi, T., and Kusai, K. Aromatic ring cleavage of β-O-4 lignin substructure model dimers by lignin peroxidase of *Phanerochaete chrysosporium. FEBS Lett.*, *205*:287–292, 1986.
35. Miki, K., Renganathan, V., Mayfield, M. B., and Gold, M. H. Aromatic ring cleavage of a β-biphenyl ether dimer catalyzed by lignin peroxidase of *Phanerochaete chrysosporium. FEBS Lett.*, *210*:199–203, 1987.
36. Umezawa, T. and Higuchi, T. Aromatic ring cleavage by lignin peroxidase. *Biogenesis and Biodegradation of Plant Cell Wall Polymers* (N. G. Lewis and M. G. Paice, eds.) ACS Symposium Series, Vol. 399, 1989.

37. Kersten, P. J., Tien, M., Kalyanaraman, B., and Kirk, T. K. The ligninase of *Phanerochaete chrysosporium* generates cation radicals from methoxybenzenes. *J. Biol. Chem.*, *260*:2609–2612, 1985.
38. Haemmerli, S. D., Schoemaker, H. E., Schmidt, H. W. H., and Leisola, M. S. A. Oxidation of veratryl alcohol by the lignin peroxidase of *Phanerochaete chrysosporium*. *FEBS Lett.*, *220*:149–154, 1987.
39. Eriksson, K.-E., Grunewald, A., Nilsson, T., and Vallander, L. A scanning electron microscopy study of the growth and attack on wood by three white-rot fungi and their cellulase-less mutants. *Holzforschung.*, *34*:207–213, 1980.
40. Ruel, K., Barnoud, F., and Eriksson, K.-E. Micromorphological and ultrastructural aspects of spruce wood degradation by wild-type *Sporotrichum pulverulentum* and its cellulase-less mutant. *Holzforschung.*, *35*:157–171, 1981.
41. Ruel, K., Barnoud, F., and Eriksson, K. E. Ultrastructural aspects of wood degradation by *Sporotrichum pulverulentum*. *Holzforschung.*, *38*:61–68, 1984.
42. Murmanis, L., Highley, T. L., and Palmer, J. G. An electron microscope study of Western Hemlock degradation by the white-rot fungus *Ganoderma applanatum*. *Holzforschung.*, *38*:11–18, 1984.
43. Otjen, L. and Blanchette, R. A. A discussion of microstructural changes in wood during decomposition by white rot basidiomycetes. *Can. J. Bot.*, *64*:905–911, 1986.
44. Bland, D. E., Foster, R. C., and Logan, A. F. The mechanism of permanganate and osmium tetroxide fixation and the distribution in the cell wall of *Pinus radiata*. *Holzforschung.*, *25*:137–142, 1971.
45. Parameswaran, N. and Liese, W. Ultrastructural localization of wall components in wood cells. *Holz als Roh-und Wekstoff.*, *40*:145–155, 1982.
46. Messner, K., Srebotnik, E., and Stachelberger, H. Electron microscopical indication of the demethylation of lignin in wood by osmium tetroxide. *Proc. 3rd Int. Conf. Biotechnology in the Pulp and Paper Industry*, Stockholm, 1986, pp. 17–19.
47. Blanchette, R. A. Ultrastructural aspects and X-ray microanalysis of selective lignin degradation by white-rot fungi. *Proc. 3rd Int. Conf. Biotechnology in the Pulp and Paper Industry*, Stockholm, 1986, pp. 7–10.
48. Nyhlen, L. and Nilsson, T. Combined T.E.M. and UV-microscopy on delignification of pine wood by *Phlebia radiata* and four other white-rotters. *Proc. Int. Symp. Lignin Enzymic and Microbial Degradation*, Paris, 1987, pp. 277–282.
49. Messner, K., Srebotnik, E., Ertler, G., Foisner, R., Pettersson, B., and Stackelberger, H. Cell wall systems and extracellular membranous structures and ligninase of wood rotting fungi. *Proc. Int. Symp. Lignin Enzymic and Microbial Degradation*, Paris, 1987, pp. 243–248.
50. Garcia, S., Latgé, M. C., Prévost, M. C., and Leisola, M. Cytochemical study of wood degradation by *Phanerochaete chrysosporium*. *Proc. Int. Symp. Lignin Enzymic and Microbial Degradation*, Paris, 1987, pp. 271–274.
51. Evans, C. S. and Frazer, M. A. Ultrastructural aspects of lignin degradation by *Coriolus versicolor*. *Proc. Int. Symp. Lignin Enzymic and Microbial Degradation*, Paris, 1987, pp. 217–222.

52. Crawford, D. L. Biodegradation of agricultural and urban wastes. In *Actinomycetes in Biotechnology* (M. Goodfellow, S. T. Williams, and M. Modarski, eds.), Academic Press, London, 1988, pp. 433–459.
53. McCarthy, A. J. Paterson, A., and Broda, P. Lignin solubilization by *Thermomonospora mesophila*. *Appl. Microbiol. Biotechnol.*, *24*:347–352, 1986.
54. Levi, M. P. and Preston, R. D. A chemical and microscopic examination of the action of the soft-rot fungus *Chaetomium globosum* on beechwood (*Fagus sylv.*). *Holzforschung.*, *19*:183–190, 1965.
55. Siefert, K. Die chemische Veränderung der Buchenholz-Zellwand durch Moderfäule (*Chaetomium globosum* Kunze). *Holzforschung.*, *24*:185–189, 1966.
56. Eslyn, W. E., Kirk, T. K., and Effland, M. J. Changes in the chemical composition of wood caused by six soft-rot fungi. *Phytopathology*, *65*:473–476, 1975.
57. Haider, K. and Trojanowski, J. A comparison of the degradation of ^{14}C-labelled DHP and cornstalk lignins by micro- and macrofungi and bacteria. In *Lignin Biodegradation: Microbiology, Chemistry, and Potential Applications*, Vol. I (T. K. Kirk, T. Higuchi, and H.-M. Chang, eds.), CRC Press, Boca Raton, FL, 1980, pp. 111–134.
58. Ander, P., Stoytschev, I., and Eriksson, K.-E. Cleavage and metabolism of methoxyl groups from vanillic and ferulic acids by brown-rot and soft-rot fungi. *Cellulose Chem. Technol.*, *22*:255–266, 1988.
59. Kirk, T. K. and Highley, T. L. Quantitative changes in structural components of conifer wood during decay by white-rot and brown-rot fungi. *Phytopathology.*, *63*:1338–1342, 1973.
60. Wilcox, W. W. Anatomical changes in wood cells attacked by fungi and bacteria. *Bot. Rev.*, *36*:1–28, 1970.
61. Enoki, A., Tanaka, H., and Fuse, G. Degradation of lignin-related compounds, pure cellulose, and wood components by white-rot and brown-rot fungi. *Holzforschung.*, *42*:85–93, 1988.
62. Antai, S. P. and Crawford, D. L. Degradation of extractive-free lignocelluloses by *Coriolus versicolor* and *Poria placenta*. *Eur. J. Appl. Microbiol.*, *14*:165–168, 1982.
63. Kirk, T. K. and Adler, E. Methoxyl-deficient structural elements in lignin of sweetgum decayed by a brown-rot fungus. *Acta Chem. Scand.*, *24*:3379–3390, 1970.
64. Kirk, T. K. Effects of a brown-rot fungus, *Lenzites trabea*, on lignin in spruce wood. *Holzforschung.*, *29*:99–107, 1975.
65. Kirk, T. K., Larsson, S., and Miksche, G. E. Aromatic hydroxylation resulting from attack of lignin by a brown-rot fungus. *Acta Chem. Scand.*, *24*:1470–1472, 1970.
66. Fukuda, K. and Haraguchi, T. Successive change of lignin through all the stages of wood decay by brown rot fungus *Tyromyces palustris*. *Tokyo Noko Daigaku Nogakubu Enshurin Hokoku.*, *21*:115–121, 1985.
67. Crawford, D. L. and Crawford, R. L. Microbial degradation of lignin. *Enzyme. Microb. Technol.*, *2*:11–22, 1980.
68. Kaplan, D. L. and Hartenstein, R. Decomposition of lignins by microorganisms. *Soil Biol. Biochem.*, *12*:65–75, 1980.

69. Enoki, A., Takahashi, M., Tanaka, H., and Fuse, G. Degradation of lignin-related compounds and wood components by white-rot and brown-rot fungi. *Mokuzai Gakkaishi.*, *31*:397–408, 1985.
69a. Ulmer, D. C., Leisola, M. S. A., Schmidt, B. H., and Fiechter, A. Rapid degradation of isolated lignins by *Phanerochaete chrysosporium*. *Appl. Environ. Microbiol.*, *45*:1795–1801, 1983.
70. Abbot, T. P. and Wicklow, D. T. Degradation of lignin by *Cyathus* species. *Appl. Environ. Microbiol.*, *47*:585–587, 1984.
71. Wicklow, D. T., Langie, R., Crabtree, S., and Detroy, R. W. Degradation of lignocellulose in wheat straw versus hardwood by *Cyathus* and related species (Nidulariaceae). *Can. J. Microbiol.*, *30*:632–636, 1984.
72. Kamaya, Y., Nakatsubo, F., and Higuchi, T. Degradation of trimeric lignin model compounds, arylglycerol-β-syringaresinol ethers, by *Fusarium solani*. *Agric. Biol. Chem.*, *47*:299–308, 1983.
73. Norris, D. M. Degradation of ^{14}C-labelled aromatic acids by *Fusarium solani*. *Appl. Environ. Microbiol.*, *40*:376–380, 1980.
74. Sutherland, J. B., Pometto, A. L., III., and Crawford, D. L. Lignocellulose degradation by *Fusarium* species. *Can. J. Bot.*, *61*:1194–1198, 1983.
75. Sutherland, J. B., Crawford, D. L., and Speedie, M. K. Decomposition of ^{14}C-labelled maple and spruce lignin by marine fungi. *Mycologia.*, *74*:511–513, 1982.
76. Milstein, O., Trojanowski, J., Gressel, L., and Huetterman, A. Biodegradation of carbon-14 labeled synthetic lignin by *Aspergillus* species. *Microbios. Lett.*, *25*:113–117, 1984.
76a. Kadam, K. L. and Drew, S. W. Study of lignin biotransformation by *Aspergillus fumigatus* and white-rot fungi using ^{14}C-labelled and unlabelled kraft lignins. *Biotechnol. Bioeng.*, *28*:394–404, 1986.
77. Milstein, O., Vered, Y., Sharma, A., Gressel, J., and Flowers, H. M. Heat and microbial treatments for nutritional upgrading of straw. *Biotech. Bioeng.*, *28*:381–386, 1986.
78. Nilsson, T., Obst, J. R., and Kirk, T. K. Unpublished results.
79. Cowling, E. B. Comparative biochemistry of the decay of sweetgum sapwood by white rot fungi and brown rot fungi. U.S. Department of Agriculture Technical Bulletin 1258. U.S. Department of Agriculture, Washington, DC, 1961.
80. Ander, P. and Eriksson, K.-E. Lignin degradation and utilization by microorganisms. In *Progress in Industrial Microbiology*, Vol. 14 (M. J. Bull, ed.), Elsevier Scientific, Amsterdam, 1978, pp. 1–58.
81. Kirk, T. K. and Moore, W. E. Removing lignin from wood with white rot fungi and digestibility of resulting wood. *Wood Fiber.*, *4*:72–79, 1972.
82. Ander, P. and Eriksson, K.-E. Selective degradation of wood components by white rot fungi. *Physiol. Plant.*, *41*:239–248, 1977.
83. Blanchette, R. A. Screening wood decayed by white rot fungi for preferential lignin degradation. *Appl. Environ. Microbiol.*, *48*:647–653, 1984.
84. Blanchette, R. A. Selective delignification of eastern hemlock by *Ganoderma tsugae*. *Phytopathology.*, *74*:153–160, 1984.
85. Blanchette, R. A., Otjen, L., Effland, M. J., and Eslyn, W. E. Changes in structural and chemical components of wood delignified by fungi. *Wood Sci. Technol.*, *19*:35–46, 1985.

86. Meier, H. Decomposition of the cell wall by wood-destroying fungi and the submicroscopic structure of tracheids of spruce and wood fibers of birch. *Holz Roh. Werkst.*, *13*:323–338, 1955.
87. Wilcox, W. W. Changes in wood microstructure through progressive stages of decay. U.S. Forest Service, research paper, FPL-70, 1968.
88. Blanchette, R. A. Wood decomposition by *Phellinus (Fomes) pini*: A scanning electron microscope study. *Can. J. Bot.*, *58*:1496–1503, 1980.
89. Otjen, L. and Blanchette, R. A. Patterns of decay caused by *Inonotus dryophilus* (Aphyllophorales:Hymenochaetaeceae), a white-pocket rot fungus of oaks. *Can. J. Bot.*, *60*:2770–2779, 1982.
90. Otjen, L. and Blanchette, R. A. Selective delignification of aspen wood blocks *in vitro* by three white rot basidiomycetes. *Appl. Environ. Microbiol.*, *50*:568–572, 1985.
91. Otjen, L. and Blanchette, R. A. *Xylobolus frustulatus* decay of oak: Patterns of selective delignification and subsequent cellulose removal. *Appl. Environ. Microbiol.*, *47*:670–676, 1984.
92. Blanchette, R. A. Manganese accumulation in wood decayed by white rot fungi. *Phytopathology.*, *74*:725–730, 1984.
93. Kawase, K. Chemical components of wood decayed under natural conditions and their properties. *J. Fac. Agric. Hokkaido Univ.*, *52*:186–245, 1962.
94. Kaneshiro, T. Lignocellulosic agricultural wastes degraded by *Pleurotus ostreatus*. *Dev. Ind. Microbiol.*, *18*:591–597, 1977.
95. Highley, T. L., Bar-Lev, S. S., Kirk, T. K., and Larsen, M. J. Influence of O_2 and CO_2 on wood decay by heartrot and saprot fungi. *Phytopathology.*, *73*:630–633, 1983.
96. Reid, I. D. Biological delignification of aspen wood by solid state fermentation with the white rot fungus *Merulius tremellosus*. *Appl. Environ. Microbiol.*, *50*:133–139, 1985.
97. Henningsson, B., Henningson, M., and Nilsson, T. Difibration of wood by the use of a white rot fungus. Swedish Royal College of Forestry Research Notes R78, Royal College of Forestry, 1972.
98. Keyser, P., Kirk, T. K., and Zeikus, J. G. Ligninolytic enzyme system of *Phanerochaete chrysosporium*: Synthesized in the absence of lignin in response to nitrogen starvation. *J. Bacteriol.*, *135*:790–797, 1978.
99. Buswell, J. A., Mollet, B., and Odier, E. Ligninolytic enzyme production by *Phanerochaete chrysosporium* under conditions of nitrogen sufficiency. *FEMS Microbiol. Lett.*, *25*:295–299, 1984.
100. Kirk, T. K. and Chang, H.-M. Decomposition of lignin by white-rot fungi. II. Characterization of heavily degraded lignins from decayed spruce. *Holzforschung.*, *29*:56–64, 1975.
101. Kirk, T. K., Schultz, E., Connors, W. J., Lorenz, L. F., and Zeikus, J. G. Influence of cultural parameters on lignin metabolism by *Phanerochaete chrysosporium*. *Arch. Microbiol.*, *117*:277–285, 1978.
102. Bar-Lev, S. S. and Kirk, T. K. Effects of molecular oxygen on lignin degradation by *Phanerochaete chrysosporium*. *Biochem. Biophys. Res. Commun.*, *99*:373–378, 1981.
103. Reid, I. D. and Seifert, K. A. Effect of an atmosphere of oxygen on growth, respiration and lignin degradation by white-rot fungi. *Can. J. Bot.*, *60*:252–260, 1982.

104. Reid, I. D. and Seifert, K. A. Lignin degradation by *Phanerochaete chrysosporium* in hyperbaric oxygen. *Can. J. Microbiol.*, *26*:1168–1171, 1980.

105. Hatakka, A. I. and Uusi-Rauva, A. K. Degradation of ^{14}C-labelled poplar wood lignins by selected white-rot fungi. *Eur. J. Appl. Microbiol. Biotechnol.*, *17*:235–242, 1983.

106. Highley, T. L., Bar-Lev, S. S., Kirk, T. K., and Larsen, M. J. Influence of O_2 and CO_2 on wood decay by heartrot and saprot fungi. *Phytopathology.*, *73*:630–633, 1983.

107. Yang, H. H., Effland, M. J., and Kirk, T. K. Factors influencing fungal degradation of lignin in a representative lignocellulosic, thermomechanical pulp. *Biotechnol. Bioeng.*, *22*:65–77, 1980.

108. Yu, H.-S. and Eriksson, K.-E. Influence of oxygen on the degradation of wood and straw by white-rot fungi. *Sven. Papperstidn.*, *88*:R57–R60, 1985.

109. Leisola, M. S. A., Ulmer, D. C., and Fiechter, A. Problem of oxygen transfer during degradation of lignin by *Phanerochaete chrysosporium*. *Eur. J. Appl. Microbiol. Biotechnol.*, *17*:113–116, 1983.

110. Faison, B. D. and Kirk, T. K. Factors involved in the regulation of a ligninase activity in *Phanerochaete chrysosporium*. *Appl. Environ. Microbiol.*, *49*:299–304, 1985.

111. Faison, B. D. and Kirk, T. K. Relationships between lignin degradation and production of reduced oxygen species by *Phanerochaete chrysosporium*. *Appl. Environ. Microbiol.*, *46*:1140–1145, 1983.

112. Shimada, M., Nakatsubo, F., Kirk, T. K., and Higuchi, T. Biosynthesis of the secondary metabolite veratryl alcohol in relation to lignin degradation in *Phanerochaete chrysosporium*. *Arch. Microbiol.*, *129*:321–324, 1981.

113. Leisola, M. S. A., Ulmer, D. C., Waldner, R., and Fiechter, A. Role of veratryl alcohol in lignin degradation by *Phanerochaete chrysosporium*. *J. Biotechnol.*, *1*:331–339, 1984.

114. Faison, B. D., Kirk, T. K., and Farrell, R. L. Role of veratryl alcohol in regulating ligninase activity in *Phanerochaete chrysosporium*. *Appl. Environ. Microbiol.*, *52*:251–254, 1986.

115. Tonon, F. and Odier, E. Influence of veratryl alcohol and hydrogen peroxide on ligninase activity and ligninase production by *Phanerochaete chrysosporium*. *Appl. Environ. Microbiol.*, *54*:466–472, 1988.

116. Ander, P. and Eriksson, K.-E. Influence of carbohydrates on lignin degradation by the white-rot fungus *Sporotrichum pulverulentum*. *Svensk. Papperstidn.*, *18*:643–652, 1975.

117. Kirk, T. K., Connors, W. J., and Zeikus, J. G. Requirement for a growth substrate during lignin decomposition by two wood-rotting fungi. *Appl. Environ. Microbiol.*, *32*:192–194, 1976.

118. Pelczar, M. J., Jr., Gottleib, S., and Day, W. C. Growth of *Polyporus versicolor* in a medium with lignin as the sole carbon source. *Arch. Biochem.*, *25*:449–451, 1950.

119. Jeffries, T. W., Choi, S., and Kirk, T. K. Nutritional regulation of lignin degradation by *Phanerochaete chrysosporium*. *Appl. Environ. Microbiol.*, *42*:290–296, 1981.

120. Leisola, M. S. A., Thanei-Wyss, U., and Fiechter, A. Strategies for production of high ligninase activities by *Phanerochaete chrysosporium*. *J. Biotechnol.*, *3*:97–107, 1985.
121. Reid, I. D. The influence of nutrient balance on lignin degradation by the white-rot fungus *Phanerochaete chrysosporium*. *Can. J. Bot.*, 57:2050–2058, 1979.
122. Bone, D. H. and Levonen-Munoz, E. Solid state fermentation of oat straw by *Polyporus* sp. A-336 and the effect of added sugars. *Biotechnol. Lett.*, *6*:657–662, 1984.
123. Leatham, G. F. The ligninolytic activities of *Lentinus edodes* and *Phanerochaete chrysosporium*. *Appl. Microbiol. Biotechnol.*, *24*:51–58, 1986.
124. Haemmerli, S. D., Leisola, M. S. A., and Feichter, A. Polymerization of lignins by ligninases from *Phanerochaete chrysosporium*. *FEMS Microbiol. Lett.*, *35*:33–36, 1986.
125. Kirk, T. K., Schultz, E., Connors, W. J., Lorenz, L. F., and Zeikus, J. G. Influence of cultural parameters on lignin metabolism by *Phanerochaete chrysosporium*. *Arch. Microbiol.*, *117*:277–285, 1978.
126. Janshekar, H., Brown, C., Haltmeier, T., Leisola, M., and Fiechter, A. Bioalteration of Kraft pine lignin by *Phanerochaete chrysosporium*. *Arch. Microbiol.*, *132*:14–21, 1982.
127. Leisola, M., Brown, C., Laurila, M., Ulmer, D., and Fiechter, A. Polysaccharide synthesis by *Phanerochaete chrysosporium* during degradation of Kraft lignin. *Appl. Microbiol. Biotechnol.*, *15*:180–184, 1982.
128. Fenn, P., Choi, S., and Kirk, T. K. Ligninolytic activity of *Phanerochaete chrysosporium*: Physiology of suppression by NH_4^+ and L-glutamate. *Arch. Microbiol.*, *130*:66–71, 1981.
129. Fenn, P. and Kirk, T. K. Relationship of nitrogen to the onset and suppression of ligninolytic activity and secondary metabolism in *Phanerochaete chrysosporium*. *Arch. Microbiol.*, *130*:59–65, 1981.
130. Reid, I. D. Effects of nitrogen supplements on degradation of aspen wood lignin and carbohydrate components by *Phanerochaete chrysosporium*. *Appl. Environ. Microbiol.*, *45*:830–837, 1983.
131. Chua, M. G. S., Choi, S., and Kirk, T. K. Mycelium binding and depolymerization of synthetic ^{14}C-labeled lignin during decomposition by *Phanerochaete chrysosporium*. *Holzforschung.*, *37*:55–61, 1983.
132. Buswell, J. A., Ander, P., and Eriksson, K.-E. Ligninolytic activity and levels of ammonia assimilating enzymes in *Sporotrichum pulverulentum*. *Arch. Microbiol.*, *133*:165–171, 1982.
133. MacDonald, M. J., Paterson, A., and Broda, P. Possible relationship between cyclic AMP and idiophasic metabolism in the white-rot fungus *Phanerochaete chrysosporium*. *J. Bacteriol.*, *160*:470–472, 1984.
133a. Commanday, F. and Macy, J. M. Effect of substrate nitrogen on lignin degradation by *Pleurotus ostreatus*. *Arch. Microbiol.*, *142*:61–65, 1985.
134. Merrill, W. and Cowling, E. B. Role of nitrogen in wood deterioration: Amounts and distribution of nitrogen in tree stems. *Can. J. Bot.*, *44*:1555–1580, 1966.
135. Leatham, G. F. and Kirk, T. K. Regulation of ligninolytic activity by nutrient nitrogen in white-rot basidiomycetes. *FEMS Microbiol. Lett.*, *16*:65–67, 1983.

136. Freer, S. N. and Detroy, R. W. Biological delignification of ^{14}C-labeled lignocelluloses by basidiomycetes: Degradation and solubilization of lignin and cellulose components. *Mycologia*, *74*:943–951, 1982.
137. Hatakka, A., Kantelinen, A., Tervilä-Wilo., and L. Viikari. Production of ligninases by *Phlebia radiata* in agitated cultures. *Proc. Int. Symp. Lignin Enzymic and Microbial Degradation*, Paris, 1987, pp. 185–189.
138. Reid, I. D., Chao, E. E., and Dawson, S. S. Lignin degradation by *Phanerochaete chrysosporium* in agitated cultures. *Can. J. Microbiol.*, *31*:88–90, 1985.
139. Jäger, A., Croan, S., and Kirk, T. K. Production of ligninases and degradation of lignin in agitated submerged cultures of *Phanerochaete chrysosporium*. *Appl. Environ. Microbiol.*, *50*:1274–1278, 1985.
140. Gold, M. H., Kuwahara, M., Chiu, A. A., and Glenn, J. K. Purification and characterization of an extracellular hydrogen peroxide requiring diarylpropane oxygenase from the white-rot basidiomycete, *Phanerochaete chrysosporium*. *Arch. Biochem. Biophys.*, *234*:353–362, 1984.
141. Leisola, M. S. A. and Fiechter, A. Ligninase production in agitated cultures by *Phanerochaete chrysosporium*. *FEMS Microbiol. Lett.*, *29*:33–36, 1985.
142. Asther, M., Corrieu, G., Drapon, R., and Odier, E. Effect of Tween 80 and oleic acid on ligninase production by *Phanerochaete chrysosporium* INA-12. *Enzyme Microb. Technol.*, *9*:245–249, 1987.
143. Asther, M., Lesage, L., Drapon, R., Corrieu, G., and Odier, E. Phospholipid and fatty acid enrichment of *Phanerochaete chrysosporium* INA-12 in relation to ligninase production. *Appl. Microbiol. Biotechnol.*, *27*:393–398, 1988.
144. Kirk, T. K., Croan, S., Tien, M., Murtagh, K. E., and Farrell, R. L. Production of multiple ligninases by *Phanerochaete chrysosporium*: Effect of selected growth conditions and use of a mutant strain. *Enzyme Microb. Technol.*, *8*:27–32, 1986.
145. Linko, Y.-Y., Leisola, M., Lindholm, N., Troller, J., Linko, P., and Fiechter, A. Continuous production of lignin peroxidase by *Phanerochaete chrysosporium*. *J. Biotechnol.*, *4*:283–289, 1986.
146. Willershausen, H., Graf, H., and Jaeger, A. Production of ligninases in stirred tank fermentors. *Proc. Int. Symp. Lignin Enzymic and Microbial Degradation*, Paris, 1987, pp. 203–207.
147. Paszczynski, A., Huynh, V.-b., and Crawford, R. L. Comparison of ligninase-1 and peroxidase-M2 from the white-rot fungus *Phanerochaete chrysosporium*. *Arch. Biochem. Biophys.*, *244*:750–765, 1986.
148. Kirkpatrick, N. and Palmer, J. M. Semi-continuous ligninase production using foam immobilized *Phanerochaete chrysosporium*. *Proc. Int. Symp. Lignin Enzymic and Microbial Degradation*, Paris, 1987, pp. 191–195.
149. Linko, S., Zhong, L-C., Leisola, M., Linko, Y.-Y., Fiechter, A., and Linko, P. Lignin peroxidase production by immobilized *Phanerochaete chrysosporium* in repeated batch shake cultures. *Proc. Int. Symp. Lignin Enzymic and Microbial Degradation*, Paris, 1987, pp. 209–213.

150. Linko, S. Production and characterization of extracellular lignin peroxidase from immobilized *Phanerochaete chrysosporium* in a 10ℓ bioreactor. *Enzyme Microb. Technol.*, *10*:410–417, 1988.
151. Tien, M. and Kirk, T. K. Lignin degrading enzyme from *Phanerochaete chrysosporium*: Purification, characterization and catalytic properties of a unique H_2O_2-requiring oxygenase. *Proc. Natl. Acad. Sci., U.S.A.*, *81*:2280–2284, 1984.
152. Hall, P. L. Enzymatic transformations of lignin. *Enzyme. Microb. Technol.*, *2*:170–176, 1980.
153. Koenigs, J. W. Production of extracellular hydrogen peroxide and peroxidase by wood-rotting fungi. *Phytopathology*, *62*:100–110, 1972.
154. Koenigs, J. W. Production of hydrogen peroxide by wood-rotting fungi in wood and its correlation with weight-loss, depolymerization, and pH changes. *Arch. Microbiol.*, *99*:129–145, 1974.
155. Forney, L. J., Reddy, C. A., Tien, M., and Aust, S. D. The involvement of hydroxyl radical derived from hydrogen peroxide in lignin degradation by the white-rot fungus *Phanerochaete chrysosporium*. *J. Biol. Chem.*, *257*:11455–11462, 1982.
156. Forney, L. J., Reddy, C. A., and Pankratz, H. A. Ultrastructural localization of hydrogen peroxide production in ligninolytic *Phanerochaete chrysosporium* cells. *Appl. Environ. Microbiol.*, *44*:732–736, 1982.
157. Highley, T. L. and Murmanis, L. L. Determination of hydrogen peroxide production in *Coriolus versicolor* and *Poria placenta* during wood degradation. *Mater. Org.*, *20*:241–252, 1985.
158. Golovleva, L. A., Maltseva, O. V., Myasoedova, N. M., and Leontievsky, A. A. *Panus tigrinus* 144-degrading lignin. *Proc. 3rd Int. Conf. Biotechnology in the Pulp and Paper Industry*, Stockholm, 1986, pp. 28–30.
159. Green, R. V. and Gould, J. M. Substrate-induced H_2O_2 production in mycelia from the lignin-degrading fungus *Phanerochaete chrysosporium*. *Biochem. Biophys. Res. Commun.*, *117*:275–281, 1983.
160. Kelley, R. L. and Reddy, C. A. Identification of glucose oxidase activity as the primary source of hydrogen peroxide production in ligninolytic cultures of *Phanerochaete chrysosporium*. *Arch. Microbiol.*, *144*:248–253, 1984.
161. Ramasamy, K., Kelley, R. L., and Reddy, C. A. Lack of lignin degradation by glucose oxidase-negative mutants of *Phanerochaete chrysosporium*. *Biochem. Biophys. Res. Commun.*, *131*:436–441, 1985.
162. Kelley, R. L., Ramasamy, K., and Reddy, C. A. Characterization of glucose oxidase negative mutants of a lignin degrading basidiomycete *Phanerochaete chrysosporium*. *Arch. Microbiol.*, *144*:254–257, 1986.
163. Eriksson, K.-E., Pettersson, B., Volc, J., and Musilek, V. Formation and partial characterization of glucose-2-oxidase, an H_2O_2 producing enzyme in *Phanerochaete chrysosporium*. *Appl. Microbiol. Biotechnol.*, *23*:257–262, 1986.
164. Volc, J., Denisova, N. P., Nemol, F., and Musilek, V. Glucose-2-oxidase activity in mycelial cultures of basidiomycetes. *Folia Microbiol.* (Prague), *30*:141–147, 1985.

165. Paszczynski, A., Huynh, V.-b., and Crawford, R. L. Enzymatic activities of an extracellular, manganese-dependent peroxidase from *Phanerochaete chrysosporium*. *FEMS Microbiol. Lett.*, *29*:37–41, 1985.

165a. Asada, Y., Miyabe, M., Kikkawa, M., and Kuwuhara, M. Oxidation of NADH by a peroxidase of the lignin degrading Basidiomycete, *Phanerochaete chrysosporium*, and its involvement in the degradation of a lignin model compound. *Agric. Biol. Chem.*, *50*:525–529, 1986.

166. Kuwuhara, M., Ishida, Y., Miyagama, Y., and Kawakami, C. Production of extracellular NAD and NADP by a lignin-degrading fungus, *Phanerochaete chrysosporium*. *J. Ferment. Technol.*, *62*:237–242, 1984.

167. Green, R. V. and Gould, J. M. Fatty acyl-coenzyme A oxidase activity and H_2O_2 production in *Phanerochaete chrysosporium* mycelia. *Biochem. Biophys. Res. Commun.*, *118*:437–443, 1984.

168. Westermark, U. and Eriksson, K.-E. Cellobiose:quinone oxidoreductase, a new wood-degrading enzyme from white-rot fungi. *Acta Chem. Scand.* Ser. B., *28*:209–214, 1974.

169. Nishida, A. and Eriksson, K.-E. Formation, purification, and partial characterization of methanol oxidase, an H_2O_2-producing enzyme in *Phanerochaete chrysosporium*. *Biotechnol. Appl. Biochem.*, *9*:325–338, 1987.

170. Buswell, J. A. and Eriksson, K.-E. Methanol metabolism in *Phanerochaete chrysosporium*. *Proc. 3rd Int. Conf. Biotechnology in the Pulp and Paper Industry*, Stockholm, 1986, pp. 20–23.

171. Kersten, P. J. and Kirk, T. K. Involvement of a new enzyme, glyoxal oxidase, in extracellular H_2O_2 production by *Phanerochaete chrysosporium*. *J. Bacteriol.*, *169*:2195–2201, 1987.

172. Bes, B., Pettersson, B., Lennholm, H., Iversen, T., and Eriksson, K.-E. Synthesis, structure and enzymatic degradation of an extracellular glucan produced in nitrogen-starved cultures of the white-rot fungus *Phanerochaete chrysosporium*. *Biotechnol. Appl. Biochem.*, *9*:310–318, 1987.

173. Buchala, A. J. and Leisola, M. Structure of the β-D-glucan secreted by *Phanerochaete chrysosporium* in continuous culture. *Carbohydrate Res.*, *165*:146–149, 1987.

174. Axelsson, K., Bjorndal, H., and Eriksson, K.-E. An extracellular glucan produced by the rot fungus *Stereum sanguinolentum*. *Acta Chem. Scand.*, *22*:1363–1364, 1968.

175. Palmer, J. G., Murmanis, L. L., and Highley, T. L. Visualization of hyphal sheath in wood-decay hymenomycetes. II. White-rotters. *Mycologia.*, *75*:1005–1010, 1983.

176. Lundquist, K. and Kirk, T. K. *De novo* synthesis and decomposition of veratryl alcohol by a lignin-degrading basidiomycete. *Phytochemistry*, *17*:1676, 1978.

177. Kawai, S., Umezawa, T., and Higuchi, T. *De novo* synthesis of veratryl alcohol by *Coriolus versicolor*. *Wood Res.*, *73*:18–21, 1986.

178. Haemmerli, S. D., Leisola, M. S. A., Sanglard, D., and Fiechter, A. Oxidation of benzo(a) pyrene by extracellular ligninases of *Phanerochaete chrysosporium*. *J. Biol. Chem.*, *261*:6900–6903, 1986.

179. Schoemaker, H. E., Harvey, P. J., Bowen, R. M., and Palmer, J. M. On the mechanism of enzymatic lignin breakdown. *FEBS Lett.*, *183*:7–12, 1985.
180. Harvey, P. J., Schoemaker, H. E., and Palmer, J. M. Veratryl alcohol as a mediator and the role of radical cations in lignin biodegradation by *Phanerochaete chrysosporium*. *FEBS Lett.*, *195*:242–246, 1986.
181. Leisola, M. S. A., Haemmerli, S. D., Smit, J. D. G., Troller, J., Waldner, R., Schoemaker, H. E., and Schmidt, H. *Phanerochaete chrysosporium*: Characterization of ligninolytic enzymes and metabolism of veratryl alcohol. *Proc. Int. Symp. Lignin Enzymic and Microbial Degradation*, Paris, 1987, pp. 81–86.
182. Gold, M. H., Mayfield, M. B., Chang, T. M., Krisnangkura, K., Shimada, M., Enoki, A., and Glenn, J. K. A *Phanerochaete chrysosporium* mutant defective in lignin degradation as well as several other secondary metabolic functions. *Arch. Microbiol.*, *132*:115–122, 1982.
182a. Liwicki, R., Paterson, A., MacDonald, M. J., and Broda, P. Phenotypic classes of phenoloxidase-negative mutants of the lignin-degrading fungus *Phanerochaete chrysosporium*. *J. Bacteriol.*, *162*:641–644, 1985.
183. Waldner, R., Leisola, M., and Fiechter, A. Production of extracellular ligninolytic enzymes by different white-rot fungi. *Proc. 3rd Int. Conf. Biotechnology in the Pulp and Paper Industry*, Stockholm, 1986, pp. 150–151.
184. Biswas-Hawkes, D., Dodson, A. P. J., Harvey, P. J., and Palmer, J. M. Ligninases from white-rot fungi. *Proc. Int. Symp. Lignin Enzymic and Microbial Degradation*, Paris, 1987, pp. 125–130.
184a. Yamazaki, I. In *Molecular Mechanisms of Oxygen Activitation* (O. Hayaishi, ed.), Academic Press, New York and London, 1974, pp. 535–558.
185. Hata, K. Investigations on lignins and lignifications. XXXIII. Studies on lignins isolated from spruce wood decayed by *Poria subacida* B11. *Holzforschung.*, *20*:142–147, 1966.
186. Kirk, T. K. and Chang, H.-m. Decomposition of lignin by white-rot fungi. I. Isolation of heavily degraded lignins from decayed spruce. *Holzforschung.*, *28*:217–222, 1974.
187. Kirk, T. K. and Chang, H.-m. Decomposition of lignin by white-rot fungi. II. Isolation of heavily degraded lignins from decayed spruce. *Holzforschung.*, *29*:56–64, 1975.
188. Chang, H.-m., Chen, C.-L., and Kirk, T. K. The chemistry of lignin degradation by white-rot fungi. In *Lignin Biodegradation: Microbiology, Chemistry, and Potential Applications*, CRC Press, Boca Raton, FL, 1980, pp. 215–230.
189. Hiroi, T. and Tamai, A. Degradation of beech wood components by a white-rot fungus, *Grifora frondosa*. In *Recent Advances in Lignin Biodegradation Research*, Uni Publishing Co., 1983, pp. 34–43.
190. Chua, M. G., Chen, C.-L., and Chang, H.-m. ^{13}C-NMR spectroscopic study of lignin degraded by *Phanerochaete chrysosporium*. I. New structures. *Holzforschung.*, *36*:165–172, 1982.

191. Tai, D., Terasawa, M., Chen, C.-L., Chang, H.-m., and Kirk, T. K. Biodegradation of guaiacyl and guaiacyl-syringyl lignins in wood by *Phanerochaete chrysosporium*. In *Recent Advances in Lignin Biodegradation Research*, Uni Publishing Co., 1983, pp. 44–63.
192. Chen, C.-L., Chang, H.-m., and Kirk, T. K. Aromatic acids produced during degradation of lignin in spruce wood by *Phanerochaete chrysosporium*. *Holzforschung.*, *36*:3–9, 1982.
193. Chen, C.-L., Chang, H.-m., and Kirk, T. K. Carboxylic acids produced through oxidative cleavage of aromatic rings during degradation of lignin in spruce wood. *J. Wood Chem. Technol.*, *3*:35–57, 1983.
194. Nakatsubo, F., Kirk, T. K., Shimada, M., and Higuchi, T. Metabolism of a phenylcoumaran substructure lignin model compound in ligninolytic cultures of *Phanerochaete chrysosporium*. *Arch. Microbiol.*, *128*:416–420, 1981.
195. Buswell, J. A., Hamp, S., and Eriksson, K.-E. Intracellular quinone reduction in *Sporotrichum pulverulentum* by a NAD(P)H:quinone oxidoreductase. *FEBS Lett.*, *108*:229–232, 1979.
196. Tai, D., Terazawa, M., Chen, C.-L., Chang, H.-m., and Kirk, T. K. *Proc. TAPPI Research and Development Division Conference*, 1982, pp. 263–272.
197. Ander, P. and Eriksson, K.-E. Methanol formation during lignin degradation by *Phanerochaete chrysosporium*. *Appl. Microbiol. Biotechnol.*, *21*:96–102, 1985.
198. Frick, T. D. and Crawford, R. L. Mechanisms of microbial demethylation of lignin model polymers. In *Recent Advances in Lignin Biodegradation Research*, Uni Publishing Co., 1983, pp. 143–152.
199. Terazawa, M., Tai, D., Chen, C.-L., Chang, H.-m., and Kirk, T. K. Identification of the constituents of low molecular weight fractions obtained from birch wood degraded by *Phanerochaete chrysosporium*. *Proc. Int. Symp. Wood Pulping Chem.* (Tokyo), *4*:150–155, 1983.
200. Fenn, P. and Kirk, T. K. Effects of Cα-oxidation in the fungal metabolism of lignin. *J. Wood. Chem. Technol.*, *4*:131–148, 1984.
201. Kirk, T. K., Yang, H. H., and Keyser, P. The chemistry and physiology of the fungal degradation of lignin. *Dev. Ind. Microbiol.*, *19*:51–61, 1978.
202. Faix, O., Mozuch, M. D., and Kirk, T. K. Degradation of Gymnosperm (guaiacyl) vs. Angiosperm (syringyl/guaiacyl) lignins by *Phanerochaete chrysosporium*. *Holzforschung.*, *39*:203–208, 1985.
203. Kirk, T. K., Chang, H.-m., and Lorenz, L. F. Topochemistry of the fungal degradation of lignin in birch wood as related to the distribution of the guaiacyl and syringyl lignins. *Wood. Sci. Technol.*, *9*:81–86, 1975.
204. Krisnangkura, K. and Gold, M. H. Characterization of guaiacyl lignin degraded by the white-rot basidiomycete *Phanerochaete chrysosporium*. *Holzforschung.*, *33*:174–176, 1979.
205. Dodson, P. J., Evans, C. S., Harvey, P. J., and Palmer, J. M. Production and properties of an extracellular peroxidase from *Coriolus versicolor* which catalyses Cα-Cβ cleavage in a lignin model compound. *FEMS Microbiol. Lett.*, *42*:17–22, 1987.

206. Niku-Paavola, M.-L. Ligninolytic enzymes of the white-rot fungus *Phlebia radiata*. *Proc. Int. Symp. Lignin Enzymic and Microbial Degradation*, Paris, 1987, pp. 119–123.

207. Oki, T., Shinmoto, M., and Ishikawa, H. Enzymatic degradation of guaiacylglycerol-β-guaiacyl ether. *Mokuzai Gakkaishi.*, *32*:448–456, 1986.

207a. Maltseva, O. V., Myasoedova, N. M., Leontievski, A. A., and Golovleva, L. A. Characteristics of the ligninolytic system of *Panus tigrinus*. *Proc. Sov. Finn. Sem. Microb. Degradation Lignocellul. Raw Mater*, Tbilisi, Georgia, U.S.S.R., 1986, pp. 74–82.

208. Kuwahara, M., Glenn, J. K., Morgan, M. A., and Gold, M. H. Separation and characterization of two extracellular H_2O_2-dependent oxidases from ligninolytic cultures of *Phanerochaete chrysosporium*. *FEBS Lett.*, *169*:247–250, 1984.

209. Harvey, P. J., Schoemaker, H. E., Bowen, R. M., and Palmer, J. M. Single-electron transfer processes and the reaction mechanism of enzymic degradation of lignin. *FEBS Lett.*, *183*:13–16, 1985.

210. Kuila, D., Tien, M., Fee, J. A., and Ondrias, M. R. Resonance Raman spectra of extracellular ligninase: Evidence of a heme active site similar to those of peroxidase. *Biochemistry.*, *24*:3394–3397, 1985.

211. Andersson, L. A., Renganathan, V., Chiu, A. A., Loehr, T. M., and Gold, M. H. Spectral characterization of diarylpropane oxygenase, a novel peroxide-dependent, lignin-degrading heme enzyme. *J. Biol. Chem.*, *260*:6080–6087, 1985.

212. Leisola, M., Meussdorfer, F., Waldner, R., and Fiechter, A. Production and identification of extracellular oxidases of *Phanerochaete chrysosporium*. *J. Biotechnol.*, *2*:379–382, 1985.

213. Leisola, M. S. A., Kozulic, B., Meussdorfer, F., and Fiechter, A. Homology among multiple extracellular peroxidases from *Phanerochaete chrysosporium*. *J. Biol. Chem.*, *262*:419–424, 1987.

214. Renganathan, V., Miki, K., and Gold, M. H. Multiple molecular forms of diarylpropane oxygenase, an H_2O_2-requiring, lignin-degrading enzyme from *Phanerochaete chrysosporium*. *Arch. Biochem. Biophys.*, *342*:304–314, 1985.

215. Farrell, R. L. Industrial applications of lignin-transforming enzymes. *Phil. Trans. Roy. Soc. Lond.*, *A321*:549–553, 1987.

216. Troller, J., Smit, J. D. G., Leisola, M. S. A., Kallen, T., Winterhalter, K. H., and Fiechter, A. Crystallization of a lignin peroxidase from the white-rot fungus *Phanerochaete chrysosporium*. *Biotechnology.*, *6*:571–573, 1988.

217. Kirk, T. K., Tien, M., Kersten, P. J., Mozuch, M. D., and Kalynaraman, B. Ligninase of *Phanerochaete chrysosporium*. Mechanism of its degradation of the non-phenolic arylglycerol-β-aryl ether substructure of lignin. *Biochem. J.*, *236*:279–287, 1986.

218. Habe, T., Shimada, M., Umezawa, T., and Higuchi, T. Evidence for deuterium retention in the products after enzymatic C-C and ether bond cleavage of deuterated lignin model compounds. *Agric. Biol. Chem.*, *49*:3505–3510, 1985.

219. Miki, K., Renganthan, V., and Gold, M. H. Novel aryl ether rearrangement catalyzed by lignin peroxidase of *Phanerochaete chrysosporium*. *FEBS Lett.*, *203*:235–238, 1986.

220. Miki, K., Renganathan, V., and Gold, M. H. Mechanism of β-aryl ether dimeric lignin model compound oxidation by lignin peroxidase of *Phanerochaete chrysosporium*. *Biochemistry.*, *25*:4790–4796, 1986.
221. Renganathan, V., Miki, K., and Gold, M. H. Role of molecular oxygen in lignin peroxidase reactions. *Arch. Biochem. Biophys.*, *246*:155–161, 1986.
222. Enoki, A., Goldsby, G. P., and Gold, M. H. Metabolism of the lignin model compounds veratrylglycerol-β-guaiacyl ether and 4-ethoxy-3-methoxyphenylglycerol-β-guaiacyl ether by *Phanerochaete chrysosporium*. *Arch. Microbiol.*, *125*:227–232, 1980.
223. Enoki, A., Goldsby, G. P., Krishnangkura, K., and Gold, M. H. Degradation of the lignin model compounds 4-ethoxy-3-methoxyphenylglycol β-guaiacyl and vanillic acid ethers by *Phanerochaete chrysosporium*. *FEMS Microbiol. Lett.*, *10*:373–377, 1981.
224. Kawai, S., Umezawa, T., and Higuchi, T. Metabolism of a nonphenolic β-O-4 lignin substructure model compound by *Coriolus versicolor*. *Agric. Biol. Chem.*, *49*:2325–2330, 1985.
225. Umezawa, T. and Higuchi, T. Mechanism of aromatic ring cleavage of β-O-4 lignin substrate models by lignin peroxidase. *FEBS Lett.*, *218*:255–260, 1987.
226. Umezawa, T., Kawai, S., Yokota, S., and Higuchi, T. Aromatic ring cleavage of various β-O-4 lignin model dimers by *Phanerochaete chrysosporium*. *Wood Res.*, *73*:8–17, 1986.
227. Umezawa, T. and Higuchi, T. Formation of a muconate in aromatic ring cleavage of a β-O-4 lignin substructure model by lignin peroxidase. *Agric. Biol. Chem.*, *51*:2281–2284, 1987.
227a. Umezawa, T. and Higuchi, T. Aromatic cleavage of β-O-4 model dimers without prior demeth(ox)ylation by lignin peroxidase. *FEBS Lett.*, *205*:293–298, 1986.
228. Kawai, S., Umezawa, T., and Higuchi, T. Arylglycerol-γ-formyl ester as an aromatic ring cleavage product of nonphenolic β-O-4 lignin substructure model compounds degraded by *Coriolus versicolor*. *Appl. Environ. Microbiol.*, *50*:1505–1508, 1985.
229. Leisola, M. S. A., Schmidt, B., Thanei-Wyss, U., and Fiechter, A. Aromatic ring cleavage of veratryl alcohol by *Phanerochaete chrysosporium*. *FEBS Lett.*, *189*:267–270, 1985.
230. Shimada, M., Hattori, T., Umezawa, T., Higuchi, T., and Uzura, K. Regiospecific oxygenations during ring cleavage of a secondary metabolite, 3,4-dimethoxybenzyl alcohol, catalyzed by lignin peroxidase. *FEBS Lett.*, *221*:327–331, 1987.
231. Enoki, A., Goldsby, G. P., and Gold, M. H. β-Ether cleavage of the lignin model compound 4-ethoxy-3-methoxyphenylglycerol-β-guaiacyl ether and derivatives by *Phanerochaete chrysosporium*. *Arch. Microbiol.*, *129*:141–145, 1981.
232. Umezawa, T., Higuchi, T., and Nakatsubo, F. Difference in $^{18}O_2$ incorporation in oxygenative degradation of β-O-4 and β-1 lignin substructures by *Phanerochaete chrysosporium*. *Agric. Biol. Chem.*, *47*:2945–2948, 1983.

233. Umezawa, T. and Higuchi, T. Incorporation of $H_2{}^{18}O$ into the Cα but not the Cβ position in degradation of a β-O-4 lignin substructure model by *Phanerochaete chrysosporium*. *Agric. Biol. Chem.*, *48*:1917–1921, 1984.
234. Umezawa, T. and Higuchi, T. Arylglycerol formation in degradation of β-O-4 lignin substructures with and without α-O-γ bond by *Phanerochaete chrysosporium*. *Wood Res.*, *71*:25–31, 1985.
235. Umezawa, T. and Higuchi, T. Role of guaiacol in the degradation of arylglycerol-β-guaiacyl ether by *Phanerochaete chrysosporium*. *FEMS Microbiol. Lett.*, *26*:123–126, 1985.
236. Huynh, V.-b., Paszczynski, A., Olson, P., and Crawford, R. L. Transformations of arylpropane lignin models by a lignin peroxidase of the white-rot fungus *Phanerochaete chrysosporium*. *Arch. Biochem. Biophys.*, *250*:139–146, 1987.
237. Umezawa, T. and Higuchi, T. A novel Cα-Cβ cleavage of a β-O-4 lignin model dimer with rearrangement of the β-aryl group by *Phanerochaete chrysosporium*. *FEBS Lett.*, *192*:147–150, 1985.
238. Renganathan, V. and Gold, M. H. Spectral characterization of the oxidized states of lignin peroxidase, an extracellular enzyme from the white rot basidiomycete *Phanerochaete chrysosporium*. *Biochemistry*, *25*:1626–1631, 1986.
239. Sanglard, D., Leisola, M. S. A., and Fiechter, A. Role of extracellular ligninases in biodegradation of benzo(a)pyrene by *Phanerochaete chrysosporium*. *Enzyme Microb. Technol.*, *8*:209–212, 1986.
240. Hammel, K. E., Kalyanaraman, B., and Kirk, T. K. Substrate free radicals are intermediates in ligninase catalysis. *Proc. Natl. Acad. Sci., U.S.A.*, *83*:3708–3712, 1986.
241. Tien, M., Kirk, T. K., Bull, C., and Fee, J. A. Steady-state and transient-state kinetic studies on the oxidation of 3,4-dimethoxybenzyl alcohol catalyzed by the ligninase of *Phanerochaete chrysosporium* Burds. *J. Biol. Chem.*, *261*:1687–1693, 1986.
242. Hammel, K. E., Kalyanaraman, B., and Kirk, T. K. Oxidation of polycyclic aromatic hydrocarbons and dibenzo[p]-dioxins by *Phanerochaete chrysosporium* ligninase. *J. Biol. Chem.*, *261*:16948–16952, 1986.
243. Shimada, M., Habe, T., Umezawa, T., Higuchi, T., and Okamoto, T. The carbon-carbon bond cleavage of a lignin model compound, 1,2-diarylpropane-1,3-diol, with a heme-enzyme model catalyst tetraphenylporphyrinatoiron (III) chloride in the presence of *tert*-butylhydroperoxide. *Biochem. Biophys. Res. Commun.*, *122*:1247–1252, 1984.
244. Habe, T., Shimada, M., Okamoto, T., Panjipan, B., and Higuchi, T. Incorporation of dioxygen into hydroxylated product during the C-C–single bond cleavage of 1,2-bis(p-methoxyphenyl) propane-1,3-diol catalyzed by hemin. A novel model system for the hemoprotein ligninase. *J. Chem. Soc. Commun.*, *600*:1323–1324, 1985.
245. Harvey, P. J., Schoemaker, H. E., and Palmer, J. M. Mechanisms of lignin catalysis. See Ref. 22. pp. 145–150.
246. Huynh, V.-b. and Crawford, R. L. Novel extracellular enzymes (ligninases) of *Phanerochaete chrysosporium*. *FEMS Microbiol. Lett.*, *28*:119–123, 1985.

247. Glenn, J. K. and Gold, M. H. Purification and characterization of an extracellular Mn(II)-dependent peroxidase from the lignin-degrading basidiomycete *Phanerochaete chrysosporium*. *Appl. Biochem. Biophys.*, *242*:329–341, 1985.
248. Glenn, J. K., Akileswaran, L., and Gold, M. H. Mn(II) oxidation is the principal function of the extracellular Mn-peroxidase from *Phanerochaete chrysosporium*. *Arch. Biochem. Biophys.*, *251*:688–696, 1987.
249. Gold, M. H., Wariishi, H., Akileswaran, L., Mino, Y., and Loehr, T. M. Spectral characterization of Mn-peroxidase, an extracellular heme enzyme from *Phanerochaete chrysosporium*. *Proc. Int. Symp. Lignin Enzymic and Microbial Degradation*, Paris, 1987, pp. 113–118.
250. Mino, Y., Wariishi, H., Blackburn, N. J., Loehr, T. M., and Gold, M. H. Spectral characterization of manganese peroxidase, an extracellular heme enzyme from the lignin-degrading basidiomycete, *Phanerochaete chrysosporium*. *J. Biol. Chem.*, *263*:7029–7036, 1988.
251. Johansson, T. and Nyman, P. O. A manganese (II)-dependent extracellular peroxidase from the white-rot fungus *Trametes versicolor*. *Acta Chem. Scand.*, *B41*:762–765, 1987.
252. Bono, J. J., Goulas, P., Portet, N., Seris, J. L. Effect of Mn^{++} on the activity of two ligninolytic enzymes from *Phanerochaete chrysosporium* Burds. *Proc. Int. Symp. Lignin Enzymic and Microbial Degradation*, Paris, 1987.
253. Noguchi, A., Shimada, M., and Higuchi, T. Studies on lignin degradation. I. Possible role of non-specific oxidation of lignin by laccase. *Holzforschung.*, *34*:86–89, 1980.
254. Kirk, T. K. Effects of microorganisms on lignin. *Annu. Rev. Phytopathol.*, *9*:185–210, 1971.
255. Arora, D. S. and Sandhu, D. K. Laccase production and wood degradation by a white-rot fungus *Daedalea flavida*. *Enzyme Microb. Technol.*, *7*:405–408, 1985.
256. Ander, P. and Eriksson, K.-E. The importance of phenol oxidase activity in lignin degradation by the white-rot fungus *Sporotrichum pulverulentum*. *Arch. Microbiol.*, *109*:1–8, 1976.
257. Namba, H., Nakatsubo, F., and Higuchi, T. Degradation of β-1 linked dilignols by *Fusarium solani* M-13-1. *Wood Res.*, *69*:52–60, 1983.
258. Iwahara, S. Metabolism of lignin-related aromatic compounds by *Fusarium* species. In *Recent Advances in Lignin Biodegradation Research*, Uni Publishing Co., Tokyo, 1983, pp. 96–111.
259. Ishihara, T. and Miyazaki, M. Demethylation of lignin and lignin models by fungal laccase. *J. Jpn. Wood Res. Soc.*, *20*:39–41, 1974.
260. Gierer, J. and Opara, A. E. Studies on the enzymatic degradation of lignin. The action of peroxidase and laccase on monomeric and dimeric model compounds. *Acta. Chem. Scand.*, *28*:2909–2922, 1973.
261. Brunow, G., Wallin, H., and Sundman, V. A comparison of the effects of a white-rot fungus and H_2O_2 horseradish peroxidase on a lignosulphonate. *Holzforschung.*, *32*:189–192, 1978.
262. Hütterman, A., Herche, C., and Haars, A. Polymerization of water-insoluble lignins by *Fomes annosus*. *Holzforschung.*, *34*:64–66, 1980.

263. Leonowicz, A., Szklarz, G., and Wojtas-Wasilewska, M. The effect of fungal laccase on fractionated lignosulfonates (peritan Na). *Phytochemistry.*, *24*:393–396, 1985.
264. Westermark, U. and Eriksson, K.-E. Cellobiose:dehydrogenase (Quinone). *Methods in Enzymology*, *160*:463–466, 1988.
265. Dekker, R. F. H. Induction and characterization of a cellobiose dehydrogenase produced by a species of *Monilia*. *J. Gen. Microbiol.*, *120*:309–316, 1980.
266. Hiroi, T. and Eriksson, K.-E. Microbiological degradation of lignin. Influence of cellulose on the degradation of lignins by the white-rot fungus *Pleurotus ostreatus*. *Sven. Papperstidn.*, *79*:157–161, 1976.
267. Hiroi, T., Eriksson, K.-E., and Stenlund, B. Microbiological degradation of lignin. Influence of cellulose upon the degradation of calcium lignosulfonate. *Sven. Papperstidn.*, *79*:162–166, 1976.
268. Odier, E., Mozuch, M., Kalyanaraman, N., and Kirk, T. K. Cellobiose:quinone oxidoreductase does not prevent oxidative coupling of phenols or polymerization of lignins by ligninase. *Proc. Int. Symp. Lignin Enzymic and Microbial Degradation*, Paris, 1987, pp. 131–136.
269. Gold, M. H. and Cheng, T. M. Induction of colonial growth and replica plating of the white-rot basidiomycete *Phanerochaete chrysosporium*. *Appl. Environ. Microbiol.*, *35*:1223–1225, 1978.
270. Gold, M. H. and Cheng, T. M. Conditions for fruit body formation in the white-rot basidiomycete *Phanerochaete chrysosporium*. *Arch. Microbiol.*, *121*:37–41, 1979.
271. Gold, M. H., Cheng, T. M., and Mayfield, M. B. Isolation and complementation studies of auxotrophic mutants of the lignin-degrading basidiomycete *Phanerochaete chrysosporium*. *Appl. Environ. Microbiol.*, *44*:996–1000, 1982.
272. Molskness, T. A., Alic, M., and Gold, M. H. Characterization of leucine auxotrophs of the white-rot basidiomycete *Phanerochaete chrysosporium*. *Appl. Environ. Microbiol.*, *51*:1170–1173, 1986.
273. Gold, M. H., Glenn, J. K., Mayfield, M. B., Morgan, M. A., and Kutsuki, H. Biochemical and genetic studies on lignin degradation by *Phanerochaete chrysosporium*. In *Recent Advances in Lignin Biodegradation Research*, Uni Publishing Co., 1983, pp. 219–232.
274. Alic, M. and Gold, M. H. Genetic recombination in the lignin-degrading basidiomycete *Phanerochaete chrysosporium*. *Appl. Environ. Microbiol.*, *50*:27–30, 1985.
275. Johnsrud, S. C. and Eriksson, K.-E. Cross-breeding of selected and mutated homokaryotic strains of *Phanerochaete chrysosporium* K-3: New cellulase deficient strains with increased ability to degrade lignin. *Appl. Microbiol. Biotechnol.*, *21*:320–327, 1985.
276. Alic, M., Letzing, C., and Gold, M. H. Mating system and basidiospore formation in the lignin-degrading basidiomycete *Phanerochaete chrysosporium*. *Appl. Environ. Microbiol.*, *53*:1464–1469, 1987.
277. Gold, M. H., Cheng, T. M., and Alic, M. Formation, fusion, and regeneration of protoplasts from wild-type and auxotrophic strains of the white-rot basidiomycete *Phanerochaete chrysosporium*. *Appl. Environ. Microbiol.*, *46*:260–263, 1983.

278. Eriksson, K.-E. and Goodell, E. W. Pleiotropic mutants of the wood-rotting fungus *Polyporous adustus* lacking cellulase, mannanase, and xylanase. *Can. J. Microbiol.*, *20*:371–378, 1974.

279. Eriksson, K.-E., Johnsrud, S. C., and Vallander, L. Degradation of lignin and lignin model compounds by various mutants of the white-rot fungus, *Sporotrichum pulverulentum*. *Arch. Microbiol.*, *135*:161–168, 1983.

280. Kirk, T. K., Tien, M., Johnsrud, S. C., and Eriksson, K.-E. Lignin degrading activity of *Phanerochaete chrysosporium* Burds.: Comparison of cellulase-negative and other strains. *Enzyme Microb. Technol.*, *8*:75–80, 1986.

281. Raeder, U. and Broda, P. Rapid preparation of DNA from filamentous fungi. *Lett. Appl. Microbiol.*, *1*:17–20, 1985.

282. Rao, T. R. and Reddy, C. A. DNA sequences from a ligninolytic filamentous fungus *Phanerochaete chrysosporium* capable of autonomous replication in yeast. *Biochem. Biophys. Res. Commun.*, *118*:821–827, 1984.

283. Raeder, U., Thompson, W., and Broda, P. Establishing molecular genetics for *Phanerochaete chrysosporium*. *Phil. Trans. R. Soc. Lond.*, *A321*:475–483, 1987.

284. Haylock, R. A., Liwicki, R., and Broda, P. The isolation of mRNA from the basidiomycete fungi *Phanerochaete chrysosporium* and *Coprinus cinereus* and its *in vitro* translation. *J. Microbiol. Methods*, *41*:55–62, 1985.

285. Zhang, Y.-Z., Zylstra, G. J., Olsen, R. H., and Reddy, C. A. Identification of cDNA clones for ligninase from *Phanerochaete chrysosporium* using synthetic oligonucleotide probes. *Biochem. Biophys. Res. Commun.*, *137*:649–656, 1986.

286. Tien, M., and Tu, C.-P.D. Cloning and sequencing of a cDNA for a ligninase from *Phanerochaete chrysosporium*. *Nature*, *326*:520–523, 1987.

287. Maione, T. E., Javahaerian, K., Belew, M. A., Gomez, L. E., and Farrell, R. L. Activities of recombinant ligninase. *Proc. Int. Symp. Lignin Enzymic and Microbial Degradation*, Paris, 1987, pp. 177–183.

288. Samuelsson, L., Mjöberg, P. J., Hartler, N., Vallander, L., and Eriksson, K.-E. Influence of fungal treatment on the strength versus energy relationship in mechanical pulping. *Sven. Papperstidn.*, *83*:221–225, 1980.

289. Eriksson, K.-E. and Vallander, L. Properties of pulp from thermo-mechanical pulping of chips pretreated with fungi. *Sven. Papperstidn.*, *85*:R33–R38, 1982.

290. Eriksson, K.-E. and Kirk, T. K. Biopulping, biobleaching and treatment of kraft bleaching effluents with white-rot fungi. In *The Principles of Biotechnology: Engineering Considerations* (C. L. Cooney and A. E. Humphrey, eds.), In *Comprehensive Biotechnology. The Principles, Applications and Regulations of Biotechnology in Industry, Agriculture and Medicine* (M. Moo-Young, ed.), Pergamon Press, New York, 1985, pp. 271–294.

291. Bar-Lev, S. S., Kirk, T. K., and Chang, H.-M. Fungal treatment can reduce energy requirements for secondary refining of TMP. *Tappi.*, *65*:111–113, 1982.

292. Johnsrud, S. C. and Eriksson, K.-E. Pulp production from different fungal pretreated materials. *Proc. 3rd Int. Conf. Biotechnology in the Pulp and Paper Industry*, Stockholm, 1986, p. 53.
293. Johnsrud, S. C., Fernandez, N., Lopez, P., Gutierrez, Saez, A., and Eriksson, K.-E. Properties of fungal pretreated high yield bagasse pulps. *Nordic Pulp Paper J., Special Issue 2*:47–52, 1987.
294. Lundquist, K, Kirk, T. K., and Connors, W. J. Fungal degradation of Kraft lignin and lignin sulfonates prepared from synthetic ^{14}C-lignins. *Arch. Microbiol.*, *112*:291–296, 1977.
295. Kirk, T. K. and Yang, H. H. Partial delignification of unbleached Kraft pulp with ligninolytic fungi. *Biotechnol. Lett. 1*:347–352, 1979.
296. Farrell, R. L. A new key enzyme for lignin degradation. *Ann. New York Acad. Sci.*, *501*:150–158, 1987.
297. Farrell, R. Biocatalysts hold promise of better pulp quality. *Tappi.*, *67*:31–33, 1984.
298. Kirk, T. K. and Chang, H.-m. Potential applications of bioligninolytic systems. *Enzyme Microb. Technol.*, *3*:189–196, 1981.
299. Bumpus, J. A., Tien, M., Wright, D., and Aust, S. D. Oxidation of persistant environmental pollutants by a white rot fungus. *Science.*, *228*:1434–1436, 1985.
300. Ander, P., Eriksson, K.-E., Kolář, M.-C., Kringstad, K., Ranning, U., and Ramel, C. Studies on the mutagenic properties of bleaching effluents Part I. *Sven. Papperstidn.*, *80*:454–459, 1977.
301. Kringstad, K. P., Ljungquist, P. O., de Sousa, F., and Stromberg, L. M. Identification and mutagenic properties of some chlorinated aliphatic compounds in spent liquor from Kraft pulp chlorination. *Environ. Sci. Technol.*, *15*:562–566, 1981.
302. Fukuzumi, T., Nishida, A., Aoshima, K., and Minami, K. Decolourization of Kraft liquor with white-rot fungi. I. Screening of the fungi and culturing condition for decolourization of Kraft waste liquor. *Mokuzai Gakkaishi.*, *23*:290–298, 1977.
303. Eaton, D., Chang, H.-m., and Kirk, T. K. Fungal decolourization of Kraft bleach plant effluents. *Tappi.*, *63*:103–106, 1980.
304. Sundman, G., Kirk, T. K., and Chang, H.-m. Fungal decolourization of Kraft bleach plant effluent. Fate of the chromophoric material. *Tappi.*, *64*:145–148, 1981.
305. Eaton, D., Chang, H.-m., Joyce, T. W., Jeffries, T. W., and Kirk, T. K. Method obtains fungal reduction of the color of extraction-stage kraft bleach effluents. *Tappi.*, *65*:89–92, 1982.
306. Livernoche, D., Jurasek, L., Desrochers, M., Dorica, J., and Veliky, I. A. Removal of color from kraft mill waste waters with cultures of white-rot fungi and with immobilized mycelium of *Coriolus versicolor*. *Biotechnol. Bioeng.*, *25*:2055–2066, 1983.
307. Livernoche, D., Jurasek, L., Desrochers, M., and Veliky, I. A. Decolourization of a kraft mill effluent with fungal (*Coriolus versicolor*) mycelium immobilized in calcium alginate gel. *Biotechnol. Lett.*, *3*:701–706, 1982.
308. Royer, G., Desrochers, M., Jurasek, L., Rouleau, D., and Mayer, R. C. Batch and continuous decolourization of bleached kraft effluents by a white-rot fungus. *J. Chem. Technol. Biotechnol.*, *35B*:14–22, 1985.

309. Chang, H.-m., Joyce, T. W., Campbell, A. G., Gerrard, E. D., Huynh, V.-B., and Kirk, T. K. Fungal decolourization of bleach plants effluents. In *Recent Advances in Lignin Biodegradation Research*, Uni Publishing Co., 1983, pp. 257–268.
310. Joyce, J. W., Chang, H.-m., Campbell, A. G., Jr., Gerrard, E. D., and Kirk, T. W. A continuous biological process to decolorize bleach plant effluents. *Biotechnol. Adv.*, *2*:301–308, 1984.
311. Chang, H.-m., Joyce, T. W., Vasudevan, B., and Kirk, T. K. Optimizing culture parameters in the MyCoR process. *Fourth Int. Symp. Wood Pulp. Chem.*, Vol. 1, Paris, 1987, pp. 159–165.
312. Huynh, V.-B., Chang, H.-m., Joyce, T. W., and Kirk, T. K. Dechlorination of chloro-organics by a white-rot fungus. *Tappi.*, *68*:98–102, 1985.
313. Paice, M. G. and Jurasek, L. Peroxidase catalysed color removal from bleach plant effluent. *Biotechnol. Bioeng.*, *26*:477–480, 1984.
314. Kolankaya, N. and Saglam, N. Biological decolorization of waste black kraft liquor and lignin biodegradation. *Doga Seri A.*, *7*:513–517, 1983.
315. Marton, J., Stern, A. M., and Marton, T. Decolorization of Kraft black liquor with *Polyporous versicolor*, a white-rot fungus. *Tappi.*, *52*:1975–1981, 1969.
316. Eaton, D. C. Mineralization of polychlorinated biphenyls by *Phanerochaete chrysosporium*: A ligninolytic fungus. *Enzyme Microb. Technol.*, *7*:194–196, 1985.
317. Bumpus, J. A. and Aust, S. D. Biodegradation of environmental pollutants by the white rot fungus *Phanerochaete chrysosporium*: involvement of the lignin degrading system. *Bio Essays*, *6*:166–170, 1987.
318. Hammel, K. E., Kalyanaraman, B., and Kirk, T. K. Oxidation of aromatic pollutants by *Phanerochaete chrysosporium* ligninase. *Proc. Int. Symp. Lignin Enzymic and Microbial Degradation*, Paris, 1987, pp. 45–49.
319. Arjmand, M. and Sandermann, H., Jr. Plant biochemistry of xenobiotics. Mineralization of chloroaniline/lignin metabolites from wheat by the white-rot fungus *Phanerochaete chrysosporium*. *Z. Naturforsch.*, *41*:206–214, 1986.
320. Bumpus, J. A. and Brock, B. J. Biodegradation of crystal violet by the white-rot fungus *Phanerochaete chrysosporium*. *Appl. Environ. Microbiol.*, *54*:1143–1150, 1988.
321. Kirk, T. K. and Moore, W. E. Removing lignin from wood with white-rot fungi and digestibility of resulting wood. *Wood Fiber.*, *4*:72–79, 1972.
322. Baker, A. J. Effect of lignin on the *in vitro* digestibility of wood pulp. *J. Anim. Sci.*, *36*:768–771, 1973.
323. Reade, A. E. and McQueen, R. E. Investigation of white-rot fungi for the conversion of poplar into a potential feedstuff for ruminants. *Can. J. Microbiol.*, *29*:457–463, 1983.
324. Agosin, E. and Odier, E. Solid-state fermentation, lignin degradation and resulting digestibility of wheat straw fermented by selected white-rot fungi. *Appl. Microbiol. Biotechnol.*, *21*:397–403, 1985.

325. Agosin, E., Tollier, M. T., Brillouet, J. M., Thivend, D. P., and Odier, E. Fungal pretreatment of wheat straw: effects on biodegradability of cell walls, structural polysaccharides, lignin and phenolic acids by rumen microorganisms. *J. Sci. Food Agric.*, *37*:97, 1986.
326. Zadrazil, F. Screening of fungi for lignin decomposition and conversion of straw into feed. *Angew Botanik*, *39*:433–452, 1985.
327. Lindenfelser, L. A., Detroy, R. W., Ramstack, J. M., and Worden, K. A. Biological modification of the lignin and cellulose components of wheat straw by *Pleurotus ostreatus*. *Dev. Ind. Microbiol.*, *20*:541–551, 1978.
328. Zadrazil, F. Conversion of different plant waste into feed by basidiomycetes. *Eur. J. Appl. Microbiol. Biotechnol.*, *9*:243–248, 1980.
329. Zadrazil, F. and Brunnert, H. Investigation of physical parameters important for the solid state fermentation of straw by white-rot fungi. *Eur. J. Appl. Microbiol. Biotechnol.*, *11*:183–188, 1981.
330. Mudgett, R. E. and Paradis, A. J. Solid-state fermentation of natural birch lignin by *Phanerochaete chrysosporium*. *Enzyme Microb. Technol.*, *7*:150–154, 1985.
331. Hatakka, A. I. Pretreatment of wheat straw by white-rot fungi for enzymic saccharification of cellulose. *Eur. J. Appl. Microbiol. Biotechnol.*, *18*:350–357, 1983.
332. Detroy, R. W., Lindenfelser, L. A., St. Julian, G., Jr., and Orton, W. L. Saccharification of wheat-straw cellulose by enzymatic hydrolysis following fermentative and chemical pretreatment. *Biotechnol. Bioeng. Symp.*, *10*:135–148, 1980.
333. Hatakka, A. I. and Pirhonen, T. I. Cultivation of wood-rotting fungi on agricultural lignocellulosic materials for the production of crude protein. *Agric. Wastes.*, *12*:81–97, 1985.
334. Matteau, P. P. and Bone, D. H. Solid-state fermentation of maple wood. *Biotechnol. Lett.*, *2*:127–132, 1980.
335. Daugulis, A. J. and Bone, D. H. Production of microbial protein from tree bark by *Phanerochaete chrysosporium*. *Biotechnol. Bioeng.*, *20*:1639–1649, 1978.
336. Ek, M. and Eriksson, K.-E. Utilization of the white-rot fungus *Sporotrichum pulverulentum* for water purification and protein production on mixed lignocellulosic wastewaters. *Biotechnol. Bioeng.*, *22*:2273–2284, 1980.
337. Kirk, T. K. Degradation and conversion of lignocelluloses. In *The Filamentous Fungi*, Vol. 4 (J. E. Smith, D. R. Berry, and B. Kristiansen, eds.), Edward Arnold, London, 1983, pp. 266–295.
338. Wood, D. A. Useful biodegradation of lignocellulose. *Annu. Proc. Phytochem. Soc. Eur.*, *26*:295–309, 1985.
339. Wood, D. A. and Leatham, G. F. Lignocellulose degradation during the life cycle of *Agaricus bisporus*. *FEMS Microbiol. Lett.*, *20*:421–424, 1983.
340. Glasser, W. G. Potential role of lignin in tomorrow's wood utilization technologies. *Forest Prod. J.*, *31*:24–29, 1981.

341. Hoyt, C. H. and Goheen, D. H. In *Lignins: Occurrence, Formation, Structure and Reactions*, Wiley-Interscience, New York, 1971, pp. 833–863.
342. Janshekar, H. and Fiechter, A. Lignin: Biosynthesis, applications, and biodegradation. *Adv. Biochem. Eng. Biotechnol.*, *27*:119–178, 1983.
343. Goheen, D. W. In *Lignins: Occurrence, Formation, Structure and Reactions*, Wiley-Interscience, New York, 1971, pp. 797–831.
344. Goldstein, I. S. Potential for converting wood into plastics. *Science.*, *189*:847–852, 1975.
345. Kirk, T. K. The discovery and promise of lignin-degrading enzymes. In *New Horizons for Biotechnological Utilization of the Forest Resource*, Marcus Wallenberg Found. Symp. Proc. No. 2, 1985, pp. 27–42.

17

DEGRADATION OF PLANT LITTER BY AQUATIC HYPHOMYCETES

SURESH K. HASIJA and PRADEEP K. SINGHAL *Rani Durgavati University, Jabalpur, India*

I. INTRODUCTION

Aquatic fungi have been reported from a wide array of environments, ranging from marine to freshwater habitats [1]. They are represented by two major groups, namely, saprolegniales (water molds) and aquatic hyphomycetes. A significant number of ascomycetes and other mastigomycetes also occur in freshwater habitats, while basidiomycetes are poorly represented. Saprolegniales and ascomycetes are not as important in the structure and function of freshwater ecosystems as are aquatic hyphomycetes [2]. Aquatic hyphomycetes are major degraders of plant litter and serve as food for stream invertebrates, but their role in the structure and function of lakes and reservoirs is not well understood [3–6]. Aquatic hyphomycetes flourish in streams, mainly on leaf litter, which enters through riparian vegetation. Initial studies on aquatic hyphomycetes have dealt mainly with their taxonomic distribution in different freshwater habitats [7–9]. Later studies relate to their occurrence and distribution in relation to the physicochemical status of habitats [10–14]. Accordingly, aquatic fungi have been categorized into (a) winter or summer species [15–18], (b) soft and hard water species [19–22], and (c) ruderal or combative species [23]. These studies could not specify factors and mechanisms that govern the occurrence and distribution of fungi in freshwater environments. In this review we will discuss factors and mechanisms that regulate the occurrence and distribution of aquatic fungi in freshwater environments.

Kaushik and Hynes [24] revealed the important role of aquatic fungi in dynamics of streams. They found that aquatic fungi accumulated nitrogen and protein during degradation of litter. They also observed that partially degraded litter was more nutritious to stream invertebrates than was fresh litter. This observation was later confirmed by independent studies, and several fungi were found to be capable of degrading the refractory structural polymers of plant litter (e.g., lignins and celluloses) [25,26]. It was assumed that fungi could degrade structural polymers of litter, since they could grow on plant litter. But the growth of fungi on a particular substrate (litter) does not mean that they can synthesize enzymes necessary for degradation of structural polymers of litter. Some literature is now available providing experimental proof for the presence of ligninases and cellulases in aquatic hyphomycetes [25–33].

Aquatic hyphomycetes, which degrade plant litter, are considered to be a valuable and nutritious source of food for stream invertebrates [34–38]. Fungi transform nutritionally poor plant litter to a palatable and nutritious detritus by their growth on litter and breakdown of polymers of litter. However, increased food value of litter may also be due to extracellular enzymes secreted by these fungi. The mobility and digestive capability of stream invertebrates may be another important aspect [39–41]. Fungi act as intermediaries in the food chain of streams, as most streams produce an insufficient quantity of autochthonous organic matter and allochthonous organic matter is not directly available to invertebrates. The present review discusses the mode of fungal transformation of litter into an important energy source for the detritus food chain in freshwater environments.

II. OCCURRENCE AND DISTRIBUTION

Aquatic fungi occur in diverse types of freshwater environments, ranging from streams of the northern [14,16,42–44] and southern hemispheres [7,10,11] to temperate [3,6] and tropical lakes [12,13,45]. Occurrence and distribution of fungi in these habitats have been related to physicochemical variables including temperature [12,13,15,17], pH [12,20,21,46], availability of substrate [10,14,22,47], and density of stream invertebrates [21,48]. It is suggested that competition among fungi colonizing a substrate may determine the richness of fungal species in the habitat in question [23]. There is evidence that chemical variables, namely, nutrients and biochemical oxygen demand [49], and addition of some chemicals, for example, pentachlorophenol [50], can also affect the occurrence and distribution of fungi in freshwater environments. These studies, however, fail to pinpoint any one factor regulating the distribution of aquatic fungi. It may be assumed that a complicated network of interactions between water chemistry, substrate availability, and behavior of invertebrates regulates the occurrence and distribution of aquatic fungi in freshwater environments.

A. Temperature

Seasonal distribution of fungi has been attributed to temporal changes in temperature [12,13,15,16]. Fungi has been noticed in both tropical [12,13] and temperate regions [15,16]. The distinction between summer and winter fungi has also been supported by laboratory experiments [13,15]. Winter species, e.g., *Alatospora acuminata*, *Flagellospora curvula*, and *Lemonniera aquatica*, grew at a minimum temperature of 1°C with optimum at 20–25°C [15]. Summer species, e.g., *Clavatospora tentacula*, *Flagellospora pennicillioides*, and *Lunolospora curvula*, failed to grow at 1°C and required a minimum incubation temperature of 5–10°C with optimum for growth increasing to 25–30°C [15]. Exchange experiments, where leaves were incubated at colder and warmer sites and interchanged after incubation periods of 1 day and 1 week, showed that the fungi of summer assemblage (after exchange) can survive at colder temperature. They can even reduce the normal dominants of colder sites if duration of the exchange period is increased [15]. Thus, an optimum temperature is required by a fungus for initial colonization of a substrate. Once a fungus establishes on the substrate, it can tolerate unfavorable temperatures as well. It is also reported that the temperature response of an individual fungus may change when it is in association with other fungal species. For example, when *Lemmoliera curvula* and *Tricladium chaetocladium* were grown in mixed culture, the optimum temperature for sporulation was reduced to 10°C for *L. curvula* (monoculture optimum 25°C) and to 5°C in case of *T. chaetocladium* (monoculture optimum 15 – 20°C) [15,51]. Since fungi, after establishment on a substrate, can resist invasion by latter colonizers [52,53] and are in association of 5 – 9 species on the substrate [27,54], they can efficiently survive and sporulate under varying temperatures.

B. pH

Soft water streams harbor twice as many fungi as do hard water streams [19,22]. Longitudinal distribution of aquatic fungi is also affected by pH levels [18]. Bärlocher and Rosset [19] found maximum numbers of fungi at pH 6.7, with a decline in number of fungi at pH levels below 6 and above 7.5 (Fig. 1). Experiments with soft and hard water fungi show that all of them grow best at pH levels between 4 and 5.5, regardless of their origin [20]. It is not yet clear how these fungi are able to survive and grow at higher pH levels. High calcium concentration of media has been considered to maintain the growth of fungi at alkaline pH [13,20]. Misra [12] found a significant positive relationship between calcium levels and frequency percentage of aquatic fungi in some tropical alkaline ponds. Higher calcium levels of media are reported to stimulate the production of pectolytic enzymes in aquatic fungi [55,56]. Transplant experiments, where soft and hard water fungi were grown on oak leaves and then placed in either soft or hard water streams, revealed that soft as well as hard water fungi could survive in alien environments (i.e., soft water fungi in hard water and hard water fungi in soft water) [21].

C. Substrate Availability

Variations in input of litter have been related to the distribution of aquatic fungi [9,43,57]. For example, Gönczöl [47] reported two distinct groups of fungi in a Hungarian stream, which either do not intermingle or do so only to a small extent. The first group (e.g., *Dactylella aquatica* and *Monotosporella tuberculata*) belonged to a portion of stream lined by beech trees. The second group (e.g., *Tetracladium marchalianum* and *Tricladium angulatum*) was distributed in a downstream portion lined by alder trees. Later studies could not record such substrate specificity in aquatic fungi [10,17,54]. The type of substrate is related to distribution of aquatic fungi [18,53,58]. Two broad types of substrates, one with more refractory constituents, which degrade slowly (e.g., pine needles, oak leaves), the other with fewer refractory constituents, which degrade quickly (e.g., maple, hickory, aspen leaves), are generally reported. High frequency and density of fungi is normally found on quickly degrading leaves, whereas slowly degrading leaves harbor a lower frequency and density of fungi.

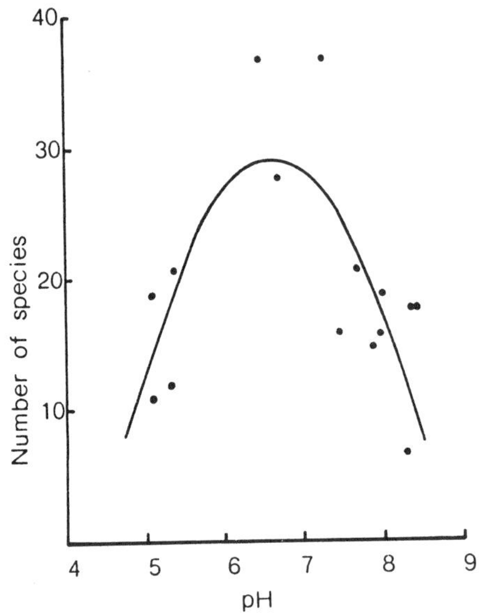

FIGURE 1. Relationship between the number of fungal species and average pH of different streams. (From Ref. 19 with permission.)

Quantity of substrate may regulate diversity of fungi in streams. Streams with heavy input of litter usually harbor more fungi than streams with less input of litter [14,19,22]. The size of substrate governs the number of colonizable niches. This follows a typical species area model. According to this model, number of fungi increases rapidly in the initial stages with increase in size of substrate, but at later stages the increase in number of fungi is gradual [59]. Thus, a larger input of litter may provide a larger number of colonizable niches and a higher number of fungi in stream. The type of litter will be a secondary factor, as prior to colonization by aquatic fungi, the litter is subjected to the activity of terrestrial microbes.

D. Density of Invertebrates

Freshwater streams have varying density of invertebrates. They perform two main functions: (a) shredding of leaf litter into smaller pieces, thereby facilitating fungal colonization on the substrate, and (b) consumption of fungal cellular material on the substrate, causing either reduction in number of colonizable niches or inhibition of some fungi from colonization. Many fungi cannot penetrate the highly lignified leaf surface and fail to colonize lignified litter. Invertebrates help fungi, by causing mechanical injury to litter, enabling entry of fungal hyphae into the leaf matrix. Higher density of shredder invertebrates has been reported during initial decay of litter [60,61]. This is important in modification of the substrate for rapid colonization by fungi. The exposed leaf matrix provides simpler polymers for the growth of fungi. Once in contact with refractory polymers of litter, there may be an induction of enzymes in fungi for breakdown of these polymers. This enables fungi to derive nutrition from the refractory polymers and to persist on the substrate. Predatory invertebrates, feeding on fungal biomass, can also consume portions of colonized substrate and may compete with fungi for resources [48]. Streams with higher densities of invertebrates may thus exhibit a higher number of aquatic fungi, immediately after the input of litter. This increase in number of aquatic fungi may be reduced subsequent to their consumption by predator invertebrates.

Three sequences in the regulation of occurrence and distribution of aquatic fungi in streams may be envisaged. (a) Optimum temperature and pH are required during the initial colonization of fungi on the substrate. (b) The size and quantity of substrate and density of invertebrates regulate the frequency and density of aquatic fungi. (c) Fungi which initially colonize and persist on the substrate may have higher frequencies along a time and space gradient. Thus, the complementary partners of unit-communities on different substrates will be expected to have a broad spectrum, whereas noncomplementary partners will have a narrow spectrum of distribution in streams.

III. DEGRADATION AND CYCLING OF LITTER

Aquatic hyphomycetes can degrade highly refractory polymers, i.e., lignins [25,31] and native celluloses [27,33,62]. It is generally observed that highly refractory litter is degraded slowly by fungi as compared to less refractory litter. Bärlocher and Kendrick [63] reported that aquatic hyphomycetes caused less loss in weight of oak leaves than with ash or maple leaves. Suberkropp et al. [28–30] also reported that oak leaves lose less weight compared to aspen leaves after inoculation with eight aquatic hyphomycetes. Such differences in ability of fungi to degrade different leaf litter at varying rates are important from the viewpoint of litter cycling and input of energy to the detritus food chain in freshwater environments.

A. Colonization on Litter

Aquatic fungi are known to colonize all types of litter which enter the stream. A recent study on decay of alder leaves illustrates the secondary status of aquatic hyphomycetes in degradation of litter [27]. The alder leaves continued to be dominated by terrestrial microbes even after one week of submersion, and aquatic hyphomycetes became dominant only after 2 weeks of submersion (Fig. 2). The ability of aquatic hyphomycetes to degrade the primary products of decomposition of mannan and xylan [31] also points towards their secondary character in degradation of litter in streams.

Swift [64] considered associations of fungal species on discrete substrates as "unit-communities." It is at this stage that fungi interact with each other and with the environment for their physiological requirements. There is evidence that a fungus, after colonization of a substrate, resists invasion by other fungi [65], and a substrate typically supports 5–9 fungi at a time during decay [27,54]. It may, therefore, be possible that pioneer colonizers selectively admit compatible species on the substrate. This can be achieved either by modification of the substrate, making it suitable for colonization by compatible fungi, or by inhibition of noncompatible fungi. Chamier et al. [27] reported a colonization model of aquatic

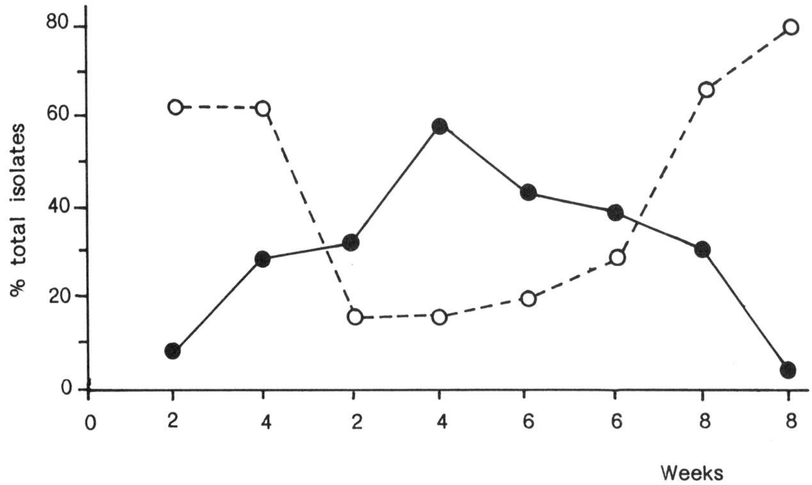

FIGURE 2. Percentage of total isolates of six dominant aquatic fungi and three dominant other fungi on submerged alder leaves at different stages of degradation in the River Bourney. ●——●, aquatic hyphomycetes; ○——○, other fungi. (From Ref. 27 with permission.)

hyphomycetes on discrete resource units (leaves) supported by field evidence. This model depicts a low number of aquatic hyphomycetes initially, which rises in time to peak, followed by a subsequent decline. According to them, aquatic hyphomycetes colonize a leaf randomly and hyphae grow at first on the surface of leaf. With time, low numbers of fungi gradually increase and their colony size develops in a nonrandom manner. These fungi finally form close-knit and overlapping colonies with their hyphae embedded in the leaf matrix. Members of this association (on leaf) do not show positive interactions, denoting that each fungus occupies an independent niche on the substrate. Aquatic hyphomycetes readily colonize lamina but are slow in colonization of ribs [27]. This is attributed to higher lignification of ribs. Terrestrial fungi, present on litter at the time of its submersion, may persist by getting suppressed during the growth of aquatic fungi. They may again dominate after the senescence of aquatic fungi (Fig. 2). Persistent terrestrial fungi reported from alder leaves are *Fusarium* sp., *Cladosporium* sp., and *Epicoccum* sp. [27]. Aquatic fungi that dominate the central period of litter degradation may be overtaken by some terrestrial fungi as well as bacteria in later stages of degradation. However, not much evidence is available on this aspect.

B. Degradation of Litter

Litter entering streams is mainly derived from riparian vegetation, which is rich in celluloses and lignins. The tissues of litter, however, vary in their proportion of structural constituents, as illustrated in Table 1. Lignins and polysaccharides are often in close physical and chemical contact as a lignocellulose complex [66]. Lignins strengthen the cell walls and render them more resistant to fungal attack [67–69]. To degrade litter, aquatic

TABLE 1. Structural Constituents in Biomass of Some Riparian Trees

Tree	Hemicellulose (%)	Cellulose (%)	Lignin (%)
Aspen	28.7	50.8	15.5
Balsam fir	24.4	42.7	28.0
Birch	37.3	38.8	19.5
Red maple	33.0	39.0	23.0
Pine	23.5	42.0	27.8
Spruce	26.5	42.0	28.6

Source: Ref. 89.

hyphomycetes should secrete enzymes capable of breakdown of lignins and celluloses. Few fungi, so far, have been reported to degrade lignins and native celluloses [25,27,31,33,62]. Aquatic fungi also have difficulty in colonization of lignified leaf surface even after secretion of ligninases and cellulases, because their hyphae cannot penetrate the leaf surface. These fungi enter into the leaf matrix after mechanical injury to litter. Such fungi may not be efficient in degradation of lignified litter even after their colonization. For example, *Flagellospora curvula*, *Articulospora inflata*, and *Tetrachaetum elegans* caused loss in weight and softening of aspen leaves after 20 days of incubation but could not cause the same degree of loss in weight or softening of oak leaves even after 30 days, though they grow well on oak leaves (Table 2) [29]. Similarly, other aquatic hyphomycetes caused negligible loss in weight of oak leaves in comparison to maple or ash leaves after 2 weeks of incubation [63], although these fungi secreted enzymes capable of depolymerization of structural polymers of the substrate. Such variations may be due to differences in the ability to degrade lignin and native cellulose or due to some extrinsic factors like temperature, nutrients in litter, or density of invertebrates.

Lignin Degradation

Lignins are degraded by three enzymes: laccase, tyrosinase, and peroxidase. Fungi have been isolated from substrates rich in lignins and are lignicolous fungi [26,70]. Very little work has been done to explore the ability of aquatic fungi to degrade lignins. Fisher et al. [25] reported ligninase activity in aquatic hyphomycetes (Table 3), employing two industrial lignins (Indulin AT and Peritan Na). They observed partial dephenolization of lignins by six fungi out of 16 tested. Zemek et al. [31] further expanded this list by testing 23 fungi out of 35 with ligninase activity (Table 3). They used water-soluble lignin derived from *Nicotiana tabacum*. Table 3 clearly demonstrates that many fungi are able to degrade lignins but that the basis is qualitative in nature. These fungi have also not been tested for different enzymes which together degrade lignin. Since different techniques are used for screening of ligninase activity in aquatic fungi, it

TABLE 2. Degradation of Oak and Aspen Leaves by Aquatic Hyphomycetes

Hyphomycete	Leaf type[a]		Maximum wt. loss (%)	Maximum ATP conc. (μg/g)[b]
Alatospora acuminata	Aspen	20	26.0	398
	Oak	20	21.4	192
	Oak	35	27.5	206
Articulospora inflata	Aspen	20	31.3	246
	Oak	20	20.6	115
	Oak	35	22.9	115
Clavariopsis aquatica	Aspen	20	27.2	483
	Oak	20	19.9	365
	Oak	35	26.1	365
Filosporella annelidica	Aspen	20	33.4	241
	Oak	20	17.4	136
	Oak	35	19.8	146
Flagellospora curvula	Aspen	20	31.9	341
	Oak	20	22.5	177
	Oak	35	26.6	177
Heliscus lugdunensis	Aspen	20	27.6	267
	Oak	20	6.5	93
	Oak	30	10.4	93
Lemonniera aquatica	Aspen	20	31.5	262
	Oak	20	21.3	153
	Oak	35	23.9	153
Tetracladium marchalianum	Aspen	20	27.1	329
	Oak	20	13.7	148
	Oak	30	16.1	192

[a]Numerals after leaf type represent days of incubation.

[b]ATP concentration represents the extent of fungal growth on leaves.

Source: Refs. 29, 30.

TABLE 3. Enzyme Activity in Aquatic Hyphomycetes

Hyphomycete	Ligninase	Pectinase	C_1	C_x	Ref.
Alatospora acuminata	+	+	+	+	31,62
A. flagellata	+	+	−	+	31
Anguillospora crassa	+	+	+	+	26,31
A. furtiva	+	0	0	0	25
A. longissima	+	+	+	+	25,26,31
A. pseudolongissima	0	+	0	+	30
Articulospora inflata	0	+	0	+	30
A. tetracladia	−	+	+	+	25−27,31
Calcalispora hiemalis	+	+	−	+	31
Clavariopsis aquatica	+	+	+	+	31,62
Clavatospora tentacula	0	0	+	0	62
Culicidospora gravida	+	+	−	+	31
Dactylella aquatica	+	+	+	+	25,31
Dendrospora erecta	+	+	−	+	31
D. tenella	−	0	0	0	25
Dimorphospora foliicola	+	0	0	0	25
Flagellospora curvula	0	+	+	+	30,62
F. penicillioides	+	+	+	+	31,62
Gyoerffyella gemallipara	+	+	−	+	31
G. rotula	+	+	−	+	25,31
G. speciosa	+	+	−	+	31
Heliscella stellata	+	+	+	+	31
Heliscus lugdunensis	+	+	+	+	26,30,31
Lateriramulosa uni-inflata	+	+	+	+	31
Lemonniera aquatica	+	+	+	+	26,27,31,62
L. centrosphaera	+	+	−	+	31
L. cornuta	+	+	−	+	31
L. terrestris	+	+	−	+	30,31
Lunulospora curvula	−	0	+	0	25,62

TABLE 3. (Continued)

Hyphomycete	Ligninase	Pectinase	C_1	C_x	Ref.
Margaritispora aquatica	+	+	–	+	31
Mycocentrospora angulata	0	+	+	+	27
Pleuropedium tricladioides	+	+	–	+	31
Pyramidospora densa	0	0	–	+	31
Scorpiosporium gracile	–	0	0	0	25
Sigmoidea prolifera	+	+	+	+	31
Taeniospora gracilis	+	+	+	+	31
Tetrachaetum elegans	+	+	–	+	27,30,31
Tetracladium marchalianum	+	+	+	+	27,31,62
T. maxilliforme	+	+	–	+	31
T. setigerum	+	+	–	+	25,31
Tricladium angulatum	+	+	–	+	31
T. giganteum	+	0	0	0	25
T. splendens	+	+	+	+	26,27,31
Varicosporium elodeae	+	+	+	+	26,31
V. giganteum	0	+	+	+	27
Volucrispora graminea	+	+	–	+	31

+ = positive activity; – = negative activity; 0 = activity not tested.

is difficult to derive any conclusion for a positive role of fungi in degradation of lignin in litter.

Cellulose Degradation

Cellulose is another refractory polymer abundantly found in litter (see Table 1). Cellulose is degraded by cellulase enzymes, namely, (a) endoglucanase or C_X (1,4,-β-D-glucan glucanohydrolase), (b) exoglucanase or C_1 (1,4,-β-D-glucan cellobiohydrolase), and (c) β-glucosidases or cellobiase. Among these, endoglucanase has been detected in a majority of fungi (Table 3). Suberkropp et al. [28,30,32] reported endoglucanase activity in 10 fungi. Zemek et al. [31] have reported C_X activity in 30 aquatic fungi (Table 3). Suberkropp and Arsuffi [30] observed higher endoglucanase activity in *Anguillospora pseudolongissima*, *Tetrachaetum elegans*, *Articulospora inflata*, and *Heliscus lugdunensis* in comparison to *Flagellospora curvula* and *Lemonniera aquatica*. Chamier et al. [27] have reported more C_X activity in *Mycocentrospora angulata* and *Tricladium splendens* and less activity in *Articulospora tetracladia*, *L. aquatica*, and *T. elegans*. These observations suggest that fungi differ in their ability to secrete this enzyme. In *T. elegans*, higher endoglucanase activity on natural substrates [30] but poor activity on artificial substrates [27] has been reported and highlights the limitation of available information in drawing any conclusion about the role of fungi in degradation of cellulose.

The ability of an organism to degrade carboxy-methyl cellulose does not mean that it can degrade native cellulose, present in microfibrils of plant cell walls. Cellulose degradation is achieved by synergistic action of C_x and C_1 enzymes. Recent studies with various forms of native cellulose indicate that some fungi can degrade this highly refractory polymer of plant litter. Different forms of native cellulose, used for assay of exoglucanase activity, are sulka floc [26], microcrystalline cellulose powder [28,32], filter paper [33,62], and avicel [27]. The information is highly inadequate and varies considerably with respect to techniques used in assay. Mostly artificial substrates for assay of exoglucanase activity have been used. The limitation of results with artificial substrate is again illustrated by the work of Jones [26], who found appreciable loss in weight of sulka floc by *H. lugdunensis* but no loss in weight of wood test blocks.

Pectin Degradation

Degradation of pectin is key in the breakdown or maceration of the leaf matrix in freshwater environments. Suberkropp and Klugg [62] concluded that one component of pectinase complex, pectin lyase, which is active up to pH 8, is predominantly responsible for initiation of leaf maceration. It is not yet known whether polygalactouronase, another component of pectinase, which is active only at acidic pH, acts synergistically with pectin lyase or independently. Pectinase activity has been reported in a number of aquatic hyphomycetes [27–32,55]. Zemek et al. [31] have reported pectolytic enzymes in 34 out of 35 fungi tested (see Table 3). Suberkropp et al. [28,32] reported pectinase activity in culture filtrates of many fungi grown on aspen and oak leaves. Chamier et al. [27] found higher activity of polygalactouronase than of pectin lyase in *L. aquatica*, *M. angulata*, and *T. elegans*, while other fungi, namely, *A. tetracladia*, *T. splendens*, and *T. marchalianum*, showed high activity of both enzymes. In a quantitative study of pectin lyase in culture filtrates of fungi grown on aspen leaves, it was found that enzymic activity varied considerably among different fungi [30]. It is likely that most aquatic hyphomycetes are able to synthesize and secrete pectinases, which are necessary to soften or macerate leaf tissues.

Mechanism of Leaf Degradation

Maximum loss in weight and maceration of leaves by fungi is generally believed to take place with maximum growth on the substrate [27–30]. Field evidence of a relationship between growth of aquatic fungi and the extent of degradation was worked out by Chamier et al. [27]. They observed growth of fungal hyphae on the leaf surface after 2 weeks of submersion (Fig. 3). By the fourth week, the hyphae were deeply embedded into the leaf matrix and the epidermis was heavily eroded. The evidence of hyphae on lamina was hardly noticeable, and the presence of hyphae on ribs was observed after 6 weeks. Laminae, however, were totally eroded, and ribs were partially damaged due to fungal growth after 8 weeks. Laboratory experiments revealed that maximum degradation of leaves can be correlated with maximum growth of fungi and endoglucanase and pectin

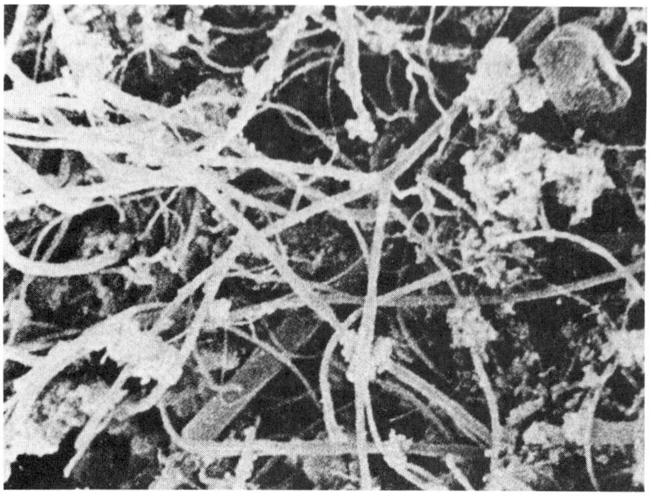

(a)

(b)

FIGURE 3. Fungal colonization of submerged alder leaves at progressive stages of degradation viewed by SEM. (a) Fungal hyphae on lamina after 2 weeks (×700). (b) Heavily eroded leaf surface with few hyphae after 6 weeks (×1000). (c) Ribs after 2 weeks' submersion showing hyphal colonization (×700). (d) Ribs after 10 weeks' submersion with little degradation (×300). (From Ref. 27 with permission.)

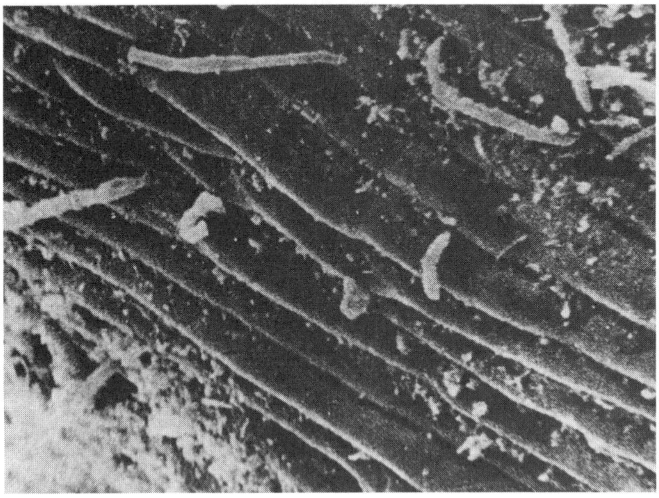

(c)

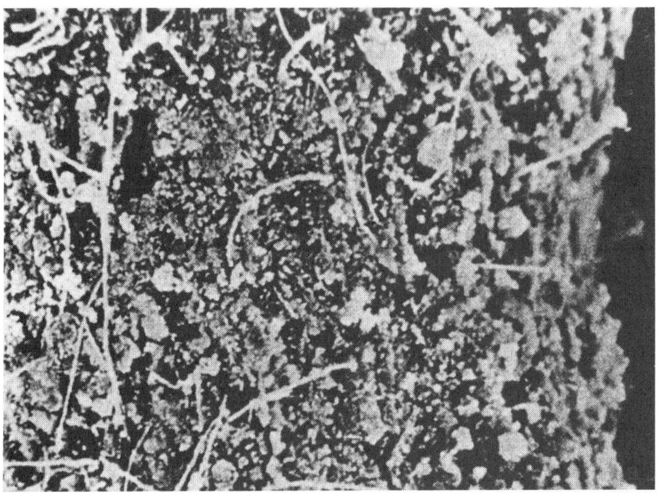

(d)

FIGURE 3. (Continued)

lyase activity by the end of this period [30]. In a comparison of different substrates, the level of enzyme activity was, however, not related to growth of the fungus and degradation of substrate [29,30,71]. There are two possible explanations for this: (a) fungus alone is not able to secrete specific enzymes, while an assemblage of five to nine fungi may provide a better and more efficient enzyme spectrum for degradation of substrates; and (b) temperature and nutrient concentration in tissues of substrate and density of invertebrates may regulate degradation by a fungus or an assemblage of fungi.

Temperature directly regulates rates of growth and metabolism in fungi. Many fungi have an optimum temperature for growth between 20 and 30°C [15,62]. Temperatures above and below this optimum may substantially affect extent of decay in a substrate. For example, *T. marchalianum* degraded hickory leaves between 15 and 30 days at 10°C, but degradation took place within 4–16 days at 18°C [32]. Field studies also show that degradation is higher in summer than winter [15]. Laboratory data, however, may not reveal a true picture of degradation by a fungus or an association of fungi in their natural habitat, as the temperature response of an individual fungus may change when in association and as a substrate have an assemblage of five to nine fungi.

Nutrient concentration of litter, especially nitrogen, also regulates degradation of substrates by fungi. Litter with a higher initial concentration of nutrients has been reported to undergo decay at a faster rate than litter with a lower nutrient concentration [67,69,72]. Higher nitrogen in tissues (of litter) supports higher growth and activity of microbes [73]. Fungi, colonizing litter, derive nitrogen through breakdown of protein polymers present in the substrate. There are indications that higher levels of nutrients in water, particularly in lentic systems, also maintain higher rates of microbial activity [72,74]. The nutrient status of streams may, however, not have an appreciable impact on fungal activity and the extent of fungal degradation. Most streams, which receive higher input of litter, may not be able to maintain a higher nutrient concentration upstream as a major proportion of nutrients will move downstream. Leached nutrients from litter may also not be retained upstream because of inefficient assimilation of nutrients by stream autotrophs.

Predator and shredder invertebrates can facilitate microbial degradation and cycling of refractory litter. Most fungi may face difficulty in colonization of lignified litter. The injury caused to litter by shredder invertebrates may expose the leaf matrix and increase litter's surface-to-volume ratio for fungal colonization. Leaves colonized by fungi represent a valuable source of food for predator invertebrates, and in the process they may significantly consume the litter. The rate of decay of litter in streams may be explicitly described by the following mathematical form:

$$\kappa \propto \frac{[M(T \times N)]\ P}{R \times S}$$

Where κ = relative decay rate of litter; M = size of microbial assemblage on litter; T = temperature; N = initial nutrient concentration in litter; P = degree of predation on fungal colonized litter; R = initial tissue refractivity of litter; and S = particle size of litter (governed by density of shredder invertebrates). This mathematical form and relationship among variables can be substantiated by substituting the relative values of each variable scaled over the range 0 to 1. Thus, in streams where temperature and initial nutrient status of litter are high along with density of shredder invertebrates, the fungal activity may be high on litter. Higher degree of predation on fungal biomass under these conditions may finally result in faster degradation and cycling of litter and a high input of energy for the detritus food chain. However, if the density of shredder invertebrates is low and temperature is suboptimal, fungi may not colonize litter efficiently. This will adversely affect the degradation and cycling of litter.

IV. AQUATIC FUNGI AS FOOD FOR INVERTEBRATES

As already discussed, undisturbed streams receive a heavy input of litter and, thus, provide substantial energy to the detritus food chain. Microorganisms colonizing this litter may determine the extent of input of nutrients and energy for the detritus food chain. Kaushik and Hynes [24] provided the first evidence in this direction. They recorded that microbes degrading the litter can increase its nitrogen and protein content. Stream invertebrates prefer this degraded litter over fresh litter as food. Several independent studies have since confirmed that litter colonized by fungi is more palatable and nutritious to stream invertebrates [29,34,39,75]. The increase in palatability and nutritive value of litter colonized by aquatic fungi may be due to the following mechanisms:

1. Fungal growth on detritus, which adds easily digestible fungal proteins, carbohydrates, and lipids on nutritionally poor leaf matrix.
2. Fungal degradation (catalysis) of litter, which transforms the refractory polymers (e.g., cellulose, lignin, hemicellulose, etc.) into simpler subunits. These subunits can be easily digested by detritivorous invertebrates.
3. Fungal extracellular enzymes (e.g., carbohydrases, proteinases, etc.), which may be ingested by invertebrates and help them to digest different polymer units or subunits of detritus.
4. Mobility and digestive ability of detritivorous invertebrates.

A. Growth on Litter

Kaushik and Hynes [24] reported that in early stages of decay, nitrogen and protein content of litter often increase. This was assumed to be the result of microbial activity, which can immobilize significant quantities of nitrogen from surrounding media. The increase in nitrogen and protein content of litter after fungal growth has been reported by many workers [24,30,68,69]. Increase in protein content could be observed only when microbial activity took place on litter, and it was generally more pronounced when fungi dominated in degradation of litter [34]. However, there are indications that fungi remain active on litter during the initial period only, while bacteria dominate later periods of decay [68,76]. These observations, together with findings of higher density of invertebrates on litter during initial decay, indicate that fungal biomass on litter is an important source of food to stream invertebrates. It is widely accepted that microbial biomass on litter represents more concentrated proteinaceous food than fresh litter for stream invertebrates [35,36]. The detritus colonized by fungi, therefore, is often termed "conditioned detritus" due to its increased food value. Conditioned detritus is reported to increase the rate of growth and survival of detritivores [77,78,80,81], and due to the presence of simpler polymers, it can be assimilated quickly and efficiently by invertebrates [40,79].

Experiments with different fungi grown on oak and aspen leaves revealed that caddisfly larvae generally fed on conditioned leaves having maximum growth of fungi. The palatability was reduced when the incubation period of fungi on litter was either curtailed or prolonged [28–30]. The changes in palatable condition of litter, however, are not fully explained by fungal growth on litter (e.g., from changes in nitrogen and ATP content of litter). The larvae showed distinct preferences for some fungi over others, and their preference for a fungus changed following a change in the substrate (Fig. 4). Caddisfly larvae preferred both oak and aspen leaves conditioned by *Flagellospora curvula*. This fungus caused twofold increase in ATP concentration of conditioned aspen leaves over conditioned oak leaves (Fig. 4). *Heliscus lungdunensis* also caused a twofold increase in ATP of aspen leaves in comparison to that in oak leaves, but caddisfly larvae did not prefer this fungus on oak leaves as they did on aspen leaves (Fig. 4). The preferential consumption of fungus according to substrate cannot be accounted for by fungal growth alone or its ability to enrich detritus with nitrogen. Since discrete substrates have an assemblage of five to nine fungal species during decay, the food of stream invertebrates will consist of a mosaic of overlapping fungal colonies on a substrate. There is evidence that detritivorous invertebrates can detect and discriminate between the leaf patches colonized by different fungi [75]. They can also consume a particular fungal mass in a mixture of colonies of different fungi growing on the litter [30]. However, the mechanisms involved in discrimination or preference of a fungus by invertebrates are not clear.

B. Degradation of Litter

Fungal degradation of litter can enhance acceptability and digestibility of conditioned litter for stream invertebrates. *Gammarus pseudolimnaeus*, a stream invertebrate, has been shown to prefer leaves and filter paper partly acid-hydrolyzed [82]. It is reported that gut fluid of *Gammarus fossarum*, another stream invertebrate, releases more reducing sugars from fungal-degraded leaves than from fresh leaves [40]. It has been demonstrated that caddisfly larvae prefer both oak and aspen leaves conditioned by the fungus *F. curvula*, which caused maximum loss in weight of leaves in comparison to litter degraded by other fungi (see Fig. 4).

C. Acquired Exoenzymes

During the growth of fungi on litter, substantial quantities of exoenzymes (e.g., carbohydrases, proteinases, ligninases, etc.) are produced. When this litter is ingested by invertebrates, they utilize acquired enzymes in digestion and assimilation of different polymer units and subunits of litter. More release of reducing sugars after ingestion of degraded litter from gut fluid of *G. fossarum*, an invertebrate, has been demonstrated [40]. The

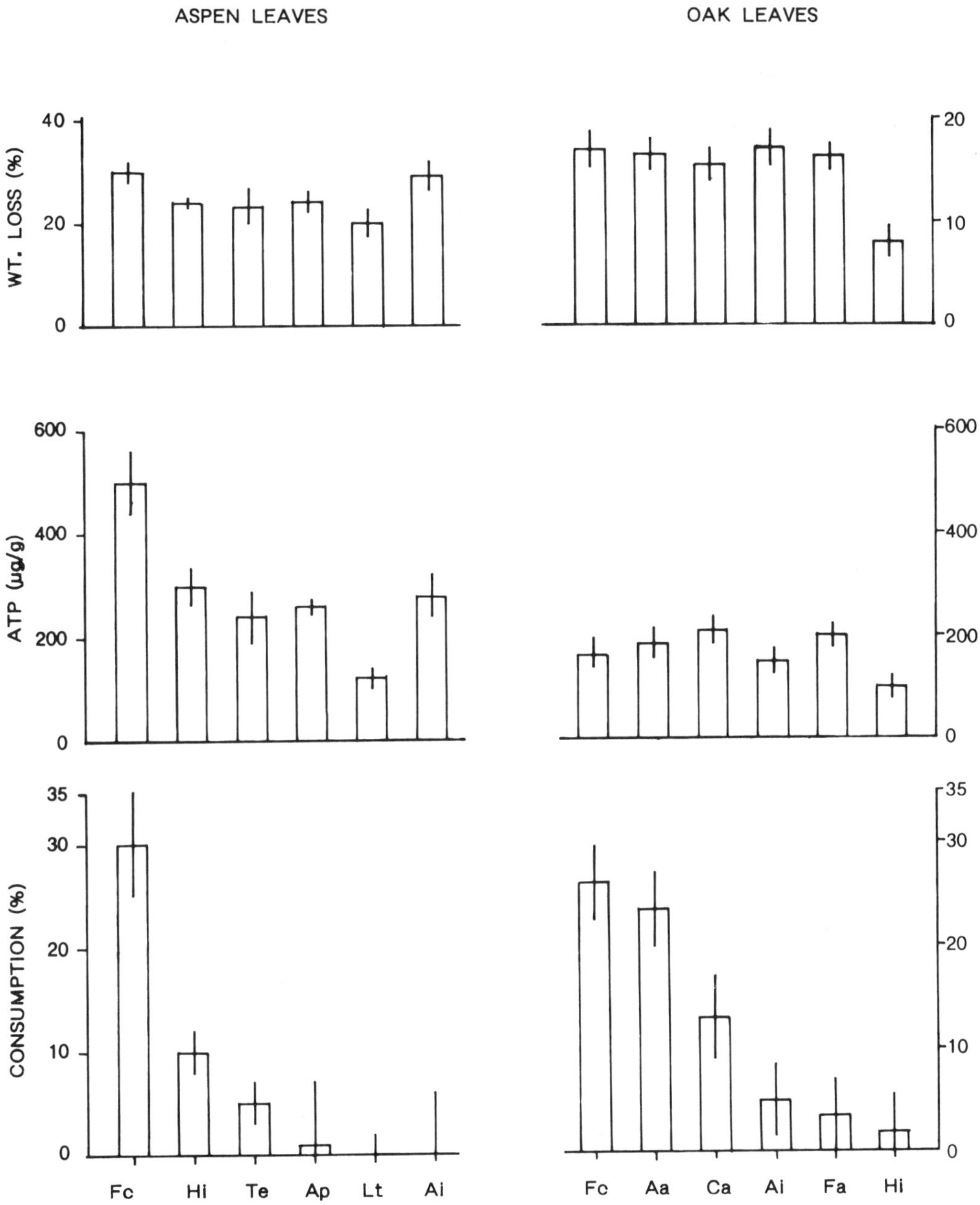

FIGURE 4. Feeding preferences of caddisfly larvae and changes in oak and aspen leaves after 15 days of incubation (ATP represents fungal growth and weight loss represents degradation). Fungal species abbreviations: Aa = *Alatospora acuminata*, Ai = *Articulospora inflata*, Ap = *Anguillospora pseudolongissima*, Ca = *Clavariopsis aquatica*, Fa = *Filosporella annelidica*, Fc = *Flagellospora curvula*, Hi = *Heliscus lugdunensis*, Lt = *Lemonniera terrestris*, Te = *Tetrachaetum elegans*. (From Ref. 29 with permission.)

use of acquired enzymes has also been demonstrated in caddisfly and stonefly [41], terrestrial isopod [83], and termites [84]. It could be the basis for weak or sporadic cellulase activity in several stream invertebrates [85,86]. It may also explain the difference in palatability of different fungi to caddisfly [29,30]. The use of acquired fungal exoenzymes by invertebrates depends upon their gut physiology. In a study of three stream invertebrates, *Tipula caloptera, Hydropsyche betteri*, and *Gammarus tigrinus*, it has been observed that the gut fluid of freshly collected *Gammarus* and *Hydropsyche* has a pH close to 7, while it is around 11 in *Tipula*. These invertebrates could not break down polysaccharides and protein polymers of unconditioned leaf powder [39]. *Gammarus* and *Hydropsyche*, however, could degrade polysaccharides following ingestion of leaf discs conditioned by *Tetracladium marchalianum* [39]. *Tipula* could not degrade polysaccharides even after ingestion of conditioned leaf discs, because the alkaline pH of its gut inactivated the acquired fungal carbohydrases.

D. Mobility and Digestive Ability of Invertebrates

The mobility of invertebrates determines the type of food and the quantity available for their consumption. Slow-moving invertebrates depend on natural input of food, which they may not be able to discriminate. On the other hand, fast-moving invertebrates can search for their food in streams and may manipulate a large proportion of their food as conditioned litter. The fast-moving invertebrates can also discriminate conditioned litter depending upon their preference of fungal species. Bärlocher and Porter [39] observed that *Gammarus*, which is a fast mover, can be selective and ingests mostly the conditioned litter. On the other hand, *Tipula*, which is a slow mover, rarely gets conditioned food and exhibits high proteinase activity. The digestion of ingested polysaccharides is facilitated by endosymbiotic bacteria present in its hindgut [41,87].

Aquatic fungi effectively transform the nutritionally poor substrate into nutritious detritus, which is important in structure and energy flow of streams because (a) production of autochthonous organic matter is poor in streams, and (b) when appreciable input of allochthonous organic matter takes place, it can be directly utilized only with the help of microorganisms, particularly aquatic fungi. The type and availability of energy to the detritus food chain will depend upon the interactions between fungal growth and catalysis on litter and mobility and digestive physiology of invertebrates. The lakes with heavy input of autochthonous input and allochthonous organic matter tend to remain unsaturated with dissolved oxygen. This does not favor the involvement of aquatic fungi in degradation of litter and input of energy to the detritus food chain in lakes and reservoirs.

V. CONCLUSION

Aquatic fungi play an important role in incorporation of allochthonous inputs to the food chain and energy flow of stream ecosystems. The occurrence and distribution of these fungi is regulated by a complicated network of interactions among environmental conditions, substrate availability, and density of predatory invertebrates. Optimum temperature and pH are required only during their establishment on the substrate, and once established the fungal species can withstand unfavorable environmental conditions. The substrate specificity is poor in aquatic fungi. The size of the substrate, however, provides a constant number of colonizable niches which are initially colonized by ruderal species, but soon replaced by competitive or persistent species. The degradation of the substrate is done efficiently by a group of complementary fungi. The degradation of litter is governed by both the ability of fungi to secrete necessary enzymes and factors like temperature, nutrient status of litter, and density of invertebrates. The palatability of litter is optimal soon after the greatest growth of fungi and degradation. The palatability of detritus and its ingestion is not the function of fungal production alone but possibly a combination of microbial production on litter and digestive physiology and mobility of detritivores. Figure 5 illustrates the role of aquatic fungi in the dynamics of stream ecosystem with mechanisms operating at each step.

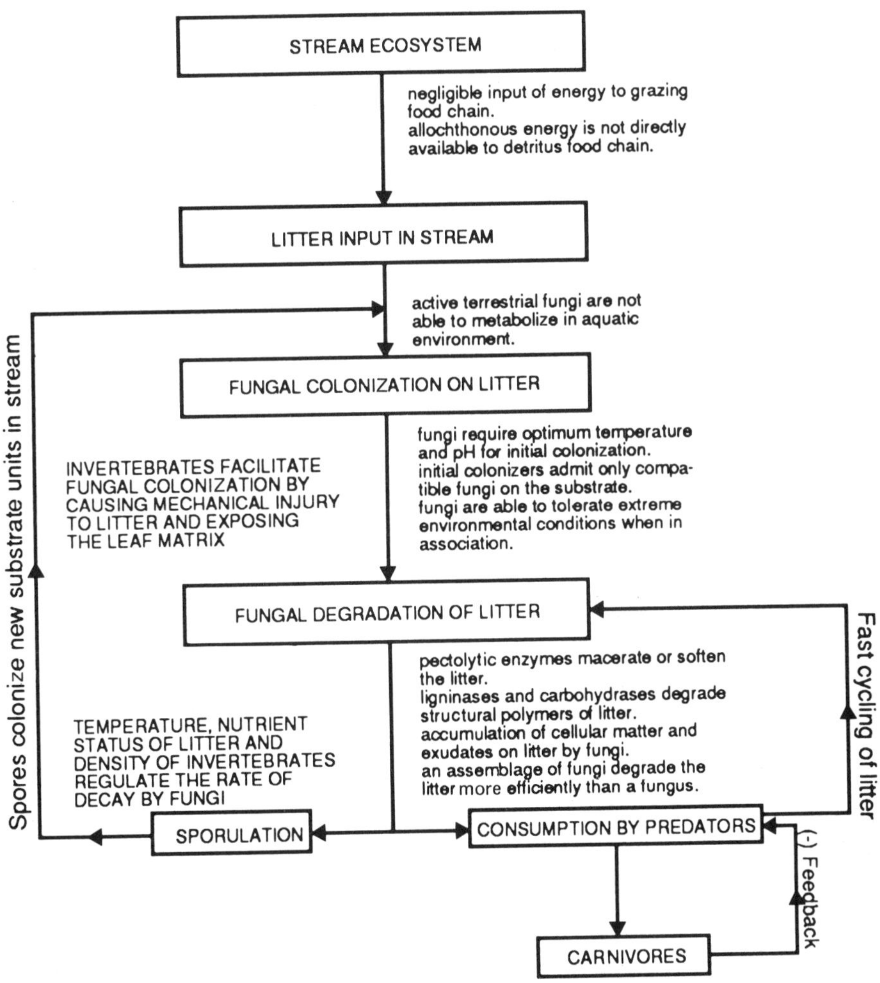

FIGURE 5. Aquatic fungi in the energy flow and dynamics of a stream ecosystem.

Much research is still needed to gain a better understanding of the role of aquatic fungi in freshwater environments. Some suggested lines are as below:

1. Recent sampling techniques [27,54,88] make it possible to enumerate the colonization patterns of aquatic fungi on litter during in situ decomposition and to enumerate the partners of the community. These techniques are different in concept and interpretation and need to be standardized for universal application and synthesis of information.
2. The mechanisms governing the co-existence of five to nine fungi on a discrete substrate are not clear, nor is how they complement each other without positive association between any two of them.
3. The techniques for assessment of different degradative enzymes in fungi vary considerably and are qualitative in nature. Not much work is available on quantitative ability of fungi to secrete these enzymes. Standardization of assay techniques and their proper details is required.
4. Stream invertebrates consume aquatic fungi, but the mechanisms involved in discrimination of different fungi by invertebrates are far from clear. Is it due to some chemical factors or to suitability of a fungus and the changes it causes to the litter?

ACKNOWLEDGMENTS

The authors express their indebtedness to G. P. Agarwal and R. C. Rajak for their helpful suggestions.

REFERENCES

1. Jones, E. B. G. (ed.). *Recent Advances in Aquatic Mycology.* Elek Science, London, 1976.
2. Dick, M. W. The ecology of aquatic phycomycetes. In *Recent Advances in Aquatic Mycology* (E. B. G. Jones, ed.), Elek Science, London, 1976, pp. 513–542.
3. Quinn, J. P. Seasonal occurrence of yeasts and other fungi in a freshwater lake. *Trans. Br. Mycol. Soc.*, *83*:53–58, 1984.
4. Godfrey, B. E. S. Growth of two terrestrial microfungi on submerged alder leaves. *Trans. Br. Mycol. Soc.*, *81*:418–424, 1983.
5. Mishra, R. R., and Tiwari, B. K. Fungi and bacteria associated with pine needles and teak leaf litter during processing in a freshwater lake. *Hydrobiologia*, *102*:123–129, 1983.
6. Park, D. On the ecology of heterotrophic microorganisms in freshwater. *Trans. Br. Mycol. Soc.*, *58*:291–299, 1972.
7. Nilsson, S. Freshwater hyphomycetes, taxonomy, morphology and ecology. *Symb. Bot. Upsal.*, *18*:1–130, 1964.
8. Ingold, C. T. An illustrated guide to aquatic and water-borne hyphomycetes (*Fungi imperfectii*) with notes on their biology. *Scientific Publication No. 30*, Freshwater Biological Association, Ambleside, 1975.

9. Webster, J., and Descals, E. Morphology, distribution and ecology of conidial fungi in freshwater habitats. In *Biology of Conidial Fungi*, Vol. 1 (G. C. Cole and B. Kendrick, eds.), Academic Press, London, 1981, pp. 295–355.
10. Aimer, R. D., and Segedin, B. P. Fluctuation in spore numbers of aquatic hyphomycetes in a New Zealand stream. *Bot. J. Linnean Soc.*, *91*:61–66, 1985.
11. Aimer, R. D., and Segedin, B. P. Some aquatic hyphomycetes from New Zealand streams. *N. Z. J. Bot.*, *23*:273–299, 1985.
12. Misra, J. K. Occurrence, distribution and seasonality of aquatic fungi as affected by chemical factors in six alkaline ponds of India. *Hydrobiologia*, *97*:185–191, 1982.
13. Hasija, S. K., and Khan, M. A. Seasonal occurrence and distribution of aquatic fungi in relation to water quality of Jabalpur lakes. In *Perspectives in Mycological Research*, Vol. 1 (S. K. Hasija, R. C. Rajak, and S. M. Singh, eds.), Today and Tomorrow Printers and Publishers, New Delhi, 1987, pp. 79–88.
14. Shearer, C. A., and Webster, J. Aquatic hyphomycete communities in the River Teign. I. Longitudinal distribution patterns. *Trans. Br. Mycol. Soc.*, *84*:489–501, 1985.
15. Suberkropp, K. Effect of temperature on seasonal occurrence of aquatic hyphomycetes. *Trans. Br. Mycol. Soc.*, *82*:53–62, 1984.
16. Iqbal, S. H., and Webster, J. Aquatic hyphomycete spora of the River Exe and its tributaries. *Trans. Br. Mycol. Soc.*, *61*:331–346, 1973.
17. Bärlocher, F., and Kendrick, B. Dynamics of the fungal population on leaves in a stream. *J. Ecol.*, *62*:761–791, 1974.
18. Triska, F. J. Seasonal distribution of aquatic hyphomycetes in relation to the disappearance of leaf litter from a woodland stream. Ph.D. thesis, Pittsburgh University, Pittsburgh, 1970.
19. Bärlocher, F., and Rosset, J. Aquatic hyphomycete spora of two Black Forest and two Swiss Jura streams. *Trans. Br. Mycol. Soc.*, *76*:479–483, 1981.
20. Rosset, J., and Bärlocher, F. Aquatic hyphomycetes: Influence of pH, Ca^{++} and HCO_3^- on growth in vitro. *Trans. Br. Mycol. Soc.*, *84*:137–145, 1985.
21. Rosset, J., and Bärlocher, F. Transplant experiments with aquatic hyphomycetes. *Verh. Internat. Verein. Limnol.*, *22*:2786–2790, 1985.
22. Wood-Eggenschwiller, S., and Bärlocher, F. Aquatic hyphomycetes in sixteen streams in France, Germany and Switzerland. *Trans. Br. Mycol. Soc.*, *81*:371–379, 1983.
23. Shearer, C. A., and Zare-Maivan, H. In vitro hyphal interactions among wood and leaf-inhabiting ascomycetes and *Fungi imperfectii* from freshwater habitats. *Mycologia*, *80*:31–37, 1988.
24. Kaushik, N. K., and Hynes, H. B. N. The fate of autumn-shed leaves that fall into streams. *Archiv. für Hydrobiol.*, *68*:465–515, 1971.
25. Fisher, P. J., Davey, R. A., and Webster, J. Degradation of lignin by aquatic and aero-aquatic hyphomycetes. *Trans. Br. Mycol. Soc.*, *80*:166–168, 1983.

26. Jones, E. B. G. Observations on the ecology of lignicolous aquatic hyphomycetes. In *The Fungal Community: Its Organization and Role in the Ecosystem* (D. T. Wicklow and G. C. Carroll, eds.), Marcel Dekker, New York, 1981, pp. 731–742.
27. Chamier, A.-C., Dixon, P. A., and Archer, S. A. The spatial distribution of fungi on decomposing alder leaves in a freshwater stream. *Oecologia, 64*:92–103, 1984.
28. Suberkropp, K., Arsuffi, T. L., and Anderson, J. P. Comparison of degradative ability, enzymatic activity and palatability of aquatic hyphomycetes grown on leaf litter. *Appl. Environ. Microbiol., 46*: 237–244, 1983.
29. Butler, S. K., and Suberkropp, K. Aquatic hyphomycetes on oak leaves: Comparison of growth, degradation and palatability. *Mycologia, 78*:922–928, 1986.
30. Suberkropp, K., and Arsuffi, T. L. Degradation, growth and changes in palatability of leaves colonized by six aquatic hyphomycete species. *Mycologia, 76*:398–407, 1984.
31. Zemek, J., Marvanová, L., Kuniák, L., and Kadlecikova, B. Hydrolytic enzymes in aquatic hyphomycetes. *Folia Microbiol., 30*:363–372, 1985.
32. Suberkropp, K., and Klug, M. J. The maceration of deciduous leaf litter by aquatic hyphomycetes. *Can. J. Bot., 58*:1025–1031, 1980.
33. Singh, N. Cellulose decomposition by some tropical aquatic hyphomycetes. *Trans. Br. Mycol. Soc., 79*:560–561, 1982.
34. Bärlocher, F. The role of fungi in nutrition of stream invertebrates. *Bot. J. Linnean Soc., 91*:83–94, 1985.
35. Bärlocher, F., and Kendrick, B. Role of aquatic hyphomycetes in the trophic structure of streams. In *The Fungal Community: Its Organization and Role in the Ecosystem* (D. T. Wicklow and G. C. Carroll, eds.), Marcel Dekker, New York, 1981, pp. 743–760.
36. Anderson, N. H., and Sedell, J. R. Detritus processing by macroinvertebrates in stream ecosystems. *Annu. Rev. Entomol., 24*:351–377, 1979.
37. Bird, G. A., and Kaushik, N. K. Coarse particulate organic matter in streams. In *Perspectives in Running Water Ecology* (M. A. Lock and D. D. Williams, eds.), Plenum Press, New York, 1981.
38. Bärlocher, F., and Kendrick, B. Hyphomycetes as intermediaries of energy flow in streams. In *Recent Advances in Aquatic Mycology* (E. B. G. Jones, ed.), Elek Science, London, 1976, pp. 435–446.
39. Bärlocher, F., and Porter, C. W. Digestive enzymes and feeding strategies of three stream invertebrates. *J. N. Am. Benthol. Soc., 5*:58–66, 1986.
40. Bärlocher, F. The contribution of fungal enzymes to the digestion of leaves by *Gammarus fossarum* koch (Amphipoda). *Oecologia, 52*:1–4, 1982.
41. Sinsabaugh, R. L., Linkins, A. E., and Benfield, E. F. Cellulose digestion and assimilation by three leaf-shredding aquatic insects. *Ecology, 66*:1464–1471, 1985.
42. Iqbal, S. H., and Webster, J. Aquatic hyphomycete spora of some Dartmoor streams. *Trans. Br. Mycol. Soc., 69*:233–241, 1977.
43. Shearer, C. A., and Webster, J. Aquatic hyphomycete communities in River Teign. II. Temporal distribution patterns. *Trans. Br. Mycol. Soc., 84*:503–507, 1985.

44. Müller-Haekel, A., and Marvanová, L. Periodicity of aquatic hyphomycetes in the sub-arctic. *Trans. Br. Mycol. Soc.*, *73*:109–116, 1979.
45. Indira, G., Mishra, R. P., and Agarwal, G. P. Some fungi from the Supatal Lake at Jabalpur (M.P.). *Nat. Acad. Sci. Letters*, *11*:65–67, 1988.
46. Hasija, S. K., and Batra, S. The distribution of *Achlya americana* (Saprolegniales) in different aquatic habitats at Jabalpur, India. *Hydrobiologia*, *61*:277–279, 1978.
47. Gönczöl, J. Ecological observations on the aquatic hyphomycetes of Hungary. I. *Acta Botanica Academial Scientiarum Hungarica*, *21*:243–264, 1975.
48. Bärlocher, F. Leaf-eating invertebrates as competitors of aquatic hyphomycetes. *Oecologia*, *47*:303–306, 1980.
49. Gupta, S., and Dubey, T. Effect of chemical contents of water on seasonal occurrence of aquatic fungi. In *Perspectives in Mycological Research*, Vol. 1 (S. K. Hasija, R. C. Rajak, and S. M. Singh, eds.), Today and Tomorrow Printers and Publishers, New Delhi, 1987, pp. 63–68.
50. Bärlocher, F., and Premdas, P. D. Effects of pentachlorophenol on aquatic hyphomycetes. *Mycologia*, *80*:135–137, 1988.
51. Webster, J., Moran, S. T., and Davey, R. A. Growth and sporulation of *Tricladium chaetocladium* and *Lunulospora curvula* in relation to temperature. *Trans. Br. Mycol. Soc.*, *67*:491–549, 1976.
52. Bärlocher, F. Conidium production from leaves and needles in four streams. *Can. J. Bot.*, *60*:1487–1494, 1982.
53. Chamier, A.-C., and Dixon, P. A. Pectinase in leaf degradation by aquatic hyphomycetes: The field study. *Oecologia*, *52*:109–115, 1982.
54. Shearer, C. A., and Lane, L. C. Comparison of three techniques for the study of aquatic hyphomycete communities. *Mycologia*, *75*:498–508, 1983.
55. Chamier, A.-C., and Dixon, P. A. Pectinase in leaf degradation by aquatic hyphomycetes: The enzymes and leaf maceration. *J. Gen. Microbiol.*, *128*:2469–2483, 1982.
56. Chamier, A.-C., and Dixon, P. A. Effects of calcium-ion concentration on leaf maceration by *Tetrachaetum elegans*. *Trans. Br. Mycol. Soc.*, *81*:415–418, 1983.
57. Akridge, R. E., and Kochn, R. D. Amphibious hyphomycetes from the San Marcos River in Texas. *Mycologia*, *79*:228–233, 1987.
58. Suberkropp, K., and Klug, M. J. Fungi and bacteria associated with leaves during processing in a woodland stream. *Ecology*, 57:707–719, 1976.
59. Wildman, H. G. Fungal colonization of resources in soil—an island biogeographical approach. *Trans. Br. Mycol. Soc.*, *88*:291–297, 1987.
60. Hill, B. H. The breakdown of macrophytes in a reservoir wetland. *Aquat. Bot.*, *21*:23–31, 1985.
61. Byren, B. A., and Davies, B. R. The influence of invertebrates on the breakdown of *Potamogeton pectinatus* L. in a coastal marina (Zandvlei, South Africa). *Hydrobiologia*, *137*:141–151, 1986.
62. Suberkropp, K., and Klug, M. J. Degradation of leaf litter by aquatic hyphomycetes. In *The Fungal Community: Its Organization and Role in the Ecosystem* (D. T. Wicklow and G. C. Carroll, eds.), Marcel Dekker, New York, 1981, pp. 761–776.

63. Bärlocher, F., and Kendrick, B. Fungi and food preferences of *Gammarus pseudolimnaeus*. *Arch. Hydrobiol.*, *72*:501–516, 1973.
64. Swift, M. Species diversity and the structure of microbial communities in terrestrial habitats. In *The Role of Terrestrial and Aquatic Organisms in Decomposition Process* (J. M. Anderson and A. Macfadayen, eds.), Blackwell Sci. Publ., London, 1976, pp. 185–222.
65. Bärlocher, F., and Schweizer, M. Effects of leaf size and decay rate on colonization by aquatic hyphomycetes. *Oikos*, *41*:205–210, 1983.
66. Sarakanen, K. V., and Ludwig, C. H. *Lignins, Occurrence, Formation, Structure, and Reactions*. Wiley-Interscience, New York, 1971.
67. Godshalk, G. L., and Wetzel, R. G. Decomposition of aquatic angiosperms. II. Particulate components. *Aquat. Bot.*, *5*:301–327, 1978.
68. Benner, R., Moran, M. A., and Hodson, R. E. Biogeochemical cycling of lignocellulosic carbon in marine and freshwater ecosystems: Relative contributions of procaryotes and eucaryotes. *Limnol. Oceanogr.*, *31*:89–100, 1986.
69. Wilson, J. O., Buchsabum, R., Valiela, I., and Swain, T. Decomposition in salt marsh ecosystems: Phenolic dynamics during decay of litter of *Spartina alterniflora*. *Mar. Ecol. Prog. Ser.*, *29*:177–187, 1986.
70. Jones, E. B. G. Lignicolous and algicolous fungi. In *Recent Advances in Aquatic Mycology* (E. B. G. Jones, ed.), Elek Science, London, 1976, pp. 1–50.
71. Butler, S. K. The degradation of oak leaves by aquatic hyphomycetes. M.S. thesis, New Mexico State University, Las Cruces, 1984.
72. DeBusk, T. A., and Dierberg, F. E. Effect of nitrogen and fiber content on the decomposition of the water hyacinth (*Eichhornia crassipes* [Mart.] Solms.). *Hydrobiologia*, *118*:199–204, 1984.
73. Brinson, M. M. Decomposition and nutrient exchanges in an alluvial swamp forest. *Ecology*, *58*:601–609, 1977.
74. Brock, T. C. M. Aspects of decomposition of *Nymphoides peltata* (Gmel.) O. Kuntze (Menyanthaceae). *Aquat. Bot.*, *19*:131–156, 1984.
75. Arsuffi, T. L. and Suberkropp, K. Leaf processing capabilities of aquatic hyphomycetes: Interspecific differences and influence on shredder feeding preferences. *Oikos*, *42*:144–154, 1984.
76. Gaur, S., Singhal, P. K., and Hasija, S. K. Involvement of fungi in biodegradation of *Eichhornia crassipes* (Mart.) Solms. In *Fungal Ecology in Relation to Human Welfare* (P. Rama Rao and C. Manoharachary, eds.), Hyderabad, 1990 (in press).
77. Rossi, L., Fano, E. A., and Basset, A. Sympatric evolution of the trophic niche of two detritivorous isopods, *Asellus aquaticus* and *Proasellus coxalis*. *Oikos*, *40*:208–215, 1983.
78. Sutcliffe, D. W., Carrick, T. R., and Willoughby, L. G. Effects of diet, body size, age and temperature on growth rates in the amphipod *Gammarus pulex*. *Freshwater Biol.*, *11*:183–214, 1981.
79. Golladay, S. W., Webster, J. R., and Benfield, E. F. Factors affecting food utilization by a leaf shredding aquatic insect: Leaf species and conditioning time. *Holarctic Ecol.*, *6*:157–162, 1983.
80. Fano, E. A., Rossi, L., and Basset, A. Fungi in diet of three benthic invertebrate species. *Bolletino di Zoologia*, *49*:29–34, 1982.
81. Kostalos, M. and Seymour, R. L. Role of microbial enriched detritus in the nutrition of *Gammarus minus* (Amphipoda). *Oikos*, *27*:512–516, 1976.

82. Bärlocher, F., and Kendrick, B. Leaf conditioning by microorganisms. *Oecologia, 20*:359–362, 1975.
83. Hassal, M., and Jennings, J. B. Adaptive features of gut structure and digestive physiology in the terrestrial isopod *Philoscia muscorum* (Scopoli). *Biol. Bull., 49*:348–364, 1975.
84. Martin, M. M., and Martin, J. S. Cellulose digestion in the midgut of the fungus-growing termite *Macrotermes natalensis*: The role of acquired digestive enzymes. *Science, 199*:1453–1455, 1978.
85. Monk, D. C. The distribution of cellulase in freshwater invertebrates of differing feeding habits. *Freshwater Biol., 6*:471–475, 1976.
86. Bjarnov, N. Carbohydrases in *Chironomus, Gammarus* and some Trichoptera larvae. *Oikos, 23*:261–263, 1972.
87. Klug, M. J., and Kotarski, S. Bacteria associated with the gut tract of larval stages of aquatic cranefly *Tipula abdominalis* (Diptera: Tipulidae). *Appl. Environ. Microbiol., 40*:408–416, 1980.
88. Shearer, C. A., and Webster, J. Aquatic hyphomycete communities in the River Teign. III. Comparison of sampling techniques. *Trans. Br. Mycol. Soc., 84*:509–518, 1985.
89. Ladisch, M. R., Lin, K. W., Voloch, M., and Tsao, G. T. Process considerations in the enzymatic hydrolysis of biomass. *Enzyme Microbiol. Technol., 5*:82–102, 1983.

18

IMPORTANCE OF WOOD DECAY FUNGI IN FOREST ECOSYSTEMS

LYNNE BODDY *College of Cardiff, University of Wales, Cardiff, Wales*

I. INTRODUCTION

Currently about one third of the land surface of the earth is covered by forest, and in forest ecosystems 90% of the biomass is woody. Even in savanna, where only a few trees are present, 80% of the biomass is woody. Thus, it is clear that wood represents a considerable global resource (Table 1). In fact, wood constitutes the largest reservoir (80% of total) of organic carbon (and hence potential energy) in the biosphere, of which a very large proportion occurs in tropical forests.

Until recently, our attention to wood decay has largely centered on the economic losses it causes in buildings and in the timber supply industries. Now, however, our view is broadening, and there is an increasing awareness of the amenity value of trees, rapid disappearance of forests, and hence also of tree diseases. Further, wood decomposition is now regarded as being an integral process in ecosystem functioning, and the use of wood decay fungi in the production of food and industrially important chemicals is becoming increasingly recognized.

This chapter, after briefly introducing the wood decay organisms and their decay abilities, will describe the involvement of these fungi in the death and decay of standing trees; their role in the carbon, energy, and nutrient cycling in forest ecosystems; how to control these fungi and manipulate them to serve our interests in forest systems; and, finally, the role of wood decay fungi as food sources for man and animals.

II. WOOD DECAY ORGANISMS

Extensive discussion of the enzymatic abilities and roles of different organisms in wood decomposition is beyond the scope of this chapter (see Ref. 1). Here I seek only to put the contribution of different groups of organisms into perspective.

Although some fungi are individually capable of effecting complete decomposition of wood, total decay is usually brought about by a wide variety

TABLE 1. Net Primary Production, Biomass and Decomposition in Terrestrial Biomes

	Biomass				
	Area (10^9 ha)	Range (t ha^{-1})	Mean (t ha^{-1})	Above-ground perennial parts (%)	Roots (%)
Tundra and alpine	0.8	1–30	6	12	75
Desert and semi-desert scrub	1.8	1–40	7	2–40	57–87
Extreme desert (rock, sand, ice)	2.4	0–2	0.2		
Temperate grassland	0.9	2–50	16	0	83
Tropical savanna	1.5	2–150	40	60	28
Woodland and shrubland	0.9	20–200	60		
Northern coniferous forest (taiga)	1.2	60–400	200	71	22
Temperate forest:					
deciduous	0.7	60–600	300	74	25
evergreen	0.5	60–2000	250		
Tropical seasonal forest	0.8	60–600	350		
Tropical rainforest	1.7	60–800	450	74	18
Swamps and marshes	0.2	30–500	150		
Cultivated land	1.4	4–120	10		

Sources: Refs. 70, 115–117.

Net primary production		Nonwoody litter dynamics		
Range ($t\ ha^{-1}\ yr^{-1}$)	Mean ($t\ ha^{-1}\ yr^{-1}$)	Input ($t\ ha^{-1}\ yr^{-1}$)	Standing crop ($t\ ha^{-1}$)	95% turnover time (yr)
0.1–4.0	1.4	1.5	44	100
0.1–2.5	0.9			
0–0.1	0.03			
2–15	6.0	7.5	5	2
2–20	9.0	9.5	3	1
2.5–12	7.0			
4–20	8.0	7.5	35	14
6–25	12.0	11.5	15	4
6–25	13.0			
10–25	16.0	30	5	0.5
10–35	22.0			
8–60	30.0			
1–40	6.5			

of organisms and abiotic factors, such as abrasion and fire. Mycelial fungi are, in many terrestrial environments, the major agents of wood decay since their mycelial form and enzymatic capacities are especially suitable for invasion of solid, bulky, spatially determinate resource units. Those fungi, which are termed "wood decay" fungi, are generally regarded as able to cause three main types of decay—white rot, brown rot and soft rot—although this is an oversimplification which masks a range of different decay characteristics.

In white rot the wood typically becomes bleached and all chemical components are decomposed. It is largely brought about by Basidiomycotina, although it is becoming increasingly apparent that xylariaceae and other Ascomycotina and imperfect Ascomycotina are white-rot fungi [2–4, G. S. Griffith and L. Boddy, unpublished]. Brown rot is a decay, again brought about by Basidiomycotina, in which the wood is discolored brown and becomes friable and cubically cracked. The simple carbon compounds, cellulose and hemicelluloses are degraded but lignin is not, although it is modified. It is the brown- and white-rot fungi that are responsible for most wood decomposition in forests, including standing trees and felled and fallen timber. Brown-rot fungi are predominantly associated with conifers and white-rot fungi with angiosperms (Table 2). The total number of currently recognized brown-rot fungal species comprise only about 6% of known wood-rotting Basidiomycotina [5].

Like brown-rot fungi, soft-rot fungi utilize simple carbon compounds, cellulose and hemicelluloses, but do not appear to break down lignin appreciably; they result in much more localized decay than brown-rot fungi. It is quantitatively important only under fluctuating or water-saturated moisture regimes. Typically, this type of decay is caused by imperfect Ascomycotina, although again its occurrence in other groups may be more widespread than is currently appreciated.

Other microorganisms, including bacteria and nondecay fungi, utilize simple carbon compounds found in living wood cells or made available by other decay organisms. They are generally unable to utilize the lignocellulose complex of the wood cell wall. Although some bacteria can degrade the latter, this is probably only of significance in water-saturated wood [e.g., 6,7] or in the presence of high concentrations of preservatives [8]. Some of these nondecay fungi do result in economic loss of timber as, although they do not cause mechanical weakness, they often leave the wood stained and hence less valuable.

Few animals have the enzymic capacity to break down lignin or for complete cellulolysis, although some contain gut symbionts or employ exosymbiotic co-digestion strategies which can cope with the problem [9,10]. Their major role in wood breakdown results from comminution, i.e., reduction in particle size brought about by feeding. Some can cause considerable economic loss, for example, woodworm (*Anobium punctatum*) in buildings, shipworms (Teridinidae) in marine timbers, and termites in wood in the tropics. They also have considerable effects on fungi in that they act as vectors, alter microenvironmental conditions and alter the decomposer community as a result of communition and direct grazing on fungi (see also below). With regard to alterations in the decomposer community, comminution tends to select for unicellular as opposed to mycelial organisms [11], and in wood invaded by invertebrates there is a shift from a flora dominated by wood decay fungi to one more characteristic of a generalized "soil" microflora [10].

TABLE 2. Brown-Rot Fungi on Conifers and Hardwoods in North America

		No. of wood-rotting fungi	No. of brown-rot fungi	Brown-rot fungi (%)
Conifers				
Ponderosa pine	*Pinus ponderosa*	228	53	23.3
Western true firs	*Abies* spp.	216	49	22.7
Spruces	*Picea*	352	66	18.8
Douglas fir	*Pseudotsuga menziesii*	241	54	22.4
Western red cedar	*Thuja plicata*	169	38	22.5
Western hemlocks	*Tsuga heterophylla*	205	50	24.4
Arizona junipers	*Juniperus* spp.	32	10	31.3
Angiosperms				
Aspen	*Populus tremuloides*	260	25	9.6
Western alders	*Alnus rubra*	178	14	7.9
Western maples	*Acer* spp.	128	9	7.0
Western birches	*Betula* spp.	102	12	11.8
Southwestern oaks	*Quercus* spp.	48	4	2.7

19% of North American polypores are brown-rot fungi

Brown-rot fungi in pileate polypores	37 of 44 primarily on conifers
Brown-rot fungi in the genus *Poria*	23 of 27 primarily on conifers
Totals	60 of 71 primarily on conifers

85% of brown-rot polypores occur primarily on conifers

Source: Ref. 5.

III. DEATH AND DECAY OF STANDING TREES

Commercial losses of standing timber result from white and brown rot decay of central, nonliving tissues—heart rot, and from death and decay of sapwood by active pathogenic mechanisms. Also, attached branches die or are killed—sometimes termed natural pruning—and subsequently decompose. Although the latter does not result in large economic loss, as such wood is not used for timber, it is important in recycling essential mineral nutrients.

A. Heart Rot

For many years, heart rot was thought to be the main or only cause of decay in standing trees. Nonetheless, while we now know that there are several types of decay, heart rot is still a major cause of economic loss and deterioration in high forest and mature amenity trees. However, in many commercial forest plantations trees are grown and harvested on short rotation and the incidence of heart rot is greatly reduced.

While a wide range of microorganisms has been isolated from the hearts of living trees [e.g., 12,13], it is apparently only a few Ascomycotina and Basidiomycotina that cause significant decay (Table 2, Fig. 1). However, deterioration of heartwood can also result from a condition known as wetwood, in which, as the name suggests, the central nonliving tissues are water saturated. This condition has been found in a range of trees, including *Abies* spp., *Populus* spp., and *Salix* spp., both in apparently healthy trees and in association with wounds, branch stubs, butt rot, and mistletoe cankers [e.g., 14,15]. The etiology is probably diverse; in some cases there is evidence that it has resulted from the physiological activity of the tree [e.g., 17], whereas in other cases the causal role has been attributed to bacteria [e.g., 17,18].

With regard to decay caused by fungi, two classes of heart rot are often recognized: butt rot, which spreads upwards from the root or root collar, and top rot, which spreads downwards from the crown. Individual fungal species generally cause one or the other of these conditions (Table 3, Fig. 1), which probably indicates fundamentally different establishment strategies [1]. Thus, fungi causing butt rot may be present in soil as mycelium, which following colonization may then spread via root contact or migratory cords and rhizomorphs, or may be present as spores. Fungi causing top rot can presumably only colonize as spores. Whatever the source of inoculum, the fungus must ultimately gain access to the heartwood. The presence of heart rot is often associated with large wounds, and in the crown the generally accepted mode of establishment is via spores arriving at wounds large enough to expose dead heartwood. This idea has led to tree surgery practices whereby pruning of branches which do not contain heartwood is regarded as being "safe," in that this will not allow establishment of heartrot fungi, whereas pruning of larger branches is not "safe." Where the presence of heart rot in standing trees is associated with large wounds, it is probably a reasonable assumption that this was the mode of establishment.

Heart rot is, however, not always associated with large wounds, and other modes of establishment are indicated (Table 4, Fig. 2). For example, in the crown *Stereum gausapatum* often establishes from latent propagules in the sapwood of intact oak (*Quercus* spp.) branches, and from there can enter the heartwood of branches or the main trunk, where it causes the characteristic decay known as "pipe rot" [19,20]. The indian paint fungus, *Echinodontium tinctorum*, can establish cryptically in twig stubs (<1 cm diameter) of western hemlock (*Tsuga heterophylla*). These stubs become embedded in the trunk as the tree increases in girth and can act as foci for colonization of heartwood under suitable conditions [21,22].

With regard to butt rot, where wounds large enough to expose heartwood in the root system probably seldom occur, entry may therefore be largely dependent on mechanisms of active pathogenesis (see below). For example, both *Heterobasidion annosum* and *Armillaria* spp., which cause butt rot in a range of tree species, invade living roots and probably subsequently invade the heartwood from there.

The fungi considered so far destroy or devalue timber, however, some wood decay fungi can increase its value. In particular, *Fistulina hepatica*, the beef steak fungus, stains the heartwood of oak to produce a timber known as brown oak, which may be 10 times more valuable than uncolonized wood [23]. While the fungus causes brown rot, the decay is very slow to develop, and the wood becomes stained well before losing mechanical strength.

The foregoing is anthropocentric, but with regard to the tree, heart rot may be advantageous, and it has been suggested that the mobilization and trapping of nutrients in a rotten center may be of adaptive value [24,25].

B. Active Pathogenesis

Colonization of trees by active pathogenic mechanisms requires establishment of a sufficient inoculum base before extension into functional sapwood. Some fungi achieve this inoculum base internally through heart rot and others adopt one of three types of strategy: ectotrophic root infection, wound infection, and canker production [1]. Some of the more important examples of fungi that adopt these strategies and the timber losses that they cause are briefly given below.

Ectotrophic Pathogens

Ectotrophic pathogens produce extensive superficial mycelium over the root surface prior to invading the root, which provides a way of establishing a sizeable inoculum base from which outer host cells, including cambium, can be killed. Subsequently, death of outer tissues results in alleviation of stressful internal environmental conditions, leaving the way open to colonization of interior sapwood. *Heterobasidion annosum* and the "honey agarics" (*Armillaria* species group) have attracted the most attention because of their widespread occurrence and economic significance. *Heterobasidion annosum* causes killing of individual roots and small trees of *Pinus* spp. in particular. Other species such as larch (*Larix* spp.), spruce (*Picea* spp.) and Douglas fir (*Pseudotsuga menziesii*) do not appear to be as susceptible to killing but are subject to butt rot by the fungus. The mycelium of *H. annosum* is apparently incapable of spreading very far in soil, and infection of trees is usually via root contact or *de novo* establishment resulting from basidiospore infection of cut stumps (Fig. 3).

The honey agarics are of worldwide distribution and cause damage to a wide range of angiospermous and gymnospermous trees and shrubs (Fig. 4). Typically, root penetration is by rhizomorphic infective organs, which then dedifferentiate to produce subcortical fans of white mycelium which spread in the cambial region. Decay then develops centripetally from this peripheral zone of occupation. Understanding of the biology and pathology of *Armillaria* has been hampered by lack of taxonomic rigor, and it is now becoming apparent that there are numerous different biological species whose pathogenicity and host ranges differ. For example, in the United Kingdom and western Europe *Armillaria mellea sensu stricto* has a wide host range among angiosperms and can also attack certain conifers, particularly on replanted sites [26,27]. Typically infection foci are small, but large foci can develop when susceptible woody species are planted close together in orchards, vine plantations, and birch (*Betula*) stands on light soil [27,28].

Armillaria ostoyae chiefly attacks resinous conifers, and in *Pinus pinaster* stands in the Landes de Gascoyne it causes losses as high as

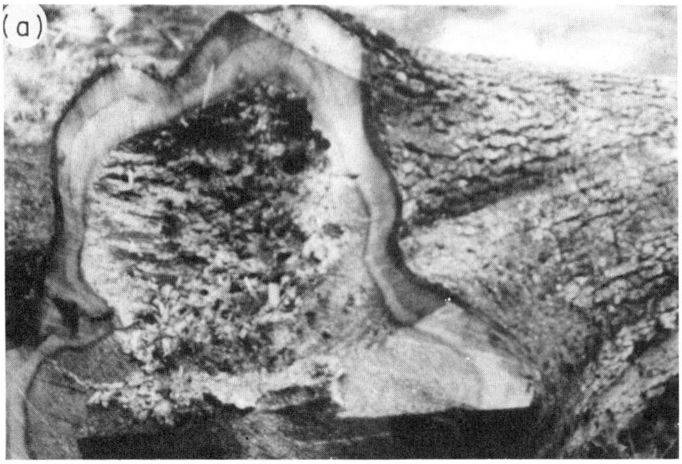

FIGURE 1. Various white and brown heart rots in coniferous and angiospermous trees. (a) Butt of a 150-year-old "veneer" oak (*Quercus robur*) from the Forest of Dean, Gloucester, United Kingdom, extensively decayed by *Grifola frondosa*, which caused a flaky white rot. (b) Brown cubical rot caused by *Tyromyces stipticus* in spruce (*Picea abies*). (c) Brown butt rot in elm (*Ulmus* sp.) due to *Rigidoporus ulmarius*. (d) Stump of a wind-broken Sitka spruce (*Picea sitchensis*) butt-rotted by *Sparassis crispa*. The brown-rotted wood is extensively covered by white mycelium. (From Ref. 1. Photographs courtesy of R. B. Pearce.)

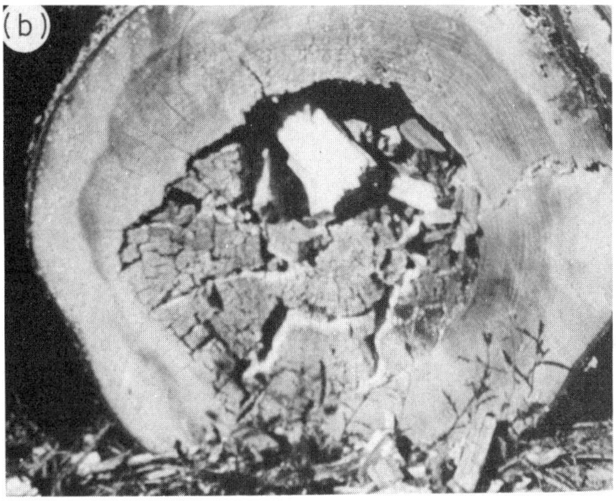

FIGURE 1. (Continued)

TABLE 3. Characteristics of Some Important Heart Rot Fungi

Decay fungus	Position of rot	Appearance	Common tree species
Armillaria species	Butt	White, stringy to spongy with dark pseudosclerotial plates	Ash (*Fraxinus*), beech (*Fagus*), Sitka spruce (*Picea sitchensis*), Norway spruce (*Picea abies*), western hemlock (*Tsuga heterophylla*), western red cedar (*Thuja plicata*), Lawson cypress (*Chamaecyparis lawsoniana*)
Daedaleopsis confragosa	Top	White rot	Alder (*Alnus glutinosa*), willow (*Salix*)
Fistulina hepatica	Top	'Brown oak'—initially rich brown, often as streaks, finally red-brown	Oak (*Quercus*), sweet chestnut (*Castanea sativa*)
Fomitopsis cytisina	Butt	White, felt-like	Ash (*Fraxinus*)
Ganoderma adspersum/applanatum	Butt and top	White mottled decay at first with brown reaction zone; later white and spongy	Beech (*Fagus*), horse chestnut (*Aesculus hippocastanum*)
Ganoderma pfeifferi	Butt and top	Yellow, mottled	Beech (*Fagus*)
Ganoderma lucidum	Butt	White rot	Oak (*Quercus*)
Ganoderma resinaceum	Butt	Leathery, white rot	Oak (*Quercus*)
Grifola frondosa	Butt	White wood with orange colored zone lines. Reddish-brown mycelial sheets in hollow trunks	Oak (*Quercus*)
Heterobasidion annosum	Butt	White pocket rot; sometimes fibrous, stringy	Beech (*Fagus*), birch (*Betula*), sitka spruce (*Picea sitchensis*), Norway spruce (*Picea abies*), western hemlock (*Tsuga heterophylla*), western red cedar (*Thuja plicata*)
Inonotus dryadeus	Butt	Dark brown water-soaked invasion zone. Then yellowish stripes: finally soft white	Oak (*Quercus*)

Inonotus hispidus	Top	Yellow; dark, gummy reaction zone	Ash (*Fraxinus*), plane (*Platanus* sp.), sycamore (*Acer pseudoplatanus*), walnut (*Juglans regia*)
Inonotus radiatus	Top	White, flaky rot	Alder (*Alnus glutinosa*)
Laetiporus sulphereus	Top	Brown cubical rot, with mycelial sheets in cracks	Cherry (*Prunus*), oak (*Quercus*), sweet chestnut (*Castanea sativa*), yew (*Taxus baccata*), willow (*Salix*)
Lyophyllum ulmarium	Top	Brown rot; wood becomes separated along the annual rings	Elm (*Ulmus*)
Meripilus giganteus	Butt	White, spongy	Beech (*Fagus*)
Phaeolus schweinitzii	Butt	Brown cubical rot with mycelial sheets	Sitka spruce (*Picea sitchensis*)
Phellinus pomaceus	Top	White, crumbly rot	Cherry (*Prunus*)
Phellinus weirii	Butt	Yellow discoloration initially; finally brown, fibrous or crumbly	Western red cedar (*Thuja plicata*), and other conifers
Pholiota squarrosa	Butt	Soft, pale brown rot; yellowish brown at first	Ash (*Fraxinus*), beech (*Fagus*), poplar (*Populus*)
Pleurotus ostreatus	Top	White, flaky rot	Beech (*Fagus*), horse chestnut (*Aesculus hippocastanum*), poplar (*Populus*)
Polyporus squamosus	Top	White to yellow spongy rot with sheets of mycelium	Elm (*Ulmus*), sycamore (*Acer pseudoplatanus*), walnut (*Juglans regia*)
Rigidoporus ulmarius	Butt	Brown cubical rot	Elm (*Ulmus*)
Sparassis crispa	Butt	Yellow-brown rot	Sitka spruce (*Picea sitchensis*)
Stereum gauspatum	Top	Pipe rot	Oak (*Quercus*)
Tyromyces stipticus	Top	Reddish-brown cubical	Norway spruce (*Picea abies*)
Ustulina deusta	Butt	Light colored decay with numerous dark zone lines	Beech (*Fagus*), lime (*Tilia*), elm (*Ulmus*)

Sources: Refs. 1, 30, and 118.

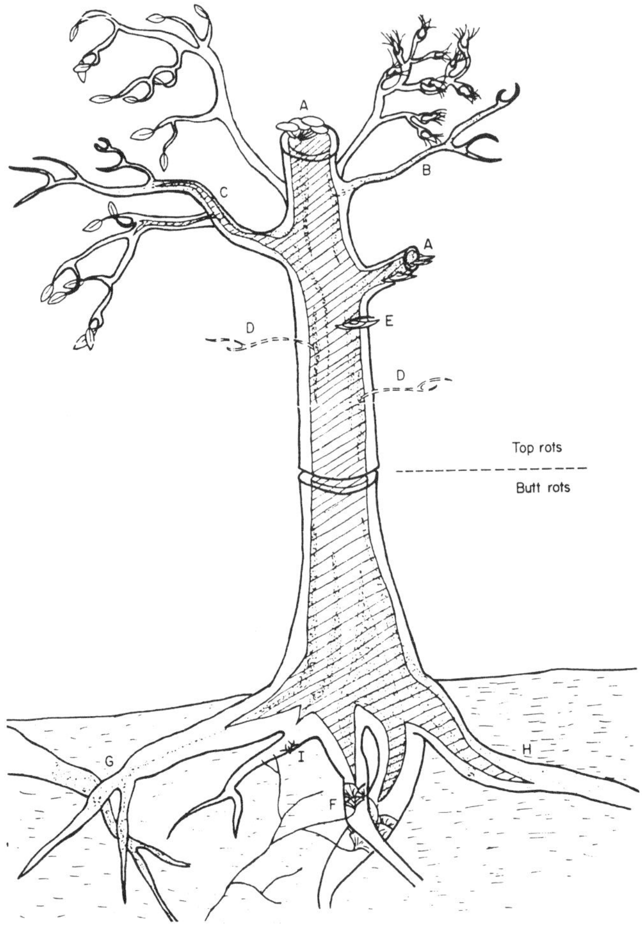

FIGURE 2. Routes of entry and exit of heart rot fungi in an idealized (from the fungal rather than the forester's viewpoint!) tree. Stippled regions represent decay columns, hatched zone represents heartwood. (A) Large wounds containing heartwood brought about by pruning or breakage. (B) Dead attached branch lacking heartwood, from which decay established in the sapwood can, under some circumstances, spread into heartwood, e.g., with *Stereum gausapatum*. More usually decay is limited at the base of such branches. (C) Intact attached branches containing heartwood in which decay establishes first in sapwood and then may enter heartwood, extending from there into the heartwood of living forks, e.g., *S. gausapatum*. (D) Branchlets, in which decay fungi become established before heartwood formation, whose stubs become increasingly buried deeper in the wood as the girth of the tree increases. When the buried stubs become enclosed in heartwood they may act as foci from which heart rot is established, e.g., *Echinodontium tinctorium*. Buried branch stubs may also provide exit routes from the heartwood, allowing fruiting on the outside of the tree (E). (F) Some butt rot fungi are able to spread from tree to tree by growing through soil in the form of mycelial cords or rhizomorphs and may become established in heartwood following initial "ectotrophic" mycelial

TABLE 4. Relationship Between Entry Court and Incidence of Infection and Volume of Decay Caused by Heart Rot Fungi in Oak (*Quercus* spp.) in the Central Hardwood Region of the United States

Infection court	Infections (%)	Decay volume (%)
Fire scars	24.5	40.1
Dead branch stubs	22.0	13.5
Branch bumps	13.1	11.1
Insect wounds	9.9	4.5
Branch stumps	9.2	9.5
Mechanical injuries	5.5	4.2
Open branch stub scars	5.4	6.9
Damaged or dead tops	5.3	6.5
Roots	3.8	3.1
Woodpecker injuries	1.3	0.8

Source: Ref. 119.

several hundred million trees per year [28]. It sometimes produces large foci in angiosperm woodlands with favorable soil conditions and colonizes suppressed oak (*Quercus* spp.) trees in evenaged plantations [29]. Both *A. ostoyae* and *A. mellea* are capable of infecting stumps by means of basidiospores, but *A. ostoyae* also forms extensive rhizomorph networks. Like *A. ostoyae*, *A. bulbosa* also forms extensive networks of rhizomorphs but is less actively pathogenic and generally only colonizes stressed or overmature trees. *Armillaria tabescens* did not cause death in any of Rishbeth's pathogenicity tests, and there is no record of it killing standing trees in the United Kingdom.

---

spread. Depicted here is the behavior of certain rhizomorph-forming species of *Armillaria* which produce "fans" of flattened rhizomorphs and white mycelium under the bark of infected roots. (G) Grafts or intimate root contacts can facilitate transmission from decayed to undecayed roots and are particularly important for fungi such as *Heterobasidion annosum* which are unable to grow as mycelium through soil. (H) Some fungi may be able to establish in root sapwood in a manner equivalent to (C) top rots, prior to establishment in the heartwood. (I) In the case of *Phaeolus schweinitzii* it has been suggested that basidiospores in soil may act as inoculum from which decay can be established in roots predisposed to infection by *Armillaria*. (From Ref. 1.)

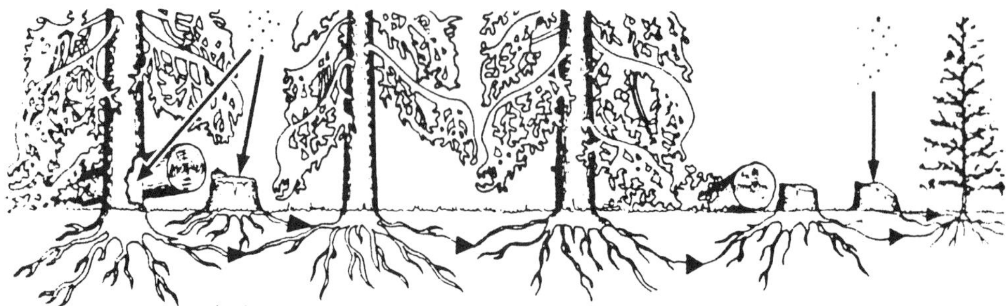

FIGURE 3. Infection sites of *Heterobasidion annosum* in conifer plantations. Stages shown left to right are as follows: (1) Thinning. Spore infections develop in wounds and on recently exposed stump surfaces. (2) Continuous. Vegetative spread occurs through root contacts between standing trees. (3) Final felling. Decay occurs in the stumps partly due to saprotrophic survival of infections already established in the previous crop, and to fresh colonization of newly exposed stump surfaces by spores. (4) Regeneration. Infections established in stumps from the previous stand are transmitted to young trees via root contacts. (From Ref. 121.)

Other ectotrophic pathogens include *Ganoderma* spp., *Inonotus tomentosus*, *Phellinus weirii*, *Rigidoporus lignosus*, *Phellinus noxious*, and several xylariaceous Ascomycotina. *Ganoderma* spp. are found worldwide and again colonize by active pathogenesis and also cause butt rot and top rot [30–32]. They are particularly damaging in the tropics where they colonize economically important plantation crops such as rubber (*Hevea brasiliensis*), tea (*Camellia sinensis*), cocoa (*Theobroma cacao*), and oil palm (*Elaeis guineensis*). Infection is probably via root contacts, with stumps of previously infected trees being important foci. *Inonotus tomentosus* causes "stand-opening disease" in white spruce (*Picea glauca*) in Saskatchewan and Manitoba [33,34]. Typically, dead and dying trees are found in patches of 0.4 ha or more surrounded by healthy trees. *Phellinus weirii* spreads from tree to tree via root contacts and, in addition to causing butt rot, causes serious root disease (laminated root rot) of conifers in the northwest United States and southwest Canada. It causes estimated losses of 4.4×10^6 m^3/yr [35] with the greatest losses occurring in second growth Douglas fir.

Rigidoporus lignosus, *P. noxious*, and the Ascomycotina *Ustulina deusta* cause economically significant diseases of rubber [1,3,36]. *Rigidoporus lignosus* also attacks teak (*Tectona grandis*) and *U. deusta* causes commercial loss of tea.

Wound Infection

Fungi that adopt this strategy establish themselves in wounded tissue and then extend pathogenically into living wood. Two economically important examples are *Stereum sanguinolentum* in conifers and *Chondrostereum purpureum* in angiosperms [1]. *Stereum sanguinolentum* causes extensive infection of sapwood and heartwood, usually associated with major wounds.

(a)

(b)

FIGURE 4. Honey fungus, *Armillaria* spp. (a) Fruit bodies, probably of *A. mellea sensu stricto*, clustered around the base of a tree stump. (b) Fruit bodies, which in having a wide cap and swollen stem base are probably of *A. bulbosa.* (c) Network of old flattened rhizomorphs produced under the bark of an elm log. (From Ref. 1. Photographs (a and b) courtesy of J. Webster.)

(c)

FIGURE 4. (Continued)

Chondrostereum purpureum causes the damaging disease known as "silverleaf" in which leaves on infected stems appear silvery. It is particularly troublesome on rosaceous fruit trees, especially *Prunus* and certain *Pyrus* varieties, but also on a range of other angiosperms. It appears to colonize preferentially wounds made in the spring and is apparently only able to colonize if it enters within 30 days [37].

Canker Production

Cankers are localized patches or strips of necrotic bark which are often caused by pathogenic fungi. As far as loss of timber is concerned, they are only really of significance if they cause decay of underlying wood or cause very extensive damage to the cambium resulting in decline in tree growth and vigor, or death. In these terms the most important canker pathogen is probably *Hypoxylon mammatum*, which causes annual losses of millions of dollars of aspen (*Populus tremuloides*) in the lake states of Michigan, Minnesota, and Wisconsin [38]. The cankers that it causes are typically centered on dead branch stubs, which may act as infection foci, and high incidence has occurred following extensive branch damage due to periodic circada (*Magicicada septendecim*) [39–41], where the wood becomes decayed below the dead bark.

C. Natural Pruning

Branches that fail to keep pace with the upward and outward spread of the canopy, like heartwood, become redundant sinks for nutrients. These become colonized by wood decay fungi [e.g., 1,42] and are eventually shed, a process sometimes called natural purning [43], resulting in mobilization of nutrients (see below).

IV. ROLE OF FUNGI IN ECOSYSTEM FUNCTIONING

The continuing flow of energy through forest ecosystems depends upon maintenance of nutrient supplies to autotrophic green plants. Since there is only a limited input of nutrients from outside, a balanced cycle which makes nutrients available again is essential for continued functioning of the ecosystem. The crucial role of wood decay fungi lies in the fact that they are the major agents of decomposition of bulky, lignocellulosic material, and hence also mediate transfer of mineral nutrients "locked up" in organic tissues back into the available soil pool. They may also play a role in soil organic matter formation.

The relative importance of wood decay fungi to ecosystem functioning varies between ecosystems, depending on the amount of woody material available for recycling (Table 5). Also, within any one area their relative importance may change because forests are static neither in space nor time. Thus, bare ground will, under suitable conditions, become covered successively by herbaceous vegetation and then by woody forms, ultimately forming climax forest communities. Even in climax forest changes occur, and they are often unevenly aged and patchy because of localized loss and reestablishment of the canopy. In plantation forestry, stands are often evenly aged and much less species diverse, although again changes occur with time.

A. Carbon and Energy Cycling

The aboveground input of litter to the forest floor in mature forests ranges between about 3500 and 5500 kg/ha/yr, and of this about 20% is twigs and 10% branches (Table 5) [44–46]. This is probably an underestimate of the amount of wood entering the decomposer system as considerable decay often occurs prior to fall [45–48]. For example, in one year of a study Boddy and Swift [45] found that over half of the branches falling to the forest floor had lost over half of their original weight. A more realistic estimate of the aboveground input of wood to the decomposer system is about 40% of total input [49]. Wood also enters the decomposer system from below ground. The amount is difficult to estimate but is probably as large as that entering from above ground. Vogt, Grier, and Vogt [50] showed that the input of fine root biomass was 2–3 times greater than the input of leaf fall in *Abies amabilis* forest.

The extent to which woody litter will accumulate on the forest floor depends on the input to the decomposer system, the extent to which decay occurred prior to fall, and the rate of turnover after fall. At early seral stages, there may be an imbalance between input and standing crop of woody litter, but in mature forests these processes will be balanced. However, on a smaller scale, there will be build-up and decline of woody standing crop on the forest floor associated with cyclical changes of gap formation and regeneration. Variation in standing crop on the forest floor also results from the considerable spatial heterogeneity of wood input; the smaller twig component is most homogeneously distributed and whole tree trunks most patchily distributed [51–53]. Correspondingly, estimates of standing crop of wood on the forest floor are quite variable and, for example, range from about 200 g m^{-2} to over 600 g m^{-2} in temperate beech (*Fagus sylvatica*) and oak (*Quercus*) woodlands.

TABLE 5. Aboveground Standing Crop and Input of Nutrients and Biomass in Woody and Nonwoody Litter in Temperate Woodlands

	Forest floor (kg/ha)			Mean residence time on forest floor (yr)	
	Biomass	Nitrogen	Phosphorus	Biomass	Nitrogen
Warm temperate:					
Broadleaf deciduous	11,480	163	11.9	2.7	4.8
Broadleaf evergreen	19,148	60	3.8	3.1	1.0
Needleleaf evergreen	20,026	362	25.1	4.6	13.8
Cold temperate:					
Broadleaf deciduous	32,207	624	50.2	10.2	19.1
Broadleaf evergreen	13,900	200	10.5	3.9	—
Needleleaf evergreen	44,574	504	44.9	17.9	32.8

Source: Refs. 50 and 59.

The decay rate of wood is influenced by a number of factors, which operate via their effect on the wood decay fungi. Size is particularly important, since large diameter components take much longer to decompose than those of smaller diameter. For example, large logs of hemlock (*Tsuga heterophylla*) can take over 300 years to decompose completely [54]. Branches and twigs usually take no more than 20 years and often much less [1]. Decay rate varies considerably between tree species, conifers tending to decay more slowly than angiosperms [55,56]. Prevailing microclimatic conditions, particularly temperature and water content, are also major determinants of decay rate [57].

While the decay rates cited above are quite slow, decomposition can be rapid: Chapela et al. [58] found that beech logs having volumes over 1000 cm^3 had sometimes completely decayed within 5 years. These differences could not simply be due to microclimate or resource quality but were probably related to the composition of the community of wood decay fungi. Few studies have examined this possibility, but it is now clear that community structure considerably affects decomposition rates (L. Boddy and E. M. Owens, unpublished).

B. Mineral Nutrient Cycling

Nonwoody components contribute the majority of mineral nutrients to the forest decomposer system (Table 5) [50–59]. Also, nutrients are more rapidly released from nonwoody tissues as they decompose more rapidly. However, by virtue of its relatively slow decay rate, wood not only forms

Mean residence time on forest floor (yr)	Nonwoody literfall (kg/ha/yr)			Woody literfall (kg/ha/yr)		
Phosphorus	Biomass	Nitrogen	Phosphorus	Biomass	Nitrogen	Phosphorus
4.3	4,236	36	3.8	891	2.6	0.8
2.2	6,484	55	3.7	—	—	—
11.0	4,432	28	2.7	1,107	2.5	0.2
11.1	3,854	43	4.6	1,046	3.7	0.2
—	3,590	—	—	—	—	—
22.1	3,144	26	3.2	602	1.1	0.1

a large standing crop of biomass but also a large reservoir of mineral nutrients, which may act as a buffer when systems are perturbed, e.g., following felling [60,61].

Nutrient cycling within forests has received a lot of attention [62–69], but there are relatively few data on the release of nutrients from wood [10,56,59–61], and there appears to be a wide variation in nutrient status at different stages of decay, and different elements show different patterns of release (Fig. 5, Table 6). Absolute concentrations of potassium decrease relatively rapidly, presumably as a result of leaching. Second to this is magnesium, which is again subject to leaching. Calcium is lost relatively more slowly, presumably reflecting greater importance as a structural component.

At early stages of decomposition, wood acts as a sink for nitrogen and phosphorus. The absolute concentrations of these elements increase as they are translocated into wood by mycelium, cords, and rhizomorphs during colonization [59] and are immobilized in the increasing fungal biomass until much of the nitrogen and phosphorus in the wood has been converted into fungal mycelium. It is not certain at which stage mineralization and release occurs, but it has been suggested that it is not until the carbon: nutrient ratio of the wood approaches that of the mycelium [70]. The carbon:nutrient ratio decreases as decay proceeds, since carbon is lost from wood, as respired carbon dioxide, but mineral nutrients remain (Table 6). This idea that mineral nutrients are released when carbon: nutrient ratios become equal is probably an oversimplification as exit structures such as sporophores and mycelial cords are produced before the

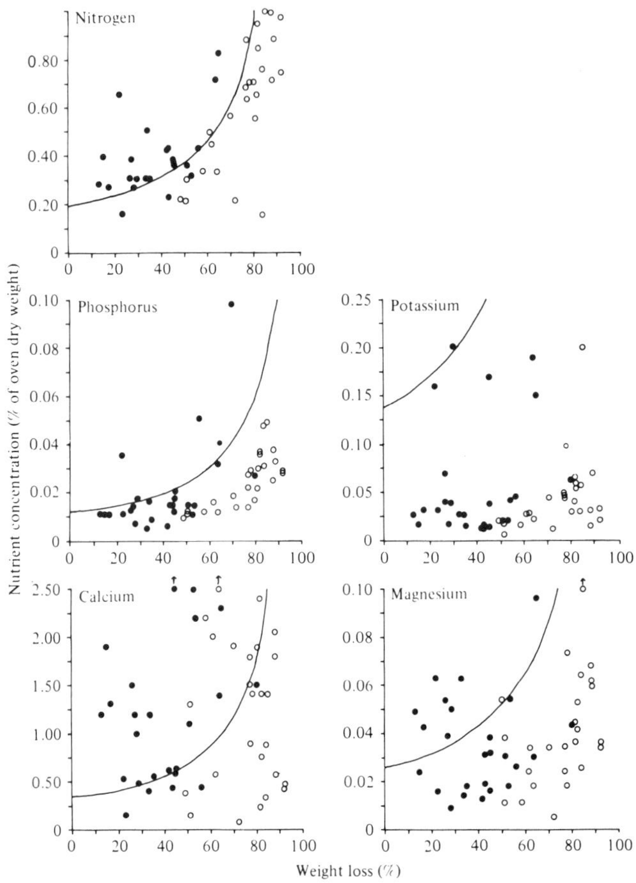

FIGURE 5. The relationship between mineral nutrient concentration (as % of oven dry weight) and weight loss (% of original dry weight) of uninvaded (●) and invaded (○) angiosperm branch wood collected from the litter layer. The line on each graph indicates the predicted change in elemental concentration as decomposition proceeds, assuming no net gain or loss. (From Refs. 10,61.)

carbon:nutrient ratio reaches this level, and these structures result in removal of nitrogen and phosphorus from the wood [59]. A major watershed in mobilization and release of nitrogen from wood appears to occur when animals invade [10,61].

Another important role of cord-forming wood decay fungi is that they are able to translocate nutrients rapidly (sometimes at rates in excess of 1 m/hr) [J. M. Wells, L. Boddy, and J. Dighton, unpublished] and over long distances. As such they may be responsible for large shifts in spatial location of nutrients.

TABLE 6. C:Nutrient Ratios for Undecayed Wood Together with Calculated Values for Decayed Wood on the Assumption that There has Been No Net Gain or Loss of the Mineral Nutrient Element

	RD ($g\ cm^{-3}$)	Weight loss (%)	C:N	C:P	C:K	C:Ca	C:Mg
Undecayed wood	0.560	0	247:1	3643:1	343:1	135:1	1787:1
Decayed wood	0.400	29	175:1	2587:1	243:1	96:1	1288:1
Decayed wood	0.300	46	133:1	1967:1	185:1	73:1	965:1
Decayed wood	0.200	64	89:1	1311:1	123:1	49:1	643:1
Decayed wood	0.100	82	44:1	655:1	62:1	24:1	322:1
Fungal mycelium	—	—	35:1	505:1	115:1	60:1	485:1
Insecta	—	—	6:1	52:1	67:1	157:1	235:1

Sources: Refs. 10, 61, 120.

C. Humus Formation

Humus is soil organic matter in which the original plant and animal remains are no longer recognizable and contains three different fractions: humic acid, fulvic acid, and humin. It is important for good soil structure, water-holding capacity, and because of the base exchange capacity which aids retention of nutrients in soil. The mechanisms of synthesis of humus molecules are uncertain, but the modified products of lignin degradation are a source of the aromatic core of the molecules of the different fractions (for further details see Refs. 71,72). Thus it may be that wood decomposition is a major pathway for humus synthesis in woodlands, and the visual character of wood (especially following brown rot) at terminal stages of decay is suggestive of this. However, Allison [73] has asserted that wood is a poor source of humus. Certainly, white rot Basidiomycotina tend to produce a bleached, relatively infertile form of humus [74], and the central role of fungi has usually been associated with non-Basidiomycotina [72].

V. MANIPULATION OF THE FOREST ECOSYSTEM

The management practices adopted in woodlands and plantation forestry can radically affect the role and relative significance of wood decay fungi within the ecosystem. If we can understand the effects of management practices on the relationships between fungus and tree and between fungus and ecosystem function, then it may be possible to lessen the impact of heart rot and pathogenic fungi. Three main aspects of management practices can be recognized as influencing the incidence of decay: replanting, wounding/thinning/pruning, and imposition of stress (for further discussion see Ref. 1).

A. Replanting

Plantation forestry involves replanting a site with a new crop following felling of the previous stand. Since felling is usually performed well before the tree becomes overmature, the incidence of top rot will often be reduced. On the other hand, the risk of infections by pathogenic and butt rot fungi may increase as these fungi may be carried over in soil or stumps from one crop to the next. This may be particularly problematical where the newly planted trees have similar or identical genotypes, which will hence be more susceptible to rapid spread of infection. Infection of trees in new stands is often associated with residual tree stumps, which act as infection foci from which pathogens may spread to adjacent trees. These stumps may have already been infected at the time of felling or may have been infected subsequently from the soil or cut surface. Indeed, *Heterobasidion annosum* root rot is a disease largely attributable to methods of plantation forestry.

Control or eradication of disease could obviously be effected by removal of stumps, and there is evidence that incidence of disease due to *Armillaria* spp. and *Phellinus weirii* is reduced following removal [75,76]. However, stump extraction is costly and greatly disturbs the site, hence other control methods are often used. Fumigation is sometimes used as a means of eradication of pathogens already established in stumps. Successful results have been obtained against *Armillaria* spp. and *P. weirii* using a variety of fumigants, including methyl bromide, carbon disulfide, Vorlex, chloropicrin, Vapam, and Di-trapex, which were poured into holes in the tops of stumps [75–78]. However, further research to determine cost effectiveness and safety are necessary before this approach can be adopted on a large scale [75].

Some tree pathogens develop in stumps after felling, and a variety of procedures have been developed to prevent their establishment. Physical and chemical barriers have been applied to the stump surface, e.g., Rishbeth [79] used creosote to control *H. annosum*, but the problem with these methods is that the barrier is eventually liable to break down.

Wounding/Thinning/Pruning

Wounding can be accidental (thinning or extraction damage) or deliberate (pruning). As already indicated, wounds can act as foci for colonization by heart rot fungi or pathogens, and therefore care should be taken to limit such damage. Losses resulting from pruning can be largely avoided by sensible timing and placement of cutting. Limbs should be cut close to the trunk so as to avoid leaving a stub, and pruning larger limbs containing heartwood should be avoided. It is also better to remove limbs during the growing season when vigorous growth is likely to seal off the wound. However, as mentioned above, colonization by *Chondrostereum purpureum* is favored at the beginning of the growing season. Chemical protectants and sealants have been employed to prevent colonization of pruning wounds, but these sometimes do more harm than good, particularly if they are phytotoxic or produce microenvironmental conditions conducive to the pathogen [80,81]. Biological control agents have also been applied to pruning wounds (see below).

Stress

Imposition of stress can predispose trees to colonization. The major causes of stress, other than those that result from replanting and wounding, involve site selection and stand density. Poor soils are frequently used, resulting in diminished or stunted growth, but even more damaging are sites subject to intermittent imposition of stress, particularly drought stress. For example, in the United Kingdom following the 1976 drought many trees became temporarily or permanently debilitated by bark diseases. On *Acer pseudoplatanus*, diamond bark canker developed often very patchily, being particularly prevalent on trees planted on soils which were normally very moist [82]. Also "beech-bark" disease and strip cankers associated with *Hypoxylon* and *Diatrype* spp. developed on beech (*Fagus sylvatica*) trees, again often associated with edaphic factors [83].

Densely packed stands may allow rapid spread of pathogens, particularly those which colonize via soil or root contacts. In such stands, trees become suppressed and subsequently colonized and infection foci may develop, as was demonstrated in the studies on suppressed *Quercus* trees where large patches of *Armillaria bulbosa* developed on one site and *A. ostoyae* developed on another [29].

B. Biological Control

A number of direct and indirect "biological control" methods have been developed to reduce incidence of tree pathogens, and ironically some of these actually involve the use of wood decay species! Direct biological control involves inoculating fungi which prevent colonization by a pathogen or oust it if it is already present. The greatest success story here has been the control of *H. annosum* in pine (*Pinus*) and Norway spruce (*Picea abies*) by the white rot fungus *Phlebia gigantea* [84–86]. This involves inoculation of stumps, immediately after felling, with a conidial suspension of *P. gigantea*, which then become established as mycelia. Protection is at least partly achieved by the hyphal interference antagonism of *H. annosum* by *P. gigantea* hyphae.

Following surface inoculation of cut stumps, there is inevitably a delay before the biological control agent becomes established in the roots, and hence such procedures are likely to be less effective against root-invading pathogens. Direct inoculation of roots presents practical difficulties, but one possibility is the use of saprotrophic cord-forming Basidiomycotina such as *Hypholoma fasciculare*, *Phanerochaete velutina*, and *Steccherinum fimbriatum*. Dowson, Rayner, and Boddy [87,88] have shown that these fungi can be successfully established in woodland soil by directly seeding with colonized wood blocks. They compete directly with root pathogens for available dead woody resources and are able actively to replace *Armillaria* spp. from dead wood (Table 7).

True decay fungi are obviously unsuitable as biological control agents of pruning wounds, since they may become established in the tree. Non-decay organisms have been suggested as a possibility, but since they often initiate community development which culminates with decay fungi, they may also lead to problems [1]. Some success has been reported in control of *Chondrostereum purpurem* by *Trichoderma viride* [89] and in controlling Basidiomycotina in *Acer rubrum* with *Trichoderma harzianum* [90].

TABLE 7. Outcome of Interactions Between Cord-Forming Fungi and *Armillaria* Species in Soil

	Armillaria bulbosa			*Armillaria cepestipes*			*Armillaria mellea*		
	Block covered (%)[a]	Depth penetrated (mm)	No. viable rhizomorphs[b]	Block covered (%)[a]	Depth penetrated (mm)	No. viable rhizomorphs[b]	Block covered (%)[a]	Depth penetrated (mm)[a]	No. viable rhizomorphs[b]
Hypholoma fasciculare	100	2	0	100	2	0	100	2	0
Phallus impudicus	50	<1	0	80	<1	0	—	—	—
Phanerochaete laevis	70	4	0	100	5	0	100	5	0
Phanerochaete velutina	100	2	0	100	2	<1	100	2	0
Steccherinum fimbriatum	50	<1	<1	100	<1	3	5	<1	—

[a]Mean values from triplicates. The data relate to colonization by cord formers.

[b]Controls without cord formers possessed numerous viable rhizomorphs.

Source: Ref. 122.

Indirect biological control involves altering conditions in stumps such that colonization by appropriate control agents is promoted. Forty percent ammonium sulfamate (AMS or "Amcide") provides an effective treatment for angiosperm stumps [91–94]. It successfully prevents regrowth of stumps and shifts the pattern of colonization towards highly combative decay species, which cause rapid and intense decay and hence remobilize nutrients and remove the stump as a possible focus for pathogenic infections. Punter in 1963 screened a range of chemicals for use against *H. annosum* in conifers other than *Pinus* spp. Sodium nitrite was most effective, but since it is highly toxic, other chemicals such as urea, disodium octoborate, and borax are preferable.

VI. WOOD DECAY FUNGI AS FOOD SOURCES AND AS DETERMINANTS OF ANIMAL ACTIVITY

Although some wood decay fungi are poisonous and many have inedible fruit bodies, e.g., many of the Aphyllophorales are tough and leathery, numerous others are edible and some are prized as great delicacies. For example, in China the "eggs" of the "stinkhorn," *Phallus impudicus*, are apparently eaten [96]. In Estonia *Pleurotus citrinopileatus* is a favorite [97], and several *Agaricus* spp. are commonly eaten in the United Kingdom and western Europe [98]. Several wood decay fungi are grown commercially, notably the shiitake, *Lentinula edodes*, and the oyster mushroom, *Pleurotus ostreatus*. Currently, cultivation of *L. edodes* is the single largest bioconversion process utilizing wood, with the Orient producing over $1 billion worth annually [99–102].

Not only do fungal sporocarps serve as food sources for man, but also they are used by other vertebrates and a wide variety of invertebrates (see Ref. 10). Furthermore, they improve the nutritional quality of the food of wood-inhabiting invertebrates and affect their behavior. Consequently, wood decay fungi also play an important role in the overall balance of decomposer organisms within forest ecosystems. A large number of mycetophagous insects use basidiocarps of white rot fungi for food and breeding grounds. Among the most common are Ciid beetles, of which 74 species have been found in sporocarps (mainly Polyporaceae) in North America [103]. Further, the sexual and asexual spores of many fungi are dispersed by feeding activities. For example, *P. impudicus* attracts flies by a smell of rotting flesh, and the sticky spores are then transmitted elsewhere on the bodies and legs of the flies.

Wood also serves as a food source for invertebrates, and prior colonization by wood decay fungi is often a prerequisite or, at least, a distinct advantage. Wood decay fungi improve wood as a food resource for invertebrates in three main ways: by enzymatic softening, destruction of allelopathic compounds, and improvement of nutritional quality. As mentioned earlier, few animals have the enzyme capacity to break down the lignocellulose complex, but even if they do, unless they can gain access to the wood or remove it to their nests they will not be able to utilize it as a food resource. The significance of the softening effect of wood decay fungi has seldom been demonstrated unequivocally, since it goes hand in hand with improvement of nutritional quality. However, in an experiment in which logs covering a range of densities were placed on the floor of a woodland in the north of England, over 75% of those invaded by larvae of *Tipula flavolineata* had densities below 0.3 gcm^{-3} [10]. Prior softening by fungi seems to be a prerequisite to successful nesting of woodpeckers. Of over 90 nest trees (of various tree and woodpecker species) examined in the United States, the heartwood of all had been softened by fungal activity, mainly by the white rot polypore *Spongipellis pachyodon* [104]. The woodpeckers were presumably able to identify suitable trees by sounding with their beaks.

Some tree species are initially repellent or unpalatable to animals because they contain compounds which have allelopathic effects. Fungal enzyme activity often results in the destruction of these compounds [e.g., 105], thus allowing invertebrate invasion. Most of the well-documented examples concern termites [10].

There is considerable evidence that partially decayed wood provides a higher quality food resource for invertebrates than undecayed wood [10]. These effects may be complex but include release of assimilable organic molecules by catabolism of polymers, the transformation of specific molecules to palatable form, and particularly the process of enrichment of mineral nutrient elements (see above). Thus, as decomposition proceeds, the carbon:nutrient ratio decreases (see Table 6), and an animal eating wood decayed by fungi will have a diet much richer in mineral nutrients. Further, the relative nutrient content of mycelium is even higher, and many animals are fungivorous, feeding directly on mycelium or ingesting it along with particles of wood. Indeed, in the case of the mutualistic association between higher termites and the white rot basidiomycetes of the genus *Termitomyces*, the fungus produces nutrient-rich bodies, termed mycotêtes, on its mycelium which are directly fed on by the termites [106]. That fungal decay of wood directly affects woodboring invertebrates is illustrated by a study of Fisher [107] on the Death Watch beetle, *Xestobium rufovillosum*. He found that the length of time taken by larvae to complete their life cycle in oak and willow timber and the rate of disintegration due to larval boring was related to the extent of fungal decay. In undecayed wood, larvae usually died. In wood slightly decayed by *Phellinus cryptarum* the life cycle took 14 to 24 months, whereas in well-decayed wood the life cycle was completed in 10 to 17 months.

In addition to the nutritional effects described above, wood decay fungi can influence the behavior of animals as a result of production of attractive and repellant chemicals. The best-documented examples again relate to termites. For example, the mycelium of *Gloeophyllum trabeum* and wood decayed by it contain termite trail-following, arrestant, and feeding stimulants which attract species of *Reticulitermes*, *Heterotermes*, *Coptotermes*, and *Kalotermes* [108–111]. By contrast, wood decayed by the white-rot fungus *Ganoderma applanatum* was toxic to *Microtermes edentatum* [112], and wood decayed by *Bjerkandera adusta*, *Coriolus versicolor*, and *Stereum sanguinolentum* was toxic to *Reticulitermes santonensis* [113]. The intricate relationship of attractant chemistry to the physiological state of the fungus was demonstrated by Boddy, Coates, and Rayner [114], who noted that fungus gnats of the genus *Bradysia* were attracted to zones of intraspecific antagonism in cultures of wood-rotting basidiomycetes and xylariaceous fungi. These zones consisted of dying and lysed hyphae, which suggests that the production of specific attractants is a product of dying mycelium.

VII. CONCLUSION

In terms of the use of forests and forest products, fungi can have both beneficial and detrimental effects. The former includes the crucial role that they play in maintaining productivity of the ecosystem, biological control of tree pathogens and sources of food for man and other animals, while detrimental effects are largely concerned with loss of timber. The aims of applied mycological research in forests are to increase beneficial effects and reduce detrimental effects. In the past such attempts have been rather hit or miss, as it is not until recently that we have begun to obtain a sound understanding of fungal ecology and, in particular, the ways in which fungi become established in wood. It is only with such knowledge that we can hope artificially to establish beneficial fungi where they are absent and to eradicate or prevent establishment of fungi in situations where they cause detrimental effects.

Development of sound management practices are leading to a trend away from planting vast areas of monoculture and away from harvesting/pruning methods which provide colonization courts and bases for build-up of detrimental fungi. The increasing awareness of the role of cord-forming fungi in nutrient cycling and as aggressive competitors of tree pathogens has yet to be capitalized on, but ongoing laboratory and field trials look promising for the future. Another beneficial role for fungi likely to increase rapidly in the immediate future lies in the production of edible fruit bodies. Commercial-scale production of shiitake and the oyster fungus is already big business in several countries and is quickly extending elsewhere to meet the increasing demand.

REFERENCES

1. Rayner, A. D. M., and Boddy, L. *Fungal Decomposition of Wood: Its Biology and Ecology*. John Wiley, Chichester, 1988.
2. Merrill, W., French, D. W., and Wood, F. A. Decay of wood by species of Xylariaceae. *Phytopathology, 54*:56–58, 1984.
3. Rogers, J. D. Xylariaceae: Systematic, biological and evolutionary aspects. *Mycologia, 71*:1–42, 1979.
4. Sutherland, J. B., and Crawford, D. L. Lignin and glucan degradation by species of the Xylariaceae. *Trans. Br. Mycol. Soc., 76*: 335–337, 1981.
5. Gilbertson, R. L. Wood-rotting fungi of North America. *Mycologia, 72*:1–49, 1980.
6. Boutelje, J. B., and Bravery, A. F. Observations on the bacterial attack of piles supporting a Stockholm building. *J. Inst. Wood Sci., 4*:45–57, 1968.
7. Holt, D. M., Jones, E. B. G., and Furtado, S. E. J. Bacterial breakdown of wood in aquatic habitats. *Brit. Wood Pres. Assoc. Ann. Conv., 2*:1979.
8. Eaton, R. A., and Dickinson, D. J. The performance of copper-chrome-arsenic treated wood in the marine environment. *Mater. Org. Beih., 3*:521–529, 1976.

9. Martin, M. M. The role of ingested enzymes in the digestive processes. In *Invertebrate-Microbial Interactions* (J. M. Anderson, A. D. M. Rayner, and D. W. H. Walton, eds.), Cambridge University Press, Cambridge, 1984, pp. 155–172.
10. Swift, M. J., and Boddy, L. Animal-microbial interactions during wood decomposition. In *Invertebrate-Microbial Interactions* (J. M. Anderson, A. D. M. Rayner, and D. W. H. Walton, eds.), Cambridge University Press, Cambridge, 1984, pp. 89–131.
11. Hanlon, R. D. G., and Anderson, J. M. Influence of macroarthropod feeding activities on microflora in decomposing oak leaves. *Soil Biol. and Biochem.*, *12*:255–261, 1980.
12. Wagner, W. W., and Davidson, R. W. Heartrots of living trees. *Bot. Rev.*, *20*:61–134, 1954.
13. Shigo, A. L., and Hillis, W. E. Heartwood, discoloured wood and microorganisms in living trees. *Ann. Rev. Phytopathol.*, *11*:197–222, 1973.
14. Hartley, C., Davidson, R. W., and Crandall, B. S. Wetwood, bacteria and increased pH in trees. *U.S. For. Serv., For. Prod. Lab. Report*, *2215*:1–34, 1961.
15. Wilcox, W. W. Some physical and mechanical properties of wetwood in white fir. *For. Prod. J.*, *18*:27–31, 1968.
16. Worrall, J. J., and Parmeter, J. R. Formation and properties of wetwood in white fir. *Phytopathology*, *72*:1209–1212, 1982.
17. Zeikus, J. G., and Ward, J. C. Methane formation in living trees: A microbial origin. *Science*, *184*:1181–1183, 1974.
18. Preece, T. F. Watermark disease of the cricket bat willow. *Forestry Commission Leaflet*, *20*:1–9, 1977.
19. Toole, E. R. Rot entrance through dead branches of southern hardwoods. *For. Sci.*, *7*:218–226, 1961.
20. Boddy, L., and Rayner, A. D. M. Population structure, intermycelial interactions and infection biology of *Stereum gausapatum*. *Trans. Br. Mycol. Soc.*, *78*:337–351, 1982.
21. Etheridge, D. E., and Craig, H. M. Factors influencing infection and initiation of decay by the Indian paint fungus (*Echinodontium tinctorium*) in western hemlock. *Can. J. For. Res.*, *6*:299–318, 1976.
22. Etheridge, D. E., Craig, H. M., and Farris, S. H. Infection of western hemlock by the Indian paint fungus via living branches. *Dep. Environ. Can. For. Serv. Bimon. Res. Notes*, *28*:3–4, 1972.
23. Cartwright, K. St. G. A re-investigation into the cause of "brown oak," *Fistulina hepatica* (Huds.) Fr. *Trans. Br. Mycol. Soc.*, *21*:68--83, 1937.
24. Janzen, D. H. Why tropical trees have rotten cores. *Biotropica*, *8*:110, 1976.
25. Dickinson, T. A., and Tanner, E. V. J. Exploitation of hollow trunks by tropical trees. *Biotropica*, *10*:231–233, 1978.
26. Rishbeth, J. Species of *Armillaria* in southern England. *Plant Pathology*, *31*:9–17, 1982.
27. Rishbeth, J. Armillaria: Resources and hosts. In *Developmental Biology of Higher Fungi* (D. Moore, L. A. Casselton, D. A. Wood, and J. C. Frankland, eds.), Cambridge University Press, Cambridge, 1985, pp. 87–101.

28. Guillaumin, J. J., Lung, B., Romagnesi, H., Marxmuller, H., Mohammed, C., Lamoure, D., Durrieu, G., Berthelay, S., and Savin, V. Taxonomy and biological cycles of the French *Armillaria* species. In *Proceedings of the Sixth International Conference on Root and Butt Rots of Forest Trees* (G. A. Kile, ed.), CSIRO, Melbourne, 1984, pp. 43–56.
29. Thompson, W., and Boddy, L. Decomposition of suppressed oak trees in even-aged plantations. II. Colonisation of tree roots by cord and rhizomorph producing basidiomycetes. *New Phytol.*, *93*: 277–291, 1983.
30. Phillips, D. A., and Burdekin, D. A. *Diseases of Forest and Ornamental Trees*. MacMillan, London and Basingstoke, 1982.
31. Varghese, G. Parasitic diseases of oil palm (*Elaeis guineensis*) with particular reference to pathological problems of this crop in Malaya. *Malayan Agriculturist*, *6*:3–14, 1965.
32. Varghese, G., and Chew, P. S. Pathological significance and problems of control of *Ganoderma* spp. parasitic to plantation crops in Malaysia. In *Proceedings of the Sixth International Conference on Root and Butt Rots of Forest Trees* (G. A. Kile, ed.), CSIRO, Melbourne, 1984, pp. 296–304.
33. Whitney, R. D. Studies in forest pathology. XXIV. *Polyporus tomentosus* Fr. as a major factor in opening disease of white spruce. *Can. J. Bot.*, *40*:1631–1658, 1962.
34. Whitney, R. D. Artificial infection of small spruce roots with *Polyporus tomentosus*. *Phytopathology*, *53*:441–443, 1963.
35. Nelson, E. E., Martin, N. E., and Williams, R. E. Laminated root rot of western conifers. *USDA For. Ser. Forest Insect and Disease Leaflet*, *159*:1–6, 1981.
36. Nicole, M., Nandris, D., and Geiger, J. P. Pathotypes of *Rigidoporus lignosus* and *Phellinus noxius* isolated from different hosts in West Africa. In *Proceedings of the Sixth International Conference on Root and Butt Rots of Forest Trees* (G. A. Kile, ed.), CSIRO, Melbourne, 1984, pp. 184–197.
37. Dye, M. H., and Wheeler, P. J. Wound dressings for the prevention of silver leaf in fruit trees caused by *Stereum purpureum* (Pers.) Fr. *New Zealand J. Agric. Res.*, *11*:874–882, 1968.
38. Marty, R. The economic impact of *Hypoxylon* canker on the Lake States resource. *U.S. For. Serv. Gen. Tech. Rep.* N.C.I., 21–26, 1972.
39. Anderson, N. A., Ostry, M. E., and Anderson, G. W. Hypoxylon canker of aspen associated with *Saperda inornata* galls. *U.S. For. Serv. Res. Note*, NC-214, 1–3, 1976.
40. Manion, P. D. Two infection sites of *Hypoxylon mammatum* in trembling aspen (*Populus tremuloides*). *Can. J. Bot.*, *53*:2621–2624, 1975.
41. Ostry, M. E., and Anderson, N. A. Infection of trembling aspen by *Hypoxylon mammatum* through Circada oviposition wounds. *Phytopathology*, *73*:1092–1096, 1983.
42. Rayner, A. D. M., and Boddy, L. Population structure and the infection biology of wood decay fungi in living trees. *Adv. Pl. Pathol.*, 5:119–160, 1986.

43. Peace, T. R. *Pathology of Trees and Shrubs.* Clarendon Press, Oxford, 1962.
44. Bray, J. R., and Gorham, E. Litter production in forests of the world. *Adv. Ecol. Res.*, *2*:101–157, 1964.
45. Boddy, L., and Swift, M. J. Wood decomposition in an abandoned beech and oak coppiced woodland in south-east England. I. Patterns of wood-litter fall. *Holarctic Ecol.*, *6*:320–332, 1983.
46. Harmon, M. E., Franklin, J. F., Swanson, F. J., Sollins, P., Gregory, S. V., Lattin, J. D., Anderson, N. H., Cline, S. P., Aumen, N. G., Sedell, J. R., Lienkaemper, G. W., Cromack, K. Jr., and Cummins, K. W. Ecology of coarse woody debris in temperate ecosystems. *Adv. Ecol. Res.*, *15*:133–302, 1986.
47. Swift, M. J., Healey, I. N., Hibberd, J. K., Sykes, J. M., Bampoe, V., and Nesbitt, M. E. The decomposition of branch-wood in the canopy and floor of a mixed deciduous woodland. *Oecologia*, *26*:139–149, 1976.
48. Christensen, O. The state of decay of woody litter determined by relative density. *Oikos*, *42*:211–219, 1984.
49. Swift, M. J. The ecology of wood decomposition. *Sci. Prog. Oxf.*, *64*:179–203, 1977.
50. Vogt, K. A., Grier, C. C., and Vogt, D. J. Production, turnover and nutrient dynamics of above- and below-ground detritus of world forests. *Adv. Ecol. Res.*, *15*:303–377, 1986.
51. Healey, I. N., and Swift, M. J. Aspects of the accumulation and decomposition of wood in the litter of a coppiced beech-oak woodland. In *Organismes du sol et production primarie, Proc. 4th Coll. Soil Zool.*, INRA, Paris, 1971, pp. 417–430.
52. Christensen, O. Estimation of standing crop and turnover of dead wood in a Danish oak forest. *Oikos*, *28*:177–186, 1977.
53. Swift, M. J., Boddy, L., and Healey, I. N. Wood decomposition in an abandoned beech and oak coppiced woodland in south-east England. II. The standing crop of wood on the forest floor with particular reference to *Tipula flavolineata* and other animals. *Holarctic Ecol.*, *7*:218–228, 1984.
54. Grier, C. C. A *Tsuga heterophylla–Picea sitchensis* ecosystem of coastal Oregon: Decomposition and nutrient balances of fallen logs. *Can. J. For. Res.*, *8*:198–200, 1978.
55. Spaulding, P., and Hansbrough, J. R. Decay of logging slash in the northeast. *U.S. Dept. Agric. Tech. Bull.*, *876*, 1944.
56. Gosz, J. R., Likens, G. E., and Bormann, F. H. Nutrient release from decomposing leaf and branch litter in the Hubbard Brook Forest, New Hampshire. *Ecol. Monogr.*, *43*:173–191, 1973.
57. Boddy, L. The micro-environment of basidiomycete mycelia in temperate deciduous woodlands. In *The Ecology and Physiology of the Fungal Mycelium* (D. H. Jennings and A. D. M. Rayner, eds.), Cambridge University Press, Cambridge, 1984, pp. 261–289.
58. Chapela, I. H., Boddy, L., and Rayner, A. D. M. Structure and development of fungal communities in beech logs four and a half years after felling. *FEMS Microbiol. Ecol.*, *53*:59–70, 1988.
59. Dighton, J., and Boddy, L. Role of fungi in nitrogen, phosphorus and sulphur cycling in temperate forest ecosystems. In *Fungi*,

Nitrogen, Phosphorus and Sulphur (L. Boddy, R. Marchant, and D. J. Read, eds.), Cambridge University Press, Cambridge, 1989, pp. 269–298.

60. Ausmus, B. Regulation of wood decomposition rates by arthropod and annelid populations. In *Soil Organisms as Components of Ecosystems* (U. Lohm and T. Persson, eds.), *Ecol. Bull.*, *25*:180–192. Stockholm, Swedish Natural Science Research Council, 1977.
61. Swift, M. J. The role of fungi and animals in the immobilisation and release of nutrient elements from decomposing branch-wood. In *Soil Organisms as Components of Ecosystems* (U. Lohm and T. Persson, eds.), *Ecol. Bull.*, *25*:193–202. Stockholm, Sedish Natural Science Research Council, 1977.
62. Ovington, J. D. Quantitative ecology and the woodland ecosystem concept. *Adv. Ecol. Res.*, *1*:103–192, 1962.
63. Duvigneaud, P., and Denaeyer-De Smet, S. Biological cycling of minerals in temperate deciduous forests. In *Analysis of Temperate Forest Ecosystems* (D. E. Reichle, ed.), Springer-Verlag, New York, 1970, pp. 199–225.
64. Bormann, F. H., and Likens, G. E. *Pattern and Process in a Forested Ecosystem.* Springer, New York, 1979.
65. Heal, O. W. Decomposition and nutrient release in even-aged plantations. In *The Ecology of Even-Aged Plantations* (E. D. Ford, D. C. Malcolm, and J. Atterson, eds.), Institute of Terrestrial Ecology, Cambridge, 1979, pp. 257–291.
66. Heal, O. W., Swift, M. J., and Anderson, J. M. Nitrogen cycling in United Kingdom forests: The relevance of basic ecological research. *Phil. Trans. R. Soc. Lond. B.*, *296*:427–444, 1981.
67. Swank, W. T., and Waide, J. B. Interpretation of nutrient cycling research in a management context: Evaluating potential effects of alternative strategies on site productivity. In *Forests: Fresh Perspectives from Ecosystem Analysis* (R. Waring, ed.), Oregon State University Press, 1980, pp. 137–158.
68. Cole, D. W., and Rapp, M. Elemental cycling in forest ecosystems. In *Dynamic Properties of Forest Ecosystems* (D. E. Reichle, ed.), Cambridge University Press, Cambridge, 1981, pp. 341–409.
69. Clark, F. E., and Rosswall, T. (eds.). *Terrestrial Nitrogen Cycles, Ecol. Bull. Stockh. 33*, 1981.
70. Swift, M. J., Heal, O. W., and Anderson, J. M. *Decomposition in Terrestrial Ecosystems.* Blackwell Scientific, Oxford, 1979.
71. Schnitzer, M., and Khan, S. U. *Humic Substances in the Environment.* Marcel Dekker, New York, 1972.
72. Haider, K., Martin, J. P., and Filip, Z. Humus biochemistry. In *Soil Biochemistry* (E. A. Paul and A. D. McLaren, eds.), Marcel Dekker, New York, 1975, pp. 195–244.
73. Allison, F. E. *Soil Organic Matter and Its Role in Crop Production.* Elsevier Scientific Pub. Co., New York, 1973.
74. Hintikka, V. Studies on white-rot humus formed by higher fungi in forest soils. *Comm. Inst. For. Fenn.*, *69*:1–68, 1970.
75. Thies, W. G., and Russell, K. W. Controlling root rots in coniferous forests of northwestern North America. In *Proceedings of the Sixth International Conference on Root and Butt Rots of Forest Trees* (G. A. Kile, ed.), CSIRO, Melbourne, 1984, pp. 379–386.

76. Van der Pas, J. B., and Hood, I. A. The effect of site preparation on the incidence of *Armillaria* root rot in *Pinus radiata* four years after conversion from indigenous forest in Omatoroa Forest, New Zealand. In *Proceedings of the Sixth International Conference on Root and Butt Rots of Forest Trees* (G. A. Kile, ed.), CSIRO, Melbourne, 1984, pp. 387–397.
77. Thies, V. G., and Nelson, E. E. Control of *Phellinus weirii* in Douglas-fir stumps by the fumigants chloropicrin, allyl alcohol, Vapam or Vorlex. *Can. J. For. Res.*, *12*:528–532, 1982.
78. Filip, G. M., and Roth, L. F. Stump injections with soil fumigants to eradicate *Armillariella mellea* from young-growth ponderosa pine killed by root rot. *C. J. For. Res.*, *7*:226–231, 1977.
79. Rishbeth, J. Control measures against *Fomes annosus* in Great Britain. Sect. 24, Congr. Int. Un. Forest Res. Org., Munich, 1967.
80. Mercer, P. C. Attitudes to pruning wounds. *Arbor. J.*, *3*:457–465, 1979.
81. Mercer, P. C. Phytotoxicity and fungitoxicity tests for tree wound plants. *Ann. Appl. Biol.*, *91*:199–202, 1979.
82. Bevercombe, G. P., and Rayner, A. D. M. Diamond-bark diseases of sycamore in Britain. *New Phytol.*, *86*:379–392, 1980.
83. Lonsdale, D. Some aspects of the pathology of environmentally stressed trees. *Yearbook International Dendrology Society, 1982*: 90–97, 1983.
84. Rishbeth, J. Stump protection against *Fomes annosus*. III. Inoculation with *Peniophora gigantea*. *Ann. Appl. Biol.*, *52*:63–77, 1963.
85. Kallio, T. Protection of spruce stumps against *Fomes annosus* (Fr.), Cooke by some wood-inhabiting fungi. *Acta For. Fenn.*, *117*:1–20, 1971.
86. Kallio, T., and Hallaksela, A.-M. Biological control of *Heterobasidion annosum* (Fr.) Bref. (*Fomes annosus*) in Finland. *Eur. J. For. Path.*, *9*:298–308, 1979.
87. Dowson, C. G., Rayner, A. D. M., and Boddy, L. Inoculation of mycelial cord-forming basidiomycetes into woodland soil and litter. I. Initial establishment. *New Phytol.*, *109*:335–341, 1988.
88. Dowson, C. G., Rayner, A. D. M., and Boddy, L. Inoculation of mycelial cord-forming basidiomycetes into woodland soil and litter. II. Resource capture and persistence. *New Phytol.*, *109*:343–349, 1988.
89. Hunter, T. Biological control as a method to be used against silver leaf. *Hort. Ind.*, Dec.:841, 1977.
90. Pottle, H. W., Shigo, A. L., and Blanchard, R. O. Biological control of wood hymenomycetes by *Trichoderma harzianum*. *Pl. Dis. Reptr.*, *61*:687–690, 1977.
91. Rishbeth, J. Chemical treatment and inoculation of hardwood stumps for control of *Armillaria mellea*. *Ann. Appl. Biol.*, *82*:57–70, 1976.
92. Rayner, A. D. M. Fungal colonization of hardwood stumps from natural sources. I. Non-basidiomycetes. *Trans. Br. Mycol. Soc.*, *69*:291–302, 1977.
93. Rayner, A. D. M. Fungal colonization of hardwood stumps from natural sources. II. Basidiomycetes. *Trans. Br. Mycol. Soc.*, *69*:303–312, 1977.

94. Rayner, A. D. M. Internal spread of fungi inoculated into hardwood stumps. *New Phytol.*, *82*:505–517, 1979.
95. Punter, D. The effects of stump treatments on fungal colonization of conifer stumps. Ph.D. thesis, University of Cambridge, 1963.
96. Boa, E. Edible stinkhorns? *The Mycologist*, *2*:107–108, 1988.
97. Parmasto, E. *Pleurotus citrinopileatus*, one of the favourites. *The Mycologist*, *1*:106–107, 1987.
98. Legg, A. The top twenty edible fungi. *The Mycologist*, *1*:109–125, 1987.
99. Royse, D. J., and Schisler, L. C. Mushrooms—their consumption, production and culture development. *Interdisciplinary Science Reviews*, *5*:324–332, 1980.
100. San Antonio, J. P. Cultivation of the shiitake mushroom. *Horticulture*, *16*:151–156, 1981.
101. Leatham, G. F. Cultivation of shiitake, the Japanese forest mushroom, on logs: A potential industry for the United States. *For. Prod. J.*, *32*:29–35, 1982.
102. Leatham, G. F. Growth and development of *Lentinus edodes* on a chemically defined medium. In *Developmental Biology of Higher Fungi* (D. Moore, L. A. Casselton, D. A. Wood, and J. C. Frankland, eds.), Cambridge University Press, Cambridge, 1985, pp. 403–427.
103. Lawrence, J. F. Host preference in Ciid beetles (Coleoptera: Ciidae) inhabiting the fruiting bodies of basidiomycetes in North America. *Bulletin of Museum of Comparative Zoology Harvard University*, *145*:163–212, 1967.
104. Conner, R. N., Miller, O. K., and Adkisson, C. S. Woodpecker dependence on trees infected by fungal heart rots. *Wilson Bulletin*, *88*:575–581, 1976.
105. Tattar, T. A., Shortle, W. C., and Rich, A. E. Sequence of microorganisms in constituents associated with discoloration and decay of sugar maples infected with *Fomes connatus*. *Phytopathology*, *61*:556–558, 1971.
106. Thomas, R. J. Ecological studies on the symbiosis of *Termitomyces* Heim with Nigerian Macrotermitinae. Ph.D. thesis, University of London, 1981.
107. Fisher, R. C. Studies of the biology of the death-watch beetle (*Xestobium rufovillosum* de Geer). IV. The effect of type and extent of fungal decay in timber upon the rate of development of the insect. *Ann. Appl. Biol.*, *28*:244–260, 1941.
108. Kovoor, J. Modifications chimiques provoquées par un termitide (*Microcerotermes edentatus* Was.) dans du bois de peuplier sain ou partiellement dégradé par des champignons. *Bulletin biologique de la France et de la Belgique*, *98*:491–510, 1964.
109. Ritter, F. J., and Coenen-Saraber, C. M. A. Food attractants and a pheromone as trail-following substances for the Saintoge termite. *Experimental and Applied Entomology*, *12*:611–622, 1969.
110. Amburgey, T. L., and Smythe, R. V. Factors influencing termite feeding on brown-rotted wood. *Sociobiology*, *3*:3–12, 1977.
111. Amburgey, T. L., and Smythe, R. V. Factors influencing the production of termite trail-following and arrestant stimuli by isolates of *Gloephyllum trabeum*. *Sociobiology*, *3*:13–25, 1977.

112. Becker, G. Versuche über den Einfluss von Braunfaulepilzen auf Wahl und Ausnutzung der Holznährung durch Termiten. *Material und Organismen, 1*:95–156, 1965.
113. Schultze-Dewitz, G., and Unger, W. Das Verhalten von *Reticulitermes lucifugus var. santonensis* de Feyteaud gegenüber Weibfaulem. *Beitrage zur Entomologie, 22*:487–490, 1972.
114. Boddy, L., Coates, D., and Rayner, A. D. M. Attraction of fungus gnats to zones of intrapsecific antagonism on agar plates. *Trans. Br. Mycol. Soc., 81*:149–151, 1983.
115. Whittaker, R. H. *Communities and Ecosystems*. Collier-MacMillan, London, 1975.
116. Rodin, L. E., and Bazilevič, N. I. *Production and Mineral Cycling in Terrestrial Vegetation*. Oliver & Boyd, Edinburgh, 1967.
117. Rodin, L. E., and Basilevič, N. I. World distribution of plant biomass. In *Functioning of Terrestrial Ecosystems at the Primary Production Level* (F. E. Eckardt, ed.), UNESCO, Paris, 1968, pp. 45–52.
118. Cartwright, K. St. G., and Findlay, W. P. K. *Decay of Timber and its Prevention*. HMSO, London, 1958.
119. Berry, F. H., and Beaton, J. A. Decay in oak in the central hardwood region. *USDA For. Res. Paper*, NE-242, 1972.
120. Allen, S. E., Grimshaw, H. M., Parkinson, J. A., and Quarmby, C. *Chemical Analysis of Ecological Materials*. Blackwell Scientific Publications, Oxford, 1974.
121. Stenlid, J. Biochemical and ecological aspects of the infection. *Biology of Heterobasidium annosum*. Sveriges Lantbruksuniversitet, Uppsala, 1986.
122. Dowson, C. G., Rayner, A. D. M., and Boddy, L. The form and outcome of mycelial interactions involving cord-forming decomposer basidiomycetes in homogenous and heterogenous environments. *New Phytol., 109*:423–432, 1988.

19

FUNGAL DEGRADATION OF PESTICIDES

U. D. SINGH and N. SETHUNATHAN *Central Rice Research Institute, Cuttack, India*

K. RAGHU *Nuclear Agriculture Division, Bhabha Atomic Research Centre, Bombay, India*

I. INTRODUCTION

Pesticides are used extensively for protection of food, feed, and fiber in agriculture and humans and cattle in public health. The fate and significance of pesticides in the environment have been the subject of worldwide research during the last three decades because of the concern about the side effects of many pesticides, recalcitrant organochlorines in particular, and their metabolites on the nontarget microorganisms. The fate of pesticides in the environment is influenced by several factors: physical, chemical, and biological. Biodegradation is probably the major or the only means of detoxification for many pesticides in the environment as highlighted in several reviews [1–8].

Biological transformations of pesticides in the complex and dynamic soil system are effected by bacteria, fungi, and actinomycetes. According to estimates of microbial biomass in various soils often by dissimilar methods, bacteria account for 0.01–0.40% (100–4000 kg/ha soil) of the total soil biomass, and fungi account for 0.05–0.5% (500–5000 kg/ha soil) [9]. Although in most aerated soils fungi, like bacteria, constitute a large part of the microbial biomass, the role of fungi in the detoxification of pesticides in the environment is less understood than that of bacteria. Available evidence indicates that fungi whose generic composition and size of the flora vary with soil type, pH, temperature, moisture, aeration, organic matter content, and other factors participate in the degradation of pesticides. In anaerobic environments, such as flooded soils and lake sediments, the role of fungi in the degradation of pesticides may not be of great significance.

Fungi, in general, are heterotrophic in nutrition, and neither sunlight nor oxidation of inorganic substances provides energy for growth of these microorganisms. The fungal distribution in many cases is determined by the availability of oxidizable carbonaceous substrates [9]. Initial research with microorganisms including fungi on pesticide metabolism was characterized by isolating microorganisms capable of using a compound as their only source of both carbon and energy, but later it was realized that other mechanisms like co-metabolic transformation, conjugation reactions, or the mere accumulation of a pesticide within a microorganism are important factors of microbial transformations. The complete degradation of a pesticide molecule to its inorganic parts or its fermentation into components that can be eventually used in an oxidative cycle like the Krebs cycle removes its potential toxicity from the environment [6]. Bollag [6] suggested four major possibilities for transformation or inactivation of a pesticide by microorganisms: (1) the pesticide is used as a substrate and energy, (2) the pesticide can undergo co-metabolism, i.e., organisms transform it but cannot derive energy for growth from it, (3) the entire pesticide molecule or its intermediate can be conjugated with naturally occurring compounds, and (4) the pesticide is incorporated and accumulated within the organism.

The term "co-metabolism" is also defined as transformation of a non-growth substrate in the obligate presence of a growth substrate or another transformable compound [10]. The nature and significance of microbial co-metabolism of xenobiotics has recently been reviewed by Janke and Fritsche [11]. Co-metabolic transformations seem to be common and widespread in natural habitats. Most transformations mediated by fungi probably involve co-metabolism.

Microbial transformations are catalyzed by specific enzymes: constitutive or adaptive. The principal biochemical reactions in the degradation of pesticides by fungi include alkylation, dealkylation, amide or ester hydrolysis, dehalogenation, dehydrogenation, oxidation, reduction, hydroxylation, ring cleavage, ether cleavage, condensation, and conjugate formation.

This chapter highlights the progress made in studies on the fungal degradation of selected pesticides of different groups.

II. FUNGICIDES

Fungicides probably rank next to insecticides and herbicides in terms of the amount of pesticides used globally. Fungicides are used to treat soilborne, seedborne, and airborne plant diseases caused by fungi. Hence, fungicides designed to control plant pathogenic fungi may be toxic not only to other fungi but also to other members of microbial community in the soil. Biological as well as nonbiological factors affect the detoxification of fungicides in soil and on plant surface. Fungicides are meant to act against various fungi, but a few fungi have been found to metabolize or detoxify certain fungicides. The role of fungi in the metabolism of selected fungicides is discussed in this section.

A. Organomercurial Compounds

Organomercurials have been used as fungicides for the control of plant diseases since the beginning of this century. However, there is a steady decline in their use, because of the high toxicity and environmental accumulation properties of these organomercurials.

Detoxification of organomercurials by microorganisms has been demonstrated only in a few cases. Munnecke and his group have provided positive evidence of microbial degradation of Semesan (2-chloro-4-hydroxymercuriphenol), Panodrench (cyanomethylmercuriguanidine), and Panogen (methyl mercury dicyandiamide). When Semesan was applied as an aqueous suspension to nonsterile soil, it was found to be inactivated after 2–3 weeks, whereas its activity persisted for some months in steam-sterilized soil [12]. It was also found that several species of *Penicillium, Aspergillus*, and *Trichoderma* isolated from Semesan-treated soil could grow satisfactorily on potato dextrose agar containing up to 10 ppm of this fungicide, and more significantly that Semesan was rapidly inactivated in soil that had been incubated with *Penicillium* spp. and *Trichoderma viride* [13,14]. The two other mercury compounds were unaffected. In these studies, no information regarding metabolic pathways or degradation products was offered, but the authors suggested that detoxification of the fungicide occurred by a three-step process: initial uptake of toxicant (sorption) by the fungal protoplast, metabolic breakdown of the toxicant, and finally possible utilization of suitable compounds for growth and energy purposes.

B. Organosulfur Compounds

Dithiocarbamates

On a worldwide basis [15] dithiocarbamates constitute the most important class of organic fungicides used for control of plant diseases. Two major groups within this class, monoalkyl and dialkyl derivatives, differ in their mode of action and degradation pathways [16].

Monoalkyldithiocarbamate: Fungicides in this group include maneb, zineb, mancozeb, metiram, and metham-sodium. The water-soluble compound nabam is not used as a practical fungicide but has served as a useful model for studying the action and degradation of ethylenebisdithiocarbamate (EBDC) fungicides.

Soil degradation products of nabam (disodium ethylenebisdithiocarbamate) include ethylenethiourea (ETU); 5,6-dihydro-3H-imidazo [2,1-C]-1,2,4-dithiazole-3-thione (DIDT); carbon disulfide (CS_2); hydrogen sulfide (H_2S); and carbonyl sulfide (COS) [17–19]. Although the major degradation product ETU was chemically produced from EBDC fungicides, microorganisms accelerated the formation of ETU from the intermediate breakdown product DIDT. Thus, *Aspergillus niger, Saccharomyces cerevisae*, and two bacteria readily formed ETU from DIDT [20]. Also, the conversion took place in the presence of reducing compounds like cystein, glutathione, or ascorbic acid and in enzyme extracts from *Pseudomonas fluorescens* and *Aspergillus niger* and in commercial glutathione reductase or lipoamide dehydrogenase.

Dialkyldithiocarbamate: Major agricultural fungicides of this group include the water-soluble derivatives ziram, ferbam, and thiram. Sodium dimethyl dithiocarbamate (NaDDC), a fungitoxic water-soluble salt, has no practical agricultural application but is frequently used in studies on mode of action or transformation. Microorganisms have been shown to behave differently in the degradation of sodium dimethyldithiocarbamate. Thus, NaDDC was converted to α-aminobutyric acid, not only by washed cell suspensions of *Saccharomyces cerevisiae, Hansenula anamola*, and *Bacterium coli*, but also by mycelial pellets of *Glomerella cingulata, Aspergillus niger*, and *Cladosporium cucumerinum* [21]. There is considerable literature on the persistence and degradation of thiram (tetramethylthiuram disulfide) in soil and in microbial cultures. Sisler and Cox [22] reported the decomposition of thiram to carbon disulfide by three soil fungi, *Penicillium notatum*, *Glomerella cingulata*, and *Fusarium roseum*. Thiram was degraded more rapidly in nonsterile soils than in sterile soils [23–26]. Persistence of thiram was directly related to initial concentration, and the rate of decomposition increased with decreasing concentration. In addition to a bacterium, *Pseudomonas fluorescens* [25], two pathogenic fungi, *Pythium ultimum* [23] and *Rhizoctonia solani* [26], were able to detoxify thiram in soil.

Substituted Phthalimides

Captan [*N*-(trichloromethylthio)-3a, 4,7,7a-tetrahydrophthalimide] is a substituted phthalimide fungicide used extensively in soil applications to control a broad spectrum of plant pathogenic fungi. Several studies have shown that captan is rapidly decomposed in direct reaction with thiol constituents of fungal cell [27–31]. Richmond and Somers [28] reported the complete decomposition of ^{35}S-labeled captan by the spores of *Neurospora crassa*, while in presence of yeast (*Saccharomyces pastorianus*) cells, significant amounts of tetrahydrophthalimide, thiophosgene carbonyl sulfide, and hydrogen disulfide were released from captan and folpet [30–32].

C. Organophosphorus Compounds

Pyrazophos

Pyrazophos [*O*-6-ethoxy-carbonyl-5-methylpyrazolo-(1,5-a)-pyrimidine 2-yl *O,O*-diethyl phosphorothioate] is a systemic fungicidal compound used for the excellent control of powdery mildews. In a comprehensive study de Waard [33] found that the fungus *Pyricularia oryzae*, but not *Pythium debaryanum* and *Saccharomyces cerevisiae*, converted pyrazophos into 6-ethoxy-carbonyl-2-hydroxy-5-methylpyrazolo [1,5-a] pyrimidine (PP) and the oxygen analog of pyrazophos (PO-pyrazophos).

Kitazin P, Edifenphos, and Inezin

Kitazin P, edifenphos (Hinosan), Inezin, and a number of related organic phosphate esters active against the rice blast pathogen, *Pyricularia oryzae*, have been developed in Japan.

The metabolism of edifenphos (*O*-ethyl-*S,S*-diphenyl phosphorodithioate) by mycelial cells of *Pyricularia oryzae* was studied by Uesugi and Tomizawa [34]. The fungicide undergoes des-*S*-phenylation involving one P-S linkage forming ethyl-*S*-phenyl hydrogen phosphorothioate. This is further degraded to ethyl dihydrogenphosphate and orthophosphoric acid. The other metabolite formed, namely, *O*-ethyl-*S*-(*p*-hydroxyphenyl)-*S*-phenyl phosphorodithioate, also undergoes des-*S*-phenylation to yield the same product as edifenphos. The possible degradation pathway by this fungus is shown in Figure 1.

Similar routes operated in the degradation of both Kitazin P and Inezin. An interesting difference from edifenphos is, however, that hydroxylation by *P. oryzae* occurs at the *meta* position instead of at the *para* position [35].

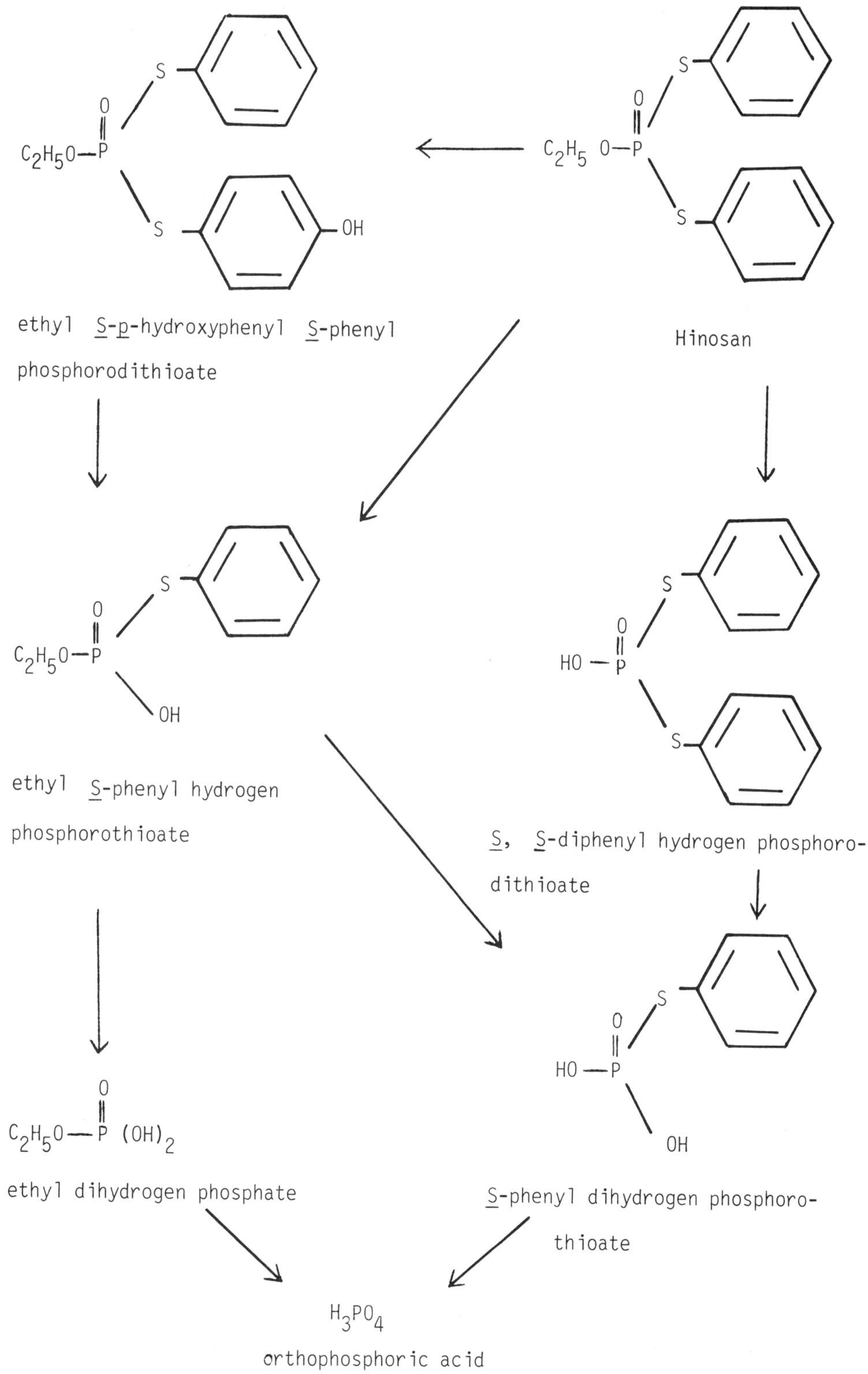

FIGURE 1. Metabolism of edifenphos by *Pyricularia oryzae*. (From Ref. 34.)

D. Aromatic Compounds

Quintozene

Quintozene (pentachloronitrobenzene, PCNB) is used as a seed and soil treatment to control several soilborne fungi. Among the organic fungicides, quintozene is the most stable, persisting for several months to one year.

A variety of fungi and actinomycetes in pure culture convert PCNB to PCA (pentachloroaniline) and PCTA (pentachlorothioanisole) [36,37]. Quintozene is reduced to PCA by a large number of soil microorganisms [1]. Nakanishi and Oku [37] confirmed the greater uptake of quintozene by sensitive fungi such as *Rhizoctonia solani* and showed that the fungicide added at 25 ppm to a liquid culture of *Fusarium oxysporum* was not only reduced to PCA but was also converted to pentachloromethyl thiobenzene.

Chloroneb

Chloroneb (1,4-dichloro-2,5-dimethoxybenzene) is highly effective against seed and/or soilborne fungal pathogens such as *Pythium* spp. and *Rhizoctonia* spp.

Hock and Sisler [38] found that *Rhizoctonia solani* demethylated chloroneb to a nontoxic metabolite 2,5-dichloro-4-methoxyphenol (DCMP). Further, Wiese and Vargas [39,40] demonstrated that some fungi (*Fusarium* spp. are most active) demethylated chloroneb to DCMP, whereas others (*Trichoderma viride* and *Mucor ramannianus*) converted DCMP to chloroneb. *Cephalosporium gramineum, Rhizoctonia solani, Mucor ramannianus*, and various species of *Fusarium* were capable of both methylation and demethylation. Thus, microbial resynthesis of chloroneb from the degradation product DCMP may account for the relative stability of chloroneb in the soil.

E. Heterocyclic Compounds

Benzimidazoles

Benomyl is a systemic fungicide extensively used in agriculture for control of a large array of plant pathogenic fungi. Carbendazim, the hydrolysis product of benomyl, is also an important systemic fungicide used against a wide range of fungal diseases.

Information concerning microbial degradation of benomyl [methyl 1-(butylcarbamoyl) benzimidazole-2-ylcarbamate] or carbendazim (methyl benzimidazole-2-ylcarbamate) is very limited. Microorganisms have been implicated in the degradation of benomyl and carbendazim based on relative persistence in sterile and nonsterile soil systems [41–43]. Microorganisms, isolated from treated soil degraded benomyl or carbendazim [41,44] to 2-aminobenzimidazole and 5-hydroxy-carbendazim. The latter compound (Fig. 2) was detected during the metabolism of benomyl by *Pellicularia sasakii, Alternaria mali* [45], and *Aspergillus nidulans* [46]. Gupta and Chatrath [47] reported the accelerated degradation of ^{14}C-labeled benomyl in sterile soil amended with *Fusarium solani*. Recently, Yarden et al. [48] isolated *Alternaria alternata* and *Bipolaris tetramera* from soil capable of degrading carbendazim.

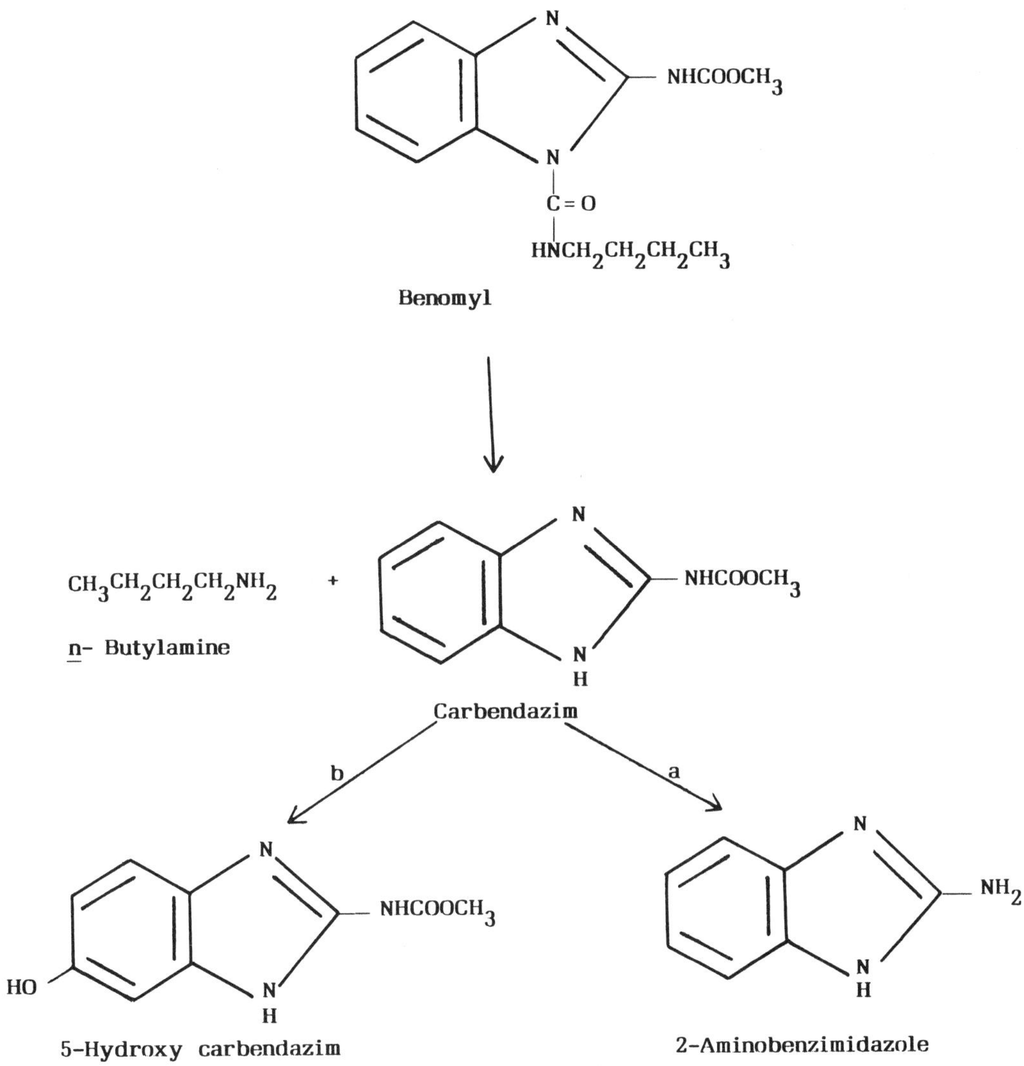

FIGURE 2. Metabolism of benomyl in soil (a) and by fungi (b). (From Ref. 187.)

Carboxylic Acid Anilides

Carboxin and mebenil are the oxathiin systemic fungicides of this group. The oxathiins are more susceptible to conversion than the benzimidazole-derived fungicides. Carboxin (2,3-dihydro-6-methyl-5-phenylcarbamoyl-1,4-oxathiin) is readily oxidized to the nonfungitoxic *S*-oxide (carboxin sulfoxide). Of several microorganisms tested, only *Ustilago maydis* could form the sulfoxide from carboxin [49]. Also, isolated mitochondria of this fungus could oxidize carboxin.

Mebenil (2-methylbenzanilide) is transformed into its *p*-hydroxyphenyl derivative by the pure cultures of *Rhizopus japonicus* [50]. In contrast, the methyl groups of the related compounds 2,5-dimethyl-3-furanilide are hydroxylated by several phycomycetous fungi, including *Rhizopus japonicus* and *Mucor* spp. [51].

Triazolylmethane

Triadimefon, triadimenol, and diclobutrazole are the important systemic fungicides of this group, having both protective and curative action. Using a replacement culture technique and ^{14}C substrates, Deas and Clifford [52] demonstrated that the three compounds were degraded by *Aspergillus niger* involving a common pathway, namely, oxidation of $-C(CH_3)_3$ to $(CH_3)_2CH_2OH$. The fungus also reduced triadimefon to triadimenol. Authors implicated both oxidative and reductive pathways in the degradation of these fungicides by *A. niger*.

III. HERBICIDES

Herbicides account for 45% of all pesticides used in developed countries, followed by insecticides (35%) and fungicides (15%). In most developing countries of the tropics and subtropics, insecticides (75%) are the most important, followed by fungicides (15%) and herbicides (7%). Hence, the majority of literature on herbicide–soil microflora interactions has been generated mostly from studies in developed countries. Yet, of the approximately 150 chemicals used as herbicides throughout the world, complete degradation pathways are known only for a few compounds. Metabolic pathways for the majority of herbicides have been only partially characterized. Herbicides are degraded in the soil by biological and nonbiological means, almost always working in conjunction with one another. Probably the very first evidence for a microbial role in the degradation of a pesticide stemmed from studies with a herbicide 2,4-D [53]. Since then a vast array of microorganisms belonging to fungi, bacteria, and actinomycetes have been implicated in the detoxification of herbicides in the soil and water environments. The role of fungi in the degradation of selected herbicides of different groups is reviewed in this section.

A. Phenoxyalkanoates

Phenoxyalkanoic acid herbicides are degraded by many bacterial and fungal isolates obtained from soil. Woodcock and his group have convincingly demonstrated the role of *Aspergillus niger* in the microbial degradation of various phenoxyalkanoates. Byrde and Woodcock [54] reported the ring hydroxylation of omega-substituted phenoxyalkanoates by *A. niger* in the *ortho* and *para* positions. Thus, phenoxyacetic acid itself was converted into a mixture of the 2-, 3-, and 4-hydroxyisomers [55]. Further, Faulkner and Woodcock [56] obtained an additional product 2-hydroxy-phenoxyacetate from 2-chlorophenoxyacetate and indicated that *A. niger* was also capable of hydrolytic dechlorination. Degradation products obtained from 2,4-D metabolism by *A. niger* were 2,5-dichloro-4-hydroxy-phenoxyacetate and 5-hydroxy derivatives [57]. The formation of phenoxyacetate indicated that the fungus effected a shift in position of a chlorine atom during ring hydroxylation. *Aspergillus niger* also cleaved the side chain (β-oxidative attack of side chain) and effected ring hydroxylation of these compounds. Another fungus, *Phytophthora megasperma* var. *sojae*, was not capable of effecting β-oxidation of *n*-butylester of 2,4-D (2,4-DB). Moreover, 3–5 times more 2,4-DB than 2,4-D was absorbed and accumulated by this fungus [58]. Recently, Sahasrabudhe and Modi [59] reported that a strain of *A. niger* was capable of dechlorinating 2,4-D prior to ring cleavage. Also, mycelial pellets of this fungus immobilized on calcium alginate dechlorinated 2,4-D for five successive cycles.

Chlorophenols are major intermediates of phenoxyalkanoate herbicides and other pesticides. They can easily form soilbound residues. Wolf and Martin [60] showed that ^{14}C-ring-labeled 2,4-D and chlorpropham (isopropyl-3-chlorophenyl-carbamate) or their transformation products were incorporated into humic-like polymers produced by fungi. An extracellular enzyme laccase from *Rhizoctonia praticola* polymerized syringic acid products and various chlorophenols [61,62]. When the laccase was incubated together with the halogenated phenols and syringic acid, two types of hybrid products were found: (1) phenols covalently bound to an orthoquinone product of syringic acid resulting in the formation of quinonoid oligomers, and (2) phenols covalently bound to decarboxylated products of syringic acid resulting in the formation of phenolic oligomers. It was concluded that all hybrid products contained only one halogenated phenol molecule and that no dehalogenation took place.

B. Phenylamides

A majority of the herbicides used in agriculture belongs to this phenylamide class. This class has been subdivided into the acylanilides, the phenylureas, and the phenylcarbamates. A common feature of all the phenylamides is that their structures are based on substituted anilines.

Acylanilides

Among the acylanilides, propanil, Karsil, dicryl, alachlor, butachlor, propachlor, and diphenamide have been used widely in weed control. Certain soil fungi could hydrolyze almost all acylanilide herbicides with concomitant formation of corresponding anilines. Several fungi including *Fusarium solani* and species of *Penicillium* and *Pulullaria* use these herbicides as a sole source of carbon and energy [63]. These organisms possess an acylamidase, which cleaves these substrates forming an aniline and an aliphatic moiety.

Propanil: *Fusarium solani* growing on propanil [*N*-(3,4-dichlorophenyl) propionamide] utilized the propionic acid formed by hydrolysis and released 3,4-dichloroaniline [64,65]. Propanil was used as a sole source of carbon and energy by this fungus [64]. Bartha and Pramer [63] showed that

3,4-dichloroaniline (DCA) and large quantities of 3,3',4,4'-tetrachloroazobenzene (TCAB) were produced in propanil-treated soil. Kaufman et al. [66] reported the formation of TCAB by *Fusarium oxysporum* growing on ^{14}C-ring-labeled 3,4-dichloroaniline. It was metabolized by amino group reaction and besides TCAB, azobenzene, triazine, and 3,4-dichloronitrobenzene were also produced. They postulated that aniline metabolism proceeded by oxidation of the amino group with successive formation of phenylhydroxylamine, nitrosobenzene, and nitrobenzene (Fig. 3). Self-condensation of the hydroxylamine or condensation with the nitrosobenzene yields azoxybenzene, whereas reaction of unchanged aniline with hydroxylamine or nitrosobenzene yields azobenzene. Another fungus, *Aspergillus niger*, metabolizes 3,4-dichloroaniline to 3,4-dichloracetaniline [67].

Synergistic interaction of two soil fungi (*Penicillium piscarium* and *Geotrichum candidum*) led to sequential transformation of propanil to TCAB [68]. *Geotrichum candidum* possessed a high concentration of peroxidase, which mediated the formation of TCAB from 3,4-dichloroaniline [69]. Another enzyme of this organism, oxidase, catalyzed azobenzene formation. These two enzymes have been used as a model system for aniline metabolism in soil [70]. Recently, Pelsy et al. [71] characterized an enzyme isolated from *Aspergillus nidulans* which could hydrolyze propanil to 3,4-dichloroaniline. This enzyme was similar to the propanil-hydrolyzing enzyme present in *Fusarium solani*, rice, and mammal.

There is evidence that acylamidases induced by propanil in two species of *Fusarium* (*F. solani* and *F. oxysporum*) differed in their substrate specificity and sensitivity to inhibitors despite their close phylogenetic relationships. The acylamidase from *F. solani* was insensitive to parathion and carbaryl, but was competitively inhibited by propachlor [64,65]. In contrast, the *F. oxysporum* acylamidase was insensitive to propachlor, carbaryl, PCMC, diuran, and chlorporpham. Interestingly, the acylamidase of *F. oxysporum* induced by PCMC (*p*-chlorophenyl methylcarbamate) could not act on PCMC but was highly active in hydrolyzing propanil [72]. PCMC also acted as a potent inhibitor of the acylamidase it induced.

Karsil and Dicryl: Sharabi and Bordeleau [73] have shown the metabolism of Karsil [*N*-(3',4'-dichlorophenyl)-2-methylpentanamide] by two species of *Penicillium* and a *Pulullaria* sp. Primary products of Karsil hydrolysis by cells and cell-free extracts of *Penicillium* spp. were identified as 2-methylvaleric acid and 3,4-dichloroaniline. The fungus metabolized the 2-methylvaleric acid to formic and acetic acids and used them as a carbon source.

The hydrolytic enzyme acylamidase, obtained from extract of *Fusarium solani* [65], hydrolyzed several acylanilides including Karsil and dicryl [*N*-(3-4-dichlorophenyl) methacrylamide]. An acylamidase from *Penicillium* sp. [73] also hydrolyzed many acylanilides, but not Karsil and dicryl. Thus, the enzymes from these two fungi differed in their specificity to the substrate.

In contrast, *Rhizopus japonicus* detoxified acylanilides not by hydrolysis as noticed with many fungi, but by hydroxylation of the aliphatic portion of Karsil and dicryl [51,74].

Propachlor, Alachlor, Metolachlor, and Butachlor: There are many reports on the decomposition of the herbicide propachlor (2'-chloro-*N*-isopropylacetanilide) by certain fungi in model experiments. According to Kaufman et al. [75], *Fusarium oxysporum* released chloride from propachlor and accumulated 2'-hydroxy-*N*-isopropylacetanilide as the major metabolite.

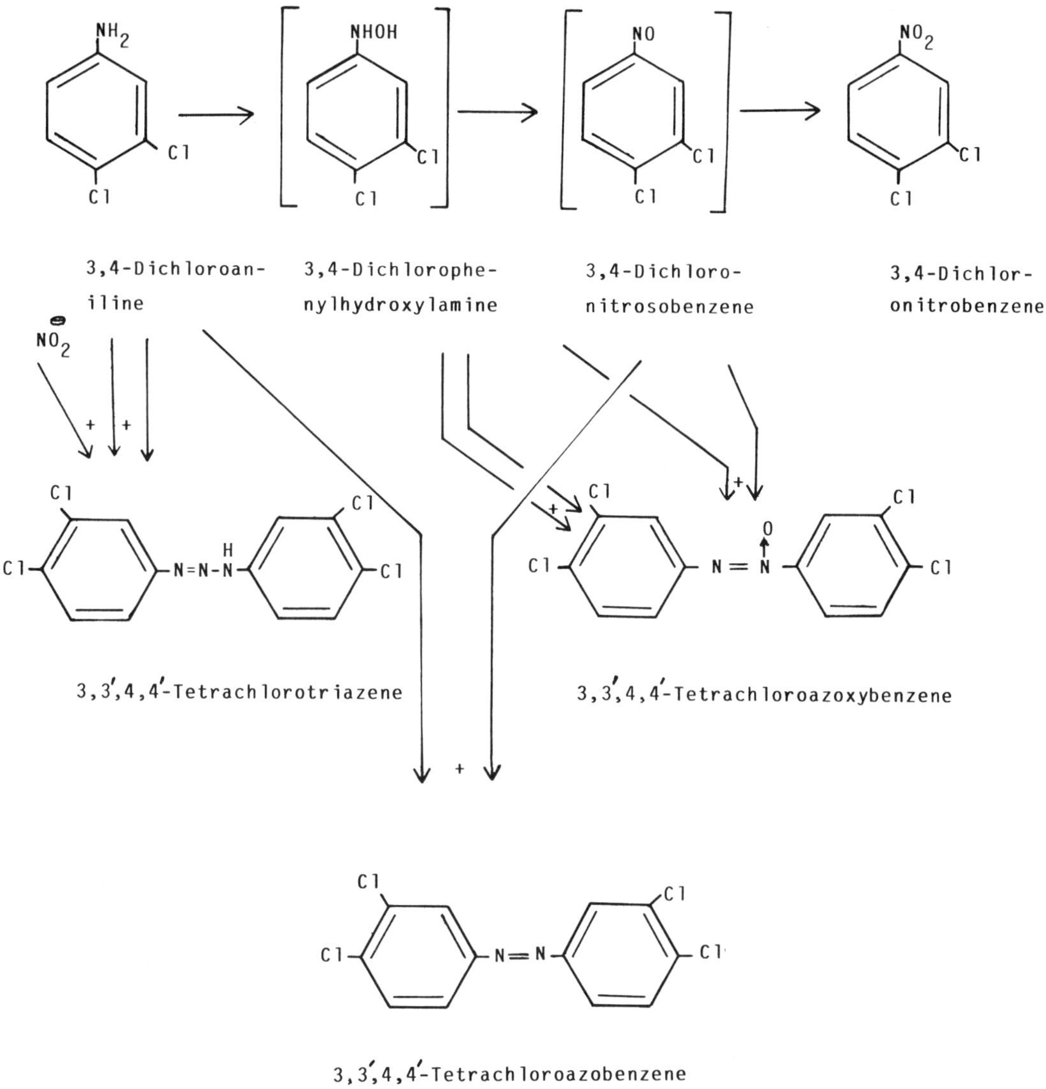

FIGURE 3. Metabolism of 3,4-dichloroaniline by *Fusarium oxysporum*. (From Ref. 187.)

Rankov and Valev [76] found that *Penicillium* (*P. lilacinum*, *P. cyclopium*, *P. purpurogenum*, and *P. sclerotiorum*), *Aspergillus* (*A. niger* and *A. flavus*), and *Trichoderma viride* accelerated the decomposition of the herbicide propachlor, leading to its detoxification.

Kaufman and Blake [77] reported the degradation of alachlor by *Fusarium oxysporum*, but no metabolites were identified. Tiedje and Hagedorn [78] demonstrated the production of chloride ions and four organic metabolites [2-chloro-2',6'-diethylacetanilide; 2,6-diethyl-*N*-(methoxymethyl)-aniline; 2,6-diethylaniline; and 1-chloroacetyl-2,3-dihydro-7-ethylindole] from alachlor by *Chaetomium globosum*. Similar metabolic studies were also conducted with Antor (H-22234) and Dual (metolachlor) herbicides using *Chaetomium globosum* in pure culture [79]. Aniline moiety was not released during degradation of metolachlor by *C. globosum*.

Chen and Wu [80] reported the conversion of butachlor [2-chloro-2',6'-diethyl-*N*-(butoxymethyl) acetanilide] by *Mucor sufui* NTU-358 to at least 8–12 metabolites. Some of these metabolites were identified: 2-chloro-2',6'-diethylacetanilide; 2-hydroxy-2',6'-diethylacetanilide; 2,6-diethylaniline; *N*-chloro acetyl-7-ethyl-2,3-dihydroindole; 2,6-diethylacetanilide; and *N*-methyl-2-chloro-2',6'-diethyl acetanilide. Lee [81] also found that butachlor was metabolized by *Chaetomium globosum* to more than 10 metabolites.

Phenylureas

There is considerable literature on the fungal degradation of herbicides belonging to the phenylurea group. Rankov [82] showed that some strains from *Aspergillus* (*A. niger* and *A. flavus*) and *Penicillium* (*P. notatum* and *P. clavigerum*) can use the herbicides monolinuron and linuron as a sole nitrogen source. *Neurospora crassa*, *Mucor* sp., *Aspergillus* sp., and *A. niger* utilized diuron and 3,4-dichloroaniline as carbon and/or nitrogen sources [83]. Schroeder [84] examined the ability of 81 fungi and 13 bacteria to decompose linuron, monolinuron, diuron, and monuron. *Aspergillus nidulans* was one of the most effective isolates and decomposed 52.8% of the linuron in culture. Geissbuhler et al. [85] also found *A. nidulans* to be the most effective isolate, which degraded more than 50% linuron [3-(3,4-dichlorophenyl)-1-methoxy-1-methylurea] in culture solution. Metabolism proceeds by successive removal of methyl groups followed by hydrolysis of phenylurea to the corresponding aniline. *Aspergillus niger* obtained from soil degraded linuron into 3,4-dichloroaniline by demethylation and subsequent hydrolysis of the amide bond [67]. Acetylation of the 3,4-dichloroaniline produced 3,4-dichloroacetaniline. Further, Glad et al. [86] reported the metabolism of linuron and its breakdown products by *Aspergillus niger*, *Trichoderma viride*, *Geotrichum candidum*, and two species of *Cladosporium herbarum* isolated from soil. Another herbicide monuron [3-(4-chlorophenyl)-1,1-dimethylurea] was converted to 4-(chlorophenyl)-urea and 3-(4-chlorophenyl)-1-1-methylurea by *Cunninghamella echinulata* [87].

The major mechanism of phenylurea detoxification involved dealkylation or *N*-dealkoxylation. *N*-methoxylation has also been described. Such reactions have been reported in vitro in pure cultures of fungi. For example, *Talaromyces wortmanii* converted metobromuron [3-(4-bromophenyl)-1-methoxy-1-methylurea] into demethylated and demethoxylated derivatives [88]. Other metabolites identified were *p*-linomophenylurea and *p*-bromoacetanilide. Similar reactions and degradation products were reported from the metabolism of linuron and monolinuron by *Aspergillus niger* [89] and chlorbromuron by *Rhizoctonia solani* [90,91].

Detailed investigations on the role of *Rhizoctonia solani* in the degradation of phenylurea herbicides have been conducted by Weinberger and Bollag [90]. Chlorbromuron [3-(4-bromo-3-chlorophenyl)-1-methoxy-1-methylurea] was metabolized to a demethylated product, 3-(3-chloro-4-bromophenyl)-1-methoxyurea. The demethylated herbicide was converted by the fungus to its demethoxylated product (the corresponding urea) and then to 3-chloro-4-bromoaniline, which subsequently disappeared from the culture medium. *Rhizoctonia solani* also attacked a large number of other phenylurea herbicides, including chloroxuron, diuron, fenuron, fluometuron, linuron, metabromuron, neburon, and siduron, despite the variations in their ring structure or functional groups.

Stepwise demethylation has been reported in the degradation of fluometuron [1-dimethyl-3-(3-trifluoromethylphenyl) urea] by *Rhizoctonia solani.* Two major metabolites were identified as 1-methyl-3-(α,α,α-trifluorotolyl) urea and (3-trifluoromethylphenyl) urea [92]. Ozaki and Hayase [93] reported the degradation of isouron by *R. solani* to 3-methylurea in liquid culture. This fungus also rapidly degraded isouron-related urea compounds by oxidative hydroxylation of the tert-butyl group at the isoxazole ring and demethylation of the urea side chain.

Phenylcarbamates

Phenylcarbamate herbicides such as barban, propham, chlorporpham, phemedipham, and swep are widely used in agriculture, because of their effectiveness at low levels of application and their general short-term persistence in soils.

Several studies indicate that the ability to hydrolyze phenylcarbamates is widespread among microorganisms. Kaufman and Blake [77] and McClure [94] have reported the growth of several species of fungi on swep (methyl-3,4-dichlorocarbanilate), propham (isopropyl phenylcarbamate), and chlorpropham (isopropyl 3-chlorophenylcarbamate). *Fusarium oxysporum* isolated from enrichment cultures from swep-treated soil released 3,4-dichloroaniline after a lag of 2–3 days [77]. Fletcher and Kaufman [95] detected 2-amino-4-chlorophenol in chlorporpham-treated soil and in cultures of *Fusarium oxysporum* as a metabolite past 3-chloroaniline. The fungus *Trichoderma viride* demethylated phenmedipham [3-methoxycarbonylaminophenyl *N*-(3'-methylphenyl) carbamate] but was unable to cleave the aromatic ring [96].

The major route of phenylcarbamate metabolism and detoxification is by hydrolysis to the free aniline, CO_2, and an alcohol. Wright and Forey [97] have demonstrated the role of *Penicillium jenseni* in the metabolism of barban (4-chlorobut-2-ynyl 3-chlorophenylcarbamate) via 3-chloroaniline. No chloride was released during the metabolism of 3-chloroaniline. The organism was unable to grow with barban or 3-chloroaniline as the sole carbon source. The barban-hydrolyzing enzyme, probably an amidase, was closely associated with the mycelial wall and was inhibited by *p*-chlorophenyl *N*-methylcarbamate (PCMC). Kaufman and Kearney [98] and Kaufman et al. [99] showed that certain methylcarbamates are competitive inhibitors of phenylcarbamate-hydrolyzing enzymes. The fact that barban hydrolysis in the *Penicillium* is also inhibited by a methylcarbamate emphasizes the similarity in microbial hydrolysis of phenylcarbamates.

C. S-Triazines

2-Chloro-s-triazine herbicides (simazine and atrazine) and 2-methylthio-s-triazine herbicides, such as simetryne, ametryne, and prometryne, have been extensively used for weed control. Numerous workers have demonstrated that s-triazines can be degraded by pure cultures of fungi and, in many cases, these compounds can act as nutrient sources for their growth. Simazine has been used as a sole or supplemental source of carbon for the growth of different species of *Aspergillus*, *Fusarium*, *Penicillium*, *Rhizopus*, and *Trichoderma* [100]. Bortels et al. [101] isolated several fungi which could use simazine only as a source of nitrogen. Recently, Shocken [102] studied the interaction of atrazine with marine higher fungi. Of the 47 isolates of marine fungi, several isolates utilized atrazine as a sole carbon or nitrogen source or mediated the removal of atrazine from the medium in shake culture in the presence of other utilizable carbon or nitrogen sources. The methylthio-s-triazine (prometryne) was used as a source of not only carbon or nitrogen but also sulfur by *Aspergillus niger*, *A. tamarii*, and *A. flavus* [103]. Wolf and Martin [104] reported that cyanuric acid (an intermediate breakdown product of s-triazines) can be completely degraded to CO_2 within 8 weeks by a soil fungus, *Stachybotrys chartarum*. Zeyer et al. [105] isolated a strain of *Sporothrix schenckii* from s-triazine herbicide-treated soils, which could degrade cyanuric acid. This fungus was able to use cyanuric acid as its sole nitrogen source, with a constant release of carbon dioxide. With the use of ^{15}N-labeled cyanuric acid, Myskow et al. [106] found that the nitrogen taken from this compound by *Aspergillus minutus* and *Pseudogymnoascus* sp. was incorporated into their proteins.

Extensive studies by Kaufman and Kearney [107] have demonstrated the role of *Aspergillus fumigatus* Freb, in the degradation of many herbicides of this group. The most common degradative mechanisms involved in the degradation of s-triazines by this fungus are dealkylation and deamination, but no ring cleavage was detected in fungal culture. This fungus can utilize simeton as the sole source of either carbon or nitrogen or both, but only the side chains of the molecule are degraded. Simazine and atrazine were also degraded by this fungus without ring cleavage [108,109]. Thus, two products were identified when this organism was grown in the presence of ^{14}C- or ^{36}Cl-labeled simazine (2-chloro-4,6-bis-ethylamino-1,3,5-triazine) [108]. One was identified as the dealkylated product 2-amino-4-chloro-6-ethylamino-s-triazine. Another metabolite, not identified, retained the chlorine atom and the intact ring structure but had lost both ethyl groups. This fungus also formed the ammelide (2,4-dihydroxy-6-amino-s-triazine) from simazine, indicating the possibility of deamination. With *A. fumigatus*, no release of chlorine atom from the chloro-s-triazine molecule has been observed. However, *Fusarium roseum* hydrolyzed atrazine (2-chloro-4-ethylamino-6-isopropylamino-1,3,5-triazine) to its corresponding hydroxy analog (2-ethylamino-4-hydroxy-6-isopropylamino-s-triazine). This indicates that microbial hydrolytic dechlorination can also occur [110]. Recently, Martinez and Castaneda [111] reported the breakdown of simazine by *Humicola* spp. and of simazine and atrazine by *Mucor* spp. in culture media inoculated with 1 g of soil.

Metamitron, a member of 1,2,4-triazinone herbicide, has been found to be degraded to desaminometamitron by *Cunninghamella echinulata* Thaxter and *Rhizopus japonicus*, by desamination [112]. Schilling et al. [113]

obtained complete transformation of metribuzin [4-amino-6-t-butyl-3-methylthio-1,2,4-triazin-5(4H)-one] by *Rhizopus japonicus* and *Cunninghamella echinulata* when incubated for 4 weeks. Degradation rate in *Aspergillus niger*, *Penicillium lilacinum*, and *Fusarium oxysporum* was less than 45%. Two major metabolites, desaminometribuzin and desaminodiketometribuzin, were produced during degradation of metribuzin by *C. echinulata*.

D. Dinitroanilines

Trifluralin (2,6-dinitro-*N*,*N*-dipropyl-4-trifluromethylaniline) is the most commonly used herbicide in the dinitroaniline series. It is a selective preemergence herbicide used for the control of grasses and broad-leaved weeds. Other dinitroanilines used as preemergence herbicides include benefin, nitralin, dinitramine, etc. Both aerobic and anaerobic degradation of these herbicides have been demonstrated [114]. However, Helling [115] suggested that as a group, the dinitroaniline herbicides are degraded more rapidly in aerobic than in anaerobic soil.

There are few reports on the involvement of fungi in the metabolism of dinitroaniline herbicides in vitro. Pure cultures of *Aspergillus fumigatus*, *Fusarium oxysporum*, and *Paecilomyces* sp. degraded dinitramine (*N'N'*-diethyl-2,6-dinitro-4-trifluromethyl-*m*-phenylenediamine) with the formation of three products, two of which were identified as the ethylamino and amino derivatives, and the third product was a benzimidazole derivative [116]. Cell-free extract of *A. fumigatus*, when supplemented with NADPH and ferrous ions, catalyzed the mono-dealkylation of dinitramine.

Kearney et al. [117] studied the metabolism of *N*-s-butyl-4-t-butyl-2,6-dinotroaniline (Amex) in soil and in a culture solution of the fungus *Paecilomyces* sp. The major metabolic products formed in soil and fungal extracts were identified as 4-t-butyl-2,6-dinitroaniline and 3-(4-t-butyl-2,6-dinitroaniline)-2-butanol, respectively.

Zayed et al. [118] investigated the microbial degradation of trifluralin by *Aspergillus carneus*, *Fusarium oxysporum*, and *Trichoderma viride*. The major metabolites were 2,6-dinitro-*N*-n-propyl-α,α,α-trifluoro-p-toluidine (III); 2,6-dinitro-α,α,α-trifluoro-p-toluidine (IV); 2-amino-6-nitro-α,α,α-trifluoro-p-toluidine (V); and 2,6-dinitro-4-trifluoromethyl phenol (VI). The degradation pathway (Fig. 4) involved initial oxidative dealkylation of the propyl groups followed by reduction of the nitro group(s).

E. Miscellaneous Herbicides

Certain herbicides that do not fall within the general classes discussed so far and susceptible to fungal degradation are discussed here.

Thiolcarbamates

S-*Ethyl* N,N-*Dipropylthiocarbamate*: EPTC (*S*-ethyl *N*,*N*-dipropylthiocarbamate) is the first of a series of thiolcarbamates developed as herbicides by Stauffer Chemical U.S.A. EPTC persisted longer in autoclaved soil than in nonautoclaved soil [119]. Mineralization of EPTC to $^{14}CO_2$ was more rapid in retreated soils than in soils not exposed to EPTC before [120,121] indicating enrichment of microorganisms capable of degrading EPTC. But the isolation and identification of EPTC degraders from soil has not been reported by these workers. Using an enrichment technique,

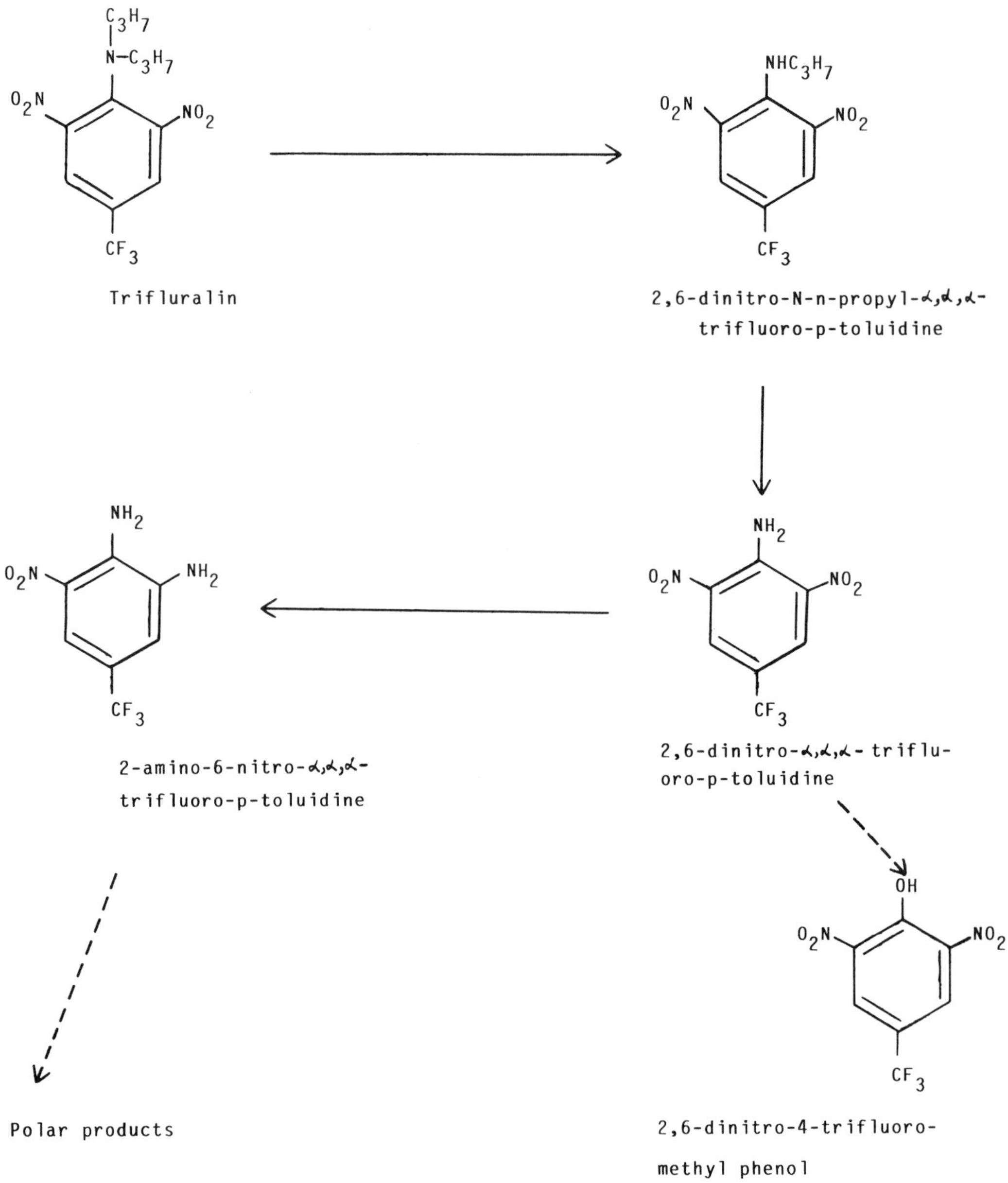

FIGURE 4. Metabolism of trifluralin by *Aspergillus carneus*, *Fusarium oxysporum*, and *Trichoderma viride*. (From Ref. 118.)

Lee [122] isolated bacteria and fungi capable of degrading EPTC from soil which had received four annual applications of the herbicide. Among the 29 fungal isolates, *Fusarium oxysporum* was the most predominant, whereas *Paecilomyces*, *Penicillium*, and *Diheterospora* were the most efficient decomposers even after 15 months storage.

Molinate: The herbicide molinate (*S*-ethyl perhydroazepine-1-carbothioate) was degraded rapidly by microorganisms [123–126]. The isolates, including three bacteria (*Mycobacterium* sp., *Flavobacterium* sp., and *Streptomyces* sp.) and the fungus *Fusarium* sp., degraded molinate. The fungus *Fusarium* sp. brings about the oxidation of azepine ring-carbon, yielding 3-OH-, 3-oxo-, 4-OH-, and 4-oxo-derivatives. These intermediates seem to be further metabolized to microbial cell constituents or carbon dioxide (Fig. 5).

Diallate: Diallate (*S*-2,3-dichloroallyl *N*,*N*-diisopropylthiolcarbamate), a thiolcarbamate herbicide, rapidly disappears from the soil by microbial action. Pure cultures of *Fusarium oxysporum* dehalogenated diallate within 20 days [77]. Cullimore and Smith [127] speculated that the closely related compound triallate may be degraded by the same diallate-degrading fungus, although degradation of triallate has not been reported in fungal cultures. Anderson and Domsch [128] found that three fungi, *Phoma eupyrena*, *Penicillium janthinellum*, and *Trichoderma harzianum*, could degrade at least 20% of the applied diallate to carbon dioxide after 10 days of incubation, but no intermediates were detected in the medium.

Pentachlorophenol

PCP (pentachlorophenol) is used as a herbicide and also as an eradicant fungicide. It is mostly degraded by bacteria, however, in a few cases certain fungi have also been implicated in its degradation. Thus, Lyr [129] reported the detoxification of PCP in vitro by certain wood-rooting fungi, such as *Trametes versicolor*, which operate by secreting laccase into the medium. Also, Konishi and Inoue [130] found that a laccase from *T. versicolor* converted PCP to quinonoid products. Cserjesi and Johnson [131] found that *Trichoderma virgatum* methylates PCP to form a stable methyl ether. Another fungus, *Fusarium* sp., has been found to degrade this biocide rapidly (25% in 4 days and 75% in 11 days) when grown in liquid malt extract medium with L-tyrosine [132].

Diquat and Paraquat (Bipyridyls)

The bipyridyls diquat (1,1'-ethylene-2,2'-bipyridylium ion) and paraquat (1,1'-dimethyl-4,4'-bipyridylium ion) have been widely used in agriculture because of their rapid defoliant action.

The soil yeast *Lipomyces starkeyi* degraded paraquat, using the compound as a source of nitrogen [133]. The ability of *Lipomyces* sp. to degrade paraquat was also reported by Anderson and Drew [134]. They have shown that this ability is an adaptive property of the organism. *Lipomyces starkeyi* can also degrade diquat [135]. Recently, Carr et al. [136] reported the in vitro degradation of ring-^{14}C- and methyl-^{14}C-paraquat (as a sole nitrogen source) by *L. starkeyi* leading to accumulation of ^{14}C-acetic acid in the extracellular medium and the evolution of $^{14}CO_2$. It is postulated that after demethylation or removal of methylamine group from the molecule, the remaining carbon skeleton is progressively degraded via a number of short-lived intermediates to acetic acid.

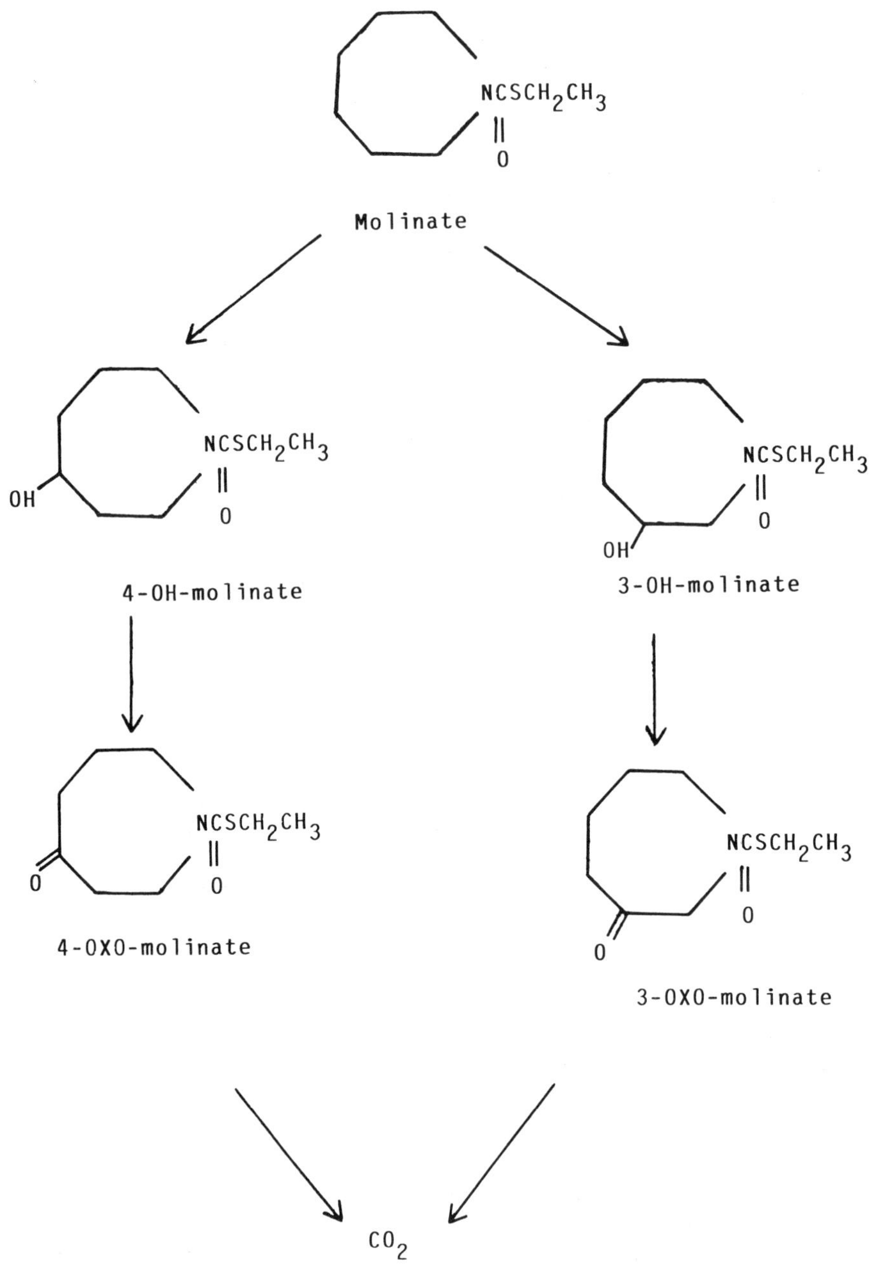

FIGURE 5. Metabolism of molinate by *Fusarium* sp. (From Ref. 126.)

IV. INSECTICIDES

In tropical countries where hot and humid environments favor the buildup of insect pests, insecticides constitute more than 60% of all pesticides used. Most of the earlier research on pesticide metabolism in soils was confined to herbicides, but in recent years considerable literature has accumulated on the metabolism of insecticides in soil and microbial cultures. The role of fungi in the degradation of insecticides in soil is emphasized in this section.

A. Chlorinated Hydrocarbons

DDT

DDT [1,1-bis(4-chlorophenyl)-2,2,2-trichloroethane] is one of the few compounds where the metabolic pathway is well elucidated. Studies, mostly with bacteria, indicated that the pathway of DDT degradation involves an initial dechlorination of the trichloromethyl group to form 1,1-dichloro-2,2 bis(4-chlorophenyl) ethane (DDD), which then undergoes further dechlorination, oxidation, and decarboxylation to form bis-(4-chlorophenyl) methane (DDM). Subsequent cleavage of one of the aromatic rings yields *p*-chlorophenylacetic acid, which may also undergo ring cleavage. The products of ring cleavage reaction are then degraded to Krebs cycle intermediates and ultimately oxidized to carbon dioxide. In some instances, DDM may be converted to 4,4'-dichlorobenzohydrol (DBH) or 4,4'-dichlorobenzophenone (DBP). DBH may undergo ring cleavage, or both DBP and DBH may undergo reductive dechlorination of the aromatic ring to form the corresponding nonchlorinated analogs, which may further undergo aromatic ring cleavage and further degradation. There are reports of degradation of DDT by fungi like *Fusarium oxysporum* [137–139], *Aspergillus flavus*, *Thanatephorus cucumeris* [140], *Mucor alternans* [141,142], and *Trichoderma viride* [143,144]. Pathways in the degradation of DDT by fungi and bacteria appeared to be almost identical.

Recently, Bumpus and Aust [145] reported a new pathway for DDT degradation (Fig. 6) by a white rot fungus, *Phanerochaete chrysosporium*. DDT was first oxidized by hydroxylation of benzylic carbon of DDT to 2,2,2-trichloro-1-1-bis(4-chlorophenyl) ethanol (dicofol), which was then dechlorinated to form 2,2-dichloro-1,1-bis (4-chlorophenyl) ethanol (FW-152). Ultimately, oxidative carbon-carbon bond cleavage led to the formation of DBP, which may then undergo aromatic ring cleavage or reductive dechlorination followed by eventual degradation to the end product, carbon dioxide. The same fungus degraded lindane and other environmental pollutants, 3,4,3',3'-tetrachlorobiphenyl, 2,4,5,2',-4',5'-hexachlorobiphenyl, 2,3,7,8-tetrachlorodibenzo-*p*-dioxin, and benzo(a)pyrene, to carbon dioxide [146]. *Phanerochaete chrysosporium* produces an extracellular lignin-degrading enzyme (ligninase) during secondary metabolism. Model studies with DDT suggested that the ability of this lignin-degrading Basidiomycete to metabolize DDT and related organochlorines is dependent on the ligninase of this fungus. The other white rot fungi, *Pleurotus ostreatus*, *Phellinus weivii*, and *Polyporus versicolor*, also mineralized DDT [145].

Aldrin and Dieldrin

Korte et al. [147] reported that cultures of *Aspergillus niger*, *A. flavus*, *Penicillium notatum*, and *P. chrysogenum* converted aldrin to

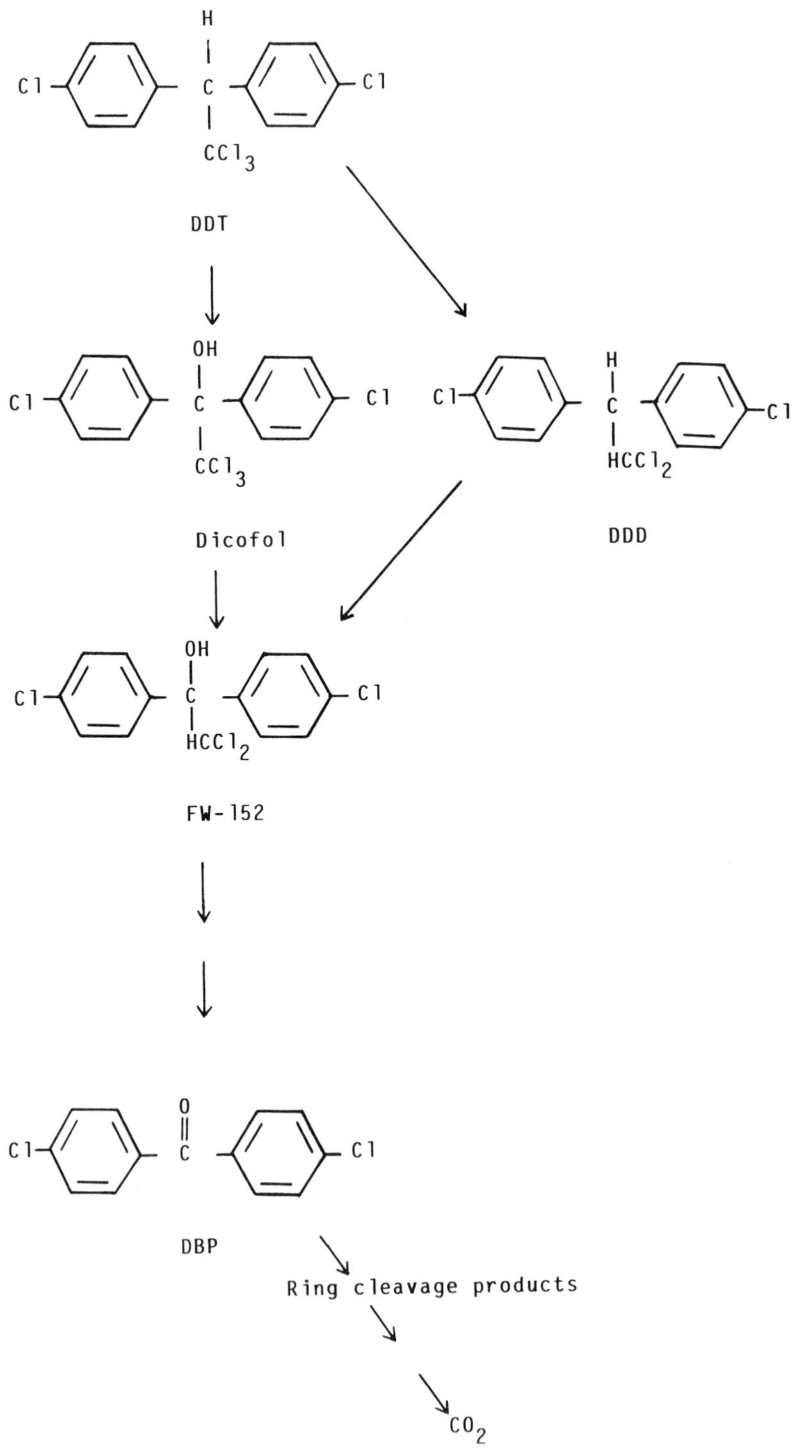

FIGURE 6. Metabolism of DDT by *Phanerochaete chrysosporium*. (From Ref. 145.)

dieldrin and four other metabolites. Tu et al. [148] found that out of 92 microbial pure cultures screened for aldrin-degrading activity, *Trichoderma* spp. were the most active fungi effecting epoxidation of aldrin to dieldrin and other metabolites. In addition, *Fusarium* sp. and *Penicillium* sp. were also effective converters of aldrin to dieldrin. The oxidative metabolism in soil leading to the formation of an epoxy-ring from the unsaturated carbon–carbon bond of the unchlorinated ring of aldrin was first described by Kiigemagi et al. [149]. Patil et al. [144] isolated 13 microorganisms including a fungus, *Trichoderma viride*, which had the ability to degrade aldrin. *Trans*-aldrindiol (6,7-*trans*-dihydroxy dihydroaldrin) was identified as one of the metabolites.

Dieldrin is a stable chlorinated hydrocarbon. However, there is evidence that certain fungi, in addition to many soil bacteria, can degrade this insecticide. Thus, Anderson et al. [141] demonstrated that *Mucor alternans* degraded 26% of the added dieldrin in 2 days. No metabolites were characterized. *Trichoderma viride* produced *trans*-aldrindiol (principal metabolite) from dieldrin [143]. An interesting aspect of dieldrin metabolism was reported by Bixby et al. [150], who demonstrated that a soil fungus, *Trichoderma kohingii*, cleaved the rings of dieldrin liberating some carbons as carbon dioxide.

Heptachlor, Endrin, and Chlordane

Microbial degradation of heptachlor has been studied by Miles et al. [151]. Heptachlor (1,4,5,6,7,8,8-heptachloro-3a,4,7,7a-tetrahydro-4,7-methanoindene) is converted to heptachlor epoxide by soil microorganisms including different species of *Rhizopus*, *Fusarium*, *Penicillium*, and *Trichoderma*. Heptachlor is chemically hydrolyzed to 1-hydroxychlordane, which is subsequently expoxidized to 1-hydroxy-2,3-epoxychlordane by microbial activity. Slow microbial degradation of β-dihydroheptachlor by *Aspergillus niger* and *Penicillium urticae* has also been reported [152].

Patil et al. [144] isolated 20 microorganisms including a fungus, *T. viride*, which degraded endrin to ketoendrin.

Iyengar and Rao [153] studied the rate of disappearance of chlordane and heptachlor by *Aspergillus niger*. The fungus could not utilize these insecticides as a sole carbon source, but it degraded up to 12.5 ppm of chlordane and heptachlor in 48 hr. Interestingly, chlordane-adapted *A. niger* could metabolize chlordane, heptachlor, and aldrin, but not HCH or DDT.

Endosulfan

Fungal degradation of endosulfan (6,7,8,9,10,10-hexachloro-1,5,5a,6,-9,9a-hexahydro-6,9-methano-2,4,3-benzodioxathiepin 3-oxide) has been demonstrated by Martens [154]. Many soil fungi produced endosulfan sulfate as the major metabolite. Other metabolic products identified were endoether, endohydroxyether, chlorendic acid, and endolacetate. El-Zorgani and Omer [155] found endosulfan diol as the major degradation product of α- and β-endosulfan by a culture of *Aspergillus niger*.

B. Organophosphorus Insecticides

Organophosphorus insecticides are mostly degraded by hydrolysis. The hydrolysis can be chemical, biological, or both. Microbial degradation involving mainly bacteria has been demonstrated for most organophosphorus insecticides. However, some reports show that a few fungi can also actively degrade these pesticides.

Malathion

Malathion [*O,O*-dimethyl-*S*-(1,2-dicarboethoxyethyl) phosphorodithioate] has been widely used because of its broad spectrum and high insecticidal activities. The hydrolysis of malathion can be both chemical and biological in nature [156,157], but most studies indicate that the major pathway of malathion disappearance in soil, water, sediments, and salt marsh environment is biologically mediated [156,158–161].

Matsumura and Boush [158] found that *Trichoderma viride* and a *Pseudomonas* sp. from a heavily contaminated soil could metabolize malathion. Soluble esterase enzymes were isolated which deethylated the carboxylic acid side chain of malathion to its monoacid and diacid derivatives (Fig. 7). In addition to the strong carboxylesterase activity, some strains of *T. viride* exhibited high demethylation activity. Similarly, Mostafa et al. [159] suggested the strong carboxyesterase activity in the degradation of malathion by *Aspergillus niger, Penicillium notatum*, and *Rhizoctonia solani.*

Another fungus, *Aspergillus oryzae*, also possessed carboxylesterase activity and metabolized malathion to β-monoacid and malathion dicarboxylic acid [162]. These intermediary products were subsequently metabolized to inorganic phosphate and thiophosphates.

Diazinon, Parathion, and Fenitrothion

Trichoderma viride readily degraded diazinon [*O,O*-diethyl-*O*-(2-isopropyl-6-methyl-4-pyrimidinyl) phosphorothioate] and parathion [*O,O*-diethyl-*O*-(*p*-nitrophenyl) phosphorothioate] [149]. Rao and Sethunathan [163] found that parathion was converted to aminoparathion by *Penicillium waksmani* isolated from a flooded acid sulfate soil.

Under aerobic conditions, a fungus *Fusarium* sp. decomposed fenitrothion [*O,O*-dimethyl-*O*-(3-methyl-4-nitrophenyl) phosphorothioate] to aminofenitrothion, formylaminofenitrothion, acetylaminofenitrothion, and 3-methyl-4-nitrophenol [164]. According to Miyamoto [165], bacteria and some fungi converted fenitrothion almost exclusively to aminofenitrothion. In a recent study, Baarschers and Heitland [166] found that *Trichoderma viride* enzymatically hydrolyzed fenitrothion and fenitrooxon with great ease in the presence of alternate nutrients. Its hydrolysis product, 3-methyl-4-nitrophenol, appeared to be subsequently degraded. Cometabolism was indicated in the fungal degradation of parent compound and 3-methyl-4-nitrophenol.

Other Organophosphorus Insecticides

Disulfoton: Bhaskaran et al. [167] demonstrated that most of the 20 fungal cultures and 10 isolates of *Streptomyces* spp. tested could use disulfoton [*O,O*-diethyl-*S*-(2-ethyl-thio-ethyl) phosphorodithioate] as a phosphorus source. Among the different microbes, species of *Aspergillus*, *Helminthosporium*, and *Streptomyces* showed maximum utilization of disulfoton as carbon as well as phosphorus sources.

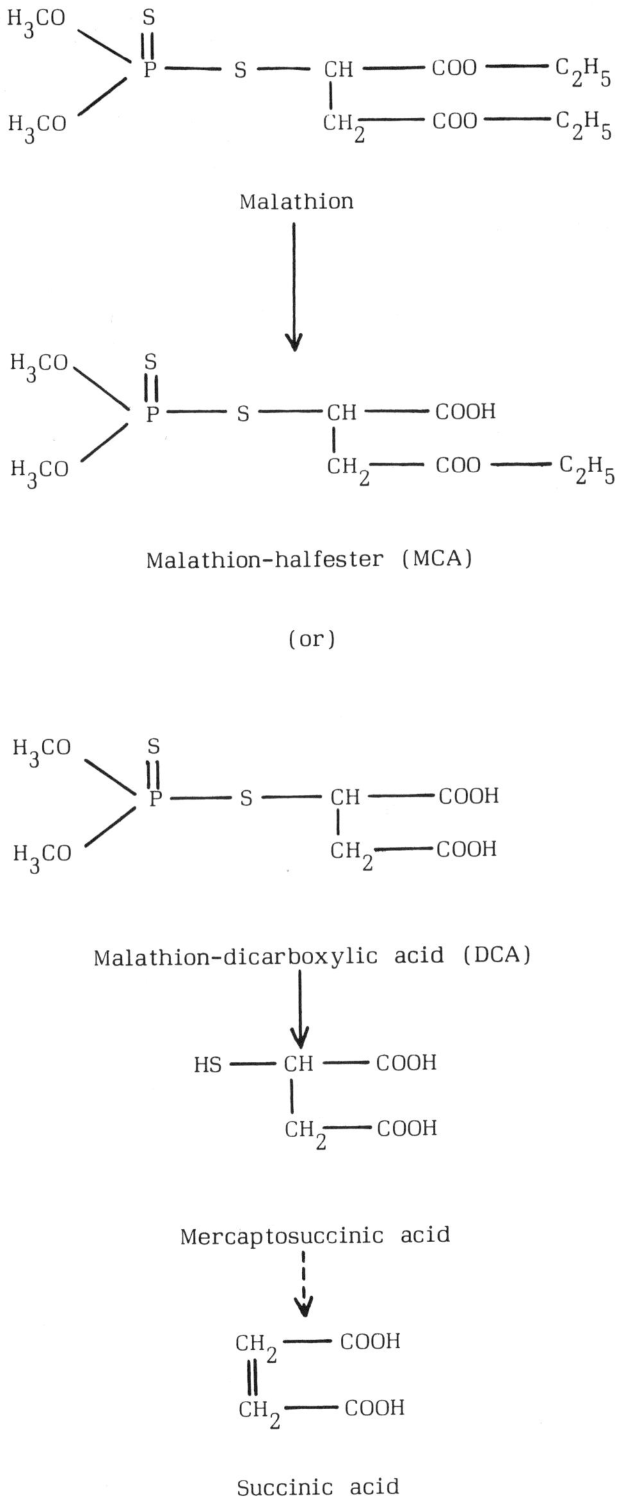

FIGURE 7. Metabolism of malathion by *Trichoderma viride*. (From Ref. 158.)

Fonofos

Flashinski and Lichtenstein [168,169] found that among the nine soil fungi tested, *Mucor plumbeus* and *Rhizopus arrhizus* were the most active in degrading fonofos (*O*-ethyl *S*-phenyl ethylphosphonodithioate). A variety of metabolites were formed from ^{14}C-ethoxy labeled and ^{14}C-ring labeled fonofos. Oxidative disulfuration to fonofoxon was followed by cleavage to *O*-ethyl ethoxyphosphonothioic acid (ETP) and *O*-ethyl ethoxyphosphonic acid (EOP). The thiophenyl moiety was oxidized to methyl phenylsulforide (MPSO) and methyl phenylsulfone ($MPSO_2$). Flashinski and Lichtenstein [170] tested the ability of *M. plumbeus, R. arrhizus*, and *Penicillium notatum* to degrade fonofos under various environmental conditions. Different nutrient sources affected the fungal degradation of fonofos. Time course studies on the metabolism of the insecticide indicated that fonofos was first absorbed by the fungal mycelium and then metabolized. This was followed by the release of the water-soluble metabolites ETP, EOP, MPSO, and $MPSO_2$. A temperature of 15 – 25°C and a pH of 6.0 – 7.0 were optimum for the degradation of fonofos by *R. arrhizus*.

Cyolane

Soil fungi (*Rhizoctonia solani, Penicillium notatum*, and *Aspergillus flavus*) and soil bacteria (two species of *Rhizobium*) metabolized cyolane [171]. Metabolites identified were mono- and diethylphosphates.

Bromophos

Trichoderma sp. and *Alternaria* sp. degraded ^{32}P-bromophos [*O*-(4-bromo-2,5-dichlorophenyl) *O,O*-dimethylphosphorothioate] to desmethyl and bidesmethyl derivatives as well as dimethyl and monomethyl phosphorothionic acids [172].

C. Carbamate Insecticides

Carbaryl

Because of the long persistence of organochlorine insecticides and high mammalian toxicity of many organophosphorus insecticides, there is a growing tendency to replace them with less toxic and less persistent carbamates. Carbaryl (1-naphthyl-*N*-methylcarbamate) is the most widely used carbamate insecticide in agriculture and public health. Microbial involvement in the degradation of carbaryl in soils has been demonstrated. Matsumura and Boush [143] isolated a strain of *Trichoderma viride* which could degrade carbaryl through hydroxylation. In recent years, microbial metabolism of carbaryl, especially with fungi, has been studied intensively by Bollag and his group. There are many indications that carbaryl is hydrolyzed not only chemically but also by biological action. Among different bacteria and fungi isolated from soil, *Fusarium solani* hydrolyzed carbaryl to 1-naphthol (Fig. 8), which was metabolized further by a constitutive enzyme system [173 – 175]. A cell-free extract from this organism converted 74% of the ^{14}C in 1-naphthol to carbon dioxide within 30 min. Another fungus, *Gliocladium roseum*, isolated from a soil treated with carbaryl for 4 weeks metabolized carbaryl by a nonhydrolytic mechanism [176]. The metabolites formed by this fungus, as a result of hydroxylation reactions, were 1-naphthyl,*N*-hydroxymethylcarbamate, 4-hydroxy-1-naphthylmethylcarbamate, and 5-hydroxy-1-naphthylmethylcarbamate. The first

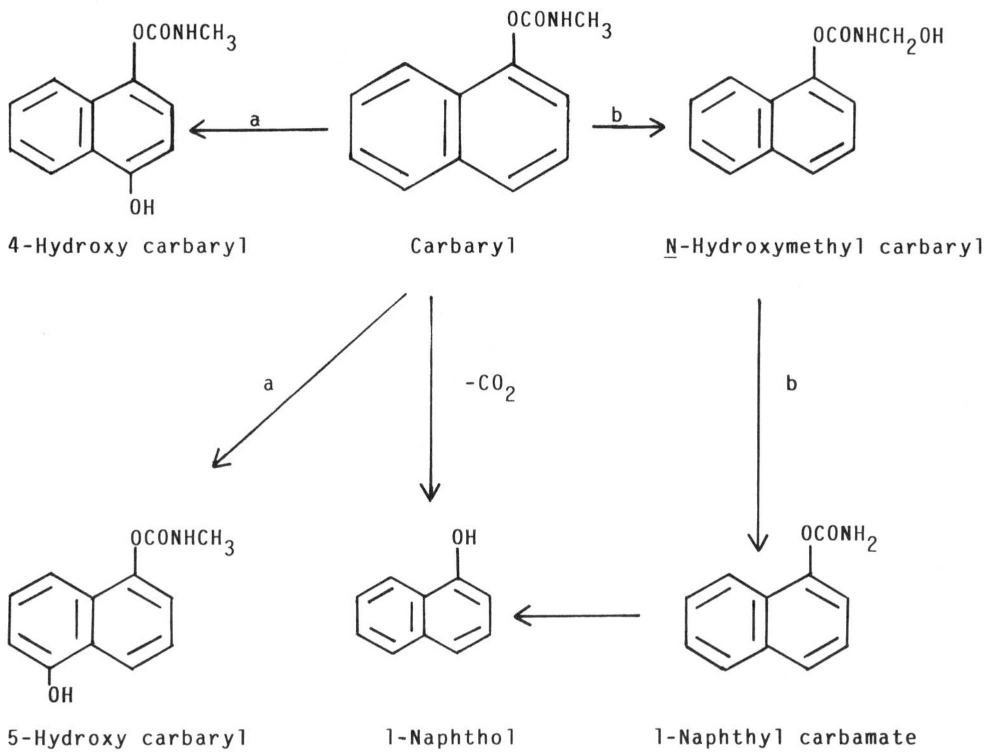

FIGURE 8. Metabolism of carbaryl by *Gliocladium roseum* (a) and *Aspergillus terreus* (b). (From Ref. 187.)

metabolite was formed by the hydroxylation of the side chain, while the remaining two metabolites were formed by hydroxylation of the aromatic ring. Most of the investigated soil fungi could metabolize carbaryl to hydroxylated derivatives; however, the products varied qualitatively as well as quantitatively with the various fungi [174]. *Penicillium* sp., *Mucor* sp., *Rhizopus* sp., and *Trichoderma viride* formed more of the ring hydroxylated metabolites, whereas other isolates, *Aspergillus terreus*, *Fusarium oxysporum*, and *Gliocladium roseum*, produced more *N*-hydroxymethyl derivatives. In another study, Rodriguez and Dorough [177] showed that certain fungal isolates (*Fusarium* sp., *Penicillium* sp., and *Aspergillus* sp.) from a carbaryl-enriched soil also produced the 4- and 5-hydroxy- and the *N*-hydroxymethyl derivatives of carbaryl and a small amount of 5-6-dihydrodihydroxy carbaryl.

Biodegradation of carbaryl has also been studied in aquatic environment. Sikka et al. [178] isolated selected yeasts, fungi, and one of the four bacteria from marine environment, which converted carbaryl and 1-naphthol to water-soluble products. Filamentous fungi (*Culcitalana* sp., *Halosphaeria* sp., *Numicola* sp., and *Aspergillus* sp.) possessed a greater ability to degrade 1-naphthol than bacteria and yeast. Hydroxylation of carbaryl, which was an important reaction in several soil fungi [174], appeared to be of minor significance in the marine fungi tested.

Degradation of 1-naphthol has been studied in detail by Bollag and his group. 1-Naphthol was completely degraded by the cell-free extracts of *Fusarium solani* [175]. After 60 min of incubation with cell-free extracts, more than 80% of ^{14}C in 1-naphthol was evolved as $^{14}CO_2$. Another fungus, *Rhizoctonia praticola*, transformed 1-naphthol to water-soluble products, and after removal of the fungal cells, the culture medium retained the ability to transform 1-naphthol [179]. *Rhizoctonia praticola* produced an extracellular phenol oxidase which polymerized phenolic and naphthalmic intermediates of many pesticides [180–182]. The ring hydroxylated carbaryl metabolites were also coupled by the fungal enzyme [182]. The extracellular enzyme responsible for the polymerizing activity was isolated and characterized as laccase [183]. Subsequent studies [61,184,185] showed that this fungal enzyme also oxidatively coupled phenolic humus constituents such as orcinol, syringic acid, vanillin, and vanillic acid. Recently, Bollag and Liu [62] demonstrated that laccase can copolymerize syringic acid products with various chlorophenols. The occurrence of similar reactions in soil may lead to the incorporation of xenobiotics into soil organic matter.

Carbofuran

Carbofuran (2,3-dihydro-2,2-dimethyl-7-benzofuranyl *N*-methylcarbamate) is a broad spectrum systemic insecticide and nematicide used as a soil treatment for control of soil insect pests.

Degradation of carbofuran occurred in both autoclaved and nonautoclaved soils but more rapidly in the latter, indicating chemical and microbial participation [186]. Recent studies have provided more convincing evidence to suggest that microorganisms, bacteria in particular, are involved in the degradation of carbofuran, especially in soils of near neutral pH and in retreated soils [187,188]. Limited information is also available on the contribution of fungi in carbofuran metabolism. Thus, Kandasamy et al. [189] isolated a *Helminthosporium* sp. which could degrade more than 70% of the carbofuran added at 20 ppm within 14 days. Other strains of *Aspergillus niger* and *Trichoderma viride* were less efficient. All these isolates formed hydroxycarbofuran as an intermediate breakdown product of carbofuran. Recently, Ampova et al. [190] found that a soil fungus, *Trichoderma viride*, from tobacco rhizosphere was resistant to high concentrations of a commercial formulation of carbofuran, Furadan, and took part in its decomposition. But no degradation products were reported.

Aldicarb

Aldicarb [2-methyl-2-(methylthio)-propionaldehyde *O*-(methylcarbamoyl) oxime] is a broad spectrum soil-applied insecticide for the control of certain insect pests, mites, and nematodes in various crops.

Limited information is available on the fungal degradation of aldicarb. Spurr and Sousa [191] showed that a soil saprophyte and plant pathogen, *Rhizoctonia solani*, could utilize aldicarb as a food source or cometabolite. Pure cultures of five soil fungi, *Gliocladium catenulatum*, *Penicillium* sp., *Rhizoctonia* sp., *Cunninghamella* sp., and *Trichoderma* sp., degraded aldicarb to organosoluble (aldicarb sulfoxide, a major metabolite) and water-soluble metabolites (aldicarb amide sulfone and aldicarb alcohol) [192]. Of the above five fungi, *Gliocladium catenulatum* was the most active, converting 65 and 35% of the aldicarb to organosoluble and water-soluble metabolites, respectively, in 21 days. Likewise, *G. catenulatum* effected maximum degradation of sulfoxide to oxime sulfoxide (14%), nitrite sulfoxide (11%), and water-soluble metabolites (28%) after 28 days of incubation.

V. MICROBIAL ADAPTATIONS AND BIOTECHNOLOGY

Edaphic and genetic processes influence the rate of microbial degradation of pesticides in soil [193]. The code for degradative enzyme synthesis is contained in DNA molecules in soil microorganisms. In bacteria, it is known that a single chromosome and movable genetic elements may alter the DNA sequences and gene expressions. However, the role of microbial genetics with reference to degradation of pesticides by fungi is less understood.

The microbial adaptation to a pesticide may occur by mutation, recombination, and through external DNA additions like gene duplication and extrachromosomal additions. Of these, extrachromosomal additions (known as ECEs—extrachromosomal elements) may play an important role in the microbial environment of the soil under pesticide pressure. The existence of ECEs has been shown in fungi, but their participation in the degradation of pesticides has yet to be studied.

Interestingly, the gene for the enzyme ligninase, produced by the Basidiomycete *Phanerochaete chrysosporium*, has been cloned and sequenced, and attempts are being made to amplify the gene expression in yeast cells by site-directed mutagenesis [199]. This is significant because ligninase has been implicated in the degradation of DDT, lindane, and chlorinated biphenyls by *P. chrysosporium* [145,146].

Any discussion on fungal metabolism without the mention of biotechnology is incomplete. The biotechnological innovations using fungi include bio-pesticides for the control of nematodes, mycoprotein-based foods, and filtration of industrial effluents.

Fungi, owing to its biomass, may serve as a filter for effluents containing pesticides which may be readily and effectively absorbed by the fungal mat. Alternatively, fungi or their enzymes (ligninase, for example) can be immobilized on suitable supports for effluents to pass through. Protoplast fusion of fungal cells for conferring varied degradative abilities is another area which merits attention.

VI. CONCLUSIONS

Fungi constitute an important part, in terms of biomass, of the microbial community in the complex and dynamic soil system. Yet, a majority of the global literature on biodegradation of pesticides is concerned with bacteria. Limited studies with fungi indicate that fungi have a remarkable ability to degrade a variety of pesticides. Many pesticide transformations enunciated by fungi have been characterized (Table 1). Both fungi and bacteria can bring about the same type of reactions. A few type reactions are enunciated exclusively by fungi.

Most studies on fungal metabolism of pesticides in pure cultures were performed in nutrient-rich media, although there are a few examples of degradation of pesticides by fungi as a sole source of carbon or nitrogen.

TABLE 1. Fungi Involved in the Degradation of Pesticides

Compounds decomposed	Degrading fungi	Metabolites detected	Ref.
Fungicides			
1. DIDT	*Aspergillus niger* and *Saccharomyces cerevisiae*	Ethylenethiourea	20
2. Sodium dimethyl dithiocarbamate	*Saccharomyces cerevisiae*, *Hansenula anamola*, *Glomerella cingulata*, *Aspergillus niger*, and *Cladosporium cucumerinum*	α-Aminobutyric acid	21
3. Thiram	*Penicillium notatum*, *Glomerella cingulata*, and *Fusarium roseum*	Carbon disulfide	22
4. Captan and folpet	*Saccharomyces pastorianus*	Tetrahydrophthalmide thiophosgene, carbonyl sulfide, and hydrogen disulfide	30,31
5. Pyrazophos	*Pyricularia oryzae*	6-Ethoxy-carbonyl-2-hydroxy-5-methyl-pyrazolo [1,5-a] pyrimidine (pp) and PO-pyrazophos	33
6. Edifenphos	*Pyricularia oryzae*	Ethyl-*S*-phenyl hydrogen phosphorothioate, ethyl dihydrogen phosphate, orthophosphoric acid, and *O*-ethyl-*s*-(*p*-hydroxyphenyl)-*S*-phenyl phosphorodithioate	34
7. PCNB	*A. niger*, *G. cingulata*, *Trichoderma viride*, *Mucor ramannianus*, and *Helminthosporium victoriae*	Pentachloroaniline (PCA) and pentachlorothioanisole (PCTA)	36
	Fusarium oxysporum	PCA and pentachloromethyl thiobenzene	37

8. Chloroneb	*Rhizoctonia solani*	2,5-Dichloro-4-methoxyphenol (DCMP)	38
	Fusarium sp.	DCMP	39,40
9. Carbendazim	*Alternaria mali* and *Pellicularia sasakii*	5-Hydroxy carbendazim	45
	Aspergillus nidulans	5-Hydroxy carbendazim	46
10. Carboxin	*Ustilago maydis*	Carboxin sulfoxide	49
11. Mebenil	*Rhizopus japonicus*	*p*-Hydroxyphenyl derivative	50
Herbicides			
1. 2,4-D and MCPA	*Aspergillus niger*	2,5-Dichloro-4-hydroxyphenoxyacetate and 5-hydroxy derivative	58
2. Propanil	*Fusarium solani*	Propionic acid and 3,4-dichloroaniline	64,65
	Penicillium piscarium and *Geotrichum candidum*	3,4-Dichloroaniline, 3,3',4,4'-tetrachloroazobenzene (TCAB), and azobenzene	68,69
	Aspergillus nidulans (enzyme isolated from the fungus)	3,4-Dichloroaniline	71
3. 3,4-dichloroaniline	*Fusarium oxysporum*	TCAB, azobenzene, triazine, and 3,4-dichloronitrobenzene	66
	Geotrichum candidum	TCAB	69
	Aspergillus niger	3,4-Dichloroacetaniline	67
4. Karsil	*Penicillium* sp. and *Pullularia* sp.	2-Methylvaleric acid, 3,4-dichloroaniline, ferric acid, and acetic acid	73
	Rhizopus japonicus	*N*-(3,4-Dichlorophenyl)-3-hydroxy-2-methyl-pentanamide	51

TABLE 1. (Continued)

Compounds decomposed	Degrading fungi	Metabolites detected	Ref.
Herbicides (Continued)			
5. Dicryl	*Rhizopus japonicus*	*N*-(3,4-Dichlorophenyl)-2,3-dihydroxy-2-methylpropionamide	74
6. Propachlor	*Fusarium oxysporum*	2-Hydroxy-*N*-isopropylacetanilide and Cl^- ion	75
7. Alachlor	*Chaetomium globosum*	2-Chloro-2',6'-diethylacetanilide, 2,6-diethyl-*N* (methoxymethyl) acitine, 2,6-diethylaniline, 1-chloroacetyl-2,3-dihydro-7-ethylindole, and Cl^- ion	78
8. Butachlor	*Mucor sufui* NTU-358	2-Chloro-2',6'-diethylacetanilide, 2,6-diethylaniline, *N*-chloroacetyl-7-ethyl-2,3-dihydroindole, 2,6-diethylacetanilide, *N*-methyl-2-chloro-2',6'-diethylacetanilide, and 2-hydroxy-2',6'-diethylacetanilide	80
9. Linuron	*Aspergillus nidulans*	3,4-Dichloroaniline	84,85
	Aspergillus niger	3,4-Dichloroaniline and 3,4-dichloroacetaniline	67
10. Monuron	*Cunninghamella echinulata*	4-(Chlorophenyl) urea and 3-(4-chlorophenyl)-1-1 methylurea	87
11. Metobromuron	*Talaromyces wortmanii*	Demethylated and demethoxylated derivatives, *p*-linomophenylurea, and *p*-bromoacetanilide	88

12.	Monolinuron	*Aspergillus niger*	Demethylated and demethoxylated derivatives, *p*-linomophenylurea, and *p*-bromoacetanilide	89
13.	Chlorbromuron	*Rhizoctonia solani*	3-(3-Chloro-4-bromophenyl)-1-methoxyurea and 3-chloro-4-bromoaniline	90
14.	Chlorbromuron	*Rhizoctonia solani*	Demethylated and demethoxylated derivatives, *p*-linomophenylurea, and *p*-bromoacetanilide	91
15.	Fluometuron	*Rhizoctonia solani*	1-Methyl-3-(α,α,α-trifluorotolyl) urea and (3-trifluoromethylphenyl) urea	92
16.	Isouron	*Rhizoctonia solani*	3-Methylurea	93
17.	Swep	*Fusarium oxysporum*	3,4-Dichloroaniline	77
18.	Chlorpropham	*Fusarium oxysporum*	2-Amino-4-chlorophenol	95
19.	Barban	*Penicillium jenseni*	3-Chloroaniline	97
20.	Simazine	*Aspergillus fumigatus*	2-Amino-4-chloro-6-ethylamino-*s*-triazine and 2,4-dihydroxy-6-amino-*s*-triazine	108
21.	Atrazine	*Fusarium roseum*	2-Ethylamino-4-hydroxy-6-isopropylamino-*s*-triazine	110
22.	Metamitron	*Rhizopus japonicus* and *Cunninghamella echinulata*	Desaminometamitron	112
23.	Metribuzin	*Cunninghamella echinulata*	Desaminometribuzin and desaminodiketometribuzin	113

TABLE 1. (Continued)

Compounds decomposed	Degrading fungi	Metabolites detected	Ref.
Herbicides (Continued)			
24. Dinitramine	*Aspergillus fumigatus*	N^3-Ethyl-2,4-dinitro-6-trifluoromethyl-m-phenylenediamine	116
		6-Amino-1-ethyl-2-methyl-7-nitro-5-trifluoromethyl benzimidazole	
		2,4-Dinitro-6-trifluoromethyl-m-phenylenediamine	
25. *N*-s-Butyl-4-t-butyl-2,6-dinitroaniline (Amex)	*Paecilomyces* sp.	4-t-Butyl-2,6-dinitroaniline; and 3-(4-t-butyl-2,6-dinitroaniline)-2-butanol	117
26. Trifluralin	*Aspergillus carneus*, *Fusarium oxysporum*, and *Trichoderma viride*	2,6-Dinitro-*N*-n-propyl-α,α,α-trifluoro-*p*-toluidine	118
		2,6-Dinitro-α,α,α-trifluoro-*p*-toluidine	
		2-Amino-6-nitro-α,α,α-trifluoro-*p*-toluidine	
		2,6-Dinitro-4-trifluoro-methylphenol	
27. Molinate	*Fusarium* sp.	3-OH-, 3-OXO-4, 4-OH-, and 4-OXO-derivatives	126
28. Diallate	*Fusarium oxysporum*	Chloride ion	77
	Phoma eupyrena, *Penicillium janthinellum*, and *Trichoderma harzianum*	Carbon dioxide	128
29. Pentachlorophenol (PCP)	*Trichoderma virgatum*, *Cephaloasca* sp., and *Penicillium* sp.	Pentachloroanisole	131

Insecticides			
1. DDT	*Trichoderma viride*	Dicofol, DDD, and DDE	143,144
	Fusarium oxysporum	DDD and DBH	137
	Aspergillus flavus ATCC 11495	DBH, DDD, and DDE	140
	Thanatephorus cucumeris	DBP, DDD, and DDE	140
	Phanerochaete chrysosporium	DDD, Dicofol, 2,2-dichloro-1-1-bis (4-chlorophenyl) ethanol [FW-152], DBP, and carbon dioxide	145
2. Aldrin	*Aspergillus niger*, *A. flavus*, *Penicillium notatum*, and *P. chrysogenum*	Epoxide and dieldrin	147
	Trichoderma sp., *Fusarium* sp., and *Penicillium* sp.	Epoxide and dieldrin	148
	Trichoderma viride	Dieldrin and trans-aldrindiol	144
3. Dieldrin	*Trichoderma viride* and *Trichoderma kohingii*	*Trans*-aldrindiol and carbon dioxide	143,150
4. Heptachlor	*Rhizopus* sp., *Fusarium* sp., *Penicillium* sp., and *Trichoderma* sp.	Heptachlor epoxide	151
5. Endrin	*Trichoderma viride*	Ketoendrin	144
6. Endosulfan	Many soil fungi	Endosulfan sulfate, endoether, endohydroxyether, chlorendic acid, and endolacetate	154
	Aspergillus niger	Endosulfan diol	155

TABLE 1. (Continued)

Compounds decomposed	Degrading fungi	Metabolites detected	Ref.
Insecticides (Continued)			
7. Malathion	*Trichoderma viride*	Malathion-half ester, malathion-dicarboxylic acid, and mercaptosuccinic acid	158
	Aspergillus niger, *Penicillium notatum*, and *Rhizoctonia solani*	Several carboxylic products, monomethyl phosphate, and thiophosphate	159
	Aspergillus oryzae	B-Monoacid and malathion dicarboxylic acid	162
8. Parathion	*Trichoderma viride*	Aminoparathion	143
	Penicillium waksmani	Aminoparathion	163
9. Fenitrothion	*Fusarium* sp.	Aminofenitrothion, formyl aminofenitrothion, acetyl aminofenitrothion, and 3-methyl-4-nitrophenol	164
	Trichoderma viride	3-Methyl-4-nitrophenol	166
10. Fenofos	*Mucor plumbeus*, *Rhizopus arrhizus*, and *Penicillium notatum*	*O*-Ethyl ethoxyphosphonothioic acid, methylphenyl sulfoxide, and methylphenyl-sulfone	168–170
11. Cyolane	*Rhizoctonia solani*, *Penicillium notatum*, and *Aspergillus flavus*	Mono and diethylphosphate	171
12. Bromophos	*Trichoderma* sp. and *Alternaria* sp.	Desmethyl and bidesmethyl derivatives, and dimethyl and monomethyl phosphorothioic acids	172

13. Carbaryl	*Fusarium solani*	1-Naphthol and carbon dioxide	173–175
	Gliocladium roseum	1-Naphthyl,*N*-hydroxymethyl carbamate, 4-hydroxy-1-naphthylmethyl carbamate, and 5-hydroxy-1-naphthylmethyl carbamate	176
	Aspergillus sp., *A. niger*, *A. terreus*, *Gliocladium roseum*, *Helminthosporium* sp., *Mucor* sp., *M. racemosus*, *Penicillium* sp. (isolate 2), *Rhizopus* sp., and *Trichoderma viride*	*N*-Hydroxymethyl carbaryl and 4- and 5-hydroxy carbaryl	174
	Fusarium roseum and *Penicillium* sp. (isolate 1)	*N*-Hydroxymethyl carbaryl	174
	Aspergillus sp., *Fusarium* sp., and *Penicillium* sp.	5,6-Dihydro-dihydroxy carbaryl, *N*-hydroxymethyl carbaryl, and 4- and 5-hydroxycarbaryl	177
	Culcitalna achraspora, *Halosphaeria mediosetigera*, and *Humicola alopallonella*	Unidentified polar metabolites	178
14. 1-Naphthol	*A. fumigatus*, *C. achraspora*, *H. mediosetigera*, and *H. alopallonella*	Unidentified water-soluble metabolites	178
	Fusarium solani	CO_2 and unidentified polar metabolites	175
	Rhizoctonia solani	Unidentified polar metabolites	179
	Rhizoctonia praticola	Dimer (4,4'-bis-1-naphthol), trimer, tetramer, and pentamer compounds of 1-naphthol	180–182

TABLE 1. (Continued)

Compounds decomposed	Degrading fungi	Metabolites detected	Ref.
Insecticides (Continued)			
15. Carbofuran	*Helminthosporium* sp., *Aspergillus niger*, and *Trichoderma viride*	Hydroxycarbofuran	186
16. Aldicarb	*Gliocladium catenulatum*, *Penicillium* sp., *Rhizoctonia* sp., *Cunninghamella* sp., and *Trichoderma* sp.	Aldicarb sulfoxide, aldicarb amide sulfone, and aldicarb alcohol	189

Rapid degradation in such nutrient-rich media implicates cometabolism in the fungal transformations of many pesticides. Such cometabolic transformations have relevance since organic sources are always present in the soil environment. In a few instances, fungal enzymes (for example, laccase from *Rhizoctonia praticola* or ligninase from *Phanerochaete chrysosporium*) involved in the biochemical transformation of pesticides have been characterized.

Our current knowledge of the biodegradation of pesticides is largely based on the research generated from studies in individual cultures of microorganisms, chiefly bacteria and, to a less extent, fungi and actinomycetes. But in the dynamic and complex soil system, these microorganisms never exist in an isolated state. A pesticide reaching the soil is attacked by the fungus, bacterium, and actinomycete in tandem or simultaneously often leading to its complete mineralization. It is possible that a fungus transforms a pesticide to an intermediate which may resist further degradation by the same fungus but is susceptible to degradation by other members of the microbial community in the soil. There are reports of synergistic interaction between a bacterium and an actinomycete leading to the accelerated degradation of an organophosphorus insecticide, diazinon (200). However, the occurrence, extent, and significance of interactions between a fungus and other members of the microbial community in the soil in relation to pesticide transformations are not well understood.

Most of the information generated so far on the genetics and biochemistry of pesticides has been concerned with aerobic bacteria. Although bacteria and fungi often employ different mechanisms in the degradation of pesticides, very little is known about the genetics and biochemistry of pesticide degradation by fungi. There is a need to develop biotechnological innovations using fungi to address problems of environmental pollutants.

ACKNOWLEDGMENTS

We thank Dr. S. Patnaik, Director, Central Rice Research Institute, Cuttack for permission to publish this manuscript, Dr. C. R. Bhatia, Associate Director, Biology Group, BARC, Bombay for his encouragement, and Dr. R. P. Singh, BGA Section, CRRI, Cuttack for his help. This project was funded, in part, by the Ministry of Environment and Forests, Government of India and PL-480 Project No. FG-IN-693.

REFERENCES

1. Menzie, C. M. *Metabolism of Pesticides*. Bureau Sport Fish Wildl., U.S. Dept. Interior, Special Scientific Report, Wildlife No. 127, 1969, p. 190.
2. Alexander, M. Microbial degradation and biological effects of pesticides in soil. In *Soil Biology*. UNESCO, Paris, 1969, pp. 209–240.
3. Alexander, M. Microbial formation of environmental pollutants. *Adv. Appl. Microbiol.*, *18*:1–73, 1974.
4. Alexander, M. Biodegradation of chemicals of environmental concern. *Science*, *211*:132–138, 1981.

5. Helling, C. S., Kearney, P. C., and Alexander, M. Behavior of pesticides in soils. *Adv. Agron.*, *23*:147–240, 1971.
6. Bollag, J. M. Microbial transformation of pesticides. *Adv. Appl. Microbiol.*, *18*:75–130, 1974.
7. Bollag, J. M. Microbial metabolism of pesticides. In *Microbial Transformations of Inactive Compounds*, Vol. 2 (J. P. Rosazza, ed.), CRC Press, Boca Raton, FL, 1982, pp. 126–167.
8. Hill, I. R., and Wright, S. J. L. *Pesticide Microbiology*. London, Academic Press, 1978, pp. 1–844.
9. Alexander, M. *Introduction to Soil Microbiology*, 2nd ed. John Wiley and Sons, New York, 1977, pp. 1–467.
10. Dalton, H., and Stirling, D. I. Cometabolism. *Phil. Trans. R. Soc. Lond. B.*, *297*:481–496, 1982.
11. Janke, D., and Fritsche, W. Nature and significance of microbial cometabolism of xenobiotics. *J. Basic Microbiol.*, *25*:603–619, 1985.
12. Munnecke, D. E., and Solberg, R. A. Inactivation of semesan in soil by fungi. *Phytopathology*, *48*:396, 1958.
13. Spanis, W., and Munnecke, D. E. Bacterial inactivation of organic fungicides. *Phytopathology*, *52*:365, 1962.
14. Spanis, W., Munnecke, D. E., and Solberg, R. A. Biological breakdown of two organic mercurial fungicides. *Phytopathology*, *52*:455–462, 1962.
15. EBDC, Fungicide Assessment Team, U.S. Dept. Agr. Assessment of ethylenebisdithiocarbamate (EBDC) fungicide uses in agriculture. I. An analysis of uses and their relationship to exposure. U.S. Dept. Agr., Washington, D.C., 1977.
16. Kaars Sijpesteijn, A., Dekhuijzen, H. M., and Vonk, J. W. Biological conversion of fungicides in plants and microorganisms. In *Antifungal Compounds*, Vol. 2 (M. R. Siegel and H. D. Sisler, eds.), Marcel Dekker, New York, 1977, pp. 91–147.
17. Moje, W., Munnecke, D. E., and Richardson, L. T. Carbonyl sulfide a volatile fungitoxicant from nabam in soil. *Nature*, *202*:831–832, 1964.
18. Munnecke, D. E. Factors affecting the efficacy of fungicides in soil. *Annu. Rev. Phytopathol.*, *10*:375–398, 1972.
19. Kaufman, D. D. Soil-fungicide interactions. In *Antifungal Compounds*, Vol. 2 (M. R. Siegel and H. D. Sisler, eds.), Marcel Dekker, New York, 1977, pp. 1–49.
20. Vonk, J. W., and Kaars, Sijpesteijn, A. Formation of ethylenethiourea from 5,6-dihydro-3*H*-imidazo-[2,1-c]-1,2,4-dithiazole-3-thione by microorganisms and reducing agents. *J. Environ. Sci. Health*, *B11*:33–37, 1976.
21. Kaars Sijpesteijn, A., Kaslander, J., and van der Kerk, G. J. M. On the conversion of sodium dimethyldithiocarbamate into its α-aminobutyric acid derivative by microorganisms. *Biochem. Biophys. Acta*, *62*:587–589, 1962.
22. Sisler, H. D., and Cox, C. E. Release of carbon disulfide from thiram by fungi. *Phytopathology*, *41*:565, 1951.
23. Munnecke, D. E., and Mickail, K. Y. Thiram persistence in soil and control of damping off caused by *Pythium ultimum*. *Phytopathology*, *57*:969–974, 1967.
24. Raghu, K., Murthy, N. B. K., and Kumarsamy, R. Degradation of thiram in soils. In *Proc. Symp. Use of Radiations and Radioisotopes in Studies of Plant Productivity*, Dept. Atomic Energy, Govt. of India. 1974, pp. 847–854.

25. Raghu, K. Soil-dithiocarbamate interactions. In *Proc. Symp. Nuclear Techniques in Studies of Metabolism: Effects and Degradation of Pesticides*, Dept. Atomic Energy, Govt. of India, 1978, pp. 13–18.
26. Chatrath, M. S., and Raju, T. N. Persistence of ^{35}S labeled thiram in soil. *Indian Phytopathol.*, *39*:312–313, 1986.
27. Lukens, R. J., and Sisler, H. D. Chemical reactions involved in the fungitoxicity of captan. *Phytopathology*, *48*:235–244, 1958.
28. Richmond, D. V., and Somers, E. Studies on the fungitoxicity of captan. IV. Reactions of captan with cell thiols. *Ann. Appl. Biol.*, *57*:231–240, 1966.
29. Richmond, D. V., and Somers, E. Studies on the fungitoxicity of captan. VI. Decomposition of ^{35}S-labeled captan by *Neurospora crassa* conidia. *Ann. Appl. Biol.*, *62*:35–43, 1968.
30. Siegel, M. R., and Sisler, H. D. Fate of the phthalimide and trichloromethylthio ($SCCl_3$) moities of folpet in toxic action on cells of *Saccharomyces pastorianus*. *Phytopathology*, *58*:1123–1128, 1968.
31. Siegel, M. R., and Sisler, H. D. Reactions of folpet with purified enzymes, nucleic acids and subcellular components of *Saccharomyces pastorianus*. *Phytopathology*, *58*:1129–1133, 1968.
32. Lukens, R. J. Thiophosgene split from captan by yeast. *Phytopathology*, *53*:881, 1963.
33. de Waard, M. A. Mechanism of action of the organophosphorus fungicide pyrazophos. *Meded. Landbouwhogeschool Wageningen*, *74-14*:1–97, 1974.
34. Uesugi, Y., and Tomizawa, C. Metabolism of *O*-ethyl-*S*,*S*-diphenylphosphorodithioate (Hinosan) by mycelial cells of *Pyricularia oryzae*. *Agric. Biol. Chem.*, *35*:941–949, 1971.
35. Uesugi, Y., and Tomizawa, C. Metabolism of *S*-benzyl-*O*-ethyl phenylphosphorothioate (Inezin) by mycelial cells of *Pyricularia oryzae*. *Agric. Biol. Chem.*, *36*:313–317, 1971.
36. Chacko, C. I., Lockwood, J. L., and Zabik, M. Chlorinated hydrocarbon pesticides: Degradation by microbes. *Science*, *154*:893–895, 1966.
37. Nakanishi, T., and Oku, H. Metabolism and accumulation of PCNB by phytopathogenic fungi in relation to selective toxicity. *Phytopathology*, *59*:1761–1762, 1969.
38. Hock, W. K., and Sisler, H. D. Metabolism of chloroneb by *Rhizoctonia solani* and other fungi. *J. Agric. Food Chem.*, *17*:123–128, 1969.
39. Wiese, M. V., and Vargas, J. M. Degradation and synthesis of chloroneb by soil microorganisms. *Phytopathology*, *62*:1121, 1972.
40. Wiese, M. V., and Vargas, J. M. Interconversion of chloroneb and 2,5-dichloro-4-methoxyphenol by soil microorganisms. *Pestic. Biochem. Physiol.*, *3*:214–222, 1973.
41. Helweg, A. Persistence of benomyl in different soil types and microbial breakdown of the fungicide in soil and agar cultures. *Tidsskr. Planteavl.*, *77*:232–243, 1973.
42. Helweg, A. Degradation and adsorption of carbendazim and 2-aminobenzimidazole in soil. *Pestic. Sci.*, *8*:71–78, 1977.
43. Sharma, S. K., and Gupta, V. K. Persistence of bavistin in soil. *Indian J. Mycol. Plant Pathol.*, *13*:381–383, 1983.
44. Fuchs, A., and Bollen, G. J. Benomyl, after seven years. In *Systemfungizide* (H. Lyr and C. Polter, ed.), Akademic-Verlag, Berlin, 1975, pp. 121–136.

45. Yasuda, Y., Hashimoto, S., Soeda, Y., and Noguchi, T. Metabolism of thiophanate-methyl by pathogenic fungi and antifungal activity of its metabolites. *Ann. Phytopathol. Soc. Japan, 39*:49–52, 1973.
46. Davidse, L. C. Metabolic conversion of methyl benzimidazole-2-yl carbamate (MBC) in *Aspergillus nidulans*. *Pestic. Biochem. Physiol., 6*:538--546, 1976.
47. Gupta, J. P., and Chatrath, M. S. Persistence of benomyl in soil. *J. Nucl. Agric. Biol., 8*:138–139, 1979.
48. Yarden, O., Katan, J., Aharonson, N., and Ben-Yephet, Y. Delayed and enhanced degradation of benomyl and carbendazim in disinfested and fungicide-treated soils. *Phytopathology, 75*:763–767, 1985.
49. Lyr, H., Ritter, G., and Banasiak, L. Detoxification of carboxin. *Zeitschr, Allg. Mikrobiol., 14*:313–320, 1973.
50. Wallnofer, P. R., Safe, S., and Hutzinger, O. Metabolism of the systemic fungicides 2-methylbenzanilide and 2-chlorobenzanilide by *Rhizopus japonicus*. *Pestic. Biochem. Physiol., 1*:458–463, 1971.
51. Wallnofer, P. R., Safe, S., and Hutzinger, O. Die Hydroxylation des Herbizids Karsil [*N*-(3',4'-dichlorophenyl)2-methylpentamide] durch *Rhizopus japonicus*. *Chemosphere, 1*:155–158, 1972.
52. Deas, A. H. B., and Clifford, D. R. Metabolism of the 1,2,4-triazolylmethane fungicides, triadimefon, triadimenol, and diclobutrazol, by *Aspergillus niger* van Tiegh. Pestic. *Biochem. Physiol., 17*:120–133, 1982.
53. Audus, L. J. The biological detoxification of 2,4-D in soils; Isolation of an effective organism. *Nature, 166*:356, 1951.
54. Byrde, R. J. W., and Woodcock, D. Fungal detoxification. 2. The metabolism of some phenoxyalkylearboxylic acids by *Aspergillus niger*. *Biochem. J., 65*:682–686, 1957.
55. Clifford, D. R., and Woodcock, D. Metabolism of phenoxyacetic acid by *Aspergillus niger* van Tiegh. *Nature, 203*:763, 1964.
56. Faulkner, J. K., and Woodcock, D. Fungal detoxification. Pt. V. Metabolism of *O*- and *p*-chlorophenoxyacetic acids by *Aspergillus niger*. *J. Chem. Soc., part IV*:5397–5400, 1961.
57. Faulkner, J. K., and Woodcock, D. Fungal detoxification. Pt. VII. Metabolism of 2,4-dichlorophenoxyacetic and 4-chloro-2-methylphenoxyacetic acids by *Aspergillus niger*. *J. Chem. Soc., Part I*:1187–1191, 1965.
58. Smith, A. E., and Phillips, D. V. Degradation of 4-(2,4-dichlorophenoxy) butyric acid (2,4-DB) by *Phytophthora megasperma*. *J. Agric. Food Chem., 24*:294–296, 1976.
59. Sahasrabudhe, S. R., and Modi, V. V. Microbial degradation of chlorinated aromatic compounds. *Microbiol. Sci., 4*:300–303, 1987.
60. Wolf, D. C., and Martin, J. P. Decomposition of fungal mycelia and humic type polymers containing carbon-14 from ring and side chain labeled 2,4-D and chlorpropham. *Soil Sci. Soc. Am. J., 40*:700–704, 1976.
61. Liu, S. Y., Minard, R. D., and Bollag, J. M. Oligomerization of syringic acid, a lignin derivative by a phenoloxidase. *Soil Sci. Soc. Am. J., 45*:1100–1105, 1981.
62. Bollag, J. M., and Liu, S. Y. Copolymerization of halogenated phenols and syringic acid. *Pestic. Biochem. Physiol., 23*:261–272, 1985.

63. Bartha, R., and Pramer, D. Metabolism of acylanilide herbicides. *Adv. Appl. Microbiol.*, *13*:317–341, 1970.
64. Lanzilotta, R. P., and Pramer, D. Herbicide transformation. I. Studies with whole cells of *Fusarium solani*. *Appl. Microbiol.*, *19*: 301–306, 1970.
65. Lanzilotta, R. P., and Pramer, D. Herbicide transformation. II. Studies with an acylamidase of *Fusarium solani*. *Appl. Microbiol.*, *19*:307–313, 1970.
66. Kaufman, D. D., Plimmer, J. R., Iwan, J., and Klingebiel, U. I. 3,3',4,4'-Tetrachloroazoxybenzene from 3,4-dichloroaniline in microbial culture. *J. Agric. Food. Chem.*, *20*:916–919, 1972.
67. Fustikova, N. S. The decomposition of linuron by fungi of the genus *Aspergillus*. *Mikrobiologiya*, *48*:57–61, 1979.
68. Bordeleau, L. M., and Bartha, R. Ecology of herbicide transformation: Synergism of two soil fungi. *Soil Biol. Biochem.*, 3:281–284, 1971.
69. Bordeleau, L. M., and Bartha, R. Biochemical transformation of herbicide-derived anilines in culture medium and in soil. *Can. J. Microbiol.*, *18*:1857–1864, 1972.
70. Bordeleau, L. M., and Bartha, R. Biochemical transformations of herbicide-derived anilines: Purification and characterization of causative enzymes. *Can. J. Microbiol.*, *18*:1865–1871, 1972.
71. Pelsy, F., Leroux, P., and Heslot, H. Properties of an *Aspergillus nidulans* propanil hydrolase. *Pestic. Biochem. Physiol.*, *27*:182–188, 1987.
72. Blake, J., and Kaufman, D. D. Characterisation of acylanilide hydrolysing enzymes from *Fusarium oxysporum* Schlecht. *Pestic. Biochem. Physiol.*, 5:305–313, 1975.
73. Sharabi, N. E., and Bordeleau, L. M. Biochemical decomposition of the herbicide *N*-(3,4-dichlorophenyl)-2-methyl pentanimide and related compounds. *Appl. Microbiol.*, *18*:369–375, 1969.
74. Wallnofer, P. R., Safe, S., and Hutzinger, O. Microbial hydroxylation of the herbicide *N*-(3,4-dichlorophenyl)-methacrylamide (dicryl). *J. Agric. Food Chem.*, *21*:502–504, 1973.
75. Kaufman, D. D., Plimmer, J. R., and Iwan, J. *Abstr. of P. 162nd* N.M. Am. Chem. Soc., Maryland, 1971.
76. Rankov, V., and Valev, B. Detoxification of the herbicide propachlor by certain microscopic fungi. *Acta Phytopathol. Acad. Sci. Hung.*, *12*:67–72, 1977.
77. Kaufman, D. D., and Blake, J. Microbial degradation of several acetamide, acylanilide, carbamate, toluidine and urea pesticides. *Soil Biol. Biochem.*, 5:297–308, 1973.
78. Tiedje, J. M., and Hagedorn, M. L. Degradation of alachlor by a soil fungus, *Chaetomium globosum*. *J. Agric. Food Chem.*, *23*:77–81, 1975.
79. McGahen, L. L., and Tiedje, J. M. Metabolism of two new acylanilide herbicides, Antor herbicide (H-22234) and Dual (metalachlor) by the soil fungus *Chaetomium globosum*. *J. Agric. Food Chem.*, *26*:414–419, 1978.
80. Chen, Y. L., and Wu, T. C. Degradation of herbicide butachlor by soil microbes. *J. Pestic. Sci.*, 3:411–417, 1978.

81. Lee, J. K. A study on degradation of butachlor by a soil fungus *Chaetomium globosum*. *J. Korean Agric. Chem. Soc.*, *21*:1–10, 1978.
82. Rankov, V. Inactivation of aresin and afalon herbicides by some soil microorganisms. *Dokl. Bolg. Akad. Nauk.*, *21*:1097–1100, 1968.
83. Alfonso, H. M. M., Martinez, V. R., Urdaminia, M., and Delos, A. Diuron decomposition by the action of microorganisms in a yellow ferrallitic soil. *Ciencias de la Agricultura*, *19*:99–104, 1984.
84. Schroeder, M. The microbial decomposition of substituted urea herbicides. *Weed Sci. Soc. Am.*, Abstr. No. 157, 81–82, 1970.
85. Geissbuhler, H., Haselbach, C., Aebi, H., and Ebner, L. The fate of *N'*-(4-chlorophenoxy)-phenyl-*N,N*-dimethyl urea (C-1983) in soils and plants. III. Breakdown in soils and plants. *Weed Res.*, *3*:277–297, 1963.
86. Glad, G., Goransson, B., Popoff, T., Theander, O., and Torstensson, N. T. L. Decomposition of linuron by fungi isolated from soil. *Swedish J. Agric. Res.*, *11*:127–134, 1981.
87. Tillmanns, G. M., Wallnofer, P. R., Engelthardt, G., Olie, K., and Hutzinger, O. Oxidative dealkylation of five phenylurea herbicides by the fungus *Cunninghamella echinulata* Thaxter. *Chemosphere*, *7*:59–69, 1978.
88. Tweedy, B. G., Loeppky, C., and Ross, J. A. Metabolism of 3-(*p*-bromophenyl)-1-methoxy-1-methylurea (metobromuron) by selected soil microorganisms. *J. Agric. Food Chem.*, *18*:851–853, 1970.
89. Borner, H. Der Abbau von Harnstoffherbiziden im Boden. *Z. PflKrankh. PflPath. PflSchutz.*, *74*:135–143, 1967.
90. Weinberger, M., and Bollag, J. M. Degradation of chlorbromuron and related compounds by the fungus *Rhizoctonia solani*. *Appl. Microbiol.*, *24*:750–754, 1972.
91. Bollag, J. M. Biochemical transformations of pesticides by soil fungi. *Critical Rev. Microbiol.*, *2*:35–38, 1972.
92. Rickard, R. W., and Camper, N. D. Degradation of fluometuron by *Rhizoctonia solani*. *Pestic. Biochem. Physiol.*, *9*:183–189, 1978.
93. Ozaki, M., and Hayase, Y. Degradation of isouron related compounds by *Rhizoctonia solani*. *J. Pestic. Sci.*, *8*:87–93, 1983.
94. McClure, G. W. Degradation of aniline herbicides by propham-adapted microorganisms. *Weed Sci.*, *22*:323–329, 1974.
95. Fletcher, C. L., and Kaufman, D. D. Hydroxylation of monochloroaniline pesticide residues by *Fusarium oxysporum* Schlecht. *J. Agric. Food Chem.*, *27*:1127–1130, 1979.
96. Bellinck, C., and Mayaudon, J. Mineralisation du phenmediphame ed derives par *Trichoderma viride*. Immobilization des residues ^{14}C. Effects sur lactivate biologique et le nombre de germes. *Rev. Ecol. Biol. Soc.*, *16*:1–7, 1979.
97. Wright, S. J. L., and Forey, A. Metabolism of the herbicide barban by a soil *Penicillium*. *Soil Biol. Biochem.*, *4*:207–213, 1972.
98. Kaufman, D. D., and Kearney, P. C. Microbial degradation of carbamate pesticide combination in soils. Div. Agric. Food Chem., ACS Abstr., A. 45, 152nd Meeting, New York, September, 1966.
99. Kaufman, D. D., Kearney, P. C., von Endt, D. W., and Miller, D. E. Methylcarbamate inhibition of phenylcarbamate metabolism in soil. *J. Agric. Food Chem.*, *18*:513–519, 1970.

100. Kaufman, D. D., Kearney, P. C., and Sheets, T. J. Microbial degradation of Simazine. *J. Agric. Food Chem.*, *13*:238–242, 1965.
101. Bortels, H., Fricke, E., and Schneider, R. Simazine decomposition by microorganisms in various soils. *Nachr. Bl. dt. Pfl. Schutzdienst.*, *19*:101–105, 1967.
102. Shocken, M. J. Interaction of the herbicide atrazine with marine higher fungi: Bioaccumulation, degradation and effects of growth. *Diss. Abs. Int.*, *B 43*:1378, 1982.
103. Murray, D. S., Rieck, W. L., and Lynd, J. Q. Utilisation of methylthio-s-triazine for growth of soil fungi. *Appl. Microbiol.*, *19*: 11–13, 1970.
104. Wolf, D. C., and Martin, J. P. Microbial decomposition of ring-^{14}C-atrazine, cyanuric acid, and 2-chloro-4,6-diamino-s-triazine. *J. Environ. Qual.*, *4*:134–139, 1975.
105. Zeyer, J., Bodmer, J., and Hutter, R. Rapid degradation of cyanuric acid by *Sperothrix schenckii.* Zentralblatt für Bakteriologie, Mikrobiologie und Hygine, I. *Abt. Oreginale*, *2*:99–110, 1981.
106. Myskow, W., Lasota, T., and Stachyra, A. Cyanuric acid—a *s*-triazine derivative as a nitrogen source for some soil microorganisms. *Acta Microbiologica Polonica.*, *32*:177–183, 1983.
107. Kaufman, D. D., and Kearney, P. C. Microbial degradation of s-triazine herbicides. *Residue Rev.*, *32*:235–265, 1970.
108. Kearney, P. C., Kaufman, D. D., and Sheets, T. J. Metabolites of simazine by *Aspergillus fumigatus.* *J. Agric. Food Chem.*, *13*: 369–372, 1965.
109. Kaufman, D. D., and Blake, J. Degradation of atrazine by soil fungi. *Soil Biol. Biochem.*, *2*:73–80, 1970.
110. Couch, R. W., Gramlich, J. V., Davis, D. E., and Funderburk, H. J., Jr. The metabolism of atrazine and simazine by soil fungi. *Proc. 5th. Weed Control Conf.*, *18*:623–631, 1965.
111. Martinez, V. R., and Castaneda, R. P. F. New microorganisms which assist in the degradation of the herbicides atrazine and simazine. *Ciencias de la Agricultura*, *20*:125–126, 1984.
112. Engelhardt, G., and Wallnofer, P. R. Microbial transformation of triazinone herbicide metamitron to desaminometamitron. *Chemosphere*, *6*:463–466, 1978.
113. Schilling, R., Engelhardt, G., and Wallnofer, P. R. Degradation of the herbicide metribuzin (Sencor) by pure cultures of *Cunninghamella echinulata* Thaxter ATCC 38447. *Chemosphere*, *14*:267–270, 1985.
114. Probst, G. W., and Tepe, J. B. Trifluralin and related compounds. In *Degradation of Herbicides* (P. C. Kearney and D. D. Kaufman, eds.), Marcel Dekker, New York, 1969, pp. 255–282.
115. Helling, C. S. Dinitroaniline herbicides in soils. *J. Environ. Qual.*, 5:1–14, 1976.
116. Laanio, T. L., Kearney, P. C., and Kaufman, D. D. Microbial metabolism of dinitramine. *Pestic. Biochem. Physiol.*, *3*:271–277, 1973.
117. Kearney, P. C., Plimmer, J. R., Williams, V. P., Klingebiel, U. I., Isensee, A. R., Laanio, T. L., Stolzenberg, G. E., and Zaylskie, R. G. Soil persistence and metabolism of *N-sec* butyl-4-*tert* butyl-2,6-dinitroaniline. *J. Agric. Food Chem.*, *22*:856–859, 1974.

118. Zayed, S. M. A. D., Mostafa, I. Y., Farghaly, M. M., Attaby, H. S., Adam, Y. M., and Mahdy, F. M. Microbial degradation of trifluralin by *Aspergillus carneus, Fusarium oxysporum* and *Trichoderma viride*. *J. Environ. Sci. Health, B 18*:253–260, 1983.
119. Sheets, T. J. Effects of soil type and time on the herbicidal activity of CDAA, CDEC and EPTC. *Weeds, 7*:442–448, 1959.
120. Obrigawitch, T., Wilson, R. G., Martin, A. R., and Roeth, F. W. The influence of temperature, moisture and prior EPTC application on the degradation of EPTC in soils. *Weed Sci., 30*:175–181, 1982.
121. Obrigawitch, T., Martin, A. R., and Roeth, F. W. Degradation of thiocarbamate herbicides in soils exhibiting rapid EPTC breakdown. *Weed Sci., 31*:187–192, 1983.
122. Lee, A. EPTC (*S*-ethyl *N,N*-dipropylthiocarbamate)-degrading microorganisms isolated from a soil previously exposed to EPTC. *Soil Biol. Biochem., 16*:529–531, 1984.
123. Imai, Y., and Kuwatsuka, S. Degradation of the herbicide molinate in soils. *J. Pestic. Sci., 7*:487–497, 1982.
124. Imai, Y., and Kuwatsuka, S. Characteristics of microflora degrading the herbicide molinate in soil. *J. Pestic. Sci., 11*:57–63, 1986.
125. Imai, Y., and Kuwatsuka, S. The mode of metabolism of the herbicide molinate by four strains of microorganisms isolated from soil. *J. Pestic. Sci., 11*:111–117, 1986.
126. Imai, Y., and Kuwatsuka, S. Metabolic pathways of the herbicide molinate in four strains of isolated soil microorganism. *J. Pestic. Sci., 11*:245–251, 1986.
127. Cullimore, D. R., and Smith, A. E. Initial studies on the breakdown of triallate. *Bull. Environ. Contam. Toxicol., 7*:36–42, 1972.
128. Anderson, J. P. E., and Domsch, K. H. Microbial degradation of the thiolcarbamate herbicide, diallate, in soil and by pure cultures of soil microorganisms. *Arch. Environ. Contam. Toxicol., 4*:1–7, 1976.
129. Lyr, H. Enzymatische detoxication chorierter phenole. *Phytopathol. Z., 47*:73–83, 1963.
130. Konishi, K., and Inoue, Y. Detoxification mechanism of pentachlorophenol by laccase of *Corialus versicolor*. *Mokuzai Gakkaishi, 18*:463–469, 1972.
131. Cserjesi, A. J., and Johnson, E. I. Methylation of pnetachlorophenol by *Trichoderma virgatum*. *Can. J. Microbiol., 18*:45–49, 1972.
132. Kwasniewska, K. The degradation of pentachlorophenol by a fungus of *Fusarium* sp. In *Trace Substances in Environment and Health*, Vol. 15 (D. D. Hamphill, ed.), 1981, pp. 392–398.
133. Baldwin, B. C., Bray, M. F., and Geoghegan, M. J. The microbial decomposition of paraquat. *Biochem. J., 101*:15, 1966.
134. Anderson, J. R., and Drew, E. A. Growth characteristics of a species of *Lipomyces* and its degradation of paraquat. *J. Gen. Microbiol., 70*:43–58, 1972.
135. Funderburk, H. H. Diquat and paraquat. In *Degradation of Herbicides* (P. C. Kearney and D. D. Kaufman, eds.), Marcel Dekker, New York, 1969, pp. 283–297.
136. Carr, R. J. G., Bilton, R. F., and Atkinson, T. Mechanism of biodegradation of paraquat by *Lipomyces starkeyi*. *Appl. Environ. Microbiol., 49*:1290–1294, 1985.

137. Engst, R., Kujawa, M., and Muller, G. Enzymatischer Abbau des DDT durch Schimmelpilze. I. Mitt. Isolierung und Identifizierung eines DDT abbauenden Schimmelpilzes. *Nahrung, 11*:401–403, 1967.
138. Engst, R., and Kujawa, M. Enzymatischer Abbau des DDT durch Schimmelpilze. 2. Mitt. Reaktionsverlauf des enzymtischen DDT-Abbaues. *Nahrung, 11*:751–760, 1967.
139. Engst, R., and Kujawa, M. Enzymatischer Abbau des DDT durch Schimmelpilze. 3. Mitt. Darstellung des 2,2-bis(*p*-chlorphenyl)-acetaldehyds (DDHO) and seine Bedeutung im Abbaucyclus. *Nahrung, 12*:783–785, 1968.
140. Subba-Rao, R. V., and Alexander, M. Bacterial and fungal co-metabolism of 1,1,1-trichloro-2,2-bis (4-chlorophenyl) ethane (DDT) and its breakdown products. *Appl. Environ. Microbiol., 49*:509–516, 1985.
141. Anderson, J. P. E., Lichtenstein, E. P., and Whittingham, W. F. Effect of *Mucor alternans* on the persistence of DDT and dieldrin in culture and in soil. *J. Econ. Entomol., 63*:1595–1599, 1970.
142. Anderson, J. P. E., and Lichtenstein, E. P. Effect of nutritional factors on DDT degradation by *Mucor alternans*. *Can. J. Microbiol., 17*:1291–1298, 1971.
143. Matsumura, F., and Boush, G. M. Degradation of insecticides by a soil fungus *Trichoderma viride*. *J. Econ. Entomol., 61*:610–612, 1968.
144. Patil, K. C., Matsumura, F., and Boush, G. M. Degradation of endrin, aldrin, and DDT by soil microorganisms. *Appl. Microbiol., 19*:879–881, 1970.
145. Bumpus, J. A., and Aust, S. D. Biodegradation of DDT [1,1,1-trichloro-2,2-bis (4-chlorophenyl) ethane] by the white rot fungus *Phanerochaete chrysosporium*. *Appl. Environ. Microbiol., 53*:2001–2008, 1987.
146. Bumpus, J. A., Tien, M., Wright, D., and Aust, S. D. Oxidation of persistent environmental pollutants by a white rot fungus. *Science, 228*:1434–1436, 1985.
147. Korte, F., Ludwig, G., and Voel, J. Unwandbung von Aldrin (^{14}C) und Dieldrin (^{14}C) durch Microorganismen, Leberhomogenate, und Moskito-larven. *Ann. Chem., 656*:135–140, 1962.
148. Tu, C. M., Miller, J. R. W., and Harris, C. R. Soil microbial degradation of aldrin. *Life Sci., 7*:311–322, 1968.
149. Kiigemagi, U., Morrison, H. E., Roberts, J. E., and Bollen, W. B. Biological and chemical studies on the decline of soil insecticides. *J. Econ. Entomol., 51*:198–204, 1958.
150. Bixby, N. W., Boush, G. M., and Matsumura, F. Degradation of dieldrin to carbon dioxide by a soil fungus *Trichoderma kohingii*. *Bull. Environ. Contam. Toxicol., 6*:491–494, 1971.
151. Miles, J. R. W., Tu, C. M., and Harris, C. R. Metabolism of heptachlor and its degradation products by soil microorganisms. *J. Econ. Entomol., 62*:1334–1337, 1969.
152. Poonawalla, N. H., and Korte, F. Metabolism of β-dihydro heptachlor ^{14}C in soil and by microorganisms. *J. Agric. Food Chem., 16*:15–16, 1968.
153. Iyengar, L., and Rao, A. V. S. P. Metabolism of chlordane and heptachlor by *Aspergillus niger*. *J. Gen. Appl. Microbiol., 19*:321–324, 1973.

154. Martens, R. Der Abbau von Endosulfan durch Mikroorganismen des Bodens. *Schriftenr. Ver. Wasser-Boden-Lufthyg.*, *37*:167–173, 1972.
155. El-Zorgani, G. A., and Omer, M. E. H. Metabolism of endosulfan isomers by *Aspergillus niger*. *Bull. Environ. Contam. Toxicol.*, *12*: 182–185, 1974.
156. Konard, J. G., Chesters, G., and Armstrong, D. E. Soil degradation of malathion, phosphorodithioate insecticide. *Soil Sci. Soc. Am. Proc.*, *33*:259–262, 1969.
157. Walker, W. W., and Stojanovic, B. J. Microbial versus chemical degradation of malathion in soil. *J. Environ. Qual.*, *2*:229–232, 1973.
158. Matsumura, F., and Boush, G. M. Malathion degradation by *Trichoderma viride* and a *Pseudomonas species*. *Science*, *153*:1278–1280, 1966.
159. Mostafa, I. Y., Bahig, M. R. E., Fakhr, I. M. I., and Adam, Y. Metabolism of organophosphorus insecticides. XIV. Malathion breakdown by soil fungi. *Z. Naturforsch*, *27b*:1115–1116, 1972.
160. Walker, W. W. Chemical and microbiological degradation of malathion and parathion in an estuarine environment. *J. Environ. Qual.*, *5*:210–216, 1976.
161. Bourquin, A. W. Degradation of malathion by salt-marsh microorganisms. *Appl. Environ. Microbiol.*, *33*:356–367, 1977.
162. Lewis, D. L., Paris, D. F., and Baughman, G. L. Transformation of malathion by a fungus, *Aspergillus oryzae*, isolated from a freshwater pond. *Bull. Environ. Contam. Toxicol.*, *13*:596–601, 1975.
163. Rao, A. V., and Sethunathan, N. Degradation of parathion by *Penicillium waksmanii* Zaleski isolated from flooded acid sulphate soil. *Arch. Mikrobiol.*, *97*:203–208, 1974.
164. Takimoto, Y., Hirota, M., Inui, H., and Miyamoto, J. Decomposition and leaching of radioactive sumithion in four different soils under laboratory condition. *J. Pestic. Sci.*, *1*:131–143, 1976.
165. Miyamoto, J. Degradation of fenitrothion in terrestrial and aquatic environments including photolytic and microbial reactions. In *Proceedings of a Symposium on Fenitrothion: The Long-Term Effects of Its Use in Forest Ecosystem* (J. R. Roberts, R. Greenhalgh, and H. K. Marshall, eds.), NRCC/CNRC No. 16073, Ottawa, Canada, 1977, pp. 105–134.
166. Baarschers, W. H., and Heitland, H. S. Biodegradation of fenitrothion and fenitrooxon by the fungus *Trichoderma viride*. *J. Agric. Food Chem.*, *34*:707–709, 1986.
167. Bhaskaran, R., Kandasamy, D., Oblisami, G., and Subramanian, T. R. Utilization of disyston as carbon and phosphorus sources by soil microflora. *Curr. Sci.*, *42*:835–836, 1973.
168. Flashinski, S. J., and Lichtenstein, E. P. Metabolism of Dyfonate by soil fungi. *Can. J. Microbiol.*, *20*:399–411, 1974.
169. Flashinski, S. J., and Lichtenstein, E. P. Degradation of Dyfonate in soil inoculated with *Rhizopus arrhizus*. *Can. J. Microbiol.*, *20*: 871–875, 1974.
170. Flashinski, S. J., and Lichtenstein, E. P. Environmental factors affecting the degradation of Dyfonate by soil fungi. *Can. J. Microbiol.*, *21*:17–25, 1975.
171. Mostafa, I. Y., Fakhr, I. M. I., and El-Zawahry, Y. A. Metabolism of organophosphorus insecticides. XV. Degradation of ^{14}C-cyolane by soil microorganisms. *Isotope Radiat. Res.*, *14*:37–42, 1982.

172. Stenersen, J. Degradation of P^{32}-bromophos by microorganisms and seedlings. *Bull. Environ. Contam. Toxicol.*, *4*:104–112, 1969.
173. Bollag, J. M., and Liu, S. Y. Degradation of Sevin by soil microorganisms. *Soil Biol. Biochem.*, *3*:337–345, 1971.
174. Bollag, J. M., and Liu, S. Y. Hydroxylations of carbaryl by soil fungi. *Nature*, *236*:177–178, 1972.
175. Bollag, J. M., and Liu, S. Y. Fungal degradation of 1-naphthol. *Can. J. Microbiol.*, *18*:1113–1117, 1972.
176. Liu, S. Y., and Bollag, J. M. Metabolism of carbaryl by a soil fungus. *J. Agric. Food Chem.*, *19*:487–490, 1971.
177. Rodriguez, L. D., and Dorough, H. W. Degradation of carbaryl by soil microorganisms. *Arch. Environ. Contam. Toxicol.*, *6*:47–56, 1977.
178. Sikka, H. C., Miyazaki, S., and Lynch, R. S. Degradation of carbaryl and 1-naphthol by marine microorganisms. *Bull. Environ. Contam. Toxicol.*, *13*:666–672, 1975.
179. Bollag, J. M., Sjoblad, R. D., Czaplicki, E. J., and Hoeppel, R. E. Transformation of 1-naphthol by the culture filtrate of *Rhizoctonia praticola*. *Soil Biol. Biochem.*, *8*:7–11, 1976.
180. Sjoblad, R., Minard, R. D., and Bollag, J. M. Polymerisation of 1-naphthol and related phenolic compounds by an extracellular enzyme. *Pestic. Biochem. Physiol.*, *6*:457–463, 1976.
181. Sjoblad, R., and Bollag, J. M. Oxidative coupling of aromatic pesticide intermediates by a fungal phenol oxidase. *Appl. Environ. Microbiol.*, *23*:906–910, 1977.
182. Bollag, J. M., Sjoblad, R. D., and Minard, R. D. Polymerisation of phenolic intermediates of pesticides by a fungal enzyme. *Experientia*, *33*:1564–1566, 1977.
183. Bollag, J. M., Sjoblad, R. D., and Liu, S. Y. Characterisation of an enzyme from *Rhizoctonia praticola* which polymerizes phenolic compounds. *Can. J. Microbiol.*, *25*:229–233, 1979.
184. Bollag, J. M., Liu, S. Y., and Minard, R. D. Cross coupling of phenolic humus constituents and 2,4-dichlorophenol. *Soil Sci. Soc. Am. J.*, *44*:52–56, 1980.
185. Bollag, J. M., Liu, S. Y., and Minard, R. D. Enzymatic oligomerisation of vanillic acid. *Soil Biol. Biochem.*, *14*:157–163, 1982.
186. Williams, I. H., Pepin, H. S., and Brown, M. J. Degradation of carbofuran by soil microorganisms. *Bull. Environ. Contam. Toxicol.*, *15*:244–249, 1976.
187. Rajagopal, B. S., Brahmaprakash, G. P., Reddy, B. R., Singh, U. D., and Sethunathan, N. Effect and persistence of selected carbamate pesticides in soil. *Residue Rev.*, *93*:1–199, 1984.
188. Ramanand, K., Sharmila, M., and Sethunathan, N. Mineralisation of carbofuran by a soil bacterium. *Appl. Environ. Microbiol.*, *54*:2129–2133, 1988.
189. Kandasamy, D., Marimuthu, J., Oblisamy, G., and Subramanian, T. R. Effect of application of insecticides on the HCN content and rhizosphere microflora of sorghum plants. *Madras Agric. J.*, *64*:302–306, 1977.
190. Ampova, G., Steffanov, D., and Gheorghieva, E. Contribution of soil fungi to the biological detoxification of Furadan and their influence on tobacco. In *Soil Biology and Conservation of the Biosphere*, Vol. 2 (J. Szegi, ed.), Akademiai Kiado, Budapest, 1984, pp. 231–237.

191. Spurr, W. H., Jr., and Sousa, A. A. Potential interactions of aldicarb and its metabolites on non-target organisms in the environment. *J. Environ. Qual.*, *3*:130–133, 1974.
192. Jones, A. S. Metabolism of aldicarb by five soil fungi. *J. Agric. Food Chem.*, *24*:115–117, 1976.
193. Kearney, P. C., and Kellog, S. T. Microbial adaptation to pesticides. *Pure Appl. Chem.*, *57*:389–403, 1985.
194. Pemberton, J. M., and Fisher, P. R. 2,4-D Plasmids and persistence. *Nature*, *268*:732–733, 1977.
195. Ghosal, D., You, I.-S., Chatterjee, D. K., and Chakrabarty, A. M. Microbial degradation of halogenated compounds. *Science*, *228*:135–142, 1985.
196. Mulbry, W. W., Karns, J. S., Kearney, P. C., Nelson, J. O., McDaniel, C. S., and Wild, J. R. Identification of a plasmid-borne parathion hydrolase gene from *Flavobacterium* sp. by southern hybridization with opd from *Pseudomonas diminuta*. *Appl. Environ. Microbiol.*, *51*:926–930, 1986.
197. Sethunathan, N., and Yoshida, T. A *Flavobacterium* sp. that degrades diazinon and parathion as sole carbon sources. *Can. J. Microbiol.*, *19*:873–875, 1973.
198. Paul, H. T., Betty, F., Chatterjee, D. K., and Chakrabarty, A. M. Genetic and molecular basis of the microbial degradation of herbicides and pesticides. In *Biotechnology for Solving Agricultural Problems* (Beltsville Symposium in Agricultural Research). Martinus Nijhoff Publishers, Dordrecht, 1986, pp. 355–368.
199. Chakrabarty, A. M. New biotechnological approaches to environmental pollution problems. *Biotech.*, *1*:67–74, 1987.
200. Gunner, H. B., and Zuckerman, B. M. Degradation of diazinon by synergistic microbial action. *Nature*, *217*:1183–1184, 1968.

Part IV

MODELING AND QUANTITATIVE APPROACHES

20

MATHEMATICAL MODELING OF VEGETATIVE GROWTH OF FILAMENTOUS FUNGI

JAMES I. PROSSER *Marischal College, University of Aberdeen, Aberdeen, Scotland*

I. INTRODUCTION

In many mycological studies the presence, absence, or nature of growth may be all that is important. Increasingly, however, it is necessary to measure growth in a quantitative manner. This may be required for purely practical or technical reasons. The specific growth rate or biomass yield coefficient may be measured to provide an absolute value or to compare growth on different media or under different environmental conditions. Such information may be of particular importance to fermentation technologists attempting to predict rates of substrate utilization or to optimize biomass production or product formation. Alternatively, quantitative measurement of growth and characterization of growth kinetics provide information on the physiology of an organism and mechanisms controlling and regulating growth.

Quantification implies the use of mathematics and of mathematical models. There exists a broad range of models, which reflects both the range of situations to which they have been applied and the range of functions they perform. Their classification will not be considered exhaustively here, but brief descriptions of the major types of models applied to fungal growth will be provided.

A. Empirical Models

Empirical models are frequently the simplest in mathematical terms. They are based solely on experimental observations and seek solely to describe experimental data quantitatively using the simplest available mathematical function. They are often derived using curve fitting techniques, e.g., linear and multiple regression. The basis for a particular function fitting the data need not be understood, and these models may therefore be termed descriptive. Thus, if hyphal length is seen to increase at a constant rate, a simple empirical model would be the equation for a straight line, and linear regression would yield a slope equal to the hyphal extension rate.

Such simple models are of enormous practical use. In our example, hyphal extension rate may be measured to determine the effect of particular compounds or environmental conditions on hyphal growth. They often represent the first stage of a modeling study and may be validated by subsequent models of greater complexity coupled to critical experimental work. In themselves, however, they provide no scientific understanding of underlying mechanisms of growth, and their use outside the range of the experimental data on which they are based is dangerous. For example, a compound may affect growth by increasing or decreasing the degree of branching with no observed effect on hyphal extension rate.

B. Mechanistic Models

Empirical models may be contrasted with mechanistic models. These consider the underlying processes controlling a system and represent them in mathematical form. Such models are equivalent to a theory, but stated mathematically rather than verbally. They are based on a set of assumptions. Some assumptions will function to simplify the system. For example, environmental conditions such as temperature and pH may be assumed to be constant and all hyphae may be assumed to extend at the same constant rate. Other assumptions are concerned with particular mechanisms proposed by the model, e.g., hyphal extenstion occurs by fusion of vesicles within the extension zone and the rate of absorption is represented by saturation-type kinetics.

Mechanistic models may serve an explanatory function, elucidating ways in which a number of interacting factors may affect the system under study. They also generate predictions which may be tested against experimental data. Such experiments must be designed to reflect as closely as possible assumptions of the first type, e.g., temperature must be constant, and to test rigorously and critically assumptions of the second type. Their development is therefore closely linked to experimental studies, and a close fit between experimental and predicted results provides support for the assumptions on which models are based.

The complexity and sophistication of mechanistic models, although allowing more detailed consideration of the system under study, often prevent analytical solution of equations on which they are based. This problem can be solved to some extent by the application of numerical approximation techniques and computer simulation languages [36]. This, however, does not allow investigation of the system under the full range of conditions and may not provide information on important relationships between components of the system.

C. Deterministic and Stochastic Models

Both mechanistic and empirical models may be deterministic or stochastic. Deterministic models of fungal growth are more common and typically consist of a set of algebraic equations or ordinary or partial differential equations. They describe relationships between dependent variables, e.g., biomass concentration, hyphal length, and an independent variable, usually time. The behavior of the system being modeled is determined solely by the equations, after provision of values for constants and initial conditions. The same sets of values will always predict the same behavior.

Stochastic models introduce statistical variability. In empirical models they may describe a body of experimental data in the form of a distribution. Other stochastic models may quantify variability in this way and then use it to predict the behavior of a system. An example will be given below in which the formation and development of a mycelial colony is described on the basis of statistical variability in properties of individual hyphae. Stochastic models are also important in describing processes, e.g., mutation, which are based on probability and, when considering small population sizes, where the behavior of individuals becomes important.

Other terms used to describe models will be defined as necessary, and a fuller classification is provided by Roels and Kossen [1]. The application of modeling within microbiology is described in a number of textbooks and reviews, e.g., Bazin [2,3] and Bazin and Prosser [4]. The distinction between the different classes of model is not always clear-cut, and models of mixed type are found, e.g., mechanistic models may contain empirical components for parts of the system for which no mechanism may be postulated.

Mathematical modeling has been applied to many areas of fungal growth, ranging from vegetative growth in laboratory culture to more applied areas associated with fungal ecology and industrial processes for production of biomass and secondary metabolites. This article will be restricted to vegetative growth of pure cultures of filamentous fungi on solid medium or in liquid culture. Although some aspects of mycelial development will be considered, models of differentiation, spore formation, etc., will not be considered.

II. THE HYPHAL EXTENSION ZONE

Extension of fungal hyphae occurs by tip growth. Material required for wall growth and membrane expansion is synthesized in a region behind the tip termed the peripheral growth zone (see below). It is transported to the tip in membrane-bound vesicles [5,6], which fuse with existing wall material in a region at the hyphal apex termed the extension zone. The length of this region may be determined as the distance from the tip to the point where hyphal diameter becomes constant or from a consideration of hyphal growth kinetics (see below).

The role of the extension zone during growth on solid medium is crucial. Growth of vegetative mycelium behind the tip is visible only through formation of branches but the rate of tip growth and hyphal extension determine the rate at which new medium may be explored and colonized. The extension zone is also the region capable of reacting to chemical or other signals involved in tropic responses. It is therefore responsible for the direction of hyphal growth and, ultimately, colony morphology and development.

The rate of hyphal extension depends on the rate at which new wall material is incorporated within the extension zone. This depends on three factors, the volume of the extension zone, or more correctly its surface area, the specific growth rate and the length of hypha supporting growth. The interaction between the second and third of these factors will be discussed below. Dependence on volume and surface area of the extension zone implies that extension zone length and hyphal diameter will show correlation with hyphal extension rate, and this has been observed experimentally [7]. Within the extension zone, the variation in hyphal diameter with increasing distance from the apex reflects hyphal tip shape, the quantitative analysis of which is important in determining the mechanism of addition of new wall material.

A. Hyphal Tip Shape

Initial attempts to model the shape of hyphal tips were based on models for growth of the characeous alga, *Nitella* [8,9]. The model considered a hypha to consist of a cylinder of constant diameter capped by a hemisphere of the same diameter. The growth of a small, initially circular area of wall material was then followed as it moved from the apex towards the cylindrical region. Growth was accompanied by a change in shape from circular to elliptical, depending on the value of the allometric coefficient, K. This defined the relative rates of growth along the meridional and longitudinal axes.

The model predicts that the rate of movement of a portion of wall material over the surface of the extension zone (dm/dt), away from the apex, is given by:

$$\frac{dm}{dt} = A \frac{r}{R}^{K}$$

where A is a constant, R is the hyphal radius in the cylindrical region, and r is the radius at any point within the extension zone, increasing from 0 to R with distance from the apex.

Green and King [8] solved their model for K = 0.5, 1, and 2 and tested predictions using an experimental model consisting of a circular rubber membrane, constrained at the rim and expanded with different types of reinforcement, reflecting different forms of anisotropic growth. These reinforcements were related to the physical properties of wall material but a more direct application for hyphal tip growth was given by Green [9]. This was based on the mechanism for hyphal growth proposed by Bartnicki-Garcia [10] in which wall material was loosened by wall-degrading enzymes prior to insertion of new wall material and subsequent rigidification.

A consistent prediction of Green's model is that the specific rate of area expansion at any point on the extension zone wall is proportional to cos ϕ, where ϕ represents the angle between the longitudinal axis and a line drawn from the base of the extension zone to that point on the wall. This arises from the assumption of hemispherical tip shape.

While this assumption may have validity for algal tip shape, fungal hyphal tips are not hemispherical. Trinci and Saunders [11] calculated the ratio of extension zone length, a, to hyphal radius at the base of the extension zone, b, for a number of fungi and found values ranging from 2.48 to 20.45. A hemispherical tip will give a value of a/b = 1, and this assumption is therefore invalid. Trinci and Saunders, therefore, developed a more general model in which tip shape was assumed to be represented as a semi-ellipsoid of revolution, with relatively large eccentricity, i.e., a >> b.

In fact, this model incorporates that of Green and King if a = b, and leads to the same relationship between specific rate of area expansion and cos ϕ. However, for increasing deviation from hemispherical shape, i.e., increasing values of a/b, the model predicts proportionality between the specific rate of area expansion and the cotangent of ϕ (Fig. 1).

The specific rate of area expansion is difficult to measure experimentally but is possible for giant sporangiophores of *Phycomyces blakesleeanus* by observing microscopically increases in distances separating starch grains placed on the tip surface [12]. Indirect estimation of variation in wall expansion with distance from the apex is provided by variation in the concentration of vesicles within the extension zone (as observed by electron microscopy) [13], and by variation in the degree of incorporation of the radiolabeled wall precursor *N*-acetyl glucosamine [14]. All of these measures show a concave upwards relationship. Data of Collinge and Trinci [13] and Gooday [14] are illustrated in Figure 2 for a > b rather than the concave downwards relationship indicated for hemispherical tip shape.

A similar approach was adopted by da Riva Ricci and Kendrick [15], in which the growth of a small area of wall material was determined by its existing area and an expansion function, θ, which varied with time and with meridional distance from the apex. A constant value of θ, implying uniform expansion over the whole tip, predicted decrease in hyphal diameter with prolonged hyphal growth. If, however, θ was assumed to decrease with increasing distance from the apex, realistic predictions were obtained, with the attainment of constant hyphal diameter. Da Riva Ricci

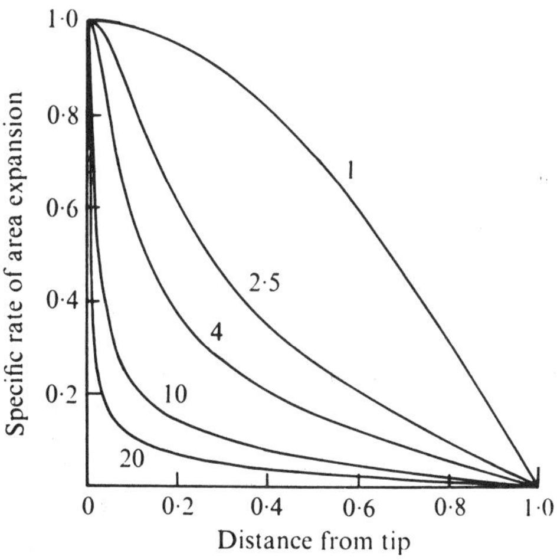

FIGURE 1. Variation in specific rate of area expansion with distance from the tip, measured along the tip surface, for selected values of the ratio a:b, where a and b represent the principal radii of curvature of a tip with semi-elliptical longitudinal cross section. All values of specific rate of area expansion are presented as proportions of the value at the apex, and distance is presented as a fraction of the total distance from the apex to the base of the extension zone. (From Ref. 11.)

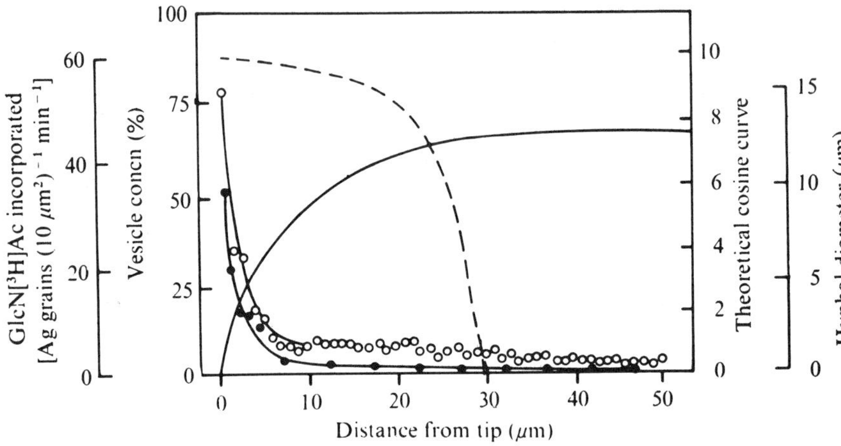

FIGURE 2. Hyphal diameter (——), vesicle concentration (○), expressed as the percentage of protoplasmic volume occupied by vesicles, of *Neurospora crassa cot*−3 and the specific rate of incorporation of *N*-acetyl glucosamine (●) of another strain of *N. crassa*, expressed as the number of silver grains per 10 μm^2 resulting from incorporation of labeled precursor in 1 min. The hypothetical cosine relationship (---) is also shown for *N. crassa cot*−3. (From Ref. 11.)

and Kendrick suggest this represents a situation in which newly incorporated wall material increases in rigidity as it moves back from the apex. This implies a different mechanism to that of Bartnicki-Garcia [10] and is relevant to subsequent models for tip growth.

B. Surface Stress Theory

The surface stress theory for microbial growth and form states that the shape adopted by microorganisms results from internal hydrostatic pressure, driven by growth and biosynthesis. This pressure is usually assumed to be uniformly distributed within a cell and is exerted on the cell wall. The cell wall then adjusts its shape to minimize stress within the wall. The shape adopted therefore reflects variation in apparent surface tension or elasticity forces within the wall and further reflects differences in mechanisms of incorporation of wall material. The theory has been developed extensively for growth of bacterial cells [16,17], but was first applied to growth of hyphal tips by Saunders and Trinci [18] and Koch [19].

Saunders and Trinci [18] compared two models for hyphal growth. The first was similar to those already described and considers expansion of an area of wall supplied with material at a rate f(r), which is a function of distance from the tip. If wall thickness, ρ, is assumed constant, then:

$$f(r) = c\rho\left(1 + \frac{1}{K}\right)Kr^{K-1}$$

where c is a constant, K is the allometric coefficient, and r is the hyphal radius at any point within the extension zone. While the allometric coefficient can be seen to depend on variation in the rate of supply of material to the tip, the equation contains no information on tip shape, i.e., tip shape is independent of f(r) and independent of variation in the rate of supply of material to the tip.

The second model, based on the surface stress theory, assumed the tip wall to be effectively elastic and capable of adjusting its shape in response to internal hydrostatic pressure to minimize surface energy within the wall. The pressure difference, P, across the wall is represented by the equation:

$$P = \alpha\left(\frac{1}{R_1} + \frac{1}{R_2}\right)$$

where α is the surface tension coefficient and R_1 and R_2 are the principal radii of curvature of the surface of the extension zone. This equation, in fact, describes the situation in a soap bubble, where P and α are constant. This implies that $R_1 = R_2$, i.e., we have a sphere, which is the shape adopted by a soap bubble. While P may be assumed constant in the extension zone, R_1 does not equal R_2, implying that α will also vary.

Saunders and Trinci [18] assumed that the extension zone had a circular transverse cross-section and calculated variation in elasticity with distance from the tip for a number of semi-elliptical tip shapes, represented by different ratios of length: radius of the extension zone (a/b). For a = b, i.e., a hemispherical tip, elasticity was found to be constant throughout the extension zone wall. With increasing values of a/b, however, elasticity, α, followed curves similar to those in Figure 1, i.e., concave upwards. Again these compare well with experimental data (Fig. 2) indicative of wall elasticity. This work suggests, therefore, that tip extension is driven by internal hydrostatic pressure, that incorporation of wall material decreases with distance from the hyphal apex, and that this decrease is reflected in decreased elasticity, which is reflected by the shape adopted by the extension zone.

Koch [19] approached this problem is a slightly different manner, based on the equation:

$$PdV = TdA$$

This equation describes the relationship between work done during an increment of growth (equal to the product of the internal hydrostatic pressure, P, and the incremental increase in volume, dV) and the energy required to create unit surface area (equal to the product of surface tension, T, and the incremental increase in area, dA). The increases in dV and dA occurring during tip growth for a small time interval, dt, were calculated by trigonometry. The resultant equation was then used to derive an expression relating the slope of the tip wall (S) to the radius of the extension zone, this relationship defining tip shape:

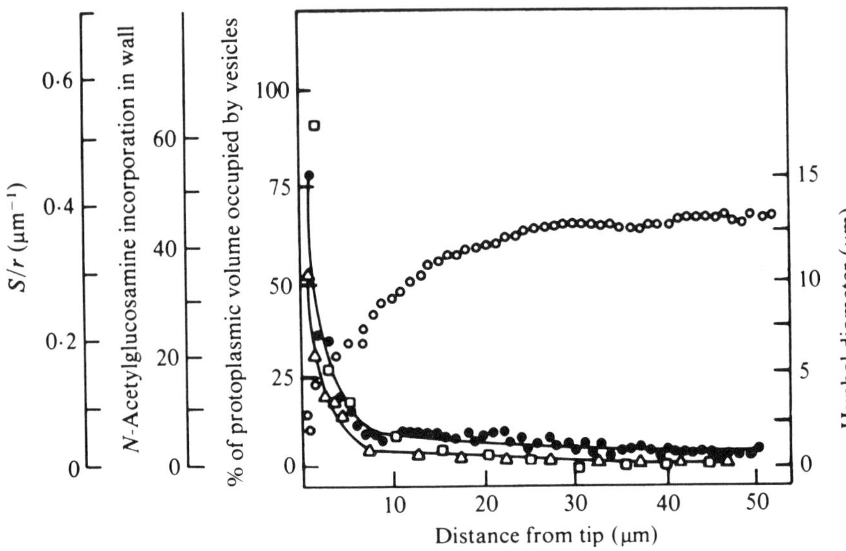

FIGURE 3. Experimental data on hyphal diameter (○), vesicle concentration (●), and specific incorporation of *N*-acetylglucosamine (△) for *N. crassa*, as in Figure 2, and values of S/r (□). (From Ref. 19.)

$$\frac{S}{r} = \frac{P}{2T}$$

T represents surface tension, which will be inversely proportional to elasticity or the indirect measures of wall elasticity discussed above. Rather than assuming the extension zone to be semi-elliptical in longitudinal cross-section, Koch measured shapes of actual tips (i.e., S/r) from which the data illustrated in Figure 2 were obtained. He then plotted S/r vs. 1/T and, as can be seen in Figure 3, found excellent fit.

This therefore provides further confirmatory evidence for the surface stress theory for growth and shape of hyphal apices. The mechanism proposed for wall growth by Bartnicki-Garcia [10] involves splitting of existing bonds within the wall and insertion of new material. This is difficult to conceive in a wall subjected to high internal hydrostatic pressure, although mechanisms to achieve this can be postulated. Recent work [6] suggests a more likely mechanism in which wall material is inserted at the tip region and then rigidifies due to formation of cross-linkages between microfibrils until complete rigidification at the base of the extension zone leads to constant hyphal diameter.

III. GROWTH OF INDIVIDUAL HYPHAE

The rate of tip extension depends on the size of the extension zone and on flux of material to the hyphal tip. Immediately following spore germination, extension of the germ tube will be supported by mobilization of material stored in the spore and biosynthesis in the germ tube itself. The latter will increase with increasing germ tube length, i.e., growth will be autocatalytic. During the initial stages of germ tube growth, therefore, hyphal length, L, increases exponentially at a constant specific rate, α:

$$L = L_o e^{\alpha t}$$

where L_0 is the spore diameter. This was first observed experimentally by Smith [20] for germ tubes of *Botrytis cinerea* with similar behavior in *Chaetomium globosum* [21] and a range of other fungi [22,23]. The period of exponential growth varies between fungi and is related to the length of the peripheral growth zone, discussed below. In *Rhizopus stolonifer* it continues until the germ tube is 40 μm in length, while sporangiophores of *P. blakesleeanus* grow exponentially until 4 mm in length.

It is relatively easy to measure the increase in germ tube length, but difficult to measure the specific rate of biomass production during exponential growth in liquid culture. It would be convenient, therefore, if α, calculated from germ tube growth on solid medium, reflected μ_m, the maximum specific growth rate on the equivalent liquid medium. Unfortunately, this is not the case due to the contribution to initial tip growth from endogenous spore reserves. This can result in values of α up to nine times greater than those of μ_m.

Following exponential growth, hyphal extension rate decreases, enters a deceleration phase, and eventually achieves a constant linear rate, K_r:

$$L = L_o + K_r (t - t_o)$$

where L_o is hyphal length a time t_o. Subsequent growth is linear and may proceed indefinitely at the same rate K_r, although there is evidence in some fungi (see Section VI.C) for increase in K_r with hyphal and mycelial development.

The change from exponential to linear hyphal extension occurs when material synthesized in distal regions is unable to reach the tip and no longer contributes to extension. Subsequent extension is therefore supported by a constant length of hypha, providing a constant amount of material.

IV. MYCELIAL GROWTH

A. Growth Kinetics

Microorganisms typically grow exponentially when nutrients are in excess, inhibitors are absent, and physiological conditions favorable for growth. Linear growth of germ tube hyphae does not allow for exponential biomass production unless hyphae can increase in diameter. This is not normally the case, but exponential growth is achieved by branch formation, which generally occurs before the linear phase of germ tube extension. Again this was first observed by Smith [20] and Plomley [21], the former observing an exponential increase in the combined lengths of parent and branch hyphae and the latter noting an exponential increase in total mycelial length.

Figure 4 illustrates mycelial growth kinetics during early growth of *Geotrichum candidum* [23]. Total mycelial length increases exponentially. The first branch is formed before linear growth of the germ tube and, after a phase of discontinuous branch production, the number of tips increases continuously and exponentially at the same specific rate as that for total mycelial length. If hyphal diameter is assumed to be constant, as is the case in these young mycelia, exponential growth is seen to result from an exponentially increasing number of hyphae, each of which extends at the same linear rate. In a sense, branch formation is equivalent to cell division in unicellular organisms. In addition, Trinci [23] demonstrated for several organisms that the specific increase in total mycelial length was equivalent to the specific growth rate in liquid culture, measured directly as biomass dry weight.

Although each branch is assumed to have the extension rate of its parent hypha, there is evidence that in *A. nidulans* [24] and *Geotrichum candidum* the parent hypha contributes to branch extension for a short

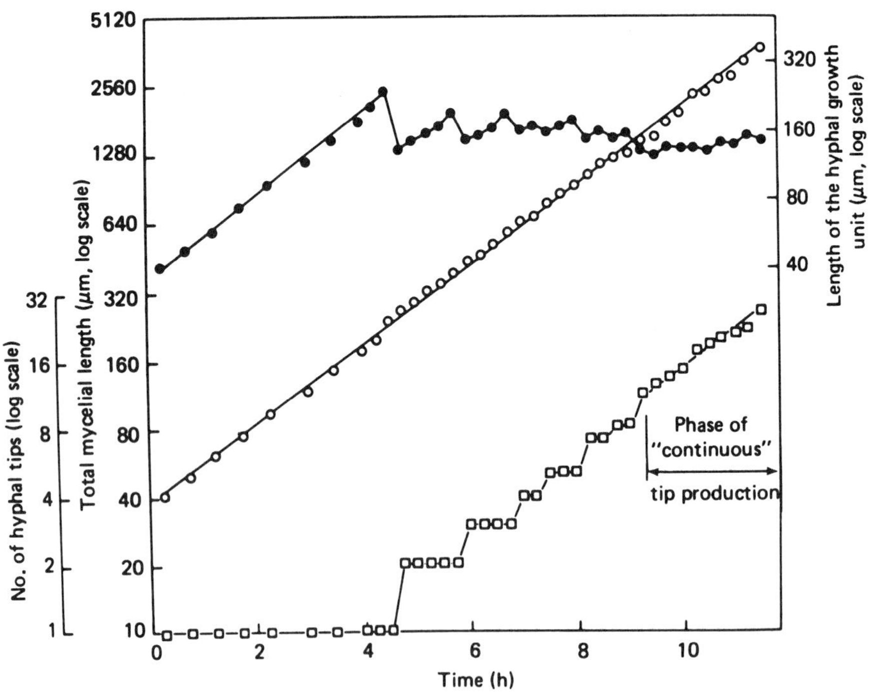

FIGURE 4. Early growth of a mycelium of *Geotrichum candidum* on solid medium, number of tips (□), total mycelial length (○), and hyphal growth unit length (●). (From Ref. 23.)

time. The kinetics of extension of branch hyphae is therefore similar to that of germ tubes, with exponential, deceleration and linear phases. There is also a degree of variation in hyphal extension rates, even in young colonies. This is important for mycelial development and will be discussed in Section VI.C.

B. A Vesicular Model

Prosser and Trinci [25] presented a model which related the growth and branching of individual hyphae to mycelial growth kinetics, incorporating recognized cellular mechanisms for growth and branching [10,13]. Vesicles containing material and enzymes required for wall expansion are produced in distal regions of the hypha and are transported to the hyphal tip at a constant rate. At the tip vesicles are absorbed at a rate represented by saturation-type kinetics. This gives rise to tip extension, the amount being related to the surface area of vesicle membrane and hyphal diameter.

Branch formation is incorporated along with consideration of septation and the duplication cycle, for filamentous fungi [26 – 29], equivalent to the cell cycle of unicellular organisms. Germ tube growth occurs until nuclear division increases the number of nuclei from four to eight. A septum is then formed, dividing nuclei equally between the newly formed apical and subapical compartments. Growth and nuclear division in the apical compartment then proceed in a similar fashion, leading to repeated nuclear division and septation.

Septation is assumed to reduce the flow of vesicles to the hyphal tip, preventing flow completely for septa without septal pores, or reducing flow for septa containing pores through which a proportion of vesicles can pass. In both cases this leads to accumulation of vesicles behind the septum and subsequent formation of a new hyphal apex and branch. The branch then grows in the manner of its parent.

This provides an example of a mechanistic model, in that it describes mycelial growth and development on the basis of cellular mechanisms for hyphal extension and branch formation. It cannot be solved analytically, and predictions were generated using a computer-based finite element model in which a hypha was considered to consist of a number of hypothetical segments or compartments. Vesicle flow was modeled as the movement of vesicles from distal to proximal compartments, this being prevented or reduced by septation. The extension zone was represented as an apical segment, receiving vesicles from the adjacent distal compartment. Within the apical segment the number of vesicles, R_V, absorbed in the time interval δt is given by the equation:

$$R_v = \frac{R_a V_0 \delta t}{V_0 + K_v}$$

where R_a is the maximum rate of adsorption, V_0 is the vesicle concentration in the apical segment, and K_V is the vesicle concentration at which the rate of absorption is half the maximum. Resultant extension, E, is given by:

$$E = (V_{0_{(t - \delta t)}} - V_{0_{(t)}}) R$$

where $V_{0_{(t)}}$ and $V_{0_{(t-\delta t)}}$ represent vesicle concentrations at times t and $t - \delta t$, and R is the increase in hyphal length resulting from absorption of a single vesicle.

The model was used to predict early mycelial growth of *G. candidum* and *A. nidulans*, and predicted results were compared with experimental data of Trinci [23]. Kinetics similar to those illustrated in Figure 4 were obtained with experimental increases in total mycelial length and number of branches, at similar specific rates, and linear extension of individual hyphae. Quantitative predictions also agreed well with experimental data, thereby providing evidence for the validity of the assumptions on which the model was based. It also indicated areas where information was lacking, thereby directing experimental work.

V. COLONY GROWTH

A colony developing from a germinated spore on solid medium rapidly achieves a circular morphology. The mechanisms involved in formation of a circular colony are discussed in Section VI. Here, the kinetics of growth of individual colonies will be discussed.

A. The Peripheral Growth Zone

The most obvious parameter for assessment of colony growth is colony radius or diameter. This is conveniently measured microscopically during early growth and, more easily, by macroscopic measurement of diameters of mature colonies over periods of hours or days. The kinetics of early colony growth reflect that of individual hyphae, with an exponential increase in radius, followed by a deceleration phase and a linear phase. The linear phase can proceed indefinitely, given sufficient medium, and the colony expands at a constant radial growth rate, K_r, equivalent to the average extension rate of hyphae at the margin of the colony. This can be represented by the simple empirical model:

$$r = r_0 + K_r t$$

where r and r_0 represent colony radii at times t and 0, respectively.

The first attempt to model growth of microorganisms on solid medium was made by Pirt [29]. He considered the mass, M, of a colony growing under conditions of nutrient excess and proposed the equation:

$$\ln M = \mu t + \ln M_0$$

where μ is the specific growth rate and M_0 the initial mass of the colony, e.g., the mass of a germinating spore. If the density (ρ) of the biomass within the colony is assumed to be constant and uniform and the colony is assumed to be disk shaped with constant height, h, then M may be related to colony radius by the equation:

$$M = \pi r^2 \rho h$$

Substitution into the previous equation provides an expression for the rate of colony radial expansion:

$$\ln r = \frac{\mu t}{2} + \ln r_0$$

i.e., the colony radius will increase at an exponential rate equivalent to the specific growth rate, μ (this takes no account of the effect of spore reserves on initial germ tube growth). Growth at the center of the colony will become limited due to the large nutrient demand created by high biomass density and restrictions imposed on nutrient supply by diffusion, or alternatively due to accumulation of inhibitory compounds or space limitation. This eventually leads to cessation of growth at the center, and subsequent colony extension results from growth in an annular ring of mycelium of constant width at the colony margin. If mycelium in this region grows at a specific rate, μ, and has a biomass Mg, then:

$$\frac{dM}{dt} = \mu \; Mg$$

Pirt assumed the cross-section of this ring to be triangular and substituting for Mg obtained the expression:

$$r = \frac{w\mu t}{2} + r_0$$

where w is the width of exponentially growing peripheral ring of mycelium. If colony height is constant, and the peripheral ring has rectangular, rather than triangular cross-section, the colony radius is seen to increase by the equation:

$$r = w\mu t + r_0$$

Comparison with the empirical model above indicates that

$$K_r = \mu w$$

Trinci [30] used the term peripheral growth zone to describe the peripheral ring of exponentially growing mycelium and suggested that its width was constant during colony development and that its specific growth rate, μ, was equivalent to specific growth rate in liquid of medium of equivalent composition.

The width, w, has been determined in three ways. The most direct is to cut the margin of a colony diagonally with a razor blade and follow subsequent extension of the colony margin. The peripheral growth zone width is the minimum length of colony margin that can be cut without affecting subsequent colony expansion. At the level of the individual hypha, therefore, w is equivalent to the maximum length of hypha supplying material for apical growth. Biosynthesis will occur in more distal regions but will not be transported to the tip at a net rate greater than the hyphal extension rate and will therefore never reach it. This view of the peripheral growth zone provides the basis for two other techniques for estimating w, these being the length of hypha associated with tropic responses and the length of hypha at which hyphal extension kinetics switch from exponential to linear.

Trinci [30] tested the validity of the peripheral growth zone concept by comparing experimentally determined values of w with those calculated from the colony radial growth rate, K_r, divided by the specific growth rate in liquid culture, μ. In all cases the two values showed close agreement within the range of experimental error, providing strong evidence for this concept of mycelial growth.

Trinci also found that w varied between different species and with variation in some, but not all environmental conditions. If w is constant, K_r provides a very convenient measure for comparing specific growth rates under different conditions, much more convenient than measurement of biomass in liquid culture. This technique, however, cannot be used if w varies with the conditions under investigation or for comparisons between species or strains.

The most obvious physical representation of the peripheral growth zone is the apical compartment of a hypha delimited by a complete septum. In hyphae containing septa with central pores, material can pass through the septum to contribute to apical growth, potentially increasing w. In aseptate hyphae, w represents the distance at which apical polarity mechanisms are insufficient to allow material synthesized in distal regions to reach the extension zone.

The peripheral growth zone concept therefore has practical use in defining the conditions under which K_r may be used to assess μ and the limitations of this technique. It also greatly increases our understanding of the control and regulation of biomass formation in mycelial organisms and the relationship between growth on solid medium and in liquid culture. It implies that strains with the same specific growth rate can have widely differing colony radial growth rates as a result of different values for w. For those with low K_r values, this must result in increased branch formation to accommodate the biomass being synthesized.

This highlights an important feature of mycelial growth, which distinguishes it from that of colony formation by unicellular organisms. By controlling and regulating the distribution of biomass between apical extension and branch formation, a mycelium can adjust to variation in nutritional conditions and maximize its ability to explore and colonize new regions. This explains the ability of filamentous fungal colonies to completely cover petri dishes while bacteria and yeasts form small, self-limiting, discrete, colonies. This morphology can be induced in filamentous fungi by conditions which lead to increased branching, effectively decreasing w and increasing depletion of nutrients in the region surrounding the colony.

The effect of branching has been incorporated by Morrison and Righelato [31]. They proposed the equation:

$$K_r = \mu\ Gk$$

where k is constant and G is the hyphal growth unit. This is determined experimentally as the ratio between total mycelial length and number of branches and can be seen in Figure 4 to reach a constant value early in colony growth. It represents the average length of hypha associated with each growing tip under conditions of unrestricted growth and provides a mechanism for temporal regulation of branch formation.

B. Application of the Logistic Equation

As stated above, growth at the center of a young colony becomes limited by a number of possible factors. The classical empirical model for growth limitation is the Verhulst-Paerl logistic equation, which has the form:

$$\frac{dN}{dt} = N\ (r - kN)$$

where N represents the number of individuals in a population, r is the intrinsic rate of increase, equivalent to maximum specific growth rate, and k is the carrying capacity of the environment, equivalent to the yield. The equation assumes that specific growth rate decreases as an inverse function of population size.

Koch [32] applied this equation to mycelial growth suggesting that specific growth rate decreased, in the above manner, with increasing distance from the colony margin. The number of individuals, N, was replaced by the amount of biomass W, occupying unit volume of space, dV, and the initial population size was replaced by S, the initial mycelial mass occupying space dV. He also replaced time, t, by T − t, where T is the time that growth was initiated and t the time a particular volume of space had been occupied. These modifications led to the equation:

$$\frac{dW}{dt} = \frac{SKe^{r(T - t)}}{K - S + Se^{r(T - t)}}\ dV \qquad T - t > 0$$

K represents the carrying capacity, i.e., the maximum biomass density possible. By further assuming a colony to have cylindrical symmetry and lag and exponential periods of growth to be negligible and restricting growth to two dimensions, Koch derived the expression:

$$W = \int_0^T \frac{Ske^{r(T - t)}}{K - S + Se^{r(T - t)}} \, 2\pi ha^2 t\,dt$$

where h is colony height and a represents colony radial growth rate, equivalent to K_r. Expressions for three-dimensional growth were also derived, equivalent to growth in liquid medium in the form of pellets. The model was solved using numerical approximation techniques and generated predictions regarding changes in mycelial mass, W, with time.

The key factor in determining the kinetics of growth is K/S, the ratio of final, or maximum, to initial mycelial density. W increases exponentially with time for values of K/S between $10^2 - 10^4$ until $K/S \simeq e^{rT}$, after which the increase is curvilinear. There is a phase of cube root kinetics following exponential growth in three dimensions, while colony growth in two dimensions follows square root kinetics when growth is limited by circular expansion of the colony.

Koch applied this model to data of Deppe [33] on colony growth of five fungi growing on membrane surfaces. Colony mass increased exponentially, with constant radial expansion, and was described best by values of $K/S > 100$. This high value follows from the need for surface colonization to enable growth above the surface and from the need to supply aerial hyphae with nutrients from the solid medium. Comparison with data of Trinci [34] on growth kinetics of pellets of *A. nidulans* in liquid culture implied a value of $K/S \simeq 1$. Here linear expansion was associated with cube root kinetics for total biomass. The small value for K/S reflects the fact that medium surrounding the pellet is simultaneously colonized by hyphae from the whole pellet surface, leaving virtually no unoccupied space which may act as a source of nutrients for subsequent growth.

This empirical model is important in providing a unified description of mycelial growth in liquid culture and on solid medium. It predicts the different kinetics observed experimentally, linear colony expansion, exponential, square-root and cube-root kinetics for biomass, and indicates the conditons under which each will occur. It also provides a target for more sophisticated mechanistic models for colony and pellet formation.

VI. MYCELIAL AND COLONY DEVELOPMENT

The early development of a mycelial colony is a well-ordered and regulated process, as indicated by the colony of *Coprinus sterquilinus* [35] illustrated in Figure 5. Hyphae tend not to cross and are oriented radially towards the colony margin, where they are evenly spaced. The growth and pattern formation exhibited by such a colony depends on four factors, discussed in detail by Prosser [36]: the rate and direction of hyphal extension and the temporal and spatial distribution of branches. The control of hyphal

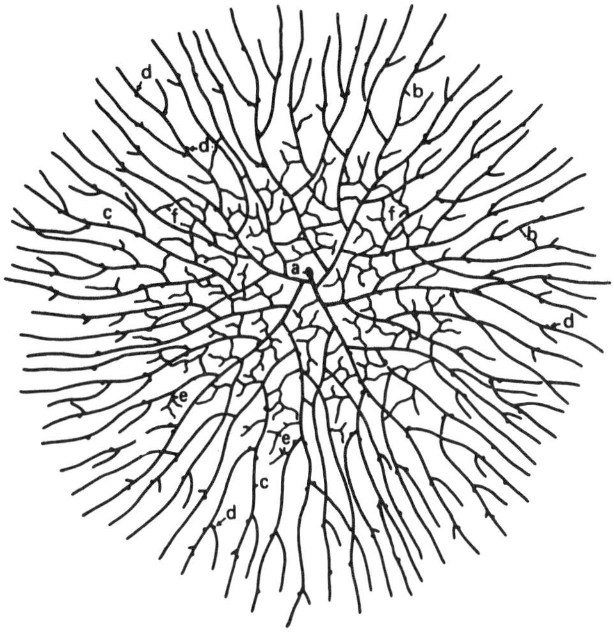

FIGURE 5. A colony of *Coprinus sterquilinus* grown on solid medium showing positions of basidiospores (a), leading hyphae (b), clamp connections (c,d), meandering lateral hyphae (e), and anastomoses (f). (From Ref. 35.)

extension rate has been discussed above. The factors which determine the direction of hyphal growth are less well understood, but the negative autotropic effects observed during mycelial growth appear to result from gradients in oxygen concentration around growing hyphae [37,38]. The reduction in extension rate resulting from reduced oxygen concentration results in an avoidance reaction, the mechanism for which is unknown but must reside in the hyphal extension zone.

The kinetics of branch formation are related to the hyphal growth unit, G, defined above as the average length of hypha associated with a growing tip. This applies only if hyphal diameter is constant throughout the mycelium, and the work of Robinson and Smith [39–40] and Smith and Robinson [42] demonstrates the importance of considering the volume, rather than the length, of the hyphal growth unit.

In young mycelia, however, hyphal diameter is generally constant, and the specific growth rate, μ, and hyphal growth unit length are related by the equation:

$$E = \mu G$$

where E represents the mean hyphal extension rate and is calculated as [7]:

$$E = \frac{2(H_t - H_0)}{B_0 + B_t}$$

where H_0 and H_t represent hyphal lengths at time 0 and 1 hr later and B_0 and B_t the respective number of hyphal tips. As discussed by Bull and Trinci [44] and Prosser [45], E, μ, and G may be independently affected by environmental conditions and paramorphogens that affect the degree of branching. The interrelationships between these factors are important in understanding variation in branching patterns and quantitative aspects of mycelial development.

Finally, the positioning of branches within hyphae, as indicated above, is associated with septum formation, and is believed to result from accumulation of vesicles behind septa, as modeled by Prosser and Trinci [25]. There is also evidence that other factors influence the site of branch initiation, e.g., branch formation in hyphae of *G. candidum* within an oxygen gradient occurs more frequently on the side of the hypha experiencing the greater concentration of oxygen [37].

A. Stochastic Models

Cohen [46] was the first to consider growth of individual hyphae and the prediction of branching patterns in young mycelia. Growth was assumed to occur by apical extension, the rate and direction of growth being determined partly by a "local density field" surrounding a tip. Extension rate was reduced as crowding increased and ceased at a certain maximum level of hyphal density, in a manner similar to the model of Koch [17] described in Section V.B. Branch formation was modeled stochastically, the probability of branch initiation at any point on a hypha being determined by the minimum interbranch distance, to some extent analogous to the hyphal growth unit, and the apical compartment length. Branch angle was determined as for the direction of hyphal extension, and the model provided a facility for biasing both hyphal extension and branching. The model was simulated by application of growth and branching rules based on the above assumptions, with branch formation determined by the outcome of a random trial.

Predictions generated by the model consisted of mycelia which bore many similarities to those observed experimentally. Quantitative comparisons were not made with experimental data, but hyphal growth was oriented away from the colony center, and the model predicted well the qualitative effects of centers of attraction or repulsion within the medium on mycelial growth.

Cohen's [46] model introduced stochastic processes into branch formation and positioning but otherwise treated mycelial growth in a deterministic manner. Hutchinson et al. [47] determined the statistical variability in the properties of individual hyphae of *Mucor hiemalis* growing on solid media and incorporated this information into a stochastic model for colony growth. Three properties of hyphal growth were considered: hyphal extension rate, interbranch distance (internodal length), and branch angle. For each they determined the mean and probability distribution around the mean for a population of hyphae in a growing mycelium.

Variability in hyphal extension rate was found to fit a half-normal distribution, with probability density function:

$$p\ (\chi/\delta) = (2/\pi\delta)^{\frac{1}{2}}\ \exp\ (-\chi^2/2\delta^2) \qquad (0 < \chi < \infty)$$

with a mean of 85.3 μm/hr and δ = 107. Interbranch distance followed a gamma distribution, with shape parameter δ = 1.5 and scale parameter θ = 0.011:

$$p\ (\chi/\alpha) = \theta^{\alpha}\chi^{\alpha - 1}\ \exp\ [-\theta\chi/\Gamma(\alpha)] \qquad (0 < \chi < \infty)$$

Branch angle followed a normal distribution with mean μ = 50° and standard deviation s = 170:

$$p\ (\chi/\mu,\delta) = (2\pi\delta)^{-\frac{1}{2}}\ \exp\ [-(x - \mu)^2/2\delta^2] \qquad (-\infty < \chi < \infty)$$

This information was then used to construct a colony from a short length of hypha, equivalent to a germ tube. Extension of this hypha was calculated for a series of time intervals by generation of random numbers which were used to determine actual increases from the mean and probability density distribution given above. At each time interval, a similar stochastic process was used, involving the distribution of interbranch lengths, to determine whether, and where, a branch should be formed. If a branch was formed its direction of growth was determined by the mean and distribution for branch angle, again by a stochastic process, and its subsequent growth and branching were modeled by application of the same procedures.

Once a branch is formed, and the branch angle determined, a branch hypha grows in a straight line. The model does not, therefore, consider hyphal avoidance reactions, and this provides the most notable morphological difference between predicted and experimentally observed colonies. A further difference is colony size, simulated colonies increasing in size faster than those observed experimentally. This results, in part, from their unidirectional growth, and variation in the size of predicted colonies appears to depend only on variation in the number of branches formed during the first few cycles of model simulation.

The main aim of this work was, however, to demonstrate that formation of a circular colony can result solely from natural and observed variation in the properties of individual hyphae. The model provided strong evidence for this in that predicted colonies quickly adopted a circular morphology, which was maintained during subsequent growth. Mechanisms for orientated growth of hyphae within a colony and regulation of biomass distribution between hyphal extension and branch formation are not necessary for formation of a circular colony, although they may have an important influence on other features of mycelial development and morphogenesis.

Adupa [48] adopted a similar stochastic approach in modeling mycelial growth of *Penicillium claviforme*. The major addition to the model of Hutchinson et al. [47] was consideration of the direction of hyphal extension, which was varied on the basis of observed statistical variability. This greatly improved the morphology of predicted colonies in comparison to that observed experimentally for *P. claviforme*. Differences between experimental and predicted morphologies suggested a need for consideration of hyphal interference and avoidance reactions. The model confirmed the ability to predict the growth and morphology of mycelial colonies on the basis of observed variations in a limited number of properties of individual hyphae, and the extension of such models to predict the effect of changes in the values of these properties, coupled with experimental testing of predictions, would be of great value.

B. Deterministic Models

Edelstein [49] proposed a new approach to the modeling of mycelial growth in which average properties of hyphae are considered, rather than the properties of individual hyphae. This model is based on two such properties. Hyphal density, ρ, is defined as the length of hyphal filaments per unit area, and has been considered in some of the models already described. Tip density, n, is defined as the number of tips per unit area and represents a new property which emphasizes the importance of the hyphal tip in colony formation. Changes in these properties depend on both time and distance and must therefore be described by partial differential equations:

$$\frac{\partial \rho}{\partial t} = nv - d(\rho)$$

$$\frac{\partial n}{\partial t} = -\frac{\partial nv}{\partial x} + \delta(n, \rho)$$

where v is the hyphal extension rate, assumed to be constant, and ∂ represents creation of tips through branch formation and is dependent on both n and ρ. The model also introduces hyphal death (through autolysis) which occurs at a constant rate, d, dependent on ρ. The first equation states that, within a certain area of medium, the change in hyphal density will depend on tip flux (nv) and cessation of extension (d). The second equation allows for an increase in the number of tips through branching and a change in tip density due to entry to or exit from the area being considered as a result of tip growth.

Edelstein then defines a series of functions for describing dichotomous and lateral branching, tip-hypha and tip-tip anastomoses, and tip death due to overcrowding. The model therefore extends those described previously by considering in depth events occurring behind the peripheral growth zone which are not directly associated with biomass production but represent the early stages of differentiation.

In its basic form the model was subjected to phase-plane analysis, in which two dependent variables are plotted against each other for a series of time-dependent solutions. This allowed investigation of different branching strategies with regard to their advantages for continued colony growth. Dichotomous branching and tip-tip anastomoses tended to produce self-limiting, discrete nonspreading colonies with growth throughout the colony and hyphal density greatest at the center. Spreading colonies resulted from mycelia forming tip-hypha anastomoses, with uniform hyphal density and maximum tip density immediately behind the colony margin.

Edelstein and Segel [50] extended this model to consider uptake of a growth-limiting substrate from the medium, the rate of which was governed by Michaelis-Menten kinetics. Within the medium, changes in the concentration of substrate were described by simple diffusion. Changes in substrate concentration within the mycelium were also modeled, being dependent on uptake, diffusive and convective flux, and consumption for growth and maintenance. This extended model provides an explanation for rhythmic growth described below, and experimental data now exist [51] against which predictions could be tested.

Edelstein et al. [51] applied the original model to growth of *Sclerotium rolfsii* on solid media. The model incorporated lateral branching, tip-hypha anastomoses, and hyphal autolysis, all of which are features of growth of this organism, and generated predictions of hyphal and tip density as functions of distance from the colony margin. These predictions showed good qualitative agreement with experimental data on hyphal density, obtained by weighing mycelia removed from growing colonies, and also indicated that the length of the hyphal growth unit is constant within the peripheral growth zone.

This approach is therefore novel and incorporates a number of features not considered by earlier models. It is also flexible in providing a general model which can be readily modified to include the properties of specific organisms. It may also be useful in constructing models for differentiation processes within developing mycelial colonies.

An important feature of Edelstein and Segel's [50] model was its description of changes in the concentration of a limiting substrate in medium beneath the colony. This has also been considered by Georgiou and Shuler [53] in a computer model, which also includes conidiophore production and conidiation. A colony is considered to consist of four components: vegetative biomass, competent biomass, which is capable of differentiation, conidiophore biomass, and conidial biomass. The rates of change of each of these components and of colony radius and glucose and nitrate concentrations are described by seven differential equations.

The model was used to generate predictions on colony radial growth kinetics, which compared favorably with experimental data. Other predictions, e.g., those relating to conidiation, were not tested. Figure 6 illustrates the predicted gradients for glucose concentration beneath colonies at 28 and 74 hr, at which times predicted colony radii were 0.68 and 1.55 cm, respectively. This compares favorably with data of Robson et al. [51] (Fig. 7) for depletion of glucose beneath colonies of *Penicillium chrysogenum*. In both cases, glucose was depleted in medium some distance ahead of the advancing colony margin. These experimental data provide a means of testing quantitative predictions of the model of Georgiou and Shuler [53] for vegetative growth, and of validating its use for modeling subsequent differentiation processes.

C. Mycelial Differentiation

From the preceding sections it is apparent that the assumption that all hyphae within a mycelium have similar properties is not always valid. The observations of Hutchinson et al. [40] quantify the variability found in various properties of individual hyphae in young mycelia. It is also commonly observed [20,21] that parental hyphae have greater diameters and

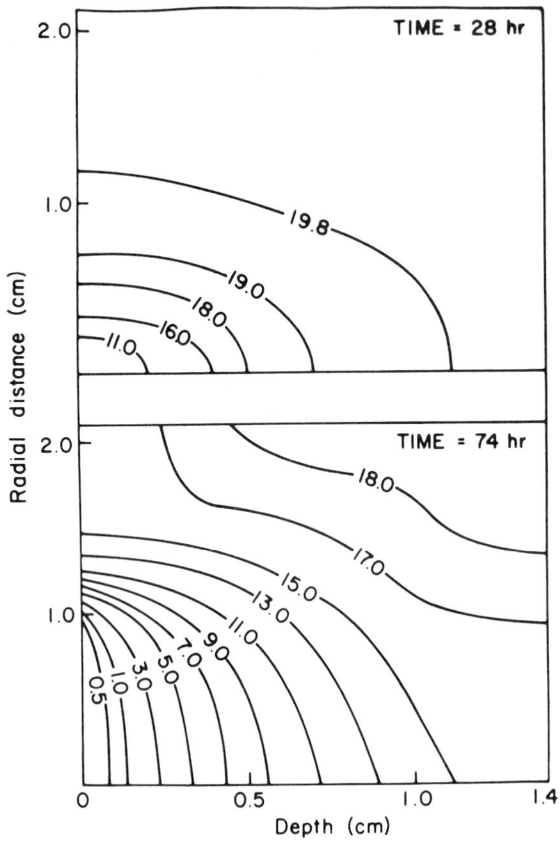

FIGURE 6. A two-dimensional glucose profile for a fungal colony growing on solid medium with glucose as the limiting substrate. (From Ref. 60.)

extension rates than the branches they subtend. For example, Butler [54] found that the extension rates of leading hyphae and primary and secondary branches of *Coprinus disseminatus* varied in the proportion 100:66:18. Based on observations of hyphae of *Neurospora crassa*, Steele and Trinci [7,43] distinguished between undifferentiated and differentiated mycelia. In the former all hyphae had similar diameters, extension zone lengths, and extension rates, as found in young mycelia. In differentiated mycelia, found at the margins of mature colonies, a hierarchy existed such that parental hyphae subtended branches in which these three properties were reduced. In addition, the extension rate of even secondary branches in differentiated mycelia was significantly greater than that of germ tube hyphae, growing under ideal conditions of substrate supply and absence of inhibitors.

The sequence of events involved in this developmental process was studied by McLean and Prosser [55]. Growth of *N. crassa* on solid medium was followed for approximately 2 days. In young mycelia, all hyphae extended at the same rate, hyphal diameter was constant and the average angle subtended by newly formed branches was 90°. At 22 hr branch angle

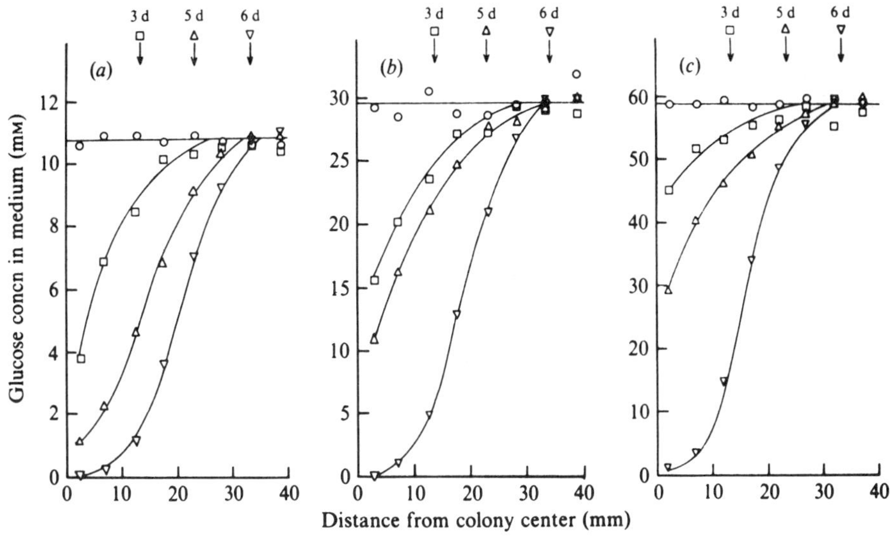

FIGURE 7. Glucose concentration of medium in uninoculated plates (○) and of medium below colonies of *Rhizoctonia cerealis* grown on medium containing (a) 10 mM glucose, (b) 25 mM glucose, or (c) 50 mM glucose for 3(□), 5(△), and 6(▽) days. Arrows indicate the position of the colony margin 3, 5, and 6 days after inoculation. (From Ref. 51.)

decreased to 63° (Fig. 8a), hyphal extension rates and diameters increased (Figs. 8b,c), and the hierarchy described above was established. This process was completed after approximately 10 hr with hyphal extension rates in the proportion 100:62:26.

Two processes appear to be operating here, and probably in other fungi. The first is a gradual increase in hyphal diameter and extension rate as the germ tube and hyphae in young mycelia develop. The second is a sudden change, in this case at 22 hr, affecting branch angle and hyphal diameter and extension rate. The most likely explanation for these changes is the production at the center of the colony, where conditions for growth will be unfavorable, of compounds which diffuse to the colony margin. At the margin they exert a paramorphogenetic effect, giving rise to radially directed growth, increased extension rate, and, possibly, reduced branch formation, thereby permitting growth at an unaltered rate.

Conclusive evidence for this mechanism is difficult to obtain but the differences in properties of hyphae in mycelia of different age must be considered by both empirical and mechanistic models of mycelial and colony growth. In addition, vegetative growth is often accompanied by formation of alternate bands of low and high hyphal density (rhythmic growth), sector formation, and spiral growth. All of these features affect colony growth rates and have implications for the modeling of vegetative growth. They also represent differentiation processes and may provide the link between modeling of vegetative growth and differentiation in fungal colonies.

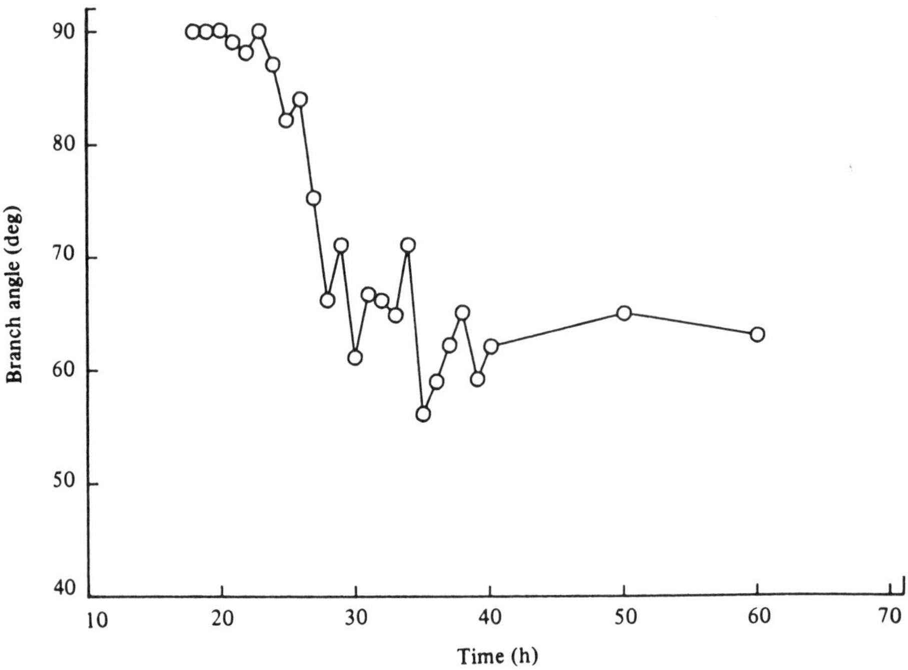

(a)

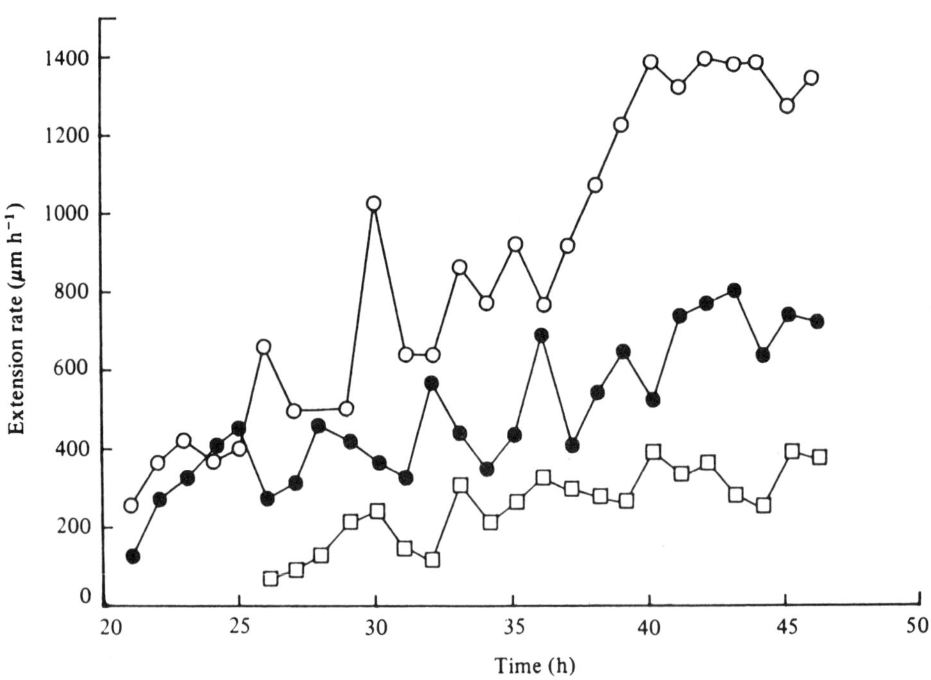

(b)

FIGURE 8. (a) Changes in branch angle, (b) hyphal extension rate, and (c) hyphal diameter of leading hyphae (○), primary branches (●), and secondary branches (□) of *Neurospora crassa*. (From Ref. 55.)

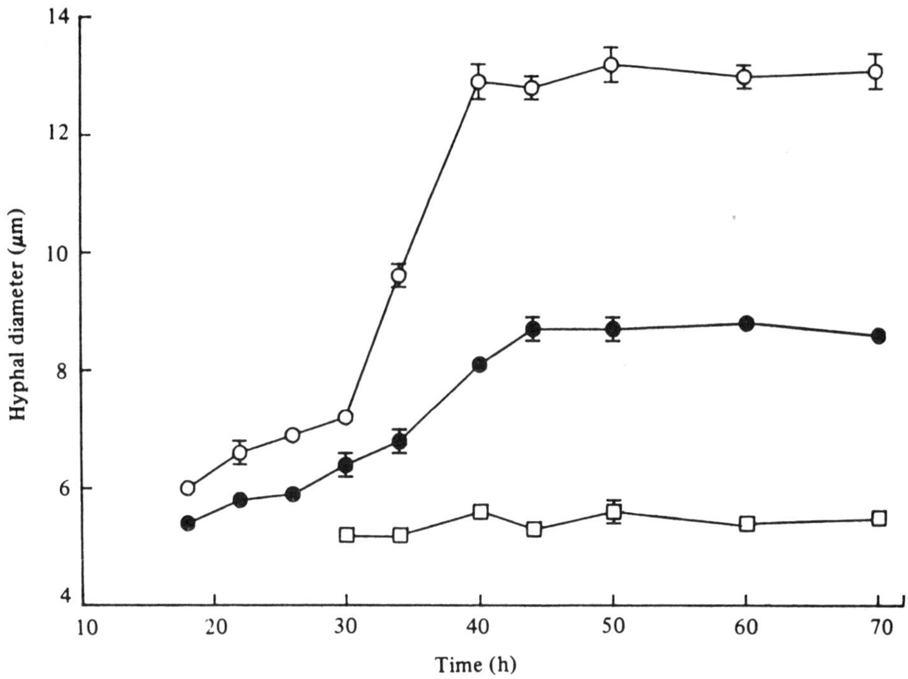

(c)

FIGURE 8. (Continued)

VII. GROWTH IN SUBMERGED CULTURE

Growth of filamentous fungi in submerged culture is similar in many respects to that of unicellular microorganisms [56,57]. Growth in batch culture involves lag, acceleration, exponential, deceleration, and stationary phases and may be used to determine maximum specific growth rate and growth yield. In continuous culture, substrate-limited growth is described by the Monod equation, and values for maximum specific growth rate, the saturation constant for growth, and maintenance coefficient have been determined experimentally.

Where differences occur, they arise through particular properties of the mycelial growth form. These may result in practical or technical problems. For example, filamentous fungi have an increased tendency for surface growth and may colonize probes and other surfaces within the culture vessel and may grow back along nutrient feed lines to contaminate medium reservoirs. Establishment of steady states in continuous culture may also be more difficult than for unicellular organisms, and at low substrate concentrations and specific growth rates growth kinetics may be complicated by differentiation processes.

A. Dispersed Growth

Despite these difficulties, liquid culture studies have provided important information on the factors controlling growth and branching in mycelia. For example, Trinci [27,28] found that the maximum specific growth rates of wild-type and spreading colonial mutants of *N. crassa* were similar, despite differences in colony radial growth rate on solid medium, with K_r values ranging from 493 to 6800 μm. These differences were shown to be due to variation in the peripheral grown zone width arising from mutations affecting hyphal diameter, apical and intercalary compartment lengths, and septal plugging.

A second example is the effect of specific growth rate on glucose-limited chemostat cultures of *G. candidum* [39–42]. Lateral branch formation increased with specific growth rate up to 0.3/hr, while apical branching only occurred during washout at dilution rates greater than the critical dilution rate. Apical polarity in continuous culture reflected that on solid medium and increased specific growth rate led to increased hyphal diameter, decreased hyphal growth unit length, but no change in hyphal growth unit volume. These studies, together with investigations of growth on solid medium, led to formulation of a model in which apical branch formation occurred upon attainment of a critical apical cell volume which was independent of glucose concentration.

B. Pellet Growth

The major difference between growth of unicells and mycelial organisms in liquid culture is the formation of pellets, consisting of aggregates of mycelia arising from aggregation of spores, entrapment of spores by young mycelia, or, rarely, aggregation of mycelia. Their morphology is variable but they are generally spherical or ellipsoidal with texture ranging from compact, with little diffusion of material into the pellet, to loose or fluffy, with adequate supply of nutrients to the entire pellet from the surrounding medium. Whitaker and Long [58] and Metz and Kossen [59] review the properties of pellets and factors controlling their production, and Wittler et al. [60] suggest that they consist of four distinct layers. An outer layer of growing hyphae surrounds a layer where hyphae show signs of autolysis. Within this is a third layer which contains cytoplasm, while the center of the pellet is hollow and contains no hyphae. These observations were made on pellets of *Penicillium chrysogenum*. The thickness of the four layers varied during growth and with varying growth conditions, and the general structure may be different in other organisms.

The simplest empirical model for pellet growth is represented by the equation:

$$M^{1/3} = M_0^{1/3} + kt$$

where M and M_0 represent total biomass at time t and 0, and k is a constant. Thus, pellet growth follows cube root kinetics. This is based on the experimental observations of many workers [61–63].

Pirt [64] adopted an approach similar to that for growth in two dimensions on solid medium (see Section V.A). Growth immediately following spore germination was exponential, but as pellet size increased, growth at the center became limited by reduced supply of nutrients and eventually ceased. Subsequent growth was restricted to a peripheral shell (equivalent to the peripheral growth zone) of width w, within which growth was exponential at a specific rate, μ. This model may be represented by the equation:

$$\frac{dm}{dt} = \mu m_g$$

where m represents the mass of a single pellet and m_g is the mass of the peripheral shell. If the pellet is assumed to be spherical with radius $r \gg w$, then

$$\frac{dr}{dt} = w\mu$$

or

$$r = w\mu t + r_0$$

where r is the pellet radius at t = 0. If the pellet is assumed to have constant and uniform biomass density, ρ, then substitution for r and r_0 yields the equation:

$$M^{1/3} = M_0^{1/3} + \left(\frac{4}{3}\pi\rho n\right)^{1/3} \mu w t$$

where M_0 is the initial total dry weight and n the number of pellets.

This model therefore predicts exponential growth followed by cube root kinetics and may also be used to calculate the pellet diameter at which growth at the center of the pellet will become limited by nutrient supply. This involves consideration of the yield of biomass (Y) on the substrate, which is likely to first limit growth and its apparent coefficient of diffusion (D') into the pellet. If S_m is the concentration of this substrate in the external medium, the critical radius (R_c) at which growth at the center becomes limited is given by:

$$R_c = \sqrt{\frac{6D'YS}{\mu}m}$$

If the limiting substrate is assumed to be oxygen, R_c values calculated from this equation agree well with those measured experimentally for pellets of *P. chrysogenum*.

The rheology of pellet cultures is similar to that of unicells and therefore provides advantages for industrialists over dispersed mycelial growth although they do result in heterogeneity within the biomass, as the central region will not be growing. This has led to attempts by others to model the transport and utilization of oxygen within pellets.

Aiba and Kobayashi [65] described the specific rate of respiration in a pellet as a hyperbolic function of dissolved oxygen concentration:

$$q'_{O_2} = q_{O_2}\left(\frac{K_m + \bar{C}}{c}\right)\left(\frac{C}{K_m + C}\right)$$

where C and $\bar{C}$ represent concentrations of oxygen within the pellet and in the surrounding medium, respectively, q'_{O_2} and q_{O_2} are the respective respiration rates, and K_m is a constant. The oxygen balance within the pellet is given by:

$$D\left(\frac{d^2C}{dr^2} + \frac{2}{r}\frac{dC}{dr}\right) = 2\rho\, q_{O_2}\left(\frac{C}{K_m + C}\right)$$

where r is pellet radius and ρ represents pellet density. This equation was used to determine the relative rates of respiration within the pellet and in the surrounding medium ($\bar{q}_{O_2}/q_{O_2}$), and predictions agreed well with experimental data of Yano et al. [66] in comparison with their earlier model.

The model of Kobayshi et al. [67] utilizes diffusion equations with boundary conditions in which the concentration of oxygen is held constant at the pellet surface and zero at the center. Respiration rate is dependent on oxygen concentration and distance from the pellet surface and an effectiveness factor is defined, representing the proportional decrease in respiration rate resulting from oxygen limitation.

Experimental results for pellet growth of *Aspergillus niger* were compared with model predictions for three situations: (a) uniform and maximum respiratory activity throughout the pellet, (b) maximum respiratory activity dependent on mycelial age, and (c) adaptation of mycelia within the pellet to low concentrations of oxygen. Although all three situations gave reasonable fit to experimental data, the third situation provided the best description of pellet respiration and was considered the best model for oxygen transfer into pellets.

The physical effects of agitation on pellet growth have been modeled by several workers. These can lead to fragmentation of pellets, which may cause hyphal damage and act as centers for formation of new pellets. Taguchi [68] proposed two types of effect. In the first, material is removed by breakage of material from the pellet surface, while the second involves complete rupture of the pellet. The two situations are presented by the following two equations:

$$\frac{dD_p}{dt} = -k_c\,(ND_i)^{5.5} D_p^{5.7}$$

where D_p represents pellet diameter, N is impellor speed, D_i is impellor diameter, and K_c is a constant, and:

$$\frac{dn}{dt} = -\alpha n \ (D_p^{3.2} N^{6.65} D_i^{8.75})$$

where n represents the number of nondisrupted pellets and α is a constant.

Edelstein and Hadar [69] presented a model which considered the size distribution of the pellet population. The basic experimental data were obtained from cultures of *Sclerotium rolfsii*, which produced spherical pellets with a compact core surrounded by looser region. The model placed pellets and hyphal fragments into compartments on the basis of their size. Pellet diameter was assumed to increase at a constant rate, and this increase in size led to movement between compartments representing different size classes. Growth of pellets in a particular size class was counteracted by fragmentation which was assumed to be proportional to pellet radius, and pellet damage or death was also incorporated. Hyphal fragments arising from pellets were placed in the compartment representing the smallest size class and subsequently developed to form new pellets.

Predictions were made of increase in total biomass with time and development of pellet size distributions. Biomass production was found to be greater than that predicted by models which do not consider fragmentation, due to the larger relative surface area for nutrient absorption presented by smaller particles. Pellet size distributions showed qualitative agreement with experimental data and were considered good supportive evidence for the approach adopted and the description of hyphal damage.

VIII. CONCLUSIONS

This article has considered only mathematical models for vegetative growth of filamentous fungi. Many other aspects of their biology have been modeled, e.g., spore germination, phototropism, spore formation, product formation, and some will form the subject of other articles in this volume.

A range of modeling approaches has been adopted. Simple empirically based models are of use in describing aspects of growth and are of practical use in determining the effects of environmental factors or strain variability on these aspects. The major advantage is their simplicity, and the biological basis for the relationships which they express need not be fully understood for them to retain their utility.

In a number of cases, however, mechanistic models have validated and explained empirical models, thereby allowing greater confidence in their use, particularly under conditions for which they have not been specifically tested. Mechanistic models are seen to be more complex, and the validity of assumptione on which they are based is determined by comparison of their predictions with experimental data. This requires provision of values for constants within the model. It is frequently impossible to determine such values, but mechanistic models can be used to determine the values to which the system is most sensitive and can thereby direct experimental work.

There is no universally "correct" modeling technique for all situations. The different modeling approaches fulfill different functions and should be used accordingly. The requirements of an industrialist wishing to predict the rate of production of biomass in an industrial fermentation are very different from those of a physiologist investigating mechanisms of hyphal extension and branch formation. Mathematical models have increased our knowledge and understanding of vegetative growth and branching, and their correct application, coupled to appropriate and related experimental work, should improve further our understanding and description of quantitative aspects of mycelial growth.

REFERENCES

1. Roels, J. A., and Kossen, N. W. F. On the modelling of microbial metabolism. *Progr. Ind. Microbiol.*, *14*:95–203, 1978.
2. Bazin, M. J. (ed.). *Mathematical Models in Microbiology*. CRC Press Inc., Boca Raton, FL, 1982.
3. Bazin, M. J. (ed.). *Mathematical Models in Microbiology*. Academic Press, London, 1983.
4. Bazin, M. J., and Prosser, J. I. (eds.). *Physiological Models in Microbiology*. CRC Press Inc., Boca Raton, FL, 1988.
5. Gooday, G. W. The hyphal tip. In *Fungal Differentiation* (J. E. Smith, ed.), Marcel Dekker Inc., New York, 1983, pp. 315–356.
6. Wessels, J. G. H. Cell wall synthesis in apical hyphal growth. *Int. Rev. Cytol.*, *104*:37–79, 1986.
7. Steele, G. C., and Trinci, A. P. J. The extension zone of mycelial hyphae. *New Phytol.*, *75*:583–587, 1975.
8. Green, P. B., and King, A. A mechanism for the origin of specifically oriented textures in development with special reference to *Nitella* wall texture. *Aust. J. Biol. Sci.*, *19*:421–437, 1966.
9. Green, P. B. Morphogenesis of the cell and organ axis-biophysical models. *Brookhaven Symp. Biol.*, *25*:166, 1974.
10. Bartnicki-Garcia, S. Fundamental aspects of morphogenesis. In *Microbial Differentiation. 23rd Symposium of the Society for General Microbiology* (J. M. Ashworth and J. E. Smith, eds.), Cambridge University Press, Cambridge, 1973, pp. 245–267.
11. Trinci, A. P. J., and Saunders, P. T. Tip growth of fungal hyphae. *J. Gen. Microbiol.*, *103*:243–248, 1977.
12. Castle, E. S. The topography of tip growth in a plant cell. *J. Gen. Physiol.*, *41*:913–926, 1958.
13. Collinge, A. J., and Trinci, A. P. J. Hyphal tips of wild-type and spreading colonial mutants of *Neurospora crassa*. *Arch. Microbiol.*, *99*:353–368, 1974.
14. Gooday, G. W. An autoradiographic study of hyphal grwoth of some fungi. *J. Gen. Microbiol.*, *67*:125–133, 1971.
15. da Riva Ricci, D., and Kendrick, B. Computer modelling of hyphal tip growth in fungi. *Can. J. Bot.*, *50*:2455–2462, 1972.
16. Koch, A. L. The surface stress theory of microbial morphogenesis. *Adv. Microbial Physiol.*, *29*:301–367, 1983.
17. Koch, A. L. How bacteria grow and divide in spite of internal hydrostatic pressure. *Can. J. Microbiol.*, *31*:1071–1084, 1985.
18. Saunders, P. T., and Trinci, A. P. J. Determination of tip shape in fungal hyphae. *J. Gen. Microbiol.*, *110*:469–473, 1979.

19. Koch, A. L. The shape of hyphal tips of fungi. *J. Gen. Microbiol.*, *128*:947–951, 1982.
20. Smith, J. H. On the early growth of the individual fungus hypha. *New Phytol.*, *23*:65–78, 1924.
21. Plomley, N. J. B. Formation of the colony in the fungus *Chaetomium*. *Aust. J. Biol. Sci.*, *12*:53–64, 1959.
22. Trinci, A. P. J. A kinetic study of the growth of *Aspergillus nidulans* and other fungi. *J. Gen. Microbiol.*, *57*:11–24, 1969.
23. Trinci, A. P. J. A study of the kinetics of hyphal extension and branch initiation of fungal mycelia. *J. Gen. Microbiol.*, *81*:225–236, 1974.
24. Fiddy, C., and Trinci, A. P. J. Mitosis, septation, branching and duplication cycle in *Aspergillus nidulans*. *J. Gen. Microbiol.*, *97*:169–184, 1976.
25. Prosser, J. I., and Trinci, A. P. J. A model for hyphal growth and branching. *J. Gen. Microbiol.*, *11*:153–164, 1979.
26. Fiddy, C., and Trinci, A. P. J. Nuclei, septation, branching and growth of *Geotrichum candidum*. *J. Gen. Microbiol.*, *97*:185–192, 1976.
27. Trinci, A. P. J. The duplication cycle and development in moulds. In *The Filamentous Fungi*, Vol. 3 (J. E. Smith and D. R. Berry, eds.), Edward Arnold, London, 1978, pp. 132–163.
28. Trinci, A. P. J. The duplication cycling and branching in fungi. In *Fungal Walls and Hyphal Growth* (J. H. Burnett and A. P. J. Trinci, eds.), Cambridge University Press, Cambridge, 1979, pp. 319–358.
29. Pirt, S. J. A kinetic study of the mode of growth of surface colonies of bacteria and fungi. *J. Gen. Microbiol.*, *47*:181–197, 1967.
30. Trinci, A. P. J. Influence of the peripheral growth zone on the radial growth rate of fungal colonies. *J. Gen. Microbiol.*, *67*:325–344, 1971.
31. Morrison, K. B., and Righelato, R. C. The relationship between hyphal branching, specific growth rate and colony radial growth rate in *Penicillium chrysogenum*. *J. Gen. Microbiol.*, *81*:517–520, 1974.
32. Koch, A. L. The kinetics of mycelial growth. *J. Gen. Microbiol.*, *89*:209–216, 1975.
33. Deppe, C. S. Protein degradation in *Schizophyllum commune*. Ph.D. thesis, Harvard University, Cambridge, MA, 1974.
34. Trinci, A. P. J. Kinetics of growth of mycelial pellets of *Aspergillus nidulans*. *Arch. Mikrobiol.*, *73*:353–367, 1970.
35. Buller, A. H. R. *Researches on Fungi*, Vol. 4. Longmans Green, London, 1931.
36. Prosser, J. I. Growth kinetics of mycelial colonies and aggregates. In *Developmental Biology of Ascomycetes* (N. Read and D. Moore, eds.), Cambridge University Press, London, 1991.
37. Robinson, P. M. Autotropism in fungal spores and hyphae. *Bot. Rev.*, *39*:367–384, 1973.
38. Robinson, P. M. Chemotropism in fungi. *Trans. Br. Mycol. Soc.*, *61*:303–313, 1973.
39. Robinson, P. M., and Smith, J. M. Morphogenesis and growth kinetics of *Geotrichum candidum* in continuous culture. *Trans. Br. Mycol. Soc.*, *66*:413–420, 1976.

40. Robinson, P. M., and Smith, J. M. Development of cells and hyphae of *Geotrichum candidum* in chemostat and batch culture. *Trans. Br. Mycol. Soc.*, *72*:39–47, 1979.
41. Robinson, P. M., and Smith, J. M. Apical branch formation and cyclic development in *Geotrichum candidum*. *Trans. Br. Mycol. Soc.*, *75*:233–238, 1980.
42. Smith, J. M., and Robinson, P. M. Development of somatic hyphae of *Geotrichum candidum*. *Trans. Br. Mycol. Soc.*, *74*:159–165, 1980.
43. Steele, G. C., and Trinci, A. P. J. Morphology and growth kinetics of hyphae of differentiated and undifferentiated mycelia of *Neurospora crassa*. *J. Gen. Microbiol.*, *91*:362–368, 1975.
44. Bull, A. T., and Trinci, A. P. J. The physiology and metabolic control of fungal growth. *Adv. Microbial Physiol.*, *15*:1–84, 1977.
45. Prosser, J. I. Hyphal growth patterns. In *Fungal Differentiation* (J. E. Smith, ed.), Marcel Dekker, New York, 1983, pp. 357–396.
46. Cohen, D. Computer simulation and biological pattern generation processes. *Nature*, *216*:246–248, 1967.
47. Hutchinson, S. A., Sharma, P., Clarke, K. R., and MacDonald, I. Control of hyphal orientation in colonies of *Mucor hiemalis*. *Trans. Br. Mycol. Soc.*, *75*:177–191, 1980.
48. Adupa, R. L. Studies on the growth of *Penicillium claviforme* (Bainier). Ph.D. thesis, University College of North Wales, Bangor, 1981.
49. Edelstein, L. The propagation of fungal colonies: A model for tissue growth. *J. Theor. Biol.*, *104*:187–210, 1983.
50. Edelstein, L., and Segel, L. A. Growth and metabolism in mycelial fungi. *J. Theor. Biol.*, *98*:187–210, 1983.
51. Robson, G. D., Bell, S. D., Kuhn, P. J., and Trinci, A. P. J. Glucose and penicillin concentrations in agar medium below fungal colonies. *J. Gen. Microbiol.*, *133*:361–352, 1961.
52. Edelstein, L., Hadar, Y., Chet, I., Henis, Y., and Segel, L. A. A model for fungal growth applied to *Sclerotium rolfsii*. *J. Gen. Microbiol.*, *129*:1873–1881, 1983.
53. Georgiou, G., and Shuler, M. L. A computer model for the growth and differentiation of a fungal colony on solid substrate. *Biotechnol. Bioeng.*, *28*:405–416, 1986.
54. Butler, G. M. Growth of hyphal branching systems in *Coprinus disseminatus*. *Ann. Bot. N.S.*, *25*:341–352, 1961.
55. McLean, K. M., and Prosser, J. I. Development of vegetative mycelium during colony growth of *Neurospora crassa*. *Trans. Br. Mycol. Soc.*, *88*:489–495, 1987.
56. Righelato, R. C. The kinetics of mycelial fungi. In *Fungal Walls and Hyphal Growth* (J. H. Burnett and A. P. J. Trinci, eds.), University Press, London, 1979, pp. 385–401.
57. van Suijdam, J. C., and Metz, B. Influence of engineering variables on the morphology of filamentous molds. *Biotechnol. Bioeng.*, *23*:111–148, 1981.
58. Whitaker, A., and Long, P. A. Fungal pelleting. *Process Biochem.*, *11*:27–31, 1973.
59. Metz, B., and Kossen, N. W. F. The growth of molds in the form of pellets—A literature review. *Biotechnol. Bioeng.*, *19*:718–799, 1977.

60. Wittler, R., Baumgartl, H., Lubbers, D. W., and Schugerl, K. Investigation of oxygen transfer into *Penicillium chrysogenum* pellets by microprobe measurements. *Biotechnol. Bioeng.*, *28*:1024–1036, 1986.
61. Emerson, R. The biology of water molds. In *Aspects of Synthesis and Order in Growth* (D. Runick, ed.), Academic Press, New York, 1955, pp. 171–208.
62. Machlis, L. Factors affecting the lag phase of growth of the filamentous fungus, *Allomyces macrogynous*. *Ann. J. Bot.*, *44*:113–119, 1957.
63. Marshall, K. C., and Alexander, M. Growth characteristics of fungi and actinomycetes. *J. Bacteriol.*, *80*:412–416, 1960.
64. Pirt, S. J. A theory of the mode of growth of fungi in the form of pellets in submerged culture. *Proc. R. Soc. London Ser.*, *166*:369–373, 1966.
65. Aiba, S., and Kobayashi, K. Comments on oxygen transfer within a mold pellet. *Biotechnol. Bioeng.*, *12*:583–588, 1971.
66. Yano, T., Kodama, T., and Yamada, K. Fermentation studies on the aerobic fermentation. VIII. Oxygen transfer within a mold pellet. *Agric. Biol. Chem.*, *25*:580–584, 1961.
67. Kobayashi, T., van Dedem, G., and Moo-Young, M. Oxygen transfer into mycelial pellets. *Biotechnol. Bioeng.*, *15*:27–45, 1973.
68. Taguchi, H. The nature of fermentation fluids. *Adv. Biochem. Eng.*, *1*:1–30, 1971.
69. Edelstein, L., and Hadar, Y. A model for pellet size distributions in submerged mycelial cultures. *J. Theor. Biol.*, *105*:427–452, 1983.

21

MODELING GROWTH AND DISPERSAL OF FUNGI IN NATURAL ENVIRONMENTS

GUY R. KNUDSEN *University of Idaho, Moscow, Idaho*

JAMES P. STACK* *Texas A&M University, College Station, Texas*

I. INTRODUCTION

A. Rationale

Fungi are adapted in various ways for dispersal and colonization of substrates. Their filamentous growth form facilitates colonization of complex matrices (e.g., soil and marine habitats) and the exploitation of complex substrates (e.g., plant, animal, and microbial organic matter). Fungal hyphae play a significant role in active dispersal primarily over short distances, from one microsite to another within a small three-dimensional space, although specialized structures such as rhizomorphs may allow movement over several meters. Hyphae can also be the propagules of long distance passive dispersal, for example, on farm equipment, by wind, in plant seeds, or on insects and other animals. Fungi produce spores of various sizes and shapes, some of which are well adapted to specific mechanisms of dispersal, e.g., splash dispersal, adherence to insects, survival in bird and insect intestinal tracts, and survival and reproduction in dung.

Although detailed studies of spore liberation have increased our understanding of the mechanics of dispersal, the significance of dispersal in the life history of a given species and to community structure in a defined ecosystem is understood for only a few fungi. Quantitative modeling can be used as an approach to determining this significance. The ability to predict the incidence and dispersal of fungi in natural environments would benefit many areas in research and management of fungi. In this review we will examine some of the different ways that mechanisms of fungal growth and dispersal have been modeled. First, however, we will briefly consider four areas where we consider this methodology to be especially useful:

(1) Plant pathology. For plant pathologists, dispersal of fungal spores and other propagules is important in determining the epidemiology of plant diseases they cause. Accurate models of spore dispersal are needed in order to understand how disease epidemics develop. Similarly, studies of spore dispersal gradients provide useful information about movement of

* *Present affiliation: EcoScience Laboratories, Inc., Amherst, Massachusetts*

fungal plant pathogens and consequent disease spread. Also, numerous fungal species continue to be evaluated as biological control agents against plant pathogens, although there have been few success stories for the biological control of soilborne plant pathogens. One hypothesis for these failures is that a thorough colonization of the soil by the control agent has not been achieved. Accurate, validated models could help determine which fungal characteristics are important for dispersal and/or colonization in natural soils, and perhaps indicate which factors could be manipulated to achieve maximum colonization.

(2) Pollution control. There is a great deal of interest in the use of microorganisms for environmental waste management through degradation or detoxification of industrial and agricultural wastes [1–3]. Although bacteria have been isolated and/or manipulated to detoxify certain recalcitrant molecules, their limited dispersal capability in soil restricts their usefulness. In contrast, because of their ability for active dispersal via hyphal growth in complex matrices, fungi may be well equipped for use as agents for decontamination of toxic waste sites. Our understanding of the molecular biology of fungi and the ability to genetically manipulate them have increased substantially in the last few years. Also, fungi are often less sensitive than bacteria to changes in temperature and moisture, permitting them to be active over a wider range of environmental conditions. With appropriate delivery systems, fungi may be able to colonize some contaminated sites much more effectively than bacteria.

(3) Risk assessment for release of genetically engineered fungi. Recombinant DNA technology offers the potential to modify fungal traits including antibiotic production, hyperparasitic activity, and substrate utilization, potentially improving the performance of fungi as biocontrol or pollution control agents. Currently, genetically engineered fungi are ready for field trials to evaluate their performance. However, the potential impact of recombinant microbes on agroecosystems is still largely unknown and remains an impediment to scientific and commercial development [4,5]. Fungi account for a large part of the total microbial population in most cultivated soils [6], and dispersal of genetic information in soil fungal communities potentially can occur through vegetative growth, spore dispersal, and through the exchange or transfer of nuclear or plasmid DNA through hyphal anastomosis. If a recombinant genotype confers a selective advantage in a particular habitat, there might be a shift in the genetic makeup of the soil population [7]. By quantifying dynamics of vegetative growth, sporulation, and ecological interactions, models can be used to assess the potential for dissemination of recombinant fungi from a site of introduction, thus serving as a means to both formulate and test hypotheses about "risks" associated with environmental release of recombinant fungi.

(4) Medical and veterinary epidemiology. The incidence of human mycoses in the United States is increasing, according to medical mycologists [8,9]. There are several possible reasons for this observation, including an increase in the number of people in an immunocompromised condition, and more sophisticated detection and diagnostic procedures. Whatever the reason, elucidating and modeling the growth and dispersal of the causal fungi might lead to a better understanding of fungal development in animal host tissues and dissemination throughout the host organism. Such information would contribute to the development of predictive models and the study

of the epidemiology of human mycoses. Where this might be of greatest use is in the study of allergenic fungi and fungi that cause respiratory mycoses and toxicoses. Spores of some fungi are capable of inducing respiratory dysfunction, acting as either specific or nonspecific allergens, or as inducers of pulmonary mycotoxicoses [10]. Also, many fungi elaborate toxic secondary metabolites in foods and feed. For example, under grain storage conditions, the level of contamination often increases as a direct result of fungal growth.

B. Types of Predictive Models

Predictive models are tools for hypothesis testing, building of theories, and practical application. The philosophy and methods of modeling biological systems have been described extensively elsewhere [11–13] (see also Chapters 19, 21, and 22 in this volume), and we will primarily limit this review to describing applications. Briefly, however, we will paraphrase Jeger's definition [14] of a mathematical model as a way to represent a system as a set of mathematical equations describing the functional dependence of one or more quantities on time, space, and each other. The model provides a quantitative description of complex interactions that would otherwise be difficult to account for.

Mathematical models are frequently described as empirical or descriptive (derived by statistically fitting mathematical equations to experimental data) vs. mechanistic or explanatory (elucidating fundamental mechanisms). In practice, a single model often contains components of both types. Jeger [15] pointed out that complex mechanistic models are usually based on empirical relationships derived for a lower level, then linked together in a logical structure. Thus, a model that is "explanatory" at one level derives from a lower level "descriptive" model. Frequently, models are also categorized as analytic vs. simulation, the relative merits of which were considered by Jeger [15]. Analytical models usually consist of a series of differential or algebraic equations. They are sometimes considered to be oversimplifications of complex processes, with limited ability to account for highly stochastic (probabilistic) processes or to accurately deal with systems having many state variables. Despite these potential limitations, powerful analytic models have been developed to describe biological systems. Simulation models have been developed in order to account for the intrinsic stochasticity and/or large number of state variables in complex systems, and are especially amenable to implementation using digital computers. Teng [16] reviewed the various interrelationships between systems analysis and simulation and other model types, and noted difficulties in comparing approaches even among modelers.

II. BASIC FUNGAL GROWTH MODELS

The general fungal life cycle can be divided into germination, colonization of a substrate, sporulation, and dispersal. At some level, vegetative growth over short distances may be considered one component of dispersal. Growth and dispersal are influenced by the nature of the environment in which they occur, e.g., aquatic, marine, or terrestrial, and the potential for growth or dispersal in natural environments varies among fungal genera, species, and even strains. There has been a substantial amount of work mathematically modeling the physiology and growth of bacteria, but relatively less effort relevant to growth and physiology of fungi. As Da Riva Ricci and Kendrick pointed out, "The average agaric is made up of miles of mycelium. And yet we know remarkably little about how hyphae grow" [17]. Early modeling efforts usually involved using models derived for bacteria or modifying them to fit mycelial growth [18]. More recently, mathematical models have appeared that specifically apply to mycelial growth of fungi [17,19–23]. Models of mycelial growth, using empirical and/or mechanistic approaches, consider various aspects of growth including mycelial branching patterns, sporulation, spore dispersal, and germination.

At the cellular level, a fungal growth model might consider the regulatory mechanisms that determine how individual hyphal cells increase in size and synthesize cellular constituents. Components include the chemical reactione required to synthesize enzymes, protoplasm, and cell walls, the availability of raw materials and energy to carry out these synthetic processes, and environmental effects that govern their rate. Bartnicki-Garcia [24] considered fundamental aspects of morphogenesis of filamentous fungal hyphae, where vegetative growth occurs only at the tip of each cylindrical hyphal filament. The tip elongates, and a septum forms to yield a cell with a distal growing tip. When branching occurs, the growing tip bifurcates and both branches elongate in separate strands. Bartnicki-Garcia made the assumption that a hyphal tip consists of a cylinder with a hemispherical apical dome wall. The cytoplasm in this apical dome region consists primarily of apical vesicles (*Spitzenkörper*), which apparently play a role in carrying materials and enzymes for subsequent plasmalemma and cell wall synthesis. The orderly growth of the cell wall was described as "a harmonious balance between the processes of wall synthesis and enzymatic wall lysis" [24].

Bartnicki-Garcia [24] formulated a general model of fungal cell wall growth, based on the assumption that growth results from the cumulative action of minute hypothetical growth units. He pointed out that cell wall growth units may have a physical correspondence to the amount of growth resulting from the discharge of a discrete number of apical vesicles. Thus, the shape and diameter of a hyphal segment may be determined by the spatial distribution of these cell wall growth units and by their relative synthetic and lytic activities. Interacting environmental factors, including temperature and depletion of enzyme substrates or precursors, would impact on hyphal morphology through this mechanism.

Da Riva Ricci and Kendrick also derived a mathematical model to simulate growth of the hyphal tip [17]. Their model assumed that the fungal cell wall exists in two conditions: "set" (inelastic, incapable of further deformation) and "unset" (elastic, able to deform in response to pressure differentials). Thus, in their model, hyphal tip elongation results from a dynamic balance between the amounts of cell wall in each of these two states.

Environmental factors all have minimum and maximum limiting effects, between which lies a range known as the limits of tolerance [25]. Quantification of these limits is necessary in order to model and predict environmental effects on growth and dispersal. Selection of appropriate environmental factors to measure depends largely on the knowledge and intuition of the investigator. Certainly, of the abiotic environmental factors affecting populations of terrestrial plants, animals, and microbes, moisture and temperature have been the most studied. Studies of mycelial growth rates on agar, where water potential was amended by addition of solutes, have generally indicated a progressive decrease in hyphal extension as water potential was decreased. This is presumably the result of reduced turgor in hyphal cells, as cell turgor provides the internal force for cell expansion [26].

Numerous models have been derived to describe growth of fungi on agar or in broth [17,19–23]; this subject is reviewed elsewhere in this volume (see Chapter 19). Most of these models assume a continuous and unlimited supply of substrate being encountered by an exponentially increasing population of hyphal tips, at least during certain defined stages of colony development. Trinci [21,22] and Trinci and Saunders [23] derived two useful parameters to describe growth: the "hyphal growth unit," defined as the total hyphal length divided by the number of hyphal tips, and the "peripheral growth zone," defined as a zone of constant proportion within which exponential growth occurs. The hyphal growth unit and the peripheral growth zone are unique to fungal species and conditions of substrate and environment.

Mycelial fungi are branching systems, and as such have certain developmental patterns in common with other branching systems including rivers, trees, and animal circulatory and nervous systems. Mathematical and computer simulation models have been developed to describe the formation of branching patterns in several of these systems [27,28]. Gull [29] defined parameters applicable to fungal branching systems. Spores of *Thamnidium elegans* were germinated on agar and resultant mycelium photographed after a short incubation period. Gull used Leopold's branching terminology [30] to quantify branch orders and lengths. A logarithmic relationship was observed between branch order number and the average length of hyphae; i.e., each hypha was longer than hyphae it produced, by a factor he called the length ratio. A second parameter, branch ratio, was defined as the ratio of numbers of hyphae in adjacent branch orders. For example, a system in which second order hyphae of mean length 10 mm produce an average of four 2-mm first order hyphae each would have parameters of branch ratio = 4 and length ratio = 5. Gull determined that these two parameters were independent of the order of the system looked at, thus concluding that branching had a definite recurrent pattern that was under some form of control. Similarly, he observed that the value for hyphal growth unit (sensu Trinci) was constant for third and fourth order branching systems.

III. GROWTH AND DISPERSAL ON VEGETATION AND IN SOIL

As growth models derived in vitro are extrapolated to natural environments, numerous difficulties of verification and validation become apparent, not the least of which is simply the problem of locating and identifying individual fungal organisms. In some cases, indirect or statistical methods have helped overcome some of these difficulties. We will describe several attempts at modeling mycelial growth of fungi on vegetation surfaces and in soil.

Quinn and Fugimoto [31] observed and photographed growth and sporulation of the powdery mildew fungus *Erysiphe polygoni* on bean leaves, and quantified parameters including time between formation of a hyphal segment and branching of that segment, mean hyphal length per branch, rate of germ tube formation, branch angle, and rate of conidiophore production. They then incorporated these parameters into a model developed previously by Quinn and Powell [32] to describe growth and sporulation of *Oidium begoniae*. The *Erysiphe* model, written in Fortran, contains graphics routines to depict hyphal growth patterns.

Lapp and Skoropad [33] looked at a different aspect of growth: the formation of appressoria on surfaces. They used a mathematical model based on a modification of the Gompertz equation to predict the number of generations and the number of appressorial formations in a population of conidia of *Colletotrichum graminicola*.

Less is known about dispersal of microbes in soil than about the mechanisms of dispersal aboveground. By volume, soil consists primarily of inorganic particles and the pore spaces between them. From 30 to 50% of the soil volume consists of pore space, with clay soils having more total pore space but smaller diameter pores. The spatial distribution of soil pore space largely constrains patterns of fungal growth in soil, while rates and extent of growth are also determined by organic matter content and the liquids and gases that occupy pores. Because the soil matrix acts as a filter for suspended particles, microbial dispersal in soil may often be limited to distances of only a few centimeters [34–38]. However, bacteria are known to move passively in soil water: extensive lateral movement and vertical penetration to depths of several meters have been observed in water-saturated soils [39–41]. Although dispersal of fungal propagules by moving water in soil has rarely been well quantified, reports that flagellated zoospores generally moved only small distances through soil under their own locomotion suggest limited active movement by soil fungi [42–44]. Within fungal species, different propagules will have different dispersal capabilities in different environments. For example, in a silt loam soil, zoospores of *P. cryptogea* [42,45] or *P. megasperma* [44,45] were unable to disperse from the point of introduction, whereas hyphae were able to grow at least 2.0 centimeters in the same soil [44]. As would be expected, different forms of the same propagule will vary in dispersal abilities. As a moving front of water passes through soil, motile zoospores will travel farther than encysted zoospores [38]. Thus, the general models of dispersal need to be validated for individual species, propagules, and environments.

Movement of nonzoosporic soil fungi over short distances occurs primarily through initial hyphal growth and colonization of nearby substrates, followed by subsequent growth from the new substrates. Sporulation may occur at this time, which aids dispersal over intermediate or longer distances via movement of soil water, soil particles, soil animals, or aerial entrainment and dissemination of the spores themselves.

There is still relatively little quantitative data available to predict the extent of dispersal, substrate colonization, and sporulation of fungi introduced into soil. A major difficulty has been detection of hyphae growing in natural, three-dimensional substrates. Parr et al. [36] attempted to quantify mycelial development of *Trichoderma viride* in a glass bead system and found a significant effect of bead size on growth patterns. Working with *Streptomyces*, Ruddick and Williams [46] monitored growth in soil using electron microscopy, fluorescent microscopy, soil stripping, and maceration techniques. They found that periods of active growth were discontinuous both in space and time, making detection difficult.

Despite inherent difficulties, some models describing dynamics of soil fungi have been developed using direct and indirect approaches. Indirect approaches include modeling of root infections by fungal phytopathogens, modeling rhizosphere size and dynamics, and quantification of soil colonization by introduced fungi.

The ability of plant pathogenic fungi in soil to attack plant roots depends on contact of a moving plant root or fungal hypha with its stationary or moving counterpart. Thus, fungi capable of occupying a large portion of the soil volume may be better pathogens. Garrett [47] discussed the relationship between pathogen propagule size and number in terms of inoculum potential, which he defined as "the energy of growth of a parasite available for infection of a host at the surface of the host organ to be infected. Inoculum potential may be increased in either or both of two ways: (a) by increase in the number of infecting units or propagules of the fungus per unit area of root surface and . . . (b) by increase in the nutritional status of such units."

Gilligan [48,49] proposed a model to predict the distance that *Gaeumannomyces graminis* var. *tritici* (causal agent of take-all of wheat and barley) could grow from colonized millet seeds and infect wheat roots. He placed colonized seeds in a two-dimensional, horizontal plane through which the wheat roots passed vertically, and estimated that *G. graminis* var. *tritici* grew about 11.6 mm from inoculum particles to reach a root. His rooting medium was an artificial, sterile, sand-soil mixture; presumably in natural soils competition would be more intense and nutrients available for growth would be more limiting.

Wilkinson et al. [50] took a statistical approach to modeling the distance that *G. graminis* var. *tritici* can grow through soil and infect a plant host. The pathogen grows from decaying fragments of colonized plant material (e.g., remnant stems and roots of the previous year's crop) and infects wheat or barley roots. Using the method described by Thompson [51], they used the distribution of the distance from a wheat root to the *n*th nearest neighbor (inoculum particle), assuming a random distribution of particles around the root. For different inoculum particle sizes and densities, they then determined an estimated mean distance for infection (EDI). EDI values in two different silt loam soils ranged from about 1.7 mm for smaller particles (0.25 – 0.50 mm diameter) to 3.9 mm for larger particles (0.5 – 1.0 mm) and 11.3 mm for the largest particles (1 – 2 mm). They observed that soil fumigation with methyl bromide resulted in greater EDI values for the smaller inoculum particle sizes, presumably due to effects of nutrient competition from the associated microbiota when particles had only a small food base. However, they observed that EDI values did not differ significantly between two soils that were conducive or suppressive to take-all disease, respectively. Thus, they concluded that antagonistic effects resulting in suppressiveness were independent of EDI.

In his review of inoculum potential, Baker [52] stated, "The inoculum density of a given soil-borne pathogen usually is not greater than that needed for reasonably efficient positioning in the three-dimensional space to cover the available infection courts." Bruehl [53] commented that this statement is questionable, since living forms, unless restrained, reproduce in excess. There may be adaptive ecological advantages to efficiency in colonization of substrates. Nedwell and Gray [54] summarized some of the advantages for soil organisms that are able to efficiently spread and disperse, and also to change their growth pattern in response to the presence of different substrates.

As mentioned above, most models of fungal growth on agar or in broth have assumed a continuous and unlimited supply of substrate being encountered by an exponentially increasing population of hyphal tips, at least during certain defined stages of colony development. However, growth of a fungus in soil represents a substantially different type of substrate–fungus interaction and does not meet those assumptions. In natural soil the physical and nutritional substrates are discontinuous. Unlike growth on agar or in broth, growth in natural soil requires traversing soil volumes depleted of utilizable substrate and potentially housing inhibitory substances and/or predator populations. Competition between saprophytic fungi may in some instances be a competition for soil microsites containing usable substrates, thus conferring an advantage to an effective "exploratory" mechanism. Hyphal growth in soil can be considered as an exploration phase (i.e., a "quest" for a utilizable substrate) followed by an exploitation phase (colonization of the substrate).

During the initial stage of colony development in soil, fungal hyphae are obtaining their energy from either endogenous reserves or from substrate contained within the soil volume in which they are residing. As the colony develops, the number of hyphal tips and the distance of the tips from the energy source increases. During the exploration of the soil matrix by hyphae, hyphal tips may encounter inhibitory substances, lytic factors, new substrates, or neutral sites. The fungus must be able to bridge gaps between utilizable substrates. While traversing the voids between soil aggregates or between utilizable substrates, hyphal branch frequency is much reduced relative to the branch frequency observed on agar. Thus, during this stage of colony development, increase in biomass or number of hyphal tips over time would be less than exponential.

When a substrate is encountered, branch frequency increases in order to maximize the colonization process. This stage of growth may be exponential depending upon the amount of available energy and the level of competition for that energy. During this intense colonization process, growth is no longer oriented away from the origin, allowing for biomass increase with no increase in the radial extension of the colony. Thus, growth of *Trichoderma harzianum* in sterile soil was observed to change from radial vegetative growth to sporulation only, when hyphae encountered a food source (wheat seeds) in the soil (G. R. Knudsen and D. J. Eschen, unpublished). Similarly, mycelial networks of two basidiomycetes, *Hypholoma fasciculare* and *Phanaerochaete velutina*, were observed to change from a sparse, radial pattern to a more diffuse growth pattern following contact with wood blocks [55].

These substrate-dependent changes in growth form are in some ways remarkably similar to searching behavior patterns that have been observed for insects foraging for dispersed food sources. Murdie and Hassel [56] analyzed searching behavior of a parasitic wasp, *Diaeretiella rapae* (Curt.), searching for aphid hosts, and of the common housefly (*Musca domestica* L.) searching for sugar droplets in an enclosed arena. They mapped the paths taken by the insects, then in the latter example analyzed distributions of direction (angle) changes as the fly took a "random walk." The mean angle was approximately zero, but when the fly encountered a drop of sugar water, the standard deviation of the distribution increased significantly, resulting in a more comprehensive search of the immediate area around the food source. For food sources with an aggregated spatial distribution, such activity would increase the likelihood of successful encounters.

Similarly, attempts to model vegetative growth patterns of fungi in natural environments will need to consider how branching and growth patterns may change in response to localized nutrient sources or other environmental factors.

Whatever the medium or matrix (e.g., agar, fermentation broth, or natural soil), the growth habit of fungi is largely determined by the source and concentration of carbon and nitrogen, i.e., the source and amount of available energy. It is probable that fungal species with high carbon and nitrogen utilization efficiencies will have a greater capacity for growth in soil when other factors are not limiting. In a comparison of mycoparasites, it was observed that strains of *Thielavia terricola* were capable of greater growth into soil from a substrate-impregnated carrier than strains of *Trichoderma* or *Gliocladium* [57]. In a similar test, strains of *Gliocladium catenulatum* varied greatly in their ability to grow into soil from a substrate-carrier [58]. The genus *Phytophthora* has species (e.g., *P. parasitica* and *P. cactorum*) with little ability to grow through nonsterile soil [59], species (*P. palmivora* and *P. erythroseptica*) capable of limited growth [60,61], and species (*P. cinnamomi* and *P. megasperma*) with moderate growth ability in nonsterile soil [44,62,63]. Again, different propagules of a single fungal species may have different dispersal capabilities in different environments, and different forms of the same propagule may vary in dispersal abilities.

Paustian and Schnurer developed physiological-based models of fungal growth in soil [64]. These models predicted biomass increase and the increase in total hyphal length as a function of the input of carbon and nitrogen, assuming a steady-state conversion into fungal biomass. In validation experiments, they determined that for *Trichoderma harzianum*, biomass increase could be predicted based on the amount of available carbon and nitrogen, as well as soil moisture. The model generated information relative to the driving forces of fungal development in soil. However, it did not predict how the hyphae were arranged in three-dimensional space within the soil volume. Understanding the spatial arrangement of the fungal hyphae and the colony morphology would improve our understanding of the mechanics of dispersal.

The most common substrates for fungi in soil are plant materials (e.g., roots, root exudates, decomposition products) and other fungi. From what is known about the spatial distribution of plant roots and soilborne plant pathogens, we can presume that the spatial distribution of utilizable substrates is aggregated to some degree. Since the distribution of utilizable substrates is not continuous or uniform, the chance of a hyphal tip encountering such a substrate is less than absolute. This, combined with potential competition from other microbes, ensures that ability to utilize the substrate is also less than absolute. For these reasons, encounters between soilborne fungi and utilizable substrates are dependent upon the occurrence of a series of events that can best be represented by probability distributions. Simulation models have been developed to account for this intrinsic stochasticity in natural soil. One simulation model predicts total hyphal length and the spatial distribution of hyphae, as reflected in the proportion of target propagules (sclerotia of *Aspergillus flavus* and *Phymatotrichum omnivorum*) contacted by the fungal biocontrol agents *Gliocladium virens* and *Gliocladium roseum* [65,66]. To simulate energy-dependent growth in the model, the total number of nodes that can be "grown" during one simulation is predetermined. It can be calculated how much biomass will result from a known amount of substrate. Prior to simulation, the total number of nodes that are available for colony development is set, independent of where in the colony they are used. Validation experiments include photography and digitizing of fungal growth in soil, and strategic placement of targets within a soil volume to determine the subsequent proportion of the targets contacted. The model was proposed as a tool to investigate combined effects of biocontrol agent density and distribution, target propagule density and distribution, and environment on the probability of contact with the target propagule. Thus, the model could be incorporated into a larger model simulating the epidemiology of mycoparasitism.

A statistical model based on maps of fungal colony development in soil was developed by Knudsen and Eschen [67], who quantified radial growth rates for the biocontrol fungus *Trichoderma harzianum* in soil. *T. harzianum* was formulated in sodium alginate granules [68], then placed beneath nylon mesh nets in pasteurized soil. Periodically, the extent of radial growth was observed microscopically and mapped on graph paper. A logistic model was used to describe radial growth over a 14-day period under different environmental conditions, then slopes of the logit curves were regressed against soil temperature, soil moisture, and amount of bran additive in the granules. Finally, the best-fit models were used to predict the proportion of randomly distributed sclerotia of *Sclerotinia sclerotiorum* that would be contacted and colonized over time by hyphae of *T. harzianum* in nonsterile soil. Perhaps due in part to microbial competition in the nonsterile soil, the initial model's predictions sometimes varied significantly from experimental observations.

Reynolds et al. [69] developed a simulation model for spread of *Phytophthora cinnamomi* through Fraser fir nursery beds. A series of regression equations was used to generate values for several model parameters. Lesion development on Douglas-fir seedlings and/or disease incidence in the seedbed was used as an indication of movement of the pathogen. As demonstrated by Kenerley, however, using disease symptoms as an indicator of pathogen movement in a similar *P. cinnamomi*–Douglas-fir nursery system underestimated the actual spread of the pathogen [70]. By direct isolation from nursery soils, they were able to detect the presence of the pathogen prior to symptom development in the seedlings.

In deriving simulation models for infection of Douglas-fir seedlings by *Fusarium oxysporum*, Bloomberg [71,72] incorporated quantitative data on temperature effects on lesion development [73], and on pathogen distribution in soil [74]. One application of his model was to explore the relative predominance of infections caused by lesion spread (i.e., hyphal growth) compared to new infections through root tips (a function of propagule distribution).

IV. AERIAL DISPERSAL OF FUNGI

Although a permanent microflora probably doesn't exist in the atmosphere [75], atmospheric conditions nonetheless play an important role in the dispersal, deposition, and survival of many fungi. Dispersal of airborne fungal propagules involves take-off into the airstream, downwind movement and dispersion in air currents, and deposition on surfaces (e.g., leaves) where subsequent death or survival and possible colonization will occur.

Many factors affect dispersal and survival, including aerosol drift, dispersion and deposition, epiphytic survival, and possible reentrainment of organisms, washoff to soil, and further dispersal by insects and other animals or farm machinery. Lacey [75] and Aylor [76] have provided comprehensive reviews of mechanisms by which spores of fungi are liberated, transported through the atmosphere, and deposited.

Much of the literature describing physical models of particle transport comes from research on atmospheric transport of air pollutants. In plant pathology, another primary motivation for development of aerial spore dispersal models has been the need to understand how disease epidemics develop. Empirically derived models have been developed from measurement of spore dispersal using a variety of spore sampling devices; recent comparisons have been made of different models to describe spore dispersal or disease gradients [77], and McCartney and Fitt [78] reviewed the construction of physical models of spore dispersal within and above crops.

Various models have been developed to describe dispersion of atmospheric particles released from continuous or noncontinuous sources. Gaussian plume models predict concentrations of particles in a downwind plume and are probably the most commonly used dispersion models [79–81]. They assume that turbulence causes airborne particles to randomly disperse, so that after any given amount of time the plume will show a scatter around its origin. This dispersion is characterized by a bivariate Gaussian (normal) distribution. Variations on Gaussian plume models specify the dispersion parameters as functions of downwind distance and atmospheric stability [81,82].

Although Gaussian dispersion models are relatively easy to manipulate mathematically, one disadvantage lies in the assumption of a continuous source and uniform wind speed. For some experimental field trials involving dispersal of fungal spores, a higher level of precision may be required than is provided by Gaussian models. Especially if the particle source is small and discontinuous in time, minor variations in wind speed and direction over short time periods can significantly affect predicted dispersal patterns. Puff diffusion models calculate the position of a particle cloud at successive time steps and offer an improvement over simple Gaussian plume models for predicting this type of event, but one disadvantage lies in the increasing computational difficulty as the time step length is decreased or wind vectors change.

More generally, a disadvantage of most empirically based statistical models is their relative inflexibility for prediction of dispersal patterns under conditions different from those for which they were derived [77]. A variety of alternative, theoretical models have been derived based entirely or in part on the underlying physical principles of particle dispersal. So-called random walk models take the approach of following a discrete number of individual particles released from a source, thus producing a representative picture of the entire plume as it diffuses in space and time. Time is also treated as a discrete value, meaning that the system is examined at regular intervals, for example, every second or fraction of a second. Movement of particles by advection or turbulence, as well as other physical influences, may vary at each time step. The structure of these models makes them appropriate for implementation on digital computers. Several random walk models have been developed for estimating dispersion from a continuous point source, including more complicated models based on a Markov chain principle of correlation between component particle velocities at successive time steps [83,84]. Knudsen [85] developed a random walk simulation model to predict short-range aerosol dispersal of microbes in field trials.

Other published microbial dispersal models have variously emphasized a statistical approach (including Gaussian plume modeling), or incorporation of physical principles, or some combination of the two approaches. Legg and Powell [86] modeled transport and deposition of *Lycopodium* spores with some success, predicting spore concentrations downwind from a constant line source. However, their model was less successful in predicting gradients of *Erysiphe graminis* conidia released from a strip of powdery mildew-infected barley in an extensive, uniform field. Aylor [87] modified Legg and Powell's model to account for passive liberation of spores by wind and its effect on the subsequent transport and deposition of spores within the crop. Aylor's model describes the decrease in concentration of *Erysiphe graminis* spores observed downwind of a source of infection within a barley crop.

Aylor and Ferrandino [88] measured vertical profiles of wind speed and aerial concentration of urediniospores of *Uromyces phaseoli* above a diseased bean field. Aerial concentrations of single and clusters of spores were determined separately, and upward flux above canopy calculated for each. They compared upward flux (i.e., escape) for singlets, doublets, and triplets; it was greater for singlets than doublets than triplets. Relative proportions of these clusters escaping the canopy agreed with the proportions of each released by the crop, as assessed by numbers caught on sticky glass slides.

Numerous physical forces act on airborne particles, including gravity (which affects sedimentation rate), temperature and atmospheric humidity (affecting survival as well as evaporative size change in aerosol droplets), solar radiation, and rainfall. The terminal velocity (sedimentation rate) of small falling particles is often estimated using Stokes' Law, which describes the relation between the size and terminal velocity of a smooth spherical object falling in a fluid medium. For microorganisms suspended in air, assuming the density of the medium is negligible compared to the density of the falling object, Stokes' Law simplifies to:

$$V_t = 0.0121\ r^2$$

where V_t is the terminal velocity of the falling object (m/sec) and r is the radius of a spherical aerosol droplet in micrometers. However, particles that are nonspherical (e.g., many fungal spores, or fungi resuspended on "rafts" of plant epidermis) have more frictional drag than spheres of the same volume and density. Boundary layers affect particle movement around surfaces such as plant stems and leaves. Movement of airborne fungal propagules through plant canopies is different than movement in open air, since plant canopies effectively filter microbial aerosols due to impaction on stems and leaves. Canopies also have the effect of slowing the wind so that sedimentation can occur. Because small particles have a lower impaction efficiency, they are less likely to penetrate plant canopies and will therefore disperse greater distances over dense vegetation [89,90].

Because fungal spores and mycelial fragments are living organisms and not inert particles, prediction of their dispersal in space is complicated by the concurrent need to predict their mortality. The death rate of airborne microbes or microbes on surfaces is a function both of cellular physiology [91] and of environmental factors including temperature, relative humidity, and solar radiation. Ultraviolet (UV) radiation can be lethal to exposed microbes, with the most damaging wavelengths being in the range of about 265 nm. However, other short wavelengths, even up to the visible range, may have deleterious effects. Rotem et al. [92] exposed spores of *Peronospora tabacina*, *Uromyces phaseoli*, and *Alternaria solani* in the field to full or partial sunlight and measured the fungicidal effect. Temperature was found to be an important factor in mortality of detached but not attached sporangia. Conidia of *A. solani* were killed after 5 days of exposure, independent of temperature, and mortality was lower when spores were shaded or protected by UV filters. Rotem et al. concluded that UV radiation rather than increased temperature was the main factor causing mortality.

Frequently, mortality of microbes during aerial dispersal has been modeled by introducing an exponential decay factor to account for death over time due to solar irradiation or dessication [93,94]. Lighthart and Frisch [95] incorporated an exponential microbial death rate into a modified Pasquill particle dispersion model [79] and predicted concentrations of viable airborne bacteria at different locations downwind from a point source. Their model assumed that the source was continuous, and meteorological conditions (wind speed, atmospheric stability class) remained constant. Lighthart and Mohr [96] also discussed the use of multiple regression models to predict survival of airborne virus particles as affected by independent variables of temperature, relative humidity, solar radiation, and time in the plume. The regression model was coupled with a model to predict dispersion and deposition of particles emitted from a continuous point source. This approach may have application to predicting dispersal of fungal spores as well.

As fungal spores are released from a point, line, or area source, deposition processes result in a decrease in spore concentration with increasing distance from the source, i.e., a dispersal gradient. Several mathematical models have been developed to describe gradients of spore dispersal or disease [97–100]. Based on a gradient model, Kampmeijer and Zadoks [101,102] developed the simulation model EPIMUL to describe spatial effects of crop distribution on progress of an epidemic.

Two of the more commonly used basic models for gradient analysis are the exponential decay model [103,104] and the power law model. The exponential model takes the form:

$$y = a\ \exp^{(-bx)}$$

where y is the number of microbes deposited, x is distance from the source, constant a reflects the source strength, and exponent b measures the steepness of the dispersal gradient. The power law model is of the form:

$$y = a\ x^{-b}$$

again, constant a and exponent b reflect source strength and gradient steepness, respectively. Either the power law model or the exponential model often provides a good fit to experimental data [77]. Fitt and McCartney [77] pointed out that because the exponential model predicts finite values of y at all distances, it is more easily incorporated into epidemic models.

Dispersal gradients may be strongly affected by weather factors including prevailing winds and turbulence, as well as by particle size. Primary dispersal gradients, produced when a source releases spores for a short time, tend to become flattened by secondary gradients resulting from subsequent generations of spores. In experimental field plots, dispersal gradients may be obscured by interplot interference [105,106] or by background contamination.

In general, validation of aerial dispersal and gradient models can be difficult and expensive. Detection of viable microbes in aerosols or deposited on vegetation or soil under field conditions requires sensitive methods and sampling at several locations. Dispersal can be evaluated in at least two ways: movement from one site to another, and successful establishment in the new site. To separate these two phenomena may require different protocols. For example, fungal propagules may be disseminated by wind or other forces but fail to become established at the site where they are introduced. Usually, the reason we are concerned with the specific distance a fungal propagule has dispersed from a given site is because we are interested whether or not that propagule survives in the new site and gives rise to a colony. Many environmental factors and genotype-dependent characteristics inherent to the organism will influence establishment, and since the potential to become established at a remote site is less than absolute, sampling at more than one location and more than one time is generally necessary. Depending upon the species and the sampling sensitivity, it may take several years to map the full extent of dispersal. The dispersal of fungal genotypes between populations of the same species may require use of sophisticated sampling and detection technology over a long period of time. Introduced organisms may remain below detection limits for several months or years.

Detecting low level concentrations of microbes in aerosols is complicated by a high dilution factor for dispersed particulates and requires sampling relatively high air flow rates for a long time [107]. Fannin [107] listed several criteria for selection of appropriate sampling methodology; these included collection efficiency, sampling sensitivity, reliability, ease of sterilization, maintenance of collected sample viability, ease of sample assay, remote operation capability, particle size discrimination, and cost. Sampling devices for aerobiology are variously based on the concepts of sedimentation (e.g., vaseline-coated slides, settle plates), filtration, impingement (e.g., AGI-30), impaction (e.g., Reynier or Anderson samplers, rotorod sampler), and cyclone scrubbing [107]. The relative merits of different sampling devices have been described elsewhere [107,108].

V. CONCLUDING REMARKS

Despite a strong theoretical basis of modeling fungal growth in vitro, there are still relatively few models of fungal growth in natural environments. There is, however, a large volume of literature in which aspects of fungal growth in soil have been at least indirectly determined; this work should prove useful in development and verification of models. Models that link the processes of growth on a substrate and dispersal between sites will greatly increase our understanding of fungal life histories. This, in turn, will provide a foundation for epidemiological studies and increase our predictive ability relative to animal and plant diseases. Such models will also be instrumental as decision-making aids for environmental managers concerned with the deployment of microorganisms for toxic waste cleanup or the biological control of plant and animal pests.

Models relevant to aerial dispersal of fungi are much more available, largely due to the large body of applicable literature on dispersal of aerial bacteria, insects, plant parts (pollen, spores), and pollutants. Also, this area of research is probably more amenable to quantification for several reasons, including the relative ease of nondestructive sampling and the discrete nature of airborne propagules. However, sampling to validate models in aerobiology is still a difficult and often expensive task.

Further developments in modeling fungal growth and dispersal in natural environments are not likely to be constrained by lack of a theoretical basis (either biological or mathematical). Nor is availability of appropriate computer hardware or software a limiting factor. However, we feel there is a major technical challenge to develop methods to reliably detect, identify and monitor fungi in air, soil, and water and on plant surfaces. Without powerful tools for validation, model development will necessarily be slowed. Advances in such diverse new technologies as molecular biology and computer-based image analysis will help provide these tools. Applications for models, we suggest, are virtually unlimited.

REFERENCES

1. Alexander, M. Biodegradation of organic chemicals. *Environ. Sci. Technol.*, *19*:106–111, 1985.
2. Chakrabarty, A. M. *Biodegradation and Detoxification of Environmental Pollutants.* CRC Press, Boca Raton, FL, 1982, p. 147.
3. Johnston, J. B., and Robinson, S. G. *Genetic Engineering and New Pollution Control Technologies.* Noyes Publications, Park Ridge, NJ, 1984, p. 580.
4. Levin, M. A., Seidler, R., Borquin, A. W., Fowle, J. R., and Barkay, T. EPA developing methods to assess environmental release. *Biotechnology*, *5*:38–45, 1987.
5. McCormick, D. Detection technology: The key to environmental biotechnology. *Biotechnology*, *4*:419–422, 1986.
6. Alexander, M. *Introduction to Microbial Ecology.* Academic Press, New York, 1977.
7. Slater, J. H. Genetic interactions in microbial communities. In *Current Perspectives in Microbial Ecology* (M. J. Klug and C. A. Reddy, eds.), American Society for Microbiology, Washington, D.C., 1984.
8. Emanuel, D. A. In *Occupational Mycoses* (A. F. DiSalvo, ed.), Lea and Febiger, Philadelphia, 1983, pp. 211–214.
9. Fraser, D. W., Ward, J. I., Ajello, L., and Plikaytis, B. D. *Amer. Med. Assn.*, *242*:1631, 1979.
10. Emanuel, D. A. In *Occupational Mycoses* (A. F. DiSalvo, ed.), Lea and Febiger, Philadelphia, 1983, pp. 201–209.

11. Maynard Smith, J. *Mathematical Ideas in Biology*. Cambridge University Press, Cambridge, 1968, p. 152.
12. Holling, C. S. The strategy of building models of complex ecological systems. In *Systems Analysis in Ecology* (K. E. F. Watt, ed.), Academic Press, New York, 1966, pp. 195–214.
13. Waggoner, P. E. Simulation of epidemics. In *Epidemics of Plant Diseases*. Springer-Verlag, New York, 1974, pp. 137–160.
14. Jeger, M. J. Modelling the dynamics of pathogen populations. In *Populations of Plant Pathogens: Their Dynamics and Genetics* (M. S. Wolfe and C. E. Caten, eds.), Blackwell Scientific Publications, Oxford, 1985.
15. Jeger, M. J. The potential of analytic compared with simulation approaches in plant disease epidemiology. In *Plant Disease Epidemiology, Vol. 1* (K. J. Leonard and W. E. Fry, eds.), Macmillan, New York, 1986, pp. 255–281.
16. Teng, P. S. A comparison of simulation approaches to epidemic modeling. *Annu. Rev. Phytopathol.*, *23*:351–379, 1985.
17. Da Riva Ricci, D., and Kendrick, B. Computer modelling of hyphal tip growth in fungi. *Canadian Journal of Botany*, *50*:2455–2462, 1972.
18. Emerson, S. The growth phase in *Neurospora* corresponding to the logarithmic phase in unicellular organisms. *Journal of Bacteriology*, *60*:221–223, 1950.
19. Prosser, J. I., and Trinci, A. P. J. A model for hyphal growth and branching. *J. Gen. Microbiol.*, *11*:153–164, 1979.
20. Trinci, A. P. J. Kinetics of growth of mycelial pellets of *Aspergillus nidulans*. *Archiv. für Microbiologie*, *73*:353–367.
21. Trinci, A. P. J. Influence of the width of the peripheral growth zone on the radial growth rate of fungal colonies on solid media. *J. Gen. Microbiol.*, *67*:325–344, 1971.
22. Trinci, A. P. J. A study of the kinetics of hyphal extension and branch initiation of fungal mycelia. *J. Gen. Microbiol.*, *81*:225–236, 1974.
23. Trinci, A. P. J., and Saunders, P. T. Tip growth of fungal hyphae. *J. Gen. Microbiol.*, *103*:243–248, 1977.
24. Bartnicki-Garcia, S. Fundamental aspects of hyphal morphogenesis. In *Microbial Differentiation (23rd Symposium of the Society for General Microbiology)* (J. M. Ashworth and J. E. Smith, eds.), Cambridge University Press, 1973, pp. 245–267.
25. Boughey, A. S. *Ecology of Populations*, 2nd ed. Macmillan, New York, 1973, p. 182.
26. Eamus, D., and Jennings, D. H. Water, turgor and osmotic potentials of fungi. In *Water, Fungi and Plants* (P. G. Ayres and L. Boddy, eds.), British Mycological Society Symposium 11, Cambridge University Press, 1986, pp. 27–48.
27. Barker, S. B., Cumming, G., and Horsfield, K. Quantitative morphometry of the branching structure of trees. *Journal of Theoretical Biology*, *40*:33–43, 1973.
28. Cohen, D. Computer simulation of biological pattern generation processes. *Nature*, *216*:246–248, 1967.
29. Gull, K. Mycelium branch patterns of *Thamnidium elegans*. *Trans. Brit. Mycol. Soc.*, *64*:321–324, 1975.
30. Thornley, J. H. M. A model of apical bifurcation applicable to trees and other organisms. *J. Theoret. Biol.*, *65*:165–176, 1977.

31. Quinn, J. A., and Fujimoto, T. T. Computer graphics simulation of growth and sporulation of *Erysiphe polygoni*. *Phytopathology, 76*: 883–888, 1986.
32. Quinn, J. A., and Powell, C. C., Jr. Growth and sporulation of *Oidium begoniae*: Development of a deterministic model. *Annals of Appl. Biol., 107*:163–178, 1985.
33. Lapp, M. S., and Skoropad, W. P. A mathematical model of conidial germination and appressorial formation for *Collectotrichum graminicola*. *Can. J. Botany, 54*:2239–2242, 1976.
34. Wallace, H. R. Dispersal in time and space: Soil pathogens. In *Plant Disease: An Advanced Treatise* (J. G. Horsfall and E. B. Cowling, eds.), Academic Press, New York, 1978, pp. 181–202.
35. Bahme, J. B., and Schroth, M. N. Spatial-temporal colonization patterns of a rhizobacterium on underground organs of potato. *Phytopathology, 77*:1093–1100, 1987.
36. Parr, J. F., Parkinson, D., and Norman, A. G. Growth and activity of soil microorganisms in glass micro-beads: II. Oxygen uptake and direct observations. *Soil Science, 103*:303–310, 1967.
37. Ruddick, S. M., and Williams, S. T. Studies on the ecology of actinomycetes in soil. V. Some factors influencing the dispersal and adsorption of spores in soil. *Soil Biol. Biochem., 4*:93–103, 1972.
38. Wilkinson, H. T., Miller, R. D., Millar, R. L. Infiltration of fungal and bacterial propagules into soil. *Soil Sci. Soc. Am. J., 45*:1034–1039, 1981.
39. McCoy, E. L., and Hagedorn, C. Quantitatively tracing bacterial transport in saturated soil systems. *Water Air Soil Pollut., 11*:467–479, 1979.
40. Bitton, G., Lahav, N., and Henis, Y. Movement and retention of *Klebsiella aerogenes* in soil columns. *Plant Soil, 40*:373–380, 1974.
41. Wollum, A. G., and Cassel, D. K. Transport of microorganisms in sand columns. *Soil Sci. Soc. Am. J., 42*:72–76, 1978.
42. Duniway, J. M. Movement of zoospores of *Phytophthora cryptogea* in soils of various textures and matric potentials. *Phytopathology, 66*: 877–882, 1976.
43. Parke, J. L., Moen, K., Rovira, A. D., and Bowen, G. D. Soil water flow affects the rhizosphere distribution of a seedborne biological control agent, *Pseudomonas fluorescens*. *Soil Biol. Biochem., 18*:583–588, 1986.
44. Stack, J. P., and Millar, R. L. Competitive colonization of organic matter in soil by *Phytophthora megasperma* f. sp. *medicaginis*. *Phytopathology, 75*:1393–1398, 1985.
45. MacDonald, J. D., and Duniway, J. M. Influence of soil texture and temperature on the motility of *Phytophthora cryptogea* and *P. megasperma* zoospores. *Phytopathology, 68*:1627–1630, 1978.
46. Ruddick, S. M., and Williams, S. T. Studies on the ecology of actinomycetes in soil, IV. *Soil Biol. Biochem., 4*:79–91, 1972.
47. Garrett, S. D. *Biology of Root-Infecting Fungi*. Cambridge University Press, London, 1956.
48. Gilligan, C. A. Modeling rhizosphere infection. *Phytopathology, 69*: 782–784, 1979.
49. Gilligan, C. A. Zone of potential infection between host roots and inoculum units of *Gaeumannomyces graminis*. *Soil Biol. Biochem., 12*:513–514, 1980.

50. Wilkinson, H. T., Alldredge, J. R., and Cook, R. J. Estimated distances for infection of wheat roots by *Gaeumannomyces graminis* var. *tritici* in soils suppressive and conducive to take-all. *Phytopathology, 75*:557–559, 1985.
51. Thompson, H. P. Distribution of distance to Nth neighbor in a population of randomly distributed individuals. *Ecology, 37*:391–394, 1956.
52. Baker, R. Inoculum potential. In *Plant Disease—An Advanced Treatise*, Vol. 2 (J. G. Horsfall and E. B. Cowling, eds.), Academic Press, New York, 1978, pp. 137–157.
53. Bruehl, G. W. *Soilborne Plant Pathogens.* Macmillan, New York, 1987, p. 368.
54. Nedwell, D. B., and Gray, T. R. G. Soils and sediments as matrices for microbial growth. In *Ecology of Microbial Communities* (M. Fletcher, T. R. G. Gray, and J. G. Jones, eds.), Soc. for Gen. Microbiol., 41st Symposium, Cambridge University Press, Cambridge, 1987, pp. 21–54.
55. Dowson, C. G., Rayner, A. D. M. and Boddy, L. Outgrowth patterns of mycelial cord-forming basidiomycetes from and between woody resource units in soil. *J. Gen. Microbiol., 132*:203–211, 1986.
56. Murdie, G., and Hassell, M. P. Food distribution, searching success and predator-prey models. In *The Mathematical Theory of the Dynamics of Biological Populations* (M. S. Bartlett and R. W. Hiorns, eds.), Academic Press, London, 1973, pp. 87–101.
57. Stack, J. P., Kenerley, C. M., and Pettit, R. E. Influence of carbon and nitrogen sources, relative carbon and nitrogen concentrations, and soil moisture on the growth in nonsterile soil of soilborne fungal antagonists. *Can. J. of Microbiol., 33*:626–631, 1987.
58. Kenerley, C. M., and Stack, J. P. Influence of assessment methods on selection of fungal antagonists of the sclerotium-forming fungus *Phymatotrichum omnivorum*. *Can. J. Microbiol., 33*:632–635, 1987.
59. Gisi, U. Investigations on the soil phase of *Phytophthora cactorum* (Leb. et Cohn) Schroet. with fluorescent optical direct observation. *Z. Pflanzenkr. Pflanzenschutz., 82*:355–377, 1975.
60. Turner, P. D. Behavior of *Phytophthora palmivora* in soil. *Plant Disease Reporter, 49*:135–137, 1965.
61. Vujicic, R., and Park, D. Behavior of *Phytophthora erythroseptica* in soil. *Trans. Br. Mycol. Soc., 47*:455–458, 1964.
62. Weste, G. Population dynamics and survival of *Phytophthora*. In *Phytophthora, its Biology, Taxonomy, Ecology, and Pathology* (D. C. Erwin, S. Bartnicki-Garcia, and P. H. Tsao, eds.), APS Press, St. Paul, MN, 1983, pp. 237–257.
63. Zentmyer, G. A., and Mircetich, S. M. Saprophytism and persistence in soil by *Phytophthora cinnamomi*. *Phytopathology, 56*:710–712, 1966.
64. Paustian, K., and J. Schnürer. Fungal growth response to carbon and nitrogen limitation: A theoretical model. *J. Soil Bio. Biochem., 19*:613–620, 1987.
65. Stack, J. P., Knudsen, G. R., and Koch, D. O. A computer simulation model to predict the dispersal of biocontrol fungi in soil. *Phytopathology, 77*:1771, 1987.
66. Stack, J. P., Knudsen, G. R., and Koch, D. O. Modeling growth of soil fungi with special reference to biocontrol of plant pathogens. *Proc. 5th Intn'l Congress of Plant Pathology*, Kyoto, Japan, 1988.

67. Knudsen, G. R., and Eschen, D. J. Hyphal growth from alginate granules containing *Trichoderma harzianum*. *Phytopathology*, *78*: 1594, 1988.
68. Walker, H. L., and Connick, W. J. Sodium alginate for production and formulation of mycoherbicides. *Weed Sci.*, *31*:333–338, 1983.
69. Reynolds, K. M., Gold, H. J., Bruck, R. I., Benson, D. M., and Campbell, C. L. Simulation of the spread of *Phytophthora cinnamomi* causing a root rot of fraser fir in nursery beds. *Phytopathology*, *76*:1190–1201, 1986.
70. Kenerley, C. M., and Bruck, R. I. Distribution and disease progress of Phytophthora root rot of fraser fir seedlings. *Phytopathology*, *77*:520–536, 1987.
71. Bloomberg, W. T. A model of damping-off and root rot of Douglas-fir seedlings caused by *Fusarium oxysporum*. *Phytopathology*, *69*: 74–81, 1979.
72. Bloomberg, W. T. Model simulations of infection of Douglas-fir seedlings by *Fusarium oxysporum*. *Phytopathology*, *69*:1072–1077, 1979.
73. Salisbury, P. J. The effect of temperature and hydrogen-ion concentration on growth of certain cultures of *Fusarium oxysporum* isolated from Douglas-fir seedlings. *Can. Dept. Agric. For. Biol. Div.*, mimeo report, 1952.
74. Bloomberg, W. T. Distribution and pathogenicity of *Fusarium oxysporum* in a forest nursery soil. *Phytopathology*, *66*:1090–1092, 1976.
75. Lacey, J. Aerial dispersal and the development of microbial communities. In *Microbial Ecology* (J. M. Lynch and N. J. Poole, eds.), Blackwell Scientific Publications, Oxford, 1979, pp. 140–170.
76. Aylor, D. E. Dispersal in time and space: Aerial pathogens. In *Plant Disease: An Advanced Treatise* (J. G. Horsfall and E. B. Cowling, eds.), Academic Press, New York, 1978, pp. 159–180.
77. Fitt, B. D. L., and McCartney, H. A. Spore dispersal in relation to epidemic models. In *Plant Disease Epidemiology*, Vol. 1 (K. J. Leonard and W. E. Fry, eds.), MacMillan, New York, 1986, pp. 311–345.
78. McCartney, H. A., and Fitt, B. D. L. Construction of dispersal models. *Adv. Plant Pathol.*, *3*:107–143, 1985.
79. Pasquill, F. The estimation of the dispersion of windborne material. *Meteorol. Mag.*, *90*:33–49, 1961.
80. Pasquill, F. *Atmospheric Diffusion*, 2nd ed. Ellis Horwood Ltd., Chichester, 1974.
81. Hanna, S. R., Briggs, G. A., and Hosker, R. P., Jr. *Handbook on Atmospheric Diffusion*, Tech. Info. Ctr., U.S. Department of Energy, DOE/TIC-11223, 1982.
82. Hanna, S. R., Briggs, G. A., Deardoff, J., Egan, B. A., Gifford, F. A., and Pasquill, F. Summary of recommendations made by the AMS workshop on stability classification schemes and sigma curves. *Bull. Am. Meteorol. Soc.*, *58*:1305–1309, 1977.
83. Hanna, S. R. Some statistics of Lagrangian and Eulerian wind fluctuations. *J. Appl. Met.*, *18*:518–525, 1979.

84. Reid, J. D. Markov chain simulations of vertical dispersion in the neutral surface layer for surface and elevated releases. *Boundary-Layer Met.*, *16*:3–22, 1979.
85. Knudsen, G. R. Model to predict aerial dispersal of bacteria during environmental release. *Appl. Environ. Microbiol.*, *55*:2641–2647, 1989.
86. Legg, B. J., and Powell, F. A. Spore dispersal in a barley crop: A mathematical model. *Agric. Meteorol.*, *20*:47–67, 1979.
87. Aylor, D. E. Modeling spore dispersal in a barley crop. *Agric. Meteorol.*, *26*:215–219, 1982.
88. Aylor, D. E., and Ferrandino, F. J. Escape of urediniospores of *Uromyces phaseoli* from a bean field canopy. *Phytopathology*, *75*:1232–1235, 1985.
89. Chamberlain, A. C. Deposition of particles to natural surfaces. In *Airborne Microbes* (P. H. Gregory and J. L. Monteith, eds.), Cambridge University Press, Cambridge, 1967, pp. 138–164.
90. Chamberlain, A. C. The movement of particles in plant communities. In *Vegetation and the Atmosphere*, Vol. 1 (J. L. Monteith, ed.), Academic Press, London, 1975, pp. 155–203.
91. Anderson, J. D., and Cox, C. S. Microbial survival. In *Airborne Microbes. Proceedings of the 17th Symposium*, Society for General Microbiology, Cambridge, 1967.
92. Rotem, J., Wooding, B., and Aylor, D. E. The role of solar radiation, especially ultraviolet, in the mortality of fungal spores. *Phytopathology*, *75*:510–514, 1985.
93. Akers, A. B., and Won, W. D. Assay of living, airborne microorganisms. In *An Introduction to Experimental Aerobiology* (R. L. Dimmick and A. B. Akers, eds.), Wiley-Interscience, New York, 1969, pp. 59–99.
94. Dimmick, R. L., and Heckly, R. J. Theoretical aspects of microbial survival. In *An Introduction to Experimental Aerobiology* (R. L. Dimmick and A. B. Akers, eds.), Wiley-Interscience, New York, 1969, pp. 347–374.
95. Lighthart, B., and Frisch, A. S. Estimation of viable airborne microbes downwind from a point source. *Appl. Environ. Microbiol.*, *21*:700–704, 1976.
96. Lighthart, B., and Mohr, J. Estimating downwind concentrations of viable airborne microorganisms in dynamic atmospheric conditions. *Appl. Environ. Microbiol.*, *53*:1580–1583, 1987.
97. Gregory, P. H. Interpreting plant disease dispersal gradients. *Annu. Rev. Phytopathol.*, *6*:189–212, 1968.
98. Mundt, C. C., and Leonard, K. J. A modification of Gregory's model for describing plant disease gradients. *Phytopathology*, *75*:930–935, 1985.
99. Roelfs, A. P. Gradients in horizontal dispersal of cereal rust uredospores. *Phytopathology*, *62*:70–76, 1972.
100. Roelfs, A. P., and Martell, L. B. Uredospore dispersal from a point source within a wheat canopy. *Phytopathology*, *74*:1262–1267, 1984.
101. Kampmeijer, P., and Zadoks, J. C. EPIMUL, a simulator of foci and epidemics in mixtures of resistant and susceptible plants, mosaics and multilines. *Simulation Monographs*, Pudoc, Wageningen, The Netherlands, 1977, p. 56.

102. Zadoks, J. C., and Kampmeijer, P. The role of crop populations and their deployment, illustrated by means of a simulator EPIMUL 97. *Ann. N.Y. Academy of Sciences, 287*:164–190, 1977.
103. Kiyosawa, S., and Shiyomi, M. A theoretical evaluation of the effect of mixing resistant variety with susceptible variety for controlling plant diseases. *Ann. Phytopathol. Soc. Jap., 38*:41–51, 1972.
104. Lindow, S. E., Knudsen, G. R., Seidler, R. J., Walter, M. V., Lambou, V. W., Amy, P. S., Schmedding, D., Prince, V., and Hern, S. Aerial dispersal and epiphytic survival of *Pseudomonas syringae* during a pretest for the release of genetically engineered strains into the environment. *Appl. Environ. Microbiol., 54*:1557–1563, 1988.
105. Jenkyn, J. F., Bainbridge, A., Dyke, G. V., and Todd, A. D. An investigation into inter-plot interactions, in experiments with mildew on barley, using balanced designs. *Ann. Appl. Biol., 92*:11–28, 1979.
106. Paysour, R. E., and Fry, W. E. Interplot interference: A model for planning field experiments with aerially disseminated pathogens. *Phytopathology, 73*:1014–1020, 1983.
107. Wolf, H. W., Skaliy, P., Hall, L. B., Harris, M. M., Decker, H. M., Buchanan, L. M., and Dahlgren, C. M. *Sampling Microbiological Aerosols*, Public Health Monograph No. 60, Public Health Service Publication No. 686, U.S. Government Printing Office, Washington, D.C., 1959, p. 53.
108. Gregory, P. H. *The Microbiology of the Atmosphere*, 2nd ed. Leonard Hill, London, 1973, p. 377.

22

STOCHASTIC MODELING OF PLANT DISEASE EPIDEMIC PROCESSES

DOUGLAS I. ROUSE *University of Wisconsin—Madison, Madison, Wisconsin*

I. INTRODUCTION

This chapter examines some probabilistic approaches to formulating mathematical models of plant pathogen population and epidemic processes. The chapter will give evidence for the proposition that stochastic models are a necessary tool for increasing our understanding of pathogen and disease dynamics. There is a vast literature on stochastic models of biological population processes, including those for single species population growth [1–8], predator-prey [4,9], and animal epidemic processes [3,5,10–12]. In support of the proposition given above, repeated referral to that literature will be made in this chapter although no attempt has been made to present an exhaustive review. The existence of such an extensive literature suggests that there is little which is, in the context of population biology, conceptually new in this chapter. Nevertheless, little of the knowledge present in that literature has made its way into the plant pathology domain. It is hoped that this chapter will stimulate the development of stochastic plant epidemic models by pointing out why development of such models may benefit our understanding of plant disease processes.

Plant disease epidemiology is a part of the science of ecology and population biology. As such, stochastic modeling techniques developed by population biologists should be regarded as potential tools of the epidemiologist. Where mathematical techniques are simply being borrowed from that literature it may be argued that the techniques are not new within the plant pathology domain except as their application results in better understanding of the biology of plant disease. Thus, papers appearing on this topic in the plant pathology literature should be evaluated on the basis of their biological contribution to our understanding of epidemic processes, not on the basis of presentation of a "new technique" for modeling population dynamics (of disease).

A. Background on Modeling

It is appropriate to begin the examination of probabilistic modeling by clarifying what a model is and why modeling is essential. Models are simplifications of reality which relate variable quantities to each other. For example, in population models the variability in population size may be related to variables such as time, spatial location, or environment. In the simplest models, population size is referred to as a dependent variable because its numerical value is determined entirely by one or more other variables, such as time, which are independent of the value of the dependent variable, in this case population size.

In more complex models there may be several dependent variables which are related to each other in a system of equations. In that case the distinction between dependent and independent variables may be less clear, since within the system a variable can be a dependent variable in one quantitative relation (equation) and an independent variable in other quantitative relations. If the system of equations is relatively simple, consisting of no more than two or three equations, then analytical mathematical techniques may be sufficient to manipulate and analyze system behavior.

If the system becomes very complex as a result of a large number of variables and equations, then a formal systems approach may be taken in which the model usually takes the form of a computer simulation. With the systems approach the physical state of the system is defined by state variables that change through time, according to rate variables. The rate variables are functions of the state variables and driving variables, where driving variables represent external forces acting on the system (e.g., environment). The systems approach pays particular attention to defining the system boundaries so that driving variables are independent of the behavior of the system (otherwise they would need to be represented in the system by rate variables governing their change in time). In simulation models of systems, the rate variables are not allowed to depend on each other directly. There are many references to systems modeling in ecology [e.g., 13,14] and agriculture [e.g., 15,16], and recently a thorough review by Teng [17] of plant disease epidemic simulation.

Even the most complex models are simplifications of reality because it is virtually impossible to include in a model all the variables that, in reality, interact with each other. A model always has associated with it, either implicitly or explicitly, a set of simplifying assumptions. These assumptions identify the ways in which the model does not mimic the variability found in reality and thus also defines the manipulated or controlled conditions under which the model is expected to function predictably.

From the above statements it can be concluded that hypotheses are models and therefore that modeling is an integral part of the modern scientific method. Experiments are designed to carefully control for "unimportant" variables as defined by the assumptions of the model. If after experimental work we find no evidence to reject the hypothesis, then we may conclude that the model is an "adequate" description of the phenomenon we are trying to understand. In addition to the variability explained by the model, there will also be variability in any experimental system that is not explained by the model. A deterministic model is usually judged to be adequate if the only variability left unexplained is random and of a low order compared with the variability explained by the model. Random variability can be defined as that variability that results from chance.

Stochastic models provide a means of incorporating this unexplained or random variability into the model explicitly, sometimes with surprising results.

Models are the way scientists quantify, describe, and explain variability. There are several ways of categorizing variability. As indicated in the last paragraph, a distinction can be made between systematic variation (that explained by the model) and random variation (that which is due to chance). In stochastic models the distinction between explained variability and unexplained variability does not exist in the same context as for deterministic models. Kurtz [18] defines three ways that randomness may arise in population processes as fluctuations due to sampling mechanisms, random motion, and/or random environment. The first two sources of randomness may be referred to as internal fluctuations, while random environmental effects are referred to as external fluctuations. Iosifescu and Tautu [5] also identify three categories of randomness. They refer to "random perturbations on a deterministic trajectory" which may be due to environmental fluctuations, "initial random conditions" where the process is deterministic but is repeated under randomly varying conditions, and "pure randomness" which might correspond to either of Kurtz's sources of internal randomness.

Chesson [9] has categorized the types or sources of random variability (due to chance mechanisms) that are important in population processes and that are biologically interpretable. Variation may occur between individuals. This is largely due to phenotypic variation ascribable to genetic differences between individuals. Stochastic modeling is widely used in population genetics to account for this kind of random variability [19,20]. The topic will not be discussed further in this chapter. Variation may also arise within individuals. This is illustrated by making the analogy between an individual and a coin. A coin may by chance come up either heads or tails when flipped. An individual spore may germinate or not by chance. Random variation may also be induced in a population due to the effects of random environmental fluctuations. A fourth kind of variation is due to spatial heterogeneity. Chesson [9] called this between-patch variability. He also recognized that variability could occur within patches due to the interaction of the other sources of variability. He discussed the role that stochastic models can play in describing these sources of variability.

B. Definition of Deterministic Model

In plant pathology the usual means of quantifying variability in pathogen or disease population processes is to develop a deterministic model. A deterministic model is one in which the values of the independent variables and parameters determine the exact value of the dependent variable. The logistic model given in Eq. 1 makes a convenient and familiar example.

$$dY/dt = r * Y * (1 - Y) \tag{1}$$

Here there are two variable quantities: Y representing disease proportion and t representing time. There are also two parameters (sometimes called constants) in the model. One is the initial condition Y_0 at time t = 0 and the other is r, the apparent infection rate. For a given value of r, Y_0, and t there can be only one value of Y. Thus, the model is deterministic.

In an experimental study the data set will exhibit variability in Y for a given time, t. Nevertheless, if Eq. 1 is used to describe that data, the resulting fitted equation will be deterministic. This is because the model-fitting procedure finds the single "best" values for the parameters. These are the values that produce the average value for Y given t. Deterministic models are justified on the basis that they describe the average value of the dependent variable given the independent variable inputs. Standard statistical techniques for fitting models, such as regression, are justified primarily because statistical theory proves these techniques will make the fitted model represent the average of the data (unless the assumptions of the statistical procedure are not met). It will be shown later in this chapter why average estimates may be inappropriate or wrong and why deterministic models may therefore be inadequate.

Models are not only used in conjunction with experimental research for the purpose of describing or explaining the results. Models can also be useful for theoretical exploration of the potential range of behavior of a system. Of interest then is the question, given a set of biological assumptions, how does the resulting model predict how population dynamics will behave? The answer to that question can then be compared with knowledge of the behavior of real populations. This approach is employed extensively in ecology [21,22]. Jeger [23] discussed this approach in the context of plant disease epidemiology and suggests it be taken more frequently. He points out that a careful comparison of the mathematical behavior of a model with what is known about the general behavior of epidemic processes can provide insight relevant to both the modeling effort and to understanding biological processes.

In this theoretical approach the model is assumed implicitly to account for all the variability that is significant with respect to population dynamics. In other words, the types of random variability described by Chesson [9] are either excluded from consideration explicitly or they are assumed implicitly to have no systematic effect on the system. In some cases, there are mathematical arguments showing that a theoretical deterministic model represents the mean of an equivalent stochastic model. However, the mean alone may be insufficient to describe the behavior of the population process. In other cases the deterministic model will not represent the mean of a similar stochastic model. The use of a stochastic model in place of a deterministic model is justified for either of these reasons.

Another problem with deterministic models is that their behavior may be very complex mathematically. For example, systems consisting of only two or three differential equations can be chaotic. A mathematically chaotic situation is said to occur when the model produces widely different and unpredictable results due to very small changes in the values of the parameters. May [24] reviews this topic and shows by example how the difference equation form of density-dependent growth exhibits chaotic behavior. He suggests that "What seems called for is some effectively stochastic description of the deterministic dynamics" of these models.

There are several ways a deterministic model may be viewed as inadequate as a description of pathogen population or epidemic growth. From a theoretical view, the model may indicate behavior that is nonsensical when compared with known biological reality. From the point of view of the empiricist, a model may be considered inadequate if either systematic departures from the data occur or the random error (unexplained variability) remaining after fitting the model is unsatisfactorily large. Within the field of plant disease epidemiology, the usual response to these problems has been to look for alternative deterministic models. This has produced two results. First, a large number of single equation density-dependent deterministic models, such as the Gompertz, Bertalanffy, Weibull, and the logistic modified in various ways, have been rediscovered (from the population biology literature) as alternative means of describing disease progression and analyzing data [25–27]. Their usefulness has been as a means of eliminating systematic errors between data and model predictions. The second approach which has been taken is to develop deterministic simulation models. This approach addresses not only the problem of systematic discrepancies between model and data but also attempts to reduce the unexplained variation by a more biologically complete description of the disease system. This is accomplished by structuring the model hierarchically in keeping with systems analysis [28]. As an example, Rouse [29] has suggested that variability in yield as a function of disease could be better explained by adding pathogen biology and disease components to plant growth simulations.

Despite the effort expended to find alternative deterministic models and develop computer simulations, there is a lot of biological variability in epidemic systems which has not been accounted for. Aylor [30] divided epidemics into component processes and then gave an estimate of how close our models came to accurately predicting the processes. In no case did he predict that we were within a factor of 2 of being able to model any individual processes. Although his "guestimates" seem high, his general conclusion cannot be disputed that our understanding of epidemic processes is far from complete.

Much of this lack of fit between models and reality simply indicates that more fundamental experimental research is needed. But at least part of this gap may be due to the conceptual basis upon which we conduct our experimentation and especially upon which we develop models. The current paradigm is that plant disease processes and their components are deterministic. The alternative paradigm is that plant disease processes are best viewed as an ensemble of probabilistic events, that is, as stochastic processes. I emphasize that this second paradigm results not only in a different perspective on model development, but just as significantly, the stochastic paradigm dramatically changes our experimental approach.

II. STOCHASTIC MODELING

In loose terms, a stochastic model is a mathematical formulation of a process that incorporates probabilistic statements. This statement seems to imply that the terms probabilistic and stochastic are synonymous. This is not quite the case since probabilistic statements can be made with reference to a single random variable, while a stochastic process is more formally defined as a set of random variables that have a common index. (There are many texts on stochastic processes, see, for example, Feller [31] or Papoulis [32].) This chapter includes discussion of models that might be viewed as probabilistic but not as stochastic given this formal definition.

A random variable is a function that assigns a real number to each element of a set of related events. For example, the proportion of flag leaves of wheat diseased on June 1 could be considered a random variable. Let Y represent that random variable. In that case, the set of different possible amounts of disease that could occur at that time is the set of events and the function, Y assigns a numerical value to each of the elements in the set, in this case the different possible amounts of disease. Notice that the random variable is formally a function which can take on all possible values of disease. Each numerical value of the random variable can be assigned a probability of occurring. The functional relationship between the realizations of the random variable and the probability assigned to each event is called the probability distribution function. It can be written as P(Y) or as P(Y = y), where y is a specific value of the random variable. With this function P(Y), we can make probabilistic statements about disease on the flag leaf on June 1, but this need not be viewed as a stochastic process. However, if we consider as random variables the disease over successive time periods, say, t = 0, t = 1, t = 2, . . . , t = n, where t represents some appropriate time scale, say weeks, then the set of random variables $Y_0, Y_1, \ldots, Y_n$ constitute a stochastic process. The index in this case is time, but it is possible to have a process that is indexed in some other way, for example, spatially. In the current example the index could be leaf position so that t = 1, t = 2, . . . , t = n represent the leaves by position in the canopy.

Many probability distribution functions can be derived as stochastic processes. For example, suppose n spores fall on a leaf. The probability of X = x lesions subsequently forming may be given by the binomial distribution

$$P(X = x/n,p) = \binom{n}{p} p^x (1 - p)^{n - x} \tag{2}$$

where the random variable X represents the number of lesions and can take on values between 0 and n with x representing a specific outcome; p = probability a spore landing on the leaf will be viable, germinate and infect. This same distribution can be derived by defining as random variables the possible events for each spore. Then each random variable has a Bernoulli distribution. These random variables are indexed by the spores and so represent (trivially) a stochastic process. The number of lesions on the leaf will still be described by the binomial distribution. The Poisson distribution can be obtained in this example if the assumption is made that spores act independently, and n is large and p is small such that n*p = L, a constant.

The Poisson distribution is also the result of a well-known stochastic process of population growth through time. This is usually referred to as the Poisson birth-death process. Rouse [27] mentions this process and illustrates its derivation in a plant epidemiological context.

There is a considerable amount of mathematical theory behind stochastic processes. Of particular concern to the application of stochastic processes is how to find the mathematical form of the probability distribution (and the properties of that distribution) starting from the indexed set of random variables. The mathematical difficulties associated with stochastic processes have often been an argument for the use of simpler deterministic models.

A. Historical Background

It is interesting to note that stochastic modeling was earliest motivated by problems posed by biologists interested in population related phenomena. To this day the large class of stochastic models referred to as branching processes are motivated to a large extent by biological applications [33]. The earliest recourse to probabilistic treatment of population processes resulted from Galton's challenge to deduce mathematical functions for the disappearance of family names. This challenge was published in 1873 as problem 4001 in the *Educational Times* [33]. The problem was not solved in final form until 1940 although almost complete solutions were obtained a decade earlier [33].

The first physical phenomenon formally described with a stochastic process was Brownian motion [34]. Curiously enough this phenomenon was named for the botanist that characterized this type of motion and proved it was of physical and not biological origin in 1828. Neither a physical explanation for, nor a mathematical treatment of, Brownian motion was forthcoming for almost a century after Brown's work. The development of a stochastic theory relevant to physics arose from the work by Einstein, among others, on this phenomenon in the early part of this century [34].

In the 1920s, medical epidemiologists began to develop stochastic models as a means of understanding and predicting epidemic behavior [10]. Significant advances in stochastic epidemic modeling took another 20 years too occur and were dependent on the further development of the mathematical theory of stochastic processes by Feller [31], Doob [35], and others. Often this theoretical development was motivated directly by consideration of biological growth processes. Simultaneously, questions raised from the study of evolution and the genetics of populations provided similar motivation for the development of probabilistic/stochastic theory.

In plant pathology, mathematical modeling is a more recent endeavor and most efforts have been directed toward deterministic models that require less mathematical sophistication. On several occasions stochastic models have been mentioned in general reviews of epidemic modeling [25–27]. Two areas of plant disease epidemiology have historically made use of probabilistic models. Both the topics of dispersal and of dose-response relations have been viewed as consisting of stochastic elements. However, only recently have the advantages of these stochastic models been explored. For example, the multiple infection transformation, presented to the plant pathology community for the first time by Gregory [36] in 1948, is based on a probabilistic argument. Nevertheless, it is used as a deterministic transformation. It is difficult to know if Van der Plank's [37] statement in 1963 that stochastic models are unnecessary for describing epidemic dynamics slowed the progress of stochastic model development. A few examples of dynamic stochastic models in plant disease epidemiology will be reviewed below.

B. Demographic Variability

Models are usually constructed with a specific goal in mind as to the source of variability they are going to explain. To illustrate some of the conceptual approaches to model construction we refer back to Eq. 1. We seek to replace the unique quantity Y in the deterministic version of this model with a function P(Y) representing the probability distribution of the random variable Y resulting from chance phenomena. The way we do this depends entirely on how we view the source of random variability.

If the variability is due directly to the population units, for example spores, and their probability of either causing disease or not, then the so-called branching process approach can be taken, in which disease develops as a sequence of births and/or deaths. In this case Y may be considered to change in discrete steps because the underlying process involves the discrete transition from state Y to state Y + 1 due to the probability of a single event occurring. This is because, regardless of the number of lesions, we can assume there is a small time interval during which only one new lesion will form. This is sometimes referred to as demographic variability and is identical to what Cesson [9] calls within-individual variation. The stochastic analog of logistic growth that arises from treating the model as a birth-death process is discussed by several authors [1,2,8,21,22,38]. Unfortunately there is no analytical solution to the process that gives the probability distribution function in closed form.

In contrast, density independent growth as modeled by the exponential equation has a stochastic analog arising from a branching process with a closed form solution. In this case, the probability of finding n individuals at time t is given by

$$f(n,t) = \frac{(n - 1)!}{(N_0 - 1)!\,(n - N_0)!}\, e^{-N_0 rt}\, [1 - e^{-rt}]^{n - N_0} \tag{3}$$

where N_0 represents the starting population size at time t = 0 and r is the probability of a birth (one new lesion) in the infinitesimal time interval dt. The mean and variance of this distribution function can be derived easily from the distribution function as

$$M = N_0 e^{rt}, \quad \sigma^2 = N_0 e^{2rt}\,(1 - e^{-rt}) \tag{4}$$

In this case the mean of the stochastic model is identical to the deterministic model for exponential growth. Since with large populations the variance-to-mean ratio will be small, it has been argued that introducing stochasticity within individuals is an unnecessary complication. It is because of this that many researchers have argued that deterministic models are adequate descriptions of population processes. In the case where we are considering an epidemic consisting of tens of millions of spores giving rise to millions of lesions exponentially, this would seem correct.

However, it has been pointed out in the literature that when populations are initially small, deviations from the mean (a measure of which is given by the variance) can have a significant effect even in the exponential model [9]. This is manifested by reduced population growth, compared with the deterministic model. It would seem possible that skewed disease progress curves fitted by models such as the Gompertz might be explained on a theoretical basis by random phenomena occurring early in the epidemic. It is interesting that attention has been given to the significant effects even small errors in estimation of disease early in an epidemic can have on the estimation of rate of disease progress, and yet no consideration has been given to examining the behavior of stochastic models that could describe that situation. Could this be an explanation for the observed correlation between estimates of initial disease and apparent infection rates?

The effect of random fluctuations when population size is initially low is clearly illustrated by competition studies with flour beetles. Park [39] found that competition between two *Tribolium* species was a function of physical environmental variables such as temperature and humidity. However, it was observed from a large number of experiments under identical environmental conditions that sometimes one species would win and sometimes the other. A satisfactory explanation was given by Neyman et al. [40]. They viewed the chance that the first species would outcompete the second as a random variable with distribution $p(n_1,n_2)$, where n_1 and n_2 are initial population numbers of each species. As Ludwig [7] has said, "The regularities which the experimenters sought could be found in the behavior of $p(n_1,n_2)$ as a function of n_1,n_2 and the environment, but not always in the results of single experiments."

C. Random Environmental Variability

Another kind of variability is due to environmental fluctuations. To address the question of variable environment on population growth, the parameters of the deterministic model given in Eq. 1 can be made functions containing random variables described by some appropriate distribution. In most cases the random component of the environment is assumed to act as "white noise." As an example, Kiester and Barakat [41] solved the following form of the logistic model:

$$\frac{dN}{dt} = r(t)*N*\left(1 - \frac{N}{K}\right) \tag{5}$$

where r(t) was taken to be a Gaussian random process (white noise). In this formulation of the logistic model, N is taken to represent absolute population size, not porportion (as in proportion disease), and K is defined as the carrying capacity. In this case the explicit solution of the model is given by

$$f(N) = \frac{[N(1 - Nk^{-1})]^{-1}}{\tau_y (2\pi)^{\frac{1}{2}}} \exp\left\{-\frac{1}{2\sigma_y^2}\left[\log\left(\frac{N}{1 - NK^{-1}}\right) - \mu_y\right]^2\right\} \tag{6}$$

where f(N) is the probability density function for $0 < N < K$. Kiester and Barakat also derive the equation for the moments of the distribution and point out that the mean of the stochastic logistic process is not equivalent to the deterministic model. Similar variations on the logistic model have been investigated by others with similar results [42–44]. As with other nonlinear population processes (e.g., logistic growth), the means of stochastic epidemic models do not in general reduce to standard deterministic models [2,10].

Kurtz [18] summarized the conditions under which appropriately derived deterministic models will provide adequate limiting cases of stochastic models. First, if the population is large, the internal random fluctuations are negligible. Second, if the random environmental fluctuations are much faster than the population dynamics, they average out. When these conditions are not met, deterministic models will be inadequate and populations should be modeled as stochastic processes.

III. STOCHASTIC MODELS OF PLANT DISEASE EPIDEMIC PROCESSES

Two statements have been made concerning mean values. First, the distributions arising from stochastic models may not have, as a mean, functions that correspond to deterministic models with analogous biological assumptions. The reason is because the underlying deterministic relationship between dependent and independent variables is nonlinear. Second, with reference to empirical research, it was stated earlier that mean values may be misleading. The reason is that the estimated mean of a distribution may be a poor predictor of the frequency of occurrence of biologically significant events. Several good examples exist in the plant pathology literature that illustrate these statements.

A. Dose-Response Relationships

Bacterial

Suppose we are interested in the relationship between pathogen population size and disease for bacterial brown spot on beans caused by *Pseudomonas syringae* pv. *syringae*. From controlled inoculation studies with several plant pathogenic bacteria on different plant species, it is known that the probability of disease is a nonlinear function of bacterial dose [45]. This relationship can be approximated by the probit function, conceptually representing the probability of disease given bacterial dose [46]. This can be written symbolically as P(d/n), where d represents disease and n represents dose. To apply this model to the field it is necessary to estimate n. A standard way of doing this has been to estimate the average number of bacteria per leaflet by washing and dilution plating of bulked samples of leaflets. However, it has been observed that in the field bacterial population size of *P. syringae* on individual leaflets is lognormally distributed [47]. As a result, there is a tendency for large bacterial population sizes to be present on a relatively small number of leaflets while the majority of leaflets will have relatively small bacterial populations. The significance of this is that disease occurs on individual leaflets, and therefore it is the frequency of those leaflets with relatively high population size that is important, not the mean.

Thus, in the field, a better relationship between disease and bacterial population size can be obtained by accounting for the relative frequencies of bacteria per leaflet. This can be done by multiplying P(d/n) by P(n) for each n and then summing the products obtained for all n. This process is described mathematically bh the following expression:

$$P(d) = \sum_{n} P(d/n) * P(n) \tag{7}$$

In the bacterial disease example, a good relationship was found between disease and bacterial population size when this model was used. With P(d/n) given by the probit function and P(n) given by the lognormal distribution function, the probability distribution of disease, P(d), was expressed by the cumulative normal distribution [48].

Use of this model requires information about the frequency of bacterial populations on leaves that cannot be obtained from a bulked sample of leaflets. Conceptually, the stochastic model provides a framework that requires a different experimental approach—one in which data are not collected for the purpose of obtaining a good estimate of the mean alone while attempting to reduce variability between treatments. Rather, the experimental approach consistent with a random model is to collect data that gives a proper representation to the actual range of random variability present in the biological system. This variability is then incorporated into the model explicitly.

Why couldn't the arithmetic mean be used as an estimate of the frequency of leaflets with high bacterial population sizes? There are several reasons. First, in this example it was found that the variance as well as the mean was variable from one field plot to another. Thus, the mean alone was not sufficient to parameterize the distribution. A further complication in this example is that the lognormal distribution is skewed, resulting in a difference between the mean and median of the distribution.

Even if the variance were constant, the relationship between dose and response given by the probit function would cause estimates obtained by the arithmetic mean to be biased. This is because the expected value of the response function is not equivalent to the response of the expected value of the dose. This fact is in turn due entirely to the fact that the probit dose response function is nonlinear.

Nonlinearity

In general, whenever the biological relationship between independent and dependent variables is nonlinear, an estimate of the dependent variable obtained from the mean value of the independent variable will not be equivalent to the estimated mean of the dependent variable obtained after evaluating the function for individual values of the independent variable. Mathematically, the mean is obtained from the first moment of the frequency distribution and is called the expectation of the distribution. The notation for the expected value (i.e., mean) is E[X], where X is the random variable of interest. Thus, the statement above about nonlinear functions can be represented by the following expression:

$$E[f(X)] \neq f(E[X]) \tag{8}$$

The conceptual implications of nonlinearity have been thoroughly discussed [49]. In the context of insect predation a numerical example is given by Rabbinge et al. [50]. They take the simple rate relationship dP/dt = c*V, where dP/dt is predation rate, V is velocity of insect movement, and c is a rate constant. They compare this linear equation to the nonlinear model dP/dt = c*SQRT(V), where SQRT means square root of V for both deterministic and stochastic varsions of the models. Using CSMP, a computer simulation language, they simulate these two models. They make V a uniform random variable and generate values for V using a random number generator during each time step of the model. Since E[SQRT(V)] ≠ SQRT(E[V]), the stochastic version of the nonlinear model gives different results than the deterministic version, while the linear model gives the same result for both versions.

Several other examples considering both dose and response as random variables, and therefore where Eq. 7 applies, can be found in the epidemiological literature. In relating dose to response it may be possible to generalize the above bacterial example to most plant pathogen systems in the field. In the field, the dose is almost never uniformly distributed with respect to sampling units. Almost all recent studies of spatial pattern indicate that pathogen populations representing initial inoculum and/or secondary inoculum are dispersed in an aggregated manner. Therefore the probability of disease is conditional on the value of dose in each of the sampling units and is not conditional on the mean dose across sampling units.

Soilborne Disease

Gilligan [51] has developed a model that utilizes the concept represented by Eq. 7 for soilborne diseases. He observes that:

$$P\left\{\begin{array}{l}\text{host}\\ \text{is}\\ \text{infected}\end{array}\right\} = P\left\{\begin{array}{l}\text{propagule}\\ \text{occurs in}\\ \text{pathozone}\end{array}\right\} * P\left\{\begin{array}{l}\text{propagule is}\\ \text{capable of}\\ \text{infecting host}\end{array}\middle|\begin{array}{l}\text{propagule}\\ \text{occurs in}\\ \text{pathozone}\end{array}\right\}$$

He derives the probability density function for the probability the host remains uninfected from a geometric argument that accounts for the distance of the propagule from the center of the pathozone. Recently, Brassett and Gilligan [52] extended this result to obtain a discrete probability model for polycyclic infection by soilborne pathogens.

Fungicide Residue

The effect of fungicide residues on disease development represents another example where dose-response relationships have been considered stochastically. In this case it is the amount of residue on an individual leaf that determines whether that leaf will be protected from infection — not necessarily the average amount of fungicide residue in the field. Thus the distribution of fungicide concentration on individual leaves is important. Bruhn and Fry [53] observed that chlorothalonil " . . . deposited on leaves was highly variable throughout the potato canopy. The relative frequency distribution of residue levels was skewed; there were greater frequencies of low than high concentrations." They found that the gamma distribution adequately described the frequency data for each of four separate strata in the plant canopy. They also observed that the values of the distribution parameters were a function of environmental conditions, plant-related factors, and method of application. They conclude that "fungicide models that ignore the variability in fungicide residue levels will overestimate the impact of fungicides on pathogen development." In a subsequent paper they incorporate this variability explicitly into a stochastic model of the dynamics of fungicide residues over time [54].

Milgroom et al. [55] have conducted similar experiments for metalaxyl residues in potato foliage. They observed that residue variability could be described by a censored lognormal distribution. The impact of environmental factors on residue levels was investigated. These results were used to develop a model for the effects of metalaxyl on epidemics of potato late blight [56] and also to investigate fungicide resistance management strategies [57].

A stochastic modeling approach can provide a means of explaining observed distributions of disease. For example, numbers of powdery mildew lesions on individual leaves of wheat were observed to be fitted by the negative binomial distribution [58]. Rouse [59] presented a plausible stochastic model for powdery mildew development by assuming that number of spores, N, landing on leaves of constant area, a, were Poisson distributed with mean aL. The number of lesions, X, conditional on N, was assumed to be binomially distributed, with p = the probability that a single spore infects. Then the $P(X) = \sum_N P(X/N) * P(N)$ results in a Poisson distribution with mean aLp. By allowing the area of leaves, a, to vary randomly, the final distribution for number of lesions is given by the following form of the negative binomial distribution:

$$P(X = x) = \binom{x + k - 1}{x} \left(\frac{Lp}{Lp + 1}\right)^x \left(\frac{1}{Lp + 1}\right)^k \quad (9)$$

This explanation for observed lesion distribution allows the parameter Lp to be interpreted as the combined effect of sporulation and infection efficiency on disease development.

B. Dispersal

Dispersal is a random process, and several models relevant to dispersal of fungal pathogens employ random variables. Consider the process of take-off, whereby a fungal spore becomes airborne. Early work on *Helminthosporium maydis* showed that in the field a wind speed above the canopy of approximately 1 m/sec was needed to entrain spores [60]. When the actual force required to detach spores was estimated experimentally, it was found that wind speeds of 5 m/sec were required [61]. The discrepancy was explained by the fact that an average wind speed of 1 to 3 m/sec above a canopy gave rise to a frequency distribution of wind gusts that included gusts above 5 m/sec [62]. Thus, it was not the average wind speed but the frequency of fast gusts of wind that directly determined spore liberation. This information was used to modify a spore dispersal model, although the model itself is not stochastic [63].

Strand and Roth [64] developed a model for spread and intensification of dwarf mistletoe based on combining conditional probabilities. This model illustrates the thought process used in combining probabilities. Like many of the other models which utilize a probabilistic approach to model construction, it is not a stochastic model as defined in this chapter. This is because the results of the model are deterministic. The model does not specify a range of outcomes with assigned probabilities for a given set of independent variables.

The process of transport has commonly been modeled as a Gaussian plume [65]. In this model the location of a spore relative to mean flight path downwind from a source is given as a function of time since release by a normal distribution. This is justified by the similarity of spore dispersal processes to diffusion. There are several reviews of diffusion processes in ecology which are relevant to dispersal [e.g., 66,67]. Two other papers in the plant pathology literature have used probabilistic models to describe the dispersal process. Waggoner [68], in a general analysis of chance of infection as a function of weather, space, and time, viewed the probability that a unit area in a field would become infected as the sum of the products (Eq. 7) of the probability of infection given that j spores were deposited in that unit area, and the probability that j spores were deposited. Minogue and Fry [69] used a double geometric distribution to describe dispersal in a disease gradient model. It must be assumed that they used Monte Carlo simulation to obtain their results although it is not explicitly stated in the paper.

C. Stochastic Simulation of Epidemic Processes

There are few examples of stochastic simulation models of plant epidemic processes. Several epidemic simulation models use distribution functions and randomly sample from these distributions with the aid of random number generators [70–73]. Shrum [72] uses the gamma distribution and the normal distribution as submodels for dispersal in EPIDEMIC. His model is not stochastic, however, because he makes no attempt to determine variability in simulated outcomes. Rather, the distribution functions are used deterministically. Teng [73] used the uniform distribution to make number of spores liberated a random variable. He also used the normal distribution to describe the random process of fungal penetration of the host. Similarly to Shrum, he does not generate variable outcomes from his simulation with a single set of input data.

Eisensmith et al. [70] created a stochastic model for spore germination using what they called a parallel boxcar train approach. In their model the time for each spore to complete the germination process is a random variable. They looked at the normal, lognormal, and gamma distributions as plausible distributions of waiting time. They also included the effect of temperature on germination in the model. Due to the nonlinear relationship between temperature and spore germination, modeling germination of a population of spores by incorporating the stochastic behavior of individuals should give biologically more correct results than a deterministic model.

In one of the only truly stochastic treatments of epidemic processes using computer simulation, Sall [71] introduced random environmental effects into her grape powdery mildew model. Her model used the logistic equation modified to account for latent period sensu Vander Plank [37] to describe powdery mildew dynamics. This basic model is modified by including a grape vine growth submodel and by making the apparent infection rate, r, a function of temperature [74]. In the stochastic version of the model, random variation is introduced into the temperature relation. Predictions from the simulation model were expressed as probabilities based on the distribution of simulation results generated by running the simulation repeatedly. For example, based on 1976 weather data, there was a 25% chance of disease severity above 20.7%. The mean level of infection was consistently 1 − 2% higher in the stochastic model compared to the deterministic version. This was due to the nonlinearity of the relationships in the model. This was statistically significant although probably not practically important.

In the powdery mildew model, Monte Carlo simulation was used to generate the distribution of outcomes. The Monte Carlo technique is conceptually straightforward as it consists of simply running the model over and over. Each time the model is executed a new random sample is taken from the distribution programmed into the model. A random number generated in this way is called a random deviate. The trick to Monte Carlo simulation is to have a good algorithm for obtaining the random deviates from given distributions. Subroutines for obtaining random deviates can be found in several numerical analysis textbooks [e.g., 75].

A problem with Monte Carlo techniques is that they use a lot of computer time. As a result, alternative methods may be used. One such approach is called compound simulation [76]. With this method each nonlinear function in the model is divided into segments that can be approximated with linear functions. Since stochastic and deterministic results coincide for each linear segment, the simulation can calculate a value just once for each segment each time step.

One of the significant points Sall [71] makes is that stochastic models make informative predictive tools for disease management. It is exactly in the management setting where knowing the degree of variability associated with a prediction is most important. Methods for optimizing control measures when disease or pest dynamics are described by a stochastic model can be found in the literature [e.g., 12].

IV. CONCLUSIONS

Stochastic models have several advantages when compared with deterministic models. They account for random variability explicitly, by using frequency distribution functions to incorporate that variability into the model. This gives stochastic models greater predictive ability compared with similar deterministic models. Second, stochastic models provide a means of reducing bias when nonlinear functions are part of the system.

Finally, stochastic modeling suggests a different approach to data collection. The stochastic paradigm says there is a significant amount of random variability in biological systems. This variability can be accounted for if it can be quantified. Although data collection will necessarily be more intensive, the benefit of a better understanding of the system may be worth it. The emphasis on frequency data rather than means in conjunction with the notion that biological systems may have nonlinear components is a major departure from the concept of a linear analysis of variance approach to experimental design.

Within plant pathology very little use of stochastic modeling has been made. Nevertheless "It is not too daring to anticipate also in plant epidemiology a great potential of stochastic models . . . " [25].

REFERENCES

1. Bailey, N. T. *The Elements of Stochastic Processes with Application to the Natural Sciences.* John Wiley, New York, 1964.
2. Bartlett, M. S. *Stochastic Population Models in Ecology and Epidemiology.* Methuen, London, 1960, p. 90.
3. Bartlett, M. S., and Hiorns, R. W. (eds.). *The Mathematical Theory of the Dynamics of Biological Populations.* Academic Press, New York, 1973.
4. Goel, N. S., and Richter-Dyn, N. *Stochastic Models in Biology.* Academic Press, New York, 1974, p. 269.
5. Iosifescu, M., and Tautu, P. *Stochastic Processes and Applications in Biology and Medicine II. Models.* Springer-Verlag, Berlin, 1973, p. 337.
6. Ludwig, D. *Stochastic Population Theories, Lecture Notes in Biomathematics*, Vol. 3. Springer-Verlag, Heidelberg, 1974.
7. Ludwig, D. Comparison of some deterministic and stochastic population theories. In *Studies in Mathematics, Vol. 16, Studies in Mathematical Biology, Part II: Population and Communities* (S. A. Levin, ed.), Mathematical Association of America, 1978, p. 624.
8. Ricciardi, L. M. Stochastic population theory: Birth and death processes. In *Mathematical Ecology: An Introduction; Biomathematics*, Vol. 17 (T. G. Hallam and S. A. Levin, eds.), Springer-Verlag, New York, 1986, p. 457.
9. Chesson, P. Predator-prey theory and variability. *Annu. Rev. Ecol. Syst.*, *9*: 323–347, 1978.
10. Bailey, N. T. J. *The Mathematical Theory of Epidemics.* Griffin, London, 1957.
11. Bartlett, M. S. Deterministic and stochastic models for recurrent epidemics. In *Proc. Third Berkeley Symposium on Mathematical Statistics and Probability*, 1956.
12. Wickwire, K. H. Optimal isolation policies for deterministic and stochastic epidemics. *Math. Biosci.*, *26*: 325–346, 1975.
13. Patton, B. C. (ed.). *Systems Analysis and Simulation in Ecology*, Vol. 1–4. Academic Press, New York, 1971.
14. de Wit, C. T., and Goudriaan, J. *Simulation of Ecological Processes.* PUDOC, Wageningen, 1978, p. 175.
15. Dent, J. B., and Blackie, M. J. *Systems Simulation in Agriculture.* Applied Science Publishers, London, 1979, p. 180.
16. Keulen, H. van, and Wolf, J. (eds.). *Modelling of Agricultural Production: Weather, Soils, and Crops.* PUDOC, Wageningen, 1986, p. 479.
17. Teng, P. S. A comparison of simulation approaches to epidemic modeling. *Annu. Rev. Phytopathol.*, *23*: 351–379, 1985.
18. Kurtz, T. G. Relationships between stochastic and deterministic population models. In *Biological Growth and Spread* (W. Jager, H. Rost, and P. Tautu, eds.), Springer-Verlag, Berlin, 1980.

19. Crow, J. F., and Kimura, M. *An Introduction to Population Genetics Theory.* Harper and Row, New York, 1970.
20. Maruyama, T. *Lecture Notes in Biomathematics, Vol. 17: Stochastic Problems in Population Genetics.* Springer-Verlag, New York, 1977, p. 246.
21. May, R. M. *Stability and Complexity in Model Ecosystems*, 2nd ed. Princeton University Press, Princeton, NJ, 1974, p. 265.
22. Smith, M. J. *Models in Ecology.* Cambridge University Press, Cambridge, 1974, p. 146.
23. Jeger, M. J. The potential of analytic compared with simulation approaches to modeling in plant disease epidemiology. In *Plant Disease Epidemiology: Population Dynamics and Management* (K. J. Leonard and W. E. Fry, eds.), MacMillan, New York, 1986, pp. 255–281.
24. May, R. M. Simple mathematical models with very complicated dynamics. *Nature, 261*:459–467, 1976.
25. Kranz, J. (ed.). *Epidemics of Plant Disease.* Springer-Verlag, New York, 1974, p. 170.
26. Madden, L. V. Quantification of disease progression. *Protection Ecology, 2*:159–176, 1980.
27. Rouse, D. I. Construction of temporal models: I. Disease progress of air-borne pathogens. In *Advances in Plant Pathology, Vol. 3: Mathematical Modelling of Crop Disease* (C. A. Gilligan, ed.), Academic Press, London, 1985, pp. 67–105.
28. Goodall, D. W. The hierarchial approach to model building. In *Critical Evaluation of Systems Analysis in Ecosystem Research and Management* (G. W. Arnold and C. T. de Wit, eds.), PUDOC, Wageningen, 1976.
29. Rouse, D. I. The use of plant growth models to predict yield loss due to disease. *Annu. Rev. Phytopathol., 26*:183–201, 1988.
30. Aylor, D. E. A framework for examining inter-regional aerial transport of fungal spores. *Agric. For. Meteorol., 38*:263–288, 1986.
31. Feller, W. *An Introduction to Probability Theory and Its Applications*, Vol. 2. Wiley Interscience, New York, 1966.
32. Papoulis, A. *Probability, Random Variables and Stochastic Processes.* McGraw Hill, New York, 1965.
33. Kendall, D. G. Branching processes since 1873. In *Studies in the History of Statistics and Probability*, Vol. II (M. Kendall and R. L. Plackett, eds.), Charles Griffin and Co., London, 1977, pp. 383–404.
34. Brush, S. G. A history of random processes, I. Brownian movement from Brown to Perrin. In *Studies in the History of Statistics and Probability*, Vol. II (M. Kendall and R. L. Plackett, eds.), Charles Griffin & Co., London, 1977, pp. 347–382.
35. Doob, J. L. *Stochastic Processes.* Wiley-Interscience, New York, 1953.
36. Gregory, P. H. The multiple-infection transformation. *Ann. Appl. Biol., 35*:412–417, 1948.
37. Van der Plank, J. E. *Plant Diseases: Epidemics and Control.* Academic Press, London, 1963, p. 349.
38. Pielou, E. C. *An Introduction to Mathematical Ecology*, 2nd ed. Wiley-Interscience, New York, 1977.
39. Park, T. Experimental studies of interspecies competition. II. Temperature, humidity, and competition in two species of Tribolium. *Physiol. Zool., 27*:117–238, 1954.

40. Neyman, J., Park, T., and Scott, E. Struggle for existence. The Tribolium model: Biological and statistical aspects. In *Proc. 3rd Berkeley Symp. Math. Stat. Prob.*, Vol. 6, 1956, pp. 41–79.
41. Kiester, A. R., and Baraket, R. Exact solutions to certain stochastic differential equation models of population growth. *Theor. Popul. Biol.*, *6*:199–216, 1974.
42. Levins, R. The effect of random variations of different types on population growth. *Proc. Nat. Acad. Sci.*, *62*:1062–1065, 1969.
43. Lewontin, R. C., and Cohen, D. On population growth in a randomly varying environment. *Proc. Nat. Acad. Sci.*, *62*:1056–1060, 1969.
44. May, R. M. Stability in randomly fluctuating versus deterministic environments. *Am. Naturalist*, *107*:621–650, 1973.
45. Ercolani, G. L. Two hypotheses on the etiology of response of plants to phytopathogenic bacteria. *J. Gen. Microbiol.*, *75*:83–95, 1973.
46. Finney, D. J. *Probit Analysis*. Cambridge University Press, Cambridge, 1947, p. 318.
47. Hirano, S. S., Nordheim, E. V., Arny, D. C., and Upper, C. D. Lognormal distribution of epiphytic bacterial populations on leaf surfaces. *Appl. Environ. Microbiol.*, *44*:695–700, 1982.
48. Rouse, D. I., Nordheim, E. V., Hirano, S. S., and Upper, C. D. A model relating the probability of foliar disease incidence to the population frequencies of bacterial plant pathogens. *Phytopathology*, *75*:505–509, 1985.
49. West, B. J. *An Essay on the Importance of Being Nonlinear, Lecture Notes in Biomathematics No. 62*, Springer-Verlag, Berlin, 1985, p. 204.
50. Rabbinge, R., Ward, S. A., and VanLaar, H. H. (eds.). *Simulation and Systems Management in Crop Protection*. PUDOC, Wageningen, 1989, p. 119.
51. Gilligan, C. A. Probability models for host infection by soilborne fungi. *Phytopathology*, *75*:61–67, 1985.
52. Brassett, P. R., and Gilligan, C. A. A discrete probability model for polycyclic infection by soil-borne plant parasites. *New Phytol.*, *109*:183–191, 1988.
53. Bruhn, J. A., and Fry, W. E. A statistical model of fungicide deposition on potato foliage. *Phytopathology*, *72*:1301–1305, 1982.
54. Bruhn, J. A., and Fry, W. E. A mathematical model of the spatial and temporal dynamics of chlorothalonil residues on potato foliage. *Phytopathology*, *72*:1306–1312, 1982.
55. Milgroom, M. G., McCulloch, C. E., and Fry, W. E. Distribution and temporal dynamics of metalaxyl in potato foliage. *Phytopathology*, *78*:555–449, 1988.
56. Milgroom, M. G., and Fry, W. E. A model for the effects of metalaxyl on late blight epidemics. *Phytopathology*, *78*:559–565, 1988.
57. Milgroom, M. G., and Fry, W. E. A simulation analysis of the epidemiological principles for fungicide resistance management in pathogen populations. *Phytopathology*, *78*:565–570, 1988.

58. Rouse, D. I., MacKenzie, D. R., Nelson, R. R., and Elliott, V. J. Distribution of wheat powdery mildew incidence in field plots and relationship to disease severity. *Phytopathology, 71*:1015–1020, 1981.
59. Rouse, D. I., Nelson, R. R., and MacKenzie, D. R. A stochastic model of horizontal resistance based on frequency distributions. *Phytopathology, 70*:951–954, 1980.
60. Aylor, D. E., and Lukens, R. J. Liberation of Helminthosporium maydis spores by wind in the field. *Phytopathology, 64*:1136–1138, 1974.
61. Aylor, D. E. Forcerequired to detach conidia of Helminthosporium maydis. *Plant Physiol., 55*:99–101, 1975.
62. Shaw, R. H., Ward, D. P., and Aylor, D. E. Frequency of occurrence of fast gusts of wind inside a corn canopy. *Appl. Meteor., 18*:167–171, 1979.
63. Aylor, D. E. Modeling spore dispersal in a barley crop. *Agric. Meteorol., 26*:215–219, 1982.
64. Strand, M. A., and L. F. Roth. Simulation model of spread and intensification of western dwarf mistletoe in thinned stands of ponderosa pine saplings. *Phytopathology, 66*:888–895, 1976.
65. Aylor, D. E. Dispersal in time and space: Aerial pathogens. In *Plant Disease: An Advanced Treatise*, Vol. 2 (J. G. Horsfall and E. B. Cowling, eds.), Academic Press, London, 1978, pp. 159–180.
66. Okubo, A. *Diffusion and Ecological Problems: Mathematical Models.* Springer-Verlag, New York, 1980, p. 254.
67. Othmer, H. G., Dunbar, S. R., and Alt, W. Models of dispersal in biological systems. *J. Math. Biol., 26*:263–298, 1988.
68. Waggoner, P. E. Weather, space, time, and chance of infection. *Phytopathology, 52*:1100–1108, 1962.
69. Minoque, K. P., and Fry, W. E. Models for the spread of disease: Model description. *Phytopathology, 73*:1168–1173, 1983.
70. Eisensmith, S. P., Rabbinge, R., and Zadoks, J. C. Development of a stochastic spore germination model. *Neth. J. Plant Pathol., 91*:137–150, 1985.
71. Sall, M. A. Uses of stochastic simulation: Grape powdery mildew example. *Z. Pflanzenk, Pflanzensch., 87*:397–403, 1980.
72. Shrum, R. Simulation of wheat stripe rust (*Puccinia striiformis* West) using EPIDEMIC, a flexible plant disease simulator. In *Progress Report No. 347*, Pennsylvania State College of Agriculture, 1975.
73. Teng, P. S., Blackie, M. J., and Close, R. C. Simulation of the barley leaf rust epidemic: Structure and validation of BARSIM-I. *Agricul. Syst., 5*:85–103, 1980.
74. Sall, M. A. Epidemiology of grape powdery mildew: A model. *Phytopathology, 70*:338–342, 1980.
75. Press, W. H., Flannery, B. P., Teukolsky, S. A., and Vetterling, W. T. *Numerical Recipes: The Art of Scientific Computing.* Cambridge, Cambridge University Press, 1986, p. 818.
76. Fransz, H. G. *The Functional Response to Prey Density in an Acarine System*, Simulation Monographs, PUDOC, Wageningen, 1974, p. 149.

23

COMPUTER-AIDED APPROACHES TO THE EPIDEMIOLOGY AND MANAGEMENT OF FUNGAL DISEASES

VERN J. ELLIOTT *Agricultural Research Station, United States Department of Agriculture, and North Carolina State University, Oxford, North Carolina*

I. INTRODUCTION

In the past, computer applications in plant pathology would have focused on a few areas centered around simulation models of plant diseases. Due to the increasing power of computers and software, coupled with the increasing computer literacy of plant pathologists, the horizons are broadened considerably. Simulation modeling of plant disease epidemics still constitutes a major thrust, but new developments like expert systems are appearing. The underlying purpose of these efforts is to gain a better understanding of diseases and to improve the options available for disease management.

In 1978, Jeffers [1] discussed the role of computers in systems analysis and emphasized that although computers are an integral tool of systems analysis, computer use does not distinguish systems analysis from other types of investigations. Likewise, although the applications in this chapter rely on computers, it is the approach to the problem which is important rather than computer use per se.

The wide availability of microcomputers is also having an impact on farm management. A survey of U.S. farmers in 1984 indicated that quantitative applications were the top ranking uses for on-farm computers [2]. These quantitative applications included accounting, financial decision aids, and production decision aids. Farmers are becoming well versed in quantitative skills of data manipulation and may develop into applied systems analysts [3]. The need for computer-based disease management tools will most likely increase, and some of those management tools will be discussed in this chapter.

A. Purpose of Models in Disease Management

Disease management involves those aspects of crop management which are carried out primarily to reduce the economic impact of disease. The type of management response for disease control varies with the perceived threat of disease. Control measures can be applied in direct response to the threat of disease, such as following a period of weather conducive to infection, or on a schedule during the part of the growing season when disease is generally progressing. Control measures can also be applied based on historic indications such as is done with seed treatments or crop rotation. Sometimes control measures are applied without specific indication, with the justification of minimizing the risk. These measures are often seen as "insurance" treatments. This situation probably developed because, although high risk situations can usually be identified, situations of marginal to minimal risk are more difficult to recognize. Hence a safe approach was taken primarily due to a lack of information as to the true risk. Models can help to visualize this risk.

Disease management decisions rely heavily on information about the present state of the system [4]. Models have the potential to improve the information known about a particular crop or disease system and, moreover, to predict the consequences of specific actions. Models allow the more complex aspects of a management system to be quickly investigated and tracked with a goal of improving decision making in terms of timing, materials, and economic consequences. The changing picture of farm economies in many parts of the world mandate that costs of disease control be more carefully controlled.

B. Purpose of Models in Research

The use of models is fundamental to the scientific method. Models, even complex computer simulation models, are hypotheses. Models serve to limit the system under study and provide a tangible means of testing. Jeger [5] discusses the use of models and distinguishes between analytical and simulation models. Although this chapter is concerned primarily with models ranging from simple forecast criteria to complex simulation models, it should be realized that analytical models aid in the development of simulation models and can be incorporated into them. Although the underlying empirical structure of many forecasting or simulation models may limit their usefulness in the development of theory [5], they are nonetheless useful in exploring strategies of disease management and in gaining an understanding of processes occurring at higher levels of integration and over greater time and distance scales than can be approached experimentally.

II. SIMULATION MODEL APPROACHES

A. Background

The term simulation model will be used in a broad sense to mean a model which tries to mimic or simulate all or part of a real system. Although the term simulation model often refers to a complex representation of the entire epidemic, it can also refer to a model which simulates a narrower part of an epidemic such as an infection cycle. A synopsis of the modeling process will be presented here to help the reader formulate a critical view of simulation models particularly with reference to factors influencing the use of the models.

A considerable amount has been written concerning the simulation modeling of plant disease. The purpose of this chapter is not to discuss the methodology of systems analysis and modeling but rather to illustrate the overall development process and the uses of models in the epidemiology and management of fungal disease. Procedures for building simulation models vary with the intended use of the model as well as the philosophy of the modeler. Simulation modeling is both an art and a science [6]. Selected material will be cited, and the reader should find these references helpful in their own right as well as a source of further references to satisfy more detailed inquiries.

B. Capturing an Epidemic in a Computer Program

As with so much in nature, every epidemic is unique. This is not surprising when we consider that epidemic progress is under the influence of a variety of factors. Some of these include the amount and timing of initial infections, varying weather conditions, planting dates, cultivar differences, and the timing and the extent of control practices. In studying nature, we often try to look for common patterns and then deal with the remaining variations as neatly as possible. Studying the nature of complex systems like epidemics is difficult, and the advent of computers was welcome.

Computers became available to engineers and physical scientists in the 1940s and 1950s and eventually moved into biological and agricultural sciences in the 1960s and 1970s. Epidemiologists, following suit, recognized the potential of computers to aid in the understanding of epidemics. Computers and available programming languages in this era dealt primarily with numeric operations, and hence it was necessary to formulate a numeric representation of epidemics.

The experience of other disciplines in dealing with complex problems resulted in the evolution of systems analysis. Systems analysis is a structured, organized approach to dealing with complex systems. Seem and Haith [3] discuss the application of systems analysis to epidemiology. The process begins with a statement of objectives about the intended use of the model. With these objectives in mind, the researcher then sets boundaries or limits on the problem. The limitations imposed at this stage may result in future problems if the model is used for unintended purposes. The system is then divided into logical, workable components. With biological systems, because of their wide-ranging complexities, it is necessary to limit the system to the key components and the interaction of those components. Again, the limits set at this point will restrict uses of the model.

For modeling purposes, the fungal infection cycle is broken into discrete steps corresponding to the different stages of the infection cycle. Infection, latent period, spore production, and spore dispersal are generally used as divisions of the infection cycle, but there appears to be no consistent rationale for dividing the cycle [6]. Waggoner hypothesizes that this approach may be due to the plant pathologist's long preoccupation with the morphologically distinct components of the pathogen's life cycle [7].

The next task is to develop submodels for each of these components. These models define how the components behave over time (temporal models) or space (spatial models) under the influence of specific controlling variables like temperature, moisture, fungicide, or even complex variables like host resistance. Hau et al. [8] present an informative discussion of the procedures involved in developing component submodels. This stage of development often involves data collection experiments in which individual components such as infection are studied under controlled conditions. It is through these submodels that environmental inputs are made which drive the model.

These submodels are then linked together in a fashion mimicking the actual epidemic. Teng [6] points out that almost without exception, the infection cycle has been the repeating subunit for simulations. Time progression is incorporated as discrete steps usually representing 24-hour or shorter time periods. The manner in which driving variables such as temperature enter this time cycle has a direct bearing on the data collection needed for model execution. In some instances, component models are used alone to predict certain events like infection.

The simulation model is then translated to a programming language. The structure and logic of the model is evident in the program (or should be). This is an important justification for publishing or otherwise making the source code available. The size and complexity of the program directly determines the minimum computer resources required and hence are related to the ways in which the model can be implemented.

The simulation model to this point involves much work and even more thought. However, it is very important to maintain the proper perspective of current computer simulation models. We have attempted to represent an epidemic that is a very complex, ever changing, and in many ways poorly understood *biological* system as a well-delimited *numerical* system with important components and interactions explicitly defined. This approach has been successful, but limits initially set for models should be kept in mind. The computer simulation model has been erected as a paradigm of the plant disease epidemic but it should be remembered that pathogens that do not necessarily recognize the constraints set during the modeling process.

C. The Problem of Validation

Although the modeling might appear to be complete at this point, the most important and often the most neglected task remains. It still must be determined if the model *simulates* the real system. Validation is important because it determines the acceptibility of model results to the modeler and more importantly to those plant pathologists not oriented to the simulation approach. Teng [9] presents an extensive review including some of the philosophy behind validation of computer models. Several considerations of that review follow. In general, the task can be subdivided into *verification* and *validation*. Verification of the model is the process of determining that the model has been accurately translated into computer code and that no programming errors or misrepresentative logic have been included. Validation of the model is the process of determining that the model faithfully represents the system under study. The approach to determining the validity of a simulation model depends primarily on the purpose of the model [9–11].

Caswell [10] distinguishes between models used for predicting the behavior of a systems and models used to gain insight into the operation of a system. Predictive models are validated by testing the accuracy of the prediction compared to the behavior of the actual system. If the model is predicting disease development, then the predicted disease would be compared to actual epidemics. Procedures for validating disease prediction models generally involve some type of visual or statistical comparison to actual epidemics [6,9,11]. Models used to gain insight into the system are, in actuality, scientific hypotheses and are assumed to be valid as long as they cannot be proven to be invalid. Considerations for testing these models are discussed in the literature [9–11].

Validation of a model may be carried out at the submodel level. This method is more amenable to controlled testing, and validation can be carried out earlier in the modeling process. Testing the whole model can be inefficient because it can be very difficult to decide what constitutes acceptable behavior [10]. However, Teng points out that since a system is generally considered to be more than just a sum of its part, validation of the whole model is mandated [9].

Welch et al. [11] discuss some important points concerning the validation of pest management models. The value of a management model lies in its overall utility. In evaluating the utility of a management model to a producer, several considerations are important [11]. Will the model increase profits relative to current practices? Is the probability of disastrous results negligible? Is the model easy to understand? Is the model compatible with other management activities? The model need not accurately predict the pathogen's behavior as long as no monetary loss occurs. In this situation, a model is considered valid if the risk of failure is less than some accepted level. This view of validation is not particularly satisfying to a research scientist, but Welch et al. [11] caution that valid management models may be discarded because inappropriate criteria were used for evaluation.

D. Implementation of Models

The role of models in disease management is to increase the knowledge of what is occurring in the field. One primary use of models has been in the forecasting of disease development. The principles involved as well as details on forecasting systems have been discussed in a number of recent works [12–15].

Computerization of many of these forecast systems is not a prerequisite to their use, but has occurred as appropriate delivery systems were developed. A partial review of the development of potato lateblight forecasting in the United States illustrates this point [16]. In this case, the forecasting system which started as a manually calculated forecast was subsequently implemented on a mainframe computer as a dial-up service. Microcomputer-based versions were eventually developed, and finally a microprocessor-based, in-field, data acquisition unit programmed to generate a spray recommendation was developed and made commercially available. Dedicated microcomputer forecast systems have been developed for other diseases as well [17].

Although it is difficult to predict future developments in this area, it is likely that forecasting systems will become oriented to the desk top microcomputer rather than to dedicated field computers. Many farms already have microcomputers, and the level of on-farm computer resources can be expected to rise as with other types of businesses. Moreover, the rapidly increasing power of desktop microcomputers offers greater possibilities for computer-aided disease management than is possible with a field unit. Field based units are generally less powerful. Since online weather services (as well as other numerous pest management resources) are already available [18], the need for dedicated weather monitoring may decrease. Additional weather data could be supplied by additional electronic data acquisition units placed in the field, but disease management models would still run on the microcomputer.

Simulation models, particularly the more complex models, have been used to develop and evaluate improved methods of disease management. The utility of simulation models for this purpose is well illustrated by Fry and Fohner [13]. Using a potato lateblight simulation model, they were able to evaluate the value of different fungicide scheduling programs under a variety of different disease situations. Conducting similar experiments in the field would be difficult because factors of weather and inoculum availability would be difficult to regulate and measure.

Simulation models are also valuable in developing management strategies such as disease resistance. For example, a simulation model was used by Mundt and Brophy [19] to investigate various factors related to using host mixtures for disease management. They were able to test host mixtures on spatial scales which could not readily be approached in field experimentation.

Simulation models have much potential value as research tools and can complement laboratory, controlled environment, and field studies. As Hau et al. [8] stated, properly validated simulation models offer a fourth dimension in experimentation. Their ultimate use will depend on how faithfully they represent the real situation.

III. EXPERT SYSTEM APPROACHES

A. Background

Expert systems represent a rapidly expanding subarea of the artificial intelligence research being conducted in computer science. An expert system is a unique class of computer program that uses knowledge, facts, and reasoning techniques to solve problems that would normally require the abilities of a human expert [20]. A human expert is recognized as an individual who, through education and experience, has gained the knowledge and insight to solve difficult problems. The demand for a particular type of human expert often exceeds the supply. Latin et al. [21] point out the growing demands being placed on extension staff and foresee that expert systems may help reduce some of the workload. The intent of an expert system is to make the knowledge and problem-solving abilities of the expert more widely available. An expert system is intended to serve as an extension of a human expert in providing information or decision support to the program user.

Although there is no standard design, an expert system basically consists of a *knowledge base* and an *inference engine* and usually includes a natural language *user interface*. A knowledge base is a set of facts and rules about a given area of knowledge. This area of knowledge is termed the *domain*. The inference engine is the control structure of the program and governs the manner in which information from the user (assertions) are evaluated against the knowledge base in reaching a goal or answer. The inference engine is the "thinking" part of the expert system. The natural language interface allows the user to easily and clearly communicate with the expert system. More elaborate systems may include features to explain decisions, maintain the knowledge base, access other programs, or interface with the expert for purposes of building and validating the knowledge base.

Expert systems have certain attributes that separate them from other computer programs. Expert systems contain extensive, specific knowledge; use search techniques; support heuristic analysis; infer new knowledge from existing knowledge; use symbolic processing; and can explain the reasoning used in problem solving [22].

Another unique feature of expert systems, or perhaps of the programming languages and techniques used, is that the data or knowledge is separate from the control structure of the program. (In a conventional program written in a language like FORTRAN or Pascal, the data is in part integrated with the control structure.) The separation of data from the control structure in an expert systems allows the data to be readily modified even during execution. Another consequence of this separation is that the knowledge base can be removed and replaced with a completely different knowledge base. An expert system emptied of its knowledge is termed an *expert system shell.* These shell systems can be used to speed the development process and have been partly responsible for the rapid expansion of expert system applications.

Expert systems have applications ranging from decision support to control systems for complex operations such as power plants. Martin and Oxman [20] describe over 100 expert systems supporting a wide variety of domains (agriculture not included), and the list probably has grown considerably since their compilation. Applications to agriculture and in particular to plant pathology are emerging and current systems under development include disease diagnosis [23,24] and disease and crop management [25–27]. These expert systems will be discussed later.

B. Inside a Hypothetical Expert System

Expert systems technology has developed outside of biological sciences, and hence much of the terminology and basic structure may be foreign to biologists save those few who have already ventured through the gate. A hypothetical expert system for disease management will be described and some terms defined to give the reader some insight into the basic characteristics of expert systems.

The domain for this hypothetical expert system would be all the facts and rules that a plant pathologist would use in making decisions about when a fungicide application was needed. This knowledge must be represented in a form and structure that the expert system can use. One structure, a rule-based structure, will be demonstrated in our expert system. An example of possible rules is given below:

```
rule 0:  if temperature was more than 15°C
         if rh greater than 90%
         then infection period occurred

rule 1:  if last application more than 18 days ago
         then fungicide deposit low

rule 2:  if rainfall since last application more than 2 inches
         then fungicide low
```

rule 3: if over 130 growing degree days from last application
if total growing degree days less than 1500
then significant new foliage present

rule 4: if total growing degree days more than 2000
then crop near harvest

rule 5: if infection period occurred
if fungicide deposit low
then recommend fungicide

rule 6: if fungicide low
if significant new foliage
then recommend fungicide

These rules reflect the rules an expert would use in deciding when a fungicide application is necessary. Some rules like rule 3 incorporate rules-of-thumb. The rules are easily read and understood in their relationship to the problem. Each rule contains one or more *conditions* (*if* part) and a *conclusion* (*then* part). Notice that some rules contain more than one condition which must be met before the consequence of the rule can be inferred. Conditions for satisfying some rules, such as rules 5 and 6, may require the evaluation of other rules. Rules 5 and 6 also generate an action or recommendation.

The knowledge base also contains a list of facts or assertions. In our hypothetical expert system we might enter certain assertions about the current problem like temperature = 23, relative humidity = 93%, and 20 days since last application. (The weather data could also be supplied as part of an automated system interfaced to the expert system.)

The second part of our expert system is the inference engine. The programming details of an inference engine vary depending on the structure of the knowledge base and the logic used in solving the problem. The most common methods used are *backward-chaining* and *forward-chaining* logic. Backward-chaining starts with the goal and then tries to satisfy the conditions for that goal. The alternative strategy of forward-chaining starts with assertions and tries to work towards a goal. In our rule-based example, we will illustrate backward-chaining logic. The inference engine also contains facilities for searching and pattern matching. (Programming languages like PROLOG contain built-in facilities for searching and pattern matching.) The inference engine could also contain facilities for explaining decisions. These are basically implemented by maintaining a list of the rules which were used in solving the query.

To illustrate the operation of our expert system, we will ask the question: Is a fungicide application recommended? The program starts by finding a conclusion which matches the query. In this case the program might select the conclusion part of rule 5 and then try to solve the condition parts of the rule. To solve the condition "infection period occurred," the program then looks at rule 0. The program would check the list of assertions or prompt the user to supply information on temperature and wetness. If the conditions of rule 0 were satisfied, the rule would *fire* and the conclusion, "infection period occurred," would be added to the list of assertions.

The inference engine would then reevaluate rule 5 and satisfy the infection part of the condition. The program would then seek to satisfy the condition about low fungicide until rule 5 could be fully satisfied, at which time the rule would fire and result in a "yes" answer to the query and the program halts. (The inference engine will usually continue backtracking until a conclusion is reached or no changes can be made to the list of assertions.)

Hopefully, this simplified view of operation will help to demystify expert systems. The process of building a working expert system is considerably more involved.

C. Building Expert Systems

Several distinct roles or tasks are involved in the development of an expert system. The skills of a domain expert, a knowledge engineer, and a programmer are generally needed. Although one individual might fill these roles, it is more likely and often more desirable that each role is filled by a separate individual [28]. Recent software advancements in expert systems have, however, resulted in expert system shells and development tools that are helping to reduce the need for knowledge engineers and programmers [20, 22]. Nonetheless, the tasks still remain.

The domain expert provides the *heuristic knowledge.* Heuristic knowledge includes all the judgments, rules-of-thumb, and hunches which the expert has developed through problem-solving experience. A plant pathologist with experience in a particular area such as disease diagnosis or disease management would be a good example of a domain expert. A farmer who successfully manages disease problems might be another example. Often, more than one expert might be used depending on the difficulty of the problem area. In terms of crop management, this approach would be valuable. The effective use of expert system methods in plant pathology will depend on how well our knowledge can be represented.

The knowledge engineer is responsible for structuring the knowledge of the domain expert. Knowledge engineering is a newly developing field and involves its own methodology [28,29]. The role of knowledge engineer has developed in response to the difficulty of trying to represent the problem-solving knowledge of the expert in a manner which is useful to the expert system program. The knowledge engineer must elucidate from the expert all the facts and rules used in making decisions and then structure that knowledge in a manner suited to the expert system being developed. This is a crucial process on which proper behavior of the expert system depends [28]. Advances in software development tools are working toward the goal of allowing the expert to directly build the knowledge base without the assistance of a knowledge engineer [22]. These tools will help develop effective knowledge representations.

Rules, as illustrated in the above example, are frequently used knowledge representations. Other methods like frames, semantic nets, scripts, and hybrid structures can be used but are generally beyond our scope [22]. Frames, however, are particularly useful for representing knowledge about objects or patterns. Weather patterns, in particular infection periods, represent patterns and can be effectively represented as frames [26].

The programmer is responsible for programming the knowledge base, the inference engine, and the user interface. Although traditional programming languages such as FORTRAN or Pascal can be used, nonprocedural languages such as LISP, PROLOG, and OPS5 are generally more suited to the task. Advances in software have resulted in expert system shells and development tools which greatly simplify the programming of expert systems [20,22].

D. Examples of Expert Systems

One of the first expert systems developed in plant pathology was a soybean disease diagnosis system, PLANT/ds [24]. This rule-based system helps the user diagnose the problem based on symptoms as well as additional information like weather conditions and stage of the growing season. Another diagnostic system, M:5, has been developed to diagnose muskmelon disorders and recommend control measures [23]. This system is also rule-based and was developed with the aid of a shell system and operates on a microcomputer.

Expert systems are also emerging as tools for disease management. An expert system for fungicide management is under development [27]. This system considers diverse factors relating to the crop, weather, pathogen, and fungicide history to provide the user with advice on application intervals and postinfection treatments. An expert system, EPINFORM, is being developed to forecast Septoria blotch and stripe rust on wheat and generate estimates of expected yield loss [25]. This rule-based system models disease progression and then determines the effect of the diseases on yield. The system is being tested for accuracy and seems to generate good estimates when Septoria blotch and stripe rust were the dominant stresses. The authors report that loss estimates were generally too low in other cases because the system could not account for all additional stress factors.

The potential for wider applications is exhibited by a more comprehensive expert system for orchard management [26]. This expert system, POMME, covers numerous aspects of apple orchard maintenance. The system advises on pest problems such as apple scab, cedar apple rust, and San Jose scale. Several other insects and diseases are also considered in making recommendations. It also deals with abiotic stresses of drought and winter injury.

E. Potential Applications in Plant Pathology

Problems which arise frequently and require the attention of an expert for timely and effective solutions are good areas for expert system development. Many aspects of disease management require the same decisions to be made repeatedly during the cropping season. Fungicide scheduling is a good example and expert systems could be very useful for a wide variety of situations. Because of their ability to use rules-of-thumb, expert systems can potentially incorporate many factors other than weather conditions.

Other disease management decisions are made on an annual basis but are made more or less independently by each manager. Cultivar selection, crop rotations, and planting dates are examples of such decisions. Expert systems should prove useful for these problems of disease management. Given the increasing number of constraints being placed on crop rotation due to economic necessity and soil conservation requirements imposed by the 1985 U.S. farm bill, an expert system for crop rotation would certainly help farmers optimize their decisions.

Numerous unsolved problems of a diverse nature face agriculture in general, and the ability of expert systems to create a fusion across disciplines may hold some solutions [30].

In the past, applications of expert systems were limited by the cost of development and the large computers needed for execution. As mentioned earlier, advancements in the expert system software have resulted in expert system shells and development tools which facilitate the process of expert systems development and are reducing the amount of expertise required as well as the time and money needed.

IV. CONCLUSIONS

Considerable progress has been made in our understanding of fungal disease epidemiology and the ability to forecast diseases, but more efforts are needed to increase the implementation of this technology. Computer-aided approaches have much to offer toward increasing the effectiveness of plant disease management because much of the process can be automated.

Expert systems have several advantages over simulation models as described above. They are able to incorporate uncertain information and deal with rule-of-thumb relationships which are difficult to incorporate into simulation models. Expert systems are also user friendly due to natural language interfaces, and this should aid in implementation. It is also possible to merge expert systems with simulation models of various sorts. The POMME expert system incorporates an apple scab simulation model to improve recommendations [26].

The demand for computerization in agriculture is increasing [31], and the technologies discussed in this chapter should help meet that need.

REFERENCES

1. Jeffers, J. N. R. *An Introduction to Systems Analysis: With Ecological Applications*. University Park Press, Baltimore, 1978, p. 198.
2. Novak, J. L., and Stegelin, F. E. U.S. farmer's attitudes towards microcomputer uses. *Comput. Electron. Agric.*, 3: 81–89, 1988.
3. Seem, R. C., and Haith, D. A. Systems analysis in epidemiology. In *Population Dynamics and Management*, Vol. 1, *Plant Disease Epidemiology* (K. J. Leonard and W. E. Fry, eds.), Macmillan, New York, 1986, pp. 232–252.
4. Strandberg, J. O. Disease and pathogen detection for disease management. In *Population Dynamics and Management*, Vol. 1 (K. J. Leonard and W. E. Fry, eds.), Macmillan, New York, 1986, pp. 153–179.
5. Jeger, M. J. The potential of analytic compared with simulation approaches to modeling in plant disease epidemiology. In *Population Dynamics and Management*, Vol. 1 (K. J. Leonard and W. E. Fry, eds.), Macmillan, New York, 1986, pp. 255–281.
6. Teng, P. S. A comparison of simulation approaches to epidemic modeling. *Annu. Rev. Phytopathol.*, *23*: 351–379, 1985.
7. Waggoner, P. E. Computer simulation of epidemics. In *How Disease Develops in Populations, Plant Disease an Advanced Treatise*, Vol. 2 (J. G. Horsfall and E. B. Cowling, eds.), Academic Press, New York, 1978, pp. 203–222.
8. Hau, B., Eisensmith, S. P., and Kranz, J. Construction of temporal models: II. Simulation of aerial epidemics. In *Mathematical Modelling*

of *Crop Disease*, *Advances in Plant Pathology*, Vol. 3 (C. A. Gilligan, ed.), Academic Press, London, 1988, pp. 31–65.

9. Teng, P. S. Validation of computer models of plant disease epidemics: A review of philosophy and methodology. *Z. Pflanzenkr. Pflanzenschutz.*, *88*:49–63, 1981.
10. Caswell, H. The validation problem. In *Systems Analysis and Simulation Ecology* (B. C. Patten, ed.), Academic Press, New York, 1976, pp. 313–325.
11. Welch, S. M., Croft, B. A., and Michels, M. F. Validation of pest management models. *Environ. Entomol.*, *10*:425–432, 1981.
12. Fry, W. E. *Principles of Plant Disease Management*, Academic Press, New York, 1982, p. 378.
13. Fry, W. E., and Fohner, G. R., Construction of predictive models: I. Forecasting disease development. In *Mathematical Modelling of Crop Disease*, Vol. 3 (C. A. Gilligan, ed.), Academic Press, London, 1988, pp. 161–178.
14. Jones, A. L. Role of wet periods in predicting foliar diseases. In *Population Dynamics and Management*, Vol. 1 (K. J. Leonard and W. E. Fry, eds.), Macmillan, New York, 1986, pp. 87–100.
15. Shrum, R. D. Forecasting epidemics. In *How Disease Develops in Populations*, Vol. 2 (J. G. Horsfall and E. B. Cowling, eds.), Academic Press, New York, 1978, pp. 223–238.
16. MacKenzie, D. R., and Smilowitz, Z. Automated pest management. In *Advances in Potato Pest Management* (J. H. Lashomb and R. Casagrande, eds.), Hutchison Ross Pub. Co., Stroudsburg, PA, 1981, pp. 261–267.
17. Jones, A. L., Fisher, P. D., Seem, R. C., Kroon, J. C., and Van deMotter, P. J. Development and commercialization of an in-field microcomputer delivery system for weather-driven predictive models. *Plant Dis.*, *68*:458–463, 1984.
18. Plug into computers! *Ag. Consultant*, *44*:9–13, 1988.
19. Mundt, C. C., and Brophy, L. S. Influence of number of host genotype units on the effectiveness of host mixtures for disease control: A modeling approach. *Phytopathology*, *78*:1087–1094, 1988.
20. Martin, J., and Oxman, S. *Building Expert Systems*. Prentice Hall, Englewood Cliffs, NJ, 1988, p. 457.
21. Latin, R. X., Miles, G. E., and Rettinger, J. C. Expert systems in plant pathology. *Plant Dis.*, *71*:866–872, 1987.
22. Rolston, D. W. *Principles of Artificial Intelligence and Expert Systems Development*. McGraw-Hill, New York, 1988, p. 257.
23. Latin, R. X., Miles, G. E., and Rettinger, J. C. M5: An expert system for diagnosing muskmelon disorders. *Phytopathology*, *77*:1775, 1987.
24. Michalski, R. S., Davis, J. H., Bisht, V. S., and Sinclair, J. B. A computer-based advisory system for diagnosing soybean diseases in Illinois. *Plant Dis.*, *67*:459–463, 1983.
25. Caristi, J., Scharen, A. L., Sharp, E. L., and Sands, D. C. Development and preliminary testing of EPINFORM, an expert system for predicting wheat disease epidemics. *Plant Dis.*, *71*:1147–1150, 1988.
26. Roach, J., Virkar, R., Drake, C., and Weaver, M. An expert system for helping apple growers. *Comput. Electron. Agric.*, *2*:97–108, 1988.

27. Travis, J. W., Hickey, K. D., and Rajotte, E. J. An expert system to aid in the application of a sterol-inhibiting fungicide in controlling apple scab. *Phytopathology*, 77:1691, 1987.
28. Hart, A. *Knowledge Acquisition for Expert Systems.* McGraw-Hill, New York, 1986, p. 180.
29. Kidd, A. L. *Knowledge Acquisition for Expert Systems.* Plenum Press, New York, 1987, p. 194.
30. Doluschitz, R., and Schmisseur, W. E. Expert systems: Applications to agriculture and farm management. *Comput. Electron. Agric.*, *2*:173–182, 1988.
31. Holt, D. A. Computers in production agriculture. *Science*, *228*: 422–427, 1985.

24

VIDEO TECHNIQUES TO MEASURE FUNGAL PLANT DISEASES

M. V. WIESE *University of Idaho, Moscow, Idaho*

DONNA M. GERTEN *Agricultural Research Station, National Clonal Germplasm Repository, United States Department of Agriculture, Corvallis, Oregon*

I. INTRODUCTION

A. Measuring Plant Diseases

Plant diseases are a concern because they adversely affect important crop and ornamental plants causing stress, injury, damage, and/or economic loss [1–3]. Plant diseases arise from a number of diverse causes. Bacteria, viruses, nematodes, fungi, and mycoplasmas are well-known biological pathogens that incite plant disease. In addition, nonbiological agents such as toxicants, pollutants, and imbalances in temperature, light, moisture, and nutrients also inflict plant stress and damage. Among the complex array of stresses and abnormalities caused by biological pathogens, disorders caused by fungi have been most investigated. The rusts and smuts of cereal grains, for example, are highly visible and were among the first diseases to be recognized and characterized. Today, plant diseases caused by fungi still comprise the bulk of our attention and the largest portion of our plant pathological literature.

Plant pathologists are proficient at recognizing plant diseases and uncovering their cause. They are less able, however, to measure or quantify the extent of the disease-induced injury, damage, and/or economic loss. In this context, pathologists have few proven methods to efficiently and definitively measure either disease or disease impact. To meet the demand for this information, many scientists resort to providing "a best estimate" of disease incidence, severity, and impact.

Usually empirically derived and accompanied by limited supporting data, our best estimates (or guesses) of disease incidence, severity, and impact have profound significance. Governments, institutions, and businesses demand and use such information to set market and product priorities, prescribe research and extension activity, and justify budget requests. We especially request and use disease measurements as justification to develop and precribe disease controls such as the use of resistant varieties, disease-limiting cultural practices, and chemical pesticides that protect against or eradicate plant pathogens. It is also important to define threshold levels for plant diseases beyond which the damage imposed begins to exceed the cost of its control.

B. Disease Incidence, Severity, and Impact

Measuring plant disease and disease impact is difficult. To quantify a plant disease and the damage and/or economic loss it imposes, careful measurements over both space and time may be necessary. We usually attempt first to quantify disease incidence, then move to measure disease severity and, subsequently, strive to relate these measurements to statements of damage and/or loss.

In measuring disease, two of the most useful and meaningful variables are incidence and severity [2]. Disease incidence is a measure of how many host units among the total are diseased, i.e., harbor symptoms and/or signs of disease. The percentage of smutted versus normal wheat heads in a field or the percentage of scabby versus symptomless apples in a barrel are measures of disease incidence. With an estimate of disease incidence in place, we can concentrate further on the diseased portion of the population in an attempt to quantify disease severity, i.e., the symptoms, signs, and/or damage thereon. Here, for example, we may first distinguish if the smutted wheat heads are completely or only partially destroyed, how many spots appear on each diseased apple, the size of the spots, what portion of the apple surface area is altered by scab, or what portion of the total sample of diseased apples is marketable.

In the end, a credible measure of disease is required for interpreting and describing disease impact and prescribing controls [3–8]. Occasionally, the measure of disease (incidence and/or severity) is related directly to loss. For example, if all spotted apples cannot be sold but all symptomless apples can, we have a ready assessment of loss. However, if smutted wheat is discounted in price, if wheat plants developing in the field with smutted heads are able to compensate and produce more tillers and heads than healthy plants, if the investment in fungicide treatment and a crop rotation sequence to control smut prove only partially effective, if spores of the causal fungus remain in the field to affect subsequent crops, etc., we may have several variables to evaluate, measure, and interpret before arriving at a credible estimate (or measurement) of loss.

C. Methodology Is the Key

Methodologies for measuring disease and disease loss are developing but still immature. Compare, for example, our ability to measure crop yields versus our ability to measure yield loss. Compare our ability to derive a mean versus our ability to derive an environmental impact statement. In each case, both measurements are important, but the latter is a much more complex and challenging activity.

To measure fungal plant diseases, we must observe them, and to observe them, we rely first on our eyes. After careful observation and study of a disease, we may be equipped to develop a diagrammatic scale for rating its severity. Over the years, several of our best known and most troublesome fungal plant diseases have been so characterized and quantified. Today, series of diagrams and/or photographs are available to aid observers in grading, rating, or categorizing a variety of diseases and disease symptoms (leaf spotting, tissue discoloration, fruit damage, etc.) [9,10]. One of the earliest diagrammatic scales of a plant disease was developed in 1892 to describe five levels (frequencies) of rust pustules on a wheat leaf surface. By comparing the standard diagrams with actual diseased (rusted) leaves, an estimate of disease severity (percent leaf surface covered by pustules) is obtained.

More recently, photographs and video images have come into use to supplement and, in some cases, replace, quantitative diagrams of disease symptoms. Today, photographic and especially video techniques permit us to see and interpret disease incidence and severity more rapidly, accurately, and efficiently. Nearly 100 years ago, we began to use the "eye" of the camera to supplement our own. Today, we may use complex arrays of polished lenses to capture images of diseased plants on photographic film, on magnetic (video) tape, and in electronic memory. It behooves us, therefore, to continue to apply these new technologies to the science and art of disease measurement.

What follows is a brief review of the status and application of video techniques to disease detection and measurement. The discussion focuses on microcomputer-assisted video image analysis (VIA) as a tool to detect and measure fungal plant diseases. The model disease chosen for this discussion is foot rot (eyespot), a relatively conspicuous and important disease of winter wheat in the Pacific Northwest of the United States caused by the soilborne fungus *Pseudocercosporella herpotrichoides*.

II. VIDEO IMAGE ANALYSIS (VIA)

We create, interpret, and analyze images using our eyes and brain. Just as our eyes present an image for interpretation by our brain, VIA involves a video camera presenting an electronic image for analysis and interpretation by a computer. Thus, VIA is an electronic technique whereby analog images from a video camera are digitized (numerically represented) and analyzed by a computer [11–13]. A typical VIA system is composed of a video camera linked with a CRT monitor, linked in turn to a computer equipped with a digitizer and a printer. The transmitted video image is analyzed using software that first segments the image area into numerous subunits called pixels. Each pixel is identified by coordinates on a rectangular grid. The software then senses, categorizes, and records the intensity and/or quality of transmitted light within each pixel. In an additional step, the pixels comprising the image are sorted on an incremental gray scale (ranging from black to white). The incremental levels of gray between black and white on the scale may vary from a few to hundreds depending on the sophistication of the VIA system. Via the computer and its image-digitizing software, individual gray levels or ranges of gray levels can be selected, sorted, and displayed. In an important final step, the computer can total the number of pixels comprising each selected gray level (or range) and calculate the proportion of the total image area it represents.

Within the software package, the digitizing routine is the workhorse. It not only segments the analog video image into numerous pixels, it permits the image to be reprocessed by sequentially or incrementally accessing and displaying only those pixels falling within the specified level or range of gray. Once video images are digitized, the proportional image area occupied by a specified subset of grey values (density slice) is readily calculated. Thereafter, this subset of gray values can be displayed on the monitor and analyzed visually or electronically for correspondence to features of interest in the original image.

III. ANALYZING VIDEO IMAGES OF FUNGAL PLANT DISEASES

The evolution of VIA for detection and measurement of plant disease stemmed in part from studies of aerial photographs using microdensitometers. These earlier activities proved that the symptoms of several plant diseases, stresses, and disorders were discernable from photographs as well as from direct observation of the affected plants. Over the years, there have been many successes using aerial color and black-and-white photographs to detect plant diseases in the field [14–25]. However, the success of photographic disease detection depends on a high degree of symptom specificity and resolution. When the only resolvable symptom is generic (such as wilting or chlorosis, for example), the exact cause of a disorder cannot easily be interpreted from photographs. Later it was found that color infrared film offered a different and often better alternative for detecting plant diseases than using regular color or black-and-white photographs. The use of color infrared photography for disease and stress detection is now popular and commonplace since diseased (stressed) plants reflect more infrared energy than do healthy (unstressed) plants [21,22,24–27]. Even with its limitations, the science of remotely detecting and quantifying plant diseases flourished with the acceptance of microdensitometer interpretation of aerial color, black-and-white, and color infrared photographs [28]. More recently, however, video images and VIA are bringing additional applications and enthusiasm to the science of plant disease detection and measurement.

The first applications of VIA to measurements of plant disease were spawned in the 1970s as part of earlier research in remote sensing, i.e., the detection of forms and objects on the earth's surface from aircraft or satellites. In the United States and elsewhere, remote sensing research grew with our interest in space exploration. This effort spurred the development of sophisticated multispectral scanning equipment and computer hardware and software systems especially designed for interpreting photographic and video images. Prior to these developments, aerial photographs and microdensitometers were the accepted tools for quantifying disease incidence and severity [27,29]. As early as 1975, however, Eyal and Brown [30] used VIA to measure the proportional area of wheat leaves covered by the lesions and dark pycnidia of *Septoria tritici*. In 1981, Nilsson [31–34] began to measure the symptoms of several other fungal plant diseases using a Leitz Texture Analyzing System (Ernst Leitz-Wetzlar GmbH, Wetzlar, West Germany) capable of discerning 256 shades of gray. Greaves et al. [17] in 1983 used a similar system (Quantemet 720 Image Analysis Computer) to measure barley yellow dwarf, an important viral disease affecting wheat, from aerial photographs. In 1983, Lindow [35] and Lindow and Webb [36] began using a microcomputer to measure the area of lesions on tomato leaves caused by *Alternaria solani* and the darkened lesions of *Ascochyta pteridium* on leaves of bracken fern. In

1984, Bronson and Klittich [37] developed software for microcomputer-assisted video image measurement of several foliar plant diseases. These and other video pioneers in plant pathology made it clear that many plant disease symptoms were sufficiently differentiable from healthy plant tissues to be captured in video images made directly from plant material or indirectly from photographs.

With the rapid improvement of microcomputers and compact video equipment in the 1980s, video image analysis became less complex and costly. Today $5,000 packages of video hardware and software have the imaging, digitizing, and analytical power of former $50,000 systems. Current VIA equipment is compact, portable, and easy to operate. Except for lacking the memory capacity to perform some complex image manipulation and processing routines, today's compact and inxpensive VIA equipment has wide applicability and accpetance in many areas of science and engineering.

IV. A CASE STUDY: WHEAT FOOT ROT

Foot rot is an important disease of winter wheat throughout most of Europe and North America [38]. The disease is especially damaging in the Pacific Northwest where in late stages it causes marked lodging and yield loss in fields of winter wheat. The disease also affects but is less consequential in barley, spring wheat, and other graminaceous hosts. Early diagnostic symptoms of foot rot are tan, elliptical, or eye-shaped lesions on the base of wheat stems at or slightly above the soil line. Under favorable environmental conditions, these lesions grow in size and depth, penetrate successive layers of leaf sheaths, and invade the culm (stem). Such lesions often girdle the stem and become sunken and brittle. They so stress and mechanically weaken the stem that all plant parts above the lesion may prematurely ripen and become conspicuously white or bleached. In the field, this process leads to the formation of conspicuous "white heads" among the greener, normally maturing heads of healthy plants. More dramatic still is the lodging that occurs when diseased and weakened stems break over at the lesion site near the soil line in response to the mechanical forces of wind and rain. Large contiguous areas of lodged wheat may occur in fields with severe and widespread foot rot development.

In 1983, Gerten and Wiese [39] assembled and used an inexpensive VIA system to analyze aerial photographs of maturing winter wheat fields to accurately determine the incidence of lodging. They verified the relationship between lodging and foot rot by manually measuring foot rot incidence and severity in the same fields. The objective of this study was to devise an accurate and efficient estimate of foot rot incidence and severity and to derive an estimate of the impact of foot rot on wheat yields over a large production area.

In this study [39], aerial photographs (35-mm color and color infrared transparencies) of lodged maturing wheat fields were taken at various altitudes to produce photographs of different scale. Eventually, photographs with a scale of 1:7500 were selected for analysis because they offered the best combination of large area and lodging resolution [40]. Each such photograph (35-mm transparency) was illuminated on the surface of a light table. Mounted immediately above the photograph was an RCA TC2000 black-and-white video camera (RCA/Closed-Circuit Video Equipment, Lancaster, PA 17604) equipped with a closeup lens and an automatic gain control. The video camera transmitted the field image from the photograph to an Apple IIe microcomputer. The computer was equipped with a CRT monitor, two disk drives, a printer, a light pen, 192K memory and a DS-65 Digisector video digitizer card (Micro Works, Del Mar, CA 92014). Each video image was digitized (by means of the analog-to-digital conversion mechanism of the digitizing card) to a matrix of 65,536 pixels. A software routine was written to locate each pixel on a rectangular grid and to identify its light intensity within one of 64 gray levels (ranging from black to white). With the original photograph scale being 1:7500, the area represented by each pixel was equivalent to 7.32×10^{-5} ha. The software program, derived from BASIC and machine language, enabled the microcomputer to function as a comprehensive image processor.

Each VIA began by situating a wheat field photograph (Fig. 1) on the light table beneath the video camera. Next the analog image from the video camera trained on the photograph was displayed on the CRT monitor (Fig. 2). The computer was then instructed to digitize the analog image and display the frequency of pixels in each of 64 gray levels (Fig. 3). Pixel distribution was displayed in a 64 gray level histogram and was standardized by exposure and/or contrast adjustments either to the camera or digitizer. In each case, the operator adjustments served to maximize the contrast and clarity of the lodged areas in each photograph.

Once the operator maximized the resolution of the lodged wheat areas within the video image, a density slicing software routine was employed. In this step, the spatial arrangement of pixels comprising only the lodged wheat areas was displayed on the monitor (Fig. 4). Here the gray levels selected for display were adjusted as necessary to achieve the best visual representation of the lodged wheat within each photograph. In a final step, the percent of the entire image area occupied by the pixel slice best representing the areas of lodged wheat was calculated and displayed.

Using this procedure, the area of lodged versus erect wheat was measured in seven selected winter wheat fields in eastern Washington and northern Idaho. As a standard for comparison, the lodged areas within the same photographs were manually traced (blackened) onto frosted acetate (Mylar) sheets. These manually prepared photointerpretations (black-and-white transparencies) were then subjected to VIA and calculations of percent lodged area were made. When either the original aerial photographs or the standard acetate photointerpretations were used, sorting the brighter (whiter) pixels from lodged field areas and the darker (blacker) pixels from erect wheat areas was readily accomplished. However, while the lodged percentages measured from the acetate sheets and original photographs were similar, they were not equivalent. For example, in one series of paired measurements, the manually prepared acetate images of lodged wheat areas ranged from 4.4 to 37.9% (0.5 to 7.6 ha) per field while the same lodged areas measured directly from aerial photographs ranged from 1.3 to 32.2% (0.2 to 6.4 ha) per field. These differences were significant from one another at $p = 0.01$ according to the F-max test.

Within the photographed fields, manual measurements of the incidence of foot rot lesions on wheat stems and final grain yield showed 9% more foot rot lesions (Table 1) and 1389 to 3416 kg/ha lower grain yields (Table 2) in lodged wheat relative to erect wheat areas. Foot rot was significantly more severe ($p = 0.005$) in lodged wheat than in erect wheat areas. In the final analysis, the lodging-foot rot complex was shown to reduce yields by 138 to 796 kg/ha per field. Overall yield impact of the

FIGURE 1. An original color-infrared aerial photograph (35-mm transparency with a scale of 1:7500) of a lodged winter wheat field. Note light lodged areas versus dark areas of upright wheat.

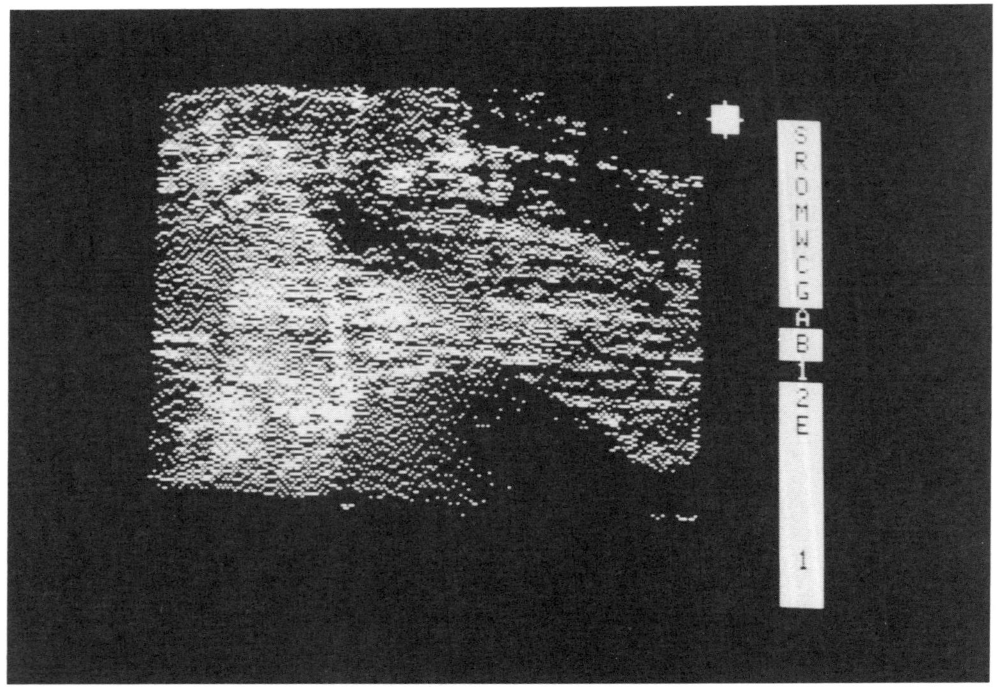

FIGURE 2. The original photograph of a lodged wheat field (Fig. 1) is reproduced on the CRT monitor using a video camera and a computer that converts the transmitted analog image to a digitized image.

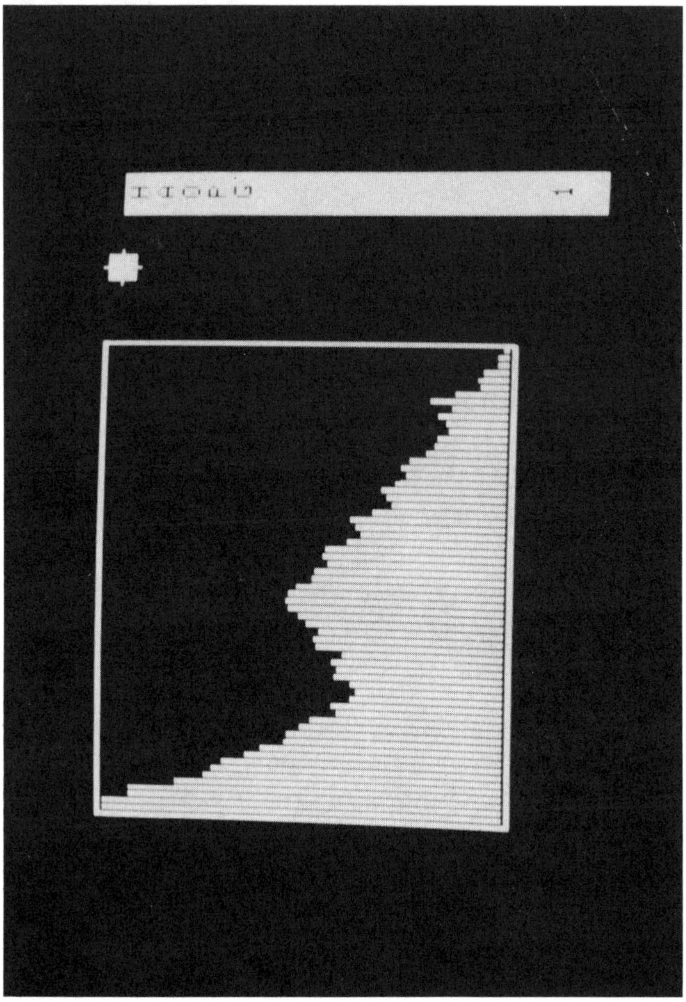

FIGURE 3. A histogram displays the relative frequency of pixels (comprising the digitized image, Fig. 2) that fall within each of 64 different gray values ranging from black (left) to white (right).

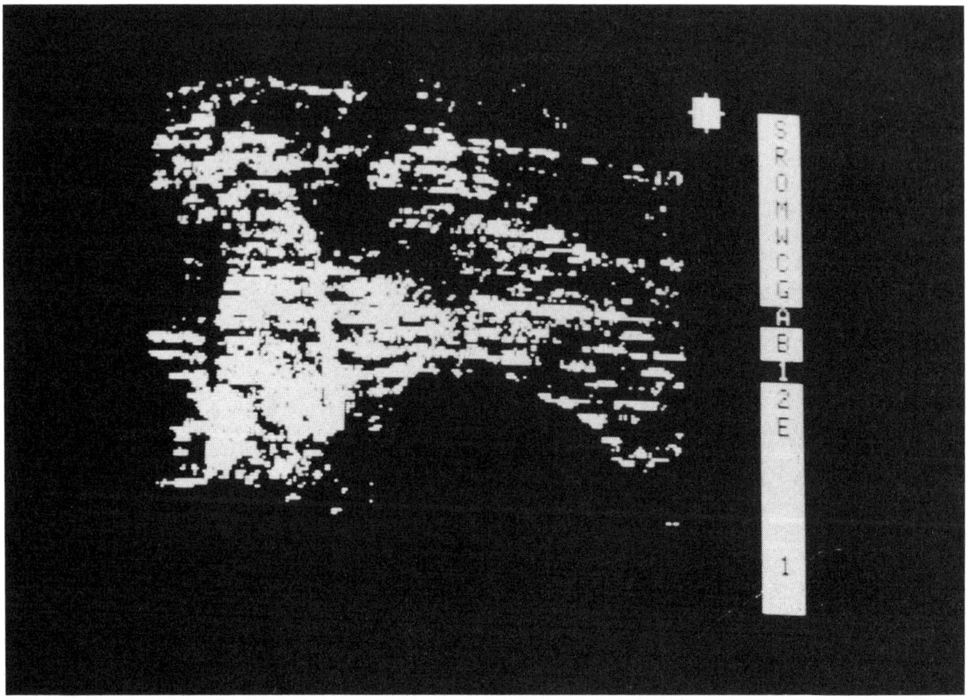

FIGURE 4. By manually selecting the range of gray values (gray density slice from the histogram in Fig. 3) that best represent the lodged wheat area, the operator displays those pixels on the CRT monitor.

TABLE 1. Foot Rot Incidence[a] and Severity on Stems Collected from Adjacent Lodged and Erect Areas in Each of Seven Winter Wheat Fields

	Severe lesion[b] (%)	Moderate lesion[c] (%)	Uninfected (%)
Erect	67	16	17
Lodged	77	9	14

[a]Mean percentage based on a minimum sample of 117 (maximum 210) erect or lodged stems from each of three locations in each of seven fields. Differences between erect and lodged percentages significant at $p = 0.005$ according to log linear model analysis.

[b]Encompassing greater than 50% of stem circumference.

[c]Encompassing less than 50% of stem circumference.

TABLE 2. Grain Yield in Adjacent Lodged and Erect Winter Wheat Areas in Each of Seven Fields

Field	Yield[a] (kg/ha)		Yield difference[b] (kg/ha)
	Erect	Lodged	
1	4376	959	3416
2	2976	703	2273
3	4565	1977	2589
4	4407	2159	2249
5	3065	1657	1408
6	4206	2817	1389
7	4893	2486	2407

[a]Mean yield of cleaned grain based on five samples one square meter in size within lodged and adjacent erect wheat areas.

[b]Yield differences significant at $p = 0.05$ according to paired t-test.

lodging-foot rot complex in each field was calculated by multiplying total lodged hectares per field by the yield difference between erect and lodged areas and dividing by total wheat hectares. Thus

$$I = LD/A$$

where: I = yield impact (kg/ha)
L = lodged area (ha)
D = yield difference between erect and lodged areas (kg/ha)
A = total field area (ha)

Yield impact estimates calculated from VIA measurements of original photographs were lower than those determined from the manually prepared acetate standards by 41 to 293 kg/ha.

V. SHORTCOMINGS OF VIDEO MEASUREMENTS

The computer-controlled digital VIA system described herein and used to measure lodging (as an expression of foot rot) incidence in winter wheat was inexpensive, rapid, accurate, and easily operated. However, one shortcoming of the above VIA procedure arose because the margin of lodged and unlodged wheat was not easily resolved. Lodged wheat areas in the center of photographs often were lighter than those toward the edges of the photographs. This presumably is why the lodging percentages calculated directly from photographs were consistent underestimates compared to lodging percentages prepared from acetate standards. Some of this underestimation may have been due to imperfectly resolved density slicing. Also, lodged wheat areas in aerial photographs, while distinct and easily discernible to the eye, were not all represented within the same definitive set of gray values (density slice). The lodged areas near the edges of the photographs were darker than those towards the center and, therefore, were incompletely resolved during density slicing.

Repeated VIA measurements made directly from photographs, although quite reproducible, were slightly more variable than measurements made from the manually drawn acetate standards. However, the margin of additional accuracy obtained from the acetate preparations was overshadowed by the simplicity and rapidity of the direct VIA procedure. Manually interpreting and graphically preparing the acetate standards is labor intensive, time consuming, and impractical when numerous measurements or numerous samples are of interest.

A video camera equipped with shading and blanking compensators perhaps would have given greater stability to the signal from the vidicon tube and provided a more consistent light intensity from image center to edge. In this same context, a color camera and a color spectral digitizer would offer greater utility, potential, and flexibility. A special imaging camera also may have avoided some parallax. Although manually drawn and time consuming, the Mylar black-and-white photo interpretations serve well as standards for area comparison and interpretation regardless of the video equipment used.

Another limitation of the VIA procedure described herein is failure to resolve visible color differences once the color or color-infrared images are reduced to black-and-white images on the CRT. For example, dark green leaf tissue and prominent disease symptoms such as dark brown lesions may not be differentiated when reproduced in shades of gray. The resolution of such symptoms, which are highly visible when reproduced in color but indistinguishable when reproduced in black and white, may necessitate reproduction using near infrared or far red radiation or reproduction using color filters to modify the reflectance of the two-color interface.

Another concern deals with interpreting the impact of foot rot on wheat yield. Specifically, the proportional yield loss attributable to lodging per se as opposed to that attributable to the combination of foot rot and lodging was not determined. Lodged but otherwise healthy wheat sometimes yields less than standing wheat. Yield reductions due to lodging can be significant especially if the lodging occurs early during grain development rather than later during grain maturation. One study [41,42] reported that the percentage of erect versus lodged culms severely infected with foot rot accounted for 47% versus 79% of the observed yield reduction, respectively.

VI. CONCLUSIONS

Video image analysis (VIA) should find many uses in plant pathology, in quantitative epidemiology, and in evaluating biological and other plant disease controls. Fungal plant diseases that produce definitive symptoms are ready candidates for measurement using VIA. Others whose symptoms are less definitive may require additional video configuration and/or ground truth for satisfactory interpretation and measurement.

Foot rot of winter wheat caused by the soilborne fungus *Pseudocercosporella herpotrichoides* is an important disease of winter wheat in the Pacific Northwest of the United States. The use of VIA to measure foot rot–induced lodging in wheat fields is just one of many real and potential VIA applications to plant disease measurement. Many studies have demonstrated that microcomputer-assisted VIA can be used to efficiently quantify both the incidence and severity of prominent disease symptoms on individual plants and plant parts or over large plant or crop areas. VIA can be applied to live video images or to images stored on magnetic tape or photographic film. Aerial photographs of diseased plants, crops, and fields are easily generated and stored for VIA interpretation at any later or convenient time. Likewise, large plant or crop areas are surveyed easily and at low cost using aerial photography and VIA compared to slower and more labor-intensive ground surveys.

REFERENCES

1. Chester, K. S. Plant disease losses: Their appraisal and interpretation. *Plant Dis. Reptr.* (Suppl.), *193*:189–362, 1950.
2. Horsfall, J. G., and Cowling, E. B. Pathometry: The measurement of plant disease. In *Plant Disease: An Advanced Treatise*, Vol. 2 (J. Horsfall and E. G. Cowling, eds.), Academic Press, New York, 1978, pp. 119–136.
3. James, W. C. Assessment of plant disease and losses. *Annu. Rev. Phytopathol.*, *12*:27–48, 1974.
4. James, W. C., and Teng, P. S. The quantification of production constraints associated with plant diseases. In *Applied Biology*, Vol. 4 (T. H. Coaker, ed.), Academic Press, New York, 1979, pp. 210–267.
5. Large, E. C. Measuring plant disease. *Annu. Rev. Phytopathol.*, *4*:9–28, 1966.
6. Romig, R. W., and Calpouzos, L. The relationship between stem rust and loss in yield of spring wheat. *Phytopathology*, *60*:1801–1805, 1970.
7. Wiese, M. V. Crop management by comprehensive appraisal of yield determining variables. *Annu. Rev. Phytopathol.*, *20*:419–432, 1982.
8. Zadoks, J. C., and Schein, R. D. Disease and crop loss assessment. In *Epidemiology and Plant Disease Management.* Oxford University Press, New York, 1979, pp. 237–281.
9. Dixon, G. R., and Doodson, J. K. Assessment keys for some diseases of vegetable, fodder, and forage crops. *J. Natl. Inst. Agric. Bot.*, *23*:299–307, 1971.
10. James, W. C. *A Manual of Assessment Keys for Plant Diseases.* Amer. Phytopath. Soc., Can. Dept. of Agric. Pub. No. 1458, St. Paul, MN, 1971.
11. Castleman, K. R. *Digital Image Processing.* Prentice-Hall Inc., Englewood Cliffs, NJ, 1979, p. 429.
12. Schowengerdt, R. A. *Techniques for Image Processing and Classification in Remote Sensing.* Academic Press, New York, 1983, p. 249.
13. Green, W. B. *Digital Image Processing: A Systems Approach.* Van Nostrand Reinhold Co., New York, 1983, p. 192.
14. Blasquez, C. H., Elliott, R. A., and Edwards, G. J. Vegetable crop management with remote sensing. *Photogramm. Eng. Rem. Sens.*, *47*:543–547, 1981.

15. Basu, P. K., Jackson, H. R., and Wallen, V. R. Estimation of pea yield loss from severe root rot and drought stress using aerial photographs and a loss conversion factor. *Can. J. Plant Sci.*, *58*:159–164, 1978.
16. Colwell, R. N. Determining the prevalence of certain cereal crop diseases by means of aerial photography. *Hilgardia*, *26*:223–286, 1956.
17. Greaves, D. A., Hooper, A. J., and Wolpole, B. J. Identification of barley yellow dwarf virus and cereal aphid infestations in winter wheat by aerial photography. *Plant Pathol.*, *32*:159–172, 1983.
18. Haglund, W. A., and Jarmin, M. C. Aerial photography for detection and identification of Fusarium wilt of peas. *Plant Dis. Reptr.*, *62*:570–572, 1978.
19. Hill, J. H., Epstein, A. H., McLaughlin, M. R., and Nyvall, R. F. Aerial detection of tobacco ringspot virus-infected soybean plants. *Plant Dis. Reptr.*, *57*:471–472, 1973.
20. Meyer, M. P., and Calpouzos, L. Detection of crop diseases. *Photogramm. Eng.*, *34*:554–556, 1968.
21. Philpotts, L. E., and Wallen, V. R. IR color for crop disease identification. *Photogramm. Eng.*, *35*:1116–1125, 1969.
22. Toler, R. W., Smith, B. D., and Harlan, J. C. Use of aerial color infrared photography to evaluate crop disease. *Plant Dis.*, *65*:24–31, 1981.
23. Wallen, V. R., and Jackson, H. R. Aerial photography as a survey technique for the assessment of bacterial blight of field beans. *Can. Plant Dis. Surv.*, *51*:163–169, 1971.
24. Wallen, V. R., and Jackson, H. R. Model of yield loss determination of bacterial blight of field beans utilizing aerial infrared photography combined with field plot studies. *Phytopathology*, *65*:942–948, 1975.
25. Wallen, V. R., and Philpotts, L. E. Disease assessment with IR color. *Photogramm. Eng.*, *37*:443–446, 1971.
26. Clark, R. V., Galaway, D. A., and Paliwal, Y. C. Aerial IR photography for disease detection in field plots of barley, oats, and wheat. (abstr.) *Phytopathology*, *71*:867, 1981.
27. Jackson, H. R., Hodgson, W. A., Wallen, V. R., Philpotts, L. E., and Hunter, J. Potato late blight intensity levels as determined by microdensitometer studies of false-color aerial photographs. *J. Biol. Photogr. Assoc.*, *39*:101–106, 1971.
28. Lillesand, T. M., and Kiefer, R. W. *Remote Sensing and Image Interpretation.* John Wiley and Sons, New York, 1979, p. 612.
29. Jackson, H. R., and Wallen, V. R. Microdensitometer measurements of sequential aerial photographs of field beans infected with bacterial blight. *Phytopathology*, *65*:961–968, 1975.
30. Eyal, Z., and Brown, M. B. A quantitative method for estimating density of Septoria tritici pycnidia on wheat leaves. *Phytopathology*, *66*:11–14, 1976.
31. Nilsson, H. E. Application of remote sensing methods and image analysis at macroscopic and microscopic levels in plant pathology. In *Crop Loss Assessment*, Minnesota Agric. Expt. Sta., Misc. Publ. 7, 1980, pp. 76–84.
32. Nilsson, H. E. Remote sensing of oil seed rape infected by Sclerotinia stem rot and verticillium wilt. Uppsala, Swedish Univ. of Agric. Sciences, Research Information Center, 1985, Jordbruk 33, p. 33.

33. Nilsson, H. E. Remote sensing of 2-row barley infected by net blotch disease. Uppsala, Swedish Univ. of Agric. Sciences, Research Information Center, 1985, Jordbruk 34, p. 101.
34. Nilsson, H. E. Remote sensing of 6-row barley infected by barley stripe disease. Uppsala, Swedish Univ. of Agric. Sciences, Research Information Center, 1985, Jordbruk 36, p. 49.
35. Lindow, S. E. Estimating disease severity of single plants. *Phytopathology, 73*:1576–1581, 1983.
36. Lindow, S. E., and Webb, R. R. Quantification of foliar plant disease symptoms by microcomputer-digitized video image analysis. *Phytopathology, 73*:520–524, 1983.
37. Bronson, C. R., and Klittich, W. M. Phytoscan 83: A computer program for quantitative disease assessment. (abstr.) *Phytopathology, 74*:871, 1984.
38. Wiese, M. V. *Compendium of Wheat Diseases*, 2nd ed. APS Press, American Phytopathological Soc., St. Paul, MN, 1987, p. 112.
39. Gerten, D. M., and Wiese, M. V. Microcomputer-assisted video image analysis of lodging in winter wheat. *Photogramm. Eng. and Rem. Sens., 53*:83–88, 1987.
40. Budanov, V. E. Lodging of cereals and its detection with the aid of aerial photography. *Zashch. Rastenii, 4*:32–33, 1979.
41. Scott, P. R., and Hollins, T. W. Effects of eyespot on yield of winter wheat. *Ann. Appl. Biol., 78*:260–279, 1974.
42. Scott, P. R., and Hollins, T. W. Prediction of yield loss due to eyespot in winter wheat. *Plant Pathol., 27*:125–131, 1978.

INDEX